Periodic Table of the El

	IB	IIB	IIIA	IVA	VA	VIA	VIIA	0
								2 4.0026 — He −272.2 (26 atm) −268.9 0.93
			5 10.811 2.0 B 2300 2550 0.80	6 12.01115 2.5 C 3570 3470 s. 0.77	7 14.0067 3.0 N −210 −196 0.74	8 15.9994 3.5 O −219 −183 0.74	9 18.9984 4.0 F −218 −188 0.72	10 20.183 — Ne −249 −246 1.12
			13 26.9815 1.5 Al 660 2327 1.25	14 28.086 1.8 Si 1414 2355 1.17	15 30.9738 2.1 P 44 280 1.10	16 32.064 2.5 S 119 445 1.04	17 35.453 3.0 Cl −101 −34 0.99	18 39.948 — Ar −189 −186 1.54
58.71 1.8 Ni	29 63.54 1.9 Cu 1083 2582 1.17	30 65.37 1.6 Zn 420 907 1.25	31 69.72 1.6 Ga 30 1983 1.25	32 72.59 1.8 Ge 958 2700 1.22	33 74.9216 2.0 As 814 (36 atm) 633 s. 1.21	34 78.96 2.4 Se 217 688 1.17	35 79.909 2.8 Br −7.3 58.8 1.14	36 83.80 — Kr −157 −153 1.69
106.4 2.2 d	47 107.870 1.9 Ag 961 2193 1.34	48 112.40 1.7 Cd 321 767 1.41	49 114.82 1.7 In 157 2000 1.50	50 118.69 1.8 Sn 232 2362 1.41	51 121.75 1.9 Sb 630 1325 1.41	52 127.60 2.1 Te 450 1390 1.37	53 126.9044 2.5 I 114 184 1.33	54 131.30 — Xe −112 −107 1.90
195.09 2.2 t	79 196.967 2.4 Au 1063 2660 1.34	80 200.59 1.9 Hg −38.9 356.9 1.44	81 204.37 1.8 Tl 304 1457 1.55	82 207.19 1.8 Pb 328 1755 1.54	83 208.980 1.9 Bi 271 1560 1.52	84 (210) 2.0 Po — — 1.53	85 (210) 2.2 At — — —	86 (222) — Rn −77 −62 —

151.96 — u	64 157.25 1.1 Gd 1312 3000 1.61	65 158.924 1.2 Tb 1356 2800 1.59	66 162.50 — Dy 1407 2600 1.59	67 164.930 1.2 Ho 1461 2600 1.58	68 167.26 1.2 Er 1497 2900 1.57	69 168.934 1.2 Tm 1545 1727 1.56	70 173.04 1.1 Yb 824 1427 1.70	71 174.97 1.2 Lu 1652 3327 1.56
(243) — m	96 (247) — Cm — — —	97 (247) — Bk — — —	98 (249) — Cf — — —	99 (254) — Es — — —	100 (253) — Fm — — —	101 (256) — Md — — —	102 (253) — No — — —	103 (257) — Lw — — —

CHEMISTRY
A Survey of Fundamentals

Chemistry
A Survey of Fundamentals

Lawrence P. Eblin
Ohio University

HARCOURT, BRACE & WORLD, INC.
New York / Chicago / San Francisco / Atlanta

Library of Congress Catalog Card Number: 68-12976

Printed in the United States of America

Preface

The purpose of this text is to present the basic concepts of chemical science together with carefully selected topics in descriptive chemistry. Some of the descriptive material is presented for the purpose of illustrating its logical derivation from basic principles. Some of it is presented because it is interesting and important in itself—for example, the treatment of compounds of biochemical significance in Chapters 33, 34, and 35.

Calculus has not been used in this book, but it is assumed that the student is competent in arithmetic and has some knowledge of elementary algebra. Significant figures, exponential numbers, and units of measurement are reviewed in the Appendix for ready reference. Sample problems and solutions are employed in the presentation of each new type of calculation throughout the text.

One of my strongest convictions is that chemical literacy is based on the ability to read and write chemical equations. Consequently, formulas and equations are introduced early in the text. By the time the student has completed his study of the second chapter, he will know how to perform calculations based on chemical formulas, how to determine molecular and atomic weights, and how to calculate empirical and molecular formulas.

The mole concept is emphasized as soon as the student has been prepared for it by a study of atoms, molecules, and ions. This concept is used consistently in solving problems based on chemical equations, titrations, and Faraday's Law. Mole fractions are employed in the treatment of Dalton's Law of Partial Pressures and Raoult's Law. Otherwise, the concentrations of solutions are expressed in terms of molarity or molality.

When stoichiometry is first considered, it is emphasized that calculations based on chemical equations assume that one of the reactants is completely consumed. It is pointed out that many reactions appear to stop before the conversion is complete. Frequent opportunities occur thereafter for reinforcing the concept of reversibility. This prepares the student for the treatment of chemical equilibrium in Chapters 15, 16, and 17. The elementary principles of chemical thermodynamics and kinetics are presented in Chapter 17, with emphasis on the concepts of free energy, entropy, and activation energy. This is an informal presentation that does not require the use of calculus. It enables the student not only to predict that a chemical reaction will proceed in one direction rather than the other, but also to calculate, in advance, the equilibrium constant for a proposed reaction from its ΔG value. The student also learns how the equilibrium constant can be calculated for any stated temperature if it is already known for some other temperature.

Electronegativity is considered when covalent bonding is first discussed. This makes it possible to present the concept of oxidation state understandably at an early point in the course. Oxidation and reduction are defined in terms of changes in oxidation numbers when they are first mentioned. A later chapter is devoted to an extended consideration of oxidation-reduction and electrochemistry. Here the student learns not only to predict the direction of oxidation-reduction reactions but also to calculate equilibrium constants from electrode potentials.

The descriptive chemistry of the elements is based on the eighteen-column periodic table. Each group of representative elements is given separate treatment. Chapters on hydrocarbons and oxygen derivatives of hydrocarbons precede this treatment so that the descriptive chemistry of the elements is an integrated treatment covering both their organic and their inorganic compounds. Further background is provided in a chapter devoted to the Arrhenius, Brønsted-Lowry, and Lewis concepts of acids and bases.

Complexes of the transition metals are considered in a separate chapter. Consideration is given here to isomerism in coordination compounds, to the stability of complexes, and to the ligand field theory, which is now regarded as offering the most satisfactory explanation of bonding in complexes of transition metals.

One of the most important reasons for studying chemistry is the pathway it opens for an effective attack on the problems of biology. For this reason chapters on carbohydrates, lipids, and proteins are included. Since the emphasis on structure in the earlier chapters includes consideration of the various types of isomerism, the student is prepared for a really meaningful treatment of carbohydrate structures and other topics of biochemical significance.

Many of the features of my briefer book *The Elements of Chemistry* (Harcourt, Brace & World, 1965) have been incorporated into the present text. At the end of each chapter are a list of the important terms introduced in the chapter, with their definitions; a set of exercises; and a list of references, each with a brief description. Answers for approximately half the numerical exercises are in the Appendix; answers for all the exercises are provided in a separate booklet available to instructors using the text. I have also prepared a laboratory manual, *Chemistry: A Survey of Laboratory Techniques and Procedures* (Harcourt, Brace & World, 1968), which follows the organization of the text.

I have written this text in a style that a student should easily understand even though he may never have had any previous exposure to chemistry. The content and organization of the book should make it equally stimulating to the student who already knows, from experience, that exploring the chemical world can be great intellectual fun. At the end of the course all students will realize, I hope, that many gaps remain in our chemical knowledge. They should also realize that we know a great deal we do not yet understand.

Many have helped, directly or indirectly, in the writing of this book. My greatest personal indebtedness is to my wife, Geraldine, who typed the manuscript with exemplary diligence and painstaking care, and to my son, James, who searched each chapter carefully for obscurities and inconsistencies. I am deeply grateful to Ohio University, and its Department of Chemistry, for a sabbatical leave during which I devoted full time to writing.

LAWRENCE P. EBLIN

Contents

Appendix 639

Index 655

CHEMISTRY
A Survey of Fundamentals

1

The Foundations of Chemistry

1-1 The Concept of Elements

One of man's earliest important achievements was chemical in nature—the discovery of fire and how to control it for cooking his food and for keeping himself warm. His continuing progress toward the goal of becoming master of the earth has involved his growing capacity for utilizing *materials*—the different kinds of matter, such as wood, water, air, and stone.

Early chemical arts included the tanning of leather, the making of glass, the extraction of dyes and medicines from plants, and the winning of such metals as iron and copper from their ores. These chemical arts developed independently of one another, by the slow method of trial and error, over many centuries. Discoveries in one of the arts meant little to those engaged in the other arts. There was no set of general principles to provide the intellectual command that characterizes the chemical arts today.

It was only three hundred years ago that Robert Boyle proposed the powerful idea that all materials are composed of a small number of *elements*. The strength of this concept lies in its usefulness for understanding the transformation of one material into another. Before people accepted the concept of elements, they thought that the *properties* of different materials were only incidental qualities and that to change a material into another material, one merely divested it of certain properties and bestowed other properties upon it. This type of mastery over materials has never come to pass. We now believe that each material has a definite set of properties and that we can change these properties only by transforming one material into one or more others.

Alchemy was the medieval art concerned with efforts to perform such profound transformations as the conversion of base metals into gold and silver. The efforts failed because the alchemists were attempting a difficult feat—the *transmutation* of one element into another—which was not accomplished until the present century. Nevertheless, the alchemists discovered three of our most important acids (nitric, sulfuric, and hydrochloric) and five elements not previously known (zinc, bismuth, antimony, arsenic, and phosphorus). Alchemy evolved into chemistry as honest and intelligent thought pervaded the art. Robert Boyle was an alchemist.

1-2 Heating Metals in Air

When man first used metals, he must have observed that many of them undergo a permanent change of character when heated in air. Iron, when heated "red hot," becomes covered with an iridescent film which changes to a brownish-black scale, called "hammer-scale," when it cools. When lead is heated in air, it melts and then becomes covered with a scum which can be recovered as a yellow powder when cold. Prolonged heating completely transforms such metals as iron and lead. For example, pure lead is converted entirely into the yellow powder (called "litharge").

The alchemists gave the name *calcination* to the transformation a metal undergoes when it is heated in air. The name was derived from *calx*, the alchemists' term for the hammer-scale, litharge, or other substance produced by the transformation. When they speculated about the nature of calcination, they concluded correctly that it must be related to the burning of combustible materials such as wood. They knew that when wood burns it disappears and ashes remain and that when a metal undergoes complete calcination it likewise disappears and the calx remains.

When we consider the quantitative aspects of the question, the burning of wood and the calcination of a metal do not seem to be so similar after all. The burning of wood produces an ash that weighs much less than the wood, but the calx produced from a metal weighs more than the metal itself. Jean Rey, a French physician, argued more than three hundred years ago that the increase in weight upon calcination comes from the air in which the metal is heated. The discovery of different kinds of "air," a hundred and fifty years later, completed the proof that Rey's supposition was correct.

Jan Baptista van Helmont, a Belgian physician, noted as long ago as 1610 that both the burning of charcoal and the fermentation of grapes produced something that seemed to be air but was clearly different from ordinary air, for it would extinguish the flame of a burning candle. (This substance is now known as carbon dioxide.) He also obtained impure samples of other "airs" with different properties. Van Helmont invented the word *gas* for these new kinds of "air." Thus, we are now able to restrict the meaning of the word *air* to that mixture of gases that forms the atmosphere of our earth.

1-3 Oxygen and Nitrogen

Other observers noted that air undergoes a reduction in volume when any substance burns in it or when an animal breathes it (Figure 1-1). These observa-

Figure 1-1

Experiment to show that air is a mixture

A cylinder is inverted over a floating candle that is burning. The water rises to take the place of the oxygen (that portion of the air involved in combustion). When the oxygen of the residual air is reduced to a certain percentage, the candle goes out. Most of the carbon dioxide formed by the burning of the candle dissolves in the water.

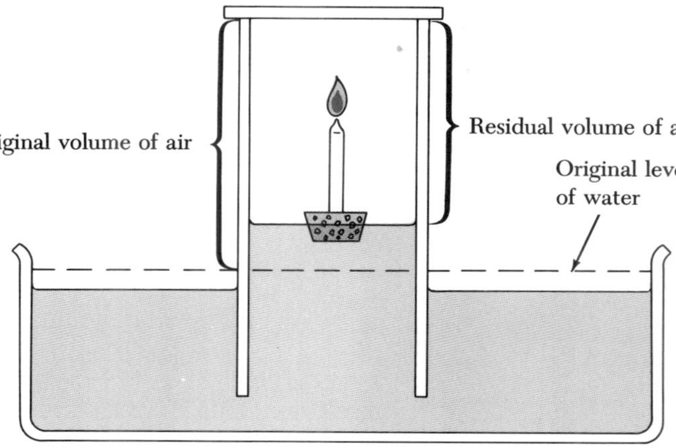

tions suggested that air is not a simple substance but a mixture of different gases. In 1774 Antoine Laurent Lavoisier, a French chemist, showed what happens when a metal is heated in air. To help solve the problem, he used a balance with which he could make weighings to the nearest 0.0005 gram.

Lavoisier placed a weighed sample of tin in a glass retort, which he sealed and weighed and then heated. The weight of the sealed retort did not change, but the tin itself (covered with a calx) gained in weight. The weight gained by the tin came from the air that was in the retort when it was sealed, for the weight of the empty retort did not change.

Lavoisier tried, by heating the calx, to remove that part of air which had combined with the tin, but he could not bring this about. However, by strongly heating the calx of mercury with a large burning-glass, Joseph Priestley, in England, succeeded in liberating a gas (Figure 1-2). He collected the gas over water and found that it supported combustion better than did ordinary air. Priestley found that mice could live in it much longer than in an equal volume of ordinary air, and he even ventured to breathe the gas himself, noting that "my breast felt peculiarly light and easy for some time afterwards." He soon made a trip to Paris and told Lavoisier of his experiments.

Lavoisier listened to Priestley with interest and repeated the experiments. He heated mercury in a retort and produced the calx, noting that part of the air in the retort was consumed. Then he heated the calx more strongly and found that the volume of air generated was equal to the volume that had previously disappeared. He also noted that the gas remaining after part of the air had been

Figure 1-2

Schematic diagram of apparatus used by Priestley in his discovery of oxygen

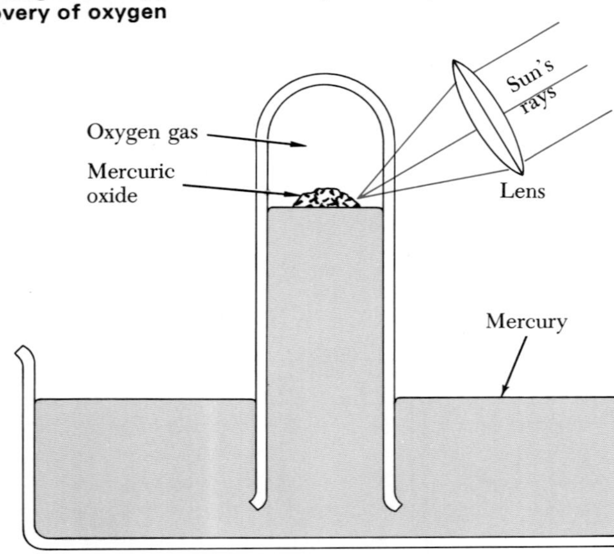

removed by combining with mercury was unfit for respiration and unable to support combustion. Yet the gas produced by strongly heating the calx was better than air in both respects. Lavoisier gave the name *azote* ("without life") to that portion of ordinary air that was incapable of supporting combustion. In the French language the gas is still known as azote, but in English it is known as nitrogen, from its occurrence in niter (which, in chemistry, is called potassium nitrate).

Lavoisier gave the name *oxygen* ("acid former") to the active component of air, the gas which supports combustion and respiration and combines with metals when they undergo calcination. His assumption that this element is a constituent of all acids was incorrect.

1-4 Different Kinds of Substances

The experiments of Priestley and Lavoisier involved two opposite kinds of chemical reaction. In *direct union* or *combination*, two or more different substances combine to form one substance, as mercury and oxygen combine to form the red oxide of mercury. We may say that the mercury oxide has been *synthesized* by the combination of the chemical elements of which it is composed. In *decomposition*, one substance produces two, or even more than two, simple substances. In the case of the mercury oxide, the decomposition produces the two chemical elements which compose it, mercury and oxygen; but there are many reactions of the decomposition type which produce substances that are not chemical elements.

The term *substance* is used to designate any material of which all specimens have the same properties and composition. Robert Boyle first defined the word *element* to mean a substance that could not be decomposed. Decomposable substances were called *compounds*. As already implied in our discussion of the experiments of Priestley and Lavoisier, compounds—but not elements—can be obtained by chemical synthesis. As a working definition we may say that an *element* is a nondecomposable, nonsynthesizable substance. Any substance that is not an element is a compound composed of elements; i.e., it may be produced from or decomposed into elements.

Many common materials are *mixtures*, not pure substances at all; i.e., they have a variable composition. It is possible, although not always convenient, to separate the components of a mixture from each other without resorting to chemical change. For example, when a mixture of sand and sugar is treated with water, only the sugar dissolves. Some materials, such as air or a mixture of sugar and water, are *homogeneous* mixtures; i.e., it is not obvious to the eye that they are composed of more than one substance. Homogeneous mixtures are usually referred to as *solutions*. Other materials, such as a sand-and-sugar mixture, are composed of different kinds of particles large enough to be individually observed and are known as *heterogeneous* mixtures.

The different kinds of matter, which we call materials, appear to be numberless. It would be difficult to understand the nature of our physical surroundings if the whole problem had not been simplified by the introduction of the ideas of mixtures, compounds, elements, combination, decomposition, etc. The study of chemistry has been aided also by the discovery that there are only a few more than one hundred chemical elements composing the hundreds of thousands of chemical compounds now known. These elements can readily be classified into families whose members are remarkably similar. If you know something about sodium, you also know something about potassium. If you are familiar with the chemistry of bromine, it is remarkable how much you know about chlorine and iodine. Yet there is something true of each chemical element that is true of it alone—each element has a "personality."

1-5 Physical and Chemical Properties

The *properties* of a material are the characteristics that distinguish it from other materials. These characteristics are usually classified as *physical* and *chemical* properties. The physical properties of a material are those that may be observed or measured without transforming it into other materials. Common physical properties include color, odor, taste, and relative heaviness. Some physical properties requiring a special test or measurement are relative hardness, electrical and thermal conductivity, tensile strength, ductility, melting point, boiling point, and solubility (the capacity a material has for dissolving in water, alcohol, or some other solvent).

Most substances that are liquid at ordinary temperatures become gaseous if sufficiently heated and solid if sufficiently cooled. Similarly, most substances that are solid at ordinary temperatures become liquid when heated to sufficiently high temperatures and gaseous at still higher temperatures. Most substances that are gaseous at ordinary temperatures become liquid if cooled sufficiently and solid if cooled to still lower temperatures.

We speak of the liquid, solid, and gaseous conditions as the three *states of matter*. A *solid* object has a size and a shape that are characteristic of that particular object; a *liquid* specimen has a volume that does not change unless the pressure or temperature changes, and it takes the shape of its container; a *gaseous* specimen assumes the shape of its container and occupies all the space available to it.

Some of the important physical properties of substances refer to their tendency to change from one state to another. Thus, a *volatile* liquid is a liquid that readily changes into gas—a change known as *vaporization* or *evaporation*. Water is characterized by its freezing point, 0° on the Celsius scale (0°C); a solid (ice) is formed when water is cooled to this temperature. The change of state from liquid to solid is known as *freezing*, *solidification*, or *crystallization*. The conversion of solid to liquid is called *melting* or *fusion*. The conversion of gas to liquid is called *condensation*.

The physical properties of a material are often a convenient means of identifying it. Two materials are identical if, under the same conditions, each of them has all the properties of the other and has no properties not possessed by the other. If one of the materials is a pure substance, the other is simply another specimen—another "piece"—of the same substance. If one of the materials is a mixture of two or more substances, the other is a mixture of the same substances in the same proportions. If this were not so, *some* of the properties would certainly be different.

It is possible—indeed it is common—for a material to be subjected to physical conditions that temporarily change some of its properties without causing it to lose its identity. Consider the substance known as common salt, the compound to which the chemical name sodium chloride has been given. It is a white, salty-tasting, odorless, crystalline solid, soluble in water. When the salt is pulverized into a finer state of subdivision, these characteristics are not altered. But when it dissolves in water it seems to disappear entirely. However, evaporation of the water leaves a residue which is still the same white, salty-tasting, odorless, crystalline solid. The salt has not ceased to be salt at any time during this series of events; i.e., its composition has not been altered. The changes that have occurred are merely physical—they are *not* chemical.

When sugar is heated, a profound change takes place: an odor develops and a black residue is obtained which does not have the sweet taste of sugar and will not dissolve in water. The sugar has undergone a permanent and fundamental change which has completely destroyed its identity: its composition has been drastically altered, and the properties of sugar are no longer evident. New substances with different properties have been produced. In brief, a *chemical reaction* has occurred.

Thus, substances possess not only physical properties but other characteristics which are appropriately called chemical properties. These include chemical stability (toward heat, light, shock, etc.) and behavior toward other substances under varying conditions. Briefly stated, the chemical properties of a substance refer to its ability to undergo chemical transformation. Thus, we say that nitroglycerin is highly explosive, gasoline is combustible, iron has the property of being converted into "iron rust" upon contact with moist air, grape juice has the property of producing alcohol by fermentation, etc.

The chemical properties of a substance are a conclusive means of identifying it. For example, the burning of diamond and the burning of graphite produce the same invisible gas, carbon dioxide, thus proving that diamond and graphite are both carbon (even though their physical properties are certainly not the same).

1-6 Some Important Elements

The discovery that air is a mixture of oxygen and nitrogen and the proof that water is a compound of oxygen and hydrogen set the stage for the development of modern chemistry. Since oxygen was discovered there has been steady progress toward an understanding of the nature of matter. Lavoisier's first list of the elements, compiled in 1789, included seven metals known even in prehistoric times: gold, silver, copper, tin, lead, mercury, and iron. Other elements on his list, such as oxygen, hydrogen, and nitrogen, had been discovered only recently.

Figure 1-3

Relative abundance of elements in the earth's crust

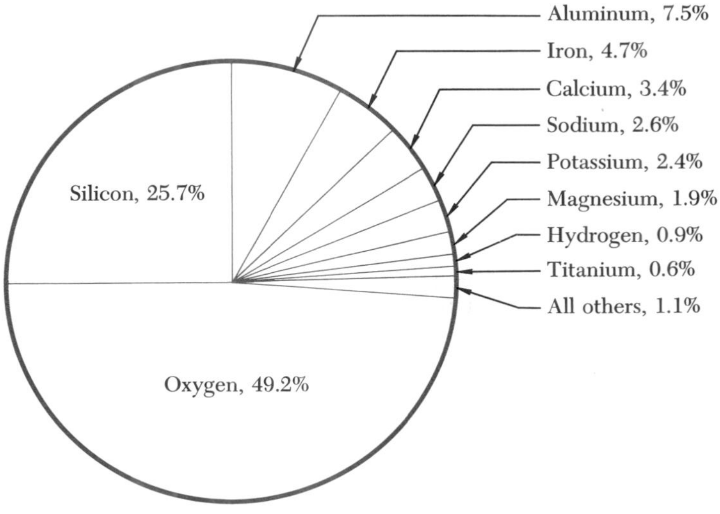

Since Lavoisier's time the list of elements has been extended continuously, and the present number stands at 103 certain, 104 probable.

Sir Humphry Davy used electric current, in 1807 and 1808, to obtain specimens of five metals never before seen by man—sodium, potassium, calcium, strontium, and barium. The invention of the Kirchhoff-Bunsen spectroscope in 1859 made possible the discovery of four elements occurring only in small amounts—rubidium, cesium, thallium, and indium. (The spectroscope is an instrument in which light is separated into its different wavelengths by being passed through a prism. Each element in the light source contributes its own characteristic set of wavelengths.) With the use of the spectroscope, helium was discovered on the sun. Several years later helium was discovered on the earth.

On the earth the most abundant element is oxygen. It constitutes almost half of the earth's crust, including the oceans, and the atmosphere (Figure 1-3). A little more than half of the remainder consists of silicon. The next most abundant elements are the six metals named aluminum, iron, calcium, sodium, potassium, and magnesium. These eight elements constitute approximately 98 percent of our environment. The remainder consists of hydrogen, titanium, and small quantities of other elements.

1-7 Oxides of Metals and Nonmetals

Many of our most important compounds consist of oxygen and one other element. Such compounds are called *oxides*, and many of them can be obtained by simply burning the element in air or in pure oxygen. The union of oxygen with another element to produce an oxide is called *oxidation*. (Oxidation has now assumed a more technical meaning which includes other processes.)

Water (hydrogen oxide) is an example of an oxide that is a liquid at ordinary temperatures. Gaseous oxides include carbon dioxide, carbon monoxide, and sulfur dioxide. Some oxides, such as zinc oxide and magnesium oxide, are white solids. Others are colored solids. For example, manganese dioxide is black and mercuric oxide is red.

About three fourths of the elements are classified as *metals*, characterized by so-called metallic properties. These properties include high conductivity for heat and electricity, ductility, malleability, and the peculiar reflective power commonly called "metallic luster." Most of the remaining elements do not have these properties and are called *nonmetals*; some examples are oxygen, sulfur, chlorine, and nitrogen. A few of the elements are not easily classified as metals or nonmetals and are known as *metalloids*; some examples are silicon, germanium, and arsenic.

Lavoisier noted that those oxides of nonmetals which are soluble in water form solutions that are *acidic* and that oxides of metals commonly form solutions that are *alkaline* or *basic*. (The properties which characterize acids and bases will be discussed in a later chapter.) So far as chemists are concerned, the decision to classify an element as a metal or as a nonmetal rests as much on whether its oxide forms an acid or a base as it does on its physical properties.

1-8 Atoms and Molecules

We are so accustomed to assuming the existence of atoms that we are likely to forget that the atomic theory was fully developed only a century and a half ago. True, certain Greek philosophers of 2500 years ago, notably Leucippus and Democritus, concluded that all kinds of matter are composed of invisibly small, separate, and indestructible particles which they called *atoms* (Greek: *atomos*, indivisible). It is difficult to determine just what led these philosophers to this view of the physical world. Perhaps consideration of such phenomena as evaporation and condensation was the reason. The poet Lucretius, a Roman contemporary of Julius Caesar, described evaporation as the departure of individual "atoms" of the liquid, like individual bees from a hive. Clearly the atomistic philosophers made no distinction between atoms and molecules.

Not all philosophers accepted the atomistic view of the structure of matter, and this theory was consequently forgotten until the revival of learning in the fifteenth and sixteenth centuries. Among those responsible for its revival were Robert Boyle and Sir Isaac Newton. In fact, Newton ascribed an atomistic character to light—he spoke of "corpuscles" of light. The corpuscular theory of light fell into disuse until its rebirth in the quantum theory of the twentieth century, but the atomic theory of matter has remained with us since its revival.

The Greeks had no concept of chemical elements. Because there seemed to be an infinite number of different materials, they assumed that the number of different kinds of atoms was unlimited. As we have seen, Robert Boyle and others developed the idea that the many different materials are composed of only a few chemical elements. Over the ensuing century this idea led in turn to another fundamental idea—that there are only as many different kinds of atoms as there are chemical elements. One kind of atom could combine with other kinds to form many different kinds of *molecules* (originally referred to as "compound atoms"). This revision of the revived atomic theory was largely the work of John Dalton, who, in 1808, applied the theory to the explanation of the fundamental chemical laws which had been developed mainly since the days of Boyle.

1-9 The Law of Constant Composition

More than three centuries ago, when Jean Rey was recording his observations on the calcination of metals in air, he noted that "the weight of the metal increased from the beginning to the end, but when the metal is saturated, it can take up no more air." In other words, 1 gram of oxygen combines with definite amounts of different metals. It has been found experimentally that 1 part by weight of oxygen combines with 1.520 parts of magnesium, 4.086 parts of zinc, etc. Even if magnesium oxide is prepared in several different ways, 1 part of oxygen still combines with 1.520 parts of magnesium. Likewise, regardless of the method used to prepare zinc oxide, 1 part of oxygen combines with 4.086 parts of zinc.

The point to all this is that two elements uniting do not unite in fortuitous quantities but in definite proportions. The *Law of Constant Composition* states that the proportion by weight of the elements present in any pure compound is always the same. This law was not discovered by any one man but evolved as chemistry developed. In fact, the law was tacitly accepted by many chemists before it was ever explicitly stated. One of the important contributions of Dalton's atomic theory was to provide an explanation for the Law of Constant Composition.

1-10 The Law of Simple Multiple Proportions

The same elements may combine in different proportions to form two or more different compounds. It is well known that carbon and oxygen combine to give two different compounds—carbon monoxide and carbon dioxide. Carbon dioxide is 27.27 percent carbon and 72.73 percent oxygen by weight; carbon monoxide is 42.88 percent carbon and 57.12 percent oxygen. A comparison of these percentages reveals nothing significant. If, however, we divide the percentage of oxygen by the percentage of carbon (i.e., grams of oxygen per 100 grams of oxide by grams of carbon per 100 grams of oxide) for each of these compounds, we obtain the number of grams of oxygen combined with 1 gram of carbon (or the number of pounds of oxygen combined with 1 pound of carbon, etc.).

$$\text{For carbon dioxide, } \frac{72.73 \text{ g oxygen}}{27.27 \text{ g carbon}} = 2.66 \text{ g oxygen per g carbon}$$

$$\text{For carbon monoxide, } \frac{57.12 \text{ g oxygen}}{42.88 \text{ g carbon}} = 1.33 \text{ g oxygen per g carbon}$$

These calculations show that in carbon dioxide 1 part by weight of carbon is combined with 2.66 parts by weight of oxygen and that in carbon monoxide 1 part by weight of carbon is combined with 1.33 parts by weight of oxygen. Clearly, in carbon dioxide twice as much oxygen is combined with a given amount of carbon as in carbon monoxide.

If we divide the percentage of carbon by the percentage of oxygen, we obtain the number of grams of carbon combined with 1 gram of oxygen in each of the two cases.

$$\text{For carbon dioxide, } \frac{27.27 \text{ g carbon}}{72.73 \text{ g oxygen}} = 0.375 \text{ g carbon per g oxygen}$$

$$\text{For carbon monoxide, } \frac{42.88 \text{ g carbon}}{57.12 \text{ g oxygen}} = 0.750 \text{ g carbon per g oxygen}$$

Thus, in carbon dioxide 1 part by weight of oxygen is combined with 0.375 part by weight of carbon, and in carbon monoxide 1 part by weight of oxygen is combined with 0.750 part by weight of carbon. In carbon dioxide only half as much carbon is combined with a given amount of oxygen as in carbon monoxide.

This example illustrates the *Law of Simple Multiple Proportions:* if there exists a series of two or more compounds containing the same elements, there is a ratio of small whole numbers between the different weights of one element that are combined with a chosen weight of another element. This law can be further illustrated by the five oxides of nitrogen, in which the number of grams of oxygen combined with 1 gram of nitrogen are 0.571, 1.142, 1.713, 2.284, and 2.855. If each of these numbers is divided by the smallest of the set, the results are 1, 2, 3, 4, and 5.

1-11 Dalton's Atomic Theory

Dalton offered his chemical adaptation of the atomic theory as an explanation of these fundamental laws of chemical combination. His assumptions were (1) that all the atoms of a particular element are alike and have the same weight and (2) that chemical combination is the chemical bonding of a definite number of atoms of each of the combining elements to form one molecule of the resulting compound.

This theory offers a simple explanation for what initially appears to be a set of highly complicated facts. Thus, we picture the two oxides of carbon as having the formulas CO and CO_2, where C is the symbol representing one carbon atom, O represents one oxygen atom, and the subscript "2," attached to the symbol for oxygen in the second formula, represents the fact that there are two atoms of oxygen combined with one carbon atom. There is twice as much oxygen per unit weight of carbon in the second oxide as there is in the first, simply because there are two atoms of oxygen per atom of carbon in the second oxide but only one atom of oxygen per atom of carbon in the first.

The Law of Simple Multiple Proportions was predicted by Dalton himself. It is difficult to separate his development of the atomic theory from his proposal of the Law of Simple Multiple Proportions. There is no ready explanation for the fact that chemists failed to notice this law before Dalton stated it, for the relationship was inherent in experiments that were made long before Dalton's publications. The Law of Constant Composition had been tacitly assumed since about 1630; the idea of elements was proposed by Boyle in 1661; and the existence of atoms had been presumed by many of the chemists.

Between 1802 and 1807 Dalton gave a number of examples of the Law of Simple Multiple Proportions from chemical analyses made by himself and others. In this same period he published his atomic theory.

1-12 The Law of Definite Proportions

Another law dealing with weight relations in chemical reactions is the *Law of Definite Proportions*, which states that in any chemical reaction there is a definite ratio between the weight of each reactant or product and the weight of any other reactant or product (Figure 1-4). As an illustration of this law, when

Figure 1-4

The Law of Definite Proportions

In any chemical reaction, the ratio of the amounts of any two substances, whether reactants or products, remains the same even though the amounts themselves are different.

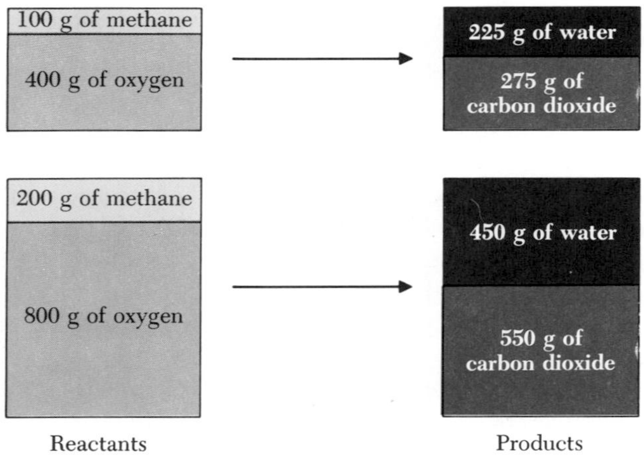

methane (the chief component of natural gas) is burned, the products are carbon dioxide and water:

$$\begin{array}{cccc} \text{Methane} + & \text{oxygen} \rightarrow & \text{carbon dioxide} + & \text{water} \\ 100\text{ g} & 400\text{ g} & 275\text{ g} & 225\text{ g} \end{array}$$

In this reaction, the weight of oxygen involved is 4 times the weight of methane being burned; the weight of carbon dioxide formed is 2.75 times the weight of methane burned; and the weight of water produced is 2.25 times the weight of methane consumed. As indicated by the numbers written beneath each reactant and product above, the burning of 100 grams of methane requires 400 grams of oxygen and produces 275 grams of carbon dioxide and 225 grams of water.

This law is a necessary corollary to the Law of Constant Composition and is readily understood in terms of atoms and molecules:

1. A molecule of any substance consists of definite numbers and kinds of atoms.
2. A definite number of molecules of each reactant is needed to provide the number of atoms of each kind that are to constitute the molecules of the reaction products.
3. Since each molecule of each reaction product consists of definite numbers and kinds of atoms, a definite number of molecules of each product will be obtained.

1-13 Mass, Weight, and Density

The two basic concepts of physical science are matter and energy. *Matter* is recognized by three characteristics: it occupies space, possesses mass, and exhibits inertia.

The *mass* of any selected portion of matter, e.g., a solid object or the liquid in a bottle or the gas inside a flask, is a measure of the amount of matter present. Ordinarily, the mass of a portion of matter is determined by the familiar process called weighing. When we determine the *weight* of an object, we measure the attractive force acting between the object and the earth. The magnitude of this force depends upon the mass of the object, the mass of the earth, and the distance between the center of the object and the center of the earth. Thus, unlike the mass, which is constant, the weight of an object varies from one point to another on the earth's surface.

The instrument with which we determine the weight of a specimen of matter in the laboratory is a chemical balance (Figure 1-5). When we weigh the specimen, we balance its attraction to the earth against the attraction which the earth has for a set of "weights." For example, a 1-gram "weight" is an object which is known to have a mass of 1 gram. The amount of matter present in such an object is 1 gram, regardless of where the object is. The attraction between it and the earth changes with altitude, and therefore its weight changes with altitude. However, since the earth's attraction for a "weight" with a mass

Figure 1-5

The balance

When the pointer is at the midpoint of the scale, the mass of the object is equal to the mass of the weights.

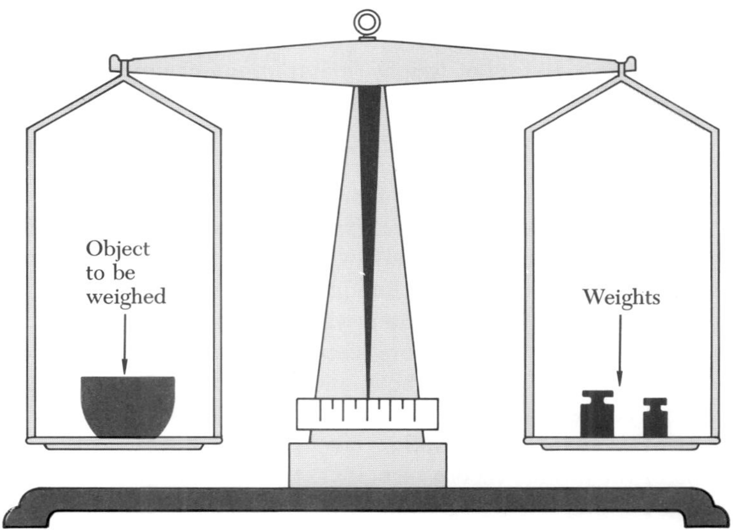

of 1 gram varies with altitude in the same manner as its attraction for any other 1-gram object, the weight of an object as determined by a chemical balance is a measure of its mass. Chemists usually refer to the mass of a specimen as its weight.

The *density* of a substance is the mass per unit volume. The density is obtained by weighing a specimen and measuring its volume. The density is, by definition, the weight (mass) divided by the volume:

$$D = \frac{M}{V}$$

where D is the density, M is the mass, and V is the volume. In scientific laboratory work the mass is usually expressed in grams and the volume in milliliters, so that the density is expressed in grams per milliliter (g/ml). The densities of gases are most often expressed in grams per liter (g/liter).

Since the volume of a given mass of a substance changes with temperature, the density of a substance must always be stated in terms of a specified temperature. For gases it is also necessary to specify the pressure, since the volume of a given mass of a gas is highly dependent on the pressure.

1-14 The Law of Conservation of Mass

When faith in magic was widespread, many believed that matter could be created and destroyed. In nature it appeared that such phenomena as the growth of a tree involved creation of matter and that matter disappeared when a candle was burned. Almost 2500 years ago, though, Democritus stated that things cannot be formed from nothing and cannot be reduced to nothing.

In 1774, when Lavoisier heated tin in a closed vessel, he found that the weight of the whole system was the same after the calcination of the tin as before the reaction; i.e., the whole system neither gained nor lost weight even though the tin and oxygen had undergone a chemical change. The principle that there is no detectable gain or loss of weight (mass) in any chemical reaction is now accepted by all scientists. The *Law of Conservation of Mass* states that in all physical and chemical processes the total mass of the substances involved remains unchanged.

Lavoisier is often given credit for first demonstrating the Law of Conservation of Mass, but it had already been explicitly stated in 1756 by the Russian chemist Mikhail Lomonossoff. The law was subjected to rigorous experimental tests as recently as 1906, when Hans Landolt at Berlin examined fifteen different reactions and was unable to detect any variation in weight that could not be attributed to experimental error. More recently, in connection with the development of atomic energy, it has been shown that under certain conditions mass and energy are interconvertible. Except for these special processes, the Law of Conservation of Mass is completely valid.

In terms of the atomic theory of matter, the Law of Conservation of Mass is true because the atoms involved in any physical or chemical process retain their identity. Even in a chemical process, the molecules of the products contain the same atoms that were originally present in the molecules of the reactants. Each individual atom retains its original mass, so the sum of the masses of the molecules of the products is the sum of the masses of the molecules of the reactants.

1-15 Conclusion

The chemical laws we have discussed are explained by the atomic theory—and this is one of the basic reasons for our belief that atoms exist. The atom is not the simple thing that Dalton believed it to be, but our purpose here has been to show why virtually all chemists believed in the *existence* of atoms long before anything was known about their structure.

New Terms

Alchemy: the medieval art that aspired to transmute base metals into gold, discover a universal cure for disease, and prolong life indefinitely. Chemistry evolved from alchemy.

Atom: one of the unit particles of elements, of which molecules are composed.

Chemistry: the science concerned with the composition of substances and the transformations they undergo.

Compound: a distinct substance with a fixed set of properties, composed of two or more constituent elements chemically combined in definite proportions.

Condensation: the change from the gaseous to the liquid state.

Decomposition: a chemical reaction in which one substance produces two or more simple substances.

Density: the ratio of the mass of a specimen of matter to its volume.

Direct union reaction (combination): a chemical reaction in which two or more different substances combine to form one substance.

Elements: the simpler substances into which a compound may be decomposed.

Freezing (solidification, crystallization): the change from the liquid state to the solid (crystalline) state.

Gaseous state: the state of matter in which a substance takes the shape of its container and fills all parts of the container to the same degree.

Liquid state: the state of matter in which a substance takes the shape of its container but has a volume of its own. The molecules are free to move among themselves but do not have the tendency to separate from one another that gas molecules do.

Mass (of a body): the amount of matter in a body.

Materials: the different kinds of matter of which the universe is composed. Some materials are mixtures, e.g., soil and paint; some are pure substances, e.g., water and sulfur.

Matter: something that occupies space, possesses mass, and exhibits inertia.

Melting (fusion): the change from the solid to the liquid state.

Metalloids: chemical elements which have both metallic and nonmetallic properties.

Metals: chemical elements which, when not chemically combined with other elements, have a peculiar luster and are good conductors of heat and electricity. Every metal forms at least one basic oxide.

Mixture: a material composed of two or more ingredients that are not chemically combined. Some mixtures are homogeneous; some are heterogeneous.

Molecule: one of the smallest particles of a substance, consisting of one or more atoms of each constituent element.

Nonmetals: chemical elements which lack the luster and conductivity that are characteristic of metals. Most nonmetals form at least one acidic oxide.

Oxide: a compound containing oxygen and one other element.

Properties: the characteristics of a material that distinguish it from other materials. These characteristics are usually classified as physical and chemical properties.

Solid state: the state of matter in which a substance has no perceptible tendency to flow but has sufficient rigidity to maintain a definite shape independent of the shape of any container in which it is placed.

Solutions: homogeneous mixtures.

Substance: a material that has a fixed composition and a definite set of properties because it has only one ingredient. As used in chemistry, a substance is pure. It may be a compound or an elementary substance, but its molecules are all alike.

Transmutation: the conversion of one element into another. In ordinary physical and chemical processes, there is no transmutation of elements. Radioactivity involves transmutation.

Vaporization (evaporation): the change from the liquid to the gaseous (vapor) state.

Weight: as commonly used in chemical practice, weight is a measure of the mass of a specimen of matter, as determined with a chemical balance. Technically, the weight of a body is the force with which it is attracted toward the earth.

The Laws of Chemical Combination

Law of Conservation of Mass: as the law applies in chemical reactions, it states that the total mass of the products of a reaction is the same as the total mass of the reactants.

Law of Constant Composition: the proportion by weight of the elements present in any pure compound is always the same.

Law of Definite Proportions: in any chemical reaction there is a definite ratio between the weight of each reactant or product and the weight of any other reactant or product.

Law of Simple Multiple Proportions: if there exists a series of two or more compounds containing the same elements, there is a ratio of small whole numbers between the different weights of one element that are combined with a chosen weight of another element.

Exercises

1-1 Distinguish between the meanings of the following paired terms: materials and substances; alchemy and chemistry; gas and air; oxygen and nitrogen; synthesis and decomposition; elements and compounds; solutions and heterogeneous mixtures; physical and chemical properties; physical and chemical changes; metals and nonmetals; atoms and molecules; mass and weight; density and weight; freezing and melting.

1-2 What facts of nature are embodied in the following generalizations?
(a) Law of Constant Composition
(b) Law of Simple Multiple Proportions
(c) Law of Definite Proportions
(d) Law of Conservation of Mass

1-3 Name several materials which have been discovered or synthesized within the present century and several which have been known since prehistoric times.

1-4 Name some materials whose uses depend on the following properties: electrical conductivity, low density, high electrical resistance, ease of stretching, difficulty of stretching, combustibility, noncombustibility, high melting point, solubility in water, insolubility in water, hardness, tensile strength.

1-5 Describe the differences among the three states of matter.

1-6 What is implied when we say that air is a mixture but water is a compound?

1-7 Classify the following properties as chemical or physical: color, flammability, compressibility, porosity, solubility.

1-8 Classify the following changes as chemical or physical: melting of ice, burning of coal, growth of a tree, digestion of food, production of light by an incandescent lamp, production of light by a burning candle.

1-9 Classify the following materials as elements, compounds, solutions, or heterogeneous mixtures: oxygen, water, air, mercury, milk.

1-10 Explain why a chemical change usually involves one or more physical changes.

References

Holton, G., and Roller, D. H. D., *Foundations of Modern Physical Science*, Addison-Wesley, Reading, Mass., 1958. The fundamental concepts of physics and chemistry presented in historical perspective.

Kesselman, B., "The Skeptical Chemist," *Chemistry*, **39** (1), 8 (1966). Robert Boyle's search for truth.

Leicester, H. M., *The Historical Background of Chemistry*, Wiley, New York, 1957. The development of chemical concepts from ancient times through the present.

Read, J., *Through Alchemy to Chemistry*, Harper & Row, New York, 1963 (paperbound). The rise and spread of alchemy and the evolution of chemistry.

Seaborg, G. T., and Valens, E. G., *Elements of the Universe*, Dutton, New York, 1962. Outgrowth of an educational-television series on the chemical elements.

Walton, H. F., "The Humanistic Values of Science," *Chemistry*, **38** (12), 8 (1965). Scientific exploration as an artistic enterprise.

Weeks, M. E., and Leicester, H. M., *Discovery of the Elements*, 6th ed., Chemical Education, Easton, Pa., 1956. Accounts of the discoveries of the chemical elements, with biographical sketches of the discoverers.

2

Formulas and Equations

2-1 The Concept of Molecules

In our discussion of the atomic theory we used the word *molecule* to designate one of the smallest particles of a substance and stated that a molecule consists of definite numbers and kinds of atoms. The atomistic philosophers made no clear distinction between atoms and molecules because they did not know about the chemical elements. The distinction between atoms and molecules became important when the concept of chemical elements was accepted.

Study of the physical behavior of gases led to the development of the *molecular theory*. (This theory evolved independently of Dalton's atomic theory.) Robert Boyle, whom we have mentioned for his part in the development of the concept of chemical elements, also investigated the behavior of gases in response to changes in the pressure applied to them. From his work and that of others there appears to be no limit to the expansibility a gas exhibits when the confining pressure is reduced more and more. It does not seem reasonable that a substance could be present in every part of a given space and also be present in every part of a space a thousand times as great. According to the molecular theory a gas is composed of particles called molecules, whose size is very small compared to the spaces between them. When the confining pressure is reduced and the gas expands, the molecules remain the same size, but there is more space between them (Figure 2-1).

Since every gas becomes a liquid when cooled sufficiently and every liquid turns to solid when cooled to its freezing point, we know that not only gases but at least some liquids and solids also consist of molecules.

One of the fundamental postulates of the molecular theory of gases is the assumption that the molecules are in rapid, random motion and that the pressure a gas exerts is due to the collisions of its molecules with the walls of the container. A further assumption is that the average velocity of the molecules in-

Figure 2-1

The molecules of a gas

When a gas expands, the molecules remain the same size, but there is more space between them.

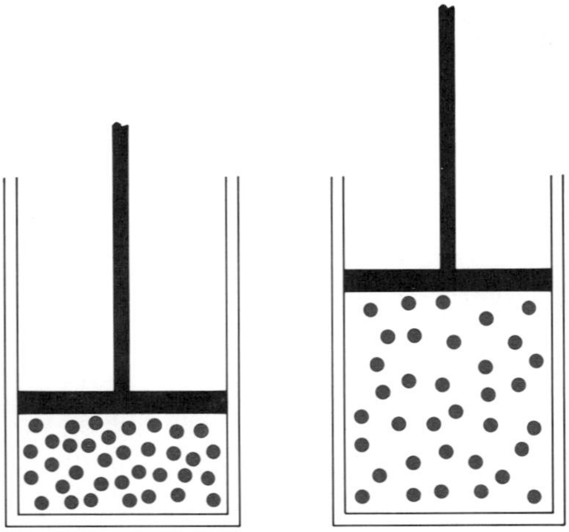

creases as the temperature increases. Since motion of the molecules is such an important aspect of the total concept, the molecular theory is usually called the *kinetic-molecular theory* or simply the *kinetic theory.*

From the standpoint of the kinetic theory, a *molecule* can be defined as a minute, uncharged particle of a substance, which moves about as a whole so that its component parts do not become detached during the motions. A molecule of a compound consists of more than one kind of atom (Figure 2-2). A molecule of an element consists of one or more atoms of the same kind (Figure 2-3).

Figure 2-2

Molecules of two compounds

A molecule of a compound consists of more than one kind of atom.

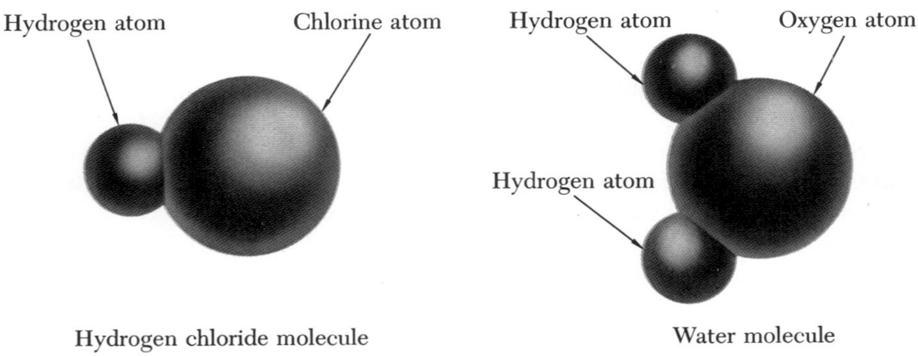

Figure 2-3

Molecules of two elements

A molecule of an element consists of one or more atoms of the same kind.

Argon molecule (one argon atom) — Hydrogen molecule (two hydrogen atoms)

2-2 Chemical Symbols and Molecular Formulas

The elements of which a compound is constituted and the number of atoms of each element in one of its molecules are indicated by its *molecular formula.* The molecular formula consists of the *symbols* of the constituent elements with subscripts to indicate the number of atoms of each element present in one molecule. For example, the molecular formula for table sugar is $C_{12}H_{22}O_{11}$. This formula means that one sugar molecule contains 12 carbon atoms, 22 hydrogen atoms, and 11 oxygen atoms.

Obviously, chemical formulas mean nothing unless the symbols of the elements can be recognized. Although 103 elements are now known, many of them are scarce and do not occur in substances with which we are familiar. The names and symbols for some of the more common metals and nonmetals are tabulated in Tables 2-1 and 2-2. Note that the symbols are often the initial letter or first two letters of the English names but that some symbols are derived from the Latin names. The symbol for one of the elements, tungsten (W), is derived from its German name (*Wolfram*). The first letter of a chemical symbol is always capitalized, and the second letter is never capitalized.

Formulas for compounds do not always contain subscripts. The compound formed by combining hydrogen with chlorine consists of molecules containing one atom of each element. The formula is written HCl, meaning H_1Cl_1. The subscript is always understood to be "1" if no subscript is written. Water contains two hydrogen atoms for every oxygen atom, and for this reason its formula is written H_2O (meaning H_2O_1).

There is no strict rule concerning the sequence in which symbols should appear in a formula. However, custom dictates that the formula for water should be writen H_2O rather than OH_2; the symbol for oxygen should appear last in the formula for any oxide; the symbol for chlorine should be last in the formula for any chloride; and the symbol for the metal should appear first in the formula for a compound consisting of a metal and a nonmetal.

The formulas of some compounds are indicated rather clearly by their names. For example, the name carbon dioxide refers to the compound whose

Table 2-1

Common elements having symbols derived from their English names

Metals		Nonmetals and metalloids	
Aluminum	Al	Arsenic	As
Barium	Ba	Boron	B
Bismuth	Bi	Bromine	Br
Cadmium	Cd	Carbon	C
Calcium	Ca	Chlorine	Cl
Chromium	Cr	Fluorine	F
Cobalt	Co	Hydrogen	H
Magnesium	Mg	Iodine	I
Manganese	Mn	Nitrogen	N
Molybdenum	Mo	Oxygen	O
Nickel	Ni	Phosphorus	P
Titanium	Ti	Silicon	Si
Zinc	Zn	Sulfur	S

Table 2-2

Elements having symbols derived from their Latin names

English name	*Latin name*	*Symbol*
Antimony	Stibium	Sb
Copper	Cuprum	Cu
Gold	Aurum	Au
Iron	Ferrum	Fe
Lead	Plumbum	Pb
Mercury	Hydrargyrum	Hg
Potassium	Kalium	K
Silver	Argentum	Ag
Sodium	Natrium	Na
Tin	Stannum	Sn

formula is CO_2, and the name carbon monoxide refers to the compound with the formula CO.

The formula for an elementary substance indicates the number of atoms present in one molecule of the substance. Thus, the formula for elementary hydrogen, H_2, indicates that a molecule of hydrogen consists of two hydrogen atoms; the formula for argon gas, Ar (meaning Ar_1), indicates that an argon molecule consists of a single argon atom.

2-3 Chemical Composition from Formulas

The molecular formula for a compound states the number of each kind of atom in one molecule. If the relative weights of the constituent atoms are known, the quantitative composition of the compound can be calculated from the formula.

The weight of a molecule is, of course, the sum of the weights of the atoms it contains. For example, when we know that a sulfur atom has twice the mass of an oxygen atom, we can quickly figure that a molecule of sulfur dioxide, SO_2, has four times the weight of an oxygen atom. From this it follows that half, or 50 percent, of the weight of a sulfur dioxide molecule can be ascribed to the sulfur atom, and the other half must be attributed to the two oxygen atoms. From its chemical formula sulfur dioxide is thus found to be 50 percent sulfur and 50 percent oxygen by weight.

The formula for sulfur trioxide is SO_3. From the formula we can compute that the weight of the molecule is five times the weight of an oxygen atom, so that the compound is two fifths, or 40 percent, sulfur and three fifths, or 60 percent, oxygen. In order to make this calculation, however, we must know that a sulfur atom is twice as heavy as an oxygen atom.

2-4 Atomic and Molecular Weights

In 1811 an Italian physicist, Amedeo Avogadro, postulated that equal volumes of gases, under identical conditions of temperature and pressure, contain equal numbers of molecules. Scientists were unwilling to accept this hypothesis until 1858, when an Italian chemist, Stanislao Cannizzaro, provided extended and convincing arguments in its behalf. *Avogadro's Principle* enables us to compare the weights of individual molecules of different gases by comparing the weights of equal volumes of the gases. For example, having found by experimental measurement that the weight of 1 liter of the gas called sulfur dioxide is (approximately) four times the weight of 1 liter of methane gas at the same temperature and pressure, we can conclude that the weight of a sulfur dioxide molecule is (approximately) four times the weight of a methane molecule.

After we have thus determined the relative weights of individual molecules of different compounds containing a particular element, we can determine the relative weight of one atom of that element by applying a principle first clearly stated by Cannizzaro. A molecule always contains a whole number of atoms of each constituent element. In a large number of compounds containing a particular element, it is probable that there will be at least one compound that has only one atom of the element in each molecule. *Cannizzaro's Principle* states that the weight of one atom of an element is the smallest weight due to that element in one molecule of any of its compounds (Figure 2-4).

Clearly, there should be universal agreement as to how the relative weights of atoms and molecules are to be expressed. Once a standard is agreed to, the weight of one atom of an element relative to the arbitrarily chosen scale is referred to as the *atomic weight* of the element; the sum of the relative weights

Figure 2-4

Cannizzaro's Principle

The weight of one atom of an element is the smallest weight due to that element in one molecule of any of its compounds. In these hydrogen compounds, the weight of one atom of hydrogen is the weight due to the one hydrogen atom in hydrogen chloride. (In each case the small balls represent hydrogen atoms.)

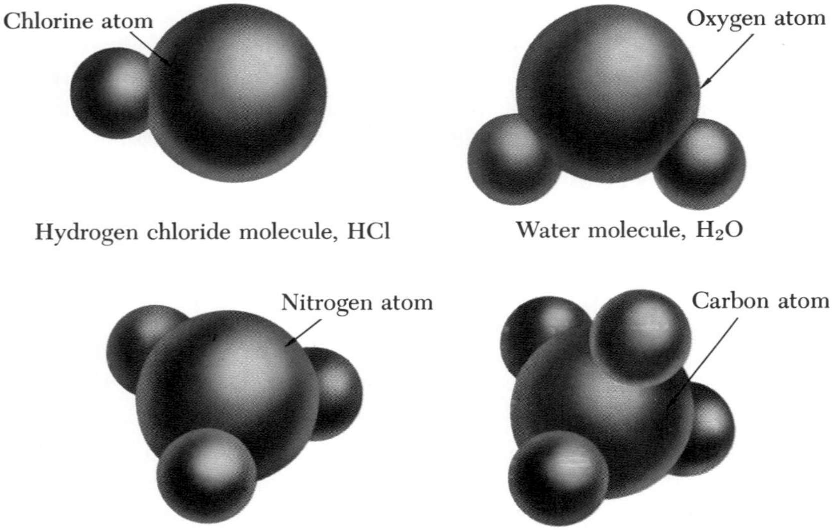

Hydrogen chloride molecule, HCl

Water molecule, H_2O

Ammonia molecule, NH_3

Methane molecule, CH_4

of the atoms in one molecule of a compound is referred to as the *molecular weight* of the compound. The standard now in use (which will be described in Section 2-5) was adopted in 1961.

The molecular weight of oxygen gas, relative to the standard now in use, is approximately 32 atomic mass units (amu). Since the density of methane is half that of oxygen, the molecular weight of methane is approximately 16 amu, and since the weight of a liter of sulfur dioxide is approximately twice the weight of a liter of oxygen (and approximately four times the weight of a liter of methane), the molecular weight of sulfur dioxide is approximately 64 amu. Chemical analysis of sulfur dioxide shows that this compound is approximately 50 percent sulfur and 50 percent oxygen by weight. Therefore, neither the atomic weight of sulfur nor the atomic weight of oxygen can be greater than 32 amu. If, however, we determine the molecular weights of a large number of sulfur and oxygen compounds, we find that 32 amu is the least weight due to sulfur in the molecular weight of any sulfur compound and that 16 amu is the least weight due to oxygen in the molecular weight of any oxygen compound. The atomic weights of sulfur and oxygen are, therefore, approximately 32 and 16 amu, respectively.

Table 2-3

Approximate atomic weight of chlorine

Compound containing chlorine	*Approximate molecular weight* (amu), *experimental*	*Percent of chlorine (by chemical analysis)*	*Weight of chlorine in one molecule of compound* (amu)
1	36.5	97.24	35.5
2	154	92.19	142
3	135	52.51	70.9
4	64.5	54.96	35.4
5	119	89.10	106

The conclusion that 16 amu is the atomic weight of oxygen is especially important: it means that there are two oxygen atoms in one molecule of oxygen gas. Therefore, the formula for oxygen gas is O_2. There is another form of gaseous oxygen, called ozone, whose density is 1.5 times the density of ordinary oxygen. Therefore, its molecular weight is 48 amu, and its formula is O_3.

From the data presented in Table 2-3 you can conclude, as an illustration of Cannizzaro's Principle, that the atomic weight of chlorine is approximately 35.5 amu. By dividing the weight of chlorine in one molecule of each compound by 35.5 amu, you can find the number of chlorine atoms in one molecule of each compound (1, 4, 2, 1, and 3, respectively).

2-5 Isotopes and the Atomic Weight Scale

One of the most stimulating discoveries of the early twentieth century was that all atoms of an element do not have the same mass. In 1913, Sir J. J. Thomson and F. W. Aston of Cambridge, England, discovered that neon gas, an elementary substance thought to consist of identical atoms, actually consists of two kinds of atoms, whose masses are approximately 20 and 22 amu. The accepted atomic weight of neon, 20.2 amu, was thus demonstrated to be simply an average of the masses of the individual isotopes of neon. *Isotopes* are atoms of the same element which differ from one another in mass.

By 1920 the pioneering work of Thomson and Aston had been developed into a precise physical method for determining the masses of atoms. In this method the atoms of the element being investigated are given electrical charges, and those with different masses but the same charge travel in different paths as they pass through the magnetic field of an instrument called the *mass spectrograph.* When the charged particles, called *ions*, fall upon a photographic film

(Figure 2-5), they strike the film at different places according to their masses. With this instrument, it has been discovered that nearly all the elements have isotopes. The chemical properties of the isotopes of an element are practically identical, and for this reason the different isotopes of a particular element usually occur together, mixed in almost the same proportion in all the compounds of the element. Nevertheless, since the atomic weight of an element as determined, for example, by application of Cannizzaro's Principle is only an average, precisely determined atomic weights will not always be the same for samples obtained from different sources. It is desirable, therefore, to choose one particular isotope of one of the elements as the standard for atomic and molecular weights.

In 1961 chemists and physicists agreed that the most abundant of the isotopes of carbon should be the standard to which all other isotopic weights, atomic weights, and molecular weights would be compared and that the standard value assigned to it should be *exactly* 12 amu. By definition, 1 amu is $\frac{1}{12}$ the mass of the carbon-12 atom. All the isotopes of hydrogen, carbon, oxygen, and chlorine that occur in significant amounts are listed in Table 2-4. The atomic weight is, by definition, the weighted average of the isotopic weights in each case. You will note that the individual isotopic weights are only slightly more or less than whole numbers. The integer which approximates the isotopic weight is called the *mass number* of the isotope. A particular isotope of an element is usually designated by its mass number, as when we write oxygen-18 or O^{18}.

Figure 2-5

The mass spectrograph

Gas molecules admitted to the evacuated space are given electrical charges. The ions emerge into a magnetic field which deflects the heaviest ions least.

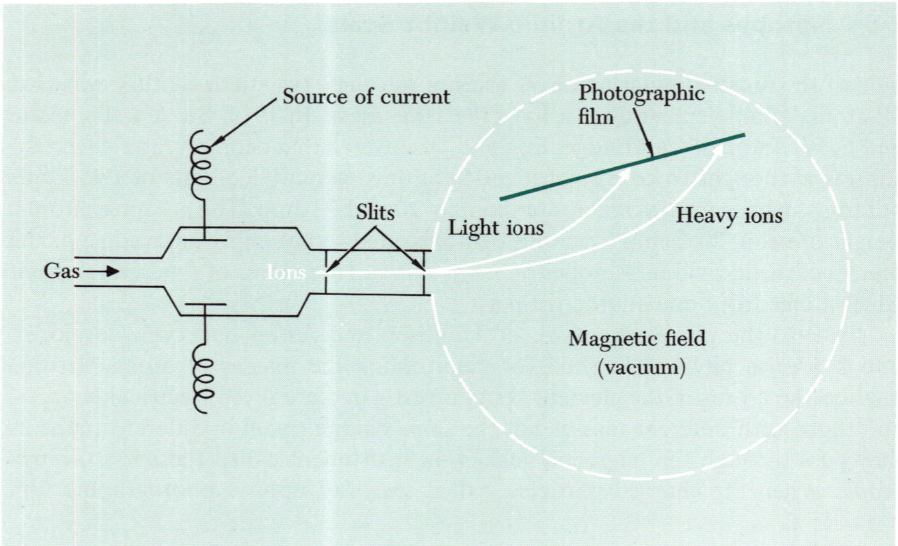

Table 2-4

Isotopic weights and atomic weights of four elements

Element	*Mass number*	*Isotopic weight* (amu)	*Abundance* (%)	*Atomic weight* (amu)
Hydrogen	1	1.0078	99.98	1.00797
	2	2.0142	0.02	
Carbon	12	12.0000	98.9	12.01115
	13	13.0034	1.1	
Oxygen	16	15.9950	99.76	15.9994
	17	16.9992	0.04	
	18	17.9992	0.20	
Chlorine	35	34.967	75.4	35.453
	37	36.966	24.6	

2-6 Percentage Composition from Chemical Formulas

Having described the scale now being used for expressing the relative weights of atoms and molecules, let us return to the interpretation of chemical formulas in terms of relative weights. As an example, the formula CO for carbon monoxide tells us that a molecule of this compound consists of one carbon atom and one oxygen atom. Since the atomic weight of carbon is 12.0 amu and the atomic weight of oxygen is 16.0 amu (values are rounded off), the molecular weight of carbon monoxide is 28.0 amu. For every 12.0 pounds of carbon in carbon monoxide there are 16.0 pounds of oxygen; for every 12.0 grams of carbon there are 16.0 grams of oxygen; etc. It makes no difference what units of weight are employed. In terms of common fractions, carbon monoxide is

$$\frac{12.0}{28.0} \text{ carbon and } \frac{16.0}{28.0} \text{ oxygen}$$

In terms of decimal fractions, carbon monoxide is $12.0 \div 28.0 = 0.429$ carbon and $16.0 \div 28.0 = 0.571$ oxygen. Composition is usually expressed in terms of percentages. Therefore, we say that carbon monoxide is 42.9 percent carbon and 57.1 percent oxygen.

The formula CO_2 states that a molecule of carbon dioxide contains one carbon atom and two oxygen atoms. Therefore, in carbon dioxide, for every 12.0 pounds of carbon there are 32.0 pounds of oxygen; for every 12.0 grams of carbon there are 32.0 grams of oxygen; etc. The sum of the atomic weights in the formula is 44.0 amu. Thus, the molecular weight of carbon dioxide is 44.0 amu. In every 44.0 parts by weight of carbon dioxide there are 12.0 parts by weight of carbon and 32.0 parts by weight of oxygen. The composition of carbon dioxide is

$$\frac{12.0 \text{ g carbon}}{44.0 \text{ g } CO_2} = 0.273 \text{ g carbon per g } CO_2 = 27.3\% \text{ carbon}$$

$$\frac{32.0 \text{ g oxygen}}{44.0 \text{ g } CO_2} = 0.727 \text{ g oxygen per g } CO_2 = 72.7\% \text{ oxygen}$$

2-7 Determination of Empirical Formulas

If you do not know the number of atoms of each element present in one molecule of a compound, you can still write its *empirical formula* (sometimes called the *simplest formula*). The empirical formula expresses only the *relative* number of atoms of each element in the compound. The first step in determining the true or molecular formula for a compound is to ascertain its empirical formula. This requires that an analysis be performed to determine the chemical composition of the compound. It also requires that the atomic weights of the constituent elements already be known.

EXAMPLE Analysis of carbon monoxide shows that it is 42.9% carbon and 57.1% oxygen. What is its empirical formula?

Calculating the empirical formula from the percentage composition is essentially the reverse of calculating the percentage composition from the formula. The percentage composition given for carbon monoxide means that each 100 amu of carbon monoxide is made up of 42.9 amu of carbon and 57.1 amu of oxygen. By dividing each percentage by the atomic weight of the corresponding element, we find the relative number of each kind of atom:

$$\text{For carbon, } \frac{42.9 \text{ amu}}{12.0 \text{ amu/atom}} = 3.58 \text{ atoms}$$

$$\text{For oxygen, } \frac{57.1 \text{ amu}}{16.0 \text{ amu/atom}} = 3.58 \text{ atoms}$$

$$\text{The ratio is } 3.58{:}3.58 = 1{:}1$$

Therefore, the empirical formula is CO.

EXAMPLE Analysis of carbon dioxide shows that it is 27.3% carbon and 72.7% oxygen. What is its empirical formula?

We first divide each percentage by the atomic weight of the corresponding element. Then we ascertain the ratio of the two quotients to obtain the relative number of each kind of atom:

$$\text{For carbon, } \frac{27.3 \text{ amu}}{12.0 \text{ amu/atom}} = 2.27 \text{ atoms}$$

$$\text{For oxygen, } \frac{72.7 \text{ amu}}{16.0 \text{ amu/atom}} = 4.54 \text{ atoms}$$

The ratio is $2.27:4.54 = 1:2$

Therefore, the empirical formula is CO_2.

EXAMPLE Ethylene is 85.6% carbon and 14.4% hydrogen. What is its empirical formula?

$$\text{For carbon, } \frac{85.6 \text{ amu}}{12.0 \text{ amu/atom}} = 7.13 \text{ atoms}$$

$$\text{For hydrogen, } \frac{14.4 \text{ amu}}{1.01 \text{ amu/atom}} = 14.3 \text{ atoms}$$

The ratio is $7.13:14.3 = 1:2$

Therefore, the empirical formula is CH_2 (or H_2C).

EXAMPLE Ethane is 79.9% carbon and 20.1% hydrogen. What is its empirical formula?

$$\text{For carbon, } \frac{79.9 \text{ amu}}{12.0 \text{ amu/atom}} = 6.66 \text{ atoms}$$

$$\text{For hydrogen, } \frac{20.1 \text{ amu}}{1.01 \text{ amu/atom}} = 19.9 \text{ atoms}$$

The ratio is $6.66:19.9 = 1:3$

Therefore, the empirical formula is CH_3 (or H_3C).

EXAMPLE Butane is 82.7% carbon and 17.3% hydrogen. What is its empirical formula?

$$\text{For carbon, } \frac{82.7 \text{ amu}}{12.0 \text{ amu/atom}} = 6.89 \text{ atoms}$$

$$\text{For hydrogen, } \frac{17.3 \text{ amu}}{1.01 \text{ amu/atom}} = 17.1 \text{ atoms}$$

The ratio is $6.89:17.1 = 1:2.5 = 2.5$ H atoms per C atom

Therefore, the empirical formula is C_2H_5 (or H_5C_2).

The same procedure can be used to calculate the empirical formula for a compound consisting of three or more elements.

EXAMPLE Potassium chlorate is 31.91% potassium, 28.93% chlorine, and 39.16% oxygen. What is its empirical formula?

We divide the percentages by the corresponding atomic weights:

$$\text{For potassium, } \frac{31.91 \text{ amu}}{39.10 \text{ amu/atom}} = 0.816 \text{ atom}$$

$$\text{For chlorine, } \frac{28.93 \text{ amu}}{35.45 \text{ amu/atom}} = 0.816 \text{ atom}$$

$$\text{For oxygen, } \frac{39.16 \text{ amu}}{16.00 \text{ amu/atom}} = 2.45 \text{ atoms}$$

Then we divide the smallest of the quotients into each of the others:

$$\frac{0.816 \text{ chlorine atom}}{0.816 \text{ potassium atom}} = 1 \text{ chlorine atom for each potassium atom}$$

$$\frac{2.45 \text{ oxygen atoms}}{0.816 \text{ potassium atom}} = 3 \text{ oxygen atoms for each potassium atom}$$

Therefore, the empirical formula is $KClO_3$.

We have now illustrated how empirical formulas are calculated from percentage composition. Every chemical compound must be subjected to chemical analysis before its empirical formula can be determined. As we have seen, if the formula is known it can be used to calculate the percentage composition; but the formula cannot be ascertained initially if the percentage composition has not been determined beforehand by analysis.

The sum of the atomic weights of the atoms in a formula can always be referred to correctly as the *formula weight* of the substance represented. Only if the formula is the molecular formula of the substance can the sum of the atomic weights be called the molecular weight of the substance.

Formulas are sometimes written with some of the symbols enclosed in parentheses. The subscript following a parenthesis applies to all of the enclosed symbols. For example, the formula $Ca(OH)_2$ implies the same composition for calcium hydroxide as if the formula were written CaO_2H_2. The advantage of writing the formula as $Ca(OH)_2$ is that it reminds us of the similarity between this compound and certain other compounds, such as sodium hydroxide, NaOH, and potassium hydroxide, KOH. In calculating the formula weight of calcium hydroxide, we must remember to find the sum of the atomic weights of the atoms enclosed within parentheses and then to double this sum before adding the atomic weight of calcium:

$$\begin{aligned} &40.1 \text{ amu} + 2(16.0 \text{ amu} + 1.0 \text{ amu}) \\ &\quad = 40.1 \text{ amu} + 2 \times 17.0 \text{ amu} \\ &\quad = 40.1 \text{ amu} + 34.0 \text{ amu} \\ &\quad = 74.1 \text{ amu} \end{aligned}$$

2-8 Determination of Molecular Formulas

The true or molecular formula gives the *actual* number of atoms of each element in one molecule of a compound. Once you determine the empirical formula, you can calculate the molecular formula if you know the molecular weight of the compound. Molecular weights can be ascertained by various experimental procedures. In Section 2-4 we discussed how the relative weights of the molecules of two gases may be compared by determining the weights of equal volumes of the gases at the same temperature and pressure. We will review this procedure in the following example.

EXAMPLE The density of O_2 gas at a certain temperature and pressure is 1.30 grams per liter, and the density of gas X at the same temperature and pressure is 1.95 grams per liter. What is the molecular weight of gas X?

The ratio of the two densities is

$$\frac{1.95\ \text{g/liter}}{1.30\ \text{g/liter}} = 1.50$$

so that the molecular weight of gas X is 1.50 times the molecular weight of O_2, which is 32.0 amu. Thus, the molecular weight of gas X is

$$1.50 \times 32.0\ \text{amu} = 48.0\ \text{amu}$$

Now that you know the molecular weight of gas X, suppose you want to calculate its molecular formula. You would first have to analyze the gas to determine its chemical composition and then find its empirical formula. Analysis shows that acetylene is 92.26 percent carbon and 7.74 percent hydrogen. Analysis of benzene shows that it also is 92.26 percent carbon and 7.74 percent hydrogen. The empirical formulas of acetylene and benzene, therefore, must be the same.

To find the relative number of carbon and hydrogen atoms, we divide each of these percentages by the atomic weight of the corresponding element:

$$\text{For carbon, } \frac{92.26\ \text{amu}}{12.01\ \text{amu/atom}} = 7.68\ \text{atoms}$$

$$\text{For hydrogen, } \frac{7.74\ \text{amu}}{1.008\ \text{amu/atom}} = 7.68\ \text{atoms}$$

$$\text{The ratio is } 7.68\!:\!7.68 = 1\!:\!1$$

Therefore, the empirical formula is CH (or HC).

Benzene and acetylene are quite dissimilar in many of their properties. For example, benzene is a liquid and acetylene is a gas; acetylene can be generated

Figure 2-6

Two compounds with the same empirical formula

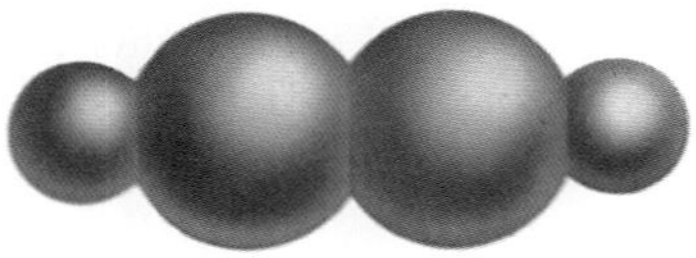

Acetylene
Molecular formula, C_2H_2
Empirical formula, CH

Benzene
Molecular formula, C_6H_6
Empirical formula, CH

by allowing water to come into contact with calcium carbide, but benzene cannot be prepared by such a simple procedure.

Since acetylene and benzene do have unlike properties, it should not be surprising that their molecular formulas are different although their empirical formulas are identical (Figure 2-6). Since their empirical formula is CH, each must have a molecular weight that is a multiple of 13 amu (the sum of the atomic weights of carbon and hydrogen); i.e., the possible molecular weights of these compounds are 13 amu, 26 amu, 39 amu, 52 amu, 65 amu, 78 amu, 91 amu, etc. The molecular weight of acetylene has been found to be 26 amu, and that of benzene, 78 amu. The molecular formulas of benzene and acetylene are, therefore, C_2H_2 and C_6H_6, respectively. Most of the experimental methods for determining molecular weights do not give precise results, but this does not really matter, for even an approximate value enables us to know the correct molecular weight if we already know the empirical formula. For example, if the molecular weight of benzene were experimentally found to be 77 amu, we would know that the correct molecular weight is 78 amu, a multiple of 13 amu (the sum of the atomic weights in the empirical formula).

2-9 Why Empirical Formulas Are Used

There are certain compounds whose molecular weights have not yet been determined. For them the only formulas we have are empirical formulas. There are certain other compounds whose molecular weights are variable, even though they have a definite percentage composition. For them, also, the only formulas we can write are empirical formulas.

The compounds known as *salts* do not consist of molecules at all; therefore, they do not have molecular formulas but only empirical formulas. The atoms of the constituent elements react in a definite ratio to form a salt, but the salt exists in an ionic rather than a molecular form. (Salts and ions are discussed more fully in later chapters.)

Many compounds have molecular formulas that are the same as their empirical formulas. For example, CO_2 is both the molecular and empirical formula for carbon dioxide. When you see a formula like H_2O_2 or C_2H_6 or $C_6H_{12}O_6$, you know that you are looking at a molecular formula.

2-10 Chemical Equations

In any chemical reaction the original substances, *reactants*, are converted into other substances, *products* (Figure 2-7). A chemical reaction can be described by a *chemical equation*, a statement employing chemical formulas instead of words. For example, the burning of carbon is a reaction between carbon (C) and oxygen gas (O_2), and the product of the reaction is carbon dioxide gas (CO_2). This reaction can be represented by the following chemical equation:

$$C + O_2 \rightarrow CO_2$$

The arrow is read "react to give." Since atoms are neither created nor destroyed in a chemical reaction, the same number of each kind of atom must be repre-

Figure 2-7

The rearrangement of atoms in a chemical reaction

The reactants, the O_2 molecule and the two H_2 molecules, contain the same atoms as the products, the two H_2O molecules. The balanced chemical equation shows this relationship.

sented on both sides of a chemical equation. In the equation above, there are two oxygen atoms and one carbon atom on each side of the arrow, but the arrangement of the atoms before and after the reaction differs.

Carbon dioxide is the product formed by burning carbon monoxide. Initially, the equation for this reaction is written simply with the formulas for the reactants and products:

$$CO + O_2 \rightarrow CO_2$$

This equation is called a *skeleton equation*. In this example the skeleton equation is *unbalanced* because the number of oxygen atoms on the left-hand side of the equation does not equal the number on the right-hand side. The equation must be *balanced*, i.e., it must be adjusted to show that the number of atoms of each element is the same on both sides.

If the formula for oxygen were O instead of O_2, the skeleton equation that we would write for the combustion of carbon monoxide would already be balanced. But one O_2 molecule has twice as many oxygen atoms as are needed to convert one molecule of CO into one molecule of CO_2. Thus, we have enough oxygen atoms in one O_2 molecule to convert two CO molecules into two CO_2 molecules. Therefore, we balance the equation as follows:

$$2CO + O_2 \rightarrow 2CO_2$$

In a balanced equation the numbers written in front of the formulas are called *coefficients*. The "2" before CO in the equation above means *two* CO molecules, and the "2" before CO_2 means *two* CO_2 molecules. The coefficient for O_2 is understood to be "1". The equation is now balanced, for if you count each kind of atom you will find two carbon atoms and four oxygen atoms on each side of the equation.

Balancing equations comes easily with practice, but knowing the formulas of reactants and products is frequently more difficult. The steps in writing a chemical equation are as follows:

1. Write the formula(s) of the reactant(s) with + signs between them:

$$CO + O_2$$

2. Place an arrow after the reactants, and to the right of the arrow write the formula(s) of the product(s):

$$CO + O_2 \rightarrow CO_2$$

3. Balance the equation by inserting the coefficients that are needed in front of the formulas, i.e., adjust the number of molecules so that the number of atoms of each element is the same on each side of the equation:

$$2CO + O_2 \rightarrow 2CO_2$$

   In making this adjustment, never change the formula of any substance, i.e., do not add or remove subscripts.

Eventually you will be able to predict the formulas of the products of many chemical reactions. However, you should first learn to balance equations in which the formulas of all starting materials and products are given.

New Terms

Atomic mass unit: one twelfth the mass of the carbon-12 atom.

Atomic weight: the weight of an average atom of an element on the scale $C^{12} = 12$ exactly. The atomic weight is the weighted average of the isotopic weights.

Avogadro's Principle: equal volumes of gases, under identical conditions of temperature and pressure, contain equal numbers of molecules.

Cannizzaro's Principle: the weight of one atom of an element is the smallest weight due to that element in one molecule of any of its compounds.

Chemical equation: a statement representing a chemical reaction, employing chemical formulas instead of words.

Chemical formula: any collection of symbols and subscripts used to represent a pure substance. Empirical formulas and molecular formulas are examples of chemical formulas.

Empirical formula (simplest formula): the combination of symbols and subscripts used to represent the relative number of atoms of the constituent elements in a compound; e.g., the empirical formulas for hydrogen peroxide and benzene are HO and CH, respectively (but their molecular formulas are H_2O_2 and C_6H_6).

Formula weight: the sum of the atomic weights of the atoms in a formula. If the formula is the molecular formula of a substance, the formula weight is also its molecular weight.

Ions: charged particles, especially atoms or molecules which have acquired electrical charges.

Isotopes: atoms of the same element which differ from one another in mass.

Isotopic weight: the weight of a particular isotope on the scale $C^{12} = 12$ exactly.

Mass number: the integer that most nearly approximates the mass of an isotope on the atomic weight scale.

Molecular formula: the combination of symbols and subscripts representing the actual number of atoms of each element in one molecule of a substance; e.g., the molecular formulas for elementary oxygen and for water are O_2 and H_2O, respectively.

Molecular weight: the weight of one molecule of a substance on the scale $C^{12} = 12$ exactly. The molecular weight is the sum of the atomic weights of the constituent elements, taking into account the number of atoms of each element in one molecule of the substance.

Symbol: a letter or a pair of letters used to represent an element or one atom of an element.

Exercises

2-1 How many atoms are in one molecule of table sugar, $C_{12}H_{22}O_{11}$? In one molecule of carbon dioxide? In 100 molecules of carbon monoxide?

2-2 What is the formula of the compound whose name is nitrogen dioxide? Dinitrogen oxide? Dinitrogen pentoxide?

2-3 A liter of sulfur dioxide weighs approximately the same as 4 liters of methane. How does the weight of a molecule of methane compare with the weight of a molecule of sulfur dioxide?

2-4 In 1 molecular weight of each of four compounds of element X, the weights of element X are found to be 28, 42, 70, and 84 amu. Is it correct to infer that the atomic weight of element X is 28 amu? Why or why not?

2-5 The least weight of nitrogen in a molecular weight of any known nitrogen compound is seven eighths as much as the least weight of oxygen in a molecular weight of any oxygen compound. If it has already been established that the atomic weight of oxygen is approximately 16 amu, what is the atomic weight of nitrogen?

2-6 The standards formerly used for atomic weights include $H = 1$, $O = 100$, $O = 16.000$. Calculate the atomic weight of chlorine on each of these scales, consulting Table 2-4 for the atomic weights of these three elements according to the present standard.

2-7 The weight of a liter of oxygen gas under standard conditions of temperature and pressure (0°C at sea level) is 1.43 grams. What is the weight of a liter of ozone under the same conditions? Of a liter of a gas whose molecular weight is 44?

2-8 Hydrogen is a diatomic gas, H_2, and helium is a monatomic gas. Consult the table on the front endpapers of this book for the atomic weights of both elements. How does the molecular weight of hydrogen compare with that of helium? How does the weight of a liter of hydrogen compare with the weight of a liter of helium?

2-9 Air is a mixture that contains approximately four molecules of nitrogen for each molecule of oxygen. A nitrogen molecule weighs seven eighths as much as an oxygen molecule. Find the average molecular weight of the gases in the air.

2-10 On the basis of what you learned in Exercise 2-9, is oxygen heavier or lighter than air? Is nitrogen heavier or lighter than air?

2-11 How would the density of a mixture of carbon monoxide and nitrogen compare with the density of either of the pure gases?

2-12 Calculate the percentage composition of uranium hexafluoride, UF_6; limestone, $CaCO_3$; methane, CH_4; ethyl alcohol, C_2H_6O; formaldehyde, CH_2O; carbon disulfide, CS_2; potassium chlorate, $KClO_3$.

2-13 Calculate the empirical formulas of the compounds with the following percentage compositions:

(a) calcium, 20.0%; bromine, 80.0%
(b) manganese, 56.4%; sulfur, 43.6%

(c) potassium, 26.58%; chromium, 35.35%; oxygen, 38.07%
(d) potassium, 28.7%; hydrogen, 1.5%; phosphorus, 22.8%; oxygen, 47.0%
(e) sodium, 32.85%; aluminum, 12.85%; fluorine, 54.30%

2-14 Compute the molecular formulas of the following compounds:

Empirical formula	*Molecular weight* (amu)
HO	34
CH_2O	30
CH_2O	150
CH_2	28
CH_2	84
CN	52

2-15 Balance the following equations (direct union reactions):

(a) $H_2 + O_2 \rightarrow H_2O$
(b) $H_2 + Cl_2 \rightarrow HCl$
(c) $H_2 + Br_2 \rightarrow HBr$
(d) $C + O_2 \rightarrow CO_2$
(e) $S + O_2 \rightarrow SO_2$
(f) $SO_2 + O_2 \rightarrow SO_3$
(g) $Na + O_2 \rightarrow Na_2O_2$
(h) $Cu + O_2 \rightarrow CuO$
(i) $Mg + O_2 \rightarrow MgO$
(j) $Hg + O_2 \rightarrow HgO$
(k) $Al + O_2 \rightarrow Al_2O_3$
(l) $P + O_2 \rightarrow P_4O_{10}$
(m) $N_2 + H_2 \rightarrow NH_3$
(n) $P_4O_6 + O_2 \rightarrow P_4O_{10}$
(o) $Sn + Cl_2 \rightarrow SnCl_4$
(p) $PCl_3 + Cl_2 \rightarrow PCl_5$
(q) $Al + Cl_2 \rightarrow AlCl_3$
(r) $CO + O_2 \rightarrow CO_2$
(s) $Ag + S \rightarrow Ag_2S$
(t) $NO + O_2 \rightarrow NO_2$

2-16 Balance the following equations (decomposition reactions):

(a) $H_2O \rightarrow H_2 + O_2$
(b) $HgO \rightarrow Hg + O_2$
(c) $KClO_3 \rightarrow KCl + O_2$
(d) $HClO \rightarrow HCl + O_2$
(e) $NH_3 \rightarrow N_2 + H_2$
(f) $NH_4OH \rightarrow NH_3 + H_2O$
(g) $NH_4NO_2 \rightarrow N_2 + H_2O$
(h) $NH_4NO_3 \rightarrow N_2O + H_2O$
(i) $NH_4Cl \rightarrow NH_3 + HCl$
(j) $Ag_2O \rightarrow Ag + O_2$
(k) $HNO_3 \rightarrow H_2O + NO_2 + O_2$
(l) $HNO_2 \rightarrow H_2O + NO + O_2$
(m) $Cu(NO_3)_2 \rightarrow CuO + NO_2 + O_2$

(n) $NaNO_3 \rightarrow NaNO_2 + O_2$
(o) $PCl_5 \rightarrow PCl_3 + Cl_2$
(p) $HClO_3 \rightarrow HClO_4 + ClO_2 + H_2O$
(q) $NaClO \rightarrow NaCl + NaClO_3$
(r) $N_2O_4 \rightarrow NO_2$
(s) $N_2O \rightarrow N_2 + O_2$
(t) $CaCO_3 \rightarrow CaO + CO_2$
(u) $Cu(OH)_2 \rightarrow CuO + H_2O$
(v) $H_2O_2 \rightarrow H_2O + O_2$

2-17 Balance the following equations (these reactions are called displacement reactions and are discussed in a later chapter):

(a) $Zn + HCl \rightarrow ZnCl_2 + H_2$
(b) $Al + HCl \rightarrow AlCl_3 + H_2$
(c) $Fe + CuSO_4 \rightarrow FeSO_4 + Cu$
(d) $Mg + H_2SO_4 \rightarrow MgSO_4 + H_2$
(e) $Na + HOH \rightarrow NaOH + H_2$
(f) $Al + H_2SO_4 \rightarrow Al_2(SO_4)_3 + H_2$
(g) $Zn + H_3PO_4 \rightarrow Zn_3(PO_4)_2 + H_2$
(h) $NaI + Cl_2 \rightarrow NaCl + I_2$
(i) $NaBr + Cl_2 \rightarrow NaCl + Br_2$
(j) $NaI + Br_2 \rightarrow NaBr + I_2$

2-18 Balance the following equations (these reactions are called neutralization reactions and are discussed in a later chapter):

(a) $NaOH + HCl \rightarrow NaCl + HOH$
(b) $KOH + H_2SO_4 \rightarrow K_2SO_4 + HOH$
(c) $Ca(OH)_2 + HNO_3 \rightarrow Ca(NO_3)_2 + HOH$
(d) $Ba(OH)_2 + H_2SO_4 \rightarrow BaSO_4 + HOH$
(e) $Al(OH)_3 + HCl \rightarrow AlCl_3 + HOH$
(f) $Al(OH)_3 + H_2SO_4 \rightarrow Al_2(SO_4)_3 + HOH$
(g) $NaOH + H_3PO_4 \rightarrow Na_3PO_4 + HOH$
(h) $Ca(OH)_2 + H_3PO_4 \rightarrow Ca_3(PO_4)_2 + HOH$
(i) $Al(OH)_3 + H_3PO_4 \rightarrow AlPO_4 + HOH$
(j) $Th(OH)_4 + H_3PO_4 \rightarrow Th_3(PO_4)_4 + HOH$

2-19 Balance the following equations (these reactions, unions of water with oxides, are discussed in a later chapter):

(a) $CO_2 + H_2O \rightarrow H_2CO_3$
(b) $P_4O_{10} + H_2O \rightarrow H_3PO_4$
(c) $P_4O_{10} + H_2O \rightarrow HPO_3$
(d) $P_4O_6 + H_2O \rightarrow H_3PO_3$
(e) $As_4O_6 + H_2O \rightarrow HAsO_2$
(f) $SO_2 + H_2O \rightarrow H_2SO_3$
(g) $SO_3 + H_2O \rightarrow H_2SO_4$
(h) $N_2O_3 + H_2O \rightarrow HNO_2$
(i) $N_2O_5 + H_2O \rightarrow HNO_3$
(j) $Na_2O + H_2O \rightarrow NaOH$
(k) $CaO + H_2O \rightarrow Ca(OH)_2$

(l) $La_2O_3 + H_2O \rightarrow La(OH)_3$
(m) $K_2O + H_2O \rightarrow KOH$
(n) $CrO_3 + H_2O \rightarrow H_2CrO_4$

References

Benson, S. W., *Chemical Calculations*, 2nd ed., Wiley, New York, 1963 (paperbound). An introduction to the use of mathematics in chemistry.

Labbauf, A., "The Carbon-12 Scale of Atomic Masses," *J. Chem. Educ.*, **39** (6), 282 (1962). Relationship of the former scale to the new scale, which became effective January 1, 1962.

Nier, A. O. C., "The Mass Spectrometer," *Sci. Am.*, March 1953, p. 68. An account of the modern version of the mass spectrograph and its many uses.

Walker, R. A., and Johnston, H., *The Language of Chemistry*, Prentice-Hall, Englewood Cliffs, N.J., 1967 (paperbound). Deals with reading, writing, and understanding chemical equations.

Wichers, E., "Why the Carbon-12 Scale?" *Chemistry* **37** (3), 12 (1964). An explanation by the chairman of the Commission on Atomic Weights of the International Union of Pure and Applied Chemistry.

Winderlich, R. (translation by R. E. Oesper), "History of the Chemical Sign Language," *J. Chem. Educ.*, **30** (2), 58 (1953). Old and new symbolism for representing elements and compounds.

3

The Periodic Table

3-1 Families of Elements

The chemical elements vary in their properties, as do the compounds they form. If there were no system for classifying the elements into a relatively small number of groups or families, the study of their individual properties would be exceedingly laborious, and the knowledge which one could retain of the chemistry of the elements would be an unsystematized mass of seemingly unrelated facts.

In the early development of chemistry, it became apparent that certain elements could be grouped together on the basis of their similarities. For example, the metals lithium, sodium, and potassium bear such close resemblance that anyone acquainted with their properties would easily note that they are much more like one another than any one of them is like any of the other elements. They are considered to be members of the same family of elements, the alkali metals. Again, chlorine, bromine, and iodine are nonmetals which resemble one another to a significant degree and can be considered members of the same family, the halogens.

3-2 Döbereiner's Triads

In 1829 Johann Döbereiner, a German chemist, noted an apparent relationship between the properties of several elements and their atomic weights. He arranged elements with similar properties into groups of three, or *triads*, in order of increasing atomic weights and found that the atomic weight of the second approximated the average of the atomic weights of the other two. Thus, the atomic weight of sodium (23) approximately equals the average of the atomic weights of the elements it so closely resembles, lithium (6.9) and potassium (39). Döbereiner's discovery encouraged chemists to look for further relationships among elements.

3-3 Newlands' Octaves

After the publication in 1858 of Cannizzaro's important paper (see Chapter 2), reliable atomic weights became available, so attempts to correlate them with other properties of elements were more successful. In 1865 John Newlands, an English chemist, arranged the known elements according to increasing atomic weight and found that among those of lowest atomic weights every element strongly resembles the eighth element beyond it. The following is an example of his arrangement:

Li	Be	B	C	N	O	F
6.9	9.0	10.8	12.0	14.0	16.0	19.0
Na	Mg	Al	Si	P	S	Cl
23.0	24.3	27.0	28.1	31.0	32.1	35.5

Although the properties of each element in a row differ decidedly from those of the other elements in the same row, there is an unmistakable gradation in properties as the atomic weight increases within the row. But the most remarkable feature of the arrangement is revealed in this set of facts:

1. Sodium bears a strong resemblance to lithium in its physical and chemical properties. Both are highly reactive metals of low density and form compounds with similar formulas: LiOH, NaOH; LiCl, NaCl; etc.
2. Magnesium and beryllium are similar. They have approximately the same density and form compounds with analogous formulas: $Be(OH)_2$, $Mg(OH)_2$; $BeCl_2$, $MgCl_2$; etc.
3. Aluminum and boron form compounds with similar formulas ($AlCl_3$, BCl_3; etc.), even though the similarities of their properties are not as marked, aluminum being a metal and boron a nonmetal.
4. Carbon and silicon are both nonmetals and form many compounds having comparable formulas: CCl_4, $SiCl_4$; CH_4, SiH_4; CO_2, SiO_2; etc. Many of the properties of these compounds with similar formulas are quite different, however.
5. The nonmetals nitrogen and phosphorus form compounds with similar formulas and properties: HNO_3, HPO_3; NH_3, PH_3; etc.
6. The chemistry of sulfur resembles the chemistry of oxygen in many ways. Similarities exist between H_2S and H_2O, CS_2 and CO_2, etc.
7. Chlorine and fluorine are members of a family of nonmetals, the halogens, which resemble one another in many respects.

The limitation of Newlands' scheme is that this *law of octaves* does not apply much beyond the two rows illustrated.

3-4 Mendeleev and the Periodic Law

Dmitri Mendeleev, in Russia, and Julius Lothar Meyer, in Germany, were the first chemists to arrange all the then-known elements in a concise *periodic table.*

In 1869 Mendeleev published a table with the elements listed in horizontal rows and vertical columns, in order of increasing atomic weights.

The success of Mendeleev's table was due to the emphasis he placed on the repetition of both physical and chemical properties at regular intervals. He was so confident about the accuracy of the arrangement that he boldly insisted that (1) where the elements did not appear in the correct order according to their atomic weights, the accepted atomic weights were in error; and (2) some of the places in his table should be left vacant, to be occupied later by elements yet to be discovered.

Using his table, Mendeleev predicted the properties of several elements not known then. Later, when they were discovered, they fitted neatly into the periodic table, and their properties proved to be as predicted. Mendeleev's observation that properties of elements recur periodically throughout the sequence of elements is called the *Periodic Law.*

Although deserving much credit for pointing out independently the periodic variation of the properties of the elements, Meyer did not pursue the problem in the thoroughgoing manner of Mendeleev.

Mendeleev's work stimulated the search for additional elements and renewed investigation of the elements that were "out of order" in his table. In most cases errors were found in the accepted atomic weights, as Mendeleev had predicted. However, there are still several instances in which the increasing atomic-weight sequence is violated. For example, tellurium must be regarded as the fifty-second element and iodine as the fifty-third, even though the atomic weight of tellurium (127.60) is greater than that of iodine (126.90). The properties of iodine resemble those of chlorine and bromine so closely that iodine is properly situated beneath them in the table, and the properties of tellurium likewise require that it be placed beneath sulfur and selenium in the periodic table.

The reversal of tellurium and iodine in the periodic table can no longer be attributed to inaccurately determined atomic weights. We are forced to acknowledge a more fundamental characteristic of the atom than atomic weight. In Chapter 2 we mentioned the discovery of isotopes in the early years of the present century. It so happens that the heavier isotopes of tellurium and the lighter isotopes of iodine are most abundant, thus making the average atomic weight of the naturally occurring mixture of isotopes greater for tellurium than for iodine.

It is now known that nearly all elements have isotopes. The fact that isotopes exist and that the different isotopes of an element have chemical properties that are virtually identical means that the atomic weight of an element, being only an average of the isotopic weights, cannot be the important factor that Mendeleev thought it to be in determining the properties of an element. To understand the feature of atomic structure that determines the chemical behavior of the elements, we must consider the nature of the subatomic particles of which atoms are composed.

3-5 Electrons and X Rays

The first of the subatomic particles to be discovered was the *electron*. Knowledge of the electron began with the study of electrical discharges through a so-called cathode-ray tube (Figure 3-1). This device consists of a glass tube with wires sealed into both ends, each wire attached to a metal disc. Through a glass side-arm the air can be drawn out with a pump. When the wires are connected to a source of high-voltage electricity, the metal discs act as electrodes, between which there is a discharge of electricity. A greenish-yellow fluorescent glow appears on the glass wall of the tube. If an obstacle is placed between the cathode (negative terminal) and the glass wall opposite, a sharp shadow is cast on the wall. The rays responsible for the fluorescence are thus found to originate at the cathode; for this reason they are called *cathode rays*.

If a pinwheel is placed in the path of the cathode rays, it rotates, indicating that the rays consist of material particles. Their deflection toward the positive side of an electric field indicates that they possess a negative charge. By measuring the amount of their deflection in electric and magnetic fields of known strengths, Sir J. J. Thomson in 1897 determined the ratio of charge to mass for the cathode-ray particles. He found that the ratio is the same in every instance, even though different metals are used as cathode material. Thus, the negatively charged particles issuing from the cathode came to be regarded as a species of subatomic particle occurring in all atoms. The name *electron* was given to this elementary unit of negative electric charge.

Meanwhile, in late 1895, the German physicist Wilhelm Konrad Röntgen was studying the behavior of a cathode-ray tube in a darkened room when he happened to notice a greenish glow from a square of cardboard lying on a nearby bench. This cardboard had been coated with crystals of a fluorescent compound in connection with a totally different experiment. On this particular occasion Röntgen had covered the cathode-ray tube with black paper so that no light could enter or leave it. He concluded that the glow of the fluorescent crystals was caused by some kind of ray coming from the cathode-ray tube,

Figure 3-1

The cathode-ray tube

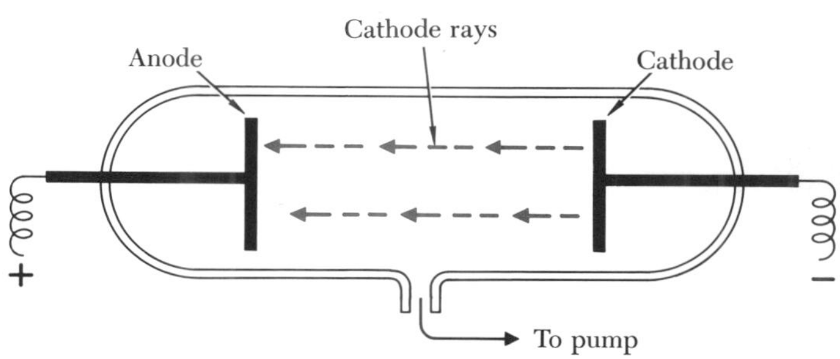

going through the black paper, and traveling to the bench where the crystals lay—even though, to the human eye, the room was utterly dark.

Röntgen found that the rays would expose photographic film and that they would pass through whatever objects he would hold between the tube and the film. His findings reached a climax when he placed a sheet of photographic film beneath his wife's hand and exposed it to the rays. When the film was developed it showed the bone structure of Frau Röntgen's hand.

Since Röntgen was unable to identify the rays coming from his tube, he called them *X rays*. Within a few weeks of their discovery, they were put to use for observing broken bones in human bodies. The role they have played in the scientific study of matter has been notable.

3-6 Charge of the Electron

During the first two decades of the present century, the American physicist Robert A. Millikan designed experiments to determine the charge of an electron. He measured the quantity of electric charge on small drops of oil. The oil droplets, sprayed from an atomizer, obtain their initial charges presumably by friction. With a droplet situated between two charged plates (Figure 3-2), the gravitational pull on it can be counterbalanced by the required electric "pull." The movement of the charged droplet is observed with a telescope. When the droplet is neither rising nor falling, i.e., when it is in equilibrium, the electric "pull" is equal to the gravitational pull. The charge on the droplet can be changed to a new value by allowing a beam of X rays to pass through the gas in which the drop is suspended. Millikan found that the charge is always an integral multiple of 1.602×10^{-19} coulomb. (The coulomb is the amount of electric charge conveyed by an electric current of one ampere in one second.) It is concluded, therefore, that the charge of one electron is 1.602×10^{-19} coulomb.

Figure 3-2

Oil-drop method for determining the charge of an electron

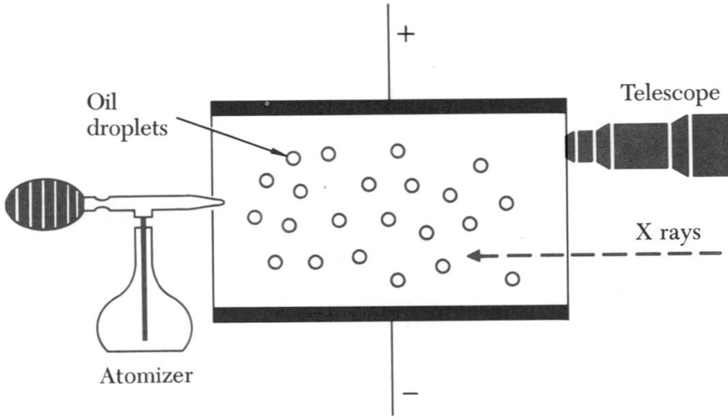

From the previously known value of the charge-mass ratio of the electron, it was now possible to calculate its mass. The result shows that it would require 1837 electrons to have the same mass as one ordinary hydrogen atom.

3-7 Rutherford's Nuclear Theory of the Atom

Normally, all matter is electrically neutral. After the discovery that all atoms contain one or more electrons, which are negatively charged, scientists naturally searched for a positively charged particle. Toward the end of the nineteenth century a number of physicists studied positively charged rays in discharge tubes containing hydrogen. Their investigations led to the discovery of the *proton*, which has exactly the same quantity of electrical charge as the electron, but positive instead of negative. (The existence of two kinds of electricity has been known since 1733. Benjamin Franklin proposed, in 1747, that the two varieties be referred to as *positive* and *negative*, and these terms are still in use.)

If you accept the idea that atoms contain (positive) protons and (negative) electrons, you will naturally wonder next about the arrangement of these particles inside the atom. The positive and negative particles might be uniformly mixed and occupy the space we are calling "the atom," or all of the particles of one kind might be grouped together in a nucleus with the particles of the other kind entirely outside the nucleus. In 1911 Lord Rutherford performed an experiment which led him to postulate that the protons are clustered together in a small nucleus at the center of the atom (Figure 3-3).

The first strong evidence that atoms are really quite complex was obtained in 1896, when Henri Becquerel, a French physicist, discovered radioactivity. The term *radioactivity* refers to the spontaneous disintegration of certain kinds of atoms, producing one or more of three kinds of radiations, designated as alpha (α), beta (β), and gamma (γ) rays.

Figure 3-3

The nuclear concept of the atom

The nucleus has mass exceeding 99.9 percent of the total mass of the atom, and charge equal to the atomic number of the atom. The spherical region outside the nucleus contains rapidly moving electrons. The diameter of the atom is 10,000 or more times the diameter of the nucleus.

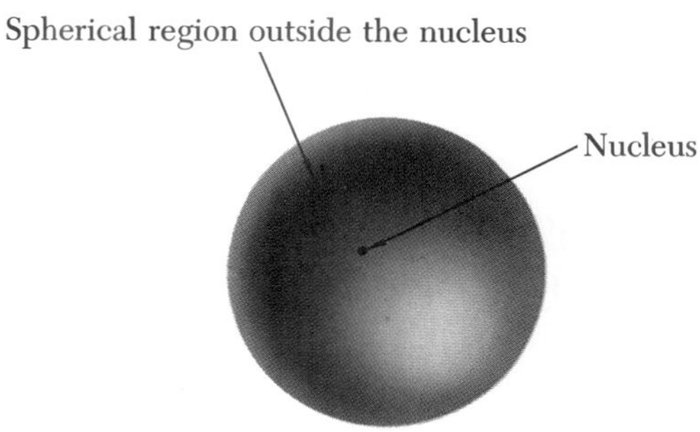

Figure 3-4

Rutherford's experiment with alpha particles and a thin gold foil

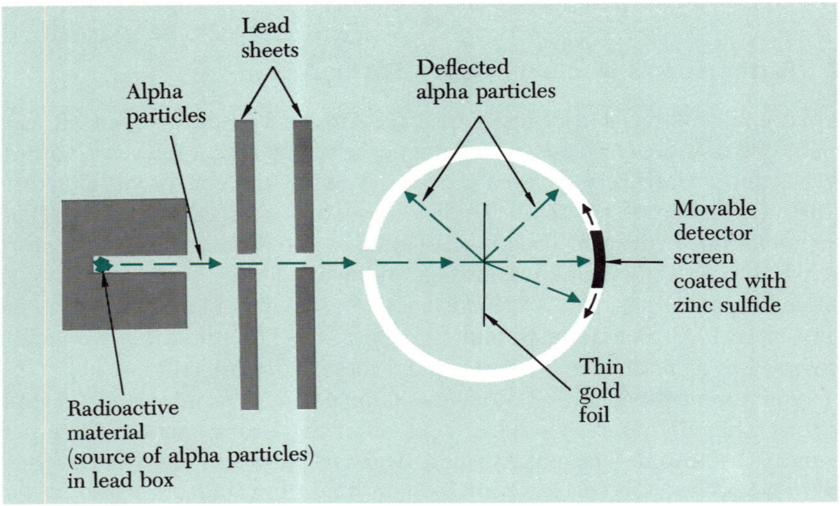

Rutherford's experiment consisted in directing the positively charged alpha particles emitted by radium (a radioactive element) through a thin gold foil and onto a movable detector screen coated with zinc sulfide (Figure 3-4). Each time an alpha particle hits the zinc sulfide screen, a momentary flash, or scintillation, can be observed. With the movable screen it is possible to observe the scintillations caused by alpha particles that have been deflected from their original direction. In Rutherford's experiment nearly all the alpha particles passed through the gold foil as if it were not there.

A gold atom is about 3 Ångström units (Å), i.e., 3×10^{-8} cm, in diameter. The gold foil that Rutherford used was about 4×10^{-5} cm thick, so the alpha particles had to travel through approximately a thousand gold atoms to reach the zinc sulfide screen—yet most of them reached the screen. From this Rutherford concluded that a gold atom must be largely "empty space." In a very few instances, an alpha particle would be deviated from its path; and occasionally (about once in 100,000 times) one of the particles would be deflected back at a large angle, indicating that it had experienced a very strong repulsive force. Rutherford concluded that there must be located within each gold atom a positively charged body capable of exerting an enormous repulsive force upon one of the positive alpha particles. This positively charged body at the center of the atom he termed the *nucleus*.

3-8 Nuclear Charge and Mass

From a mathematical analysis of the frequency with which the alpha particles are deflected, Rutherford, knowing that the alpha particle has twice the charge of a proton, calculated that the nucleus of the gold atom has the same charge as 79 protons. In the periodic table of the elements, gold is the seventy-ninth element, and it was referred to as the element with atomic number 79 even before Rutherford's research. The results of his investigation suggested that the *atomic number* of an element is something more than its serial number: it is the number of protons in the nucleus of an atom of that element.

In Rutherford's laboratory in England, the young physicist Henry Moseley, in 1913, compared the wavelengths of the characteristic X rays of a number of elements and thus arrived at values for the nuclear charges of the atoms. Moseley utilized the discovery that wavelengths of X rays can be determined by use of a crystal as a diffraction grating. He found that the wavelengths of corresponding lines in the spectra generally decrease as the atomic weights increase, but it was not possible to find any exact relationship between X-ray wavelengths and atomic weights. There was, however, an unfailing relationship between X-ray wavelengths and the ordinal position of the elements in the periodic system (Figure 3-5).

Moseley's discovery provided proof that there is in the atom a fundamental quantity which increases by regular steps as we pass from one element in the

Figure 3-5

Relationship between X-ray wavelengths and atomic numbers

Each element gives more than one line. Here the wavelengths of the two principal lines are shown to decrease regularly with the order of the elements in the periodic system.

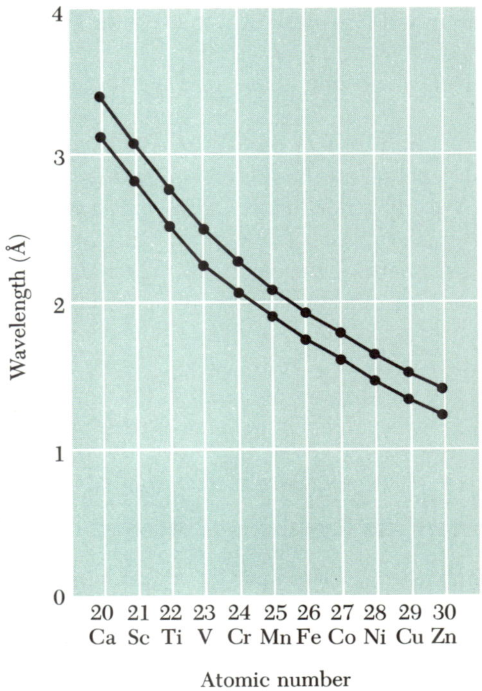

Table 3-1

Subatomic particles

Name	*Charge*[a]	*Mass*[b] (amu)
Proton	+1	1.0073
Neutron	0	1.0087
Electron	−1	0.000548

[a] The unit of charge employed here is the amount of charge of the electron itself.
[b] The mass is expressed on the atomic-weight scale, defined as C^{12} = 12 exactly.

periodic system to the next. This quantity is the *charge* of the nucleus, i.e., the number of protons in the nucleus.

The atom must also contain another kind of particle in addition to protons and electrons, because the atomic weight of an element—and the mass number of each isotope—exceeds the atomic number (except in the case of H^1). According to Rutherford's concept of the nuclear atom, the space outside the nucleus is occupied solely by electrons. The third type of subatomic particle, therefore, must be in the nucleus with the protons—and it must not have much effect on the chemical behavior of the atom. In 1932 this particle was discovered by an English physicist, James Chadwick, and was named the *neutron*. The mass of a proton is 1.0073 amu and the mass of a neutron is 1.0087 amu. The mass of an electron is only 0.000548 amu (Table 3-1).

The *mass number* of an isotope is the total number of protons and neutrons in the nucleus.

Rutherford's results may appear paradoxical in suggesting that the chemical behavior of an atom depends on the number of protons in a tiny nucleus at the center of the atom. We must not forget, however, that an atom contains electrons. In modern chemistry the behavior of an element is explained in terms of the number and arrangement of the electrons, which are outside the nucleus of the atom. The amount of charge carried by an electron is the same as that possessed by a proton; since they are oppositely charged, there are as many electrons as there are protons in a neutral atom. Thus, the atomic number of an atom is the same as the number of electrons in the neutral, isolated atom.

An *element* can now be defined as *a substance all of whose atoms have the same atomic number.*

3-9 Arrangement of the Electrons in Atoms

Conclusions regarding the arrangement of extranuclear electrons have been based on the periodic table of the elements and on studies of atomic spectra.

If an electric discharge is passed through a gas confined at low pressure in a sealed tube, the gas gives off light which is found to consist of certain wavelengths characteristic of its elements. If the gas consists of only one element, the spectral lines constitute an *atomic spectrum*. To explain the existence of line spectra and the energy atoms radiate, Niels Bohr, a Danish physicist, in 1913 pictured the atom as a tiny positive nucleus with electrons in motion around the nucleus, analogous to the movement of the planets around the sun. Although we no longer regard electron movement as orbital, the essential feature of Bohr's theory has been retained: the energy of an electron in an atom is *quantized*, i.e., the energy is limited to certain values. This means that an electron may not change its energy gradually but must shift all at once to a different *energy level*. Thus, the Bohr theory of the atom was attuned to the quantum theory of electromagnetic radiation, proposed in 1900 by the German physicist Max Planck. The different levels corresponding to large energy differences which are possible for electrons in atoms are designated by the numbers 1, 2, 3, 4, etc. (Figure 3-6). The number of the energy level, n, is also called the *principal quantum number*.

The maximum number of electrons that can exist in an energy level is $2n^2$. Thus, for the first or lowest energy level, $n = 1$, so the maximum number of electrons for this level is 2. The second energy level, $n = 2$, can contain $2(2^2)$ or 8 electrons. The maximum population for the third level is 18; for the fourth level, 32; for the fifth level, 50; etc. These energy levels are sometimes referred to as *shells*. The first energy level is frequently called the *K* shell; the second, the *L* shell; the third, the *M* shell; the fourth, the *N* shell; the fifth, the *O* shell; etc.

Suppose you were building up an atom. Each electron enters the lowest energy level available, so the single electron in a hydrogen atom (at. no. 1) would be in the *K* shell. The two electrons in a helium atom (at. no. 2) would completely fill the *K* shell. With lithium (at. no. 3), two electrons would fill the *K* shell, and the third would have to go to the next-lowest energy level, the *L* shell. Consider magnesium (at. no. 12): two electrons complete the *K* shell, the next eight electrons fill the second energy level (the *L* shell), and the last two electrons occupy the *M* shell. In Table 3-2, the first twenty elements are listed, with the number of electrons in each shell.

Figure 3-6

Energy levels of electrons in an atom

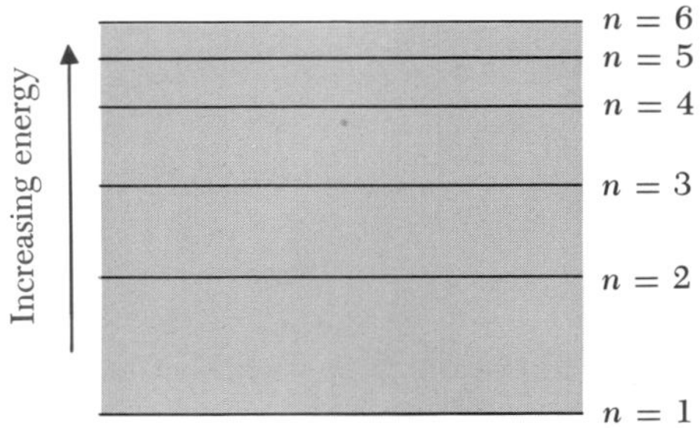

Table 3-2

Arrangement of the electrons in the atoms of the first twenty elements

Atomic number	Name	Symbol	Shell *K*	*L*	*M*	*N*
1	Hydrogen	H	1			
2	Helium	He	2			
3	Lithium	Li	2	1		
4	Beryllium	Be	2	2		
5	Boron	B	2	3		
6	Carbon	C	2	4		
7	Nitrogen	N	2	5		
8	Oxygen	O	2	6		
9	Fluorine	F	2	7		
10	Neon	Ne	2	8		
11	Sodium	Na	2	8	1	
12	Magnesium	Mg	2	8	2	
13	Aluminum	Al	2	8	3	
14	Silicon	Si	2	8	4	
15	Phosphorus	P	2	8	5	
16	Sulfur	S	2	8	6	
17	Chlorine	Cl	2	8	7	
18	Argon	Ar	2	8	8	
19	Potassium	K	2	8	8	1
20	Calcium	Ca	2	8	8	2

If you arrange these first twenty elements (including helium, neon, and argon, which were not known in Newlands' time) according to Newlands' octaves and beneath each symbol note the number of electrons in the *outermost* shell, a periodic variation is evident:

H							He
1							2
Li	Be	B	C	N	O	F	Ne
1	2	3	4	5	6	7	8
Na	Mg	Al	Si	P	S	Cl	Ar
1	2	3	4	5	6	7	8
K	Ca						
1	2						

There is only one electron in the outermost shell of lithium, sodium, and potassium—hence their resemblance. The presence of seven electrons in the outermost energy level of a chlorine atom causes it to behave much like a fluorine atom, which also has seven electrons in its outermost shell. The similarities among the other elements are accounted for in the same manner. This is the basis for the periodic law, which may be restated as follows: many of the physical and chemical properties of the elements vary periodically with their atomic numbers. Note that the outermost shells of helium and neon, members of the noble-gas family, are completely filled. Since the maximum population of the *M* shell is eighteen, you might expect argon—the next noble gas after neon—to have eighteen electrons in its outermost energy level, but argon has only eight electrons in the *M* shell. The explanation for this is considered in Chapter 4.

3-10 Periods and Groups in the Periodic Table

Several elements have been discovered since Mendeleev's periodic table was first constructed, but all of them have fitted into the table without difficulty. The imperfections that exist in any periodic table of the elements have motivated various people, including Mendeleev himself, to construct tables with new features. Julius Thomsen, a Danish chemist, in 1895 proposed a form of periodic table essentially like that shown in Figure 3-7.

In the table, the elements are arranged in order of increasing atomic number in seven rows and eighteen columns. The elements in any one row constitute a *series* or *period*; those in any one column constitute a *group*. Usually, the elements in a group consist of a *family* of elements of similar properties. For example, the Group IA elements (except hydrogen) are the alkali metals; the Group VIIA elements are the halogens; and those in Group 0 are the noble-gas elements. This periodic recurrence of similar properties is attributed to the presence of identical numbers of electrons in the outermost shells of the atoms. Restricting ourselves momentarily to the first twenty elements, we recall that the halogens, fluorine and chlorine, both have seven outer electrons; that the alkali metal atoms, lithium, sodium, and potassium, have one electron in their outermost shell; etc.

3-11 The First Five Periods

The *first period* of the table contains only the elements hydrogen, ${}_{1}H$, and helium, ${}_{2}He$. (The numbers written as subscripts preceding the symbols are the atomic numbers.) The *second period* contains eight elements, beginning with lithium, ${}_{3}Li$, and ending with neon, ${}_{10}Ne$. The *third period* also contains eight elements, beginning with sodium, ${}_{11}Na$, and ending with argon, ${}_{18}Ar$.

If the sequence of periodic similarities in electronic configuration and chemical properties that occurs among the first eighteen elements were to repeat itself once more, the next noble-gas element after argon would be element 26. Actually, element 26 is iron, and the next noble-gas element is krypton, element 36. Thus, the *fourth period* contains eighteen elements, or ten more than either the second or third period.

Figure 3-7

Periodic table of the elements

TRANSITION ELEMENTS (IIIB–VIII)

	IA	IIA	IIIB	IVB	VB	VIB	VIIB	VIII	
1	1 H Hydrogen								
2	3 Li Lithium	4 Be Beryllium							
3	11 Na Sodium	12 Mg Magnesium							
4	19 K Potassium	20 Ca Calcium	21 Sc Scandium	22 Ti Titanium	23 V Vanadium	24 Cr Chromium	25 Mn Manganese	26 Fe Iron	27 Co Cobalt
5	37 Rb Rubidium	38 Sr Strontium	39 Y Yttrium	40 Zr Zirconium	41 Nb Niobium	42 Mo Molybdenum	43 Tc Technetium	44 Ru Ruthenium	45 Rh Rhodium
6	55 Cs Cesium	56 Ba Barium	57 La* Lanthanum	72 Hf Hafnium	73 Ta Tantalum	74 W Tungsten	75 Re Rhenium	76 Os Osmium	77 Ir Iridium
7	87 Fr Francium	88 Ra Radium	89 Ac† Actinium						

* **Lanthanides**	58 Ce Cerium	59 Pr Praseodymium	60 Nd Neodymium	61 Pm Promethium	62 Sm Samarium
† **Actinides**	90 Th Thorium	91 Pa Protactinium	92 U Uranium	93 Np Neptunium	94 Pu Plutonium

The first two elements in the fourth period are members of Groups IA and IIA, respectively. Potassium, $_{19}$K, is an alkali metal, as is sodium, $_{11}$Na; and calcium, $_{20}$Ca, belongs to the same family as magnesium, $_{12}$Mg. Similarly, the last six elements in the fourth period are members of Groups IIIA, IVA, VA, VIA, VIIA, and 0, respectively. The ten elements beginning with scandium,

NONMETALS

	IB	IIB	IIIA	IVA	VA	VIA	VIIA	0
								2 He Helium
			5 B Boron	6 C Carbon	7 N Nitrogen	8 O Oxygen	9 F Fluorine	10 Ne Neon
			13 Al Aluminum	14 Si Silicon	15 P Phosphorus	16 S Sulfur	17 Cl Chlorine	18 Ar Argon
28 Ni ickel	29 Cu Copper	30 Zn Zinc	31 Ga Gallium	32 Ge Germanium	33 As Arsenic	34 Se Selenium	35 Br Bromine	36 Kr Krypton
46 Pd alladium	47 Ag Silver	48 Cd Cadmium	49 In Indium	50 Sn Tin	51 Sb Antimony	52 Te Tellurium	53 I Iodine	54 Xe Xenon
78 Pt latinum	79 Au Gold	80 Hg Mercury	81 Tl Thallium	82 Pb Lead	83 Bi Bismuth	84 Po Polonium	85 At Astatine	86 Rn Radon

63 Eu uropium	64 Gd Gadolinium	65 Tb Terbium	66 Dy Dysprosium	67 Ho Holmium	68 Er Erbium	69 Tm Thulium	70 Yb Ytterbium	71 Lu Lutetium
95 Am mericium	96 Cm Curium	97 Bk Berkelium	98 Cf Californium	99 Es Einsteinium	100 Fm Fermium	101 Md Mendelevium	102 No Nobelium	103 Lw Lawrencium

$_{21}$Sc, and ending with zinc, $_{30}$Zn, have no counterparts in the preceding period. They are known as *transition elements.*

The *fifth period* also contains eighteen elements; each element has a counterpart immediately above it in the fourth period. The first two elements in the period are members of Groups IA and IIA, and the last six belong to Groups

IIIA, IVA, VA, VIA, VIIA, and 0, respectively. The ten elements beginning with yttrium, $_{39}Y$, and ending with cadmium, $_{48}Cd$, are transition elements.

In the second and third periods there is a noticeable gradation in the properties of the elements. The first element of each period is a highly active metal; the last element is a noble gas; the next-to-last element is a highly active nonmetal; and the intervening five elements shift gradually in their properties from metallic to nonmetallic. The same is true of the A groups and noble gases in the other periods. Among the transition elements (the B groups), the differences between successive elements are much less marked than among the so-called *representative* elements (the A groups and the noble gases).

3-12 The Sixth and Seventh Periods

If the sequence of similarities that occurs among the elements of the fourth and fifth periods were to repeat itself once more, the next noble-gas element after xenon, $_{54}Xe$, would be element 72. Actually, the next noble-gas element is radon, element 86. Thus, the *sixth period* is an even longer period, consisting of thirty-two elements. As in the previous periods, the first two elements are members of Groups IA and IIA, and the last six are members of Groups IIIA, IVA, VA, VIA, VIIA, and 0, respectively. The twenty-four elements beginning with lanthanum, $_{57}La$, and ending with mercury, $_{80}Hg$, are transition elements. Of these twenty-four transition elements, only lanthanum and those beginning with hafnium, $_{72}Hf$, and ending with mercury have counterparts immediately above them in the fifth period. The fourteen elements beginning with cerium, $_{58}Ce$, and ending with lutetium, $_{71}Lu$, have no counterparts in the preceding periods. These elements are very much alike and are known collectively as the *lanthanides,* since they immediately follow lanthanum, which they closely resemble. The lanthanides are *inner transition elements.* We have followed the custom of showing the lanthanides separately because their insertion between Groups IIIB and IVB makes the table inconveniently long.

Since the third period is the same length as the second (eight elements) and the fifth period is the same length as the fourth (eighteen elements), we would expect the seventh period to contain thirty-two elements, like the sixth. In other words, the next noble-gas element after radon should be element 118. Since only the first 103 elements are now known, radon is the last of the known noble-gas elements. However, each of the elements beyond radon falls into place in the *seventh period* beneath its counterpart in the sixth period. The seventh period, like the preceding periods, begins with an alkali metal, francium, $_{87}Fr$; the second member of the period, radium, $_{88}Ra$, is the counterpart of barium. The fourteen elements beginning with thorium, $_{90}Th$, and ending with lawrencium, $_{103}Lw$, are counterparts of the lanthanides and are listed beneath them. These elements are known as the *actinides,* since they immediately follow actinium, $_{89}Ac$. They constitute the second inner transition series.

3-13 Relationships Among the Elements

Elements in the different groups exhibit important features:

1. Members of a chemical family are found in the same column. For example:
   a. The noble gases, also known as the rare gases (and formerly as the inert gases), are all found in Group 0.
   b. Group VIIA, immediately preceding the noble gases, contains the halogens, the most active nonmetals.
   c. The elements of Group IA, whose atomic numbers immediately follow those of the noble-gas elements, are the alkali metals, the most active metals.
2. The physical and chemical properties of the elements in a group show a more or less regular gradation with increasing atomic number. For example:
   a. The melting points of the alkali metals (on the Celsius scale): $_{3}Li$, 180°; $_{11}Na$, 98°; $_{19}K$, 63°; $_{37}Rb$, 39°; $_{55}Cs$, 29° (Figure 3-8).
   b. The radii of the atoms of the elements of Group VIA (in Ångström units): $_{8}O$, 0.74; $_{16}S$, 1.04; $_{34}Se$, 1.17; $_{52}Te$, 1.37 (Figure 3-9).
   c. The solubilities of the sulfides of the metals of Group IIB (grams per 100 ml of water at 18°C): ZnS, 0.00069; CdS, 0.00013; HgS, 0.000001.
3. In general, elements in the same group form compounds with similar formulas.
   a. Consider the formulas for the highest oxides (those containing the most oxygen) of the Group VIA elements: SO_3, SeO_3, TeO_3.
   b. Consider the formulas for the highest oxides of the Group VB elements: V_2O_5, Nb_2O_5, Ta_2O_5.
4. In the A groups, metallic properties are accentuated with increase in atomic number. (Metallic properties include thermal and electrical conductivity, malleability, ductility, luster, and the ability to form positive ions.)

Figure 3-8

Melting points of the alkali metals

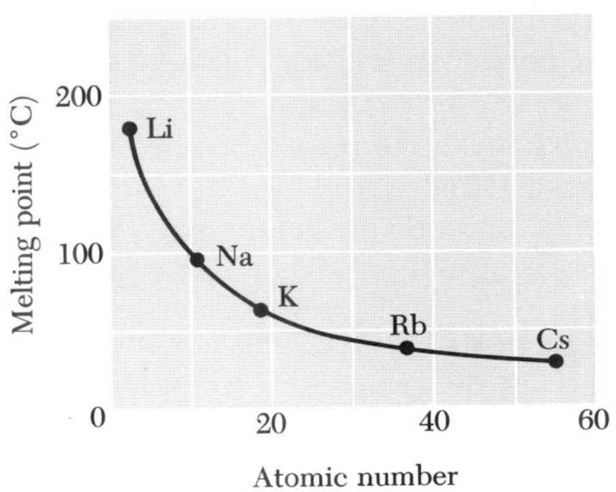

Figure 3-9

Radii of Group VIA atoms

5. In the A groups, nonmetallic properties diminish with increase in atomic number. For example, in Group VA the properties change from those of a gaseous element (nitrogen) to those of a lustrous solid (bismuth).
6. The most metallic elements are in the lower left region, and the most nonmetallic ones are in the upper right. All nonmetals are found in A groups. In Figure 3-7, a heavy line separates the metals from the nonmetals. However, most of the elements near this line are borderline elements, sometimes called *metalloids*, and are difficult to classify. For example, the commercially available forms of arsenic and selenium are sold under the labels "arsenic metal" and "selenium metal," but chemists generally agree that these elements should be classified as nonmetals (unless we take refuge under the classification of metalloids).
7. The first member of a group is usually more different from the other members of the group than the other members are different from each other. For example:
   a. Fluorine does not resemble chlorine as much as chlorine resembles bromine.
   b. Boron is considerably different from the other elements of Group IIIA; it is the only nonmetal in the group.
8. The members of an A group usually have a few points of resemblance with members of the corresponding B group. For example:
   a. Manganese in Group VIIB and chlorine in Group VIIA form acids with similar formulas and properties: permanganic acid, $HMnO_4$, and perchloric acid, $HClO_4$.
   b. The properties of titanium compounds in many instances resemble those of the corresponding silicon compounds.

Grouping similar elements into the same columns of the table simplifies the study of the elements, but the gradation of properties within each group must be remembered. In some cases certain properties of one element are quite different from those of other elements in the same group. For example, chlorine, bromine, and iodine form oxyacids (acids containing oxygen) with the formulas $HClO_3$, $HBrO_3$, and HIO_3, respectively, but fluorine does not form an oxyacid.

3-14 Prediction of "New" Elements

When Mendeleev constructed his first table, only sixty-two elements were known, so he left several blank spaces for the elements yet to be discovered. All these blanks have now been filled. Mendeleev had the insight and courage to predict the properties of the missing elements, and his predictions provided substantial help in the search for these elements. For example, the properties which he predicted in 1871 for element 32 (on the basis of its position immediately beneath silicon and to the left of arsenic) are much like the properties this element was found to have when it was discovered by the German chemist Clemens Winkler in 1886. Winkler named the element *germanium*.

None of the members of the noble-gas family were known when Mendeleev constructed his table, but when argon was discovered in the earth's atmosphere in 1894, the existence of an entire group of elements previously unknown was implied. A search for the other members of the group was initiated, and by 1898 all of the presently known noble gases had been discovered.

As recently as 1939 the spaces for elements 43, 61, 85, and 87 were blank, but since then all of these elements have been discovered. Mendeleev did not predict the existence of the transuranium elements (elements beyond uranium); he provided for only ninety-two elements. In connection with wartime research on atomic energy, elements 93 and 94 were obtained as products of nuclear reactions, and eleven transuranium elements have now been obtained by this procedure. Most of them have been obtained only in very small quantities and are too unstable to exist in nature. On the basis of periodic-table relationships, the properties of each of these "synthetic" elements have been rather successfully predicted prior to their discovery. Element 103 is lawrencium ($_{130}Lw$), which was obtained in 1961 at the Lawrence Radiation Laboratory of the University of California. Little is known about its properties, but lawrencium apparently completes the actinide series. It is expected that any elements beyond 103 will occupy the main portion of the periodic table, with element 104 situated beneath hafnium ($_{72}Hf$), etc. Already, the preparation of element 104 has been claimed by a group of Russian scientists and the name kurchatovium (symbol, Ku) assigned to it.

3-15 Value of the Table in Stimulating Research

Many investigations of the physical and chemical properties of the elements have been based on apparent discrepancies between the alleged properties and the properties expected on the basis of location in the periodic table. These investigations have sometimes resulted in the correction of serious errors.

Chemical research has frequently been based on familiarity with periodic-table relationships. For example, if a certain element is useful as a catalyst for a particular type of reaction, the catalytic action of other elements selected according to their position in the table certainly merits investigation.

The table is a useful guide in the determination of atomic weights. If an element does not form molecular compounds, application of Cannizzaro's Principle is impossible. Radium is such an element. When radium was discovered it was found to combine with chlorine in the ratio 113:35.5. If the formula for the compound is RaCl, the atomic weight of radium is 113; if the formula for the chloride is $RaCl_2$, the atomic weight of radium is 226. There is no way to determine the formula for the chloride as long as the atomic weight of radium is unknown; without the formula there is no way to know whether the atomic weight is 113 or a multiple of 113. However, in the periodic table there was a blank space which Mendeleev left directly beneath barium. The other known properties of radium indicate that it belongs in this space. With radium located in this position, an atomic weight of 226 becomes reasonable—an atomic weight of 113 would require that radium be situated near cadmium and indium, where there is no available space.

New Terms

Actinides: the fourteen elements following actinium (second inner transition series) in the periodic table.

Atomic nucleus: the very small region at the center of an atom in which all the protons and neutrons are located. The diameter of an atom is 10,000 or more times the diameter of the nucleus, but the nucleus contains more than 99.9 percent of the mass of the atom.

Atomic number: the number of units of positive charge possessed by the nucleus of an atom. The atomic number is the same as the number of protons in the nucleus, and it determines the ordinal position of the element in the periodic table.

Atomic spectrum: a pattern of sharply defined spectral lines corresponding to definite wavelengths caused by jumps of electrons between different energy levels.

Cathode rays: the invisible rays which move from the cathode (negative electrode) to the anode (positive electrode) in an evacuated tube when the electrodes are connected to a source of high voltage. Cathode rays consist of electrons.

Coulomb: the amount of electrical charge conveyed by an electric current of one ampere in one second.

Electron: a subatomic particle with a mass of 9.1×10^{-28} g (which is $\frac{1}{1837}$ of the mass of a hydrogen atom) and a negative charge of 1.602×10^{-19} coulomb. On the atomic-weight scale the mass of an electron is 0.000548. The charge of an electron is the unit usually employed to express the charges of subatomic particles and ions.

Element: a substance in which all the atoms have the same atomic number.

Energy level: one of a set of states of energy that may be assumed by an electron in an atom.

Group: the set of chemical elements appearing in any one column of the periodic table (Group IA, Group IIA, Group IIIB, etc.).

Inner transition elements: the elements in the sixth and seventh periods of the periodic table that have no counterparts in pervious periods. The lanthanides (elements 58–71) constitute the first inner transition series; the actinides (elements 90–103), the second.

Lanthanides: the fourteen elements following lanthanum (first inner transition series) in the periodic table.

Mass number: the integer which most nearly expresses the mass of an isotope on the atomic-weight scale. The mass number of an atom is the total number of protons and neutrons in its nucleus.

Neutron: a subatomic particle, devoid of electrical charge, with a mass of 1.0087 atomic mass units.

Period: the set of chemical elements appearing in any one row of the periodic table (first period, second period, etc.).

Periodic Law: if the elements are considered in the order of their atomic numbers, elements with similar properties recur periodically. In brief, most of the properties of the elements are periodic functions of the atomic numbers.

Periodic table: an arrangement of the chemical elements in rows and columns in the order of their atomic numbers, with similar elements in the same columns.

Principal quantum numbers: the numbers (1, 2, 3, etc.) used to designate the various principal energy levels which an electron may occupy in an atom.

Proton: a subatomic particle with a positive charge of 1.602×10^{-19} coulomb and a mass of 1.0073 atomic mass units.

Radioactivity: the spontaneous disintegration of certain kinds of atoms, producing one or more of three kinds of radiations: alpha, beta, and gamma rays.

Representative elements: the nontransition elements, i.e., the A groups and the noble gases.

Shell: a term commonly used in referring to the electrons of a given principal energy level, e.g., *K* shell, *L* shell, etc.

Transition elements: the B groups, i.e., the elements in the fourth and succeeding periods of the periodic table that have no counterparts in preceding periods.

X rays: the highly penetrating electromagnetic radiation emanating from the spot where cathode rays strike a metal. X rays are also known as Röntgen rays. Their wavelengths are expressed in Ångström (Å) units.

Exercises

3-1 Predict the missing properties of sodium and bromine:

Element	*Atomic radius* (Å)	*Melting point* (°C)	*Boiling point* (°C)
Lithium	1.23	180	1326
Sodium			
Potassium	2.03	63	757
Chlorine	0.99	−101	−34
Bromine			
Iodine	1.33	114	184

3-2 Given the formulas $LiNO_3$, $Mg(NO_3)_2$, AlF_3, CS_2, H_2O, $SiCl_4$, predict the formulas for the following compounds: potassium nitrate, beryllium nitrate, boron fluoride, silicon sulfide, hydrogen sulfide, carbon fluoride.

3-3 The atomic weight of argon (element 18) is 39.948, and the atomic weight of potassium (element 19) is 39.102. What can be said, on the basis of this information alone, regarding the relative abundance of the different isotopes of these two elements?

3-4 How can a piece of matter be electrically neutral if it consists of atoms that contain electrically charged particles?

3-5 State, define, describe, or explain:
(a) the Periodic Law
(b) periods, groups, families
(c) metals, nonmetals, metalloids
(d) atomic number and mass number
(e) transition elements and representative elements

3-6 Discuss the basis of Mendeleev's predictions.

3-7 Do you think that an element presently unknown may be discovered and found to have an atomic number that places it between magnesium and aluminum in the periodic table? Justify your answer.

References

Friend, J. N., *Man and the Chemical Elements*, 2nd ed., Scribner's, New York, 1961. Written from a historical and industrial viewpoint.

Hoffman, B., *The Strange Story of the Quantum*, 2nd ed., Dover, New York, 1959 (paperbound). An account of the growth of our atomic knowledge for the general reader.

Mazurs, E. G., "Ups and Downs of the Periodic Table," *Chemistry*, **39** (7), 6 (1966). Many of the different forms of periodic tables are shown.

Van Spronsen, J. W., "The Prehistory of the Periodic System of the Elements," *J. Chem. Educ.*, **36**, (11), 565 (1959). A consideration of the state of chemical theory at the time the Periodic Law was taking shape.

4

The Electronic Structure of Atoms

4-1 Electron Orbitals in Atoms

The periodicity that results when the elements are arranged in order of increasing atomic number is due to the variations in the arrangement of the electrons in the atoms. The tendency of certain properties to "repeat" at certain intervals is correlated with repeating electronic configurations in the outer shells of the atoms. One situation that especially requires an explanation is the variation in the lengths of the periods in the table: a period of 2 elements ($_1$H and $_2$He), followed by two periods of 8 elements ($_3$Li through $_{10}$Ne and $_{11}$Na through $_{18}$Ar), then two periods of 18 elements ($_{19}$K through $_{36}$Kr and $_{37}$Rb through $_{54}$Xe), a period of 32 elements ($_{55}$Cs through $_{86}$Rn), and finally an incomplete period.

The quantum theory of the atom has undergone extensive modification in the last half-century. The original Bohr model of the atom showed each electron revolving in a definite orbit outside the nucleus, but we no longer think that an electron behaves like a planet in a fixed orbit. The flat orbits of the Bohr model have been replaced by three-dimensional orbitals. An *orbital* is a region within an atom which can be occupied by no more than two electrons. The "shape" of an orbital is portrayed by drawing a figure in such a manner that there is, say, a 95 percent probability that the electron will be within the region represented by the figure.

When each electron is in the lowest energy level available to it (Section 3-9), the atom is in the *ground state*. If any electron in the atom is excited into a higher energy level, the atom is in an *excited state*. In accounting for the arrangement of the elements in the periodic table, our first concern is with the electronic arrangement in the ground states of atoms. In the ground states of the 103 known elements, electrons inhabit four types of orbital, referred to by the four letters *s*, *p*, *d*, and *f*. (The historical origin of these one-letter names for orbitals is in

Figure 4-1

Shapes of *s* and *p* orbitals

(a) *s* orbital. (b) *p* orbital.

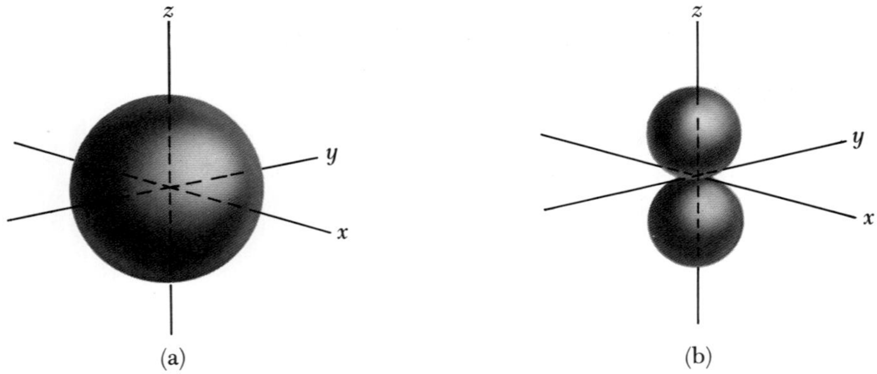

spectroscopy, where the letters are used to label different series of spectral lines designated "sharp," "principal," "diffuse," and "fundamental.")

Without attempting to describe the motion of an electron that inhabits an *s* orbital, we nevertheless can say that the shape of the orbital itself is spherically symmetrical (Figure 4-1a). This means only that there is a spherical region surrounding the nucleus which has a 95 percent probability of "containing" the electron.

A *p* orbital has approximately the shape of a dumbbell, with two spheres situated on opposite sides of the nucleus (Figure 4-1b). The cross section of a *p* orbital can be represented by a loop in the shape of a figure eight. An electron that inhabits a *p* orbital has equal probability of being in either half of it.

The shapes of *d* and *f* orbitals are considerably more complicated.

Any orbital, whatever its type, can accommodate *only two* electrons. Thus, the electronic population of a particular orbital is always one of three numbers: 0, 1, or 2. According to theory, all electrons are behaving as though they were spinning and this spinning charge sets up a tiny magnetic field. An electron can be thought to spin about its axis in a clockwise or counterclockwise direction; these two directions of spin give rise to magnetic moments in opposite directions. Two electrons can inhabit the same orbital only if they have opposite "spins."

4-2 Shells and Subshells of Electrons

When atoms of the known elements are in their ground states, all electrons are contained within seven shells designated by the principal quantum numbers (see Section 3-9). The average distance from the nucleus of electrons in the second shell is greater than for those in the first; for those in the third it is greater than for those in the second; etc. The energy of electrons in successive shells increases (i.e., the electrons are held less tightly) as the distance between the shell and the nucleus increases. However, all the electrons in a given energy level, with the exception of the *K* shell, do not have exactly the same energy. Actually, each level consists of several orbitals grouped together; i.e., each shell contains one or more *subshells*. Electrons in the same subshell have the same energy, but electrons in different subshells do not have the same energy (Figure 4-2).

Each shell contains one *s* orbital. The *K* shell contains only this one orbital, designated 1*s* (where 1 is the principal quantum number, *n*). Strictly speaking, the *K* shell is not divided into subshells; "the 1*s* orbital," "the 1*s* subshell," and "the *K* shell" are expressions that have the same meaning.

The second (*L*) shell contains two subshells: one consisting of one *s* orbital, designated 2*s*; and a second subshell consisting of three *p* orbitals, each designated 2*p*. Each *p* orbital is perpendicular to the other two so that, considered together, they are distributed about the nucleus in a symmetrical manner.

The third (*M*) shell contains three subshells consisting of 3*s*, 3*p*, and 3*d* orbitals. There are five *d* orbitals, which are spatially distributed so that the *d* subshell is symmetrical with respect to the nucleus.

Figure 4-2

Energy of electrons in different subshells

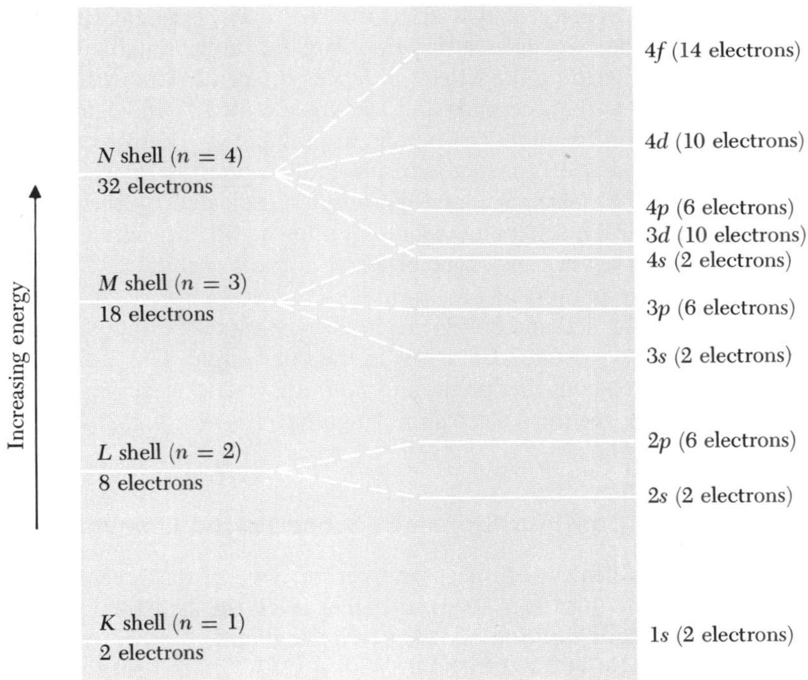

The fourth (N) shell contains four subshells designated $4s$, $4p$, $4d$, and $4f$. An f subshell consists of seven f orbitals symmetrically distributed about the nucleus.

The energy of an electron in a given orbital in a particular shell is lower (i.e., the electrons are held more tightly) than that of an electron in an orbital of the same type in a shell with a higher number, i.e., $1s < 2s < 3s < 4s < 5s < 6s < 7s$. Within a given shell the s electrons have the lowest energy and the f electrons have the highest energy, i.e., $s < p < d < f$.

Each shell contains as many subshells as are indicated by the principal quantum number: one in the first, two in the second, three in the third, etc. Each subshell consists of an odd number of orbitals: one s orbital; three p orbitals; five d orbitals; and seven f orbitals.

As we have noted, any one orbital, regardless of its type, has a maximum capacity of two electrons. Since the nth shell contains n^2 orbitals, the total electron capacity of the nth shell is $2n^2$, i.e., 2 for the first, 8 for the second, 18 for the third, 32 for the fourth, etc.

4-3 Electronic Configurations of First Period Elements

The arrangement of the elements in the periodic table is the result of the principles of atomic structure which we considered in Section 4-2. The first period of the table contains only two elements, hydrogen and helium. In the first (K) shell of any atom there is only one subshell, consisting of a single orbital, $1s$, with a maximum capacity of two electrons. We may represent the electronic configuration of the hydrogen atom by using the abbreviated notation $1s^1$. The number "1" preceding the letter s indicates the principal quantum number ($n = 1$); s represents the s orbital; and the superscript "1" indicates that there is one electron in this orbital. The helium atom has two electrons, both in the same orbital; the notation for its electronic configuration is $1s^2$.

In Table 4-1, each electronic configuration is indicated by the abbreviated notation and also by a schematic representation in which a circle is used for each orbital. The presence of a lone electron in the $1s$ orbital of the hydrogen atom is indicated by an upturned arrow in the $1s$ circle. Since the entering helium electron fills the $1s$ orbital and has its magnetic moment in the opposite direction, it is symbolized by a downturned arrow in the same circle. The K shell, i.e., the $1s$ subshell, is complete at this point, and no more electrons can be added to it. In the next element, the third electron must go into the next higher energy level, which is the $2s$ subshell.

4-4 Electronic Configurations of Second Period Elements

Lithium is the first element of the second period. Two of its three electrons are in the first shell; the third is in the next higher level, the $2s$ orbital. In abbreviated notation its configuration is $1s^22s^1$. In beryllium, the next element, the fourth electron is also in the $2s$ orbital. With a $1s^22s^2$ configuration, the beryl-

Table 4-1

Electronic configurations of the first eighteen elements

	Element	*Abbreviated notation*	*Schematic representation* 1*s*	2*s*	2*p*	3*s*	3*p*
First period	$_{1}$H	$1s^1$	(↑)				
	$_{2}$He	$1s^2$	(↑↓)				
Second period	$_{3}$Li	$1s^22s^1$	(↑↓)	(↑)	( )( )( )		
	$_{4}$Be	$1s^22s^2$	(↑↓)	(↑↓)	( )( )( )		
	$_{5}$B	$1s^22s^22p^1$	(↑↓)	(↑↓)	(↑)( )( )		
	$_{6}$C	$1s^22s^22p^2$	(↑↓)	(↑↓)	(↑)(↑)( )		
	$_{7}$N	$1s^22s^22p^3$	(↑↓)	(↑↓)	(↑)(↑)(↑)		
	$_{8}$O	$1s^22s^22p^4$	(↑↓)	(↑↓)	(↑↓)(↑)(↑)		
	$_{9}$F	$1s^22s^22p^5$	(↑↓)	(↑↓)	(↑↓)(↑↓)(↑)		
	$_{10}$Ne	$1s^22s^22p^6$	(↑↓)	(↑↓)	(↑↓)(↑↓)(↑↓)		
Third period	$_{11}$Na	$1s^22s^22p^63s^1$	(↑↓)	(↑↓)	(↑↓)(↑↓)(↑↓)	(↑)	( )( )( )
	$_{12}$Mg	$1s^22s^22p^63s^2$	(↑↓)	(↑↓)	(↑↓)(↑↓)(↑↓)	(↑↓)	( )( )( )
	$_{13}$Al	$1s^22s^22p^63s^23p^1$	(↑↓)	(↑↓)	(↑↓)(↑↓)(↑↓)	(↑↓)	(↑)( )( )
	$_{14}$Si	$1s^22s^22p^63s^23p^2$	(↑↓)	(↑↓)	(↑↓)(↑↓)(↑↓)	(↑↓)	(↑)(↑)( )
	$_{15}$P	$1s^22s^22p^63s^23p^3$	(↑↓)	(↑↓)	(↑↓)(↑↓)(↑↓)	(↑↓)	(↑)(↑)(↑)
	$_{16}$S	$1s^22s^22p^63s^23p^4$	(↑↓)	(↑↓)	(↑↓)(↑↓)(↑↓)	(↑↓)	(↑↓)(↑)(↑)
	$_{17}$Cl	$1s^22s^22p^63s^23p^5$	(↑↓)	(↑↓)	(↑↓)(↑↓)(↑↓)	(↑↓)	(↑↓)(↑↓)(↑)
	$_{18}$Ar	$1s^22s^22p^63s^23p^6$	(↑↓)	(↑↓)	(↑↓)(↑↓)(↑↓)	(↑↓)	(↑↓)(↑↓)(↑↓)

lium atom has a completed 2*s* orbital. The remaining elements of this series have electrons in the next higher level, the 2*p* orbitals. Remember there are three of these, each with a capacity for two electrons.

In the boron atom, one electron is located in a 2*p* orbital, so the configuration may be represented as $1s^22s^22p^1$. The carbon atom has two electrons in 2*p* orbitals, the configuration being $1s^22s^22p^2$. In Table 4-1 you can see that the sixth electron of the carbon atom is not in the same 2*p* orbital as the fifth electron. Likewise, the seventh electron of the nitrogen atom is in the third 2*p* orbital. According to the *principle of maximum multiplicity*, electrons enter each orbital of a given subshell singly as long as this possibility exists—only after each of the orbitals has one electron does a second electron enter any orbital of that subshell. This principle is sometimes referred to as *Hund's Rule*.

In the remaining elements of this period, the 2*p* orbitals are progressively filled, so there is a $2p^4$ configuration for oxygen, $2p^5$ for fluorine, and $2p^6$ for neon. In the neon atom both the *K* shell and the *L* shell are filled to their maximum capacity. Helium and neon are the only noble gases whose outer shells are filled to capacity.

4-5 An Alternative Notation for Electronic Configurations

In the helium atom the *K* shell contains as many electrons as it contains in an atom of any other element, however great the atomic number. To emphasize the fact that elements beyond helium have the helium configuration in the *K* shell, we may modify our notation by writing (He)$2s^1$ for lithium, (He)$2s^2$ for beryllium, (He)$2s^2 2p^1$ for boron, etc. A similar procedure may be used for elements following the other noble gases, e.g., (Ne)$3s^1$ for sodium.

4-6 Electronic Configurations of Third Period Elements

The first element of the third period, sodium, has one electron in the *M* shell, i.e., in the $3s$ orbital, and its electronic configuration may be represented as $1s^2 2s^2 2p^6 3s^1$ or (Ne)$3s^1$. The successive elements in this period have the $3s$ and $3p$ orbitals progressively filled in the same manner as the elements of the second period have the $2s$ and $2p$ orbitals progressively filled (see Table 4-1). This explains the positions of these elements in the periodic table: magnesium beneath beryllium, aluminum beneath boron, etc.

The third period ends with element 18, argon, which has the same outer-shell configuration, s^2p^6, as the neon atom. Argon, like neon, is a noble gas, even though the maximum capacity of the *M* shell is eighteen rather than eight. In fact, all the noble gases except helium have eight electrons, i.e., the s^2p^6 configuration, in the outer shell.

4-7 The Aufbau Principle

In considering electronic configurations we have assumed that each entering electron goes into the lowest energy level in which a vacancy exists. This is the *Aufbau Principle* (German: *Aufbau*, building up). With certain exceptions each successive element has the configuration of the element preceding it, plus one electron.

In the first three periods it is sufficient to use two general principles: (1) electrons in each successive shell (*K, L, M, N*, etc.) have increasing average energies; (2) electrons in orbitals of a given shell have increasing average energies in the order *s, p, d, f*. However, in the higher shells the orbitals overlap in energy. For instance, the $3d$ orbitals are of higher energy than the $4s$ (see Figure 4-2); the $4d$ orbitals are of higher energy than the $5s$; and the $4f$ orbitals are of higher energy than either the $6s$ or $5p$. In Figure 4-3 the energy associated with each type of orbital is represented diagrammatically.

Figure 4-3

Approximate relative energy of electrons in subshells

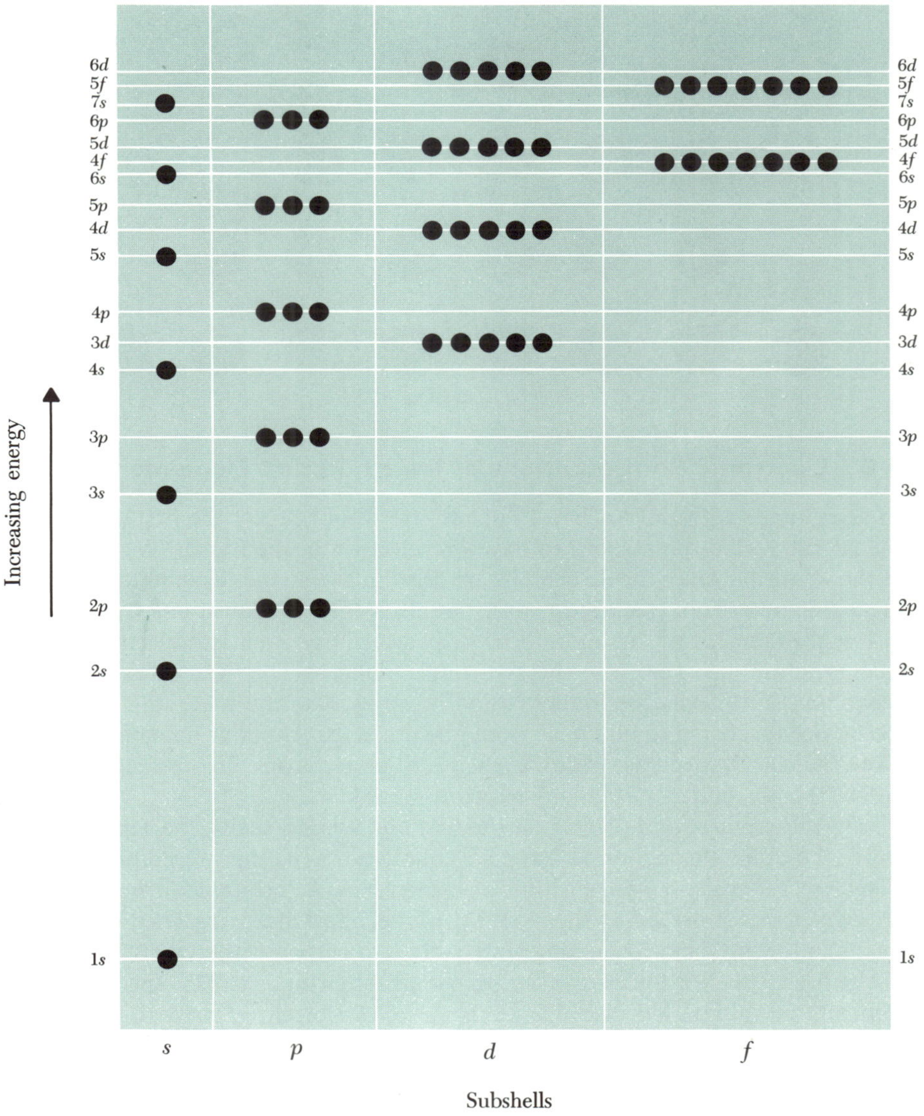

A mnemonic chart for the order of filling of orbitals is provided in Figure 4-4. The orbitals are presented in the same order as in Figure 4-3. In using this idealized chart we assume that each successive element has the same configuration as the element preceding it, with one electron added. Exceptions are pointed out.

Figure 4-4

Approximate order of entry of electrons into subshells

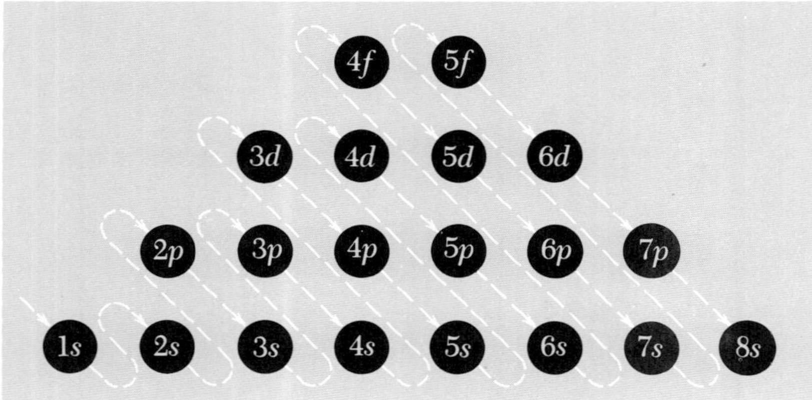

4-8 Electronic Configurations of Fourth Period Elements

In accordance with the Aufbau Principle, we would expect the first element of the fourth period, potassium, to have one electron in the fourth (N) shell, i.e., in the $4s$ orbital, since the $3p$ orbitals are filled in argon, the element immediately preceding potassium (see Figure 4-4). The electronic configuration of potassium is $1s^2 2s^2 2p^6 3s^2 3p^6 4s^1$ or, more briefly, (Ar)$4s^1$. The twentieth electron of the calcium atom also goes into the $4s$ orbital, filling it. The twenty-first element, scandium, shows why the fourth period is longer than the preceding periods—the $3d$ orbitals begin to fill up, as would be predicted from our mnemonic chart. Note that the $3d$ orbitals, with a capacity of ten electrons, are completely filled before the $4p$ orbitals obtain any electrons (Table 4-2).

Since there are no d orbitals in the first and second shells, the ten elements from scandium through zinc have no counterparts in the preceding periods. They are referred to as transition elements. The $4s$ orbital is being filled in the two elements preceding them, and the $4p$ orbitals are being filled in the elements following them (beginning with $_{31}$Ga and ending with $_{36}$Kr). The filling of the fourth shell is interrupted to permit the $3d$ orbitals to fill. The transition elements of this period, $_{21}$Sc through $_{30}$Zn, are often referred to as the $3d$ elements.

In Table 4-2 note the discrepancy between the configurations of chromium ($_{24}$Cr) and copper ($_{29}$Cu) and Figure 4-4. The $3d^5 4s^1$ arrangement gives to the chromium atom a *half-filled* $3d$ subshell, a more stable arrangement than the expected $3d^4 4s^2$. In the case of $_{29}$Cu the $3d^{10} 4s^1$ arrangement gives a filled $3d$ subshell and thereby a filled M shell. The stability of filled and half-filled p and d subshells is a principle that is frequently encountered.

In Figure 4-5 the electronic configurations of the elements are presented in abbreviated form. As you follow the order of entry of electrons with in-

Table 4-2

Electronic configurations of fourth period elements

Element	*Abbreviated notation*	*Schematic representation of outer orbitals*				
		3*s*	3*p*	3*d*	4*s*	4*p*
$_{19}$K	(Ar)$4s^1$	(↑↓)	(↑↓)(↑↓)(↑↓)	○○○○○	(↑)	○○○
$_{20}$Ca	(Ar)$4s^2$	(↑↓)	(↑↓)(↑↓)(↑↓)	○○○○○	(↑↓)	○○○
$_{21}$Sc	(Ar)$3d^14s^2$	(↑↓)	(↑↓)(↑↓)(↑↓)	(↑)○○○○	(↑↓)	○○○
$_{22}$Ti	(Ar)$3d^24s^2$	(↑↓)	(↑↓)(↑↓)(↑↓)	(↑)(↑)○○○	(↑↓)	○○○
$_{23}$V	(Ar)$3d^34s^2$	(↑↓)	(↑↓)(↑↓)(↑↓)	(↑)(↑)(↑)○○	(↑↓)	○○○
$_{24}$Cr	(Ar)$3d^54s^1$	(↑↓)	(↑↓)(↑↓)(↑↓)	(↑)(↑)(↑)(↑)(↑)	(↑)	○○○
$_{25}$Mn	(Ar)$3d^54s^2$	(↑↓)	(↑↓)(↑↓)(↑↓)	(↑)(↑)(↑)(↑)(↑)	(↑↓)	○○○
$_{26}$Fe	(Ar)$3d^64s^2$	(↑↓)	(↑↓)(↑↓)(↑↓)	(↑↓)(↑)(↑)(↑)(↑)	(↑↓)	○○○
$_{27}$Co	(Ar)$3d^74s^2$	(↑↓)	(↑↓)(↑↓)(↑↓)	(↑↓)(↑↓)(↑)(↑)(↑)	(↑↓)	○○○
$_{28}$Ni	(Ar)$3d^84s^2$	(↑↓)	(↑↓)(↑↓)(↑↓)	(↑↓)(↑↓)(↑↓)(↑)(↑)	(↑↓)	○○○
$_{29}$Cu	(Ar)$3d^{10}4s^1$	(↑↓)	(↑↓)(↑↓)(↑↓)	(↑↓)(↑↓)(↑↓)(↑↓)(↑↓)	(↑)	○○○
$_{30}$Zn	(Ar)$3d^{10}4s^2$	(↑↓)	(↑↓)(↑↓)(↑↓)	(↑↓)(↑↓)(↑↓)(↑↓)(↑↓)	(↑↓)	○○○
$_{31}$Ga	(Ar)$3d^{10}4s^24p^1$	(↑↓)	(↑↓)(↑↓)(↑↓)	(↑↓)(↑↓)(↑↓)(↑↓)(↑↓)	(↑↓)	(↑)○○
$_{32}$Ge	(Ar)$3d^{10}4s^24p^2$	(↑↓)	(↑↓)(↑↓)(↑↓)	(↑↓)(↑↓)(↑↓)(↑↓)(↑↓)	(↑↓)	(↑)(↑)○
$_{33}$As	(Ar)$3d^{10}4s^24p^3$	(↑↓)	(↑↓)(↑↓)(↑↓)	(↑↓)(↑↓)(↑↓)(↑↓)(↑↓)	(↑↓)	(↑)(↑)(↑)
$_{34}$Se	(Ar)$3d^{10}4s^24p^4$	(↑↓)	(↑↓)(↑↓)(↑↓)	(↑↓)(↑↓)(↑↓)(↑↓)(↑↓)	(↑↓)	(↑↓)(↑)(↑)
$_{35}$Br	(Ar)$3d^{10}4s^24p^5$	(↑↓)	(↑↓)(↑↓)(↑↓)	(↑↓)(↑↓)(↑↓)(↑↓)(↑↓)	(↑↓)	(↑↓)(↑↓)(↑)
$_{36}$Kr	(Ar)$3d^{10}4s^24p^6$	(↑↓)	(↑↓)(↑↓)(↑↓)	(↑↓)(↑↓)(↑↓)(↑↓)(↑↓)	(↑↓)	(↑↓)(↑↓)(↑↓)

creasing atomic number, i.e., across a period, you will notice that only those subshells are given in which electrons have just entered. Once a subshell is completed, it is omitted from the notation (except for the *s* subshells in the transition and inner-transition series). In Table 4-3 the electronic configurations of the elements are given in full.

4-9 Electronic Configurations of Fifth Period Elements

Figure 4-4 indicates that after $_{36}$Kr, (Ar)$3d^{10}4s^24p^6$, the next two electrons will enter the 5*s* orbital. Thus, $_{37}$Rb has a (Kr)$5s^1$ arrangement, and $_{38}$Sr has a (Kr)$5s^2$ configuration. Consequently, these two elements fall beneath $_{19}$K and $_{20}$Ca in the periodic table. The entire fifth period is essentially a repetition of the fourth. The elements $_{39}$Y through $_{48}$Cd comprise the second transition series; they are the 4*d* elements. After $_{48}$Cd the 5*p* orbitals fill, and the period ends with $_{54}$Xe, another noble-gas element.

A comparison of the configurations listed in Table 4-3 with those expected on the basis of Figure 4-4 reveals that there is one more electron in the 4*d*

Figure 4-5

Table showing periodicity in electron configurations of the elements

1 H $1s^1$								
3 Li $2s^1$	4 Be $2s^2$							
11 Na $3s^1$	12 Mg $3s^2$							
19 K $4s^1$	20 Ca $4s^2$	21 Sc $3d^14s^2$	22 Ti $3d^24s^2$	23 V $3d^34s^2$	24 Cr $3d^54s^1$	25 Mn $3d^54s^2$	26 Fe $3d^64s^2$	27 Co $3d^74s^2$
37 Rb $5s^1$	38 Sr $5s^2$	39 Y $4d^15s^2$	40 Zr $4d^25s^2$	41 Nb $4d^45s^1$	42 Mo $4d^55s^1$	43 Tc $4d^65s^1$	44 Ru $4d^75s^1$	45 Rh $4d^85s^1$
55 Cs $6s^1$	56 Ba $6s^2$	57 La* $5d^16s^2$	72 Hf $5d^26s^2$	73 Ta $5d^36s^2$	74 W $5d^46s^2$	75 Re $5d^56s^2$	76 Os $5d^66s^2$	77 Ir $5d^76s^2$
87 Fr $7s^1$	88 Ra $7s^2$	89 Ac† $6d^17s^2$						
			* **Lanthanides**	58 Ce $4f^15d^16s^2$	59 Pr $4f^36s^2$	60 Nd $4f^46s^2$	61 Pm $4f^56s^2$	62 Sm $4f^66s^2$
			† **Actinides**	90 Th $6d^27s^2$	91 Pa $5f^26d^17s^2$	92 U $5f^36d^17s^2$	93 Np $5f^46d^17s^2$	94 Pu $5f^67s^2$

subshell (and one less in the 5*s*) than expected in the series $_{41}$Nb through $_{47}$Ag. The difference between the energies of 4*d* and 5*s* orbitals is not great (see Figure 4-3). We have here a delicate balance between the tendency of all competing orbitals to become filled or half-filled as soon as possible and the tendency of an entire inner shell to become completely filled as soon as competing tendencies permit.

									2 He $1s^2$
			5 B $2p^1$	6 C $2p^2$	7 N $2p^3$	8 O $2p^4$	9 F $2p^5$	10 Ne $2p^6$
			13 Al $3p^1$	14 Si $3p^2$	15 P $3p^3$	16 S $3p^4$	17 Cl $3p^5$	18 Ar $3p^6$
28 Ni $3d^8 4s^2$	29 Cu $3d^{10}4s^1$	30 Zn $3d^{10}4s^2$	31 Ga $4p^1$	32 Ge $4p^2$	33 As $4p^3$	34 Se $4p^4$	35 Br $4p^5$	36 Kr $4p^6$
46 Pd $4d^{10}$	47 Ag $4d^{10}5s^1$	48 Cd $4d^{10}5s^2$	49 In $5p^1$	50 Sn $5p^2$	51 Sb $5p^3$	52 Te $5p^4$	53 I $5p^5$	54 Xe $5p^6$
78 Pt $5d^9 6s^1$	79 Au $5d^{10}6s^1$	80 Hg $5d^{10}6s^2$	81 Tl $6p^1$	82 Pb $6p^2$	83 Bi $6p^3$	84 Po $6p^4$	85 At $6p^5$	86 Rn $6p^6$

63 Eu $4f^7 6s^2$	64 Gd $4f^7 5d^1 6s^2$	65 Tb $4f^9 6s^2$	66 Dy $4f^{10}6s^2$	67 Ho $4f^{11}6s^2$	68 Er $4f^{12}6s^2$	69 Tm $4f^{13}6s^2$	70 Yb $4f^{14}6s^2$	71 Lu $4f^{14}5d^1 6s^2$
95 Am $5f^7 7s^2$	96 Cm $5f^7 6d^1 7s^2$	97 Bk $5f^9 7s^2$	98 Cf $5f^{10}7s^2$	99 Es $5f^{11}7s^2$	100 Fm $5f^{12}7s^2$	101 Md $5f^{13}7s^2$	102 No $5f^{14}7s^2$	103 Lw $5f^{14}6d^1 7s^2$

4-10 Electronic Configurations in the Sixth Period

In the sixth period the entering electrons of the first two elements, $_{55}$Cs and $_{56}$Ba, appear in the 6*s* orbital, thus following the pattern of the preceding series. In this period, however, a complication develops which has no parallel in any preceding period. According to Figure 4-4, the filling of the 6*s* orbital

Table 4-3

Electronic configurations of the elements

Atomic number	Element	1(K)	2(L)		3(M)			4(N)				5(O)				6(P)			7(Q)
		s	*s*	*p*	*s*	*p*	*d*	*s*	*p*	*d*	*f*	*s*	*p*	*d*	*f*	*s*	*p*	*d*	*s*
1	H	1																	
2	He	2																	
3	Li	2	1																
4	Be	2	2																
5	B	2	2	1															
6	C	2	2	2															
7	N	2	2	3															
8	O	2	2	4															
9	F	2	2	5															
10	Ne	2	2	6															
11	Na	2	2	6	1														
12	Mg	2	2	6	2														
13	Al	2	2	6	2	1													
14	Si	2	2	6	2	2													
15	P	2	2	6	2	3													
16	S	2	2	6	2	4													
17	Cl	2	2	6	2	5													
18	Ar	2	2	6	2	6													
19	K	2	2	6	2	6		1											
20	Ca	2	2	6	2	6		2											
21	Sc	2	2	6	2	6	1	2											
22	Ti	2	2	6	2	6	2	2											
23	V	2	2	6	2	6	3	2											
24	Cr	2	2	6	2	6	5	1											
25	Mn	2	2	6	2	6	5	2											
26	Fe	2	2	6	2	6	6	2											
27	Co	2	2	6	2	6	7	2											
28	Ni	2	2	6	2	6	8	2											
29	Cu	2	2	6	2	6	10	1											
30	Zn	2	2	6	2	6	10	2											
31	Ga	2	2	6	2	6	10	2	1										
32	Ge	2	2	6	2	6	10	2	2										
33	As	2	2	6	2	6	10	2	3										
34	Se	2	2	6	2	6	10	2	4										
35	Br	2	2	6	2	6	10	2	5										
36	Kr	2	2	6	2	6	10	2	6										
37	Rb	2	2	6	2	6	10	2	6			1							
38	Sr	2	2	6	2	6	10	2	6			2							
39	Y	2	2	6	2	6	10	2	6	1		2							
40	Zr	2	2	6	2	6	10	2	6	2		2							
41	Nb	2	2	6	2	6	10	2	6	4		1							
42	Mo	2	2	6	2	6	10	2	6	5		1							
43	Tc	2	2	6	2	6	10	2	6	6		1							
44	Ru	2	2	6	2	6	10	2	6	7		1							
45	Rh	2	2	6	2	6	10	2	6	8		1							
46	Pd	2	2	6	2	6	10	2	6	10									
47	Ag	2	2	6	2	6	10	2	6	10		1							
48	Cd	2	2	6	2	6	10	2	6	10		2							
49	In	2	2	6	2	6	10	2	6	10		2	1						
50	Sn	2	2	6	2	6	10	2	6	10		2	2						
51	Sb	2	2	6	2	6	10	2	6	10		2	3						
52	Te	2	2	6	2	6	10	2	6	10		2	4						

Atomic number	Element	1(*K*)	2(*L*)		3(*M*)			4(*N*)				5(*O*)				6(*P*)			7(*Q*)
		s	*s*	*p*	*s*	*p*	*d*	*s*	*p*	*d*	*f*	*s*	*p*	*d*	*f*	*s*	*p*	*d*	*s*
53	I	2	2	6	2	6	10	2	6	10		2	5						
54	Xe	2	2	6	2	6	10	2	6	10		2	6						
55	Cs	2	2	6	2	6	10	2	6	10		2	6			1			
56	Ba	2	2	6	2	6	10	2	6	10		2	6			2			
57	La	2	2	6	2	6	10	2	6	10		2	6	1		2			
58	Ce	2	2	6	2	6	10	2	6	10	1	2	6	1		2			
59	Pr	2	2	6	2	6	10	2	6	10	3	2	6			2			
60	Nd	2	2	6	2	6	10	2	6	10	4	2	6			2			
61	Pm	2	2	6	2	6	10	2	6	10	5	2	6			2			
62	Sm	2	2	6	2	6	10	2	6	10	6	2	6			2			
63	Eu	2	2	6	2	6	10	2	6	10	7	2	6			2			
64	Gd	2	2	6	2	6	10	2	6	10	7	2	6	1		2			
65	Tb	2	2	6	2	6	10	2	6	10	9	2	6			2			
66	Dy	2	2	6	2	6	10	2	6	10	10	2	6			2			
67	Ho	2	2	6	2	6	10	2	6	10	11	2	6			2			
68	Er	2	2	6	2	6	10	2	6	10	12	2	6			2			
69	Tm	2	2	6	2	6	10	2	6	10	13	2	6			2			
70	Yb	2	2	6	2	6	10	2	6	10	14	2	6			2			
71	Lu	2	2	6	2	6	10	2	6	10	14	2	6	1		2			
72	Hf	2	2	6	2	6	10	2	6	10	14	2	6	2		2			
73	Ta	2	2	6	2	6	10	2	6	10	14	2	6	3		2			
74	W	2	2	6	2	6	10	2	6	10	14	2	6	4		2			
75	Re	2	2	6	2	6	10	2	6	10	14	2	6	5		2			
76	Os	2	2	6	2	6	10	2	6	10	14	2	6	6		2			
77	Ir	2	2	6	2	6	10	2	6	10	14	2	6	7		2			
78	Pt	2	2	6	2	6	10	2	6	10	14	2	6	9		1			
79	Au	2	2	6	2	6	10	2	6	10	14	2	6	10		1			
80	Hg	2	2	6	2	6	10	2	6	10	14	2	6	10		2			
81	Tl	2	2	6	2	6	10	2	6	10	14	2	6	10		2	1		
82	Pb	2	2	6	2	6	10	2	6	10	14	2	6	10		2	2		
83	Bi	2	2	6	2	6	10	2	6	10	14	2	6	10		2	3		
84	Po	2	2	6	2	6	10	2	6	10	14	2	6	10		2	4		
85	At	2	2	6	2	6	10	2	6	10	14	2	6	10		2	5		
86	Rn	2	2	6	2	6	10	2	6	10	14	2	6	10		2	6		
87	Fr	2	2	6	2	6	10	2	6	10	14	2	6	10		2	6		1
88	Ra	2	2	6	2	6	10	2	6	10	14	2	6	10		2	6		2
89	Ac	2	2	6	2	6	10	2	6	10	14	2	6	10		2	6	1	2
90	Th	2	2	6	2	6	10	2	6	10	14	2	6	10		2	6	2	2
91	Pa	2	2	6	2	6	10	2	6	10	14	2	6	10	2	2	6	1	2
92	U	2	2	6	2	6	10	2	6	10	14	2	6	10	3	2	6	1	2
93	Np	2	2	6	2	6	10	2	6	10	14	2	6	10	4	2	6	1	2
94	Pu	2	2	6	2	6	10	2	6	10	14	2	6	10	6	2	6		2
95	Am	2	2	6	2	6	10	2	6	10	14	2	6	10	7	2	6		2
96	Cm	2	2	6	2	6	10	2	6	10	14	2	6	10	7	2	6	1	2
97	Bk	2	2	6	2	6	10	2	6	10	14	2	6	10	9	2	6		2
98	Cf	2	2	6	2	6	10	2	6	10	14	2	6	10	10	2	6		2
99	Es	2	2	6	2	6	10	2	6	10	14	2	6	10	11	2	6		2
100	Fm	2	2	6	2	6	10	2	6	10	14	2	6	10	12	2	6		2
101	Md	2	2	6	2	6	10	2	6	10	14	2	6	10	13	2	6		2
102	No	2	2	6	2	6	10	2	6	10	14	2	6	10	14	2	6		2
103	Lw	2	2	6	2	6	10	2	6	10	14	2	6	10	14	2	6	1	2

is followed by the occupancy of the 4*f*, 5*d*, and 6*p* orbitals, respectively. The energy levels 4*f* and 5*d* are so close (see Figure 4-3) that the chart is not completely reliable at this point, for in $_{57}$La, the fifty-seventh electron enters a vacant 5*d* orbital instead of a vacant 4*f* orbital. Thereafter, however, the 4*f* orbitals fill with almost uninterrupted regularity, followed by the filling of the 5*d* and then the 6*p* orbitals (see Table 4-3). Thus, elements 57–80 constitute the third transition series, and elements 58–71, the lanthanides, are the first inner transition series. The lanthanides are sometimes referred to as the 4*f* elements. The other elements of the third transition series (lanthanum and elements 72–80) are the 5*d* elements.

4-11 Electronic Configurations in the Seventh Period

The seventh period begins with the filling of the 7*s* orbital, as expected. In the remaining elements of this period electrons enter the 6*d* and 5*f* orbitals in much the same manner as they entered the 5*d* and 4*f* orbitals. The difference between the energies of the 6*d* and 5*f* orbitals is so slight that there is some uncertainty as to the actual configurations of several of the actinides. In general, the actinides (the second inner transition series) are the 5*f* elements and are thus the analogues of the 4*f* elements, the lanthanides.

On the basis of Figure 4-5 (see also Table 4-3), we would expect elements 104–112, if ever obtained, to complete a fourth transition series, i.e., to be the 6*d* elements. We may also speculate that elements 113–118 would be 7*p* elements and that the 8*s* subshell included for good measure in Figure 4-4 would not be occupied until we reached elements 119 and 120.

New Terms

Aufbau Principle: the principle that each element has the electron configuration of the preceding element plus one electron that enters the lowest-energy orbital in which a vacancy exists.

Excited state: any state of an atom in which at least one electron is in an energy level higher than the lowest possible.

Ground state: the lowest energy state of an atom, i.e., the state in which the electrons are in the lowest-energy orbitals available.

Orbital: a region within an atom which can be occupied by electrons. Any orbital can accommodate only two electrons.

Principle of maximum multiplicity (Hund's Rule): the principle according to which no orbital of a given subshell contains two electrons unless every other orbital of that subshell contains one electron.

Subshell: a set of like orbitals within the same shell; e.g., the 2*p* subshell consists of three *p* orbitals in the second (*L*) shell.

Exercises

4-1 State, define, describe, or explain:

(a) orbitals, subshells, shells, principal quantum numbers
(b) ground state and excited state of an atom

4-2 What is the maximum number of electrons in the third shell? The fourth shell? The outermost shell?

4-3 What is the electronic configuration characteristic of the noble gases? The halogens? The alkali metals?

4-4 Compare the electronic configurations of elements of Groups VIIA and VIIB, IA and IB, IVA and IVB, 0 and VIII.

4-5 Show the electronic configurations of elements with the following atomic numbers: 7, 14, 21, 36, 55, 65, 79, 85, 100.

4-6 Use the periodic table to find:

(a) the atomic number of lead
(b) the element whose electronic configuration is $1s^22s^22p^63s^23p^63d^{10}4s^24p^65s^2$
(c) the two elements most like selenium
(d) the two elements most like gadolinium
(e) the highest oxidation state for tungsten
(f) whether KCl should be ionic or covalent

4-7 Without the periodic table, ascertain:

(a) which two elements should most resemble strontium
(b) the atomic numbers of the halogens
(c) how many alkali metals exist
(d) whether phosphorus is a metal or a nonmetal

4-8 Do you think hydrogen should be grouped with the elements of Group IA or with those of Group VIIA? What are the arguments for and against your choice?

4-9 What are the electronic structures characteristic of transition elements? Inner transition elements?

References

Becker, C., "Geometry of the *f* Orbitals," *J. Chem. Educ.*, **41** (7), 358 (1964). Photographs of models.

Campbell, J. A., "Interpreting Electronic Structures," *J. Chem. Educ.*, **26** (9), 477 (1949). An interpretation of the periodicity of chemical properties in terms of atomic orbitals available for bonding; description of an *s, p, d, f* demonstration board.

Friedman, H. G., Choppin, G. R., and Feuerbacher, D. G., "The Shapes of the *f* Orbitals," *J. Chem. Educ.*, **41** (7), 354 (1964). Illustrations that are difficult to find in journals or textbooks.

Glasstone, S., *Sourcebook on Atomic Energy*, 3rd ed., Van Nostrand, Princeton, N.J., 1967. A very readable and authoritative book. Chapter 2 deals with fundamental particles, and Chapter 4 considers the structure of the atom.

Moeller, T., *The Chemistry of the Lanthanides*, Reinhold, New York, 1963 (paperbound). Electronic configurations of lanthanide and actinide elements are presented on pages 14 and 100, respectively.

Pearson, R. G., *Chem. Eng. News*, June 29, 1959, p. 73. Photographs of models of *d* orbitals.

Quagliano, J. V., *Chemistry*, 2nd ed., Prentice-Hall, Englewood Cliffs, N.J., 1963. Chapter 12 deals with the periodic system and atomic structure.

5

The Formation of Ions and Molecules

5-1 Formation of Ions

As we have previously stated (Section 2-9), salts consist of ions instead of molecules. When we compare the formulas for different salts, we note that the ions have different amounts of charge. Sodium chloride (NaCl) consists of Na^+ ions and Cl^- ions; calcium chloride ($CaCl_2$) is composed of Ca^{++} ions and Cl^- ions; sodium sulfide (Na_2S), of Na^+ ions and S^{--} ions; and calcium sulfide (CaS), of ions which are both *doubly charged*, Ca^{++} and S^{--}.

Why does a calcium ion have twice the charge of a sodium ion, and why does chlorine form an ion whose charge is only half as great as that of a sulfide ion? To answer such questions, we need to know how atoms become ions.

A negative ion is formed when one or more electrons from some outside source join and become a part of an atom. A negative charge is imparted because the species formed contains more electrons than there are protons in the nucleus (Figure 5-1).

$$\underset{\text{chlorine atom}}{Cl} \quad + \quad \underset{\text{one electron}}{e^-} \quad \rightarrow \quad \underset{\text{chloride ion}}{Cl^-} \text{ (a mononegative ion)}$$

$$\underset{\text{sulfur atom}}{S} \quad + \quad \underset{\text{two electrons}}{2e^-} \quad \rightarrow \quad \underset{\text{sulfide ion}}{S^{--}} \text{ (a binegative ion)}$$

A sulfide ion has twice the charge of a chloride ion because the sulfur atom acquires two extra electrons in becoming a sulfide ion, while the chlorine atom acquires only one extra electron. This difference exists because a chlorine atom has seven electrons in its outermost shell (one less than eight), while a sulfur atom has only six electrons in its outermost shell (two less than eight). Both the sulfide ion and the chloride ion have eight electrons in their outermost shell—the same number of electrons that neon and argon atoms have in their outermost shells. Neon and argon are two of the noble-gas elements, which do

Figure 5-1

Negative ions

Atoms of nonmetals form negative ions.

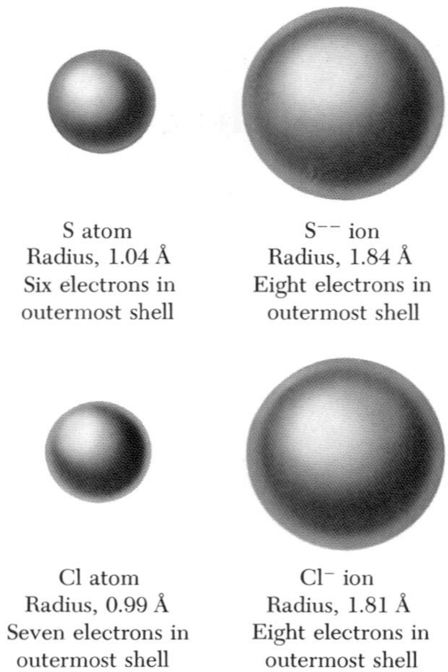

not readily form either negative or positive ions. In general, an atom of a nonmetal, in forming a negative ion, will gain enough electrons so that the outermost shell of the ion produced will have a noble-gas structure. Then all the *s* and *p* orbitals of the outermost shell are filled. This generalization is sometimes referred to as the "octet rule," since the *s* and *p* orbitals have a total capacity of eight electrons.

A positive ion is formed when an atom loses one or more electrons (Figure 5-2). The atom is neutral before the electrons are lost; consequently, the loss of electrons will convert an atom into a positive ion:

$$\underset{\text{sodium atom}}{Na} \rightarrow \underset{\text{sodium ion}}{Na^{+}} \text{ (a monopositive ion)} + \underset{\text{one electron}}{e^{-}}$$

$$\underset{\text{calcium atom}}{Ca} \rightarrow \underset{\text{calcium ion}}{Ca^{++}} \text{ (a bipositive ion)} + \underset{\text{two electrons}}{2e^{-}}$$

By losing the two electrons in its outermost shell, calcium becomes a doubly charged positive ion. Sodium becomes an ion with one positive charge when it loses the single electron in its outermost shell. Neither atom loses electrons spontaneously, but either may serve as the "outside source" of electrons required in the formation of such negative ions as Cl^{-} and S^{--}.

Figure 5-2

Positive ions

Atoms of metals form positive ions.

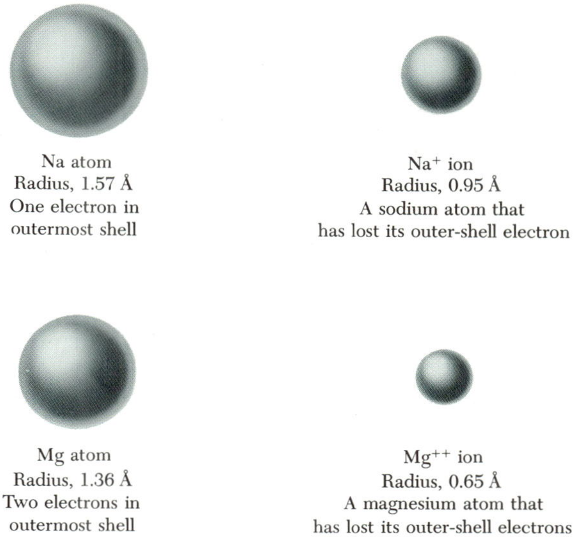

5-2 Ionization Energy

The smaller the number of electrons in the outermost shell of a neutral atom, the greater the tendency of the atom to form a positive ion by permitting electrons to be taken away from it. The energy required to remove an electron from a neutral atom can be measured experimentally and with considerable accuracy. The conversion of neutral particles into ions is called *ionization*. The voltage required to remove an electron from an atom to form an ion is called the *ionization potential*, and the energy expended in removing it is often expressed in electron volts. To be consistent with practices described in Chapter 6 (where the mole concept is discussed), we shall refer to the amount of energy required to remove one electron from each atom in a *mole* of atoms, which is most conveniently expressed in kilocalories (kcal). (A mole of atoms is 6.023×10^{23} atoms. The weight of a mole of atoms is W g, where W is the atomic weight of the element. For example, the weight of a mole of calcium atoms is 40.08 g.) To express the ionization energy in kilocalories per mole, we can utilize the fact that 1 electron volt in ionization potential is equivalent to 23.06 kcal per mole.

The energy required to remove one electron from an atom is called its *first ionization energy;* the energy required to remove a second electron is called the *second ionization energy;* etc. The first ionization energies of the first twenty elements are listed in Table 5-1. Note that there is a general increase in ionization energy as the atomic number increases from lithium (at. no. 3) to neon

Table 5-1

First ionization energies of the first twenty elements (kcal/mole)

H 313							He 566
Li 124	Be 215	B 191	C 261	N 335	O 314	F 402	Ne 497
Na 118	Mg 176	Al 137	Si 188	P 254	S 239	Cl 300	Ar 363
K 100	Ca 140						

(at. no. 10) and from sodium (at. no. 11) to argon (at. no. 18). This is due to the increase in nuclear charge, for the higher the nuclear charge (atomic number), the greater the attractive force exerted by the nucleus (positive) upon the electron (negative), thus making it more difficult to remove the electron from the atom. We have already stated that Li and Na have only one electron in the outermost shell; these elements, of the first twenty, are the ones from whose atoms an electron is most easily removed. Conversely, fluorine and chlorine have seven electrons in the outermost shell; it is more difficult to remove an electron from one of their atoms than from any of the other elements except the noble-gas elements. The extra stability of half-filled subshells is indicated by the ionization energies of nitrogen and phosphorus, which are higher than those of the elements that come immediately before and after them.

Figure 5-3

First ionization energies of metals of Group IIA

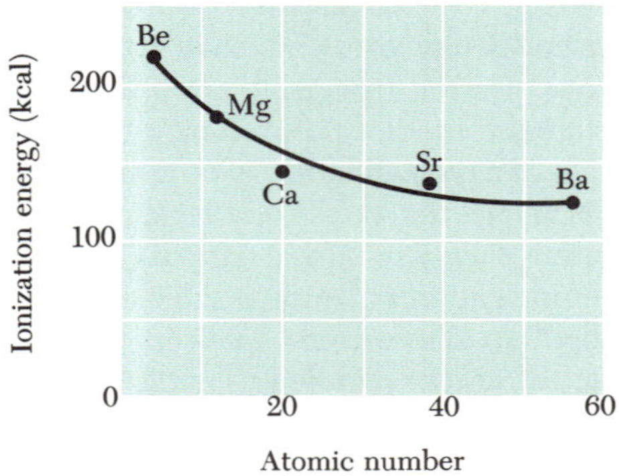

When we compare any one of the first twenty elements with the elements listed beneath it, we find in every instance a decrease in ionization energy with increasing atomic number. This can be attributed to the increased distance between the nucleus and the electron that is being removed. The first ionization energies of the metals of Group IIA, in kilocalories per mole, are ${}_4$Be, 215; ${}_{12}$Mg, 176; ${}_{20}$Ca, 140; ${}_{38}$Sr, 132; ${}_{56}$Ba, 120 (Figure 5-3).

5-3 Electron Affinity and Ionic Bonding

The energy released when an atom acquires an electron, forming a negative ion, is called the *electron affinity*. Unfortunately, this cannot be determined by direct measurement, and in most instances the values estimated from indirect measurements are not truly reliable. The values for the three similar halogens, chlorine (at. no. 17), bromine (at. no. 35), and iodine (at. no. 53), are 87, 82, and 76 kcal per mole, respectively. Here again we see a trend: the further the outermost shell is from the nucleus, the less the electron affinity.

Since the nonmetals have high electron affinities and the metals have very low electron affinities, only the nonmetals tend to form negative ions by accepting electrons from other atoms. In other words, atoms which need only one or two electrons to form an outermost shell of eight can readily acquire the necessary electrons from other atoms. The "other atoms" from which the electrons are most easily removed are metals of low ionization energy, i.e., atoms with only one or two electrons in the outermost shell. Elements such as lithium, sodium, and potassium readily react with such elements as fluorine and chlorine to form *ionic compounds*.

Electronic symbols—symbols which show the number of electrons in the outermost shell of an atom—may be written for the elements involved in such reactions. In Table 5-2, electronic symbols are given for the first twenty elements.

Table 5-2

Electronic symbols for the first twenty elements

$\mathrm{H}\cdot$							$\mathrm{He}:$
$\mathrm{Li}\cdot$	$\dot{\mathrm{Be}}\cdot$	$\underset{\cdot}{\dot{\mathrm{B}}}\cdot$	$\cdot\underset{\cdot}{\dot{\mathrm{C}}}\cdot$	$:\underset{\cdot}{\dot{\mathrm{N}}}\cdot$	$:\underset{\cdot\cdot}{\dot{\mathrm{O}}}\cdot$	$:\underset{\cdot\cdot}{\overset{\cdot\cdot}{\mathrm{F}}}\cdot$	$:\underset{\cdot\cdot}{\overset{\cdot\cdot}{\mathrm{Ne}}}:$
$\mathrm{Na}\cdot$	$\dot{\mathrm{Mg}}\cdot$	$\underset{\cdot}{\dot{\mathrm{Al}}}\cdot$	$\cdot\underset{\cdot}{\dot{\mathrm{Si}}}\cdot$	$:\underset{\cdot}{\dot{\mathrm{P}}}\cdot$	$:\underset{\cdot\cdot}{\dot{\mathrm{S}}}\cdot$	$:\underset{\cdot\cdot}{\overset{\cdot\cdot}{\mathrm{Cl}}}\cdot$	$:\underset{\cdot\cdot}{\overset{\cdot\cdot}{\mathrm{Ar}}}:$
$\mathrm{K}\cdot$	$\dot{\mathrm{Ca}}\cdot$						

With these symbols, the formation of sodium chloride, calcium chloride, and calcium oxide is represented as follows:

$$\underset{\text{sodium atom}}{\text{Na}\cdot} \quad + \quad \underset{\text{chlorine atom}}{\cdot\underset{\cdot\cdot}{\overset{\cdot\cdot}{\text{Cl}}}:} \quad \rightarrow \quad \underset{\text{sodium ion}}{\text{Na}^{+}} \quad + \quad \underset{\text{chloride ion}}{\left[:\underset{\cdot\cdot}{\overset{\cdot\cdot}{\text{Cl}}}:\right]^{-}} \qquad \text{(NaCl, sodium chloride)}$$

$$\underset{\text{calcium atom}}{\dot{\text{Ca}}\cdot} \quad + \quad \underset{\text{two chlorine atoms}}{2\ \cdot\underset{\cdot\cdot}{\overset{\cdot\cdot}{\text{Cl}}}:} \quad \rightarrow \quad \underset{\text{calcium ion}}{\text{Ca}^{++}} \quad + \quad \underset{\text{two chloride ions}}{2\left[:\underset{\cdot\cdot}{\overset{\cdot\cdot}{\text{Cl}}}:\right]^{-}} \qquad (\text{CaCl}_2\text{, calcium chloride})$$

$$\underset{\text{calcium atom}}{\dot{\text{Ca}}\cdot} \quad + \quad \underset{\text{oxygen atom}}{\cdot\underset{\cdot\cdot}{\dot{\text{O}}}:} \quad \rightarrow \quad \underset{\text{calcium ion}}{\text{Ca}^{++}} \quad + \quad \underset{\text{oxide ion}}{\left[:\underset{\cdot\cdot}{\overset{\cdot\cdot}{\text{O}}}:\right]^{--}} \qquad \text{(CaO, calcium oxide)}$$

In the ionic compound NaCl each sodium "atom" is really an ion, for it bears a positive charge, and each chlorine "atom" is really an ion also, for it bears a negative charge. In any sample of the compound large enough to see or work with, there is a large number of ions—but the number of Na^+ ions is the same as the number of Cl^- ions. The formula NaCl (sometimes written Na^+Cl^-) is intended to convey the information that the two kinds of ions are present in equal number in any sample of the compound, large or small. The formula for any ionic compound is only an empirical formula.

You must always realize, then, that the formula for an ionic compound does not represent a molecule. Ionic compounds do not consist of molecules. In a crystal of an ionic compound, the ions are held together by virtue of their charges. Each ion is an independent particle, and when the compound is heated to its melting point, the ions are free to move about. But in the solid crystal they are lined up in an orderly pattern (in all three dimensions), and each ion is attracted by all its oppositely charged neighbors. There are no individual bonds between particular ions—there is no pairing of atoms. When we speak of *ionic bonding*, we are referring to the type of chemical bonding in which atoms are converted into oppositely charged ions rather than united to form a molecule by sharing a pair of electrons.

Sodium metal, composed of sodium atoms, is a soft metal with a silvery appearance (when a freshly cut surface is observed), which corrodes rapidly in air because of the formation of sodium oxide on the surface and which reacts rapidly with water, evolving hydrogen gas. Chlorine is a poisonous gas of a light greenish-yellow color, which corrodes metals rapidly through formation of chlorides. Yet their compound, sodium chloride, which contains sodium ions and chloride ions, is a white solid that is necessary in our diet and is used as table salt. This crystalline compound illustrates the profound difference between the properties of atoms and the properties of their corresponding ions. Because atoms and ions are so different, you must carefully differentiate between them—a sodium ion is written Na^+, and a sodium atom is written Na (without the charge sign).

5-4 Covalent Bonds

Nonmetals do not ordinarily form positive ions and metals do not form negative ions. Even some of the metals are reluctant to form positive ions, and some nonmetals practically never form negative ions. In order for a given pair of elements to form an ionic compound, there must be a significant difference between their electron affinities and between their ionization energies. This limits the number of possible ionic compounds, but there are many other compounds that do not contain ions. These *nonionic compounds* consist of molecules composed of atoms that are usually held together by *covalent bonds.* A covalent bond consists of a pair of electrons shared by two atoms. The simplest example of a covalent bond is the shared electron pair which holds the two hydrogen atoms together in the hydrogen molecule, H_2:

$$\mathrm{H}\cdot \;+\; \cdot\mathrm{H} \rightarrow \mathrm{H}:\mathrm{H}$$

Since each hydrogen atom has one electron in its *K* shell (the *K* shell is filled when it has two electrons in it), the electrons from two hydrogen atoms can form a pair that is shared equally by the two nuclei. With the two electrons held jointly by the nuclei, the two atoms are bonded together very strongly. A relatively large amount of energy, 104 kcal, is required to break the bonds in all the molecules of a mole (i.e., 2.016 g or 6.023×10^{23} molecules) of H_2; and, likewise, when two hydrogen atoms combine to form an H_2 molecule, 104 kcal are evolved per mole of H_2 formed.

If the two electrons of the covalent bond in the H_2 molecule are counted for each atom, each hydrogen atom has the same electronic arrangement as a helium atom. When two chlorine atoms, each with seven electrons in its outermost shell, share one pair of electrons between them (Figure 5-4),

$$:\overset{\cdot\cdot}{\underset{\cdot\cdot}{\mathrm{Cl}}}\cdot \;+\; \cdot\overset{\cdot\cdot}{\underset{\cdot\cdot}{\mathrm{Cl}}}: \;\rightarrow\; :\overset{\cdot\cdot}{\underset{\cdot\cdot}{\mathrm{Cl}}}:\overset{\cdot\cdot}{\underset{\cdot\cdot}{\mathrm{Cl}}}:$$

each atom has the stable electronic structure of an argon atom. Thus, in forming molecules by sharing electrons—as in forming ions by gaining or losing electrons—atoms tend toward an outer-shell electron arrangement like that of the nearest noble-gas element. Our explanation of covalent bond formation is therefore basically another application of the tendency of atoms to fill the outer *s* and *p* orbitals.

The formula for water is H_2O rather than OH because an oxygen atom, with only six electrons in its outermost shell, must share a pair of electrons with each of *two* hydrogen atoms if it is to attain a noble-gas structure:

$$2\mathrm{H}\cdot \;+\; \cdot\overset{\cdot\cdot}{\underset{\cdot}{\mathrm{O}}}: \;\rightarrow\; \begin{array}{l}\mathrm{H}:\overset{\cdot\cdot}{\underset{\cdot\cdot}{\mathrm{O}}}: \\ \quad\;\mathrm{H}\end{array}$$

Figure 5-4

Formation of a covalent bond

$$:\ddot{\underset{..}{Cl}}\cdot \;+\; \cdot\ddot{\underset{..}{Cl}}: \;\rightarrow\; :\ddot{\underset{..}{Cl}}:\ddot{\underset{..}{Cl}}: \text{ (not to scale)}$$

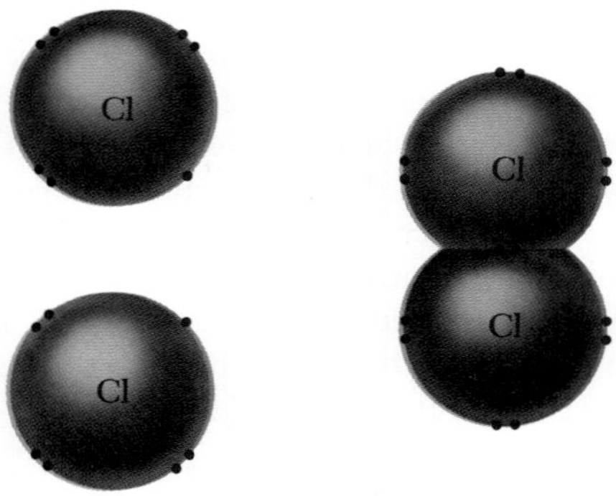

A nitrogen atom, with only five electrons in its outermost shell, shares a pair of electrons with each of *three* hydrogen atoms, thus giving it a noble-gas structure; this explains the fact that the formula for ammonia is NH_3:

$$3H\cdot + \cdot\dot{\underset{\cdot}{N}}: \rightarrow H:\overset{H}{\underset{H}{\ddot{\underset{..}{N}}}}:$$

The formula for methane is CH_4 in consequence of the presence of only four electrons in the outermost shell of an isolated carbon atom:

$$4H\cdot + \cdot\dot{\underset{\cdot}{C}}\cdot \rightarrow H:\overset{H}{\underset{H}{\ddot{\underset{..}{C}}}}:H$$

In some molecules two atoms are chemically bonded by sharing *two* pairs of electrons. For example, in the CO_2 molecule, $\ddot{\underset{..}{O}}::C::\ddot{\underset{..}{O}}$, there is such a *double bond* between the carbon atom and each oxygen atom. In the N_2 molecule, $:N:::N:$, there is a *triple bond* between the two nitrogen atoms.

A covalent bond is sometimes established between two atoms of which one furnishes *both* of the shared electrons. This type of bond is called a *coordinate covalent bond* or, more briefly, a *coordinate bond*. Such a bond accounts for the formation of the compound NH_3BF_3 by an "addition reaction" occurring between ammonia and boron trifluoride.

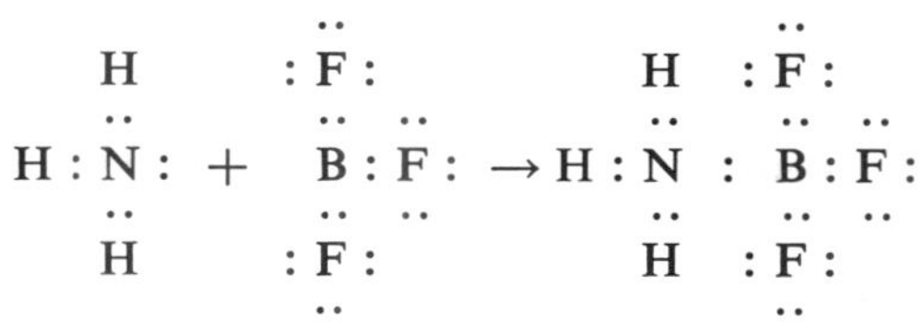

The formation of this compound is due to the presence of the unshared pair in the outermost shell of the nitrogen atom in NH_3 and the vacancy in the outermost shell of the boron atom in BF_3.

It is customary to indicate a covalent bond in the structural formulas of compounds by drawing a line to represent the shared pair of electrons. Thus, the H_2 molecule can be represented as being made up of two covalently bonded hydrogen atoms by the formula H—H. Ordinarily, structural formulas written in this manner do not include unshared electrons. For example, the formula for chlorine would be Cl—Cl. If the unshared electrons are included, the formula is an *electronic formula* whether the bonding electrons are represented by a line or by a pair of dots. There is no difference in meaning between $:\ddot{\underset{..}{Cl}}—\ddot{\underset{..}{Cl}}:$ and $:\ddot{\underset{..}{Cl}}:\ddot{\underset{..}{Cl}}:$.

With lines representing covalent bonds, the structural formulas (line formulas) for ammonia, methane, and the addition compound formed between ammonia and boron trifluoride are as follows:

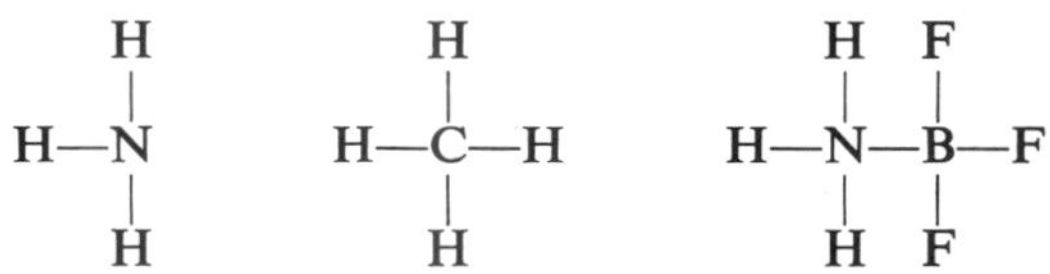

ammonia methane NH_3BF_3

The bond between the nitrogen atom and the boron atom in NH_3BF_3 is like any other covalent bond since all electrons are identical. Therefore, there is no real need for indicating that both electrons are furnished by the nitrogen atom. However, this is sometimes done in the following manner:

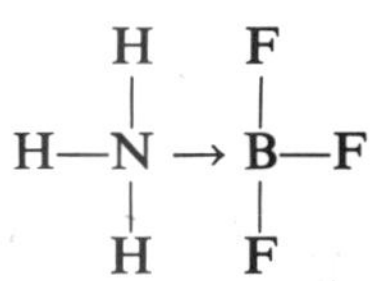

A double bond between atoms can be represented by two lines. Thus, the formula for carbon dioxide can be written O═C═O. Triple bonds are represented by three lines, as in the N_2 molecule, N≡N.

5-5 Electronegativity

When a pair of electrons is shared by two unlike atoms, such as a hydrogen atom and a chlorine atom,

$$\mathrm{H}\cdot + \cdot\overset{\cdot\cdot}{\underset{\cdot\cdot}{\mathrm{Cl}}}: \rightarrow \mathrm{H}:\overset{\cdot\cdot}{\underset{\cdot\cdot}{\mathrm{Cl}}}:$$

they are not shared equally. The electrons, on the average, spend more time in the vicinity of the chlorine nucleus than in the vicinity of the hydrogen nucleus. The cause of this unequal sharing is the greater attraction of the chlorine atom for the bonding electrons in comparison with the hydrogen atom. This tendency to attract shared electrons is called *electronegativity*. Although it is difficult to measure, electronegativity has effects which are important in determining the properties of covalent compounds. For example, HCl is considered a *polar compound* because chlorine is more electronegative than hydrogen. The two bonding electrons, one from each atom, come from atoms which are initially neutral, but in the HCl molecule the covalent bond has a partial ionic character. The center of positive charge in the molecule is not at the same point as the center of negative charge; i.e., the chlorine end of the molecule is somewhat negative, and the hydrogen end of the molecule is positive to the same degree (Figure 5-5). When placed in an electric field, such polar molecules tend to turn so that their negative ends are oriented toward the positive plate and their positive ends are oriented toward the negative plate (Figure 5-6). When there is no electric field the molecules are not aligned in any one direction.

The electronegativity values of the first twenty elements, as evaluated by procedures suggested by American chemist Linus Pauling, are listed in Table 5-3. Among the representative elements, electronegativity increases with atomic number in each row and decreases with atomic number in each column. Values are not yet available for the noble-gas elements.

Figure 5-5

A polar molecule

In the HCl molecule (a), the electrons are not shared equally since chlorine is more electronegative than hydrogen. As represented by (b), the molecule is relatively positive at the hydrogen end and negative at the chlorine end.

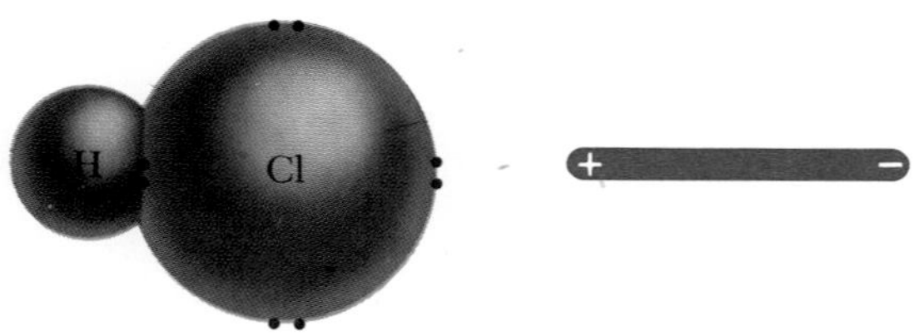

(a) (b)

Figure 5-6

Polar molecules in an electric field

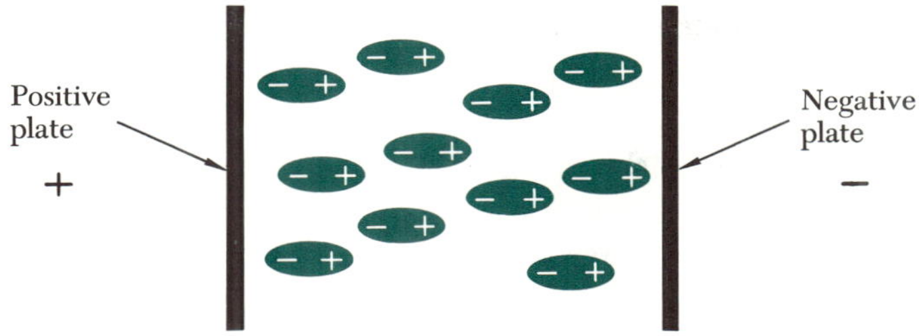

5-6 The Concept of Oxidation State

You have now learned that certain compounds are ionic and others are covalent. Sodium and potassium form positive ions, each with a charge of +1 because each atom loses one (and only one) electron to form an ion. A chlorine atom can gain one (but only one) electron to form a negative ion with a charge of −1. The difference between the electronegativities of chlorine and sodium permits chlorine to remove an electron from sodium. In an H_2 molecule neither atom has removed an electron from the other—instead, they share an electron pair because their electronegativities are identical. The same situation prevails in a Cl_2 molecule. In HCl molecules, chlorine, being more electronegative than hydrogen, acquires more than an equal share of the bonding electrons. Thus, even in HCl molecules the *tendency* of chlorine atoms to form negative ions is evident. Since the molecule is polar, we may say that the covalent bond has *partial ionic character.* Since the extent to which a covalent bond has partial ionic character depends on the relative electronegativities of the two atoms

Table 5-3

Electronegativity values of the first twenty elements

H 2.1							He —
Li 1.0	Be 1.5	B 2.0	C 2.5	N 3.0	O 3.5	F 4.0	Ne —
Na 0.9	Mg 1.2	Al 1.5	Si 1.8	P 2.1	S 2.5	Cl 3.0	Ar —
K 0.8	Ca 1.0						

involved, the ionic character varies from one compound to another. Without concerning ourselves now with the different degrees of partial ionic character in different bonds, let us note merely that the charge which the chlorine atom *tends* or appears to have in HCl or CCl_4 or PCl_3 is -1, the same charge that actually exists on the Cl^- ion in the ionic compound NaCl. This is the basis for the assignment of an arbitrary *oxidation number* to an atom in a molecule.

If a compound is actually and truly ionic, the oxidation number of each ion is, by definition, the same as its charge. Thus, in NaCl the oxidation number of the sodium ion is $+1$, and we can say that the element sodium is in the *oxidation state* $+1$; the oxidation number of the chloride ion is -1, and we can say, therefore, that the element chlorine is in the oxidation state -1. In $CaCl_2$ chlorine is in the oxidation state -1, but calcium is in the oxidation state $+2$. In MgO the oxidation state of magnesium is $+2$, and the oxidation state of oxygen is -2.

If a compound is covalent, the oxidation number arbitrarily assigned to each atom is the same as if the compound were ionic. In the covalent compounds HCl, CCl_4, and PCl_3, chlorine is in the same oxidation state (-1) as in the ionic compounds NaCl and $CaCl_2$. In HCl hydrogen is in the oxidation state $+1$, just as sodium is in the $+1$ state in NaCl. Likewise, we say that in CCl_4 carbon is in the $+4$ state, although carbon is not present as an ion with a $+4$ charge (and never is in *any* compound). Similarly, the oxidation number of the phosphorus atom in PCl_3 is $+3$.

To determine which of the two atoms sharing a pair of electrons has the *positive* oxidation number, we must know the relative electronegativities of the two. Let us determine, for example, which element in $SiCl_4$ is in the positive oxidation state. Consulting Table 5-3, we note that silicon is considerably less electronegative than chlorine, so silicon is regarded as being in the $+4$ state and chlorine in the -1 state. Can it also be predicted whether the compound is ionic or covalent? Although compounds formed of elements having great electronegativity differences are ionic, it is difficult to know just how great the difference must be. In two covalent compounds composed of different pairs of elements, the degree of partial ionic character may differ even if the electronegativity differences are equal.

The concept of oxidation state is a highly arbitrary one, but much of its usefulness depends on this very fact. For example, you do not need to know whether $SiCl_4$ is ionic or covalent in order to deduce the correct oxidation numbers of silicon and chlorine in this compound. Conversely, oxidation numbers are useful for predicting the formulas for the products of chemical reactions.

5-7 Assignment of Oxidation Numbers

The following general rules for determining oxidation states are arbitrary, but they are justified by their usefulness and are consistent with the concept of relative electronegativities:

1. The oxidation state of any free or uncombined element is zero, regardless of the number of atoms in one molecule. For example, in O_2, O_3, P_4, S_8, Ne, and Mg, the oxidation state is zero.
2. The sum of the oxidation numbers of all the atoms in the formula for a compound is zero. By definition a compound is neutral, i.e., the oxidation numbers add up to zero whether the compound is ionic or molecular. Calcium chloride, $CaCl_2$, is composed of Ca^{++} ions and Cl^- ions; any specimen of the compound is neutral because there are twice as many Cl^- ions as Ca^{++} ions. In carbon dioxide (CO_2), which is composed of neutral molecules, each oxygen atom is regarded as having an oxidation number -2 because oxygen forms O^{--} ions, as in such ionic oxides as CaO. The oxidation number of the carbon atom in carbon dioxide is, therefore, $+4$. There are no O^{--} ions in CO_2, and our decision to regard the oxidation state of carbon as $+4$ in CO_2 does not mean that C^{4+} ions are present.
3. The sum of the oxidation numbers of all the atoms in an ion is the same as the charge on the ion. For example, in a sulfate ion, SO_4^{--}, where the oxidation number of each oxygen atom is -2, we must regard the oxidation number of the sulfur atom as $+6$, since $+6 + 4(-2) = -2$, the actual charge of the sulfate ion. In the ammonium ion, NH_4^+, where the oxidation number of each hydrogen atom is $+1$, the oxidation number of the nitrogen atom is -3, since $-3 + 4(+1) = +1$, the charge of the ammonium ion.

Usually, in the case of a ternary compound, two of the three elements will have oxidation numbers about which there is little if any doubt, and we can calculate the oxidation state of the third element from the values of the other two. For example, in $MgSO_4$, if we know that magnesium forms Mg^{++} ions and that the oxidation state of oxygen here is -2, we may represent the oxidation state of sulfur by x and write the equation

$$+2 + x - 8 = 0$$

Therefore, $x = +6$. This answer is identical with the value obtained above for the oxidation number of the sulfur atom in SO_4^{--} ion because $MgSO_4$ is composed of Mg^{++} and SO_4^{--} ions.

5-8 Oxidation States and the Periodic Table

The periodicity of the oxidation states, one of the useful relationships associated with the periodic table, can be summarized as follows:

1. Elements of Group IA always have the oxidation number $+1$ in their compounds, with the exception of hydrogen, which is in the -1 state in hydrides of active metals, e.g., NaH, CaH_2.

2. Elements of Group IIA always show the oxidation number +2 in their compounds.
3. The oxidation state +3 is the only important one for the elements of Group IIIA, with the exception of thallium, which also forms thallium(I) compounds.
4. The maximum oxidation number for elements of Group IVA is +4, but tin(II) and lead(II) compounds are important, and the oxidation numbers of the elements of this group vary from +4 to −4.
5. Nitrogen forms compounds in every integral oxidation state from +5 to −3, but the most important oxidation states for the other elements in Group VA are +3 and +5.
6. The most important oxidation number for oxygen is −2, but the other members of Group VIA exhibit oxidation numbers +4 and +6, as well as −2, in many of their compounds.
7. The highest oxidation number shown by members of Group VIIA is +7, and the lowest is −1. Fluorine never has a positive oxidation number.

New Terms

Chemical bonding: a general term referring to the mechanism whereby atoms form either molecules or ions. The union of elements to form compounds and the union of like atoms to form molecules of an element are both included.

Coordinate bond: a covalent bond in which both electrons of the shared pair are furnished by the same atom.

Covalent bonding: the type of chemical bonding in which electron pairs are shared by the atoms that are thus bonded together.

Electron affinity: the energy released when an electron is added to a neutral, isolated atom, forming a negative ion.

Electronegativity: the tendency of a covalently bonded atom to attract the bonding electron pair.

Electronic symbol: the chemical symbol for an element with dots, small circles, or other suitable markings to indicate the number of electrons in the outermost shell; e.g., $:\overset{\cdot\cdot}{\underset{\cdot\cdot}{Cl}}\cdot$, $\overset{\circ\circ}{\underset{\circ\circ}{{}^{\circ}_{\circ}Cl}}{}^{\circ}$, $\overset{\times\times}{\underset{\times\times}{{}^{\times}_{\times}Cl}}{}^{\times}$, or $|\overline{\underline{Cl}}{}^{\times}$ for a chlorine atom and $:\overset{\cdot\cdot}{\underset{\cdot\cdot}{Cl}}:^{-}$, $\overset{\circ\circ}{\underset{\circ\circ}{{}^{\circ}_{\circ}Cl{}^{\circ}_{\circ}}}{}^{-}$, $\overset{\times\times}{\underset{\times\times}{{}^{\times}_{\times}Cl{}^{\times}_{\times}}}{}^{-}$, or $|\overline{\underline{Cl}}|^{-}$ for a chloride ion.

Ion: an atom, or a group of atoms, in which the number of electrons and the number of protons are not the same.

Ionic bonding: the type of chemical bonding in which electrons have been transferred from atoms of low ionization potential to atoms of high electron affinity, forming positive and negative ions.

Ionization: the conversion of neutral particles into ions.

Ionization energy: the energy required to remove an electron from an atom.

Ionization potential: the voltage required to remove an electron from an atom.

Mole: a general term referring to 6.023×10^{23} chemical units. A mole of atoms is 6.023×10^{23} atoms. A mole of molecules is 6.023×10^{23} molecules. A mole of ions is 6.023×10^{23} ions. A mole of electrons is 6.023×10^{23} electrons. (The mole concept is discussed in Chapter 6.)

Octet rule: the tendency of atoms, when they combine, to form complete octets of electrons in their outermost shells. There are numerous exceptions to this "rule."

Oxidation number: an arbitrary number assigned to each atom in an ion or molecule so that the sum of the oxidation numbers of all the atoms is the charge of the ion or is zero for a molecule.

Oxidation state: a term used to indicate the oxidation number of each atom of an element. For example, iron is in the same oxidation state (+3) in Fe_2O_3 as in $FeCl_3$.

Polar compound: a compound consisting of polar molecules. In a polar molecule, which contains one or more polar bonds, the positive and negative centers of charge do not coincide. A polar bond is a covalent bond between two atoms which differ in electronegativity.

Exercises

5-1 Of what ions are the following compounds composed: LiI, KBr, Na_2S, CaO, Al_2S_3, CaF_2?

5-2 Using electronic symbols, indicate the formation of the following ionic compounds: KF, MgS, Ca_3P_2, Na_2S, Li_3P, CaF_2.

5-3 Describe the chemical bonding involved in the formation of the following molecules: F_2, H_2, HF, H_2O, NH_3, CH_4, CCl_4.

5-4 Using the periodic table and the table of electronic configurations, predict whether stable compounds with the following formulas will be found to exist: Mg_3O_2, Mg_2O_3, PO, KCl_2, RbBr, CoNe, BaSe, NaI.

5-5 Why do the ionization energies of the alkali metals decrease with increasing atomic number?

5-6 Why does nitrogen have an ionization energy greater than oxygen?

5-7 Explain why a calcium ion has a charge equal to two protons and a sodium ion has a charge equal to one proton.

5-8 State, define, describe, or explain:
(a) ionic bonding, covalent bonding, chemical bonding
(b) oxidation number, oxidation state

5-9 Write line formulas for NH_3, BF_3, and NH_3BF_3.

References

Companion, A. L., *Chemical Bonding*, McGraw-Hill, New York, 1964 (paperbound). Covalent, ionic, and metallic bonding are described.

Ehret, W. F., "The Role of Electrons in Interatomic Relations," *J. Chem. Educ.*, **25** (5), 291 (1948). A brief treatment of the formation of ionic and covalent bonds.

Gray, H. B., *Electrons and Chemical Bonding*, Benjamin, New York, 1965 (paperbound). Molecular orbital theory is used to describe bonding in molecules.

Kieffer, W. F., "The Use of Electron Structure in Interpreting Chemical Reactions," *J. Chem. Educ.*, **25** (10), 537 (1948). A discussion of the arrangements of electrons in atomic orbitals and the combinations of orbitals in bonding.

Pauling, L., *The Chemical Bond*, Cornell University Press, Ithaca, New York, 1967 (paperbound). A brief introduction to modern structural chemistry, intended especially for the use of students.

Sanderson, R. T., "Principles of Chemical Bonding," *J. Chem. Educ.*, **38** (8), 382 (1961). Twenty-five generalizations relating chemical bonding to atomic structure.

6

The Mole Concept and Stoichiometry

6-1 Moles of Atoms

Individual atoms are so minute that the smallest piece of matter visible to the human eye contains more of them than one person could count in a lifetime. Although we cannot work with atoms and molecules individually, it becomes important to know when we are dealing with equal numbers of them, for comparisons between equal numbers of atoms or molecules are often more useful than comparisons between equal weights of substances.

If we know the atomic and molecular weights of the substances with which we are working, it is easy to know when we are dealing with equal numbers of atoms or molecules. For example, the atomic weights of carbon and oxygen are 12.011 and 15.999, respectively. If we take 12.011 g of carbon and 15.999 g of oxygen, we have weights that are in the same proportion as the atomic weights; therefore, the number of carbon atoms taken is the same as the number of oxygen atoms. When the weight in grams is numerically equal to the atomic weight of an element, that amount is called a *gram-mole*, or simply a *mole*, of atoms. Therefore, 12.011 g of carbon is 1 mole of carbon atoms, and 15.999 g of oxygen is 1 mole of oxygen atoms.

Convention has established the gram as the unit of weight for defining molar amounts. This is, of course, arbitrary and merely a matter of convenience. Consistency in such matters facilitates communication among individuals with similar interests. However, other units of weight are sometimes used for defining molar amounts, especially in the chemical industries, where English units are still in use. Thus, a *pound-mole* of carbon atoms is 12.011 lb of carbon, and a pound-mole of oxygen atoms is 15.999 lb of oxygen.

A mole of carbon atoms is a very large but definite number of carbon atoms, and a mole of oxygen atoms is that same number of oxygen atoms. This number is referred to as *Avogadro's Number*, N, and is known to be 6.023×10^{23}. We can then say that N atoms of any element constitute a mole of atoms of that element or that N atoms of an element weigh W grams, where W is the atomic weight of that element. The numerical value of Avogadro's Number depends on two completely arbitrary decisions: the choice of numerical scale that has been made for atomic weights and the use of the gram as the unit of weight.

In Sections 14-3 and 14-4 we shall see how the numerical value of Avogadro's Number was obtained.

6-2 Moles of Molecules

The molecular formula for carbon dioxide, CO_2, means that a molecule of carbon dioxide consists of one carbon atom and two oxygen atoms, so that the molecular weight of carbon dioxide is 44.009, i.e., 12.011 + 2(15.999). If we take 44.009 g of carbon dioxide, we have N molecules of carbon dioxide; i.e., we have a mole of carbon dioxide molecules. One mole of carbon dioxide molecules contains 1 mole of carbon atoms and 2 moles of oxygen atoms.

The molecular weight of methane, CH_4, is 16.043, i.e., 12.011 + 4(1.008). In 16.043 g (1 mole) of methane the number of methane molecules is the same as the number of carbon dioxide molecules in 44.009 g (1 mole) of carbon dioxide. Again, the number is 6.023×10^{23}, Avogadro's Number; 1 mole of methane molecules contains 1 mole of carbon atoms and 4 moles of hydrogen atoms.

In a mole of any substance composed of molecules and weighing G grams, where G is the molecular weight of the substance, there are N molecules.

6-3 Moles of Ions

The compounds known as salts consist of ions instead of molecules. Since their formulas do not represent molecules, the sum of the atomic weights can only be considered a formula weight. Nevertheless, the number of grams numerically equal to the formula weight is termed a mole of the compound. A mole of the ionic compound FeS does not consist of N molecules, but it does consist of a mole of Fe^{++} ions and a mole of S^{--} ions. The formula FeS represents a chemical unit just as surely as if it were a molecule; in this instance, the chemical unit is a pair of ions.

We can generalize and say that 1 mole contains Avogadro's Number of formula units of any species, whether the unit be an atom, a molecule, an ion, or an assemblage of ions (Figure 6–1).

6-4 Weight Relations in Chemical Reactions

From the balanced equation

$$2CO + O_2 \rightarrow 2CO_2$$

we can see that two molecules of carbon monoxide react with one molecule of oxygen, producing two molecules of carbon dioxide. This implies, of course,

Figure 6-1

The mole as a chemical unit

One mole contains Avogadro's Number of formula units. The cubes are represented in the relative sizes needed to contain the quantities indicated.

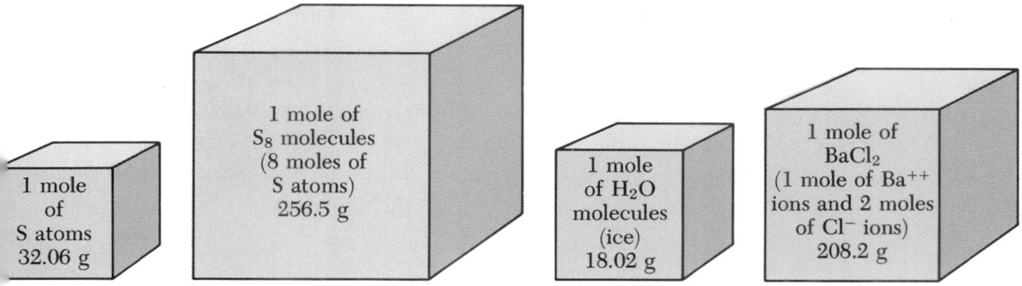

that 200 molecules of carbon monoxide react with 100 molecules of oxygen, producing 200 molecules of carbon dioxide; i.e., any number, n, of carbon monoxide molecules will react with half that number, $\frac{1}{2}n$, of oxygen molecules to form n carbon dioxide molecules. Therefore, 2 moles of carbon monoxide molecules react with 1 mole of oxygen molecules, producing 2 moles of carbon dioxide molecules, because 1 mole of each of these substances is 6.023×10^{23} molecules.

Since a molecule of carbon monoxide consists of one carbon atom (atomic weight, 12.011) and one oxygen atom (atomic weight, 15.999), the molecular weight of carbon monoxide is 28.010; thus, 1 mole of carbon monoxide equals 28.010 g, and 2 moles equal 56.020 g. For oxygen gas, O_2, we see that a mole of oxygen molecules is

$$2 \times 15.999 \text{ g} = 31.998 \text{ g}$$

A mole of carbon dioxide molecules is 44.009 g (12.011 g + 31.998 g), and 2 moles equal 88.018 g. It inevitably follows that 56.020 g of carbon monoxide react with 31.998 g of oxygen, producing 88.018 g of carbon dioxide. All we need to get such information from a balanced chemical equation is a table of atomic weights.

Thus, the quantitative relations inherent in balanced equations can be used to calculate the weights of reactants and products in chemical reactions, as illustrated in the following examples.

EXAMPLE How many moles of hydrogen gas (H_2) are needed to produce 10 moles of ammonia gas (NH_3)?

First we write the balanced equation:

$$N_2 + 3H_2 \rightarrow 2NH_3$$

This tells us that 3 moles of H_2 are needed to produce 2 moles of

NH_3, or 1.5 moles of H_2 for each mole of NH_3. Thus, 10 moles of NH_3 require 1.5 moles $\times$ 10 = 15 moles of H_2.

EXAMPLE How many grams of nitrogen are needed to produce 10 moles of ammonia?

The equation (see the first example) tells us that 1 mole of N_2 is needed to produce 2 moles of NH_3, or 0.5 mole of N_2 for each mole of NH_3. Thus, 10 moles of NH_3 require 0.5 mole $\times$ 10 = 5 moles of N_2. In the table of atomic weights we find that the atomic weight of nitrogen (rounded off to three digits) is 14.0, so 1 mole of N_2 is 28.0 g. Thus, the required 5 moles of N_2 weigh 5 $\times$ 28.0 g = 140 g.

EXAMPLE How many grams of ammonia will be produced from 100 g of hydrogen?

The atomic weight of hydrogen is 1.008, so the weight of 1 mole of H_2 is 2 $\times$ 1.008 g = 2.016 g.

$$\frac{100 \text{ g}}{2.016 \text{ g/mole}} = 49.6 \text{ moles of } H_2$$

The equation (see the first example) indicates that each mole of H_2 produces $\frac{2}{3}$ or 0.667 mole of NH_3, so 49.6 moles of H_2 will produce

$$0.667 \times 49.6 \text{ moles} = 33.1 \text{ moles of } NH_3$$

The weight of 1 mole of NH_3 is 14.0 g + 3.0 g = 17.0 g. Thus, the weight of the 33.1 moles of NH_3 is 33.1 $\times$ 17.0 g = 563 g.

EXAMPLE How many moles of water will be produced if 10 moles of sulfuric acid are neutralized with aluminum hydroxide?

From the balanced equation

$$2Al(OH)_3 + 3H_2SO_4 \rightarrow Al_2(SO_4)_3 + 6H_2O$$

it is apparent that 3 moles of H_2SO_4 produce 6 moles of H_2O, or 1 mole of H_2SO_4 produces 2 moles of H_2O. Thus, 10 moles of H_2SO_4 will produce 2 moles $\times$ 10 = 20 moles of H_2O.

EXAMPLE In the reaction of aluminum hydroxide with sulfuric acid, how many grams of aluminum sulfate will be produced from 10 moles of sulfuric acid?

The equation (see the preceding example) indicates that 3 moles of H_2SO_4 are required to produce 1 mole of $Al_2(SO_4)_3$. Thus, 10 moles of H_2SO_4 will produce $\frac{10}{3}$ moles of $Al_2(SO_4)_3$. The atomic weights of Al, S, and O are 26.98, 32.06, and 16.00, respectively. Therefore, the weight of 1 mole of $Al_2(SO_4)_3$ is (2×26.98) g + $3[32.06 + (4 \times 16.00)]$ g = 53.96 g + (3×96.06) g = 53.96 g + 288.18 g = 342.1 g. The weight of $\frac{10}{3}$ moles of $Al_2(SO_4)_3$ is

$$\tfrac{10}{3} \times 342.1\ \text{g} = 1140\ \text{g}$$

The fact that aluminum sulfate is an ionic instead of a molecular compound causes no difficulty in the solution of this problem. However, some weight-relation problems make specific reference to ions.

EXAMPLE How many grams of H_3O^+ ions are produced when 100 g of HCl gas dissolve in water?

From the equation

$$HCl + H_2O \rightarrow H_3O^+ + Cl^-$$

it is apparent that 1 mole of HCl molecules produces 1 mole of H_3O^+ ions. From the atomic weights we see that 1 mole of HCl molecules weighs 1.008 g + 35.453 g = 36.461 g.

$$\frac{100\ \text{g}}{36.461\ \text{g/mole}} = 2.75\ \text{moles of HCl}$$

Thus, 2.75 moles of H_3O^+ ions are produced; 1 mole of H_3O^+ ions weighs 3.02 g + 16.00 g = 19.02 g. Thus, 2.75 moles of H_3O^+ ions weigh

$$2.75 \times 19.0\ \text{g} = 52.3\ \text{g}$$

6-5 Incomplete Reactions

The type of chemical calculation illustrated in the previous section is often called *stoichiometry*. The basic concept in the stoichiometric approach to chemical problems is that the chemical equation, correctly interpreted, gives the correct relationship between the weights of the reacting substances.

We assume, in making a stoichiometric calculation, that *one* of the reactants is completely consumed. Indeed, if all reactants are present in the molar amounts required by the equation, we assume that all of the reactants are completely transformed. However, some reactions appear to stop before the conversion of reactants to products is complete. This is the case if the reaction is *reversible* under the prevailing conditions, for example, if N_2 and H_2 gases are brought together they react in accordance with the equation

$$N_2 + 3H_2 \rightarrow 2NH_3$$

to produce ammonia. However, at temperatures that permit N_2 and H_2 to react, NH_3 decomposes to give N_2 and H_2. Thus, when the reaction indicated by the above equation is occurring, a reverse reaction indicated by the following equation is occurring simultaneously:

$$2NH_3 \rightarrow N_2 + 3H_2$$

Each reaction produces the substances that are the reactants in the other reaction; so if none of the substances is permitted to escape, neither reaction actually stops, and the equation can be written as follows:

$$N_2 + 3H_2 \rightleftharpoons 2NH_3$$

A state of *chemical equilibrium* is established in which the rate of the forward reaction equals the rate of the reverse reaction. In any such case of incomplete reaction, we cannot compute the weights of the products from the weights of the reactants unless we know how nearly the reaction has approached completion.

New Terms

Avogadro's Number: the number of atoms in a mole (i.e., a gram-mole) of atoms, the number of molecules in a mole of molecules, the number of ions in a mole of ions, the number of electrons in a faraday of electricity, or the number of whatever chemical units one may wish to designate in a mole of such units. The number, N, is 6.023×10^{23}.

Chemical equilibrium: the state which prevails when a chemical reaction is occurring at the same rate as its opposite.

Gram-mole of atoms: the quantity of an element the weight of which in grams is numerically equal to the atomic weight of the element; e.g., since the atomic weight of sodium is 22.9898, a gram-mole of sodium atoms is 22.9898 g.

Mole of atoms: an incomplete expression which should be taken to mean a gram-mole of atoms unless otherwise stated; e.g., a mole of sodium atoms is 22.9898 grams, not 22.9898 pounds, ounces, kilograms, etc.

Pound-mole of atoms: the quantity of an element the weight of which in pounds is numerically equal to the atomic weight of the element; e.g., a pound-mole of sodium atoms is 22.9898 lb.

Reversible reaction: a reaction in which the products are able to interact to produce the starting materials.

Stoichiometry: computation based on the weight relations implicit in chemical equations.

Exercises

6-1 What is the weight of 1 mole of sulfur atoms? Of 2.5 moles of carbon atoms? Of 0.003 mole of tungsten atoms? Of 4.2×10^{-13} mole of iron atoms?

6-2 How many moles of nitrogen atoms are there in 100 g of nitrogen atoms? In 100 g of nitrogen molecules? In 1000 g of potassium nitrate, KNO_3?

6-3 What is the weight of 1 mole of carbon monoxide molecules? Of 5 moles of water molecules ? Of 4×10^{-6} mole of SO_4^{--} ions?

6-4 How many moles of nitrogen molecules are there in 500 g of N_2?

6-5 In a mixture of 2 kg each of H_2 and CO, how many moles of molecules are there? How many moles of atoms?

6-6 How many moles of Al^{+++} ions are there in 1 kg of $Al(NO_3)_3$? In 1 kg of $Al_2(SO_4)_3$?

6-7 How many moles of aluminum sulfate are there in 1 kg of $Al_2(SO_4)_3$?

6-8 (a) How many moles of oxygen are required for the combustion of 3 moles of methane? The equation for the reaction is

$$CH_4 + 2O_2 \rightarrow CO_2 + 2H_2O$$

(b) How many moles of oxygen are necessary to burn 100 g of methane?
(c) How many grams of oxygen are needed to burn 4 moles of methane?
(d) How many grams of oxygen are needed to burn 50 g of methane?
(e) What weight of water is produced if the weight of carbon dioxide obtained is 200 g?

6-9 Given that the atomic weight of mercury is 200.59, calculate the weight of an average mercury atom and express the answer in grams.

6-10 Express the value of 1 amu in grams. How many atomic mass units make 1 g?

References

Hildebrand, J. H., and Powell, R. E., *Principles of Chemistry*, Macmillan, New York, 1964, Chapter 21, "Avogadro's Number." An outline of various methods for determining the number of molecules in a mole.

Kieffer, W. F., *The Mole Concept in Chemistry*, Reinhold, New York, 1962 (paperbound). The many uses of the mole concept in chemical arithmetic.

Roberts, E., *A Programmed Sequence on Exponential Notation*, Freeman, San Francisco, 1962 (paperbound). Developed by Chemical Education Material Study (CHEM Study).

Sienko, M. J., *Stoichiometry and Structure*, Benjamin, New York, 1964 (paperbound). Chapters 6 and 7 are concerned with the mole and calculations based on chemical equations.

Young, J. A., *Chemical Concepts*, Prentice-Hall, Englewood Cliffs, N.J., 1963, pp. 41–72. Programmed instruction. Chemical arithmetic based on the mole concept.

7

Oxygen and the Oxides

7-1 Physical Properties of Oxygen

Oxygen is a colorless, odorless, and tasteless gas, which constitutes one fifth of the earth's atmosphere, where it exists as O_2 molecules. It has a density of 1.429 g per liter at atmospheric pressure and 0°C (these *standard conditions* of temperature and pressure are discussed in Section 9-7). Therefore, it is about 10 percent heavier than air. The other major component of the earth's atmosphere is colorless, odorless, tasteless nitrogen gas, which consists of N_2 molecules.

Under standard conditions only 5.0 ml of O_2 gas will dissolve in 100 ml of water. Under the same conditions, only 2.3 ml of N_2 gas will dissolve. Since one fifth of the molecules in air are O_2 molecules and four fifths are N_2 molecules, the volume of air dissolving in 100 ml of water under standard conditions is

$$(0.2 \times 5.0 \text{ ml}) + (0.8 \times 2.3 \text{ ml}) = 1.0 \text{ ml} + 1.8 \text{ ml} = 2.8 \text{ ml}$$

The dissolved air is about one third O_2 molecules, i.e., 1.0 ml ÷ 2.8 ml = 0.35, an advantageous situation for plants and animals that live under water.

Like all other gases, oxygen can be liquefied, but this requires a low temperature (−119°C) and a pressure of 50 *atmospheres*, i.e., a pressure fifty times as great as the pressure exerted by the atmosphere.

Under a pressure of 1 atmosphere, liquid oxygen boils at −183°C. At −219°C, liquid oxygen freezes to form a bluish-white solid. The density of the pale blue liquid is 1.14 g per ml at its boiling point; the density of the solid is somewhat greater (1.43 g per ml at −252°C), as is true of most substances.

Oxygen gas is *paramagnetic*; i.e., it is feebly attracted into a magnetic field. The liquid and solid are more strongly attracted. Paramagnetic substances contain one or more electrons that are not paired off in the usual way. Consequently, there is a magnetic moment which results in attraction by a magnetic field. This and other considerations indicate that the electronic formula for oxygen may be $:\ddot{\underset{\cdot}{O}}:\ddot{\underset{\cdot}{O}}:$, where the single dots represent unpaired electrons (clearly violating the octet rule). The *bond length* (i.e., distance between the two nuclei) in the O_2 molecule is 1.20 Å.

7-2 Chemical Properties of Oxygen

At ordinary temperatures oxygen is only moderately active chemically, but its activity increases markedly at higher temperatures. Such elements as iron, phosphorus, and sodium slowly combine with oxygen at ordinary temperatures. At temperatures slightly above room temperature, phosphorus bursts into flame, and at higher temperatures sodium and iron burst into flame. At still higher temperatures such elements as carbon and sulfur likewise "take fire," although they are not affected at ordinary temperatures. The hazards of a pure oxygen atmosphere were dramatically and tragically demonstrated by the disastrous fire in the Apollo spacecraft which took the lives of three astronauts in January 1967.

In the oxides, the oxidation state of oxygen is −2 (except in the peroxides and superoxides). Elements that are characterized by more than one oxidation state may form more than one oxide. For example, depending upon the adequacy of the oxygen supply, carbon forms either CO or CO_2. The names usually employed for these two compounds are carbon monoxide and carbon dioxide, respectively. Similarly, the two oxides of sulfur, SO_2 and SO_3, are named sulfur dioxide and sulfur trioxide. In some cases, however, oxides are referred to by names that indicate, by the use of parenthesized Roman numerals, the oxidation state of the element with which the oxygen is combined. Thus, a good name for P_4O_6 is phosphorus(III) oxide, while P_4O_{10} can be called phosphorus (V) oxide. The oxide SnO is known by its new name as tin(II) oxide and by its old name as stannous oxide; SnO_2 can be called tin (IV)oxide, or it can be designated stannic oxide (its old name). MgO is known simply as magnesium oxide because this is the only oxide (other than the peroxide) that is formed by magnesium.

If a very hot iron wire is thrust into pure oxygen, it burns rapidly, throwing off white-hot sparks. The product, at room temperature, is a black solid, Fe_3O_4 (often called "magnetic oxide of iron"), in which the oxidation state of the iron is not at all obvious. It behaves like a mixture of iron(II) oxide, FeO, and iron(III) oxide, Fe_2O_3. Its formula is often written $FeO \cdot Fe_2O_3$. Its scientific name is iron(II,III) oxide.

Some oxides, such as oxides of chlorine, bromine, and iodine, as well as some of the oxides of nitrogen, cannot be obtained by direct union. Oxides of virtually all the chemical elements are known, although some must be prepared by indirect procedures. In the compound OF_2, the fluoride of oxygen, the oxidation state of oxygen must be +2 because fluorine is more electronegative than oxygen and is assigned the −1 state.

A compound will react with oxygen if it is composed of elements that will do so; the reaction of oxygen with such a compound produces oxides of the constituent elements. Thus, methane, the chief constituent of natural gas, reacts as follows:

$$CH_4 + 2O_2 \rightarrow CO_2 + 2H_2O$$

Certain oxides in which an element is not in its highest possible oxidation state will combine with additional oxygen. Thus, carbon monoxide can be used as a fuel gas:

$$2CO + O_2 \rightarrow 2CO_2$$

Such oxides as CO_2 and MgO do not react with oxygen because in these compounds the elements are already in the highest oxidation states possible for them, i.e., all the electrons in their outermost shells are already involved in chemical bonding.

A common *test for oxygen* is made with a glowing wood splint. If a gaseous mixture contains oxygen in high concentration, the oxygen will cause the glowing splint to burst into flame. Nitrous oxide, N_2O, will do the same thing, but its presence can often be ruled out from a knowledge of the source of the sample.

7-3 Occurrence of Oxygen in Nature

In addition to its occurrence in the elementary state in the earth's atmosphere, oxygen is a constituent of many natural compounds. As can be computed from the formula H_2O, oxygen constitutes nearly 89 percent by weight of water—the most abundant of all chemical compounds on earth. The minerals that make up the earth's solid surface, such as sandstone, limestone, and clay, are about one half oxygen and are chiefly silicates and carbonates. Oxygen is also a constituent of the compounds present in living organisms, such as proteins, fats, and carbohydrates. In fact, the human body is nearly two thirds oxygen.

Oxygen is by far the most abundant element in the earth's crust (that portion of the earth accessible for chemical analysis, including the air, the land, the lakes, and the seas). The total weight of the oxygen is approximately the same as the sum of the weights of all the other elements.

Some of the most important ores are oxides of metals. These include hematite, Fe_2O_3; bauxite, Al_2O_3; cuprite, Cu_2O; cassiterite, SnO_2; and pyrolusite, MnO_2.

7-4 Heat of Reaction

When oxygen unites with an element or a compound, heat is usually liberated. Any reaction that is accompanied by the liberation of heat is said to be *exothermic*. Once a highly exothermic reaction has begun, it usually proceeds until one of the reactants has been completely consumed.

In any exothermic reaction there is a definite relationship between the quantity of heat liberated and the quantities of materials reacting. When 1 mole of carbon atoms reacts with 1 mole of oxygen molecules, 1 mole of carbon dioxide and 94.4 kcal of heat are produced. This can be indicated in the equation for the reaction:

$$C + O_2 \rightarrow CO_2 \qquad \Delta H = -94.4 \text{ kcal}$$

By definition, *heat of reaction* is the heat *absorbed* in a reaction. In the above example the heat of reaction is −94.4 kcal. The heat of reaction is negative for any exothermic reaction.

Heat of reaction is represented by the symbol ΔH (read "delta H") because every substance is assumed to have a certain *heat content* or *enthalpy*, H, so that the heat of a reaction is simply the difference between the heat content of the products and the heat content of the reactants:

$$\Delta H_{\text{reaction}} = H_{\text{products}} - H_{\text{reactants}}$$

In an exothermic reaction the heat content of the products is less than the heat content of the reactants; the difference is the heat liberated.

Any reaction that is accompanied by the absorption of heat is said to be *endothermic*. The heat of reaction is just as definite for an endothermic reaction as for an exothermic reaction, and it can be indicated in the equation for the reaction:

$$2H_2O \rightarrow 2H_2 + O_2 \qquad \Delta H = +116 \text{ kcal}$$

In this example the heat of reaction is +116 kcal. The heat of reaction is positive for any endothermic reaction.

If a reaction can be made to proceed in either direction, the heat set free when it proceeds in one direction is the same as the heat absorbed when it goes in the other direction. This is simply an application of the Law of Conservation of Energy.

7-5 Combustion

The term *combustion* is used to denote any reaction in which both light and heat are produced. In ordinary language, it refers to *burning* or *flaming*. Ordinarily, combustion involves reaction with atmospheric oxygen. A temperature sufficiently high to maintain a flame is produced by the heat set free in the reaction.

The temperature at which a combustible substance bursts into flame is called its *kindling temperature* or *ignition point*. Substances that undergo *slow oxidation* below their kindling temperatures, such as phosphorus and oily rags, react at an ever-increasing rate as the temperature rises. If the heat is not removed as rapidly as it is produced, the kindling temperature may be reached and the material will burst into flame. Such an incident is called *spontaneous combustion*.

The kindling temperature of a substance depends on its chemical nature and on the concentration of oxygen in the gaseous mixture (the "atmosphere") in which it is burning. If, after being ignited in air, a burning substance is lowered into a vessel of pure oxygen, the burning becomes much more rapid.

Another factor affecting the kindling temperature is the physical condition of the fuel. Wood chips catch on fire at a lower temperature than large boards.

Combustion ordinarily involves union with oxygen, but this is not always the case. For example, hydrogen will burn in a chlorine atmosphere equally as well as in an oxygen atmosphere. The following equations represent the two similar events:

$$H_2 + Cl_2 \rightarrow 2HCl \qquad \Delta H = -44 \text{ kcal}$$

$$2H_2 + O_2 \rightarrow 2H_2O \qquad \Delta H = -116 \text{ kcal}$$

Originally, the term *oxidation* was used to denote only union with oxygen, but its meaning has been extended to include any instance of an increase in oxidation number. Thus, hydrogen undergoes oxidation when it combines with chlorine as well as when it combines with oxygen, for in both cases the hydrogen goes from the 0 state to the +1 state.

7-6 Importance of Oxygen

The energy required to maintain normal body temperature in an animal and to enable the animal to perform muscular work is obtained from the oxidation of certain materials in the body. The oxygen necessary for this slow but continuous oxidation is normally acquired from the air. In the process of respiration, oxygen passes into the blood in the lungs, combining with hemoglobin to form the red oxyhemoglobin. In this form the oxygen reaches the various tissues throughout the body, where it is consumed in reactions that produce water and carbon dioxide.

Oxygen stored in cylinders is used to produce enriched air (about 50 percent oxygen) for patients suffering from impaired respiration. It is also used in submarines and in high-altitude aircraft.

Power and heat for many industrial operations are obtained by the burning of coal, oil, or gas, using oxygen from the air. Atmospheric oxygen plays an important role in the decay of wood, and dissolved oxygen is involved in the disposal of sewage.

Chemical industries are using pure oxygen on an increasing scale. Even the steel industry is now using enormous quantities of oxygen to remove carbon from iron in the production of steel in open-hearth furnaces. Large quantities of oxygen are used in the cutting and welding of metals with oxyacetylene and oxyhydrogen torches. Liquid oxygen ("LOX") is used as the oxidizing agent with many rocket fuels.

7-7 Oxides of Metals and Nonmetals

In the oxides of the various elements the oxidation states of the elements combined with oxygen vary from +1, as in Cu_2O and H_2O, to +7, as in Cl_2O_7 and Mn_2O_7. Practically all oxides of nonmetals are acidic; i.e., if they dissolve in water, the solutions turn litmus red and show the other characteristics of acid solutions:

$$CO_2 + H_2O \rightarrow \underset{\text{carbonic acid}}{H_2CO_3}$$

$$P_4O_{10} + 6H_2O \rightarrow \underset{\text{phosphoric acid}}{4H_3PO_4}$$

Oxides of metals in which the metals are in their lowest oxidation states are basic; i.e., if they dissolve in water, the solutions turn litmus blue and show the other characteristics of solutions containing bases:

$$Na_2O + H_2O \rightarrow \underset{\text{sodium hydroxide}}{2Na^+ + 2OH^-}$$

$$CaO + H_2O \rightarrow \underset{\text{calcium hydroxide}}{Ca^{++} + 2OH^-}$$

In their highest oxidation states, metals characterized by variable oxidation states form acidic oxides:

$$Mn_2O_7 + H_2O \rightarrow \underset{\text{permanganic acid}}{2H^+ + 2MnO_4{}^-}$$

$$CrO_3 + H_2O \rightarrow \underset{\text{chromic acid}}{2H^+ + CrO_4{}^{--}}$$

The name *oxygen* (Greek: acid former) was given to this element by Lavoisier because he thought it to be the one element present in all acids. He was mistaken in this supposition; there are several acids that do not contain oxygen.

7-8 Acids

The first *acids* discovered were of vegetable origin. Probably acetic acid, the acid in sour wine, was the first one known. *Acid* is derived from the Latin word *acidus*, meaning "sour." Thus, so far as the origin of the name is concerned, the acids are the *sour substances*. As usually happens in the development of a scientific discipline, the meaning of the word has changed considerably since it first came into use. Compounds whose common names include the word *acid* have the following properties:

1. They have a sour taste.
2. They impart characteristic colors to complex substances known as *indicators*; e.g., they cause the vegetable dyestuff known as litmus to develop a red color.
3. They react with bases to form salts and water. For example,

$$\underset{\text{base}}{NaOH} + \underset{\text{acid}}{HCl} \rightarrow \underset{\text{salt}}{NaCl} + \underset{\text{water}}{H_2O}$$

4. They contain hydrogen, and hydrogen gas is produced when active metals, such as zinc and magnesium, react with them. For example,

$$Zn + 2HCl \rightarrow H_2 + ZnCl_2$$

5. In aqueous solutions they give rise to the *hydronium ion*, H_3O^+.

There are thousands of acids, of which the following are quite common:

1. hydrochloric acid, HCl (known commercially as muriatic acid)
2. nitric acid, HNO_3 (known to the alchemists as *aqua fortis*)
3. sulfuric acid, H_2SO_4 (once called oil of vitriol)
4. acetic acid, $HC_2H_3O_2$ (present in vinegar)
5. phosphoric acid, H_3PO_4
6. carbonic acid, H_2CO_3 (present in "carbonated" beverages)

Aqueous solutions of acids are able to conduct electric current because they contain ions that are produced when an acid dissolves in water. For example,

$$H_2O + HCl \rightarrow H_3O^+ + Cl^-$$
$$H_2O + HC_2H_3O_2 \rightleftharpoons H_3O^+ + C_2H_3O_2^-$$

If all, or virtually all, of the acid molecules react with the water, the acid is considered a *strong acid*, e.g., HCl. If only a fraction of the molecules of the acid react with the water (as indicated by the double arrow), the acid is called a *weak acid*, e.g., $HC_2H_3O_2$.

The characteristic acidic properties of an aqueous solution of an acid are due to the presence of hydronium ions, H_3O^+, in the solution. The hydronium ion can be regarded as a hydrogen ion, H^+, to which a water molecule is attached. Instead of writing H_3O^+ for the hydronium ion, it is permissible and convenient to write H^+. The greater the concentration of hydronium ions in the solution, the more acidic the solution, i.e., the more evident the acidic properties are. Thus, a solution of a weak acid is not as sour as a solution of a strong acid. In the presence of a weak acid, i.e., an acid that gives a relatively low concentration of hydronium ions, an indicator may not develop the characteristic color that is evident with strong acids.

7-9 Bases

Historically, *bases* have been regarded as "anti-acids"; i.e., they are substances which cause a solution containing an acid to lose its acidity. Substances commonly regarded as bases have the following properties:

1. They have a peculiar taste, which can be described as caustic or bitter, and they are slippery to the touch.
2. They turn litmus blue and affect other indicators in a characteristic manner.
3. They react with acids to form salts and water. For example,

$$\underset{\text{base}}{2KOH} + \underset{\text{acid}}{H_2SO_4} \rightarrow \underset{\text{salt}}{K_2SO_4} + \underset{\text{water}}{2H_2O}$$

4. Their aqueous solutions contain the *hydroxide ion*, OH^-.

There are many bases, of which the following are well known:

1. sodium hydroxide, $NaOH$ (formerly called caustic soda)
2. potassium hydroxide, KOH (formerly called caustic potash)
3. calcium hydroxide, $Ca(OH)_2$ (known commonly as slaked lime)
4. ammonium hydroxide, NH_4OH (present in ammonia solution)

Unlike the acids, some of the most important bases are ionic compounds; i.e., in the crystalline state they consist of ions, not molecules. Sodium hydroxide, for example, is a crystalline solid consisting of equal numbers of Na^+ and OH^- ions. When sodium hydroxide dissolves in water it *dissociates* (i.e., breaks apart) only in the sense that the polar water molecules attract the Na^+ and OH^- ions and break the bonds between the ions in the crystal. The resulting solution is, of course, a good conductor of electricity because of the presence of the ions.

Some bases are polar covalent compounds. When they dissolve in water they react with the water to produce ions. For example,

$$NH_3 + H_2O \rightleftharpoons NH_4^+ + OH^-$$
$$CH_3NH_2 + H_2O \rightleftharpoons CH_3NH_3^+ + OH^-$$

Usually, but not always, the covalent bases are weak bases; i.e., the aqueous solutions do not conduct electricity as well as do sodium hydroxide and the other ionic bases. Some bases that are commonly referred to as weak bases owe their apparent weakness to their slight solubility. For example, $Ba(OH)_2$ and $Ca(OH)_2$ are ionic compounds, but their saturated aqueous solutions do not contain a high concentration of OH^- ions because these compounds are not highly soluble in water.

The characteristic basic, or *alkaline*, properties of aqueous solutions of bases are due to the hydroxide ions, OH^-, in the solution. The greater the concentration of hydroxide ions in a solution, the greater the extent to which its alkaline properties are evident. An indicator that shows its "basic color" when a strong base is present may fail to do so if the base is weak, for a certain minimum concentration of hydroxide ions is required to affect the indicator.

7-10 Salts

The *salts* have been regarded historically as the compounds which, together with water, are obtained when the appropriate acids and bases interact. For example, sodium hydroxide ($NaOH$) and hydrochloric acid (HCl) interact to form water and a salt, sodium chloride ($NaCl$). The salt consists of the positive ion of the parent base and the negative ion of the parent acid. Well-known salts include the following:

1. sodium chloride, $NaCl$ (table salt, rock salt)
2. potassium iodide, KI
3. calcium carbonate, $CaCO_3$ (limestone)
4. lead arsenate, $Pb_3(AsO_4)_2$
5. silver nitrate, $AgNO_3$ (formerly called lunar caustic)

Crystalline salts were once thought to consist of molecules. Within twenty years of the discovery of X rays, crystallographers were using them to investigate the structures of crystals. These investigations indicate, beyond any reasonable doubt, that crystalline salts are composed of ions. Thus, the "electrolytic dissociation" that occurs when a salt dissolves in water consists of polar water molecules attracting the positive and negative ions and breaking the bonds between the ions in the crystal. This, not the dissociation of molecules into ions, is what produces a solution capable of conducting electricity. Confirmation of this interpretation is obtained by testing the conducting abilities of molten salts. Without exception, if a salt is heated until it melts, the resulting liquid is a conductor of electricity. Since a molten salt is liquid, the ions of which it consists are free to move. In the solid state, though, the ions are firmly held in place by the electrical forces of attraction that exist between each ion and its oppositely charged neighbors.

How did the idea originate that, for example, the calcium ions in calcium chloride have twice as much charge as the sodium ions in sodium chloride? The idea is actually based on a simple principle—the electroneutrality of solutions and ionic solids. The formulas of these two compounds are $CaCl_2$ and NaCl, respectively. The positive sodium ion, Na^+, therefore must bear the same quantity of charge as the negative chloride ion, Cl^-, but there are two of the Cl^- ions in the calcium chloride, so that the positive calcium ion must have twice the amount of charge carried by a chloride ion or a sodium ion. This is indicated by symbolizing it as Ca^{++} or Ca^{2+}.

When the charge of one of the ions of a salt is known, the charge of the other ion can be derived if the formula for the salt is known; e.g., knowing that manganese nitrate, $Mn(NO_3)_2$, gives NO_3^- ions enables us to see that the manganese ion is bipositive, i.e., Mn^{++}. Conversely, we can derive the formulas of salts from the charges of their ions, since the salts are electrically neutral (Table 7-1).

Table 7-1

Relationship between formulas of salts and the charges of their constituent ions

Positive ion	*Negative ion*	*Salt*
Ba^{++}	SO_4^{--}	$BaSO_4$
Al^{+++}	SO_4^{--}	$Al_2(SO_4)_3$
Ba^{++}	NO_3^-	$Ba(NO_3)_2$
Al^{+++}	NO_3^-	$Al(NO_3)_3$
Al^{+++}	PO_4^{---}	$AlPO_4$
Ba^{++}	PO_4^{---}	$Ba_3(PO_4)_2$
K^+	PO_4^{---}	K_3PO_4
Ag^+	S^{--}	Ag_2S

7-11 Neutralization

When H^+ and OH^- ions are brought together from separate sources, they combine to form H_2O molecules. This process, known as *neutralization*, is typical of acids and bases. In addition to water, a salt is produced in the neutralization process. For example,

$$NaOH + HCl \rightarrow NaCl + H_2O$$

The formula of each of the salts listed in Table 7-2 can be predicted from the formulas of its parent acid and parent base. Thus, we can easily write and balance the equation for any neutralization reaction.

EXAMPLE Complete and balance the equation for the neutralization of sulfuric acid with sodium hydroxide, given their formulas, H_2SO_4 and NaOH.

Since this is a neutralization reaction, the products are water and a salt. The formula for water can conveniently be written here as HOH, for each water molecule is formed from one H^+ ion (from the acid) and one OH^- ion (from the base). The formula for the salt can be determined from the charges of the positive ion of the base, Na^+, and the negative ion of the acid, SO_4^{--}. Thus, before balancing, we have

$$NaOH + H_2SO_4 \rightarrow Na_2SO_4 + HOH \quad \text{(not balanced)}$$

If you learned how to balance equations from Chapter 2, you will have no difficulty in arriving at the final result:

$$2NaOH + H_2SO_4 \rightarrow Na_2SO_4 + 2HOH$$

Table 7-2

Relationship between formulas of salts and formulas of the corresponding bases and acids

Base	*Acid*	*Salt*
NaOH	HCl	NaCl
KOH	HNO_3	KNO_3
$Ca(OH)_2$	H_2SO_4	$CaSO_4$
NH_4OH	H_3PO_4	$(NH_4)_3PO_4$
NaOH	H_2CO_3	Na_2CO_3
$Ca(OH)_2$	$HC_2H_3O_2$	$Ca(C_2H_3O_2)_2$
$Ca(OH)_2$	H_3PO_4	$Ca_3(PO_4)_2$

7-12 Anhydrides of Acids and Bases

The *anhydride* of an acid is the oxide which is formed when the acid loses the elements of water; most such oxides will react with water to form the corresponding acid. The formula for an *acidic anhydride* may readily be determined from the formula of the acid, since the oxidation state of each element is the same in the acid as in its anhydride. For example, let us determine the formula for the anhydride of silicic acid, H_2SiO_3. Since the oxidation number of silicon in H_2SiO_3 is +4, the formula for the anhydride of H_2SiO_3 is SiO_2 (silicon dioxide), for only this formula would show silicon and oxygen in oxidation states +4 and −2, respectively. By similar reasoning the formula for the anhydride of HNO_3 (nitric acid) is N_2O_5, for only this formula would show nitrogen and oxygen in the oxidation states +5 and −2, respectively.

The anhydride of a base is the oxide formed when the elements of water are removed from the base. Many *basic anhydrides* react with water to form the corresponding base. The oxidation state of each element is the same in the oxide as in the base itself. We can use this relationship to infer the formula of the oxide from the formula of the base. For example, let us determine the formula for the anhydride of the base calcium hydroxide, $Ca(OH)_2$. Since the oxidation state of calcium in $Ca(OH)_2$ is +2, the formula for its anhydride is CaO. Similarly, the formula for the anhydride of $Fe(OH)_3$ is Fe_2O_3.

The formula for the anhydride of an acid or base may easily be derived without reference to oxidation numbers by mentally removing all "H" as water. Thus, removing both H atoms as H_2O from H_2SiO_3 leaves SiO_2; removing both H atoms as H_2O from $Ca(OH)_2$ leaves CaO. This method is slightly more complicated in such cases as $Fe(OH)_3$ or H_3PO_4, for after removal of one water two molecules must be combined to remove the remaining hydrogen.

Writing equations for the decomposition of bases or *oxyacids* (acids containing oxygen, hydrogen, and a third element) to produce water and the anhydride of the base or acid is easy if we recall how the formula for the anhydride can be determined from the formula of the base or acid.

EXAMPLE Complete the equation

$$Al(OH)_3 \rightarrow$$

Since the oxidation state of aluminum in the hydroxide (base) is +3, the formula for the anhydride must be Al_2O_3. The first step in completing the equation is to write the formulas for the products:

$$Al(OH)_3 \rightarrow Al_2O_3 + H_2O \quad \text{(not balanced)}$$

Then we must balance the equation by inserting the necessary coefficients:

$$2Al(OH)_3 \rightarrow Al_2O_3 + 3H_2O$$

EXAMPLE Complete the equation

$$H_2CO_3 \rightarrow$$

Since the oxidation state of carbon in the acid is +4, the formula for the anhydride is CO_2, and the equation is completed as follows:

$$H_2CO_3 \rightarrow H_2O + CO_2$$

7-13 Predicting Formulas for Bases and Acids

Equations for the reactions of basic or acidic oxides with water must be written with caution. In the base or acid obtained in such a reaction, the element is in the same oxidation state as in the oxide. For example,

$$CaO + H_2O \rightarrow Ca(OH)_2$$
$$SO_3 + H_2O \rightarrow H_2SO_4$$

On this basis, one can generally predict the formula for the acid or base corresponding to the oxide. For example, the formula for the hydroxide corresponding to the oxide K_2O is KOH and that for MgO is $Mg(OH)_2$. For the oxide Al_2O_3 there is a hydroxide with the formula $Al(OH)_3$, but there is also a compound with the formula AlO(OH)—and Al_2O_3 is the anhydride of both of them.

There is no way of predicting that sulfur trioxide should react with water to give H_2SO_4; the formula could conceivably be H_4SO_5 (the sulfur would be in the +6 state). The formula for sulfuric acid is H_2SO_4 rather than H_4SO_5, and nitric acid is HNO_3 rather than H_3NO_4. On the other hand there are three phosphoric acids, with the formulas HPO_3, H_3PO_4, and $H_4P_2O_7$; they all have the same anhydride because phosphorus is in the +5 state in each.

7-14 Reactions of Basic and Acidic Oxides

When a basic oxide reacts with an acid, the products are water and the same salt that would result from reaction of the acid with the corresponding basic hydroxide. For example, from the equation for the reaction of calcium hydroxide with nitric acid,

$$(Ca^{++} + 2OH^-) + 2(H^+ + NO_3^-) \rightarrow (Ca^{++} + 2NO_3^-) + 2HOH$$

you can infer that the reaction of the basic oxide, CaO, with nitric acid will proceed according to the equation

$$CaO(s) + 2(H^+ + NO_3^-) \rightarrow (Ca^{++} + 2NO_3^-) + H_2O$$

In this equation we have written the formula for calcium oxide in "molecular" rather than ionic form because it is a solid, as indicated by the letter *s* in parentheses, even though it is an ionic solid.

The reaction of an acidic oxide with a base produces water and the same salt that would be obtained if the base were reacting with the acid itself. Thus, from the equation for the reaction of nitric acid with potassium hydroxide,

$$(H^+ + NO_3^-) + (K^+ + OH^-) \rightarrow (K^+ + NO_3^-) + HOH$$

we can write the equation for the reaction of potassium hydroxide with the acidic oxide N_2O_5:

$$N_2O_5 + 2(K^+ + OH^-) \rightarrow 2(K^+ + NO_3^-) + H_2O$$

The reaction of a basic oxide and an acidic oxide produces the same salt that would be obtained by a reaction between the corresponding base and acid. From the equation

$$2(K^+ + OH^-) + \underset{\text{weak acid}}{H_2CO_3} \rightarrow (2K^+ + CO_3^{--}) + 2HOH$$

you can infer that the reaction between solid potassium oxide and gaseous carbon dioxide will proceed according to the equation

$$K_2O(s) + CO_2(g) \rightarrow K_2CO_3(s)$$

A parenthetical *g* indicates that a substance is in the gaseous state.

Many oxides are not soluble enough in water to be readily identified as either acidic or basic. However, if such an oxide "dissolves" when treated with a strong acid, its basic character is thereby established:

$$MgO + 2H^+ \rightarrow Mg^{++} + H_2O$$

If an oxide goes into solution by chemical reaction when treated with a strong base, it is identified as an acidic oxide:

$$SnO_2 + 2OH^- \rightarrow SnO_3^{--} + H_2O$$

Certain oxides that are insoluble in water will go into solution when treated with either a strong acid or a strong base:

$$SnO + 2H^+ \rightarrow Sn^{++} + H_2O$$

$$SnO + 2OH^- \rightarrow SnO_2^{--} + H_2O$$

Such oxides are said to be *amphoteric*, and the behavior itself is called *amphoterism*.

7-15 Peroxides and Superoxides

Binary compounds (compounds composed of two elements) that contain oxygen in the oxidation state -1 are called *peroxides*. Sodium peroxide, Na_2O_2, is obtained by the reaction of sodium with oxygen; in order to obtain the normal oxide, Na_2O, by this process, it is necessary to heat sodium in a limited supply of air under carefully controlled conditions. Sodium peroxide is an ionic solid consisting of Na^+ and O_2^{--} ions. The peroxide ion has the electronic structure

$\left[:\ddot{\underset{..}{O}}:\ddot{\underset{..}{O}}:\right]^{--}$. Metal peroxides are salts of the weak acid hydrogen peroxide, H_2O_2, which is to say that hydrogen peroxide can be obtained by reaction of a strong acid with a metal peroxide. For example,

$$BaO_2(s) + 2H^+ \rightarrow Ba^{++} + H_2O_2$$

The heavier alkali metals (potassium, rubidium, and cesium) react with oxygen to form *superoxides* in which oxygen is in the oxidation state $-\frac{1}{2}$. An example is potassium superoxide, KO_2. These solids are colored and paramagnetic because the superoxide ion, O_2^-, contains an unpaired electron. This ion exists only in the solid state. In the presence of water, superoxides produce oxygen and hydrogen peroxide:

$$2KO_2(s) + 2H_2O \rightarrow O_2\uparrow + H_2O_2 + 2K^+ + 2OH^-$$

7-16 Preparation of Oxygen

The industrial source of oxygen is air. Since here the element is not chemically combined, its preparation necessitates only a physical separation from the other components of the atmosphere, chiefly nitrogen. To separate the oxygen from the nitrogen, air is liquefied (by compression and cooling). Since the boiling point of liquid oxygen (−183°C) is 13° higher than that of liquid nitrogen (−196°C), the nitrogen boils away from the liquid air, leaving the oxygen behind. This method of separation is called *fractional distillation* of liquid air. The oxygen and nitrogen are often stored and shipped in steel cylinders under pressures exceeding 100 atmospheres.

Laboratory methods of preparing oxygen are often chosen on the basis of convenience rather than cost of the starting materials. The most commonly used method consists of the thermal decomposition of potassium chlorate, $KClO_3$, a man-made compound consisting of potassium ions, K^+, and chlorate ions, ClO_3^-. The three oxygen atoms are covalently bonded to the chlorine atom in the chlorate ion. The white crystalline solid melts when heated to 368°C; if heated to 400°C, the chlorate decomposes, releasing oxygen and producing the chloride:

$$2KClO_3 \rightarrow 2KCl + 3O_2\uparrow$$

The same reaction occurs rapidly when a mixture of potassium chlorate and a small amount of the black powder manganese dioxide (MnO_2) is heated to only 250°C. The weight of the manganese dioxide recoverable after the decomposition of the potassium chlorate is the same as the original weight. Although manganese dioxide does not itself undergo decomposition at this temperature, it causes the decomposition of the chlorate to occur rapidly at a much lower temperature. Here, the manganese dioxide acts as a *catalyst*; its effect in causing the reaction to occur more rapidly than it otherwise would is called *catalysis*. Many industrial processes, as well as numerous laboratory operations, are successful because of the catalysts used to speed up certain reactions.

Other compounds of the salt type liberate at least part of their oxygen upon being heated; e.g., sodium nitrate loses one third of its oxygen:

$$2NaNO_3 \rightarrow 2NaNO_2 + O_2\uparrow$$

Oxides of a few of the least active metals undergo complete decomposition when heated:

$$2HgO \rightarrow 2Hg + O_2$$
$$2Ag_2O \rightarrow 4Ag + O_2$$

Oxides of some metals (metals that can have more than one oxidation state) liberate only part of their oxygen when they are heated. Such reactions usually require very high temperatures:

$$\underset{\text{lead(IV) oxide}}{2PbO_2} \rightarrow O_2 + \underset{\text{lead(II) oxide}}{2PbO}$$

$$\underset{\text{manganese(IV) oxide}}{3MnO_2} \rightarrow O_2 + \underset{\text{manganese(II, III) oxide}}{Mn_3O_4}$$

Joseph Priestley is credited with the discovery of oxygen in 1774 by heating mercury(II) oxide (Section 1-3). Karl Scheele, in Sweden, had obtained oxygen gas by the action of heat on several compounds in the previous year, but his work was not published until 1777.

When oxygen is prepared by thermal decomposition of a solid, the solid is placed in a test tube and heated. A delivery tube leads the oxygen into a collecting bottle (which is originally filled with water and inverted in a trough of water). As the bubbles of oxygen gas emerge from the end of the delivery tube, they rise to the upper part of the inverted bottle, and the level of the water recedes. In order for any gas to be thus collected by displacement of a liquid, the gas must not be highly soluble in that liquid.

Another reaction that can be employed for the generation of oxygen in a laboratory is the reaction between water and sodium peroxide. The hydrogen peroxide resulting from the reaction undergoes rapid decomposition by the catalytic effect of the OH^- ions:

$$Na_2O_2(s) + 2HOH \rightarrow H_2O_2 + 2Na^+ + 2OH^-$$
$$2H_2O_2 \rightarrow 2H_2O + O_2\uparrow$$

Since the hydrogen peroxide decomposes as fast as it forms, it seems reasonable to combine the two equations into one expression by addition. Before doing this, we must first double all coefficients in the first equation so that we have 2 moles of H_2O_2 to decompose in the second:

$$2Na_2O_2(s) + 4HOH \rightarrow 2H_2O_2 + 4Na^+ + 4OH^-$$
$$2H_2O_2 \rightarrow 2HOH + O_2\uparrow$$
$$\overline{2Na_2O_2(s) + 2HOH \rightarrow 4Na^+ + 4OH^- + O_2\uparrow}$$

The electrolysis of water is another process that can be used to generate oxygen. Only a very small fraction of water, a covalent compound, is dissociated

Figure 7-1

Electrolysis of water

The volume of hydrogen obtained at the cathode is twice the volume of oxygen obtained at the anode.

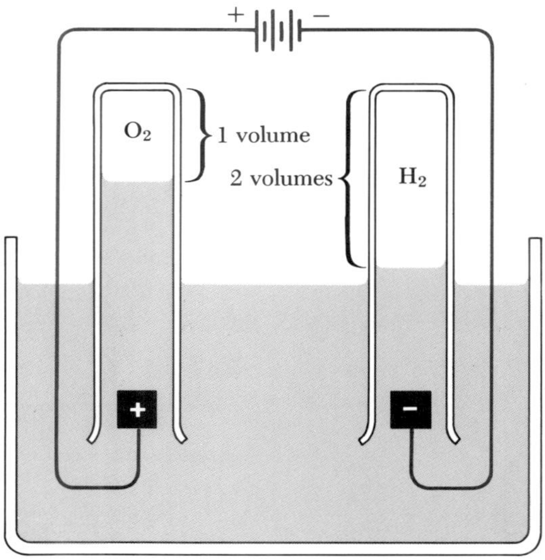

into ions, so water is essentially a nonconductor of electricity. If a small amount of sulfuric acid is dissolved in water, current passes readily through the solution. With the passage of direct current, hydrogen gas is liberated at the negative electrode and oxygen gas at the positive electrode:

$$2H_2O \rightarrow 2H_2\uparrow + O_2\uparrow$$

Since the number of H_2 molecules produced is twice the number of O_2 molecules liberated, the volume of hydrogen gas obtained is twice the volume of oxygen (Figure 7-1).

7-17 Ozone

When energy is added to ordinary diatomic oxygen under certain conditions, the triatomic molecule O_3, called *ozone*, is formed:

$$3O_2(g) + 68 \text{ kcal} \rightarrow 2O_3(g)$$

Not much O_2 can be converted to O_3 by simply adding heat, but significant quantities of ozone are produced if energy is added in certain other forms, notably electrical energy and high-energy radiation. The most satisfactory way to prepare O_3 is to pass a silent electrical discharge through O_2. An ozonizer, the apparatus for this conversion, consists of two concentric glass tubes with

Figure 7-2

The ozonizer

(a) End view of concentric tubes. (b) Side view, showing connections to induction coil generating 10,000 volts.

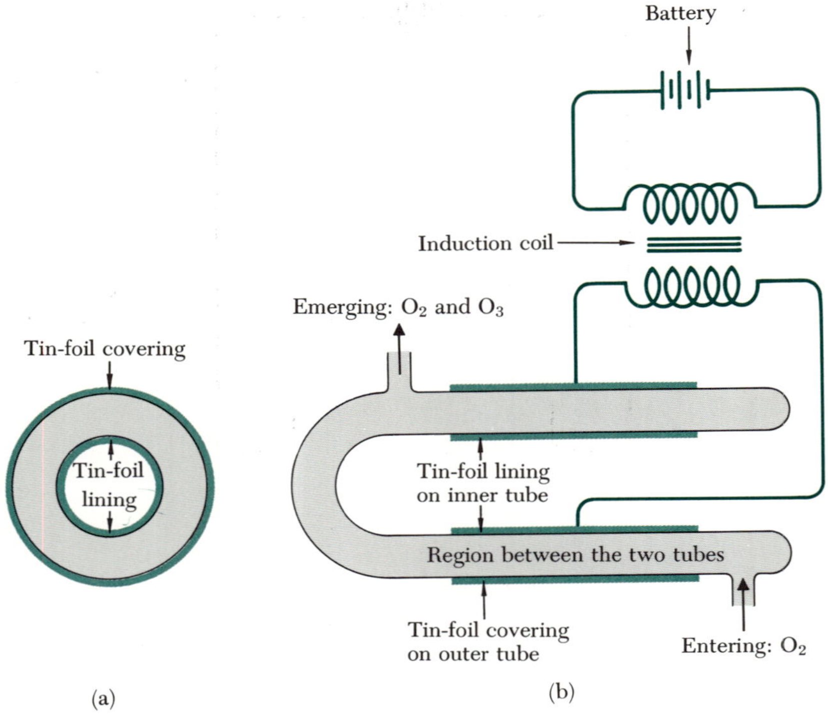

O_2 flowing through the annular space between them (Figure 7-2). A layer of tin foil is fastened to the inside of the inner tube and to the outside of the outer tube. These tin-foil layers are connected to an alternating-current source of 10,000 volts. Usually, only a small percentage of the O_2 molecules are converted to O_3 molecules as they pass through the ozonizer.

Ozone is formed to some extent by bolts of lightning and by sparking electric motors. It exists in the upper atmosphere, where it is formed by ultraviolet radiation from the sun. There it serves a useful purpose in absorbing most of the ultraviolet solar radiation, shielding living beings on the surface of the earth from fatal ultraviolet damage.

Ozone is a pale blue gas with a pungent odor. On standing, it slowly reverts to O_2. At high concentrations, O_3 tends to undergo spontaneous decomposition with explosive violence, so mixtures of O_2 and O_3 containing more than 13 percent O_3 are hazardous:

$$2O_3 \longrightarrow 3O_2 \qquad \Delta H = -68 \text{ kcal}$$

Ozone is more easily liquefied than diatomic oxygen. The normal boiling point of the deep-blue liquid, i.e., the temperature at which it boils in a vessel open to the atmosphere, is $-112°C$, or 71° higher than the boiling point of O_2. Ozone freezes to a dark-violet solid at $-250°C$.

Because of the greater internal energy of its molecules (indicated by its endothermic formation from O_2 and its exothermic reversion to the O_2 state), O_3 generally reacts with oxidizable substances more rapidly than does O_2. Ozone also reacts with some substances that are not affected by O_2. As an oxidizing agent, it is nearly as strong as elementary fluorine (the strongest one known). Ozone is used commercially for bleaching oils, waxes, flour, and paper pulp and for deodorizing air and sterilizing drinking-water. It is under investigation for possible use as an oxidizing agent for rocket fuels.

Ordinary diatomic oxygen, O_2, and the triatomic ozone, O_3, are both gases under ordinary conditions, and both are varieties of elementary oxygen. They constitute an example of the phenomenon called *allotropy* or *allotropism* (Greek: *allotropia*, variety), which is defined as the existence of an element in two or more different forms in the same physical state. Thus, O_2 and O_3 are the *allotropes* or *allotropic forms* of elementary oxygen. Only the O_3 has a special name. Both O_2 and O_3 can exist in all three states (gaseous, liquid, and solid)—an unusual situation for the allotropic forms of an element.

The O_3 molecule is not paramagnetic. Unlike the O_2 molecule, it presumably has all its electrons paired. The three atoms in the O_3 molecule are arranged in the form of an isosceles triangle in which the atoms at the base are not bonded to each other. The angle is 127°.

If the octet rule is followed, it is possible to write the electronic formula in two different ways:

Each of these formulas shows one pair of shared electrons for one of the bonds and two pairs of shared electrons for the other bond. In other words, each formula represents one bond as single and the other as double. If one bond were really single and the other one really double, we would expect, from our knowledge of single and double bonds in other molecules, that the bond lengths and strengths would be different. Actually, the central atom is the same distance, 1.26 Å, from each of the other atoms.

According to the *theory of resonance*, proposed by Linus Pauling in 1931, if two (or more) reasonable electronic structures can be written for a molecule, the "real" structure is not identical to either of them but is a *resonance hybrid* that is intermediate in character between the two. Thus, the "real" structure of the ozone molecule is not correctly represented by either of the structures shown above but is a resonance hybrid that partakes of the character of both. The two bonds are equivalent; each is more than single but less than double.

No satisfactory symbolism has been invented for representing resonance structures; i.e., no one electronic formula can represent a resonance hybrid. The two formulas written for the O_3 are referred to as *contributing structures.* It must be kept in mind that these structures do not correspond to different kinds of molecules—there is only one kind of ozone molecule, and its electronic structure is a hybrid of the two contributing structures shown.

7-18 Shapes of Triatomic Molecules and Ions

As shown above, the O_3 molecule is *angular* rather than *linear.* A useful rule for predicting the arrangement in space of the atoms in a triatomic molecule or ion is the following:

1. The arrangement is linear if the central atom has no unshared electron pairs, as in the cases of carbon dioxide and carbon disulfide:

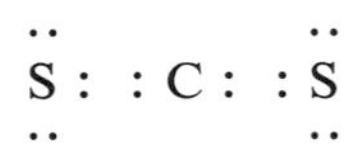

2. The arrangement is angular if the central atom has one or more unshared pairs, as in the water molecule and the chlorite ion:

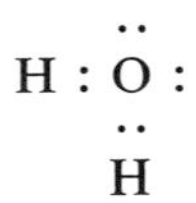

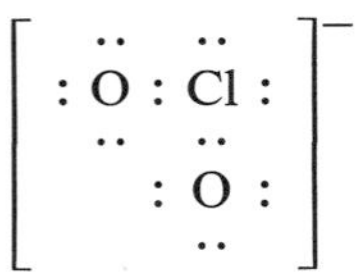

New Terms

Acid: a compound that gives rise to hydronium ions (H_3O^+) in an aqueous solution. Acids react with bases to form salts and water.

Acidic anhydride: an acidic oxide, designated as the anhydride of a specified acid; e.g., we say that SO_2 is an acidic oxide, or that SO_2 is the anhydride of sulfurous acid.

Allotropy: the existence of a free element in two or more forms in the same physical state. O_2 and O_3 are the allotropic forms of free oxygen. Graphite and diamond, both crystalline solids, are allotropes of carbon.

Amphoterism: the ability to act as an acid and also as a base.

Anhydride: the oxide formed when an acid or base loses the elements of water. The oxidation state of each element is the same in the oxide as in the acid or base of which it is the anhydride.

Base: a compound that gives rise to hydroxide ions (OH^-) in an aqueous solution. Bases react with acids to form water and salts.

Basic anhydride: a basic oxide, designated as the anhydride of a specified base; e.g., we say that CaO is a basic oxide, or that CaO is the anhydride of the base calcium hydroxide.

Catalysis: the increase in the rate of a reaction resulting from the presence of a substance, called the catalyst, which is still present in the same quantity at the end of the reaction.

Chemical resonance: the phenomenon shown by a molecule or ion to which two or more electronic structures, differing only in the disposition of electrons, can be assigned. In such instances, no single electronic formula can be assigned that conforms to the octet rule and to the observed properties of the substance.

Combustion: any chemical process accompanied by the evolution of light and heat. Commonly, combustion involves union with oxygen.

Endothermic process: a process of any kind in which heat is absorbed from the surroundings, as in the melting of ice, the evaporation of a liquid, or the union of carbon and sulfur to form CS_2. Any chemical reaction in which heat is absorbed is an endothermic reaction.

Exothermic process: a process of any kind in which heat is evolved, as in the liquefaction of a gas, the freezing of a liquid, or the burning of a candle. Any chemical reaction in which heat is evolved is an exothermic reaction.

Heat of reaction: the quantity of heat absorbed when a reaction occurs. The heat of reaction is positive for an endothermic reaction and negative for an exothermic reaction.

Indicator: a substance, natural or synthetic, which exhibits different colors in aqueous solutions, depending on the degree of acidity or alkalinity of the solutions.

Kindling temperature: the temperature at which a substance bursts into flame.

Neutralization: the process in which H_3O^+ and OH^- ions, brought together from separate sources, combine to form H_2O molecules. The reaction between an acid and a base, producing water and a salt, is a typical neutralization reaction.

Oxidation: an increase in oxidation number. (A decrease in oxidation number is *reduction.*)

Oxide: a compound containing oxygen and one other element.

Oxyacid: an acid which contains both hydrogen and oxygen, combined with a third element.

Paramagnetic: subject to being attracted into a magnetic field. (Substances subject to repulsion by a magnetic field are *diamagnetic.*)

Peroxides: binary compounds that contain oxygen in the oxidation state -1.

Resonance hybrid: a molecule or ion to which no single electronic formula can be assigned that conforms to the observed properties of the substance. Instead, the structure is a hybrid between two (or more) contributing structures.

Salt: an ionic compound consisting of the cation of a base and the anion of an acid.

Spontaneous combustion: combustion resulting from self-ignition, the heat evolved during slow oxidation being responsible for raising the temperature to the kindling temperature.

Standard conditions: atmospheric pressure and 0°C.

Strong acid: an acid which gives rise to a high concentration of ions in its solutions, enabling the solutions to serve as good conductors of electricity.

Strong base: a base which gives rise to a high concentration of ions in solution, thus making a solution of a strong base a good conductor of electricity.

Superoxides: binary compounds of oxygen in which the oxidation state of oxygen is $-\frac{1}{2}$.

Weak acid: an acid which gives rise to only a small concentration of ions in aqueous solution.

Weak base: a base which gives rise to a relatively small concentration of ions in aqueous solution.

Exercises

7-1 Compare the reactions of sodium oxide and sodium peroxide with water.

7-2 Describe in some detail a laboratory method for the preparation of oxygen, and explain why it is not used commercially.

7-3 Describe the commercial preparation of oxygen, and explain why it is not used as a laboratory method.

7-4 Why is ozone not storable?

7-5 List some compounds that will not react with oxygen, and explain why they will not.

7-6 If $KClO_3$, Na_2O_2, $NaNO_3$, and HgO all cost the same amount per kilogram (which is certainly not the case), which would be the most economical source of oxygen?

7-7 Which would give the higher temperature, a gas-air torch or a gas-oxygen torch? Why?

7-8 Illustrate the reactions of a nonmetallic oxide, a basic metallic oxide, and an acidic metallic oxide with water.

7-9 Define exothermic reaction, spontaneous combustion, allotropy, resonance, and amphoterism.

7-10 How many grams of oxygen could be obtained by complete thermal decomposition of 10 moles of silver oxide?

7-11 *Neutralization.* Write and balance the following neutralization equations:

(a) $KOH + HBr \rightarrow$
(b) $CsOH + H_2SeO_4 \rightarrow$
(c) $NH_4OH + H_3PO_4 \rightarrow$
(d) $Sr(OH)_2 + HNO_3 \rightarrow$
(e) $Fe(OH)_2(s) + H_2SO_4 \rightarrow$
(f) $Cu(OH)_2(s) + H_3PO_4 \rightarrow$
(g) $Fe(OH)_3(s) + HBrO_2 \rightarrow$
(h) $Ga(OH)_3(s) + H_3AsO_4 \rightarrow$

7-12 *Decomposition of acids and bases.* Complete and balance the following equations (in each case, water is one of the two products, and the anhydride of the acid or base is the other):

(a) $Ba(OH)_2 \rightarrow$
(b) $Fe(OH)_3 \rightarrow$
(c) $Cu(OH)_2 \rightarrow$
(d) $HNO_3 \rightarrow$
(e) $H_3PO_4 \rightarrow$
(f) $H_2CO_3 \rightarrow$
(g) $Zn(OH)_2 \rightarrow$
(h) $H_2ZnO_2 \rightarrow$
(i) $H_2SiO_3 \rightarrow$
(j) $H_4SiO_4 \rightarrow$
(k) $HMnO_4 \rightarrow$
(l) $Ga(OH)_3 \rightarrow$
(m) $Th(OH)_4 \rightarrow$

7-13 *Reactions of basic oxides with acids.* Complete and balance:

(a) $Na_2O + HCl \rightarrow$
(b) $K_2O + H_2SO_4 \rightarrow$
(c) $CaO + HNO_3 \rightarrow$
(d) $BaO + H_2SO_4 \rightarrow$
(e) $Al_2O_3 + HCl \rightarrow$
(f) $Al_2O_3 + H_2SO_4 \rightarrow$
(g) $Na_2O + H_3PO_4 \rightarrow$
(h) $CaO + H_3PO_4 \rightarrow$
(i) $Al_2O_3 + H_3PO_4 \rightarrow$
(j) $ThO_2 + H_3PO_4 \rightarrow$

7-14 *Reactions of acidic oxides with bases.* Complete and balance:

(a) $NaOH + SO_3 \rightarrow$
(b) $Ca(OH)_2 + CO_2 \rightarrow$
(c) $Al(OH)_3 + P_4O_{10} \rightarrow$
(d) $NaOH + CO_2 \rightarrow$
(e) $Ca(OH)_2 + N_2O_5 \rightarrow$
(f) $Ga(OH)_3 + SO_2 \rightarrow$
(g) $Th(OH)_4 + Mn_2O_7 \rightarrow$

7-15 *Reactions of acidic oxides with basic oxides.* Complete and balance:

(a) $CaO + CO_2 \rightarrow$
(b) $Na_2O + SO_3 \rightarrow$
(c) $Al_2O_3 + P_4O_{10} \rightarrow$
(d) $Ga_2O_3 + SO_2 \rightarrow$
(e) $BaO + SO_3 \rightarrow$
(f) $ThO_2 + P_4O_{10} \rightarrow$
(g) $CaO + Mn_2O_7 \rightarrow$
(h) $CuO + CO_2 \rightarrow$
(i) $Fe_2O_3 + CrO_3 \rightarrow$

References

Azimov, I., "The Composition of the Atmosphere," *J. Chem. Educ.*, **32** (12), 633 (1955). A discussion of the relative amounts of the different isotopic species of oxygen and nitrogen in the atmosphere.

Dole, M., "History of Oxygen," *Science*, **109**, 77 (1949). A discussion of the origin of oxygen as an element and of oxygen as a free element in the earth's atmosphere.

Farber, E., "Oxygen, the Element with Two Faces," *Chemistry*, **39** (5), 17 (1966). The life-giving and life-taking aspects of our most abundant element.

Gaidis, J. M., and Rochow, E. G., "The Catalyzed Decomposition of Potassium Chlorate," *J. Chem. Educ.*, **40** (2), 78 (1963). Report on investigations of the role played by MnO_2 and the factors that determine its effectiveness.

Miller, L. E., "Chemistry in the Stratosphere and Upper Atmosphere," *J. Chem. Educ.*, **31** (3), 112 (1954). The variation in the chemical nature of the atmosphere with altitude.

Partington, J. R., "The Discovery of Oxygen," *J. Chem. Educ.*, **39** (3), 123 (1962). A consideration of the independent discoveries of Priestley and Scheele by an eminent chemical historian.

Watt, G. W., Hatch, L. F., and Lagowski, J. J., *Chemistry*, Norton, New York, 1964, Chapter 7, "The Atmosphere." Normal and accidental components of the atmosphere are discussed.

8

The Chemistry of Hydrogen

8-1 Physical Properties

Hydrogen, at ordinary temperatures and pressures, exists as a gas which is colorless, odorless, tasteless, and only very slightly soluble in water. The density of the gas at standard conditions is 0.08987 g per liter, making it the lightest substance known. It is only one sixteenth as dense as oxygen and one fourteenth as dense as air (Figure 8-1). Since its molecular weight is about 2, or twice its atomic weight, we know that the gas is diatomic, i.e., its formula is H_2.

The H_2 molecule consists of two hydrogen nuclei, 0.75 Å apart, sharing a pair of electrons. Its nonpolar character prevents it from being appreciably soluble in water; it is, therefore, feasible to collect it and store it over water. The nonpolar character of the molecule also makes difficult the liquefaction of hydrogen. The gas cannot be liquefied at temperatures above −240°C; this temperature, therefore, is known as the *critical temperature* of hydrogen. The normal boiling point of the liquid is −253°C. Liquid hydrogen, the lightest liquid known, has a density of only 0.07 g per ml at the boiling point. Hydrogen was first liquefied in 1898 by Sir James Dewar, Scottish chemist and physicist. In 1899 Dewar found that liquid hydrogen freezes to a transparent solid at −259°C.

8-2 Chemical Properties

At suitable temperatures elementary hydrogen combines directly with many elements. The reaction with oxygen is exothermic, releasing large amounts of energy:

$$2H_2(g) + O_2(g) \rightarrow 2H_2O(g) \qquad \Delta H = -116 \text{ kcal}$$

The above equation indicates that the *heat of combustion* of H_2 gas is −58 kcal

Figure 8-1

Hydrogen, the lightest gas

Since hydrogen gas is lighter than air, it can be poured upward.

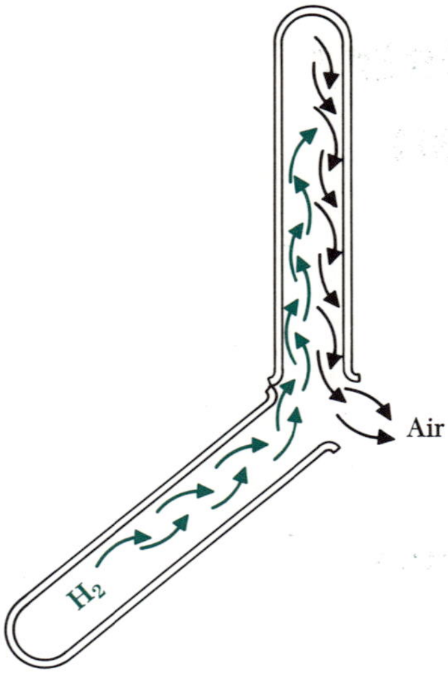

per mole. Alternatively, we may say that the *heat of formation* of gaseous H_2O from its elements is -58 kcal per mole.

The flammability of H_2 gas is one of its most dramatic characteristics. Mixtures of hydrogen with air or oxygen explode violently; therefore, this gas must be handled with caution.

In 1781 Henry Cavendish, in England, proved that hydrogen is an element by showing that water is the only product formed by burning it. He therefore is credited with the discovery of the element. Lavoisier gave the name *hydrogen* ("water former") to the element.

Figure 8-2

The oxyhydrogen torch

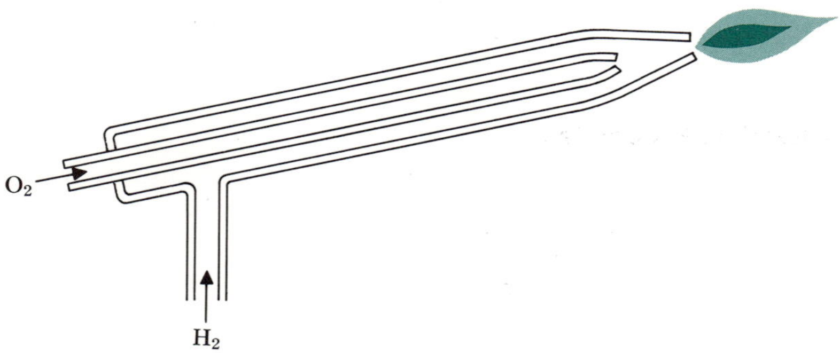

Figure 8-3

The atomic hydrogen torch

Molecular hydrogen is changed to atomic hydrogen in the electric arc. When the hydrogen burns, the heat of formation of molecular hydrogen from atomic hydrogen is added to the heat of combustion.

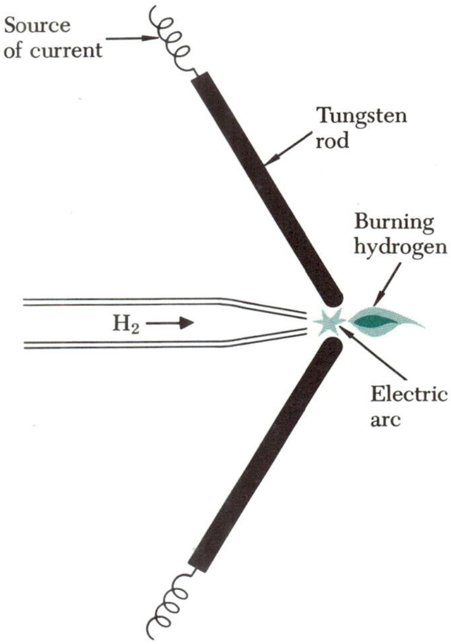

The heat of combustion of H_2 gas is utilized in the oxyhydrogen torch (Figure 8-2) to produce temperatures approximating 3000°C. In the atomic hydrogen torch (Figure 8-3)—where temperatures approaching 5000°C can be produced—H_2 molecules are dissociated by an electric arc, giving monatomic molecules ("atomic" hydrogen):

$$H_2 \rightarrow 2H \qquad \Delta H = +104 \text{ kcal}$$

The resulting H atoms impinge upon the surface of the metal to be cut or welded, where they react with oxygen. According to the principle of energy conservation, the absorption of 104 kcal in the dissociation of 1 mole of H_2 molecules implies that 104 kcal are released when 2 moles of H atoms unite to form 1 mole of H_2 molecules. This large amount of energy, in addition to the heat of combustion of H_2 gas, is liberated when monatomic hydrogen undergoes combustion in the atomic hydrogen torch. Since the reaction of H atoms with oxygen is a surface-catalyzed process, the release of large quantities of energy occurs precisely where it is most effective in melting the metal.

A chlorine atmosphere supports the combustion of hydrogen, forming hydrogen chloride gas, HCl. Mixtures of hydrogen and chlorine gases are stable in the dark but explode with considerable violence in sunlight. The reaction of fluorine with hydrogen also occurs with explosive violence, even at liquid hydrogen temperatures:

$$H_2(g) + F_2(g) \rightarrow 2HF(g) \qquad \Delta H = -130 \text{ kcal}$$

The direct union of hydrogen with nitrogen to form ammonia, NH_3, is one of the most important reactions in the chemical industry, but it is an inherently slow reaction which requires moderate heating, high pressure, and a suitable catalyst.

Hydrogen gas units directly with the very active metals (Groups IA and IIA), but these reactions require heat. Such hydrides (e.g., NaH) as those formed in this type of reaction are ionic compounds that are obtained by bubbling the H_2 gas through the molten metal:

$$2Na + H_2 \rightarrow 2Na^+H^-$$

The reaction of hydrogen with a compound frequently involves simply the addition of the hydrogen to the compound, a type of reaction called *hydrogenation.* The reactions of carbon monoxide and of ethene (ethylene) with hydrogen are examples of hydrogenation:

$$CO + 2H_2 \xrightarrow{\text{catalyst}} \underset{\text{methanol}}{CH_3OH}$$

$$\underset{\text{ethene}}{C_2H_4} + H_2 \xrightarrow{\text{catalyst}} \underset{\text{ethane}}{C_2H_6}$$

When H_2 gas is passed over the hot oxides of some of the less active metals, the hydrogen removes oxygen from the metal and liberates the metal in the free state. For example, passing H_2 gas over black copper(II) oxide liberates reddish copper metal (Figure 8-4):

$$CuO + H_2 \xrightarrow{\text{heat}} Cu + H_2O$$

Figure 8-4

Reduction of a hot metal oxide by hydrogen

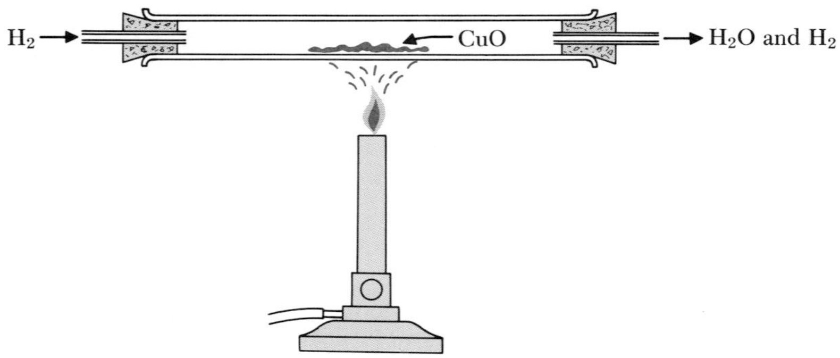

In metallurgical practice metals are commonly released from their oxide ores by heating with coke or charcoal (essentially carbon). In the metallurgy of tungsten the use of carbon is not practical for this purpose, but H_2 gas is quite satisfactory:

$$WO_3 + 3H_2 \rightarrow W + 3H_2O$$

Historically, the term *reduction* has been used to denote the liberation of a metal from its ore, and the agent used to effect the liberation of the metal has been called the *reducing agent*. The release of the metal involves a change of its oxidation state from some positive value to 0. For example, when WO_3 is reduced to tungsten metal by hydrogen gas, the oxidation number of tungsten changes from +6 to 0. Simultaneously, the oxidation state of hydrogen changes from 0 to +1. Reduction is currently defined by chemists as a downward change in oxidation state and is thus the opposite of oxidation, which is an upward change in oxidation state.

8-3 Binary Compounds of Hydrogen

In the best-known binary compounds of hydrogen, the hydrogen is in the oxidation state +1; i.e., the second element is more electronegative than the hydrogen. Such compounds include the hydrogen halides (compounds of hydrogen and halogens), water, hydrogen sulfide, ammonia, and the hydrocarbons.

Three of the hydrogen halides, HCl, HBr, and HI, are strong acids. Except at high concentrations, they react completely with water in their aqueous solutions to give hydronium ions and halide ions:

$$HCl + H_2O \rightarrow H_3O^+ + Cl^-$$
$$HBr + H_2O \rightarrow H_3O^+ + Br^-$$
$$HI + H_2O \rightarrow H_3O^+ + I^-$$

By contrast, HF is a weak acid; i.e., only a small percentage of the HF molecules are converted to ions in aqueous solution:

$$HF + H_2O \rightleftharpoons H_3O^+ + F^-$$

The elements in Group VIA form binary hydrogen compounds that are weak acids: H_2O, H_2S, H_2Se, and H_2Te. Water can be considered amphoteric—it undergoes ionic dissociation to only a slight degree, giving equal molar quantities of hydronium and hydroxide ions:

$$\mathrm{H:\underset{H}{\overset{..}{\underset{..}{O}}}:} + \mathrm{H:\underset{H}{\overset{..}{\underset{..}{O}}}:} \rightleftharpoons \left[\mathrm{H:\underset{H}{\overset{..}{\underset{..}{O}}}:H}\right]^+ + \left[\mathrm{:\overset{..}{\underset{..}{O}}:H}\right]^-$$

Nitrogen, a Group VA element, forms a binary compound with hydrogen (NH_3, ammonia) in which the nitrogen is more electronegative than hydrogen. (In general, the more electronegative element is shown last in a formula, but the formula for ammonia is customarily written NH_3 rather than H_3N.) Ammonia is a weak base, and its aqueous solution contains a low concentration of hydroxide ions produced by its incomplete reaction with the solvent:

$$\mathrm{H}:\underset{\underset{\mathrm{H}}{\cdot\cdot}}{\overset{\overset{\mathrm{H}}{\cdot\cdot}}{\mathrm{N}}}: + \mathrm{H}:\underset{\underset{\mathrm{H}}{\cdot\cdot}}{\overset{\cdot\cdot}{\mathrm{O}}}: \rightleftharpoons \left[\mathrm{H}:\underset{\underset{\mathrm{H}}{\cdot\cdot}}{\overset{\overset{\mathrm{H}}{\cdot\cdot}}{\mathrm{N}}}:\mathrm{H}\right]^{+} + \left[:\underset{\cdot\cdot}{\overset{\cdot\cdot}{\mathrm{O}}}:\mathrm{H}\right]^{-}$$

The analogous phosphorus compound (PH_3, phosphine) is an unstable, very weak base. It is obtained by indirect means rather than by direct union of phosphorus and hydrogen.

Methane and the other hydrocarbons are also obtained by indirect means, for carbon does not readily react with hydrogen. Any compound consisting of carbon and hydrogen only is known as a *hydrocarbon* (not to be confused with *carbohydrates*, which contain oxygen). The number of possible hydrocarbons is very great—it is not known how many may exist. The field of *organic chemistry* is concerned with the hydrocarbons and their derivatives. Most hydrocarbons are devoid of both acidic and basic properties, in the ordinary sense of the terms. They are stable compounds and occur in nature in petroleum and natural gas. Although hydrogen is less electronegative than carbon, formulas for the hydrocarbons are written with the C preceding the H; e.g., the formula for methane is usually written CH_4 instead of H_4C.

Binary compounds of hydrogen with the active metals of Groups IA and IIA are crystalline solids containing the *hydride ion*, H^-. The crystals are hard and have relatively high melting points. In the liquid state these compounds are good conductors of electric current, with H_2 gas liberated at the positive electrode:

$$2H^- \rightarrow H_2\uparrow + 2e^-$$

True *hydrides* are binary compounds in which hydrogen is combined with an atom less electronegative than itself. In these compounds hydrogen is in oxidation state -1, as in NaH and CaH_2. Ionic hydrides cannot exist in the presence of water, for a reaction occurs in which the hydride ion is oxidized by water, producing H_2 gas:

$$H^- + HOH \rightarrow H_2\uparrow + OH^-$$

The ionic hydrides therefore are strong reducing agents. Complex hydrides, such as lithium aluminum hydride, $LiAlH_4$, are extremely useful reducing agents.

8-4 Position of Hydrogen in the Periodic Table

Many of the most important hydrogen compounds are characterized by the $+1$ oxidation state for hydrogen. Certain ones of these yield the H^+ ion in high concentration in solution, although the compounds themselves are covalent. For example, HCl, a polar covalent compound, is a gas under ordinary conditions, but its aqueous solution contains large quantities of H^+ ion. In this respect hydrogen resembles the metals of Group IA. Of course, the hydrogen ion in aqueous solution is hydrated,

$$H^+ + H_2O \rightarrow H_3O^+$$

but probably all ions are hydrated in aqueous solution. Although the H^+ ion is never found in crystalline solids, and in spite of the fact that all binary compounds of hydrogen with more electronegative elements are covalent, placing hydrogen in Group IA of the periodic table is reasonable.

The structural difference between the hydrogen atom and the other atoms of Group IA—the fact that in the hydrogen atom the outer shell is the *only* shell—accounts for the nonexistence of the H^+ ion. If a hydrogen atom were to lose its electron completely, a completely disrobed nucleus would remain. The size of the H^+ ion would be only about $\frac{1}{10,000}$ the size of any existing ion. The intensity of the positively charged "atmosphere" around the H^+ ion would be enormous, and the H^+ ion would form a covalent bond with any atom, such as the oxygen atom in a water molecule, that has electrons available for sharing.

8-5 Uses of Hydrogen

Because of its lightness, hydrogen has been used for inflating balloons. Jacques Charles, a French physicist, made the first flight in a hydrogen-filled balloon in 1783. (The first human flight took place six weeks earlier with a hot-air balloon.) Because of the hazard associated with highly flammable hydrogen, helium gas is generally used now for inflating balloons. The "lifting power" of helium gas is almost as great as that of hydrogen, even though its molecular weight is twice as great; lifting power depends on the difference between the density of the gas and the density of the air.

Large amounts of hydrogen are used in the manufacture of synthetic ammonia and synthetic methanol and in the hydrogenation of oils, such as cottonseed oil, to form solid fats.

Hydrogen is used as a reducing agent in the metallurgy of tungsten and as a fuel in the oxyhydrogen torch and the atomic hydrogen torch. Liquid hydrogen is being developed as a rocket fuel.

8-6 Occurrence of Hydrogen in Nature

About 1 percent by weight of the earth's crust is hydrogen, but about 15 percent of all atoms contained in it are hydrogen atoms. Most terrestrial hydrogen is combined with oxygen as water. Hydrogen is a constituent of all plant and animal tissues, and large quantities of the element occur in petroleum and natural gas which are composed of hydrocarbons.

The abundance of O_2 in the atmosphere and the ease of reaction of hydrogen with oxygen explain why there is very little H_2 in the atmosphere of the earth. In addition, the earth's gravitational force is too small to prevent the escape of molecules with molecular weights as small as that of H_2. About 90 percent of the mass of the sun is hydrogen, and the spectral analysis of starlight indicates that hydrogen is the chief component of most stars. In all likelihood hydrogen is the most abundant element in the universe.

8-7 Reactions Producing Hydrogen

Reactions producing H_2 involve either reduction of hydrogen compounds in which the oxidation state of hydrogen is $+1$ or oxidation of hydrides (compounds in which the oxidation state of hydrogen is -1). In the most familiar reactions H_2 is displaced from water or an acid by a metal. Electrolytic decomposition of water produces H_2 at the negative electrode and O_2 at the positive electrode. Electrolytic decomposition of ionic hydrides in the molten state liberates H_2 at the positive electrode.

8-8 Commercial Preparation of Hydrogen

Very few of the reactions that yield H_2 are suitable for its production on the large scale required by the chemical industries. The primary consideration is cost—cost of the raw materials and the expense of the operation itself.

Electrolysis of aqueous solutions can be used as a method for manufacturing H_2 in regions where electric power is cheap. When an aqueous solution of sulfuric acid of low concentration is subjected to electrolysis, H_2 is liberated at the negative electrode and O_2 at the positive electrode. More frequently, concentrated sodium hydroxide solutions are electrolyzed with iron or nickel electrodes. Although it is easy to write the over-all equation for the process,

$$2H_2O \rightarrow 2H_2\uparrow + O_2\uparrow$$

an understanding of what happens at each electrode requires a knowledge of oxidation-reduction chemistry, the subject of Chapter 20.

The electrolytic manufacture of sodium hydroxide and chlorine from sodium chloride brine yields H_2 as a by-product:

$$2Na^+ + 2Cl^- + 2H_2O \rightarrow 2Na^+ + 2OH^- + H_2\uparrow + Cl_2\uparrow$$

The *pyrolysis* (thermal decomposition) of methane is now one of the important methods for manufacturing hydrogen gas:

$$CH_4(g) \xrightarrow[\text{catalyst}]{\text{heat}} C(s) + 2H_2(g)$$

Catalyzed thermal decompositions such as this are referred to by the term "cracking." A 1:3 mixture of CO and H_2 is obtained by the reaction of steam with methane:

$$CH_4(g) + H_2O(g) \rightarrow CO(g) + 3H_2(g)$$

Hydrogen gas can be obtained economically, mixed with carbon monoxide, by passing steam over hot coke:

$$C(s) + H_2O(g) \xrightarrow{1000^\circ C} H_2(g) + CO(g)$$

The 1:1 mixture of CO and H_2 is used as a domestic and industrial fuel under the name "water gas." The heating value of the fuel can be increased by fortifying it with gaseous hydrocarbons obtained by the "cracking" of petroleum; the fortified fuel is "carbureted water gas."

Hydrogen is also a by-product of the "coking" of coal, which involves heating coal in the absence of air to cause its "destructive distillation." The process yields coke, coal tar, and coal gas; the coal gas is about 50 percent H_2.

Hydrogen is stored and transported in thick steel cylinders at a pressure of 1000 atmospheres; the weight of the H_2 contained in one of the cylinders is less than 1 percent of the total weight. Therefore, to save transportation costs, hydrogen is used at the plant where it is produced, whenever this is possible. In spite of the cost of manufacturing calcium hydride, it is used to produce H_2 on a small scale (by its reaction with water) since it is conveniently transported in lightweight containers.

8-9 Activity Series of Metals

Certain metals, such as sodium, react vigorously with cold water, liberating H_2 gas and forming metal hydroxides.

$$2Na + 2HOH \longrightarrow 2(Na^+ + OH^-) + H_2\uparrow$$

Certain other metals, such as magnesium, react with steam but not with cold water.

$$Mg(s) + H_2O(\text{steam}) \longrightarrow MgO(s) + H_2(g)$$

Less active metals displace hydrogen from aqueous solutions of acids—some more vigorously than others—although they do not appear to react with water itself. Still other metals, such as silver, do not displace hydrogen from acids at all.

On the basis of these and other observations, the metals have been arranged in order of their chemical activity in the presence of water. The *activity series* given in Table 8-1 is an abbreviated list, including only the most familiar metals. Each metal is more active than any of the metals below it in the series. Only those at the top of the series are able to liberate hydrogen from cold water. They are followed by metals that can displace hydrogen from steam but cannot react with cold water. Any of the metals listed above hydrogen in the series are capable of displacing it from acids, but those listed below hydrogen are unable to do this.

Hydrogen gas can reduce the oxides of metals listed below chromium but cannot reduce the oxides of metals above chromium. Oxides of metals listed near the bottom of the series, such as HgO, undergo decomposition by heat alone; i.e., the metal is reduced to the free state even without the use of hydrogen.

Any metal in the series will displace any metal below it from aqueous solutions of its salts. For example, iron displaces copper from aqueous solutions of copper salts, and copper displaces silver from solutions of silver salts:

Table 8-1

Activity series of metals

Oxides	Metal	Reactions
Oxides not reduced by H_2	K	React with cold water liberating H_2; react violently with acids
	Ca	
	Na	
	Mg	React with steam or with acids, yielding H_2
	Al	
	Mn	
	Zn	
	Cr	
Oxides reduced by H_2	Fe	
	Cd	
	Co	React with acids, liberating H_2
	Ni	
	Sn	
	Pb	
	H	
	Cu	React with oxygen, giving oxides
Oxides reduced by heat alone	Hg	
	Ag	Form oxides indirectly
	Pt	
	Au	

$$Fe(s) + (Cu^{++} + SO_4^{--}) \rightarrow (Fe^{++} + SO_4^{--}) + Cu(s)$$
$$Cu(s) + 2(Ag^{+} + NO_3^{-}) \rightarrow (Cu^{++} + 2NO_3^{-}) + 2Ag(s)$$

Since the negative ions take no part in these reactions, they may be omitted from the equations:

$$Fe(s) + Cu^{++} \rightarrow Fe^{++} + Cu(s)$$
$$Cu(s) + 2Ag^{+} \rightarrow Cu^{++} + 2Ag(s)$$

8-10 The Two Forms of Molecular Hydrogen

The two nuclei and two electrons which constitute a hydrogen molecule are all spinning, each on its own axis. According to the electronic theory of chemical bonding, the electrons must spin in opposite directions, but the two nuclei can spin in either the same or opposite directions. In an *orthohydrogen* molecule, the nuclei are spinning in the same direction; in a *parahydrogen* molecule, they are spinning in opposite directions (Figure 8-5).

At ordinary temperatures the normal ratio of orthohydrogen to parahydrogen is about 3:1. As the temperature is lowered, the ratio changes, and there is a slow conversion of orthohydrogen to parahydrogen. The conversion is exo-

Figure 8-5

The two forms of molecular hydrogen

(a) Orthohydrogen, in which the nuclear spins are in the same direction. (b) Parahydrogen, in which the nuclear spins are opposed.

(a) (b)

thermic, and the heat that is given off causes liquid hydrogen to boil. Since liquid parahydrogen does not undergo a similar conversion, it is more easily shipped and stored than ordinary liquid hydrogen. Methods recently have been developed for converting orthohydrogen into parahydrogen catalytically. The use of liquid hydrogen as a rocket fuel has been made possible by this development.

8-11 The Isotopes of Hydrogen

In 1931 Harold C. Urey, an American chemist, made one of the important discoveries of this century when he found that one atom in every 5000 hydrogen atoms has a mass of 2 amu.

A difference of one or more units in atomic mass from one isotope to another in the heavy elements causes only minor variations in the properties of these isotopes (because the percentage difference in their masses is so small), but the same difference in atomic mass of the very light elements can result in an appreciable variation in properties. Since a *heavy* hydrogen atom has twice the mass of an ordinary hydrogen atom, the properties of heavy hydrogen and ordinary hydrogen differ considerably.

Usually, the different isotopes of an element are not given special names. However, the properties of heavy hydrogen are so different from those of ordinary hydrogen that Urey named it *deuterium* (from the Greek root meaning "two"). It is represented by the symbol D. Deuterium is not regarded as a new chemical element, but its discovery was more significant than the discovery of some of the new elements.

There are three isotopic varieties of the diatomic hydrogen molecule: H_2, HD, and D_2. Their melting points and boiling points are compared in Table 8-2. (The boiling point of HD is missing from the table because it cannot be determined: at temperatures approaching the boiling point, HD undergoes partial conversion to H_2 and D_2.) The density of D_2 gas is, of course, twice as great as that of H_2 gas.

The composition of a deuterium nucleus accounts for its greater mass. The nucleus of a light hydrogen atom consists only of a proton, but the nucleus

Table 8-2

Isotopic varieties of hydrogen

	H_2	HD	D_2
Melting point (°K)	13.95	16.60	18.65
Boiling point (°K)	20.38		23.6

of a deuterium atom consists of a proton and a neutron. Although the nucleus of a deuterium atom is not one of the "fundamental" particles, it has been given the special name *deuteron.*

Deuterium has proved to be a useful research tool. For example, replacing certain H atoms by D atoms in the molecules of various organic compounds aids in the investigation of the mechanisms of organic reactions and biochemical processes.

A third isotope of hydrogen with a mass of 3, *tritium* (T), has been discovered. Only very small amounts of tritium occur naturally because of its radioactive instability. The fact that it exists at all is due to the action of cosmic rays. The tritium nucleus consists of a proton and two neutrons.

Since the two heavy hydrogen isotopes have been named deuterium and tritium, the most abundant isotope, whose atomic nucleus consists of a lone proton, has been named *protium.* No special symbol, other than H, is used to represent protium.

New Terms

Activity series: a listing of chemical elements in the order of their activity in the presence of water.

Critical temperature: the highest temperature at which it is possible for a substance to exist in the liquid state, however great the pressure applied to it.

Deuterium: the hydrogen isotope of mass number 2; its symbol is D.

Deuteron: the nucleus of a deuterium atom.

Heat of combustion: the quantity of heat absorbed during the combustion of a specified quantity of an element or compound. (This definition is formulated to be in technical agreement with the definition of *heat of reaction* given in Chapter 7. The quantity of heat absorbed during combustion is, of course, negative; i.e., a combustion reaction is always exothermic.)

Heat of formation: the quantity of heat absorbed during the formation of a specified quantity (e.g., 1 mole) of a substance from its constituent elements. (If the formation of a substance from its elements is an exothermic reaction, the heat of formation of the substance is negative.)

Hydride: a binary compound in which hydrogen is combined with an atom less electronegative than itself. In all hydrides, hydrogen is in the oxidation state -1.

Hydrocarbon: a binary compound consisting of hydrogen and carbon.

Hydrogenation: a chemical reaction in which hydrogen is simply added onto another compound.

Organic chemistry: the branch of chemistry that deals with hydrocarbons and their derivatives; the chemistry of carbon compounds.

Orthohydrogen: the form of H_2 molecule in which the nuclei are spinning in the same direction.

Parahydrogen: the form of H_2 molecule in which the nuclei are spinning in opposite directions.

Protium: the hydrogen isotope of mass number 1; its symbol is H.

Pyrolysis: decomposition brought about by heat.

Reduction: a downward change in oxidation state. (Reduction is the opposite of oxidation.)

Tritium: the hydrogen isotope of mass number 3; its symbol is T.

Exercises

8-1 Write equations for the following reactions:

(a) sodium and water
(b) potassium and "heavy water" (D_2O)
(c) zinc and dilute hydrochloric acid
(d) magnesium and dilute sulfuric acid
(e) lithium hydride and water

8-2 Calculate the empirical formula for the compound that is 18.3% lithium, 71.1% aluminum, and 10.6% hydrogen.

8-3 Compute the amount of heat liberated by (a) the burning of 100 g of H_2 and (b) the burning of 100 ounces of H_2.

8-4 In the reaction

$$CH_4 + 2O_2 \rightarrow CO_2 + 2H_2O$$

which element is oxidized? Which element is reduced? What changes in oxidation state have these two elements undergone?

8-5 On the basis of the activity series of the metals, why is gold a precious metal?

8-6 What is the relation of parahydrogen to space travel?

8-7 Why has there been unusual interest in the isotopes of hydrogen?

References

Sanderson, R. T., *Principles of Chemistry*, Wiley, New York, 1963, Chapter 13, "Fundamental Interpretations of Hydrogen Compounds." A consideration of the different types of binary compounds of hydrogen.

Selwood, P. W., *Chemical Principles*, Holt, Rinehart & Winston, New York, 1964, Chapter 17, "Hydrogen." Includes consideration of the ortho and para forms of molecular hydrogen.

Urey, H. C., "Significance of Hydrogen Isotopes," *Ind. Eng. Chem.*, **26**, 803 (1934). By the American chemist who discovered deuterium.

Weeks, M. E., and Leicester, H. M., *Discovery of the Elements*, 6th ed., Chemical Education, Easton, Pa., 1956, pp. 197–205. The history of the discovery of hydrogen as an element.

9

The Nature of Gases

9-1 Principle of the Barometer

Since gases are less tangible than liquids and solids, it is not quite so obvious that they have weight. The melting of a certain weight of ice produces liquid water having the same weight; likewise, the evaporation of the water produces water vapor, i.e., gaseous water, having still the same weight.

The most familiar of all gases is the air we breathe. In 1643 Evangelista Torricelli, an Italian physicist and an associate of Galileo, obtained the first evidence that air has weight. Torricelli sealed a long glass tube at one end, poured in mercury from the other end until the tube was filled, covered the open end, inverted the tube into a dish of mercury, and then uncovered the end of the tube. The mercury fell a short distance in the tube but remained about 76 cm (approximately 30 inches) higher than the surface of the mercury in the dish. When Torricelli repeated the experiment with other tubes of various lengths and diameters the results were always the same (provided the tube was at least 76 cm long)—the mercury descended in the tube until it was 76 cm higher than the surface of the mercury in the dish.

At that time the phenomenon of suction was explained, in the Aristotelian tradition, by the phrase "nature abhors a vacuum." The Aristotelian explanation for sucking a liquid up a straw would be that the liquid rises to fill the vacuum that nature abhors. A similar explanation would be given for the mercury remaining higher in Torricelli's tube than in the dish outside. But the critical question is based on the quantitative measurements that were made. Why does the mercury in the tube stand 76 cm higher than the outside surface? Does nature's abhorrence of a vacuum not exceed 76 cm? (It must be noted that the height at which the liquid stands is greater than 76 cm if a liquid of lesser density than mercury is used.)

Figure 9-1

The barometer

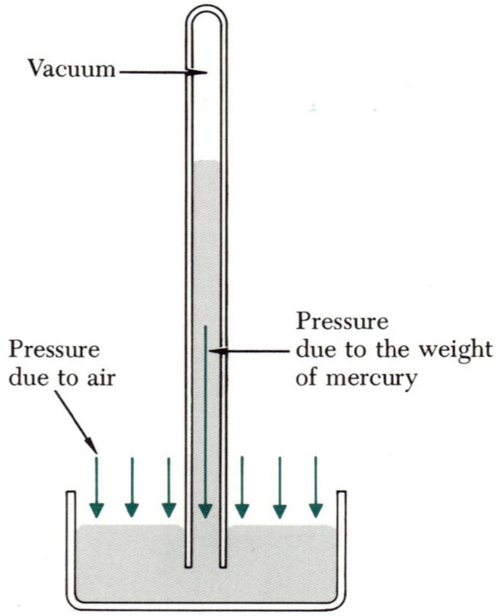

Galileo reasoned that the mercury was held up by the push of the outside air rather than by the pull of the vacuum inside the tube. The weight of all the air directly above the dish—a layer approximately 500 miles thick, we now realize—is a force acting downward upon the surface of the mercury in the dish. This air pressing down tends to cause the mercury to rise in the tube. The equilibrium position of the mercury level is such that the weight of the mercury in the tube above the level of the mercury in the dish is the same as the weight of a column of air (of the same cross-sectional area) extending from the surface of the mercury to the outer limits of the atmosphere.

Torricelli thought that the mercury column should drop on being carried up a mountain, because the amount of air above the surface of the mercury would then be less. To test Galileo's explanation he took the apparatus up a mountain and found that the height of the mercury column became progressively less with increasing altitude.

Torricelli's apparatus is the mercury *barometer*, the instrument used for measuring atmospheric pressure in scientific laboratories, weather stations, and wherever accuracy is important (Figure 9-1). The space above the mercury in the closed tube is empty space, except for a small quantity of mercury vapor, and is known as a *Torricellian vacuum.*

9-2 Units of Pressure

If the column of mercury in a barometer is 76.00 cm high and 1.000 cm^2 (square centimeter) in cross section, the volume of mercury is 76.00 cm^3 (cubic centimeters). Since the density of mercury is 13.55 g per cm^3, the weight of the column is 76.00 cm^3 × 13.55 g per cm^3, or 1030 g. In other words, the *pressure*, i.e., force per unit area, of the atmosphere is 1030 g per cm^2. If a larger tube is used, the weight of mercury in the tube is greater, but the force of this weight is distributed over a larger area, so the pressure is the same as before. For example, if the cross section of the tube is doubled, the weight of the mercury is twice as great. This doubled force is acting on an area of 2 cm^2 instead of only 1 cm^2, so the force per unit area is unchanged.

A pressure of 1030 g per cm^2 corresponds to a value of 14.7 p.s.i. (pounds per square inch). Atmospheric pressure varies not only with altitude but with changing weather conditions, chiefly because of the varying amounts of water vapor in the air, water molecules not being as heavy as O_2 or N_2 molecules. Fluctuating temperature also changes the density of the air.

When comparing barometer readings taken at different times or places, we do not have to convert the readings to actual pressure units. Even without knowing that a 76-cm reading means a pressure of 1030 g per cm^2, we realize that a 57-cm reading means a pressure $\frac{57}{76}$ or three fourths as great. It has been customary to express atmospheric pressure in terms of cm Hg (centimeters of mercury) or mm Hg (millimeters of mercury). There is increasing use of the term *torr* to designate the pressure exerted by a column of mercury 1 mm high.

The *standard atmospheric pressure* is defined as exactly 76.00 cm Hg (760.0 torrs) at sea level and at 0°C and is referred to as *1 atmosphere*. The pressures exerted by confined gases, especially those confined under high pressures, are usually expressed in atmospheres.

9-3 Gas Density

With the use of modern chemical balances and "vacuum pumps," we can make accurate direct determinations of the weights of known volumes of gases confined under known pressures. To do this we evacuate a flask, weigh it, fill it with gas, and weigh it again. The weight of the gas divided by the volume is the density of the gas. Gas densities are usually expressed in grams per liter.

The density of a gas is so dependent on pressure and temperature that density figures are of no value unless accompanied by pressure and temperature information. At ordinary pressures and temperatures, gas densities vary from a few tenths of a gram to several grams per liter. Solids and liquids are generally about a thousand times as dense as gases. This strongly suggests that the molecules of a gas must be relatively far apart.

9-4 Boyle's Law

We sometimes speak of the pressure *of* a gas and at other times of the pressure *on* a gas. Actually, the two are the same. In some situations it is more natural to speak of the pressure which the gas exerts against the walls of its container, and in other situations we focus on the pressure used to confine the gas. This difference in terminology is illustrated by the following two problems, which initially might appear to be completely different in nature.

EXAMPLE 1 If a gas exerts a pressure of 740 torrs (740 mm Hg) when confined within a 5-liter container, how much pressure will it exert when confined to a 4-liter container?

EXAMPLE 2 If the volume of a sample of gas is 850 ml when confined under a pressure of 1000 g per cm^2, what volume will it occupy under a pressure of 800 g per cm^2?

Both of these questions can be answered by applying *Boyle's Law*, which states that at constant temperature, the volume (V) occupied by a gas varies inversely with the confining pressure (P), or that the pressure exerted by a gas varies inversely with the volume within which it is confined. Mathematically, at constant temperature,

$$PV = \text{constant}$$

or

$$P_1V_1 = P_2V_2$$

The original pressure and volume of the gas are denoted by P_1 and V_1, respectively; the new pressure and volume, by P_2 and V_2, respectively. (Note that "P_1V_1" means "P_1 multiplied by V_1.") Boyle's Law is represented graphically in Figure 9-2. According to Boyle's Law, all gases behave alike. However, at low temperatures or at high pressures, deviations may occur. When the law is obeyed, the gas is said to exhibit *ideal behavior*.

Figure 9-2

Pressure-volume relationship at constant temperature, as stated by Boyle's Law

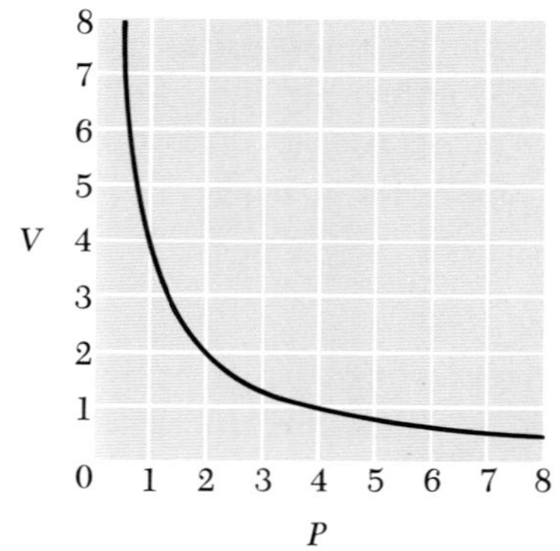

SOLUTION TO EXAMPLE 1 The volume changes to $\frac{4}{5}$ of the original volume. The pressure varies inversely; i.e., the decrease in volume causes the pressure to increase to $\frac{5}{4}$ of the original pressure. Therefore,

$$\text{New pressure} = \text{old pressure} \times \frac{5 \text{ liters}}{4 \text{ liters}}$$
$$= 740 \text{ torrs} \times \frac{5}{4}$$
$$= 925 \text{ torrs}$$

SOLUTION TO EXAMPLE 2 The pressure is reduced to $\frac{800}{1000}$ of the original pressure. Since the volume varies inversely, it increases to $\frac{1000}{800}$ of the original volume. Therefore,

$$\text{New volume} = \text{old volume} \times \frac{1000 \text{ g/cm}^2}{800 \text{ g/cm}^2}$$
$$= 850 \text{ ml} \times \frac{1000}{800}$$
$$= 1064 \text{ ml}$$

Only three digits are significant—a fact that is best expressed by writing the answer as 1.06×10^3 ml (or 1.06 liters).

9-5 Heat and Temperature

Temperature is a measure, on a suitable scale, of the degree of hotness of a substance. In order for a substance to become hotter, i.e., to reach a higher temperature, energy in the form of *heat* must be allowed to enter it. Thus temperature is a property that fixes the direction of heat flow. If heat flows from one object to another when the two are placed in contact, the one object has a higher temperature than the other. If two such bodies remain in contact for a sufficient period of time, the hotter body will "cool off" and the colder body will "warm up" until both have the same temperature.

All too often, people confuse the concepts of heat and temperature. Temperature describes hotness, i.e., the degree or intensity of heat, but heat itself is measured as a quantity of energy. *Energy* is often defined as the capacity to do work, *work* meaning motion against an opposing force. *Kinetic energy* is the energy that a moving body possesses because of its motion. *Potential energy* is the energy that a body possesses because of its position with respect to other objects. Kinetic and potential energies are encountered in such familiar forms as mechanical energy, electrical energy, chemical energy, heat, light, and sound. One form of energy can be converted into another. It is especially easy for other forms of energy to be converted into heat.

The unit of heat is the *calorie* (cal), which is the amount of heat required to raise the temperature of 1 g of water 1° on the Celsius scale. It is sometimes

Figure 9-3

The thermometer

Comparison of Celsius and Fahrenheit scales.

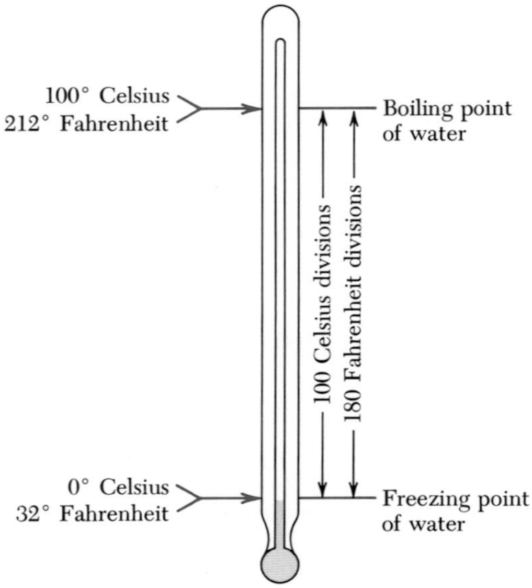

more convenient to use the *kilocalorie* (kcal, 1000 cal), which is the "big calorie," or Calorie, used in the field of nutrition.

To ascertain the temperature of a substance, we bring it into contact with a *thermometer*, the instrument used to measure temperature. The substance and the thermometer must remain in contact until they have reached *thermal equilibrium*, i.e., until they have attained the same temperature, and the temperature is then read on the thermometer scale.

The most common type of thermometer consists of a piece of small-bore glass tubing attached to a reservoir or bulb (Figure 9-3). The bulb and part of the tubing are filled with a liquid (usually mercury), which expands when heated and contracts when cooled. As the temperature of the liquid in the bulb increases or decreases, the visible thread of liquid rises or falls alongside the thermometer scale.

In order to translate the position of the liquid thread into a temperature reading, we must use a standard reference as a basis of comparison. The two reference standards are (1) the freezing point of water and (2) the boiling point of water, at normal atmospheric pressure. The two temperature scales currently in use are those introduced by a German physicist, Gabriel Fahrenheit (1686–1736), and by a Swedish astronomer, Anders Celsius (1701–1744). Arbitrarily, the freezing point of water has been assigned the numerical value of 32° on the Fahrenheit (F) scale and 0° on the Celsius (C) scale. The boiling point of water at normal atmospheric pressure has been arbitrarily fixed as 212° on the Fahrenheit scale and 100° on the Celsius scale.

To provide a thermometer with a temperature scale, we first place the thermometer bulb in a mixture of ice and water. When the liquid in the bulb is at the same temperature as the ice-water mixture, we note the position of the thread in the stem of the thermometer and mark the column either 32°F or 0°C at this point. When the liquid in the bulb is at the temperature at which water boils at normal atmospheric pressure, we mark the column 212°F or 100°C at the position of the thread. Finally, we divide the distance between the two marks into 180 equal divisions (degrees), if we are constructing a Fahrenheit thermometer, or into 100 equal divisions (degrees), if we are making a Celsius thermometer.

The conversion of a Fahrenheit temperature into a Celsius reading, or vice versa, requires a calculation. A Fahrenheit degree is $\frac{5}{9}$ the size of a Celsius degree; a Celsius degree is $\frac{9}{5}$ as large as a Fahrenheit degree. Note also that a temperature reading of 0°C would be 32°F. Therefore,

$$°F = 32 + \tfrac{9}{5}°C$$
$$°C = \tfrac{5}{9}(°F - 32)$$

9-6 Charles' Law

More than a century elapsed between Boyle's discovery of the pressure-volume relationship of gases (in 1660) and Jacques Charles' investigation of the effect of temperature changes on the volume of a gas. In 1787 Charles discovered that the volume of a gas varies in a linear manner with temperature (Figure 9-4) and that the volume is doubled when the Celsius temperature is raised from 0 to 273° with the pressure held constant. Stating it another way, the volume is halved when the Celsius temperature is lowered from 273 to 0° with the pressure held constant. If a gas were to follow this relationship indefinitely upon cooling, its volume would be reduced to 0 at −273°C! Actually, all gases have been found

Figure 9-4
Charles' Law

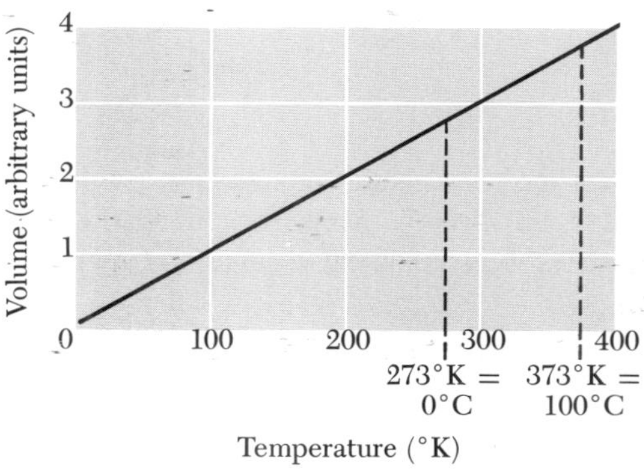

to liquefy before this temperature is reached—and the gas laws of Charles and Boyle do not apply to liquids. Nevertheless, this seemingly lowest attainable temperature has proved to be important.

In 1848 William Thomson, a British physicist later raised to peerage as Lord Kelvin, proposed a new scale of temperature with degrees the size of Celsius degrees but with $-273°C$ redesignated as 0° absolute, 0°C as 273° absolute, 273°C as 546° absolute, etc. Thus, measured on this new scale, the *absolute* Celsius scale, the volume of a gas is directly proportional to the temperature. The lowest point on the scale, most recently established as $-273.15°C$, is called *absolute zero*. The absolute Celsius scale is now known as the Kelvin scale. Temperature on the Celsius scale is converted to temperature on the Kelvin scale by adding 273.15°:

$$°K = °C + 273.15°$$

According to *Charles' Law*, the volume occupied by a fixed quantity of gas confined under constant pressure is directly proportional to the absolute temperature. Mathematically, at constant pressure,

$$V = \text{constant} \times T \text{ (where } T \text{ is in °K)}$$

or

$$\frac{V_1}{T_1} = \frac{V_2}{T_2}$$

At high pressures and at temperatures near the liquefaction point, the behavior of real gases deviates from Charles' Law. When the law is obeyed, the gas is said to exhibit *ideal behavior*. In solving problems, we shall assume ideal behavior.

EXAMPLE The volume of a gas is 453 ml at 27°C. What will be its volume at 127°C if the pressure is not permitted to change?

$$\text{New volume} = \text{old volume} \times \text{temperature factor}$$

The temperature *increases* from 300°K to 400°K; i.e., it increases to $\frac{400}{300}$ of its original temperature. Hence, the volume will also increase by the factor $\frac{400}{300}$:

$$\text{New volume} = 453 \text{ ml} \times \frac{400°K}{300°K}$$
$$= 604 \text{ ml}$$

If the volume of a gas were held constant, you would expect the pressure exerted by the gas to be directly proportional to the absolute temperature. This corollary of Charles' Law has been verified by experiment. Mathematically, at constant volume,

$$P = \text{constant} \times T \text{ (where } T \text{ is in °K)}$$

or

$$\frac{P_1}{T_1} = \frac{P_2}{T_2}$$

Table 9-1

Four scales of temperature

	°*Celsius*	°*Kelvin*	°*Fahrenheit*	°*Rankine*
Boiling point of water	100	373.15	212	671.67
Freezing point of water	0	273.15	32	491.67
Absolute zero	−273.15	0	−459.67	0

EXAMPLE If the pressure inside a tank of gas is 1.500×10^3 g per cm^2 at 77°C, what pressure will the gas exert if the tank is cooled to 20°C?

$$\text{New pressure} = \text{old pressure} \times \text{temperature factor}$$

The temperature *decreases* from 350°K to 293°K, i.e., to $\frac{293}{350}$ of its original temperature. Hence, the pressure also decreases by the factor $\frac{293}{350}$:

$$\begin{aligned}\text{New pressure} &= 1.500 \times 10^3 \text{ g/cm}^2 \times \frac{293°\text{K}}{350°\text{K}} \\ &= 1.26 \times 10^3 \text{ g/cm}^2\end{aligned}$$

If we convert −273.15°C to Fahrenheit, we find that absolute zero is −459.67°F. Those who prefer to use Fahrenheit temperatures have established an absolute Fahrenheit scale which they call the Rankine scale, in honor of William J. M. Rankine, a Scottish engineer and physicist. Temperature on the Fahrenheit scale is converted to temperature on the Rankine scale by adding 459.67°:

$$°\text{R} = °\text{F} + 459.67°$$

Since most scientific work is done with Celsius temperatures, the expression *absolute temperature* usually refers to the Kelvin scale. In fact, the abbreviation A instead of K is sometimes used to indicate Kelvin temperatures. It is better practice to use K and to refer to the corresponding temperature scale as the Kelvin scale—for both the Rankine and Kelvin scales are absolute scales of temperature.

The four scales of temperature are compared in Table 9-1.

9-7 The Combined Gas Law Equation

According to Boyle's Law, the product PV is constant as long as T is fixed; and according to Charles' Law, the ratio V/T is constant as long as P is fixed (also, the ratio P/T is constant as long as V is fixed). These relationships may be combined into one expression, the *combined gas law equation*:

$$\frac{PV}{T} = \text{constant}$$

or

$$\frac{P_1V_1}{T_1} = \frac{P_2V_2}{T_2}$$

This equation holds true for any gas showing ideal behavior as long as we are dealing with a fixed weight, i.e., a fixed number of molecules, of the gas.

Conditions (i.e., temperature and pressure) under which experiments are performed vary considerably, but the combined gas law equation permits easy standardization of results. Measured gas volumes can be compared directly with one another only if they are measured at, or recalculated to, the same temperature and pressure. One atmosphere, i.e., 76.00 cm Hg (760.0 torrs), has been chosen as the *standard pressure* for the measurement of gas volumes. The melting point of ice, 273.15°K (0°C), has been selected as the *standard temperature*. Unless stated otherwise, any specified gas volume is the volume that the gas would occupy under these *standard conditions* of temperature and pressure. Standard conditions are abbreviated as SC or as STP (for "standard temperature and pressure").

EXAMPLE If the volume of a gas is 362 ml at 25°C and 745 torrs, what would be its volume at standard conditions?

New volume = old volume × temperature factor × pressure factor

The temperature *decreases* by the factor $\frac{273}{298}$, which tends to cause the volume to *decrease* also by the factor $\frac{273}{298}$. The pressure *increases* by the factor $\frac{760}{745}$, and this tends to cause the volume to *decrease* by the factor $\frac{745}{760}$:

$$\begin{aligned}\text{New volume} &= 362 \text{ ml} \times \frac{273°\text{K}}{298°\text{K}} \times \frac{745 \text{ torrs}}{760 \text{ torrs}}\\ &= 362 \text{ ml} \times 0.916 \times 0.980\\ &= 362 \text{ ml} \times 0.898\\ &= 325 \text{ ml}\end{aligned}$$

9-8 Molar Volume of a Gas

The volume occupied by 1 mole of a gas at standard conditions is called the *molar volume* of that gas. Several values are listed in Table 9-2. The *ideal gas* (the last item in Table 9-2) is an imaginary gas that behaves in perfect accord with Boyle's Law and Charles' Law. Most real gases behave like an ideal gas under certain conditions, but no real gas behaves ideally under all conditions. The value given in Table 9-2 for the molar volume of an ideal gas is obtained from measurements made on real gases at temperatures high enough and pressures low enough for ideal behavior to be approached closely, the measured volumes being recalculated to standard conditions by use of Boyle's Law and Charles' Law.

Table 9-2

Volume of 1 mole of gas at standard conditions

Gas	*Molar volume (liters)*
Hydrogen	22.432
Nitrogen	22.403
Oxygen	22.392
Hydrogen chloride	22.246
Sulfur dioxide	21.889
Ideal gas	22.414

For the first three gases listed in Table 9-2, agreement with ideal behavior is quite close. As a practical matter, we can assume that at standard conditions the molar volume of any gas is 22.4 liters (Figure 9-5). This information can be used profitably in several types of calculation.

EXAMPLE What is the density, i.e., the weight of 1 liter, of carbon monoxide at STP?

The weight of 1 mole of CO is 12.011 g + 15.999 g, or 28.010 g. The density expected at STP is 28.010 g/mole ÷ 22.4 liters/mole = 1.25 g/liter. (The value obtained experimentally is 1.2504 g/liter.)

EXAMPLE The weight of 1 liter of chlorine gas at STP is 3.214 g. What is the molecular weight of chlorine?

Assuming that the volume of 1 mole of chlorine gas at STP is 22.4 liters, the weight of 1 mole is 22.4 liters/mole × 3.214 g/liter = 72.0 g/mole; i.e., the molecular weight of chlorine is 72.0.

The atomic weight of chlorine is 35.453, which is approximately half the value we have calculated for the molecular weight. This

Figure 9-5

Molar volume of a gas

A mole of any gas occupies a volume of approximately 22.4 liters at standard conditions.

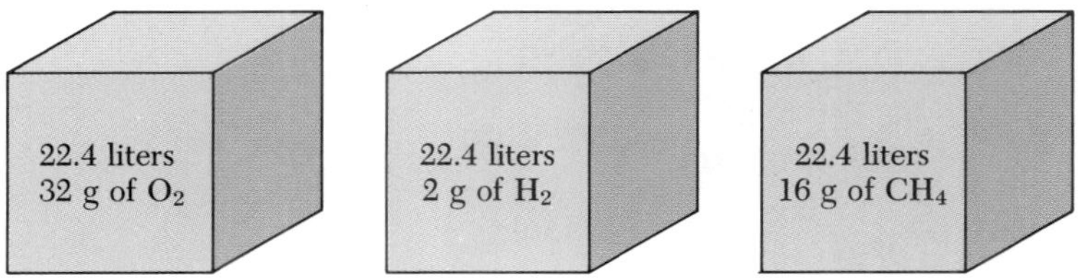

indicates beyond any reasonable doubt that the chlorine molecule consists of two chlorine atoms. The correct molecular weight therefore is exactly twice the atomic weight, i.e., 2×35.453, or 70.906, instead of 72.0. The molar volume principle does not give exactly the correct value for the molecular weight in this example, but nevertheless we have been able to arrive at the correct value by following through with the other information at our disposal. We have deliberately chosen, for this example, a gas that does not obey the gas laws well at STP.

9-9 The General Gas Law Equation

We have seen that for any gas sample in general, PV/T is a constant as long as Boyle's Law and Charles' Law are obeyed. The numerical value of the constant depends upon the quantity of gas in the sample, i.e., the number of moles. Under any specified conditions of temperature and pressure the volume occupied by 2 moles of a gas is exactly twice the volume occupied by 1 mole of the same gas. If we let n represent the number of moles, we can say that PV/nT is constant for all gases under all conditions as long as the gases show ideal behavior. This universal constant is known as the *molar gas constant* and is represented by R. The equation

$$PV = nRT$$

is known as the *general gas law equation.*

The numerical value of the gas constant R can be calculated by substituting the proper known values into the general gas law equation. Substituting $n = 1$ mole, $P = 1$ atmosphere, $V = 22.414$ liters, and $T = 273.15°\text{K}$,

$$R = \frac{PV}{nT} = \frac{(1\text{ atm})(22.414\text{ liters})}{(1\text{ mole})(273.15°\text{K})} = 0.082054\text{ liter-atm per mole per deg}$$

The numerical value of R depends on the units in which P, V, n, and T are expressed.

The general gas law equation can be used to calculate the volume of a specified quantity of gas at a given pressure and temperature, the pressure exerted by a given amount of gas, the molecular weight of a gas, and other information.

EXAMPLE Calculate the volume occupied by 5.0 g of carbon monoxide gas at 25°C and 740 torrs pressure.

$$T = (273 + 25)°\text{K} = 298°\text{K}$$

$$P = \frac{740\text{ torrs}}{760\text{ torrs/atm}} = 0.974\text{ atm}$$

$$n = \frac{5.0\text{ g}}{28.0\text{ g/mole}} = 0.18\text{ mole}$$

$$R = 0.082 \text{ liter-atm per mole per deg}$$

$$V = \frac{nRT}{P} = \frac{0.18 \text{ mole} \times 0.082 \text{ liter-atm per mole per deg} \times 298°\text{K}}{0.974 \text{ atm}}$$

$$= 4.5 \text{ liters}$$

EXAMPLE If 5.20 g of a gas occupies a volume of 1140 ml at 78°C and 780 torrs pressure, what is the molecular weight of the gas? Let x = molecular weight (so that 1 mole weighs x g). Then

$$n = \frac{5.20 \text{ g}}{x \text{ g/mole}} = \frac{5.20}{x} \text{ moles}$$

$$P = \frac{780 \text{ torrs}}{760 \text{ torrs/atm}} = 1.026 \text{ atm}$$

$$T = (273 + 78)°\text{K} = 351°\text{K}$$

$$V = \frac{1140 \text{ ml}}{1000 \text{ ml/liter}} = 1.140 \text{ liters}$$

$$R = 0.0821 \text{ liter-atm per mole per deg}$$

$$PV = \frac{5.20RT}{x}$$

$$x = \frac{5.20RT}{PV} = \frac{5.20 \text{ g} \times 0.0821 \text{ liter-atm per mole per deg} \times 351°\text{K}}{1.026 \text{ atm} \times 1.140 \text{ liters}}$$

$$= 128 \text{ g/mole}$$

9-10 Dalton's Law of Partial Pressures

In a paper published in 1803, John Dalton announced his discovery of the *Law of Partial Pressures*. According to this law, in a mixture of gases the molecules of each species exert the same pressure as they would exert if they were present alone, so that the total pressure is the sum of these "partial pressures" exerted by the different gases in the mixture (Figure 9-6):

$$P_{\text{total}} = P_1 + P_2 + P_3 + \cdots$$

The significance of this law is best understood in terms of moles and mole fractions. In any mixture the number of moles of each component is the number of grams of that component divided by its molar weight. The *mole fraction*

Figure 9-6

An example of Dalton's Law of Partial Pressures

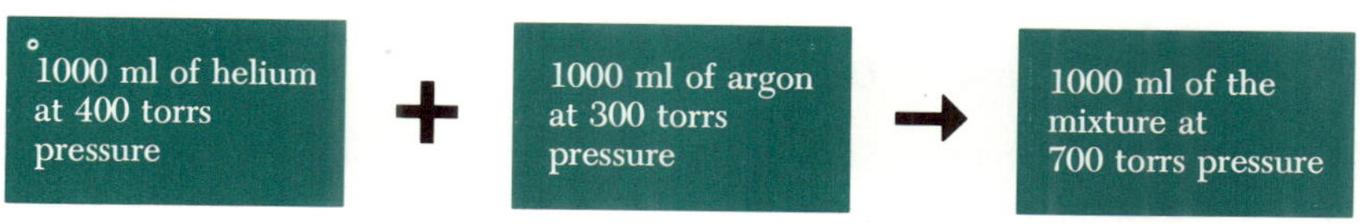

of a component of a mixture is the number of moles of that component in the mixture divided by the total number of moles of all components in the mixture. Thus, if a mixture contains n_A moles of component A and n_B moles of component B, the mole fraction X_A of component A is

$$X_A = \frac{n_A}{n_A + n_B}$$

Suppose we have a mixture of 16 g of hydrogen gas (H_2) and 56 g of carbon monoxide (CO). The molecular weights of these two gases being 2 and 28, respectively, we have

$$\frac{16\text{ g}}{2\text{ g/mole}} \text{ or 8 moles } H_2 \quad \text{and} \quad \frac{56\text{ g}}{28\text{ g/mole}} \text{ or 2 moles CO}$$

We have, then, a total of 10 moles, of which 8 are H_2 and 2 are CO. The mole fractions are $\frac{8}{10}$ or 0.8 for H_2 and $\frac{2}{10}$ or 0.2 for CO.

In the mixture for which the mole fraction of hydrogen is 0.8, it is important to realize that 0.8, i.e., 80 percent, of the molecules are H_2 molecules; similarly, since the mole fraction of carbon monoxide is 0.2, 20 percent of the molecules are CO molecules. Dalton discovered that in such a mixture 0.8 of the pressure is due to H_2 molecules and the remaining 0.2 of the pressure is exerted by CO molecules. Thus, in any gaseous mixture of A and B, the partial pressure, P_A, of component A is

$$P_A = X_A P_{\text{total}}$$

and similarly the partial pressure P_B of component B is

$$P_B = X_B P_{\text{total}}$$

As a corollary to Dalton's Law the total volume of any gaseous mixture is the sum of the partial volumes of the components of the mixture:

$$V_{\text{total}} = V_1 + V_2 + V_3 + \cdots$$

The partial volume of a component is the volume which that component would occupy if it were present alone at the total pressure of the mixture and at the same temperature. Insofar as the ideal gas laws are applicable, $V_1 = X_1 V_{\text{total}}$, $V_2 = X_2 V_{\text{total}}$, and so on.

The chemical composition of gas mixtures is frequently stated in terms of percentage by volume. Thus, dry air is stated to be 78 percent N_2, 21 percent O_2, and 1 percent Ar by volume. Mixing 78 cc of N_2, 21 cc of O_2, and 1 cc of Ar will produce 100 cc of mixture. In such a mixture 78 percent of the molecules are N_2 molecules, 21 percent are O_2 molecules, and 1 percent are Ar molecules. Therefore, 78 percent of the pressure is exerted by N_2 molecules, 21 percent by O_2 molecules, and 1 percent by Ar molecules.

9-11 Graham's Law of Diffusion Rates

In 1829 Thomas Graham, a Scottish chemist, completed an investigation of the diffusion of gases. *Diffusion* is the spontaneous spreading of a substance throughout the space available to it. The diffusibility of gases is one of their important properties. If a gas is placed in an empty container, it almost instantly pervades the entire space inside the container. If the container is already occupied by another gas, diffusion of the added gas nevertheless occurs—although more slowly, since no pressure difference exists to hasten the process. For example, if ammonia gas is liberated in a room, its odor soon pervades the room. The presence of air does not prevent the diffusion of the ammonia.

Graham compared the rates at which different gases diffuse and found that the rate depends upon the density of the gas. In a series of gases the lightest gas diffuses most rapidly, and the gas with the greatest density diffuses most slowly. For example, the density of sulfur dioxide gas is four times that of methane; methane diffuses more rapidly than sulfur dioxide—but only *twice* as fast. Oxygen has a density sixteen times that of hydrogen; hydrogen diffuses *four* times as rapidly as oxygen. Thus, we have a square-root relationship. Mathematically, *Graham's Law of Diffusion Rates* states that

$$\frac{R_1}{R_2} = \frac{\sqrt{d_2}}{\sqrt{d_1}} \quad \text{or} \quad \frac{R_1}{R_2} = \sqrt{\frac{d_2}{d_1}}$$

i.e., the relative rates of diffusion of any two gases under the same conditions are inversely proportional to the square roots of their densities.

EXAMPLE Gas A has a density of 1.24 g per liter. The density of gas B is 4.96 g per liter. Which gas will diffuse more rapidly, and in what ratio?

The ratio of the two densities is 4.96:1.24, or 4:1; i.e., gas B is four times as dense as gas A. Therefore, gas A will diffuse more rapidly than gas B by the factor $\sqrt{4}$, or 2. Gas A will diffuse twice as rapidly as gas B.

About thirty years after the Law of Diffusion Rates wes announced, Avogadro's Principle (Section 2-4) attained universal acceptance. It was then realized that the density of a gas is directly proportional to the weights of the individual molecules. Thus, Graham's Law can be stated in terms of molecular weights:

$$\frac{R_1}{R_2} = \frac{\sqrt{M_2}}{\sqrt{M_1}} \quad \text{or} \quad \frac{R_1}{R_2} = \sqrt{\frac{M_2}{M_1}}$$

EXAMPLE Which gas will diffuse more rapidly, methane (CH_4) or sulfur dioxide (SO_2), and in what ratio?

The molecular weights are 16 for methane (12 + 4) and 64 for sulfur dioxide (32 + 32). Thus, the ratio of the molecular weights is 64:16, or 4:1; i.e., the molecular weight of sulfur dioxide is four times that of methane. Thus, methane will diffuse more rapidly than sulfur dioxide by the factor $\sqrt{4}$, or 2—twice as fast.

9-12 The Kinetic Theory of Gases

The three properties of gases that are especially important are compressibility, thermal expansion, and diffusibility. In comparison with solids and liquids, gases are very easily compressed; all gases tend to behave alike in this regard, as stated by Boyle's Law. When heated, gases expand to a much greater extent than do solids or liquids; all gases tend to behave alike in this respect also, as stated by Charles' Law. All gases are characterized by diffusibility, but the rates at which different gases diffuse depend on their molecular weights. These properties of gases can be explained by the *kinetic theory*.

The kinetic theory assumes that all gases consist of very small, discrete particles called molecules, which are in constant, rapid motion. The molecules are assumed to be so small that the average distance between them is many times their diameter. The rapidly moving molecules collide with one another, and with the walls of the container, without loss of kinetic energy; i.e., a collision between a fast molecule and a slow molecule probably will result in speeding up the slow one and slowing down the fast one (although this depends upon the angle of impact), but the sum of the kinetic energies of the two molecules after the collision will be the same as the sum of the kinetic energies prior to the collision.

The kinetic energy of a moving molecule, like that of any moving object, is the energy associated with its motion. The quantity of kinetic energy such a molecule possesses is equal to the work it is capable of performing in being brought to rest and is given by the expression

$$E_k = \tfrac{1}{2}mu^2$$

where m is its mass and u is its velocity. The kinetic theory assumes that the average kinetic energy of the molecules of a gas is directly proportional to the absolute temperature. The molecules of different gases at the same temperature have the same kinetic energy. In the *ideal gas* (which has no real existence), there are *no* attractive forces between the molecules; also, the molecules of the ideal gas are infinitesimally small, so that their own volume is insignificant in relation to the volume occupied by the gas.

9-13 The Explanation of Gas Behavior

The assumptions of the kinetic theory account for the observed properties of gases and for the laws that express them:

1. *Compressibility*. If a gas consists of particles that are separated from one another by large spaces, then the reason that the gas is easily compressible is that the molecules are merely crowded closer together when the volume of the container is reduced.
2. *Diffusibility*. If the molecules of two different gases are in rapid motion and are relatively far apart, the spontaneous diffusion of each into the

other is the result of molecules of each kind moving into the spaces between the molecules of the other kind.

3. *Boyle's Law.* When the volume of a gas is reduced, the molecules have less space in which to move; they collide with the walls of the container more frequently. Since pressure depends upon the number of impacts per second, the pressure exerted by a gas is inversely proportional to the volume which it occupies.
4. *Dalton's Law.* If there is no attraction between gas molecules, each kind of molecule in a mixture of gases strikes the walls of the container the same number of times per second, and with the same force, as if it were the only kind of molecule present. Thus, the partial pressure of a gas is not changed by the presence of other gases. Each partial pressure is proportional to the number of molecules of that species present, and the total pressure exerted by the mixture of gases is the sum of the pressures exerted by the individual components.
5. *Graham's Law.* If, on the average, molecules of different gases at the same temperature have the same kinetic energy, then molecules of one kind, *A*, must have velocities that are different from the velocities of the molecules of another gas, *B*, unless they have the same masses. Mathematically, if the kinetic energy is the same for the two gases,

$$\tfrac{1}{2}m_A{u_A}^2 = \tfrac{1}{2}m_B{u_B}^2$$

and

$$\frac{{u_A}^2}{{u_B}^2} = \frac{m_B}{m_A}$$

so that

$$\frac{u_A}{u_B} = \sqrt{\frac{m_B}{m_A}}$$

For example, a methane molecule (molecular weight, 16), *A*, has a mass one fourth as great as the mass of a sulfur dioxide molecule (molecular weight, 64), *B*. If

$$\frac{m_B}{m_A} = 4$$

then

$$\frac{u_A}{u_B} = \sqrt{4} = 2$$

The methane molecules have an average velocity twice that of the sulfur dioxide molecules. Thus, the kinetic theory shows why methane diffuses twice as fast as sulfur dioxide and why, in general, different gases diffuse at rates that vary inversely with the square roots of their molecular weights or densities.

6. *Charles' Law.* The kinetic theory postulates that the absolute temperature is a manifestation of the average kinetic energy of the molecules and that doubling the absolute temperature means doubling the average kinetic energy. (Thus, absolute zero is the temperature of no kinetic energy.) The pressure exerted by a gas confined within a fixed volume is proportional to nE_k, i.e., the number of molecules per unit volume times their

average kinetic energy. Doubling the absolute temperature doubles E_k and therefore doubles the pressure which the gas exerts, provided the volume is kept constant. Alternatively, the volume must be permitted to double if the pressure is to remain the same in spite of doubled temperature.

7. *Avogadro's Principle.* If two gases are at the same temperature, then E_k (the average kinetic energy per molecule) is the same for the one gas as for the other. If the two gases are also exerting the same pressure, then nE_k (the number of molecules per unit volume multiplied by their average kinetic energy) is also the same for the two gases. Therefore, if two gases are at the same temperature and pressure, n must be the same for both. The number of molecules per unit volume is the same—i.e., equal volumes contain the same number of molecules—if gases are at the same temperature and pressure.

9-14 Gay-Lussac's Law of Combining Volumes of Gases

In the discussion of Dalton's Law of Partial Pressures we considered gases which, when mixed, do not react with each other. Of course, there are cases in which mixed gases do react. Sometimes gases react as soon as they are mixed, as when colorless NO gas comes into contact with air and reacts with the oxygen to form reddish-brown NO_2 gas:

$$2NO + O_2 \rightarrow 2NO_2$$

More commonly, a mixture of gases requires some input of energy to get the reaction started. A mixture of H_2 and Cl_2 does not react in the dark, but light of a certain frequency (in the ultraviolet range) initiates a violent reaction:

$$H_2(g) + Cl_2(g) \rightarrow 2HCl(g) \qquad \Delta H = -44 \text{ kcal}$$

A mixture of H_2 and O_2 does not react until a flame or an electric spark comes into contact with the mixture; then the two gases react with explosive violence:

$$2H_2(g) + O_2(g) \rightarrow 2H_2O(g) \qquad \Delta H = -116 \text{ kcal}$$

In 1809 Joseph Louis Gay-Lussac, French chemist and physicist, completed a series of measurements on the volumes of gases that react with each other and proposed a law now known as *Gay-Lussac's Law of Combining Volumes of Gases.* This law states that gases react in a simple numerical ratio by volume, and the volume of each gaseous product bears a simple numerical ratio to the volume of each gaseous reactant and to the volume of each other gaseous product. Thus, when Cl_2 and H_2 react, the volume of Cl_2 that reacts is the same as the volume of H_2 with which it reacts; i.e., the volume ratio is 1:1. When O_2 and H_2 react, the volume of H_2 reacting is twice the volume of O_2 with which it reacts; i.e., the volume ratio is 2:1.

In terms of Avogadro's Principle, H_2 and Cl_2 react in the volume ratio 1:1 because each molecule of H_2 reacts with one molecule of Cl_2. If a volume of

H_2 contains, say, one trillion H_2 molecules, then the volume of Cl_2 with which it reacts is the volume that contains one trillion Cl_2 molecules. According to Avogadro's Principle, one trillion Cl_2 molecules occupy the same volume as one trillion H_2 molecules. Avogadro proposed his hypothesis in 1811 to provide an explanation for the Law of Combining Volumes, which Gay-Lussac had discovered two years earlier.

9-15 Development of the Kinetic Theory

When Avogadro proposed his hypothesis, he was simply making a shrewd guess that seemed reasonable. He could not deduce his hypothesis from the kinetic theory of gases, as we have done, for the kinetic theory was not well developed at that time. The beginnings of the mathematical development of the kinetic theory were made by Daniel Bernoulli in 1738 in Switzerland, but interest in the theory did not develop until a century later. Then a detailed kinetic theory of gases developed through the efforts of many physical scientists, including Rudolf Clausius in Germany, J. J. Waterston and James Clerk Maxwell in England, Ludwig Boltzmann in Austria, and J. Willard Gibbs in America.

From the equations of the kinetic theory it is possible to calculate the average velocity of gas molecules, the average distance that a molecule travels between successive collisions with other molecules, the number of collisions that a molecule has per second, etc. For example, the average velocity of H_2 molecules at 0°C is 1.84×10^5 cm per sec (equivalent to 3780 mi per hr). At a pressure of 1 atmosphere, the average distance that a hydrogen molecule travels between collisions at 0°C is only 1.12×10^{-5} cm; the number of collisions experienced by one molecule in 1 second is 1.6×10^{10} (i.e., between ten and twenty billion).

9-16 Deviations from Ideal Behavior

Boyle's Law, Charles' Law, and the other gas laws describe the behavior of real gases satisfactorily at ordinary pressures and temperatures, but at low temperatures and high pressures the behavior of any real gas deviates markedly from ideal behavior.

At high pressures gas molecules are relatively close together, and the actual volume of the molecules themselves becomes significant compared with the free space between them. As we would expect, a gas becomes less compressible at high pressures; thereafter, doubling the pressure does not reduce the volume by half. Under extremely high pressures, the space occupied by the molecules becomes a substantial fraction of the space within which the gas is confined, and the gas resists further decrease in volume, a resistance characteristic of liquids.

At low temperatures all gases become more compressible than would be predicted by the gas laws. This phenomenon is due to attractive forces between molecules which cause them to crowd closer together. The attractive forces between nonpolar molecules are called *van der Waals forces* after Johannes van der Waals, the Dutch physicist who postulated their existence in 1873. These forces, although important, are effective only over very short distances.

Because the two factors that cause nonideal behavior in gases act in opposition, the net deviation at a particular temperature and pressure depends upon the predominant factor. This varies with different gases.

Deviation from ideal behavior, except under extreme conditions, is not serious enough to discredit the kinetic theory of gases. Indeed, we should not expect molecules to be mere mathematical points, and the very existence of liquids is evidence that there is an attraction between molecules. We should expect that the behavior of a real gas would be affected by the actual volume of the molecules and by the attractive forces between them.

The van der Waals forces of attraction between nonpolar molecules are a result of the electrical attraction between the nuclei of one molecule and the electrons of another. This is so nearly compensated for by the repulsion of nuclei by nuclei and of electrons by electrons that the net forces of attraction become significant only when the molecules are crowded close together—at temperatures low enough for the disordering effect of molecular motion to be overcome. Large molecules attract one another more strongly than do small ones because they contain more electrons.

The van der Waals forces of attraction exist because of the motion of electrons in atoms and molecules. Consider two argon atoms at a moment when they happen to be close to each other, as represented in Figure 9-7. Although the electron distribution in an argon atom is highly symmetrical when averaged over even a short period of time, we can imagine that at a particular instant there is a lack of perfect symmetry with a slight preponderance of electrons in one direction (arising through the laws of chance in the motions of electrons). At this instant, the left side of atom *A* is slightly negative, and the right side is slightly positive. The positive side of *A* attracts the electrons in its neighbor, atom *B*, leaving it similarly unsymmetrical. Thus, each atom is momentarily polar, with a consequent attraction between them. Even though the attraction between these particular atoms will persist for only a very brief period of time, any other pair of atoms has the same opportunity to develop such momentary

Figure 9-7

Van der Waals forces of attraction between two argon atoms

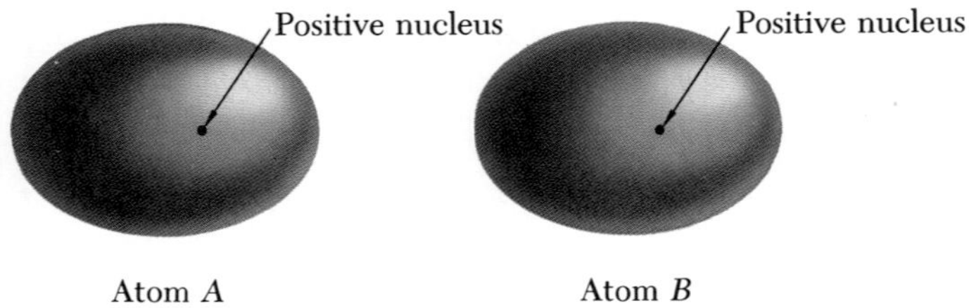

attractions between them. Consequently, there never is an instant when van der Waals attractive forces cease to exist.

One other type of attractive force, which is stronger than van der Waals forces, can exist between molecules. Polar substances deviate substantially from ideal behavior because of the permanent polarity of their molecules. The positive end of each polar molecule attracts the negative end of each of its close neighbors with sufficient effectiveness to keep the substance in the liquid state at temperatures considerably above the boiling points of substances whose only intermolecular attractive forces are of the van der Waals type. Thus, HCl, mentioned in Section 5-5 as an example of a polar compound, has a boiling point 101° higher than that of argon, even though its molecular weight is several units less than the molecular weight of argon. Even more striking is the case of H_2O, which has a boiling point 346° higher than that of neon, although its molecular weight is less than the molecular weight of neon.

New Terms

Absolute temperature: temperature measured from −273.15°C (−459.67°F), which is absolute zero. In scientific work the measurement is made in Celsius degrees; i.e., absolute temperature refers to the Kelvin scale, which is the absolute Celsius scale. The Rankine scale is the absolute Fahrenheit scale.

Absolute zero: the lowest possible temperature (−273.15°C or −459.67°F). Temperatures within a few thousandths of a degree of 0°K have been attained. In the kinetic theory of gases, absolute zero is the temperature at which the kinetic energy of the molecules of an ideal gas is zero.

Barometer: an instrument for measuring atmospheric pressure.

Calorie: the quantity of heat required to raise the temperature of 1 g of H_2O 1°C. The kilocalorie (kcal, 1000 cal) is the "big calorie," or Calorie, used in the field of nutrition.

Combined gas law equation: the equation that combines the Boyle's Law equation and the Charles' Law equation into one expression. It can be written $\frac{PV}{T} = \text{constant}$ or $\frac{P_1V_1}{T_1} = \frac{P_2V_2}{T_2}$.

Diffusion: the spontaneous spreading of a substance throughout the space available to it.

Energy: the capacity to do work.

General gas law equation: the equation that states the relationship of the volume of a gas to the pressure, the temperature, and the number of moles: $PV = nRT$.

Heat: the form of energy that passes from a body of higher temperature to a body of lower temperature.

Ideal behavior of a gas: perfect conformity with Boyle's Law and Charles' Law.

Ideal gas: an imaginary gas that behaves in perfect accord with Boyle's Law and Charles' Law.

Kinetic energy: the energy that a moving object possesses by virtue of its motion. The quantity of kinetic energy such an object possesses is equal to the work it is capable of performing in being brought to rest. This is expressed by the equation $E_k = \frac{1}{2}mu^2$, where m is the mass of the object and u is its velocity.

Molar gas constant: the constant R in the general gas law equation.

Molar volume: the volume of 1 mole. Unless stated otherwise, the molar volume of a gas is the volume of 1 mole of the gas at standard conditions. The molar volume of an ideal gas is 22.414 liters at STP.

Mole fraction (of a component of a mixture): the number of moles of a component of a mixture divided by the total number of moles of all components in the mixture.

Potential energy: the energy that a body possesses by virtue of its position; e.g., if you lift this book above the desk, you do work and thereby impart potential energy to the book.

Pressure: force per unit area.

Standard pressure: the pressure exerted by a column of mercury 760 mm (29.9 inches) high, designated 1 atmosphere (1 atm) because it is the pressure exerted by the earth's atmosphere, or 760 torrs.

Standard temperature: 0° Celsius (32° Fahrenheit).

Temperature: degree of hotness, measured on a suitable scale. The temperature of an object is not to be confused with the heat that it contains.

Thermometer: the instrument used for measuring temperature.

Torr: the pressure exerted by a column of mercury 1 mm high.

Torricellian vacuum: the space above the mercury inside a mercury barometer, devoid of all matter except a small concentration of mercury vapor.

Van der Waals forces: the weak forces of attraction that exist between nonpolar molecules, due to the effect of the atomic nuclei within one molecule upon the electrons of another. The van der Waals forces are significant only when molecules are very close together. At small distances they are balanced by forces of repulsion between outer electron shells of the molecules.

Work: motion against an opposing force.

The Gas Laws

Boyle's Law: at constant temperature, the volume occupied by a gas varies inversely with the confining pressure. It can be written PV = constant or $P_1V_1 = P_2V_2$.

Charles' Law: the volume occupied by a fixed quantity of gas confined under constant pressure is directly proportional to the absolute temperature. It can be written

$$V = \text{constant} \times T \quad \text{or} \quad \frac{V_1}{T_1} = \frac{V_2}{T_2}, \qquad \text{where } T \text{ is in °K.}$$

Dalton's Law of Partial Pressures: in a mixture of gases the molecules of each species exert the same pressure they would exert if they were present alone, so that the total pressure is the sum of these partial pressures exerted by the different gases in the mixture. It can be written $P_{\text{total}} = P_1 + P_2 + P_3 + \cdots$.

Gay-Lussac's Law of Combining Volumes of Gases: gases react in a simple numerical ratio by volume, and the volume of each gaseous product bears a simple numerical ratio to the volume of each gaseous reactant and to the volume of each other gaseous product.

Exercises

9-1 At what temperature does a Celsius thermometer give the same reading as a Fahrenheit thermometer?

9-2 The normal temperature of the human body is 98.6°F. What is this on the Celsius scale? Rankine? Kelvin?

9-3 Mercury boils at 356.9°C, and helium boils at −268.9°C. Convert these temperatures to the Kelvin and Fahrenheit scales.

9-4 Prove that Fahrenheit-Celsius conversions can be made by using the following formulas:

$$°F = \tfrac{9}{5}(°C + 40) - 40$$
$$°C = \tfrac{5}{9}(°F + 40) - 40$$

9-5 If we have 400 ml of oxygen gas at 740 torrs pressure, what volume will we have at 925 torrs (temperature remaining constant)?

9-6 If a gas has a volume of 500 ml at 25°C, what volume will it occupy at 50°C, if the pressure is not permitted to change?

9-7 If the volume of a gas is 1.492 liters at 25°C and 735 torrs, what will its volume be at standard conditions?

9-8 What is the weight of 1500 ml of carbon monoxide at STP?

9-9 If the weight of 350 ml of a gas at STP is 0.78 g, what is its molecular weight?

9-10 What is the molecular weight of a gas if 1.00 g has a volume of 437 ml at 23°C and 738 torrs?

9-11 A gaseous mixture is 50% N_2, 30% O_2, and 20% CO_2 by volume. How many molecules of each kind are contained in every 100 molecules of the mixture? What is the mole fraction of each component? What is the partial pressure of each gas if the total pressure is 750 torrs?

9-12 In a mixture of 50 g of nitrogen and 50 g of oxygen, what is the partial pressure of each gas if the total pressure is 1 atmosphere?

9-13 Arrange the following gases in order of increasing diffusion rates: N_2, Ar, O_2, He, Ne, H_2, Cl_2, NH_3.

9-14 If 10 ml of a certain gas pass through a porous wall in the same length of time as 20 ml of N_2 under the same conditions, how does the density of the gas compare with that of N_2? What is its molecular weight?

9-15 What volume of O_2 is needed to burn 10 liters of ethane, C_2H_6? (The products are water and carbon dioxide.)

9-16 If a mixture of 500 ml of H_2 and 500 ml of O_2 is ignited with an electric spark, what volume of which gas will remain unreacted?

References

Brown, E. H., "Some Early Thermometers," *J. Chem. Educ.*, **11** (8), 448 (1934). A collection of drawings of early thermometers from the time of Galileo to the time of Fahrenheit.

Drago, R. S., *Prerequisites for College Chemistry*, Harcourt, Brace & World, New York, 1966 (paperbound). Chapter 5, "Gases and the Kinetic Theory," includes a selection of solved problems and self-testing exercises.

Hildebrand, J. H., *An Introduction to Molecular Kinetic Theory*, Reinhold, New York, 1963 (paperbound). The emphasis in this book is upon the concepts, rather than the mathematical formulas, of kinetic theory.

Miller, Jr., F., *College Physics*, 2nd ed., Harcourt, Brace & World, New York, 1967. Chapters 13–16 deal with the concepts of heat and temperature and with the thermal behavior of gases.

Ruckstuhl, A., "Thomas Graham's Study of the Diffusion of Gases," *J. Chem. Educ.*, **28** (11), 594 (1951). An account of the experimental work that led to the discovery of the Law of Diffusion Rates.

Stimson, H. F., "Celsius versus Centigrade: The Nomenclature of the Temperature Scale of Science," *Science*, **136**, 254 (1962). The history of the Celsius scale, including the official 1960 definition which makes the adjective *centigrade* inexact.

Wilson, M., and the Editors of Life, *Energy*, Time, Inc., New York, 1963. A description, accompanied by drawings and photographs, of energy in its many forms.

Wilson, R. E., "Standards of Temperature," *Am. Sci.*, **41**, 271 (1953). An account of the practical problem of establishing fixed reference points for thermometry.

10

The Properties of Liquids

10-1 The Liquid State

In the liquid state a substance lacks the compressibility that is characteristic of gases. Under ordinary conditions, doubling the pressure on a gas reduces its volume by half. A similar increase in the pressure applied to a liquid has little effect on the volume, for in the liquid state most of the space is occupied by the molecules themselves.

Figure 10-1

Uniform expansion of mercury

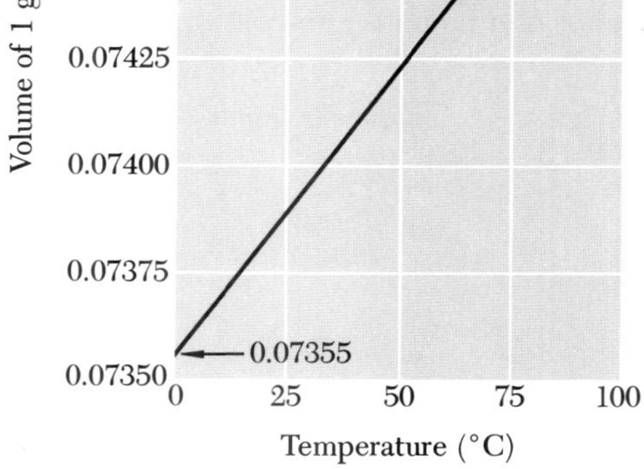

Because the molecules are so close together in the liquid state, the density of a liquid is of the order of a thousand times as great as the density of a gas under atmospheric pressure. For this reason the densities of liquids are usually expressed in grams per milliliter, while the densities of gases are expressed in grams per liter. All liquids show changes of volume when the temperature changes, but the changes are small compared with those shown by gases. Usually, but not always, liquids expand with rising temperature. The expansion is usually quite uniform (Figure 10-1), but structural complications can alter this.

Both liquids and gases are *fluids*; i.e., under the influence of gravity they readily flow. Thus, either a liquid or a gas will assume the shape of the container in which it is placed. However, while the volume of a gas is the same as that of the container, a liquid has its own volume; i.e., it occupies only the bottom portion of the container. Consequently, a liquid, unlike a gas, has a surface. If a liquid is present in a container of small cross section, such as a buret, a pipet, or a barometer tube, the surface of the liquid is called the *meniscus*. The meniscus may be either concave or convex, depending on whether the containing walls are wetted by the liquid (see Figure 10-3).

10-2 Surface Tension

The existence of a liquid is a result of the attractive forces between the molecules. These forces hold the liquid together. In substances that are difficult to liquefy, the only attractive forces are van der Waals forces. In substances that are liquid at room temperature, such as water, and in substances that are easily liquefied by moderate cooling, such as ammonia and sulfur dioxide, much stronger attractive forces exist, which are due to the polar character of the molecules.

A molecule in the interior of a liquid is acted upon equally from all directions by the intermolecular attractive forces, which are effective only at short distances. A molecule near the surface, however, is acted upon by attractive

Figure 10-2

Directions of forces on a molecule in a liquid

A molecule in the interior of a liquid is attracted equally in all directions; a molecule near the surface is attracted inward.

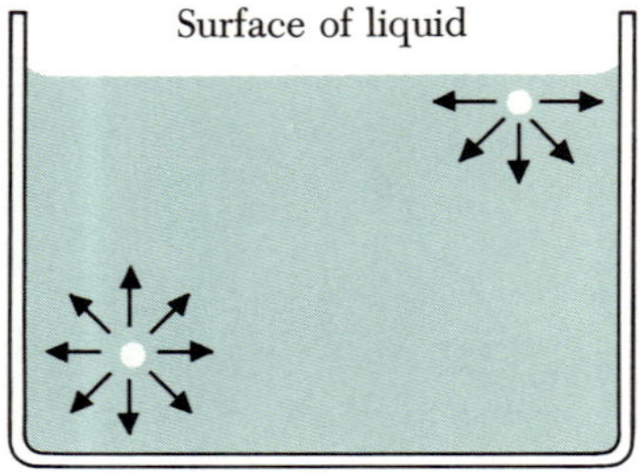

forces beneath it that are not balanced from above (Figure 10-2). The net result is an inward pull, called *surface tension*, that causes every liquid surface to become as small as possible. Thus, a free liquid drop, such as an oil drop in water, tends to be spherical, because a sphere has a minimum surface area. The spherical shapes of raindrops and teardrops are slightly distorted by the act of falling; and stationary drops on a solid surface, especially if they are composed of a dense liquid, are somewhat flattened by the action of gravity.

The magnitude of the surface tension is characteristic of a given liquid for a given temperature. With increasing temperature the surface tension decreases because the intensified disorder associated with the increased molecular motion reduces the effectiveness of the attractive forces between the molecules.

The ability of certain insects to stay afloat and move about on the surface of water is due to surface tension. Surface tension accounts for the fact that a razor blade can float on water, even though its density greatly exceeds that of water.

Capillary action is due to surface tension. If a capillary tube (a tube of small bore) is immerged in a liquid that wets the walls of the tube, the liquid adhering to the walls pulls the body of the liquid up so that the total surface of the liquid, including that adhering to the walls, is minimized as much as gravity permits (Figure 10-3a); such is the case with water in a glass capillary, for water is attracted by glass. If the liquid does not wet the walls of the capillary, the pull of surface tension causes the level of the liquid inside the tube to fall so that the total surface of the liquid, including that in contact with the walls, is diminished (Figure 10-3b); such is the case with mercury in a glass capillary—the forces of attraction between mercury and glass are not as strong as the cohesive forces within the mercury.

Figure 10-3

Capillary action

(a) Rise of a liquid that wets the wall. (b) Depression of a nonwetting liquid.

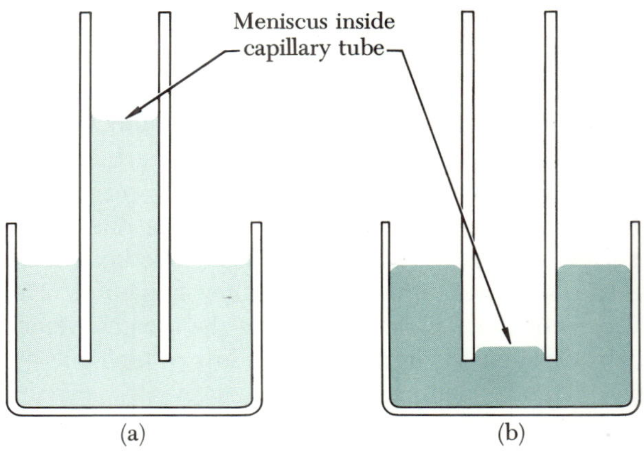

10-3 Viscosity

Although, by definition, fluids can flow, each individual fluid has a characteristic amount of internal friction or resistance to flow that is known as *viscosity*. The viscosities of any two liquids can be compared by measuring the time required for a specified volume of each to be forced through a tube of small diameter by a fixed pressure. Gravity causes a liquid to flow downward at a measurable rate through a narrow tube held vertically. Thus, the simplest apparatus used to measure the viscosities of liquids consists of a pipet and a stopwatch.

Viscosity varies greatly from materials having long pour times, such as heavy lubricating oils, to liquids of low viscosity, such as ether or benzene. There is roughly an inverse relation between viscosity and *volatility*, i.e., the tendency of a liquid to change to the gaseous state. This is because strong intermolecular forces lessen both the ability of a liquid to flow and its tendency to evaporate. Liquids, such as the heavy oils, that are composed of large molecules have high viscosities (and low volatilities) partly because the strength of the van der Waals forces increases with molecular weight and partly because of the greater difficulty of untangling the large molecules.

The viscosity of a gas increases as the temperature increases, but the viscosities of most liquids decrease with rising temperature (about 2 percent per Celsius degree) since the attraction between molecules is partially overcome by increased kinetic activity.

Viscosity is one of the frequently measured properties of blood and other animal fluids. It is an important factor in the designing of chemical engineering equipment where the cost of pumping depends greatly on the voscosity of the fluid. Thus, viscosity is relevant to practical work as well as to theoretical problems.

10-4 Evaporation

One of the most familiar properties of liquids is their ability to *evaporate*, i.e., to change from the liquid to the gaseous state. We should ask why some liquids have much greater tendency to evaporate, i.e., are more volatile, than others and why the volatility of a liquid increases with rising temperature.

The molecules of a liquid move in a random fashion. Although the average kinetic energy of the molecules is constant at any particular temperature, the individual molecules do not all have the same kinetic energy—at any particular instant the kinetic energy of some molecules is greater than others. Those molecules in the surface of a liquid that have the highest kinetic energy overcome the attractive forces exerted upon them by their neighbors. Thus, evaporation involves the escape of those molecules having the greatest velocities.

Since evaporation consists in the selective loss of high-energy molecules, the average energy of the remaining molecules continually decreases. Unless heat is supplied from an outside source, the temperature of the liquid drops. The regulation of the body temperature of many mammals is based in part on the cooling produced by evaporation of the watery liquid secreted by sweat glands.

At higher temperatures greater numbers of molecules have the necessary energy to escape from the surface. This is the reason for the increase in volatility that all liquids show with rising temperature. Since evaporation occurs at the surface of a liquid, the rate of evaporation can also be increased by increasing the surface area.

Since the attractive forces between molecules of some liquids are greater than between molecules of other liquids, volatility is a property that varies from substance to substance. The most polar substances are the least volatile. Among the nonpolar substances, the most volatile ones are those with the smallest molecular weights, for their molecules contain the fewest electrons, and the van der Waals forces are therefore quite weak.

10-5 Vapor Pressure

If a liquid is placed in a closed container, evaporation will reduce the volume of liquid to a certain point, but not beyond this point. This can be interpreted in terms of the kinetic theory. As the liquid evaporates, the molecular population above the surface increases. The molecules in the vapor phase, like the molecules of any gas, are in random motion. Some of the vapor molecules collide with the walls of the container; others collide with the surface of the liquid. Any vapor molecule colliding with the liquid will almost certainly be captured by, and again become a part of, the liquid. When the molecular population in the vapor phase attains a sufficiently high value, the rate at which molecules return to the surface of the liquid equals the rate at which molecules leave the surface to enter the vapor phase. When the rate of recondensation equals the rate of evaporation, a state of *dynamic equilibrium* prevails (Figure 10-4). After this point is reached, the relative amounts of liquid and vapor remain unchanged

Figure 10-4

Dynamic equilibrium

Evaporation in a closed vessel leads to a state of dynamic equilibrium.

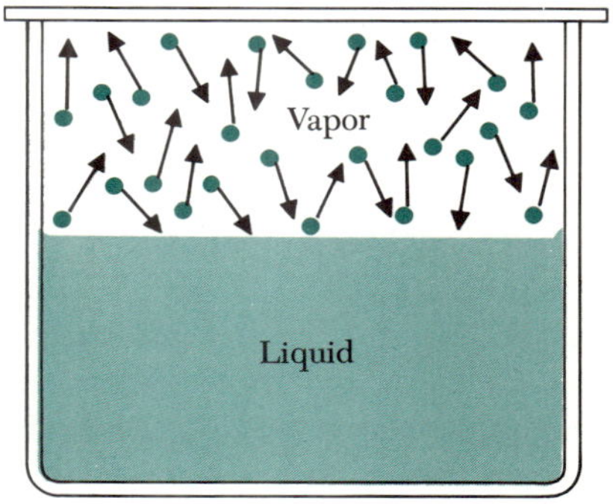

Figure 10-5

Change of vapor pressure with temperature

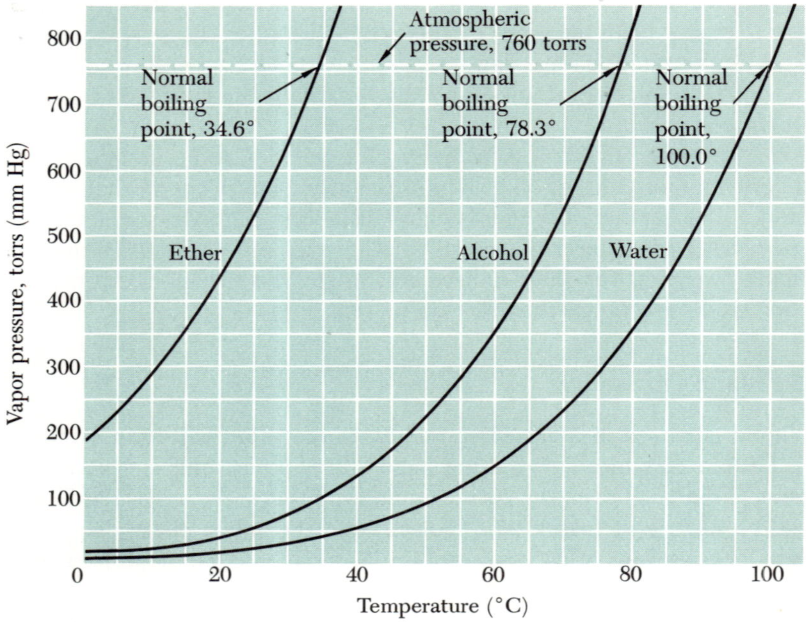

as long as the temperature is kept constant. Note that at equilibrium one molecule evaporates for every molecule that condenses, keeping the concentration of the vapor constant. The molecules in the vapor exert a definite pressure. The pressure that the vapor exerts when it is in equilibrium with its liquid is called the *vapor pressure* of the liquid.

The vapor pressure of any liquid is a measure of the escaping tendency of its molecules. Vapor pressures increase with rising temperatures, but the vapor pressures of different liquids at the same temperature differ greatly (Figure 10-5). Thus, at 25°C the vapor pressure of water is only 23.6 torrs (mm Hg), but the vapor pressure of ether is 537 torrs at the same temperature.

10-6 Boiling Point and Critical Temperature

Ordinary evaporation occurs only at the surface, but when a liquid is heated to a sufficiently high temperature, bubbles of vapor form inside the liquid and rise to the surface; i.e., the liquid *boils*. In more technical language, the phenomenon is known as *ebullition*. The temperature at which the boiling occurs is called the *boiling point*.

In order for a vapor bubble to exist beneath the surface of a liquid, the pressure exerted outward by the vapor molecules must be sufficient to offset the crushing effect of the atmosphere overlying the liquid. Thus, boiling occurs at the temperature at which the vapor pressure of the liquid is equal to the external pressure acting downward upon the surface of the liquid. Since barometric pressure varies from time to time and from place to place, the boiling point of a liquid is not always the same temperature. The *normal boiling point* of a liquid is the temperature at which its vapor pressure is 760 torrs (1 atmosphere). By controlling the external pressure, one can cause a liquid to boil either below or above its normal boiling point. Thus, the vapor pressure curve for a liquid is also a boiling point curve; i.e., from the curve, one can read the temperature at which the liquid will boil under any specified external pressure (Figure 10-5).

A liquid can be heated without boiling to a temperature above its normal boiling point if the external pressure is increased beyond 1 atmosphere. As the vapor pressure curve indicates, higher external pressures necessitate higher temperatures to effect boiling. This principle receives practical application in pressure cookers.

The vapor pressure curve for any liquid terminates at a certain temperature, called the *critical temperature*, above which it is impossible for that substance to exist in the liquid state, however great the pressure applied to it. For example, the critical temperature of hydrogen is −240°C (Section 8-1). The vapor pressure which a liquid exerts at its critical temperature is called its *critical pressure*.

The vapor pressures, molecular weights, boiling points, and critical temperatures of some common liquids are listed in Table 10-1. If a comparison is made among substances with similar molecular structures, such as CH_3Cl, CH_2Cl_2, $CHCl_3$, and CCl_4, the effect of molecular weight becomes discernible, viz., the greater the molecular weight the less the vapor pressure at a given temperature. Consequently, the higher the molecular weight the higher the boiling point and the higher the critical temperature. This is not a gravitational effect; it is a consequence of the increased number of electrons in the more massive molecules, which results in stronger van der Waals forces.

Frequently, the normal boiling point of a substance is approximately 0.6 times its critical temperature (both expressed as absolute temperatures).

Table 10-1

Properties of selected liquids

Liquid	*Molecular weight*	*Vapor pressure at 20°C* (torrs)	*Boiling point* (°K)	*Critical temperature* (°K)
Water	18	18	373	647
Alcohol	46	44	352	576
Acetone	58	185	330	508
Ether	74	442	308	467
Carbon tetrachloride	154	91	350	556

Figure 10-6

A gas collected over water

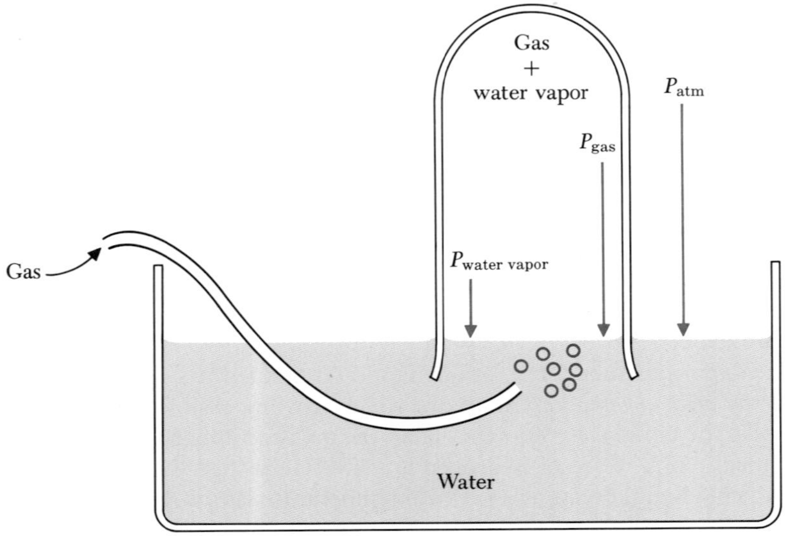

10-7 Collection of Gases over Water

A commonly used procedure for collecting a gas and measuring its volume involves the physical displacement of a liquid (Figure 10-6). A cylinder is filled with the liquid and inverted over a trough that also contains the liquid. The gas to be collected is led from a tank or generator through a tube beneath the surface of the liquid in the trough and upward into the inverted cylinder. In this way the gas is prevented from escaping into or coming in contact with the air. Before the volume of collected gas is measured, the cylinder is raised or lowered until the surface of the liquid inside the cylinder is level with the outside surface. Thus, the pressure exerted by the gas is equalized with the pressure of the atmosphere, which is ascertained by reading the laboratory barometer.

Since water is the liquid over which gases are usually collected (except such gases as ammonia and hydrogen chloride that are highly soluble in water), the collected sample contains water vapor. It is usually safe to assume that the gas is saturated with water vapor, i.e., that the partial pressure of the water vapor is identical with the vapor pressure of water, so that the rate of condensation is equal to the rate of evaporation. The pressure exerted by the gas itself is then obtained by applying Dalton's Law of Partial Pressures—the vapor pressure of water at the prevailing temperature is subtracted from the total pressure (the barometric reading). The vapor pressure of water at various temperatures is listed in the Appendix.

No vapor pressure correction is necessary when a gas is collected and measured over mercury because the vapor pressure of mercury at room temperatures is not sufficiently great to be significant.

EXAMPLE A gas is collected over water at 23°C. The barometric pressure is 753 torrs. What is the partial pressure of the gas?

The vapor pressure of water at 23°C is 21 torrs (see Appendix). The pressure exerted by the gas is 753 torrs minus the vapor pressure of water at 23°C; i.e., 753 torrs − 21 torrs = 732 torrs.

EXAMPLE A gas is collected over water at a pressure of 743 torrs and a temperature of 19°C. The measured volume is 247 ml. What will be the volume of the dry gas at standard conditions?

The pressure exerted by the gas itself is 743 torrs minus the vapor pressure of water at 19°C; i.e., 743 torrs − 16 torrs = 727 torrs. At standard conditions the pressure is greater than 727 torrs, so the pressure factor by which the measured volume should be multiplied is a fraction less than 1, i.e., $\frac{727}{760}$. The temperature at STP is less than 19°C, so the correct temperature factor is also a fraction less than 1, i.e., $\frac{273}{292}$. Therefore,

$$
\begin{aligned}
V_{\text{STP}} &= 247 \text{ ml} \times \frac{727 \text{ torrs}}{760 \text{ torrs}} \times \frac{273°\text{K}}{292°\text{K}} \\
&= 247 \text{ ml} \times 0.957 \times 0.935 \\
&= 247 \text{ ml} \times 0.895 \\
&= 221 \text{ ml}
\end{aligned}
$$

10-8 Distillation

One of the operations most commonly associated with chemistry is *distillation*, a process by which a liquid is vaporized by heating and its vapor condensed by cooling. A typical distillation apparatus for laboratory use is shown in Figure 10-7. A thermometer placed in the neck of the distilling flask registers the temperature of the vapor (the boiling point of the liquid). The vapor emerges from the flask into the inner tube of the condenser, where it is cooled and condensed to the liquid state by cold water circulating through the condenser jacket. The liquid distillate is collected in a flask or other container.

Simple distillation is effective in separating a pure liquid from impurities that do not vaporize at the boiling point of the liquid. Water is often purified by distillation. The nonvolatile mineral matter that is objectionable in water to be used in laboratory work (or in steam irons, storage batteries, etc.) is left behind in the distilling flask.

10-9 Heat of Vaporization

We have already seen that evaporation causes the temperature of a liquid to drop unless heat is supplied from an outside source. The quantity of heat that must be supplied to vaporize a stated quantity (such as 1 g or 1 mole) of a liquid,

Figure 10-7

Distillation

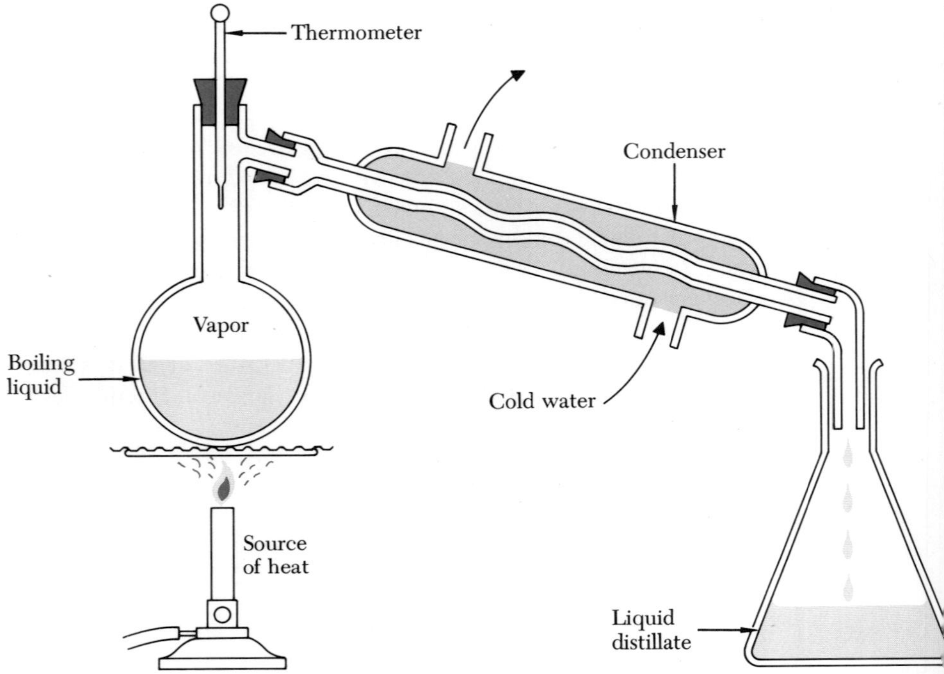

without change of temperature, is known as the *heat of vaporization* and is characteristic of the liquid. The quantity of heat required to vaporize 1 mole is called the *molar heat of vaporization*. For water, a highly polar substance, the heat of vaporization is 539 cal per g (or 9700 cal per mole) at 100°C, the normal boiling point. For carbon tetrachloride, a nonpolar compound, the heat of vaporization is only 46.4 cal per g (or 7140 cal per mole) at 77°C, the boiling point.

The heat of vaporization of a liquid varies with the temperature at which the vaporization occurs; it decreases with rising temperature until it becomes zero at the critical temperature. The heat of vaporization of water decreases from 595 cal per g at 0°C to 446 cal per g at 220°C (see Table 10-2 and Figure 10-8); it becomes zero at the critical temperature, 374°C.

According to the empirical rule proposed by Frederick Trouton in 1884, dividing the molar heat of vaporization of a liquid (in calories) by the normal boiling point (in °K) gives approximately 22 cal per mole per degree. The equation for *Trouton's Rule* is, therefore,

$$\frac{\text{Molar heat of vaporization (cal/mole)}}{\text{Normal boiling point (°K)}} = 22 \text{ cal per mole per °K}$$

Table 10-2

Heat of vaporization of water

Temperature (°C)	*Heat of vaporization* (cal/g)
0	595.4
20	584.9
40	574.2
60	562.8
80	551.1
100	538.7
120	525.6
140	511.5
160	496.5
180	480.6
200	463.8
220	446.0

Figure 10-8

Heat of vaporization of water

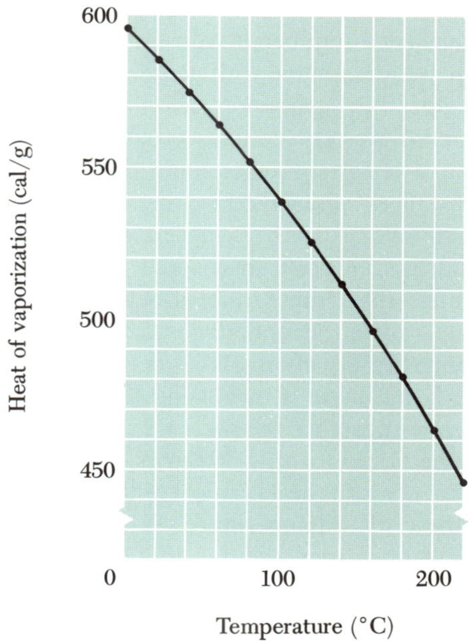

Trouton's Rule is followed fairly well by nonpolar liquids but much less well by polar compounds.

When a gas is condensed to the liquid state, a quantity of heat, known as the *heat of condensation*, is released; this quantity of heat is equal to the heat of vaporization at the same temperature. Thus, steam produces more serious burns than does boiling water because of its large heat of condensation.

The operation of mechanical refrigerators is based on the heat of vaporization of liquids. Heat energy is removed from the interior of the refrigerator by evaporation of a liquid, and the vapor is transported to the outside of the refrigerator, where it is liquefied by compression and cooling. A satisfactory refrigerant must be a substance that is readily condensed from the gaseous state to the liquid state at the operating temperature; it must also have a high heat of vaporization to produce an adequate cooling effect inside the refrigerator. One of the first refrigerants used in mechanical domestic refrigerators was ammonia, which had been used previously as the refrigerant in commercial ice plants. Sulfur dioxide was also used extensively in refrigeration until the development, since 1930, of the nontoxic Freons, which have also found wide use in aerosol sprays and other devices. The name Freon is applied to several relatively simple organic compounds containing fluorine, such as CCl_2F_2.

10-10 Liquefaction of Gases

Systematic work on the liquefaction of gases was initiated by Michael Faraday early in the nineteenth century. At that time the lowest temperatures possible were those obtained from mixtures of ice and salt. Later, with mixtures of solid carbon dioxide and ether, temperatures as low as $-110°C$ could be obtained. By 1845 Faraday liquefied a number of substances previously known only in the gaseous state, but he was unable to liquefy some of the best-known gases, such as hydrogen, oxygen, nitrogen, and carbon monoxide. Attempts to liquefy these gases by using pressures exceeding 3000 atmospheres were unsuccessful. The opinion prevailed that certain gases could not be liquefied under any circumstances; such substances were called permanent gases.

We now know that any gas can be liquefied if its temperature is sufficiently reduced. Thomas Andrews, in 1869, discovered that a gas cannot be liquefied by pressure alone unless the temperature is below a certain point, which he called the critical temperature (Section 10-6).

The pressure required to liquefy a gas is the same as the vapor pressure of the substance in the liquid state at the same temperature. Therefore, at the normal boiling point of a substance the pressure that must be applied to convert the gaseous form to the liquid form is 760 torrs. The critical pressure, i.e., the pressure required to liquefy a gas at its critical temperature, is the vapor pressure of the substance at the critical temperature. When vapor and liquid are both present at the critical temperature, the density of the vapor and the density of the liquid are identical.

A *vapor* is the gaseous form of a substance at or below its critical temperature. The term is never used to refer to a gas above its critical temperature.

The ability of a gas to become a liquid is due to the attractive forces between the molecules. Since the magnitude of these attractive forces varies from one substance to another according to differences in molecular structure, each substance has its characteristic critical temperature, just as it has its characteristic boiling point, vapor pressure at any specified temperature, etc. The intermolecular attractive forces are not dependent on temperature. Liquefaction results when a gas is cooled to a temperature sufficiently low that the kinetic energy of its molecules is no longer adequate to overcome the forces of attraction.

At high pressures gases can be liquefied at higher temperatures than would otherwise be possible because the molecules are crowded together enough for the intermolecular attractive forces to be effective. Above the critical temperature the kinetic energy of the molecules is too great to be overcome by these attractive forces, regardless of the distance between the molecules. Since the molecules of all gases have the same average kinetic energy at a given temperature (Section 9-12), the critical temperature values depend on the strengths of the attractive forces and are, in a sense, a measure of these strengths. Thus, the attractive forces between water molecules are so strong that water can exist as a liquid at any temperature below 374°C if sufficient pressure is applied; but the forces between helium molecules are so weak that liquid helium cannot exist at any temperature above 5.25°K (−267.9°C).

10-11 Mixtures of Liquids

When two liquids come into the presence of each other, they may form a homogeneous mixture (i.e., a solution), or one liquid may persist in floating atop the other with an easily observed interface between the two phases. Two or more liquids that mix with one another in all proportions to form solutions are said to be *miscible*; e.g., water and alcohol are miscible. Liquids that do not intermingle to form solutions, such as water and gasoline, are said to be *immiscible*. If two liquids, A and B, are only partially miscible, a limited amount of each will dissolve in the other. If liquid A is present in excess (i.e., more is present than can dissolve in liquid B), two layers form, and each layer is a solution. In one layer the solvent is B, and it contains as much A as can dissolve in it; in the other layer the solvent is A, and it contains as much B as can dissolve in it. Briefly stated, one layer is a saturated solution of A in B, and the other layer is a saturated solution of B in A.

Liquids that are miscible consist of molecules that are of similar character. Nonpolar liquids, such as carbon disulfide, CS_2, and carbon tetrachloride, CCl_4, are readily miscible with one another, but they will not dissolve in water because of its highly polar character. On the other hand, polar compounds, such as methyl alcohol, CH_3OH, and ethyl alcohol, C_2H_5OH, are miscible with one another and with water.

In a mixture of two similar liquids, the properties of the mixture are an average of the properties of the individual liquids. Thus, if equal volumes of two liquids are mixed together and the densities of the two liquids are 0.74 g per ml and 0.86 g per ml, one would expect the density of the mixture to be 0.80 g per ml. If this is found to be true, the mixture is an *ideal solution*, i.e., the molecules of each kind are in an environment very much like the one they were in before mixing.

The vapor pressure of an ideal solution can be calculated from its composition. The vapor pressure of each component above an ideal solution is proportional to the mole fraction of that component in the solution. According to the law announced by the French chemist François Raoult in 1886 (known as *Raoult's Law*), the partial vapor pressure of liquid A above an ideal solution in which its mole fraction is X_A is given by the relation

$$P_A = X_A P_A{}^\circ$$

where $P_A{}^\circ$ is the vapor pressure of pure liquid A at the same temperature. If the ideal solution consists of two liquids, the partial vapor pressure of liquid B above the solution is given by the relation

$$P_B = X_B P_B{}^\circ$$

and the total vapor pressure above the solution is

$$\begin{aligned} P_{\text{total}} &= P_A + P_B \\ &= X_A P_A{}^\circ + X_B P_B{}^\circ \end{aligned}$$

Since

$$X_A + X_B = 1$$

then

$$X_A = 1 - X_B$$

and

$$\begin{aligned} P_{\text{total}} &= (1 - X_B)P_A{}^\circ + X_B P_B{}^\circ \\ &= P_A{}^\circ - X_B P_A{}^\circ + X_B P_B{}^\circ \\ &= (P_B{}^\circ - P_A{}^\circ)X_B + P_A{}^\circ \end{aligned}$$

In words, the total vapor pressure above an ideal solution is a function of the vapor pressures of the two pure substances and the mole fraction of one of these, X_B. Since $P_A{}^\circ$ and $P_B{}^\circ$ are fixed quantities for a given temperature, the equation is of the form $y = mx + b$. When the total vapor pressure is plotted against the mole fraction X_B, a straight line is obtained for an ideal solution. This is shown in Figure 10-9, where the ordinate is vapor pressure and the abscissa is X_B. The limits of the diagram correspond to $X_B = 0$ (where $X_A = 1$) and $X_B = 1$ (where $X_A = 0$). At $X_B = 0$, only pure A is present, and the ordinate at $X_B = 0$ represents the vapor pressure of pure A. At $X_B = 1$, only pure B is present, and the ordinate at $X_B = 1$ represents the vapor pressure of pure B. Between these two limits the ordinates represent the vapor pressures of all possible mixtures of A and B.

The composition of the vapor above an ideal solution is different from the composition of the liquid phase itself. If we let Y_B be the mole fraction of B

Figure 10-9

Vapor pressure of miscible liquids

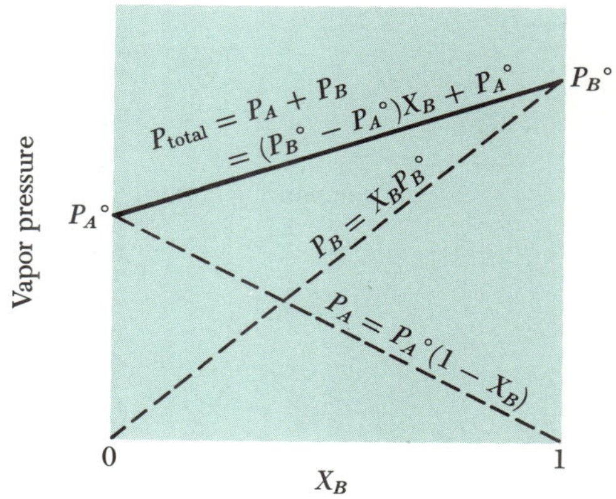

in the vapor above a solution of composition X_B, then, according to Dalton's Law of Partial Pressures,

$$Y_B = \frac{P_B}{P_{total}}$$

According to Raoult's Law, as we have seen, $P_B = X_BP_B{}^\circ$ and $P_{total} = (P_B{}^\circ - P_A{}^\circ)X_B + P_A{}^\circ$. Consequently,

$$Y_B = \frac{X_BP_B{}^\circ}{(P_B{}^\circ - P_A{}^\circ)X_B + P_A{}^\circ}$$

It is seen from this equation that the mole fraction of B in the vapor above an ideal solution of composition X_B can be calculated if the vapor pressures of the pure liquids are known. When such calculations are made for several values of X_B, the result is a vapor composition curve such as the one shown in Figure 10-10. This curve may be used to determine the composition of the vapor that is in equilibrium with an ideal solution of any specified composition. For example, if the solution has the composition $X_B = 0.3$, we move up to point c to read the total vapor pressure of the solution, and when we move over to point d we see that this total vapor pressure corresponds to a vapor composition represented approximately by $Y_B = 0.5$. The vapor is always richer in B than the solution is, B being the more volatile component.

If a liquid mixture does not obey Raoult's Law, the vapor pressure for any composition cannot be calculated reliably.

10-12 Fractional Distillation

According to Raoult's Law, the total vapor pressure in equilibrium with a solution composed of two volatile liquids is always intermediate between the vapor pressures of the two pure liquids. In such cases the vapor produced by

Figure 10-10

Liquid-vapor equilibrium for an ideal solution at constant temperature

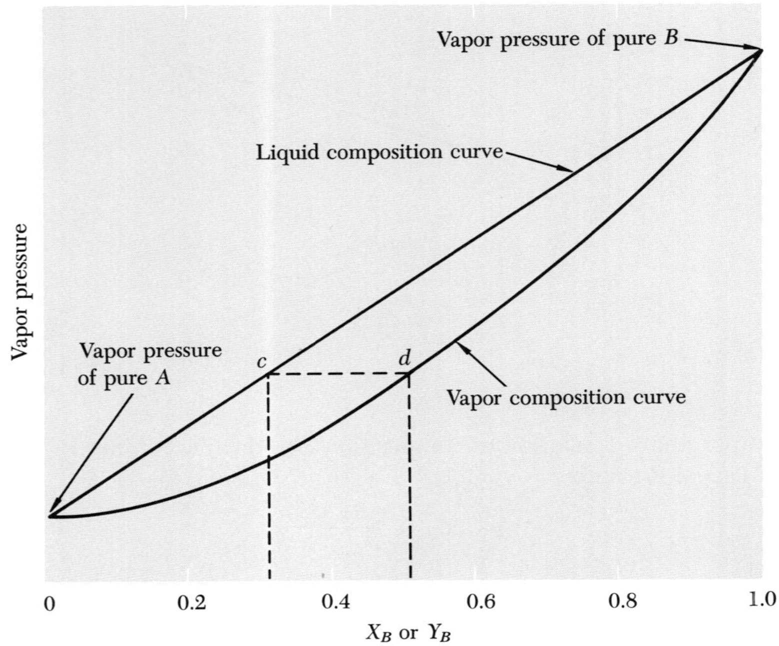

boiling the solution always contains relatively more of the more volatile component than does the solution itself. If the vapor is condensed, the resulting liquid mixture is richer in the lower-boiling component than was the original mixture. If the vapor is condensed in successive portions and each of these fractions is subjected to repeated redistillation, each component can eventually be obtained in an essentially pure state. Such a process is called *fractional distillation*. In practice, fractional distillation is accomplished by use of a fractionating column in which there is constant contact at all levels between ascending vapor and descending liquid.

Modern distillation plants are capable of separating complex mixtures, even those in which the components have only a narrow range of boiling points. Such plants are widely used in the refining of petroleum. Nitrogen, oxygen, and argon gases are obtained commercially by the fractional distillation of liquid air.

A liquid mixture does not need to obey Raoult's Law in order for its components to be separated by fractional distillation.

New Terms

Boiling point: the temperature at which a liquid boils (usually referring to the normal boiling point).

Critical pressure: the pressure required to liquefy a gas at its critical temperature; the vapor pressure of a liquid at the critical temperature.

Distillation: the process of vaporizing a liquid by the application of heat and condensing its vapor by cooling.

Dynamic equilibrium: a state of balance between opposing actions taking place at the same rate; e.g., when the rate of recondensation equals the rate of evaporation, a state of dynamic equilibrium prevails.

Ebullition (boiling): the formation of bubbles of vapor beneath the surface of a liquid.

Fractional distillation: a process by which the components of a liquid mixture are vaporized in the order of their boiling points and the liquid condensates are collected as separate fractions.

Heat of vaporization: the quantity of heat that must be supplied to vaporize a stated quantity (e.g., 1 g or 1 mole) of a liquid without changing its temperature.

Meniscus: the curved upper surface of a liquid column in a container of small cross section (e.g., a buret, pipet, or capillary tube). The meniscus may be either concave or convex, depending on whether the containing walls are wetted by the liquid.

Miscibility: the ability of two or more liquids to form a solution.

Normal boiling point: the temperature at which a liquid boils when the opposing pressure is 1 atmosphere; the temperature at which the vapor pressure of a liquid is 760 torrs.

Surface tension: the inward pull, due to intermolecular attraction, that causes a liquid surface to become as small as possible.

Trouton's Rule: dividing the molar heat of vaporization of a liquid (in calories) by the normal boiling point (in °K) gives approximately 22 cal per mole per degree.

Vapor: the gaseous form of a substance at or below its critical temperature.

Vapor pressure: the pressure that a vapor exerts when it is in equilibrium with its liquid.

Viscosity: the resistance of a fluid to flow.

Volatility: the tendency of a liquid to evaporate.

Exercises

10-1 At what temperature would water boil if the barometer reading were 730 mm? 775 mm? 760 mm?

10-2 How many calories are required to evaporate 1 kg of water at its boiling point? To evaporate 1 kg of carbon tetrachloride at its boiling point?

10-3 How well do water and carbon tetrachloride follow Trouton's Rule?

10-4 Oxygen gas generated by heating potassium chlorate is found to have a volume of 437 ml when collected over water at 23°C and 752 torrs. What volume would the gas have if it were dry at standard conditions?

10-5 Why do we say that the equilibrium between a liquid and its vapor is a dynamic equilibrium?

10-6 Summarize all the desirable properties of an ideal refrigerant.

10-7 The specific heat of iron is 0.113 cal per g per °C. This means that 0.113 cal is required to raise the temperature of 1.00 g of iron 1.00 Celsius degree. What weight of steam at 100°C must be condensed to release sufficient heat to raise the temperature of 1 kg of iron from 90°C to 100°C?

10-8 Raoult's Law applies to solutions containing nonvolatile materials, such as sugar, dissolved in volatile liquids. If a mole of sugar is dissolved in 49 moles of water, to what value will the vapor pressure of the water be reduced? How would this affect the boiling point?

10-9 The molar heat of vaporization of ammonia at refrigeration temperatures is about 5270 cal. When a mole of water freezes, the heat liberated is about 1440 cal. How many grams of ammonia must vaporize in a refrigerator that is 100% efficient to remove the heat that is released upon freezing 100 g of water?

References

Bernal, J. D., "The Structure of Liquids," *Sci. Am.*, August 1960, p. 124. An account of the author's attempt to construct a general theory of the liquid state.

Chemical Education Material Study, *Chemistry, an Experimental Science*, Freeman, San Francisco, 1963, Chapter 5, "Liquids and Solids." This chapter in the CHEMS text is concerned with all the condensed phases of matter, including solutions.

Lapp, R. E., and the Editors of Life, *Matter*, Time, Inc., New York, 1963, Chapter 4, "The Restless Surge of the Liquid State." Drawings and color photographs add to the reader's enjoyment.

11

The Solid State

11-1 Properties of Solids

In contrast to liquids, a sample of matter in the solid state is relatively hard and rigid and retains its size and shape regardless of the size or shape of the container. The densities of solids are of the same order of magnitude as the densities of liquids. Solids, like liquids, are only very slightly compressible; pressures of hundreds of atmospheres have very little effect on their volumes.

All liquids become solids when they are cooled to sufficiently low temperatures. Some substances, e.g., helium, do not solidify until they have almost reached absolute zero. Other substances, e.g., carbon, remain solids at such high temperatures that little is known about their liquid and gaseous states.

It is not difficult to understand why liquids and solids are resistant to compression, for the molecules of a substance in either of these states are in contact with one another. Solids have definite shape and lack the fluidity of liquids because the molecules have fixed positions; i.e., they are unable to move over and beyond one another to assume new positions.

11-2 Freezing and Melting

As a liquid is cooled its molecules lose kinetic energy, but the attractive forces between molecules are independent of temperature. At a sufficiently low temperature, which depends on the particular substance, the molecules cease the random motion characteristic of molecules in the liquid and gaseous states and fit themselves into a regular three-dimensional pattern. This is what happens when a liquid *freezes* to become a crystalline solid.

When the temperature of a cooling liquid is plotted over a period of time, a curve is obtained like that in Figure 11-1 if the liquid is a pure substance and if heat is removed from it at a uniform rate. As each molecule takes up its

Figure 11-1

Cooling curve for a pure liquid

The broken line represents supercooling.

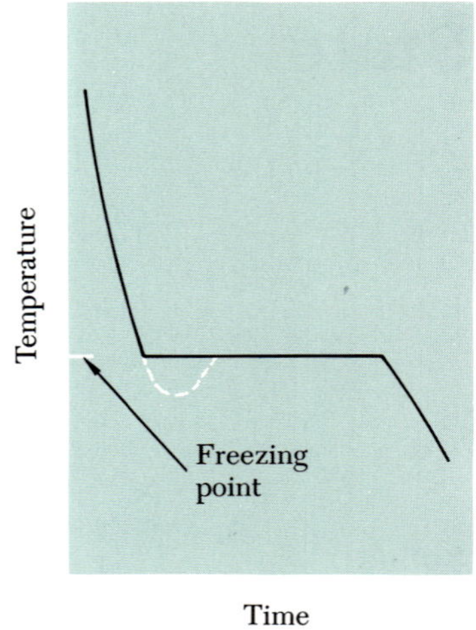

position to become part of the crystal structure characteristic of most solids, its potential energy drops (crystals are discussed later in this chapter). This decrease of potential energy is the source of the heat that is liberated during freezing. During the crystallization process, the average kinetic energy of the molecules in both the solid and the liquid phases is the same. The kinetic activity of the molecules in the solid phase is restricted to vibration; unlike the molecules in the liquid phase, they are not free to move about. The horizontal portion of the cooling curve indicates the temperature at which freezing occurs. As you can see, the temperature does not undergo any further reduction until crystallization is complete. Loss of heat without reduction of temperature is characteristic of the freezing of a liquid just as it is for the liquefaction of a gas. The heat liberated during freezing is called the *heat of solidification* or the *heat of crystallization*.

Further removal of heat after solidification is complete results in a reduction of the temperature of the solid. If the process is reversed and the solid is heated to the temperature at which it formed, the solid undergoes *melting* or *fusion* to form its liquid. The *melting point* of a substance is defined as the temperature at which the solid and liquid phases can exist in equilibrium. The melting point of the pure solid is the same as the *freezing point* of the pure liquid. Melting points are frequently used in the identification of substances.

The heat absorbed during melting is called the *heat of fusion*. It is, of course, the same quantity as the heat of crystallization. The quantity of heat required to melt 1 mole of a solid is called the *molar heat of fusion*. The heat of fusion of ice is 1440 cal per mole (or 80 cal per g) and that of sodium chloride is 7250 cal per mole. The difference between these values indicates the greater attractive forces which exist in NaCl.

The melting point of a substance is affected slightly by pressure. For example, ice skating is possible because the high pressure of the blade of the skate causes the ice to melt. This produces a thin film of water which provides lubrication, allowing the edge of the skate to move with little friction. After the skate has moved on and the pressure has been released, the water freezes.

Whether the melting point of a substance is raised or lowered by the application of pressure depends on whether the substance expands or contracts upon melting. Most liquids contract upon freezing (water is an exception). Pressure raises the melting points of most solids, but it lowers the melting point of ice.

Most cooling curves are complicated in the manner indicated by the broken line in Figure 11-1. Instead of crystallizing when the freezing point is first reached, the liquid *supercools* (as indicated by the broken line). This happens when the molecules fail to line up properly for crystallization to begin. The temperature falls below the freezing point, and the molecules continue to move about until by chance they fall into the correct pattern. When a "nucleus" of sufficient size has been built up, additional molecules rapidly attach themselves to it, and the crystal "grows." A large number of molecules are likely to crystallize out simultaneously as soon as the supercooled condition is "relieved." The heat thus liberated causes the whole system to warm up to the freezing point, and thereafter the temperature remains fixed until freezing is complete.

11-3 Vapor Pressure of Solids

Certain solids, such as the moth repellents naphthalene and dichlorobenzene, give off odors that are readily detectable. This can be explained only by recognizing that these solids vaporize directly from the solid state to the gaseous state. Like a liquid, a solid has a vapor pressure, which is the pressure exerted by the vapor that is in equilibrium with the solid. Such an equilibrium can exist only within a closed space.

The vapor pressure of a solid increases with rising temperature in much the same manner as the vapor pressure of a liquid. For example, the vapor pressure of iodine is 0.202 torr at 0°C and 45.5 torrs at 100°C. Those solids which have measurable vapor pressures at room temperature consist of molecules which are held together in the crystal by rather feeble forces. In an ionic crystal the attractive forces are greater, and the vapor pressure is low. Many solids exhibit no measurable vapor pressures, but it is a theoretical convenience to assume that every solid has a certain vapor pressure at every temperature.

The passage of a substance directly from the solid state to the gaseous state is called *sublimation*, and the vapor pressure of a solid is sometimes referred to as its *sublimation pressure*. The disappearance of snow on days when the temperature is below 0°C is due to sublimation. Moth balls slowly disappear on standing in the open because of sublimation of the naphthalene. Most familiar is the sublimation of solid carbon dioxide, well known as Dry Ice.

Table 11-1

Vapor pressure of ice and water

Temperature (°C)	*Vapor pressure* (torrs) *Ice*	*Liquid water*	*Difference*
−1.0	4.217	4.258	0.041
−0.8	4.287	4.320	0.033
−0.6	4.359	4.385	0.026
−0.4	4.431	4.448	0.017
−0.2	4.504	4.513	0.009
0.0	4.579	4.579	0
+0.2		4.647	
+0.4		4.715	
+0.6		4.785	
+0.8		4.855	
+1.0		4.926	

It is instructive to consider the vapor pressures of the solid and liquid forms of a substance in the neighborhood of the melting point. In Table 11-1 the vapor pressures of ice and of supercooled liquid water are given at intervals of 0.2° from −1 to 0°C, along with the vapor pressure of liquid water from 0 to +1°C. Note that at 0°C the vapor pressure of ice is the same as that of liquid water. The vapor pressures of the solid and liquid phases of a substance are identical at the melting point. The vapor pressure of a solid or a liquid is a measure of its tendency to escape into the gaseous state. Below 0°C liquid water and ice cannot exist in equilibrium because the supercooled water has a higher vapor pressure than ice at these temperatures. If the liquid water were in a beaker within a closed vessel which also contained a piece of ice, the liquid would escape into the vapor phase, from which it would condense in solid form onto the ice, and the beaker would eventually become empty; if ice were in the beaker with the water, the water would immediately freeze there.

Table 11-1 gives no values for the vapor pressure of ice above 0°C because this information is not available. If ice is heated above 0°C, it melts.

Above the true melting point of a substance, the vapor pressure of the (imaginary) superheated solid would be greater than that of the liquid. Below the melting point the vapor pressure of the supercooled liquid is greater than that of the solid. The melting point is the temperature at which the solid and the liquid have the same vapor pressure.

11-4 Crystallography

The first milestone on the road to understanding the nature of the solid state was the realization that most solids are composed of crystals. A *crystal* is a homogeneous piece of solid matter which has the natural shape of a polyhedron. This means that it is bounded by plane surfaces, or faces, symmetrically arranged. *Crystallography* is the science of describing precisely the external forms of crystals and the arrangement of the atoms, ions, or molecules of which they are composed. Crystallography may be said to have begun in 1669, when Nicolaus Steno noted that different crystals of quartz had identical angles between corresponding faces.

Sometimes each particle of a solid substance is a single crystal, but quite commonly, especially with metals, a single piece of the solid is an aggregate of crystals. Often a material is a collection of small solid particles, such as sand, which takes on the shape of its container just as a liquid does, but each particle may still be a single crystal. Although a crystal may have had its corners and edges rounded off sufficiently to appear noncrystalline to the eye, its crystalline structure can be observed with the aid of a microscope.

In 1784 the French mineralogist René Just Haüy was examining a large calcite crystal when he accidentally dropped it and noted that it broke along definite planes, producing small pieces having the shape of the original crystal. He surmised that the external regularity of a crystal is due to a regularity of its internal structure. Haüy's brilliant deduction, made before Dalton's atomic theory was formulated, brought increasing attention to crystallography during the early part of the nineteenth century.

There are only seven systems of crystal symmetry (Figure 11-2). The names of the seven systems are geometrical terms which refer to the relative lengths of the axes of symmetry of the crystals and the angles which the axes make with one another. An *axis of symmetry* is an imaginary line passing through the center of the crystal about which the crystal may be rotated with the result that it presents exactly the same appearance more than once during one rotation.

The seven crystal systems are as follows:

1. the cubic system, in which all angles are 90°, and all three axes are equal
2. the tetragonal system, in which all angles are 90° but one axis is longer or shorter than the others
3. the rhombic system, in which all angles are 90° but no two axes are of the same length
4. the monoclinic system, with two unequal axes perpendicular to each other and a third axis, of different length from either of the others, which is not perpendicular to either of the others
5. the trigonal or rhombohedral system, in which there are three equal axes and three equal angles not 90°
6. the hexagonal system, in which there are parallel sides perpendicular to two hexagonal faces
7. the triclinic system, with three unequal axes and three unequal angles not 90°.

Some substances can exist in more than one crystalline form. This phenomenon is known as *polymorphism*. For example, calcium carbonate occurs

Figure 11-2

The seven crystal systems

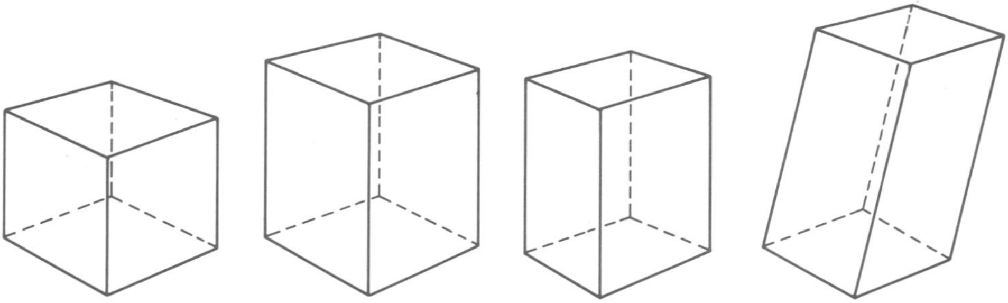

Cubic system (three equal axes at right angles).

Tetragonal system (three axes at right angles, one axis longer or shorter than the others).

Rhombic system (three axes at right angles, no two axes of the same length).

Monoclinic system (two unequal axes at right angles, third axis of still different length and not at a right angle).

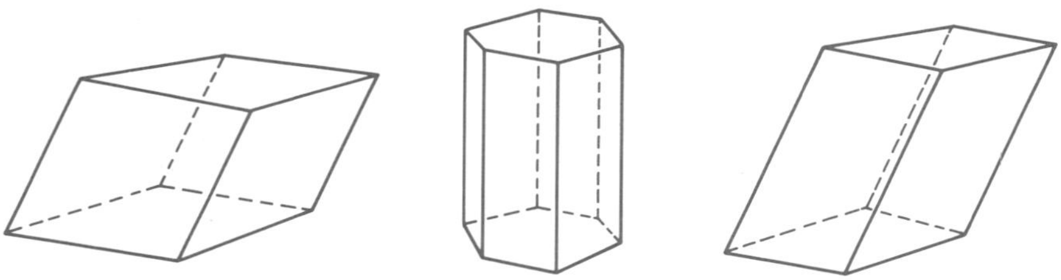

Trigonal system (three equal axes, three equal angles, no right angles).

Hexagonal system (parallel sides, perpendicular to two hexagonal faces).

Triclinic system (three unequal axes, three unequal angles, no right angles).

in nature both as calcite, with a trigonal structure, and as aragonite, with a rhombic structure. The temperature at which crystallization occurs determines which form the crystals assume. (Calcite and aragonite are typical mineralogical names which cannot be expected to have much, if any, chemical significance. The name calcite is derived from *calx*, which is Latin for "lime," and the name aragonite is derived from Aragon, the place where the mineral was first found.)

When polymorphism occurs with an element, the different crystalline modifications are called allotropes, or allotropic forms, of the element. Examples include rhombic and monoclinic sulfur, graphite and diamond, white and red phosphorus, and white and gray tin.

You will recall that oxygen exists in two allotropic forms in the gaseous state because of the existence of two kinds of molecules, O_2 and O_3. Thus, polymorphism is not the sole cause of allotropy.

Figure 11-3

Unit cell of sodium chloride

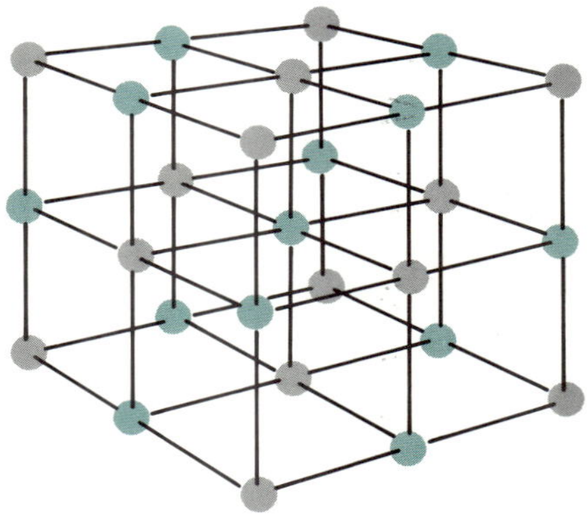

11-5 Internal Structure of Crystals

Nineteenth-century crystallographers were able to obtain a surprising amount of information about the arrangement of particles inside crystalline solids, but their only methods for doing so were indirect, based on logical deductions from external appearances. Much more exact and detailed information can be obtained by X-ray diffraction.

By mathematical analysis of X-ray diffraction patterns, X-ray crystallographers can "map" the relative positions occupied by the particles inside crystals and can determine whether the particles are atoms, ions, or molecules. If we let a point represent the position of each particle, the pattern of points which represents the arrangement of the particles in a crystal is known as a *space lattice*. In 1848 a French crystallographer, Auguste Bravais, showed that all possible space lattices can be assigned to one of fourteen types, now known as the fourteen Bravais lattices.

The space lattice is thought of as extending throughout the entire crystal, but when comparing one crystal structure with another it is necessary to con-

Figure 11-4

Structure of NaCl

The large spheres represent Cl^- ions; the small spheres, Na^+ ions.

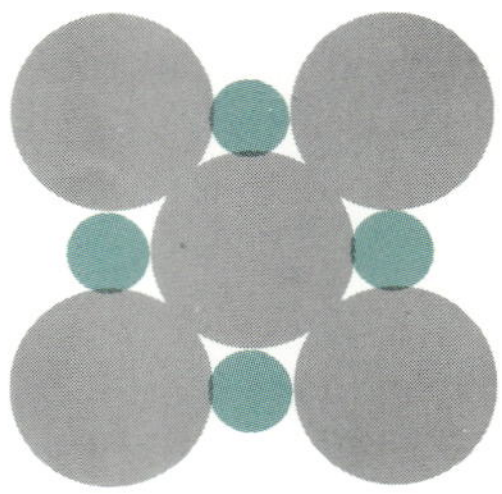

sider only enough of it to represent the arrangement. This smallest fraction of a space lattice that is the pattern for describing the whole lattice is called a *unit cell.* The unit cell of sodium chloride is shown in Figure 11-3, where blue circles represent the position of the Na^+ ions and gray circles the position of the Cl^- ions. The Na^+ and Cl^- ions have quite different sizes, as represented in the model shown in Figure 11-4.

11-6 Types of Solids

In structural chemistry, solids are classified according to the nature of the particles that compose them and the type of bonding that exists between the particles. From this point of view, there are four types of solids: ionic, molecular, network, and metallic.

Ionic solids melt, when heated to sufficiently high temperatures, to form liquids that are good conductors of electricity because they consist of ions. The crystal lattice of an ionic solid, such as Na^+Cl^-, consists of a regular array of positive and negative ions. The relative sizes of the ions, and their relative numbers as expressed in the chemical formula of the substance, determine the arrangement of the ions in an ionic crystal. The forces which hold an ionic crystal together are the strong electrostatic forces which operate between positive and negative charges. For this reason ionic solids are usually rather hard and have high melting points (and still higher boiling points). They are soluble only in polar solvents, such as water. The solutions are good conductors of electric current because they contain ions.

Molecular solids are crystalline substances consisting of molecules. The atoms within each molecule are held together by covalent bonds, but the forces holding the crystal together are the much weaker forces of intermolecular attraction. If the molecules are nonpolar, as in the case of hydrogen, H_2, or sulfur, S_8, or carbon tetrachloride, CCl_4, the only attractive forces between molecules are van der Waals forces. This type of intermolecular attraction is so weak that crystals of nonpolar substances are soft and have low melting points. In fact, many nonpolar substances are liquids or even gases at ordinary temperatures; many of those that are solid have substantial vapor pressures unless they have high molecular weights, as in the case of sulfur.

The attractive forces operating between polar molecules are appreciably stronger than those operating between nonpolar molecules, for the negative end of one molecule attracts the positive end of another. In general, substances consisting of polar molecules form harder crystals than do nonpolar substances, have higher melting and boiling points, and are less volatile. Thus, their properties are intermediate between those of the ionic substances and those of the nonpolar molecular substances. Examples include water, ammonia, sulfuric acid, and the alcohols.

The *network solids* consist of atoms united by covalent bonds into a network extending throughout the crystal. For example, in diamond (Figure 11-5) each carbon atom is located at the center of an imaginary tetrahedron and is co-

Figure 11-5

Arrangement of carbon atoms in diamond

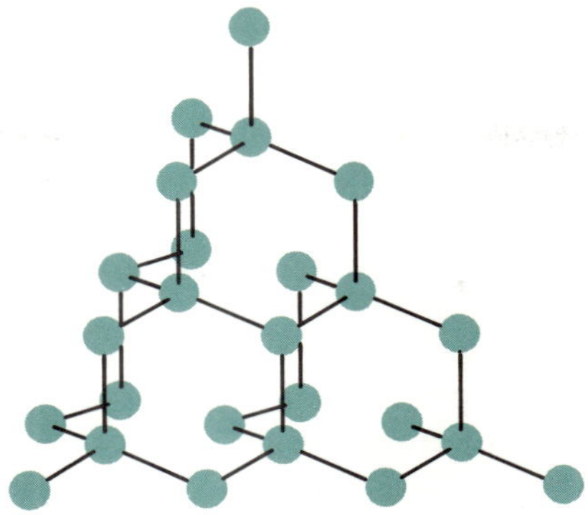

valently bonded to the carbon atom at each of the four corners. Each carbon atom is therefore both at the center of one tetrahedron and at a corner that is common to four other tetrahedra. Thus, every carbon atom in the interior of a diamond is joined by single covalent bonds to four other carbon atoms. Diamond is a hard substance with a high melting point because the bonds are quite strong, with all the bonds of the same length and strength in an interlocking arrangement. The C—C distance is 1.54 Å.

Network solids are also known as *covalent solids* or *atomic crystals.* Any solid of this type is very hard, has a high melting point, and is a nonconductor of electricity. In order for a crystal of this type to be broken, melted, or vaporized, a great number of strong bonds must be broken. Such substances are insoluble in all solvents except those that react chemically with them. Besides diamond, examples include silicon carbide, SiC (the industrial abrasive known as carborundum), and aluminum nitride, AlN.

In a *metallic solid* the lattice points are occupied by positive ions which are held together by more or less mobile electrons. Most metal atoms, in the isolated state, have only one, two, or three electrons in the outermost shell, but in a metallic crystal each atom is surrounded usually by either eight or twelve nearest neighbors. Thus, there are not enough electrons for each atom to be covalently bonded to each of its neighbors as are the carbon atoms in diamond. Instead, according to the best evidence available, the outer-shell electrons of all the atoms are free to pass from one atom to another so that they are shared by many atoms. This special type of modified covalent bonding is known as *metallic bonding.*

The characteristic properties of metals include high conductivity for heat and electricity and the luster which they have in both the solid and liquid states. All these properties are attributed to the mobility of the electrons. Most metals have high melting and boiling points, indicating that the bonding between atoms is strong. In general, they can be easily stretched or pressed into new shapes; apparently, the mobility of the electrons makes it possible for the bonds to be readily broken and replaced by new ones.

11-7 Specific Heats of Solid Elements

The quantity of heat required to raise the temperature of 1 g of a substance 1°C is called its *specific heat*. In 1819 two French chemists, Pierre Dulong and Alexis Petit, discovered that the specific heats of elements in the solid state are inversely proportional to their atomic weights. In Table 11-2, where present-day values of atomic weights and specific heats have been rounded off to three digits and their products to two digits, the molar heat capacities of the solid elements listed average 6.2 cal per degree. This is the average for most solid elements. This relationship between the atomic weight and specific heat of a solid element is known as the *Law of Dulong and Petit*. It is expressed in the following equation:

$$\text{Atomic weight} = 6.2 \div \text{specific heat}$$

Historically, this law has been useful for determining approximate values for atomic weights, for merely by determining the specific heat of a solid element one can estimate its approximate atomic weight.

EXAMPLE The specific heat of iron is determined to be 0.108 cal per g. Calculate an approximate value for its atomic weight.

$$\text{Approximate atomic weight} = \frac{6.2}{0.108} = 57$$

Table 11-2

Specific and molar heat capacities of selected solid elements

Element	*Atomic weight*	*Specific heat* (cal/g)	*Molar heat capacity* (cal/mole) (atomic weight × specific heat)
Nickel	58.7	0.105	6.2
Copper	63.5	0.0921	5.9
Zinc	65.4	0.0925	6.0
Tin	119	0.0542	6.4
Antimony	122	0.0504	6.1
Iodine	127	0.0523	6.6
Lead	207	0.0306	6.3
Bismuth	209	0.0294	6.1

To appreciate the value of the Law of Dulong and Petit, let us consider the fact that the weight of silver combining with 15.9994 g of oxygen (1 mole of oxygen atoms) is 215.74 g. What is the atomic weight of silver? If the formula for silver oxide is AgO, the answer is 215.74; if the formula for silver oxide is Ag_2O, the answer is $215.74 \div 2 = 107.87$; etc. Until the approximate atomic weight of silver is known, the formula for silver oxide cannot be known. However, the specific heat of silver is 0.0565 cal per g, so an approximate value for the atomic weight can be obtained by applying the Law of Dulong and Petit: $6.2 \div 0.0565 = 110$. This result indicates that the correct value to choose for the atomic weight of silver is 107.87 and that the formula for silver oxide is Ag_2O.

The theoretical significance of this law lies in the fact that 1 gram-atomic weight (1 mole of atoms) of every element consists of the same number of atoms (Avogadro's Number). Thus, a gram-atomic weight of one element has about the same heat capacity as a gram-atomic weight of another element because approximately the same quantity of heat is needed to increase the kinetic energy of one group of atoms as to increase the kinetic energy of an equal number of atoms of some other kind by the same amount.

A few of the elements, especially those of atomic weights below 35, do not obey the law at ordinary temperatures; but as higher temperatures are approached, all elements obey the law. This behavior is predicted in modern theoretical work based on quantum mechanics.

11-8 Sizes of Atoms and Ions in Crystals

The sizes of atoms and ions can be deduced from interatomic and interionic spacings in crystals, as measured by X-ray analysis. Such spacings are actually the distances between nuclei of atoms or ions that are nearest neighbors. For example, in sodium metal the distance between the nucleus of a sodium atom and the nucleus of each of its nearest neighbors is 3.14 Å. Since the atoms are contiguous, the radius of a sodium atom is taken to be half of this value, or 1.57 Å. What lies outside the nucleus of an atom is an electron cloud which, like the earth's atmosphere, has no theoretical termination distance; but, as a practical matter, the outermost electron of the sodium atom is thought of as being within a sphere of 1.57 Å radius almost all the time.

In general, as we proceed from left to right in any period of the periodic table, each element has a smaller atom than the element preceding it. The noble gases are an exception: their atoms are larger than the atoms of the elements of next lower atomic number. All of the atoms of the same period have the same number of electron shells. As we proceed from each element to the next, the magnitude of the electrostatic attraction between the nucleus and the surrounding electrons increases with increasing nuclear charge. All electron shells are thereby pulled in closer to the nuclei so that the radius of each atom is less than that of the atom which precedes it, as illustrated in Table 11-3.

Table 11-3

Atomic sizes of elements of the third period

Element	$_{11}$Na	$_{12}$Mg	$_{13}$Al	$_{14}$Si	$_{15}$P	$_{16}$S	$_{17}$Cl	$_{18}$Ar
Radius (Å)	1.57	1.36	1.25	1.17	1.10	1.04	0.99	1.54

The steady change in properties of the elements of the third period, from actively metallic sodium to actively nonmetallic chlorine, can be attributed to the smaller atomic size of each succeeding element. The outermost electrons are less readily removed as their distance from the nucleus decreases. On the other hand, the attraction for enough extra electrons from other atoms to complete the octet increases as the distance between the nucleus and the outermost shell decreases. The ionization energy and electron affinity are measures of these tendencies.

In each succeeding period of the periodic table, one more shell is occupied by electrons. The effect of this on atomic size is best shown by comparing atoms in the same group, since they have the same number of electrons in the outermost shell. The atomic sizes of the elements in Groups IA and VIIA are given in Table 11-4. The alkali metals show a decreasing opposition to having an electron removed as their atomic numbers, and thus their atomic sizes, increase. Similarly, the electron affinity of the halogens decreases with increasing atomic number.

A positive ion is smaller than the parent atom. The formation of a sodium ion from a sodium atom, for example, involves the removal of the outermost electron shell, and the remaining shells are drawn closer to the nucleus because the nuclear charge now exceeds the number of extranuclear electrons. The positive ions of succeeding elements in the same group are larger because of the

Table 11-4

Atomic sizes of alkali metals and halogens

Period	*Group IA (alkali metal)*	*Radius* (Å)	*Group VIIA (halogen)*	*Radius* (Å)
2	$_{3}$Li	1.23	$_{9}$F	0.72
3	$_{11}$Na	1.57	$_{17}$Cl	0.99
4	$_{19}$K	2.03	$_{35}$Br	1.14
5	$_{37}$Rb	2.16	$_{53}$I	1.33
6	$_{55}$Cs	2.35		

Table 11-5

Sizes of positive ions and their parent atoms

Atom	*Radius* (Å)	*Ion*	*Radius* (Å)	*Ion-atom ratio*
Li	1.23	Li^+	0.60	0.49
Na	1.57	Na^+	0.95	0.61
K	2.03	K^+	1.33	0.66
Rb	2.16	Rb^+	1.48	0.69
Cs	2.35	Cs^+	1.69	0.72
Be	0.89	Be^{++}	0.31	0.35
Mg	1.36	Mg^{++}	0.65	0.48
Ca	1.74	Ca^{++}	0.99	0.57
Sr	1.91	Sr^{++}	1.13	0.59
Ba	1.98	Ba^{++}	1.35	0.68

increased number of electron shells, but in each case the ion is smaller than its parent atom. This is illustrated with metals of Groups IA and IIA in Table 11-5.

Conversion of an atom to a negative ion by addition of one or more electrons results in increased interelectronic repulsion in the outermost shell and also makes the number of electrons outside the nucleus exceed the nuclear charge. Both effects contribute toward making the negative ion larger than the parent atom, as illustrated in Table 11-6.

Comparison of the size of a noble-gas atom with the sizes of positive and negative ions having the same number of electrons shows most clearly the effect

Table 11-6

Sizes of negative ions and their parent atoms

Atom	*Radius* (Å)	*Ion*	*Radius* (Å)	*Ion-atom ratio*
F	0.72	F^-	1.36	1.89
Cl	0.99	Cl^-	1.81	1.83
Br	1.14	Br^-	1.95	1.71
I	1.33	I^-	2.16	1.62
O	0.74	O^{--}	1.40	1.89
S	1.04	S^{--}	1.84	1.77
Se	1.17	Se^{--}	1.98	1.69
Te	1.37	Te^{--}	2.21	1.61

Table 11-7

Sizes of isoelectronic atoms and ions

Ion or atom with ten electrons	N^{---}	O^{--}	F^{-}	Ne	Na^{+}	Mg^{++}	Al^{+++}	Si^{4+}
Nuclear charge	7	8	9	10	11	12	13	14
Radius (Å)	1.71	1.40	1.36	1.12	0.95	0.65	0.50	0.41
Ion or atom with eighteen electrons	P^{---}	S^{--}	Cl^{-}	Ar	K^{+}	Ca^{++}	Sc^{+++}	Ti^{4+}
Nuclear charge	15	16	17	18	19	20	21	22
Radius (Å)	2.12	1.84	1.81	1.54	1.33	0.99	0.81	0.64

of increasing nuclear charge. In Table 11-7 the sizes of two sets of *isoelectronic* atoms and ions (i.e., atoms and ions having the same number of electrons) are seen to decrease markedly with increasing nuclear charge.

To observe the effect of varying the number of electrons while keeping the nuclear charge constant, we must compare the sizes of differently charged ions of the same element. This is done for four metals in Table 11-8.

New Terms

Crystal: a homogeneous piece of solid matter with the natural shape of a polyhedron. It is bounded by plane surfaces, which are symmetrically arranged.

Crystallography: the science that deals with the external forms of crystals and their internal structures (the arrangements of the constituent atoms, ions, or molecules).

Heat of fusion: the quantity of heat absorbed during melting of a stated quantity (e.g., 1 g or 1 mole) of a solid without changing the temperature. The quantity of heat required to melt 1 mole of a solid is called *molar heat of fusion.*

Table 11-8

Sizes of differently charged ions of the same element

Ion	*Radius* (Å)	*Ion*	*Radius* (Å)
Fe^{++}	0.83	Fe^{+++}	0.60
Tl^{+}	1.49	Tl^{+++}	1.05
Pb^{++}	1.32	Pb^{4+}	0.84
Ce^{+++}	1.18	Ce^{4+}	1.02

Heat of solidification (heat of crystallization): the quantity of heat absorbed during freezing, usually expressed in calories per gram or calories per mole. (This definition is formulated to be in technical agreement with the general proposition that the heat of a process should be defined as the heat absorbed when the process occurs. The quantity of heat absorbed during freezing is, of course, negative; i.e., freezing is always exothermic.) Numerically, heat of solidification is the same as heat of fusion.

Ionic solid: a solid in which the lattice points are occupied by positive and negative ions. The electrostatic forces operating between the positive and negative charges hold an ionic crystal together.

Isoelectronic: having the same number of extranuclear electrons.

Law of Dulong and Petit: the specific heats of elements in the solid state are inversely proportional to their atomic weights. It is expressed by the equation

$$\text{Atomic weight} = 6.2 \div \text{specific heat}$$

Melting point (freezing point): the temperature at which the solid and liquid phases of a substance can exist in equilibrium. At this point the solid and liquid phases have the same vapor pressure.

Metallic solid: a crystalline substance in which the lattice points are occupied by positive ions held together by more or less mobile electrons.

Molecular solid: a crystalline substance consisting of molecules. The atoms within each molecule are held together by covalent bonds. If the molecules are nonpolar, the attractions between molecules are due to van der Waals forces. If the molecules are polar, the intermolecular binding is due to the attraction between the negative end of one molecule and the positive end of another.

Network solid (covalent solid or atomic crystal): a solid which consists of atoms united by covalent bonds into a network that extends throughout the crystal.

Polymorphism: the occurrence of a substance in more than one crystalline form.

Space lattice: the pattern of points which represents the arrangement of the constituent particles in a crystal.

Specific heat: the quantity of heat (number of calories) that must be supplied to raise the temperature of 1 g of a substance 1°C.

Sublimation: evaporation of a solid.

Sublimation pressure: the name sometimes given to the vapor pressure of a solid.

Supercooled liquid: a liquid which has been cooled below its freezing point without freezing.

Unit cell: the smallest fraction of a space lattice that is the pattern for describing the whole lattice.

Exercises

11-1 Explain how snow from a light snowfall may disappear at temperatures below the freezing point of water.

11-2 Suggest a procedure for separating pure iodine from a mixture of iodine and potassium iodide.

11-3 The specific heat of ice is 0.50 cal per g per °C, and that of steam is 0.48 cal per g per °C. Calculate the amount of heat required to convert 25 lb of ice at −5°C to steam at 110°C.

11-4 Compare and explain the relative electrical conductivities of melted iodine, melted sodium, and melted sodium iodide.

11-5 Why does diamond have a small vapor pressure?

11-6 Using the Law of Dulong and Petit, calculate a value for the specific heat of each of three metals and compare this value with the experimental value to be found in reference books.

11-7 Explain why positive ions are smaller and negative ions are larger than the parent atoms.

References

Paul, M. A., King, E. J., and Farinholt, L. H., *General Chemistry*, Harcourt Brace & World, New York, 1967. Chapter 9 deals with the solid and liquid states.

Quagliano, J. V., *Chemistry*, 2nd ed., Prentice-Hall, Englewood Cliffs, N.J., 1963. Chapter 14 is concerned with the sizes of atoms and ions.

Sisler, H. H., *Electronic Structure, Properties, and the Periodic Law*, Reinhold, New York, 1963 (paperbound). A volume in the "Selected Topics in Modern Chemistry" series. Chapter 3 is concerned with the relation of properties to bond type.

12

Water

12-1 Importance of Water

Water is by far the most abundant chemical compound on earth. As a solid and liquid it covers about three fourths of the earth's surface, including the oceans, the fresh-water lakes, and the enormous quantities of ice in polar and mountainous regions.

Water is present in all living organisms; the human body is more than two thirds water. Water is essential to life processes; it is the medium through which plants take in nutrient material from the soil and in which metabolic processes occur.

So-called dry land would be wholly uninhabitable if it were completely dry. If air did not always contain water vapor, life would be impossible because of the rapid and fatal desiccation which would befall all organisms. The extent to which air is saturated with water vapor is expressed in terms of relative humidity.

Water in motion has been one of the most important agents in bringing about geological changes. Most of the surface of the earth has been sculptured by the falling rain, the flowing rivers, and the scoring glaciers.

In chemistry, water is used extensively as a solvent and as a reagent. Of all solvents water is, of course, the cheapest, but its importance is associated with its ability to dissolve a variety of materials. Consequently, chemical solutions are frequently aqueous. The chemical properties traditionally associated with acids, bases, and salts are mainly exhibited in an aqueous medium. One of the properties that distinguishes water from most other liquids is its unusual ability to dissolve electrolytes.

In the United States the amount of water used each year averages about 2500 tons per capita. One of the critical problems attending the population explosion now under way may very well be the problem of obtaining an adequate supply of fresh water. Recent advances in developing desalting processes that are economical and practical give promise that the oceans may become an important source of water for uses which cannot be served by untreated saline water.

12-2 Physical Properties of Water

Although water is a familiar substance, it is one of the most unusual compounds found in nature. Many of the physical properties of water make it appear abnormal in comparison with other compounds.

In the pure state water is a colorless, transparent liquid devoid of taste and odor. In Table 12-1 water is compared with other compounds of comparable chemical composition. Note its relatively high freezing point and boiling point.

Water has a greater density in the liquid state than in the solid state. Most solids sink in their own liquid, but ice floats on water. The density of liquid water at 0°C, the freezing point, is 0.9998 g per ml, but the density of ice at the same temperature is only 0.917 g per ml. In relation to natural phenomena it is important that water expands on freezing and that ice floats on water. If ice were heavier than water, it would sink to the bottom and expose the overlying liquid to further cooling and freezing so that rivers and lakes would eventually become completely frozen from top to bottom. The weathering of rocks is partly due to the pressure exerted by water freezing in crevices.

Most liquids become less dense with rising temperature. Above 4°C water expands with rising temperature, but below 4°C it expands with falling temperature. At 8°C the density is the same as at 0°C. In a lake of water covered with ice, the warmest water is at the bottom (since the temperature at its maximum density is 4° above the freezing point). With the coldest water near the surface, the ice sheet thickens from the top down; but the rate at which heat is lost to the air from the warmest water is appreciably reduced by the overlying layers of ice and water. Therefore, the rate of freezing is substantially restrained.

In comparison with other liquids, water has an unusually large specific heat: 1 cal per g per °C. In nature this property has a moderating effect on the temperature in regions near large bodies of water. Use is made of this property in such applications as hot-water heating systems and automobile cooling systems.

Table 12-1

Freezing points and boiling points of similar compounds

Formula	*Formula weight*	*Freezing point* (°C)	*Boiling point* (°C)
H_2O	18	0	100
H_2S	34	−86	−61
H_2Se	81	−65	−41
H_2Te	130	−51	−2

Water has an abnormally low vapor pressure and, consequently, an abnormally high boiling point. The *heat of fusion* of ice is unusually large—80 cal per g. The heat of vaporization, 540 cal per g, is also relatively high. Water does not conform to Trouton's Rule (Section 10-9); the molar heat of vaporization divided by the absolute boiling point is well in excess of 22:

$$\frac{18 \text{ g/mole} \times 540 \text{ cal/g}}{373°\text{K}} = 26 \text{ cal per mole per °K}$$

The unusual properties of water are a consequence of its molecular structure, which will be discussed at some length in the following pages of this chapter.

12-3 Water as a Standard

Since water is a relatively stable compound and can be obtained in a high state of purity with relative ease, many of its properties are used as standard references. For instance, a mixture of pure ice and pure water can easily be obtained, and the temperature of the mixture can be kept constant for a relatively long time. By definition, this temperature is 0°C. The boiling point of pure water at 1 atmosphere is 100°C; this temperature is easily measured and serves as another convenient reference point.

The standard of volume is the liter, which is defined as the volume of 1 kg of water at 4°C. Thus, the volume of 1 g of water at 4°C is 1 ml; i.e., the density of water at 4°C is 1 g per ml.

One calorie is, by definition, the quantity of heat required to raise the temperature of 1 g of water 1°C. More precisely put, to raise the temperature of 1 g of water from 14.5 to 15.5°C requires 1 cal. The specific heats of most liquids decrease somewhat with rising temperature, and this is true of water. To raise the temperature of 1 g of water from 4 to 5°C requires 1.0049 cal, but to raise it from 30 to 31°C requires only 0.9971 cal.

12-4 Structure of the Water Molecule

The determination of the composition of water was one of the most significant events in chemical history. In 1781 Henry Cavendish showed that water is the product formed when hydrogen gas burns in air; prior to this discovery, water was regarded as an element. Only a few years later, combustion in air was proved to involve union with oxygen. Thus, water was identified as a compound consisting of hydrogen and oxygen. Lavoisier determined its composition by weight, but no method was available for determining its formula. Since the atomic weights of the elements could not be ascertained, water was assumed to have the simplest imaginable formula, HO. Present-day values for atomic weights indicate that the empirical formula is H_2O, and molecular weight determinations show that H_2O is the molecular formula for steam.

In the H_2O molecule the oxygen atom shares a pair of electrons with each of two hydrogen atoms. Since oxygen is more electronegative than hydrogen

Figure 12-1

The H_2O molecule

(a) The water molecule is V-shaped, and there are two unshared pairs of electrons. (b) The angle between the O—H bonds is 105°, and each bond is polar, with each H atom carrying a partial positive charge. (c) A scale model showing the two H atoms attached to the larger O atom.

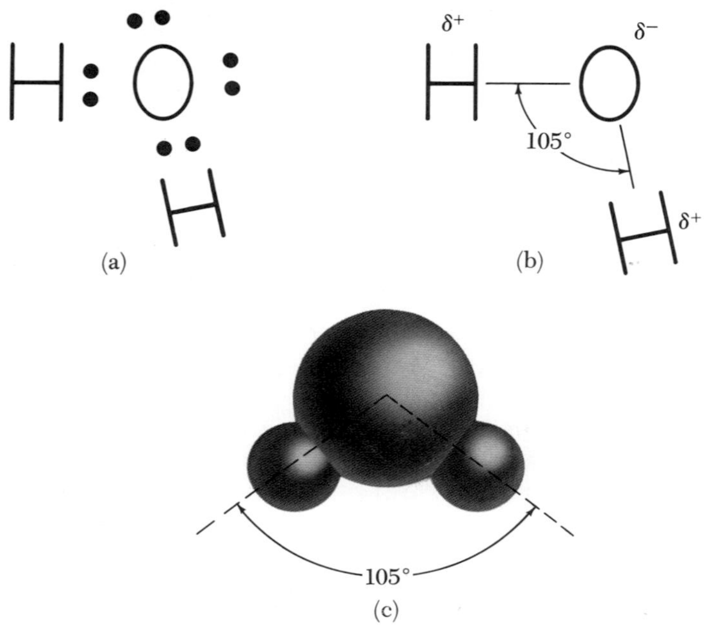

(i.e., an oxygen atom attracts electrons more strongly than a hydrogen atom), the shared electrons spend more time near the oxygen atom. Thus, each hydrogen atom has a partial positive charge, δ^+, and the oxygen atom has a partial negative charge, δ^- (Figure 12-1b). In other words, each O—H bond is polar.

Generally, atoms are held at definite angles to one another in molecules. The angle between the O—H bonds in the water molecule is an important consideration because of their polarity. If the angle were 180°, i.e., if the molecule were linear, the polarity of each bond would be offset by the polarity of the other. Just as the linear carbon dioxide molecule, $\ddot{\underset{..}{O}}=C=\ddot{\underset{..}{O}}$, is nonpolar even though it contains two polar bonds, so would the water molecule be nonpolar if the two hydrogens were on opposite sides of the oxygen atom.

Actually, the water molecule is highly polar, and this fact must be considered to understand the properties of water. In order that the molecule be highly polar, the angle between the two O—H bonds must be substantially less than 180°. The electronic configuration of the outer shell of a free oxygen atom consists of a filled *s* orbital, one filled *p* orbital, and two half-filled *p* orbitals (Table 4-1). Since the three *p* orbitals are mutually perpendicular, we could reasonably expect a 90° bond angle in any molecule of the type X_2O, such as H_2O. Experimental measurements of the polarity of water indicate that the angle is 105°. According to one theory, mutual repulsion of the two protons accounts for the angle being 15° greater than a right angle.

It must not be thought that the partial charge of a hydrogen atom in a water molecule is as great as the charge of a proton. The symbols δ^+ and δ^- are used to indicate only that partial charges are present on the atoms. In contrast, the positive ion in an ionic compound such as Na^+Cl^- or $Ca^{++}O^{--}$ has had one or more electrons almost completely removed from it, and the negative ion has almost complete possession of these electrons. Thus, the positive ions in ionic compounds have charges equal to one or more protons, and the negative ions have charges equal to one or more electrons; but in polar molecules, such as HCl or H_2O, the covalently bonded atoms have only partial charges.

The structure of the water molecule is illustrated in three different ways in Figure 12-1. The relative positions of the three atomic nuclei are indicated in all three drawings. As a matter of convenience, a water molecule is commonly symbolized as

$$\begin{array}{c} \;\;\cdot\cdot \\ H : O : \\ \;\;\cdot\cdot \\ \;\;H \end{array}$$

This representation is not intended to imply a bond angle of 90° but only that the molecule is not linear. If no electrons are shown, the formula is written as HOH or H_2O (either is to be regarded as structurally noncommittal).

12-5 Hydrogen Bonding

A water molecule acts as if it had a positive end and a negative end. The positive end of such a molecule attracts the negative end of each of its neighbor molecules. This electrical attraction constitutes a linkage between polar molecules which is important in determining the properties of a substance. Although such a linkage is far weaker than the covalent bonds which link atoms together within molecules, or the ionic bonds which hold the ions in fixed positions in a crystal of an ionic compound, it is a much stronger force than the van der Waals attraction between nonpolar molecules.

The electrostatic interaction between molecules containing an F—H, O—H, or N—H bond is substantially greater than in other cases. Molecules containing such bonds have a tendency to form linkages with other molecules containing a fluorine atom, an oxygen atom, or a nitrogen atom. This linkage is called a *hydrogen bond*, since a hydrogen atom serves as the bridge between two atoms having strong electron-attracting power.

In the vapor state, water consists of simple H_2O molecules, but in the liquid state it consists of aggregates formed by simple water molecules linking together through hydrogen bonds. In other words, water is an *associated liquid*. Liquids in which the molecules do not associate are *normal liquids*. This aggregation of water molecules can be represented by the general formula $(H_2O)_n$, where n is an integer greater than 1. If $n = 2$, the aggregate is a *dimer*, $(H_2O)_2$, with the following formula:

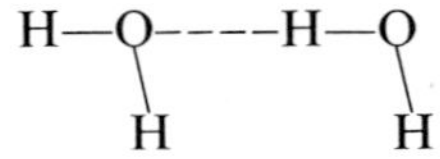

In this formula a solid line represents the covalent bond between a hydrogen and an oxygen atom within a water molecule, and the broken line indicates a hydrogen bond between atoms in different water molecules.

If $n = 3$, the aggregate is a *trimer*, $(H_2O)_3$. In general, if many molecules are associated together, the resulting aggregate is called a *polymer*. In liquid water n is variable and small.

A hydrogen bond is considerably stronger than the electrostatic attraction which exists between polar molecules with no hydrogen linkage, but a hydrogen bond between two molecules is very much weaker than a covalent bond within a molecule. For example, the energy of the covalent O—H bond within a water molecule is about 100 kcal per mole, but the energy of the hydrogen bond between two water molecules is about 3 kcal per mole. The distance between an H and an O atom that are joined by a hydrogen bond is much greater than the distance between an H and an O atom that are covalently bonded within a water molecule.

Investigations of the crystal structure of ice have revealed an arrangement of the water molecules in which each oxygen atom is surrounded symmetrically, in three dimensions, by a tetrahedral arrangement of four other oxygen atoms. Infrared spectroscopy indicates that between each two oxygen atoms is a hydrogen atom which is nearer one oxygen atom than the other. A covalent bond exists between the hydrogen atom and the nearer oxygen atom, and a hydrogen bond exists between the same hydrogen atom and the more distant oxygen atom. This arrangement results in an open structure, i.e., there are holes in it (Figure 12-2), which accounts for ice having a lesser density than liquid water. When ice melts, some of the hydrogen bonds are broken, the open structure collapses, and the water molecules pack more closely together. Therefore, the liquid occupies a smaller volume than the ice from which it was formed.

As the temperature of liquid water is raised, more hydrogen bonds break and the volume tends to decrease. However, the increasing kinetic activity of the molecules which accompanies increasing temperatures causes them to move farther apart. These two effects work toward opposite results. Above 4°C the kinetic effect predominates, and the density of water decreases with rising temperature. Below 4°C the structure collapse is more important, and the density increases with rising temperature between 0 and 4°C. Consequently, water has its maximum density at 4°C.

An input of energy is needed in the breaking of hydrogen bonds, which occurs when ice melts. This energy requirement accounts for the unusually high melting point and large heat of fusion of ice (relative to other compounds of comparable structure, such as H_2S, that do not form hydrogen

Figure 12-2

Structure of ice

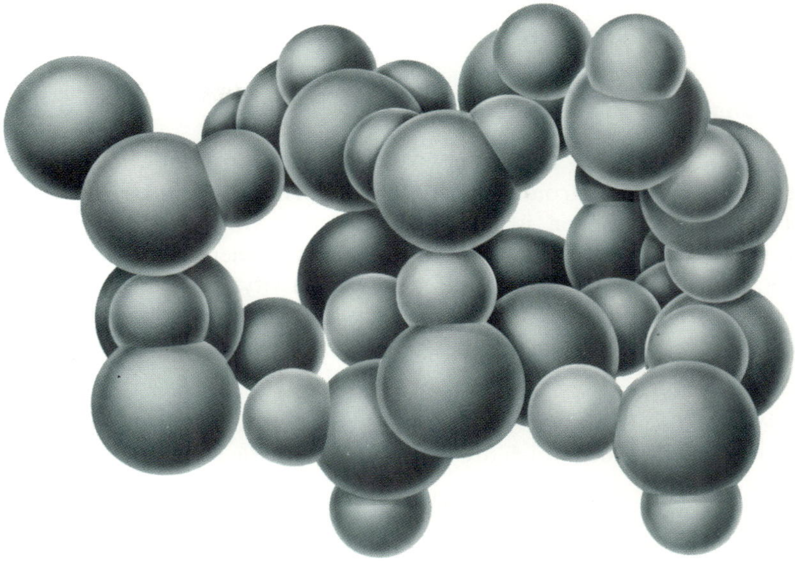

bonds). The high specific heat is partially explained by the continued breaking of hydrogen bonds with increasing temperature. Vaporization requires breaking of all remaining hydrogen bonds, so that the boiling point and heat of vaporization are also high. Failure to obey Trouton's Rule is attributed to the same cause; such failure is characteristic of associated liquids.

12-6 Water as a Solvent

Solubility is an important topic in chemistry because reactions generally occur more rapidly if the reacting substances are dissolved in an appropriate solvent. In nature many of the most important processes occur in the solution state, e.g., the absorption of nutrient salts by plants.

The ability of water to dissolve large quantities of certain hydrogen compounds is related to its tendency to form hydrogen bonds with them. Such highly soluble compounds include methyl alcohol, CH_3OH, and ammonia, NH_3, both of which form hydrogen bonds themselves. In order for any substance to dissolve in water, the water molecules must become unbonded from one another and form bonds, instead, with the particles of the dissolving substance. Consequently, nonpolar substances such as methane are insoluble in water.

The ability of water to dissolve ionic compounds is due to the nature of the water molecules. When an ionic compound is placed in contact with water, the polar water molecules attract the ions of the solid crystal and reduce the strength of the ionic bonds within the crystal. The negative (oxygen) ends of the water molecules attract the positive ions of the crystal, and the positive (hydrogen) ends attract the negative ions. The result is that the ions are separated from one another and are dispersed through the water. Ionic compounds do not dissolve in nonpolar solvents such as carbon tetrachloride.

12-7 Chemical Properties of Water

Water has little tendency to decompose by the action of heat alone. This property could be predicted from the fact that the formation of water by the union of oxygen with hydrogen is highly exothermic. Although thermally stable, water enters readily into reactions with a variety of elements and compounds.

Metals that occupy a high position in the activity series displace hydrogen from water (see Section 8-10). Fluorine is the only element that displaces oxygen from water:

$$2F_2 + 2H_2O \rightarrow 4HF + O_2$$

Oxides of the alkali metals (Group IA) combine with water, liberating heat and forming hydroxides which are highly soluble strong bases. For example,

$$Na_2O + H_2O \rightarrow 2Na^+ + 2OH^-$$

Oxides of the alkaline-earth metals (Group IIA) react less readily and give hydroxides which are much less soluble. For example,

$$CaO + H_2O \rightarrow Ca(OH)_2$$

Most other metallic oxides react with water only slowly if at all.

Oxides of nonmetals react with water to from acids. For example, carbon dioxide forms carbonic acid (a weak acid):

$$CO_2 + H_2O \rightleftharpoons H_2CO_3$$

Certain oxides of metals in which the metals show their highest oxidation states also react with water to form acids. For example, chromium(VI) oxide forms chromic acid (a strong acid):

$$CrO_3 + H_2O \rightarrow 2H^+ + CrO_4^{--}$$

In addition to the types of oxides mentioned above, certain other compounds cannot exist in the presence of water because of a chemical reaction that continues until either the water or the other compound has been completely consumed. For example, acetylene gas is prepared by such a reaction of water with calcium carbide:

$$CaC_2 + 2HOH \rightarrow Ca(OH)_2 + C_2H_2\uparrow$$

12-8 Hydrates

When an aqueous solution of a soluble salt is concentrated by evaporation, the crystals which separate frequently contain the salt and water combined in definite proportions. Such compounds are known as *hydrates*. A familiar example is $CuSO_4 \cdot 5H_2O$, commonly known as blue vitriol. Its chemical name is copper(II) sulfate 5-hydrate, or cupric sulfate pentahydrate. The dot (·) used in the formula indicates that the hydrate is formed by the chemical combination of 5 moles of H_2O with 1 mole of $CuSO_4$ and also that the 5 moles of H_2O can in turn be removed from the hydrate. The formula for the hydrate is sometimes written $CuSO_4(H_2O)_5$.

When the blue crystals of copper(II) sulfate 5-hydrate are heated, water is lost and the crystalline structure of the hydrated salt appears also to be lost. The salt that remains is a white powder, $CuSO_4$, which may appropriately be referred to as anhydrous copper(II) sulfate:

$$CuSO_4 \cdot 5H_2O \rightleftharpoons CuSO_4 + 5H_2O$$

This equation is written with a double arrow to indicate that the reaction is reversible, for addition of water to the anhydrous salt produces the hydrated salt again.

The solution obtained by dissolving copper(II) sulfate 5-hydrate in water cannot be distinguished from the solution that is formed by dissolving anhydrous copper(II) sulfate. Certainly, most salts become hydrated in aqueous solution; i.e., all ions, positive and negative, are probably hydrated in solution. It is impossible to be sure how many water molecules are attached to an ion in solution, but crystal structure investigations of hydrates outside of solution have produced some information. For example, in copper(II) sulfate 5-hydrate the copper(II) ion, Cu^{++}, is combined with four water molecules, $[Cu(H_2O)_4]^{++}$, and the sulfate ion is combined with one water molecule, $[SO_4(H_2O)]^{--}$.

Water molecules combine with metal ions by coordinate bonds, i.e., covalent bonds in which both electrons of the shared pair are furnished by the same atom (oxygen atoms in this case):

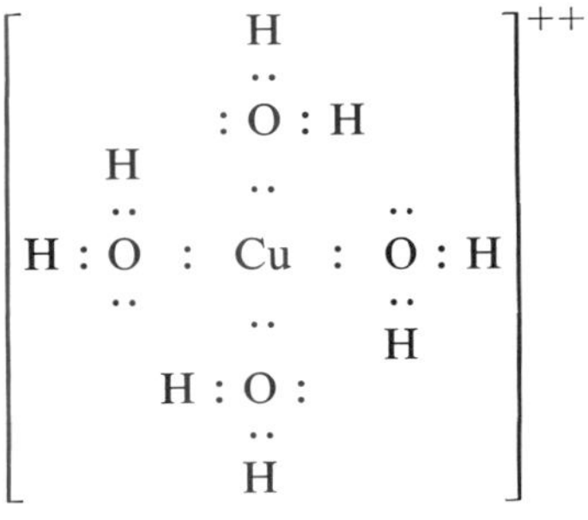

Water molecules are attached to negative ions by hydrogen bonds.

In some hydrates water molecules occupy certain positions in the crystal structure without being attached by chemical bonds to either the positive or the negative ions. Such water is called *lattice water*. An example is the hydrate $KAl(SO_4)_2 \cdot 12H_2O$, in which six H_2O molecules are lattice water molecules and the other six are attached by coordinate bonds to the aluminum ion.

Water of hydration should not be confused with the water which is taken up by an oxide in becoming a hydroxide (or given up by a hydroxide in becoming an oxide). For example, when aluminum hydroxide, $Al(OH)_3$, loses water it becomes aluminum oxide, Al_2O_3; but the crystal structure of Al_2O_3 is entirely different from that of $Al(OH)_3$, and $Al(OH)_3$ does not "contain" water in the sense that a hydrate contains water of hydration.

12-9 Efflorescence and Deliquescence

If the crystals of certain hydrates are placed in the space above the mercury in a barometer tube, the level of the mercury falls, indicating that the hydrate exerts a vapor pressure. If the vapor pressure of the crystal is greater than the partial pressure of the water in the air, the crystal will lose part or all of its water of hydration when exposed to air. This spontaneous loss of water of hydration to the atmosphere is known as *efflorescence*. Whether a hydrate is stable against efflorescence when exposed to air depends upon the humidity of the air, the nature of the hydrate, and the temperature. The vapor pressure of a hydrate increases with rising temperature. If the partial pressure of the water vapor in the air is greater than 7.8 torrs but less than 14 torrs, $CuSO_4 \cdot 5H_2O$ is stable, but $Na_2SO_4 \cdot 10H_2O$ effloresces rapidly at room temperature. If the partial pressure of the water vapor is greater than 14 torrs, both hydrates are stable; if it is less than 7.8 torrs, both effloresce.

Some salts form more than one hydrate, in which case each hydrate has its own characteristic vapor pressure. Thus, if the partial pressure of the water vapor in the air is less than 47 torrs but greater than 30 torrs, $CuSO_4 \cdot 5H_2O$ effloresces at 50°C to form $CuSO_4 \cdot 3H_2O$. If the partial pressure of the water vapor is less than 30 torrs but more than 4.5 torrs, $CuSO_4 \cdot 3H_2O$ loses water and becomes $CuSO_4 \cdot H_2O$ at 50°C. Only if the partial pressure of water vapor in the air is less than 4.5 torrs will $CuSO_4 \cdot H_2O$ change into anhydrous $CuSO_4$ at 50°C.

Mixtures of two hydrates of a given salt can be used to regulate the humidity within a closed space. If the humidity of the air is greater than the vapor pressure of the higher hydrate (e.g., $CuSO_4 \cdot 5H_2O$), the excess water will be removed by the lower hydrate (e.g., $CuSO_4 \cdot 3H_2O$). If the humidity is less than this value, water will be released by the higher hydrate.

A *hygroscopic* substance absorbs moisture from the air. If a hygroscopic substance takes up so much water vapor that an aqueous solution is formed, the substance is *deliquescent*. Any substance that deliquesces is very soluble in water, but finely divided substances, such as flour, may be hygroscopic even though they are not soluble in water and are thus incapable of deliquescence.

Hydrate formation is sometimes, but not always, involved in hygroscopic or deliquescent behavior. For example, sodium hydroxide forms no crystalline hydrate, but pellets of NaOH exposed to the air take up so much water that they form a solution. Both anhydrous calcium chloride and $CaCl_2 \cdot 6H_2O$ deliquesce. When the anhydrous salt takes up water vapor, all of it is converted to $CaCl_2 \cdot 6H_2O$ before the first drop of solution forms. The use of calcium chloride to keep down the dust on dirt roads is an application of its deliquescent property.

12-10 Treatment of Natural Water

Rain water is relatively pure except for dust particles and dissolved gases. When it falls to the ground and travels over and through the surface of the earth, many different types of material become dissolved or suspended in it. Appreciable quantities of soluble salts are present in water that has percolated through soil, and the suspended materials present in streams flowing through populous areas include organic material and bacteria.

Water for human consumption should be free of harmful bacteria. Water which passes through layers of soil and porous rock undergoes a natural filtration that removes most bacteria; therefore, water from deep wells is often safe for drinking. This water is likely, however, to contain calcium, magnesium, and iron(II) compounds which prevent the effective cleansing action of soap. Such water is known as *hard water*.

Harmful bacteria can be removed from water that is to be used for drinking by filtration through beds of sand and by chlorination. Water to be used for chemical purposes may need to be purified by distillation, but this is an expensive operation. The removal of much of the hardness of a city water supply (to render it fit for use in laundries and steam boilers) is accomplished by a process called *water softening*. For example, in a large-scale water-softening operation, sodium carbonate ("soda ash") is used to bring about the removal of Ca^{++} ions by precipitation as calcium carbonate. If the compound responsible for the hardness of the water is calcium sulfate, the following reaction occurs:

$$(Ca^{++} + SO_4^{--}) + (2Na^+ + CO_3^{--}) \rightarrow CaCO_3 \downarrow + (2Na^+ + SO_4^{--})$$

The softened water contains the same number of moles of sodium sulfate as the moles of calcium sulfate originally present; but the sodium salt formed does not interfere with the cleansing action of soap, nor does it form boiler scale.

In recent years processes have been developed for softening water by the use of ion-exchange resins. This topic will be considered in Section 29-11.

12-11 Heavy Water

Since hydrogen, as it occurs in nature, is a mixture of the two isotopes protium and deuterium, all hydrogen compounds occurring in nature are mixtures of molecules containing protium atoms and molecules containing deuterium atoms. Water that is composed of deuterium and oxygen is called *deuterium oxide* (D_2O) or, less precisely, *heavy water*. Since protium atoms are 5000 times as abundant as deuterium atoms, the properties of ordinary water are approximately the same as those of protium oxide.

Table 12-2

Physical properties of ordinary water and deuterium oxide

	Ordinary water	*Deuterium oxide*
Density at 20°C (g/ml)	0.998	1.105
Maximum density (g/ml)	1.000	1.107
Temperature of maximum density (°C)	4.0	11.6
Melting point (°C)	0.0	3.8
Boiling point (°C)	100.0	101.42
Heat of vaporization (cal/mole)	9720	9970
Heat of fusion (cal/mole)	1440	1500

Heavy water was first obtained by electrolysis of ordinary water. Since protium oxide undergoes electrolytic decomposition at a somewhat faster rate than deuterium oxide, the heavy water tends to remain behind in the electrolytic cell. If the cell contents are continuously replenished by addition of ordinary water, the residue becomes enriched in deuterium oxide.

Although heavy water and ordinary water appear identical, many of their physical properties are different, as indicated in Table 12-2. Since heavy water is about 10 percent heavier than ordinary water, it can be detected by precision methods for determining density when there is only 1 part of it in 100,000 parts of the H_2O-D_2O mixture.

Protium oxide and deuterium oxide are identical chemically except for differences in reaction rates. This property, also true of other hydrogen compounds, makes it possible to study chemical reactions and biochemical processes such as digestion and metabolism by replacing protium with deuterium in molecules of chemical reagents or in the food given to an organism. The subsequent whereabouts of the deuterium atoms can be ascertained by density determinations. Thus, deuterium can be used as a "tracer" element without recourse to radioactive isotopes.

A deuterium atom and a protium atom can both be present in the same water molecule. As a matter of fact, the heavy water present in small amounts in ordinary water is mostly HOD. One can easily see that the presence of two molecules of HOD in 10,000 molecules of H_2O is much more probable than the presence of one molecule of D_2O. The properties of pure HOD are intermediate between those of protium oxide and those of deuterium oxide.

Like ordinary water, D_2O and HOD show strong tendencies toward hydrogen bond formation.

New Terms

Associated liquid: a liquid whose molecules are linked together to form aggregates. An aggregate consisting of two molecules is a *dimer*. An aggregate consisting of three molecules is a *trimer*. If a large number of molecules are linked together, the aggregate is a *polymer*. (In *normal liquids* the molecules do not associate.)

Deliquesce: to become wet and dissolve gradually in moisture absorbed from the air.

Effloresce: to lose water of hydration to the atmosphere spontaneously.

Hard water: water in which it is not easy to obtain the cleansing action of soap. Hardness of water is due to Ca^{++}, Mg^{++}, and Fe^{++} ions.

Heavy water (deuterium oxide): water that is composed of deuterium and oxygen. Its formula is D_2O.

Hydrate: a crystalline substance that contains chemically combined water in a definite proportion.

Hydrogen bond: the bond between two molecules linked together through a hydrogen atom which serves as the bridge between two atoms having strong electron-attracting power. Hydrogen bonding is an especially strong electrostatic attraction between polar molecules containing F—H, O—H, or N—H bonds. Hydrogen bonds are much weaker than ionic or covalent bonds but much stronger than van der Waals forces.

Hygroscopic: having a tendency to absorb moisture from the air.

Exercises

12-1 How do you explain the fact that water molecules are polar? That water molecules are associated in the liquid state? That water pipes burst when water freezes inside them?

12-2 Calculate the percentage of water in Epsom salts, $MgSO_4 \cdot 7H_2O$.

12-3 Methane, ammonia, and water have molecular weights that are nearly the same: 16, 17, and 18, respectively. The boiling points of the three compounds are 112, 240, and 373°K, respectively. Why are their boiling points so different?

12-4 When borax, a hydrated salt, is heated it loses water, and the weight is reduced to 52.77 percent of the original value. The formula of the anhydrous salt that remains is $Na_2B_4O_7$. Calculate the formula of the hydrate.

12-5 A student who knew that deuterium gas has twice the density of protium gas had difficulty reconciling this fact with the statement that heavy water is only 10 percent heavier than ordinary water. How would you point out the fallacy in his reasoning?

12-6 Explain the fact that heavy water does not have its maximum density at the same temperature as ordinary water.

12-7 By means of suitable chemical equations, show that water sometimes acts as an oxidizing agent, sometimes as a reducing agent, and sometimes takes part in reactions in which it is neither.

12-8 Can water always be used to extinguish fires? Why or why not?

12-9 Calculate how many moles of each of the following hydrates would have to be dehydrated to obtain 5.00 kg of the anhydrous salt: $CuSO_4 \cdot 5H_2O$, $CuSO_4 \cdot H_2O$, $Na_2CO_3 \cdot 10H_2O$.

12-10 People who wear eyeglasses are plagued by having their glasses coated with a film of water when they enter a warm room on a cold day. Why does this happen?

12-11 Suppose 1 kg of ice (at 0°C) is placed in 1 kg of water at 40°C in a vessel which permits no heat to leave or enter. Certain physical changes will occur. When these changes have ceased and a condition of equilibrium has been reached, what will the vessel contain? What temperature will prevail?

12-12 Calculate the percentage of water of hydration in $KAl(SO_4)_2 \cdot 12D_2O$.

References

Chalmers, B., "How Water Freezes," *Sci. Am.*, February 1959, p. 114. A detailed consideration of the mechanism of formation of ice crystals, including the various shapes of snowflakes.

Chemical Bond Approach Project, *Chemical Systems*, McGraw-Hill, New York, 1964, Chapter 18, "Water." This last chapter in the CBA text is devoted entirely to the structure and chemistry of water.

Davis, K. S., and Day, J. A., *Water, the Mirror of Science*, Anchor Books (Doubleday), Garden City, New York, 1961 (paperbound). The peculiarities of water are emphasized.

Gillam, W. S., and McCutchan, J. W., "Demineralization of Saline Waters," *Science*, **134**, 1041 (1961). An account of desalination processes that may solve our impending water crisis.

Katz, J. J., "The Biology of Heavy Water," *Sci. Am.*, July 1960, p. 106. The effects on living organisms of replacing H_2O with D_2O.

Keirstead, R. E., "Water, a Basic Natural Resource," *J. Chem. Educ.*, **32** (2), 99 (1955). A discussion of man's fresh-water needs and the adequacy of the supply.

Leopold, L. B., Davis, K. S., and the Editors of Life, *Water*, Time, Inc., New York, 1966. An account of the many aspects of water with numerous color photographs.

13

Solutions

13-1 Introduction

When sugar crystals are stirred with water, a clear mixture of sugar and water is obtained. Such a homogeneous mixture, in which the individual substances cannot be seen, even with a microscope, is called a *solution*. When a solution is formed by dissolving a solid in a liquid, the solid is called the *solute*, and the liquid is called the *solvent*. The process of dissolution occurs because the solvent molecules have an attraction for the solute molecules that is comparable with the attraction the solute molecules have for one another. In the dissolved state the solute is molecularly dispersed; i.e., each particle of solute is an individual molecule. (In solutions of electrolytes some or all of the dissolved particles are electrically charged; i.e., they are ions. This will be discussed in Sections 13-6 and 13-7.)

If only a small quantity of the solute is present in comparison with the amount of solvent, the solution is said to be *dilute*. The solution is described as *concentrated* if it contains a relatively large amount of solute. Since the terms *dilute* and *concentrated* are too vague to be highly useful, it is more satisfactory to describe solutions in terms of the quantity of solute present in a given amount of solvent or solution.

13-2 Molality

One useful method of stating the concentration of a solution is to give the number of moles of solute that are present in 1 kg (1000 g) of solvent. *Molality*, usually designated by m, is the name given to this number. To calculate the molality of a solution, we divide the number of grams of solute by the number of grams in 1 mole of solute to obtain the number of moles of solute present. We then divide this number by the number of kilograms of solvent in the solution to obtain the number of moles of solute per kilogram of solvent.

EXAMPLE A solution is prepared by dissolving 1 mole of sugar in 2000 g of water. What is its molality?

The 1 mole of the solute is dissolved in 2000 g, which is 2.000 kg, of solvent. The quantity of solute per kilogram of solvent is 1 mole per 2.000 kg or 0.5 mole per kg. The solution is, therefore, 0.5 *m*; i.e., the molality of the solution is 0.5.

EXAMPLE If 25 g of sugar, $C_{12}H_{22}O_{11}$, is dissolved in 1000 g of water, what is the molality of the solution?

Since the formula for sugar is $C_{12}H_{22}O_{11}$, the formula weight is $(12 \times 12) + (22 \times 1) + (11 \times 16) = 342$. Therefore, 1 mole of sugar is 342 g. The 25 g is 25 g/(342 g/mole) or 0.073 mole. Since 0.073 mole is dissolved in 1.000 kg of water, the solution is 0.073 *m*.

EXAMPLE What is the molality of a solution obtained by dissolving 32.5 g of table salt (sodium chloride) in 1500 g of water?

When we find that the formula for the salt is NaCl, we can compute the formula weight to be 23.0 + 35.5 = 58.5; 1 mole of salt is 58.5 g, and 32.5 g is 32.5 g/(58.5 g/mole) = 0.556 mole. This quantity is dissolved in 1500 g of water, which is

$$\frac{1500\ \text{g}}{1000\ \text{g/kg}} = 1.500\ \text{kg}$$

The quantity of salt present per kilogram of water is

$$\frac{0.556\ \text{mole}}{1.500\ \text{kg}} = 0.371\ \text{mole/kg}$$

Thus, the molality of the solution is 0.371.

13-3 Effect of the Solute on the Freezing Point of the Solvent

One of the most frequently investigated aspects of solutions has been the effect the solute has on the physical properties of the solvent. In general, addition of a solute lessens the tendency of the solvent to evaporate and makes it necessary that the solution be heated to a temperature higher than the boiling point of the pure solvent in order for it to boil (unless the solute is itself a volatile substance); for the solution to freeze, it must be cooled to a temperature lower than the freezing point of the pure solvent.

It is common knowledge that solutions freeze at lower temperatures than do pure liquids. Addition of various antifreezes to the water in automobile radiators is a well-known procedure for preventing the water from freezing when it is cooled to 0°C (32°F). The extent to which the freezing point is lowered is related to the molality. In concentrated solutions the situation is complicated and somewhat unpredictable, but in dilute solutions the freezing-point depression is directly proportional to the molality. The freezing point of water, for example, is lowered at the rate of 1.86°C for each mole of solute per kilogram of water (if the solute is not an electrolyte). If 0.100 mole (34.2 g) of sugar, $C_{12}H_{22}O_{11}$, is dissolved in 1.000 kg of water, the freezing point is lowered 0.100 mole × 1.86°C/mole, or 0.186°C; i.e., the solution would begin to freeze at −0.186°C instead of at 0°C.

This example indicates a method by which the molecular weight of a substance can be determined experimentally. If ΔT_f represents the amount that the freezing point is lowered (Δ means "the change in"), m the molality, and K_f a constant characteristic of the particular solvent and known as the molal freezing-point depression constant (1.86°C/mole in the case of water), then the relationship can be stated as follows:

$$\Delta T_f = K_f m$$

From this relationship the molality of a solution can be calculated if the freezing-point lowering has been measured and the value of the constant K_f is known:

$$m = \frac{\Delta T_f}{K_f}$$

On the other hand, the molality of the solution is the number of moles of solute per kilogram of solvent. Since the number of moles of solute present is the number of grams of solute divided by its formula weight, we have the necessary information for calculating the molecular weight of the solute.

EXAMPLE A solution prepared by dissolving 9.00 g of glucose in 150 g of water has a freezing point of −0.620°C. What is the molecular weight of glucose?

From the freezing-point lowering, 0.620°C, we first calculate the molality of the solution (number of moles of glucose per kilogram of water):

$$m = \frac{0.620°\text{C}}{1.86°\text{C/mole}} = 0.333 \text{ mole}$$

For the weight (grams) of glucose per kilogram of water, we have

$$\frac{9.00 \text{ g glucose}}{150 \text{ g water}} \times 1000 \text{ g water/kg water} = 60.0 \text{ g glucose/kg water}$$

Since the solution contains 0.333 mole of glucose per kilogram of water and 60.0 g of glucose per kilogram of water, it follows that

$$0.333 \text{ mole glucose} = 60.0 \text{ g glucose}$$

and

$$1 \text{ mole glucose} = \frac{60.0 \text{ g}}{0.333} = 180 \text{ g}$$

Thus, the molecular weight of glucose is 180. (Check this against the formula, $C_6H_{12}O_6$. You should not expect results to be this good in all cases, for the laws of solutions are limited in their accuracy, just as the gas laws are.)

13-4 Elevation of the Boiling Point

The temperature of a liquid must be raised above its normal boiling point for it to boil when a nonvolatile substance is dissolved in it. The extent to which the boiling point is raised is related to the molality. As with the lowering of the freezing point, the situation is complicated and somewhat unpredictable in concentrated solutions, but the elevation of the boiling point is directly proportional to the molality in dilute solutions. The boiling point of water, for example, is raised at the rate of 0.512°C for each mole of solute per kilogram of water (if the solute is not an electrolyte). If 0.100 mole (34.2 g) of sugar is dissolved in 1.000 kg of water, the boiling point is raised 0.100 mole × 0.512°C/mole, or 0.051°C; i.e., the solution would boil at 100.051°C instead of 100.000°C.

This example indicates another method by which the molecular weight of a substance can be determined from experimental data. If ΔT_b represents the amount that the boiling point is raised, m the molality, and K_b a constant characteristic of the particular solvent and known as the molal boiling-point elevation constant (0.512°C/mole in the case of water), the relationship can be stated as follows:

$$\Delta T_b = K_b m$$

From this relationship the molality of a solution can be calculated if the boiling-point elevation has been measured and the value of the constant K_b is known:

$$m = \frac{\Delta T_b}{K_b}$$

Knowing also the composition of the solution in terms of grams of solute and grams of solvent, we have the information we need to calculate the molecular weight of the solute.

EXAMPLE A solution prepared by dissolving 3.89 g of a certain compound in 187 g chloroform boils at 62.27°C. What is the molecular weight of the solute?

Consulting Table 13-1, we find that the boiling point of pure chloroform is 61.26°C. The boiling-point elevation is the difference between the boiling point of chloroform and the boiling point of the solution:

$$\Delta T_b = 62.27°\text{C} - 61.26°\text{C} = 1.01°\text{C}$$

Table 13-1 also shows that K_b for chloroform is 3.63°C/mole. From the boiling-point elevation we now calculate the molality (the number

Table 13-1

Molal freezing-point depression and boiling-point elevation constants

Solvent	T_f(°C)	K_f(°C/mole)	T_b(°C)	K_b(°C/mole)
Benzene	5.49	5.08	80.1	2.53
Camphor	176	37.7	208	5.95
Carbon tetrachloride	−22.8	31.8	76.8	5.03
Chloroform	−63.5	4.68	61.26	3.63
Water	0.00	1.86	100.00	0.512

of moles of solute in 1 kg of solvent):

$$m = \frac{1.01°\text{C}}{3.63°\text{C/mole}} = 0.278 \text{ mole}$$

The quantity of solvent in which the 3.89 g of solute is dissolved is

$$\frac{187 \text{ g}}{1000 \text{ g/kg}} = 0.187 \text{ kg}$$

The weight of solute per kilogram of solvent is

$$\frac{3.89 \text{ g solute}}{0.187 \text{ kg solvent}} = 20.8 \text{ g solute/kg solvent}$$

Thus, the solution contains 0.278 *mole* of solute per kilogram of solvent and 20.8 *grams* of solute per kilogram of solvent. Therefore,

$$0.278 \text{ mole of solute} = 20.8 \text{ g solute}$$

and the weight of 1 mole of the solute is

$$\frac{20.8 \text{ g}}{0.278 \text{ mole}} = 74.8 \text{ g/mole}$$

Thus, the molecular weight of the solute is calculated to be 75. (We will not claim three-digit accuracy because we do not expect the laws of solutions to be completely accurate.)

13-5 Lowering of the Vapor Pressure

We saw in Section 10-11 that the vapor pressure, P, of a liquid in a solution that obeys Raoult's Law is given by the expression

$$P = XP°$$

where $P°$ is the vapor pressure of the pure liquid at the same temperature and X is its mole fraction in the solution. If a solution is prepared by dissolving a nonvolatile solid in a volatile liquid, the above expression gives the *total* vapor pressure. Therefore, at all temperatures the vapor pressure of the solution is less than the vapor pressure of the pure solvent. For example, at the normal boiling point of the pure solvent the vapor pressure of the solution is less than 1 atmosphere. This explains why a solution containing a nonvolatile solute must be heated to a temperature higher than the boiling point of the pure solvent in order for it to boil.

The same considerations explain why a solution must be cooled to a temperature lower than the freezing point of the pure solvent in order for it to begin to freeze. At the freezing point the vapor pressure of the solid is the same as the vapor pressure of the liquid (Section 11-3). At the temperature at which the pure solvent would freeze (e.g., 0°C in the case of water), the vapor pressure of the solvent component of a solution is less than the vapor pressure of the solid, even if the solute is volatile. It is not possible, therefore, to have frozen solvent in equilibrium with a solution at the freezing point of the pure solvent. If the temperature is reduced, both the vapor pressure of the solid and the vapor pressure of the liquid solvent in the solution become smaller. The vapor pressure of the solid falls off more rapidly than the vapor pressure of the liquid solvent, so that there is some lower temperature at which the vapor pressure of the solid and the vapor pressure of the liquid solvent are identical. This temperature is the freezing point of the solution.

The mole fraction of the solvent, X_A, and the mole fraction of the solute, X_B, are related by the expression $X_A = 1 - X_B$. Therefore, the Raoult's Law expression, $P_A = X_A P_A{}^\circ$, can be written in the form

$$P_A = (1 - X_B)P_A{}^\circ = P_A{}^\circ - X_B P_A{}^\circ$$

Solving this for X_B, we get

$$X_B = \frac{P_A{}^\circ - P_A}{P_A{}^\circ}$$

The numerator, $P_A{}^\circ - P_A$, is the amount that the vapor pressure is lowered by having the solute dissolved in it. This lowering of the vapor pressure can be appropriately symbolized as ΔP. Therefore,

$$\Delta P = P_A{}^\circ - P_A = X_B P_A{}^\circ$$

This equation states that the amount by which the vapor pressure of the solvent is reduced is proportional to the mole fraction of solute. Written in the form

$$\frac{\Delta P}{P_A{}^\circ} = X_B$$

it states that the *relative* lowering of the vapor pressure has the same numerical value as the mole fraction of solute. It suggests still another method for determining the molecular weight of the solute. Upon measuring the vapor pressure of the solution we can calculate the mole fraction of the solute:

$$X_B = \frac{\Delta P}{P_A{}^\circ}$$

Since we know the number of grams of solute and the number of grams of solvent used to prepare the solution, we can calculate the molecular weight of the solute.

EXAMPLE At 20°C the vapor pressure of ether, $(C_2H_5)_2O$, is 442.2 torrs. At the same temperature the vapor pressure of a solution prepared by dissolving 10.0 g of a nonvolatile organic compound in 100 g of ether is 426.0 torrs. What is the molecular weight of the solute?

Letting M_A represent the molecular weight of the solvent, W_A the number of grams of solvent in the solution, M_B the molecular weight of the solute, and W_B the number of grams of solute in the solution, the mole fraction of the solute is

$$X_B = \frac{W_B/M_B}{W_A/M_A + W_B/M_B}$$

Therefore,

$$P_A^\circ - P_A = \left[\frac{W_B/M_B}{W_A/M_A + W_B/M_B}\right] P_A^\circ$$

Rearranged, the equation reads

$$M_B = M_A\left(\frac{W_B}{W_A}\right)\left(\frac{P_A}{P_A^\circ - P_A}\right)$$

Now we can substitute the numerical values given for the different quantities in the statement of the problem. The molecular weight of ether, obtained by adding the atomic weights of its constituent elements, is

$$\begin{aligned} M_A &= 2[(12.01 \times 2) + (1.008 \times 5)] + 16.00 = 74.12 \\ M_B &= 74.12 \times \frac{10.0\text{ g}}{100\text{ g}} \times \frac{426.0\text{ torrs}}{16.2\text{ torrs}} \\ &= 74.12 \times 0.100 \times 26.3 \\ &= 195 \end{aligned}$$

Since the lowering of the vapor pressure of the solvent is proportional to the mole fraction of solute, we might expect the elevation of the boiling point and the depression of the freezing point also to be proportional to the mole fraction of the solute because these effects are due to the lowering of the vapor pressure. They are, indeed, but in dilute solutions (as you may be able to prove for yourself) molality in a stated solvent is approximately proportional to the mole fraction of solute. By working with molalities instead of mole fractions we need not know the molecular weight of the solvent. Both K_f and K_b can be evaluated empirically by determining the lowering of the freezing point and the raising of the boiling point produced by solutes of known molecular weight. The data in Table 13-1 were obtained in this way.

13-6 Electrolytes

There is a certain class of compounds that affect the freezing point of water to a greater extent than can be predicted from the equation $\Delta T_f = K_f m$. They likewise affect the boiling point to a greater extent than can be predicted from the equation $\Delta T_b = K_b m$, and the relative reduction of the vapor pressure exceeds the mole fraction of solute. These compounds, in aqueous solutions, conduct an electric current. Such substances are known as *electrolytes* and include acids, bases, and salts.

An electric current is, by definition, a movement of charged particles. In solutions of electrolytes, the charged particles are ions, i.e., atoms or groups of atoms that bear one or more positive or negative charges.

Svante Arrhenius, a Swedish chemist, proposed a theory of ionization in 1887 to explain electrolytic conduction and the abnormal effects electrolytes have upon the freezing point, boiling point, and vapor pressure of water. Arrhenius thought, like the others of his day, that everything consists of molecules. We have already considered the fact that some substances, such as salts, consist of ions instead of molecules. Since Arrhenius did not know this, he assumed that the ions present in a solution of any electrolyte originate through the dissociation of the molecules of the solute:

$$NaCl \rightarrow Na^+ + Cl^-$$
$$HCl \rightarrow H^+ + Cl^-$$
$$NaOH \rightarrow Na^+ + OH^-$$

If each molecule splits into two ions, there are two particles present instead of one molecule. In this way Arrhenius explained why 1 mole of NaCl or HCl or NaOH in 1 kg of water lowers the freezing point of water approximately *twice* the expected amount: 3.72°C instead of 1.86°C. By assuming that the dissociation gives charged particles (ions) instead of neutral atoms, Arrhenius explained the ability of these solutions to conduct an electric current.

Our present-day knowledge enables us to correct Arrhenius on two counts. (1) Salts and some hydroxides, such as NaCl and NaOH, consist of ions, not molecules, in the crystalline state; when they dissolve in water, the crystals dissociate, i.e., they break apart, but there is no dissociation of molecules. (2) Although acids, such as HCl, consist of molecules (as Arrhenius supposed), the formation of ions when they dissolve is due to an interaction with the solvent, water:

$$HCl + H_2O \rightarrow H_3O^+ + Cl^-$$

Arrhenius correctly supposed that in a solution of sodium sulfate, Na_2SO_4, there are singly charged Na^+ ions and doubly charged SO_4^{--} ions, there being twice as many of the former as of the latter:

$$Na_2SO_4(s) \rightarrow 2Na^+ + SO_4^{--}$$

In solutions of this salt the effect on the freezing point of water is three times as great as for a solution of a nonelectrolyte of the same molality. In a calcium chloride ($CaCl_2$) solution there are Ca^{++} ions and Cl^- ions:

$$CaCl_2(s) \rightarrow Ca^{++} + 2Cl^-$$

In a solution of aluminum sulfate, $Al_2(SO_4)_3$, there are Al^{+++} ions and SO_4^{--} ions:

$$Al_2(SO_4)_3(s) \rightarrow 2Al^{+++} + 3SO_4^{--}$$

In "mole language," if 1 mole of NaCl is dissolved in water, the solution contains 1 mole of Na^+ ions and 1 mole of Cl^- ions, a total of 2 moles of ions. When 1 mole of Na_2SO_4 is dissolved in water, 2 moles of Na^+ ions and 1 mole of SO_4^{--} ions, a total of 3 moles, are formed. Similarly, there are 3 moles of ions in 1 mole of $CaCl_2$ and 5 moles of ions in 1 mole of $Al_2(SO_4)_3$. Thus, if 0.0100 mole of each of these salts is dissolved in separate kilogram quantities of water, the expected freezing points are: NaCl, −0.0372°C; Na_2SO_4, −0.0558°C; $CaCl_2$, −0.0558°C; $Al_2(SO_4)_3$, −0.0930°C. (The actual freezing points are somewhat higher; i.e., the freezing point is lowered somewhat less than expected because of interionic attraction.)

13-7 Weak Electrolytes Versus Strong Electrolytes

Arrhenius was aware that some electrolytes do not conduct electric current or depress the freezing point of water as effectively as others. To explain these closely correlated facts, he made the important assumption that *only a fraction* of the molecules are dissociated into ions. He symbolized this situation by using two arrows in the expression representing the dissociation:

$$\underset{\text{hydrocyanic acid molecules}}{HCN} \rightleftharpoons \underset{\text{hydrogen ions}}{H^+} + \underset{\text{cyanide ions}}{CN^-}$$

$$\underset{\text{acetic acid molecules}}{HC_2H_3O_2} \rightleftharpoons \underset{\text{hydrogen ions}}{H^+} + \underset{\text{acetate ions}}{C_2H_3O_2^-}$$

Acids such as hydrocyanic acid and acetic acid are examples of weak acids. As with the strong acid HCl, the formation of ions actually involves reaction with water:

$$HCN + H_2O \rightleftharpoons H_3O^+ + CN^-$$
$$HC_2H_3O_2 + H_2O \rightleftharpoons H_3O^+ + C_2H_3O_2^-$$

However, we must not become so zealous about correcting the statements of Arrhenius that we lose sight of the main points in his theory:

1. Aqueous solutions of electrolytes contain ions, which are charged atoms or groups of atoms, whose motion through the solution under the influence of an applied voltage constitutes an electric current.

2. In a solution where molecules dissociate into ions, the total number of particles is greater and the effect on the freezing point (and other related properties) of the solvent is correspondingly greater.
3. Electrolytes differ in the extent to which they form ions in solution. Those which give solutions that are good conductors of electricity are *strong electrolytes* and those which give slightly conducting solutions are *weak electrolytes*.

13-8 Molarity

A chemically significant method of expressing the concentration of a solution is to state the number of moles of the solute contained in 1 liter of the solution. The *molarity* of a solution is defined as this number. A molar (M) solution, i.e., a 1 M (pronounced "one-molar") solution, of sugar in water contains 1 mole (342 g) of sugar, $C_{12}H_{22}O_{11}$, in 1 liter of solution. Note that a molar solution is *not* prepared by dissolving 1 mole of the solute in 1 liter of the solvent. Instead, the 1 mole of solute is dissolved in a quantity of solvent (usually water) substantially less than 1 liter in volume, and then the solution is diluted with more solvent until the volume becomes exactly 1 liter. This is usually done in a volumetric flask, which is a flask with a mark on its neck to indicate the point to which the flask must be filled in order to contain the specified volume.

Unless otherwise specified, we assume the solvent is water, so such an expression as 3 M HCl refers to a solution in which the solvent is water, with HCl (the solute) present in a concentration of 3 moles per liter. If there is any likelihood of ambiguity, we should refer to the solution specifically as an aqueous solution to indicate that the solvent is water.

The number of moles of solute present in a solution is directly proportional to its volume and to its concentration (expressed in moles per liter). Thus,

$$\text{Number of moles} = \text{molarity} \times \text{volume (in liters)}$$

This relationship enables us to calculate the molarity of a solution if we know the volume and the moles present. We can also use the relationship to calculate the volume of solution required to contain a designated number of moles.

EXAMPLE How many grams of sugar are needed to prepare 500 ml (0.500 liter) of 0.250 M solution?

One liter of 0.250 M solution contains 0.250 mole; 0.500 liter of 0.250 M solution contains

$$0.250 \text{ mole/liter} \times 0.500 \text{ liter} = 0.125 \text{ mole}$$

One mole of sugar, $C_{12}H_{22}O_{11}$, weighs 342 g; 0.125 mole of sugar weighs

$$342 \text{ g/mole} \times 0.125 \text{ mole} = 42.8 \text{ g}$$

EXAMPLE What is the molarity of a solution prepared by dissolving 13.2 g of NaCl in sufficient water to give 2.000 liters?

One mole of NaCl weighs 58.44 g.

$$\frac{13.2 \text{ g}}{58.44 \text{ g/mole}} = 0.226 \text{ mole of NaCl}$$

$$\frac{0.226 \text{ mole}}{2.000 \text{ liters}} = 0.113\ M$$

EXAMPLE What volume of 2.00 *M* NaOH would be needed to furnish 12.0 g of NaOH for a reaction?

One mole of NaOH weighs 40.00 g.

$$\frac{12.0 \text{ g}}{40.00 \text{ g/mole}} = 0.300 \text{ mole}$$

$$\frac{0.300 \text{ mole}}{2.00 \text{ moles/liter}} = 0.150 \text{ liter}$$

EXAMPLE How would you prepare 200 ml of 3.0 *M* HCl from 12 *M* HCl solution?

In 200 ml of 3.0 *M* HCl solution there are 3.0 moles/liter × 0.200 liter = 0.600 mole.

$$\frac{0.600 \text{ mole}}{12 \text{ moles/liter}} = 0.050 \text{ liter} = 50 \text{ ml of } 12\ M \text{ HCl}$$

Place 50 ml of 12 *M* HCl in a 200-ml volumetric flask and add enough water to bring the level of the solution up to the mark.

13-9 Titration

The simplest method for determining the concentration of a base solution is to measure the volume of an acid solution of known concentration required to neutralize it. This determination is made by a process called *titration* (Figure 13-1). A measured volume of the base solution is placed in a flask, together with a few drops of an indicator solution, and the acid solution is added from a buret until the base solution is neutralized. This point is signaled by the indicator, which changes color. The volume of acid solution added is then read from the buret. From the information thus obtained one can calculate the concentration of the base solution, provided he knows the equation for the reaction for which the indicator has signaled the end point.

EXAMPLE Neutralization of 25.00 ml of NaOH solution required 23.78 ml of 0.0492 *M* H_2SO_4 solution. What is the concentration of the base solution?

Figure 13-1

Titration of base with acid

The buret contains an acid solution of known concentration; the flask contains a base solution with an indicator.

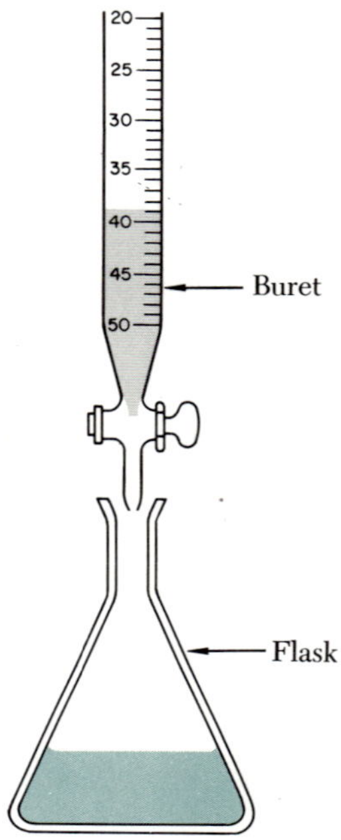

Write a balanced equation for the reaction:

$$2NaOH + H_2SO_4 \rightarrow Na_2SO_4 + 2H_2O$$

The equation shows that the number of moles of NaOH in the 25.00 ml of base solution is twice the number of moles of H_2SO_4 in the acid solution used in the titration. The number of moles of H_2SO_4 used in the titration is

$$0.0492 \text{ mole/liter} \times 0.02378 \text{ liter} = 0.001170 \text{ mole}$$

The number of moles of NaOH is twice 0.001170, or 0.002340 mole. The concentration of the NaOH solution is

$$\frac{0.002340 \text{ mole}}{0.02500 \text{ liter}} = 0.0936\ M$$

The concentration of an acid solution can likewise be determined by titrating with a base solution of known concentration. Precipitation reactions, suitably chosen, can also serve as the means for determining the concentrations of solutions by titration. In fact, any reaction is suitable for use in titration, provided the equation for the reaction is known and some means is available for indicating the end point.

New Terms

Electrolyte: a compound that conducts electricity in the molten state or in the dissolved state.

Molality of a solution: the number of moles of solute per kilogram of solvent. Molality is usually designated by *m*.

Molarity of a solution: the number of moles of solute in 1 liter of the solution. Molarity is usually designated by *M*.

Solute: in a solution, the dissolved substance.

Solvent: in a solution, the medium in which the solute is dissolved.

Strong electrolyte: any electrolyte (acid, base, or salt) that is a good conductor of electricity in the dissolved state because of the high concentration of its ions.

Titration: the process of determining the volume of a solution needed to react with a given volume of another solution (or with a given weight of some material). An important example is the addition of an acid solution to a base solution (or vice versa) to determine the concentration of the base (or acid).

Weak electrolyte: any electrolyte that is only a fair or poor conductor of electricity in the dissolved state because the substance furnishes only a few ions in solution.

Exercises

13-1 If 3 moles of a solid are dissolved in 4 moles of a chosen solvent, what is the molality of the solution? This question cannot be answered without additional information. What additional information is required?

13-2 If 300 g of sugar, $C_{12}H_{22}O_{11}$, are dissolved in 1500 g of water, what is the molality of the solution? To what temperature must it be cooled before ice can begin to form?

13-3 If the freezing point of an aqueous solution is $-0.372°C$, what is the molality of the solution? If the solution was prepared by dissolving 9.37 g of a solid in 250 g of water, what is the molecular weight of the solid?

13-4 What would be the boiling point, at 1 atmosphere, of a solution containing 100.0 g of sucrose, $C_{12}H_{22}O_{11}$, in 400.0 g of water?

13-5 The freezing point of an aqueous solution of a nonvolatile compound is $-0.742°C$. What is the boiling point of the solution?

13-6 Calculate the boiling point of a solution composed of 200 g of water, 0.0500 mole of NaCl, 0.1500 mole of Na_2SO_4, and 0.2000 mole of sucrose, $C_{12}H_{22}O_{11}$.

13-7 (a) A solution made by dissolving 10.80 g of glucose in 100.0 g of water is found to have a vapor pressure of 22.854 torrs. At the same temperature the vapor pressure of pure water is 23.109 torrs. Assuming Raoult's Law to be valid for this solution, calculate the molecular weight of glucose.
(b) The emprical formula for glucose is CH_2O. According to the result obtained in (a), what is the molecular formula of glucose?

13-8 What is the origin of the ions in an aqueous solution of Na_2SO_4? H_2SO_4?

13-9 What freezing points would you expect for aqueous solutions of the following compositions:

(a) $C_{12}H_{22}O_{11}$ in a 1.5 *m* (molal) solution
(b) NaCl in a 0.25 *m* solution
(c) Na_2SO_4 in a 2.5 *m* solution

13-10 If only 0.1% of the molecules of solute in a solution of a weak acid, HA, react with water to give H_3O^+ and A^- ions, what percentage of the HA molecules are still present as molecules?

13-11 What volume of a 0.0540 *M* solution of barium hydroxide, $Ba(OH)_2$, is required to neutralize 28.7 ml of an 0.0932 *M* solution of nitric acid?

References

Kieffer, W. F., *The Mole Concept in Chemistry*, Reinhold, New York, 1962 (paperbound). In Chapter 6, "Chemical Equivalence," the author recommends abandonment of the conventional designation of "gram-equivalent weight" and the correlated "normality" scale of concentrations.

Sienko, M. J., *Chemistry Problems*, Benjamin, New York, 1967 (paperbound). Chapter 11 is concerned with calculations involving molality, molarity, Raoult's Law, etc.

14

Ionic Reactions

14-1 Conductivity and Electrolysis

Acids, bases, and salts are known collectively as *electrolytes* because their aqueous solutions conduct an electric current. Pure water is virtually a nonconductor, but even minute quantities of an acid, base, or salt dissolved in water produce solutions through which electric currents can pass with ease. To demonstrate the electrical conductivity of an electrolytic solution, we can immerse strips of metal in the solution to serve as *electrodes*, through which the current can leave and enter the solution, and connect the electrodes to a source of current with an electric light bulb included in the circuit (Figure 14-1). If the solution is a good conductor, the bulb glows brightly; if it is a poor conductor, the bulb glows feebly; if it is a nonconductor, the light remains "off."

Either alternating current or direct current can be used to test the electrical conductivity of a solution. If direct current is used, one electrode remains positive and the other electrode remains negative during all the time that the current is "on." Under these circumstances chemical change occurs at each electrode as long as the current is "on." For example, if the current is flowing through a solution containing copper bromide, copper is plated out upon the *negative* electrode, and elemental bromine is produced at the *positive* electrode. The chemical change thus produced by the electric current is called *electrolysis*. The copper metal forms at the surface of the negative electrode because this is where the positive copper ions acquire the electrons needed to convert them to neutral atoms:

$$\underset{\text{copper ion}}{Cu^{++}} + \underset{\text{two electrons}}{2e^-} \rightarrow \underset{\text{copper atom}}{Cu}$$

The negative bromide ions lose electrons to the positive electrode and are thereby converted to neutral atoms, which combine in pairs to form bromine molecules:

Figure 14-1

Apparatus for testing conductivity of solutions

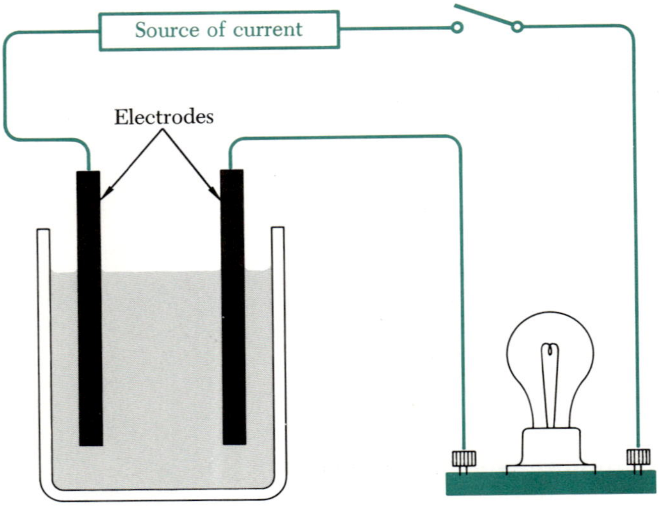

$$\underset{\text{two bromide ions}}{2Br^-} \rightarrow \underset{\text{bromine molecule}}{Br_2} + \underset{\text{two electrons}}{2e^-}$$

In an *electrolytic cell*, i.e., a cell in which electrolysis is occurring, the negatively charged electrode is called the *cathode*, and the positive electrode is called the *anode* (Figure 14-2). The ions that go to the cathode are called *cations*, and the ions that go to the anode are called *anions*. Thus, cations are positively charged ions, and anions are negative ions.

14-2 Conduction of Electricity by Pure Liquids

The movement of electrically charged particles of any kind constitutes an electric current. In metallic conduction, the charged particles are electrons. Mercury is the only important metal that is liquid at room temperature. It is, therefore, the only readily available pure liquid that is a good conductor under ordinary conditions. It would be difficult indeed to find a suitable substitute for the mercury in mercury switches.

Certain compounds conduct an electric current in the pure liquid state, but all such compounds are solids at room temperature, so they must be heated to their melting points to be able to conduct electric current. All these substances are ionic compounds, e.g., sodium chloride, NaCl. Conduction of electric current by pure liquid compounds consists of the movement of the ions of which they are composed. All compounds that are liquid at room temperature are covalent; they do not conduct electric current in the pure state because they consist of molecules, which are neutral, rather than ions, which are charged.

Figure 14-2

Electrolysis

Chemical changes occur at both electrodes.

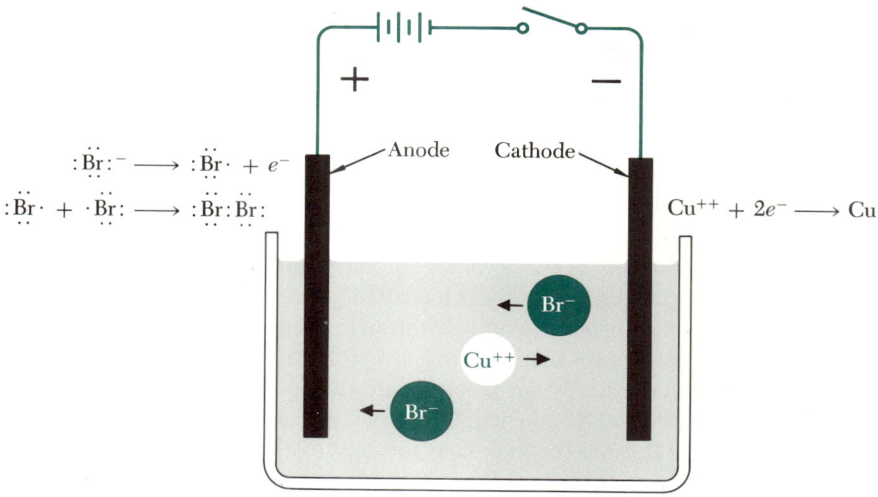

Melted sodium chloride, like the crystalline salt itself, consists of equal numbers of positive sodium ions, Na^+, and negative chloride ions, Cl^-. When the fused salt is present in an electrolytic cell, the sodium ions (cations) are attracted to the negative electrode (cathode), where they combine with electrons to form sodium atoms, i.e., metallic sodium:

$$Na^+ + e^- \rightarrow Na$$

The oxidation state of the sodium changes from +1 to 0. By definition, any change from a higher to a lower oxidation state is reduction. Note that the reduction occurs at the cathode.

Chloride ions (anions) are attracted to the positive electrode (anode), where each ion gives up an electron and becomes a chlorine atom. The chlorine atoms combine to form molecules of chlorine gas:

$$2Cl^- \rightarrow Cl_2 + 2e^-$$

The oxidation state of the chlorine changes from −1 to 0. By definition, any upward change in oxidation state is oxidation, and the electrode at which oxidation occurs is the anode.

The equation for the over-all reaction occurring in an electrolytic cell is obtained by adding the equations representing the reactions occurring at the individual electrodes. These equations may be called *electronic equations*, since the electrode reactions involve gain or loss of electrons. Before the electronic equations are combined, the number of electrons gained in the one equation must be made the same as the number lost in the other equation. Thus, in the sodium chloride example the cathode equation must be multiplied by 2:

$$\begin{array}{rr} \text{CATHODE:} & 2Na^+ + 2e^- \rightarrow 2Na \\ \text{ANODE:} & 2Cl^- \rightarrow Cl_2 + 2e^- \\ \hline \text{NET:} & 2Na^+ + 2Cl^- \rightarrow 2Na + Cl_2 \end{array}$$

14-3 Faraday's Law of Electrolysis

The idea that the charges of ions are related as small whole numbers is associated with the findings made by Michael Faraday in his investigations of electrolysis. Early in the nineteenth century it was found that the passage of electric current through aqueous solutions of acids, bases, or salts caused chemical changes to occur at the two electrodes. Electrolysis has already been defined in Section 14-1 as any chemical change brought about by electric current.

In 1832 Faraday began investigating the relationship of the *amount* of chemical change produced by an electric current to (1) the strength of the current (i.e., the *amperage*), (2) the length of time the current was passed through the electrolytic cell, and (3) the chemical identity of the electrolyte. He found that passage of current through a specified solution produces an amount of chemical change that is directly proportional to the amperage and to the time during which the current flows. For example, a 5-ampere current passing through a copper sulfate solution for 2 minutes will liberate the same amount of copper as a 1-ampere current in 10 minutes. A *coulomb* is the amount of electricity carried by a 1-ampere current in 1 second:

$$\text{Amperes} \times \text{seconds} = \text{coulombs}$$

Faraday found, then, that the amount of chemical change is directly proportional to the number of coulombs used.

Faraday investigated the relationship of the amount of electrolysis to the chemical identity of the electrolyte by passing the same current through each of several solutions contained in electrolytic cells connected in series, as shown in Figure 14-3. He found that the number of coulombs required for the liberation of 1 mole of silver atoms (107.870 g) from a silver salt will liberate only $\frac{1}{2}$ mole (31.77 g) of copper atoms from $CuCl_2$, $Cu(NO_3)_2$, $CuSO_4$, etc. This same number of coulombs (96,490) will liberate 1 mole of chlorine atoms (35.453 g) from NaCl, $CuCl_2$, etc., but only $\frac{1}{2}$ mole (16.032 g) of sulfur atoms from Na_2S, K_2S, etc. Faraday's results thus indicated that the quantity of charge carried by a chloride ion is the same as that carried by a silver ion, that a copper ion has twice the charge of a silver ion, a sulfide ion twice the charge of a chloride ion, etc.

The findings of Faraday led the Irish physicist G. Johnstone Stoney to propose the electron theory (1874). This theory assumed electricity to be "atomistic" in much the same manner as matter itself. In other words, Stoney assumed that there is a fundamental particle of negative electricity, for which

Figure 14-3

Electrolysis in cells in series

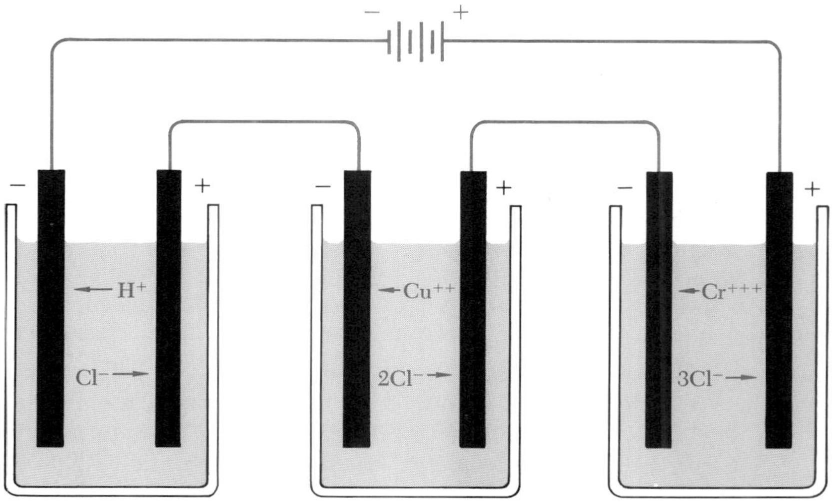

he proposed, in 1891, the name *electron*. Proof of the existence of the electron has since been obtained, as described in Chapter 3. We can now say that a silver ion, Ag^+, is a silver atom that has lost one electron; a chloride ion, Cl^-, is a chlorine atom that has gained one electron; a calcium ion, Ca^{++}, is a calcium atom that has lost two electrons; a sulfide ion, S^{--}, is a sulfur atom that has gained two electrons; etc. In electrolysis, a positive ion is converted to an atom by gaining the number of electrons required to neutralize its charge. For example,

$$\underset{\text{silver ion}}{Ag^+} + \underset{\text{one electron}}{e^-} \rightarrow \underset{\text{silver atom}}{Ag}$$

$$\underset{\text{copper ion}}{Cu^{++}} + \underset{\text{two electrons}}{2e^-} \rightarrow \underset{\text{copper atom}}{Cu}$$

The electrolytic liberation of atoms of a nonmetal at the positive electrode consists of the conversion of each negative ion to an atom by losing its "extra" electrons. For example,

$$2Cl^- \rightarrow Cl_2 + 2e^-$$
$$S^{--} \rightarrow S + 2e^-$$

Since 96,490 coulombs will liberate 1 mole of silver atoms, and since each silver atom is formed by the combination of one electron and one silver ion, it is apparent that 96,490 coulombs is the total charge of a mole of electrons. This quantity of electricity is now called a *faraday*.

Faraday's Law of Electrolysis may be stated in the following (modernized) manner: 1 faraday discharges 1 mole of singly charged ions, $\frac{1}{2}$ mole of doubly charged ions, etc. at each electrode. Thus, if 1 faraday is passed through the cells shown in Figure 14-3, there will be liberated 1 mole of H atoms, $\frac{1}{2}$ mole of Cu atoms, and $\frac{1}{3}$ mole of Cr atoms, respectively. One mole of Cl atoms will be liberated in each cell.

Knowledge of this law enables us to calculate the weights of the products obtained by electrolysis.

EXAMPLE How many moles of Cd^{++} ions will be converted to Cd atoms by 3 faradays?

Since 1 faraday will liberate 1 mole of singly charged ions but only $\frac{1}{2}$ mole of doubly charged ions, 3 faradays will convert 3 faradays $\times$ $\frac{1}{2}$ mole/faraday = 1.5 moles of Cd^{++} ions to Cd atoms.

EXAMPLE How many grams of silver will be plated out by 0.01 faraday?

One faraday will liberate 1 mole of singly charged atoms, such as Ag^+; 0.01 faraday will convert 0.01 mole of Ag^+ ions to Ag atoms. Since the weight of 1 mole of silver is 107.870 g, the weight of the silver is

$$0.01 \text{ mole} \times 107.870 \text{ g/mole} = 1.07870 \text{ g}$$

If the 0.01 mole is itself regarded as an experimentally determined quantity, it should be borne in mind that it has only one significant figure; in this case, the answer is best expressed as 1 g. If the question had stated that 0.010 faraday was used (note the two significant figures), the answer would best be expressed as 1.1 g; i.e., we would round off our answer to two digits. (See the appendix for a discussion of significant figures.)

EXAMPLE How many grams of copper metal will be plated out by passage of a 1.92-ampere current for 30.0 minutes through a $CuSO_4$ solution?

Convert time to seconds:

$$30.0 \text{ min} \times 60.0 \text{ sec/min} = 1800 \text{ sec} \quad \text{(3 significant figures)}$$

Calculate coulombs (multiply amperes by seconds):

$$1.92 \text{ amp} \times 1800 \text{ sec} = 3456 \text{ coulombs} \quad \text{(3 significant figures)}$$

Calculate faradays (1 faraday is 96,490 coulombs):

$$\frac{3456 \text{ coulombs}}{96{,}490 \text{ coulombs/faraday}} = 0.0358 \text{ faraday}$$

Calculate the weight liberated by 1 faraday:

Since Cu^{++} ion is doubly charged, 1 faraday will liberate exactly $\frac{1}{2}$ mole.

$$0.5 \text{ mole} \times 63.54 \text{ g/mole} = 31.77 \text{ g Cu}$$

Multiply faradays by grams per faraday:

$$0.0358 \text{ faraday} \times 31.77 \text{ g Cu/faraday} = 1.14 \text{ g Cu}$$

14-4 Avogadro's Number

The charge of the electron is 1.602×10^{-19} coulomb, and the value for the faraday is 96,490 coulombs. Now we can calculate Avogadro's Number:

$$N = \frac{96{,}490 \text{ coulombs/faraday}}{1.602 \times 10^{-19} \text{ coulomb/electron}}$$

$$= 6.023 \times 10^{23} \text{ electrons/faraday}$$

This is not only the number of electrons in a mole of electrons but also the number of molecules in a mole of molecules and the number of atoms in a mole of atoms. We can now define a mole as 6.023×10^{23} molecules, atoms, ions, electrons, or whatever chemical units are being considered.

14-5 Metathetical Reactions

Metathesis is the name given to the type of reaction in which two substances react to produce two new substances of the same class or classes. There are three types of metathetical reactions:

1. Two salts react, producing two salts. For example,

$$AgNO_3 + NaCl \rightarrow AgCl\downarrow + NaNO_3$$

   This reaction occurs when an aqueous solution of silver nitrate is mixed with an aqueous solution of sodium chloride. The reaction occurs because one of the products, silver chloride, is insoluble in water (the solvent); i.e., it forms a *precipitate*. This fact is indicated by a downturned arrow following the formula for silver chloride in the equation.

   In general, if neither of the possible products of the reaction between two salts is insoluble, there will be no reaction between the salts. For example,

$$NaCl + KNO_3 \rightarrow \text{no reaction}$$

   The sodium chloride solution contains Na^+ and Cl^- ions; the potassium nitrate solution contains K^+ and NO_3^- ions. When the two solutions are mixed, no reaction occurs because both of the possible products, potassium chloride (KCl) and sodium nitrate ($NaNO_3$), are soluble in water. Since no precipitate forms, the solution contains all four ions, K^+, Na^+, Cl^-, and NO_3^- (Figure 14-4).

2. A salt of a weak acid reacts with a strong acid to form a salt and a weak acid. For example,

$$NaC_2H_3O_2 + HCl \rightarrow NaCl + HC_2H_3O_2$$

Figure 14-4

Two salts that do not react

When NaCl and KNO_3 solutions are mixed, no reaction occurs.

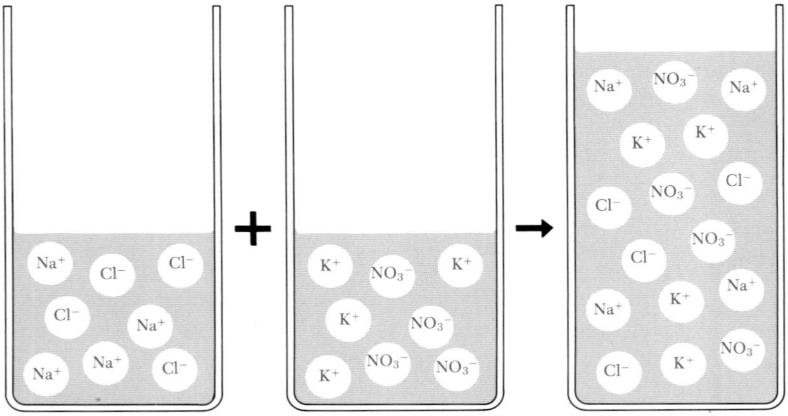

The sodium acetate solution contains Na^+ and $C_2H_3O_2^-$ ions, and the hydrochloric acid solution contains H^+ (actually, H_3O^+) and Cl^- ions. When the two solutions are mixed, the H^+ and $C_2H_3O_2^-$ ions combine to form $HC_2H_3O_2$ molecules. The reaction occurs because $HC_2H_3O_2$ is a weak acid; i.e., acetic acid molecules do not dissociate readily into ions.

Another example of this kind of metathetical reaction is the following:

$$CaCl_2 + H_2SO_4 \xrightarrow{\text{heat}} CaSO_4 + 2HCl\uparrow$$

When "concentrated" sulfuric acid, i.e., pure hydrogen sulfate, is added to solid calcium chloride and the mixture is heated, a reaction occurs. The upturned arrow indicates that hydrogen chloride is a gas, which escapes as quickly as it is formed. (HCl gas is highly soluble in water, but in this reaction water is absent.) The escape of the HCl gas permits the reaction to continue until one of the reactants is completely consumed.

A third example is the following:

$$Na_2CO_3 + 2HCl \rightarrow 2NaCl + H_2O + CO_2\uparrow$$

This reaction occurs when hydrochloric acid (the aqueous solution of hydrogen chloride) is added to solid Na_2CO_3 or to an aqueous solution of Na_2CO_3. Carbonate ions, CO_3^{--}, would be expected to combine with H^+ ions to form H_2CO_3 (carbonic acid) molecules. Perhaps they do; but carbonic acid is an unstable compound which cannot be isolated, and it decomposes as rapidly as it is formed:

$$H_2CO_3 \rightarrow H_2O + CO_2\uparrow$$

The escape of the CO_2 gas assures the continuation of the reaction.

3. A salt reacts with a base, producing a salt and a base. For example,

$$MgCl_2 + 2NaOH \rightarrow 2NaCl + Mg(OH)_2\downarrow$$

When aqueous solutions of sodium hydroxide and magnesium chloride mix, sodium chloride and magnesium hydroxide are formed. Magnesium hydroxide is of low solubility and removes itself from the solution.

A second example of this kind of metathetical reaction is

$$NH_4Cl + NaOH \rightarrow NaCl + NH_3\uparrow + H_2O$$

The combination of the NH_4^+ ions from ammonium chloride and the OH^- ions from sodium hydroxide produces an unstable base, ammonium hydroxide, which decomposes as fast as it forms:

$$NH_4OH \rightarrow NH_3\uparrow + H_2O$$

Since one of the products, NH_3, escapes as a gas, the reaction proceeds until it is complete.

14-6 Ionic Equations

Since the reactions described in the previous section consist primarily of the combination of ions to form the products, the reactions may be represented by equations in which only the participating ions are shown. For example, the metathetical reaction between silver nitrate and sodium chloride is actually a reaction in which only Ag^+ and Cl^- ions take part; the Na^+ and NO_3^- ions are *spectator ions* (Figure 14-5):

Figure 14-5

Spectator ions

In the reaction between NaCl and $AgNO_3$, Na^+ and NO_3^- are spectator ions.

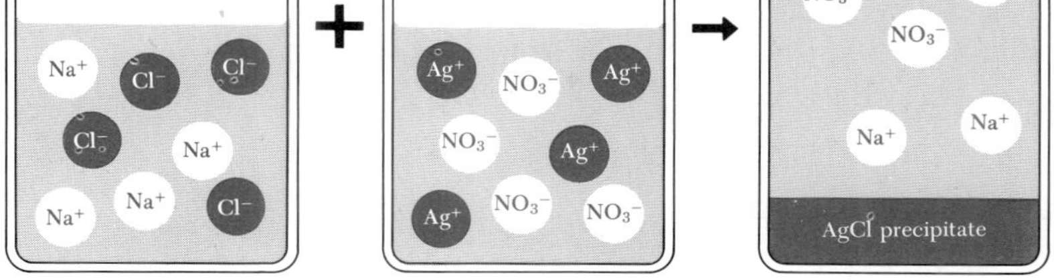

$$Ag^+ + NO_3^- + Na^+ + Cl^- \rightarrow AgCl\downarrow + Na^+ + NO_3^-$$

Since the Na^+ and NO_3^- ions do not take part in the reaction they may be omitted from the equation. Thus, instead of writing the *complete ionic equation*, we may write the *net ionic equation* as follows:

$$Ag^+ + Cl^- \rightarrow AgCl\downarrow$$

This equation represents the reaction that occurs when *any* silver salt solution is mixed with a solution of *any* ionic chloride, such as KCl, $CaCl_2$, or $MgCl_2$. We can also include HCl in the list; although HCl is a covalent substance, its aqueous solution contains a high concentration of Cl^- ions.

The reaction that occurs when sodium acetate and hydrochloric acid solutions are mixed occurs whenever any ionic acetate is mixed with any strong acid. This reaction can be represented by the net or general ionic equation

$$H^+ + C_2H_3O_2^- \rightleftharpoons HC_2H_3O_2$$

In the specific case of sodium acetate and hydrochloric acid solutions, the complete ionic equation is

$$H^+ + Cl^- + Na^+ + C_2H_3O_2^- \rightleftharpoons HC_2H_3O_2 + Na^+ + Cl^-$$

The reaction between sodium carbonate and hydrochloric acid can be represented by the complete ionic equation

$$2Na^+ + CO_3^{--} + 2(H^+ + Cl^-) \rightarrow 2(Na^+ + Cl^-) + H_2O + CO_2\uparrow$$

It can also be represented by the net or general ionic equation

$$2H^+ + CO_3^{--} \rightarrow H_2CO_3 \rightarrow H_2O + CO_2\uparrow$$

The same result would be obtained by allowing, say, potassium carbonate or calcium carbonate to react with hydrochloric acid or any other strong acid.

The precipitation of magnesium hydroxide, which occurs whenever any magnesium salt and any strong base are brought together, is represented by the general equation

$$Mg^{++} + 2OH^- \rightarrow Mg(OH)_2\downarrow$$

This represents what happens—*all* that happens—when sodium hydroxide solution and magnesium chloride solution are mixed. It also represents what happens when *potassium* hydroxide and magnesium *sulfate* are brought together. If we consider only sodium hydroxide, potassium hydroxide, magnesium nitrate, and magnesium sulfate, we see that either hydroxide may react with either magnesium salt, but the result is the same in each case:

$$2(Na^+ + OH^-) + (Mg^{++} + 2NO_3^-) \rightarrow Mg(OH)_2\downarrow + 2(Na^+ + NO_3^-)$$
$$2(K^+ + OH^-) + (Mg^{++} + 2NO_3^-) \rightarrow Mg(OH)_2\downarrow + 2(K^+ + NO_3^-)$$
$$2(Na^+ + OH^-) + (Mg^{++} + SO_4^{--}) \rightarrow Mg(OH)_2\downarrow + (2Na^+ + SO_4^{--})$$
$$2(K^+ + OH^-) + (Mg^{++} + SO_4^{--}) \rightarrow Mg(OH)_2\downarrow + (2K^+ + SO_4^{--})$$

Since the result is the same in each case, it saves space and mental effort if we simply write the one equation

$$Mg^{++} + 2OH^- \rightarrow Mg(OH)_2\downarrow$$

When magnesium nitrate and sodium hydroxide are used, the NO_3^- and Na^+ ions do not take part in the reaction. They are therefore spectator ions. Similarly, in the last example above, the SO_4^{--} and K^+ ions are spectator ions. Sodium nitrate, potassium sulfate, etc. are products of the above reactions only in the sense that if the solution is separated from the $Mg(OH)_2$ precipitate and the water evaporated, crystals of $NaNO_3$, K_2SO_4, etc. will be obtained. At the time the starting solutions are mixed, however, the only chemical event is the precipitation of magnesium hydroxide.

14-7 Solubilities of Common Ionic Solids

From the discussion in the two preceding sections it is apparent that the solubility or insolubility of an ionic solid in water is one of its most important properties. Chemical analysis is based to a considerable degree on such differences. The following are general statements regarding the solubilities of some of the most common of these compounds. A knowledge of these statements will be of value to anyone contemplating the further study of chemistry.

1. The following types of ionic compounds are generally soluble in water:
   a. all acetates, e.g., $NaC_2H_3O_2$, $Ca(C_2H_3O_2)_2$
   b. all nitrates, e.g., $AgNO_3$, $Pb(NO_3)_2$
   c. all chlorides, bromides, and iodides (except those of silver, mercury(I), and lead), e.g., NaCl, $CuBr_2$, CaI_2
   d. all sulfates (except those of barium, strontium, and lead), e.g., $MgSO_4$, $CuSO_4$, $Al_2(SO_4)_3$
   e. all alkali metal (Group IA) hydroxides (and barium hydroxide), e.g., NaOH, KOH
   f. all common salts of sodium, potassium, and ammonium, e.g., Na_2CO_3, K_3PO_4, $(NH_4)_2S$, $NaHCO_3$, KI
   g. all alkaline-earth metal (Group IIA) hydrogen carbonates ("bicarbonates"), e.g., $Ca(HCO_3)_2$, $Ba(HCO_3)_2$
2. The following inorganic ionic solids are insoluble or only very slightly soluble in water:
   a. all metal hydroxides (except those of the alkali metals and ammonium), e.g., $Mg(OH)_2$, $Fe(OH)_3$
   b. all carbonates (except those of the alkali metals and ammonium), e.g., $CaCO_3$, $PbCO_3$
   c. all phosphates (except those of the alkali metals and ammonium), e.g., $Ca_3(PO_4)_2$, $AlPO_4$
   d. all sulfides (except those of the alkali metals and ammonium), e.g., CuS, Bi_2S_3

New Terms

Anion: a negative ion.

Anode: the positive electrode in an electrolytic cell.

Cathode: the negative electrode in an electrolytic cell.

Cation: a positive ion.

Complete (total) ionic equation: an equation, representing a reaction involving ionic compounds, in which all the individual ions, including spectator ions, are represented.

Electrode: one of the terminals at which an electric current enters or leaves an electrolytic solution.

Electrolysis: chemical change brought about by electric current.

Electronic equation: an equation of the type used to represent the reactions occurring at the individual electrodes during electrolysis, in which electrons are gained (at the cathode) or lost (at the anode); e.g., $Ag^+ + e^- \rightarrow Ag$.

Faraday: the quantity of electricity equal to the total charge of a mole of monopositive or mononegative ions. It therefore consists of a mole of electrons. Its numerical value, in electrical units, is 96,490 coulombs.

Faraday's Law of Electrolysis: 1 faraday discharges 1 mole of singly charged ions, $\frac{1}{2}$ mole of doubly charged ions, etc. at each electrode in electrolytic cells.

General ionic equation: same as net ionic equation; so called because it represents all reactions of a given type.

Metathesis: a reaction in which two substances react to produce two new substances of the same class or classes.

Net ionic equation: an equation, representing a reaction involving ionic compounds, from which spectator ions are omitted.

Spectator ions: the ions which are "partners" of the ions participating in a reaction but which do not take an actual part in the reaction themselves.

Exercises

14-1 What is the total charge, in coulombs, of the following:

(a) 3 moles of electrons

(b) 2 moles of Cl^- ions

(c) 4 moles of Ca^{++} ions

(d) all the ions in 5 moles of $Al_2(SO_4)_3$

14-2 How long would it take a 1.64-ampere current to plate out 12.35 g of cadmium from $Cd(NO_3)_2$ solution?

14-3 How long must a current of 0.012 ampere pass through an aqueous NaCl solution in order to liberate 0.018 mole of Cl_2 at the anode?

14-4 A current of 2.40 amperes is passed for 24.00 hours through a cell containing a solution of zinc sulfate, $ZnSO_4$. What weight of zinc is deposited?

14-5 If a current deposits 1.75 g of copper on the cathode when passed through a copper(II) sulfate solution for 7.25 hours, what is the amperage? (Assume a steady current.)

14-6 A steady current passed through a solution of copper(II) chloride deposits 0.456 g of copper in 18.6 min. (a) What volume of chlorine gas is evolved? (Assume none of the chlorine dissolves.) (b) What weight of copper would be deposited in 1.00 hour?

14-7 *Metathetical reactions.* Complete and balance as molecular equations, as complete ionic equations, and as net ionic equations:

(a) $NH_4Cl + Ca(OH)_2 \rightarrow$
(b) $NaNO_3 + H_2SO_4 \rightarrow$
(c) $AgNO_3 + NaCl \rightarrow$
(d) $CaCl_2 + H_2SO_4 \rightarrow$
(e) $(NH_4)_2CO_3 + Sr(NO_3)_2 \rightarrow$
(f) $Ag_2SO_4 + NaI \rightarrow$
(g) $H_2S + CuSO_4 \rightarrow$
(h) $NaOH + FeCl_3 \rightarrow$
(i) $H_3PO_4 + FeCl_3 \rightarrow$
(j) $CuSO_4 + Ca(OH)_2 \rightarrow$

14-8 *Ionic equations.* Complete and balance the following equations, assuming that the product is a (neutral) insoluble solid or a (neutral) molecule of a covalent compound.

(a) $H^+ + OH^- \rightarrow$
(b) $H_3O^+ + OH^- \rightarrow$
(c) $Ag^+ + Br^- \rightarrow$
(d) $Cu^{++} + S^{--} \rightarrow$
(e) $Fe^{+++} + OH^- \rightarrow$
(f) $Ca^{++} + AsO_4^{---} \rightarrow$
(g) $Mg^{++} + SiO_3^{--} \rightarrow$
(h) $Ca^{++} + F^- \rightarrow$
(i) $Mn^{++} + OH^- \rightarrow$
(j) $Ag^+ + S^{--} \rightarrow$
(k) $H^+ + C_2O_4^{--} \rightarrow$
(l) $H^+ + BO_3^{---} \rightarrow$
(m) $H_3O^+ + NO_2^- \rightarrow$
(n) $NH_4^+ + OH^- \rightarrow$

References

Glasstone, S., *Sourcebook on Atomic Energy*, 3rd ed., Van Nostrand, Princeton, N.J., 1967. Chapter 2 includes accounts of the contributions of Faraday and Stoney.

Kieffer, W. F., *The Mole Concept in Chemistry*, Reinhold, New York, 1962 (paperbound). In Chapter 7, "Electrochemistry," the author discusses electrolysis in terms of the faraday as a mole of electrons, avoiding use of "equivalents."

Shen, S. Y., "What is the Real Value of a Faraday?" *Chemistry*, **39** (2), 8 (1966). An explanation of the discrepancies in reported values for the faraday. The best value now available is concluded to be 96,490 ± 2 coulombs.

15

Chemical Equilibrium

15-1 Reversible Reactions

Some chemical reactions are the exact opposite of others. When steam is passed over hot iron, the reaction that occurs is represented by the equation

$$3Fe + 4H_2O \rightarrow Fe_3O_4 + 4H_2$$

If the situation is reversed and hydrogen is passed over hot iron oxide, the reaction is represented by the equation

$$Fe_3O_4 + 4H_2 \rightarrow 3Fe + 4H_2O$$

Since the second reaction is the exact opposite of the first one, both can be represented by one equation:

$$3Fe + 4H_2O \rightleftharpoons Fe_3O_4 + 4H_2$$

Here the double arrows indicate, first, that the left-to-right and right-to-left reactions can both occur and, second, that both reactions can occur simultaneously if suitable conditions prevail. Arbitrarily, the substances appearing on the left are referred to as *reactants*, and those on the right are called the *products* of the reaction.

If the equation is written in the form

$$Fe_3O_4 + 4H_2 \rightleftharpoons 3Fe + 4H_2O$$

the substances appearing on the left *as now written* are considered to be the reactants, and those on the right are the products. Ordinarily, the equation is written so that the starting materials appear on the left.

15-2 The Point of Equilibrium

If hydrogen and iodine are present together in equal molar quantities in a closed container at 600°C, both elements are in the gaseous state, and a reaction occurs which is represented by the equation

$$H_2 + I_2 \rightarrow 2HI$$

The reaction continues until 80 percent of the hydrogen and iodine are converted to hydrogen iodide gas. Conversely, if pure hydrogen iodide is heated in a closed container at 600°C, a reaction occurs which is represented by the equation

$$2HI \rightarrow H_2 + I_2$$

and it continues until 20 percent of the hydrogen iodide has decomposed. Both reactions may be represented by a single equation with the use of double arrows:

$$H_2 + I_2 \rightleftharpoons 2HI$$

With the equation written in this manner, the formation of hydrogen iodide is referred to as the *forward reaction*, and the decomposition of hydrogen iodide is designated the *reverse reaction*.

Either the forward reaction or the reverse reaction will approach a state containing the same proportions of hydrogen, iodine, and hydrogen iodide. Once this state has been reached, and so long as the experimental conditions are not changed, the proportions of the components in the mixture will remain constant. This situation is referred to as a state of *chemical equilibrium*; the mixture is an *equilibrium mixture*; each component of the equilibrium mixture is an *equilibrant*.

The state of equilibrium can be approached from either direction: by starting with hydrogen iodide alone and allowing it to decompose until decomposition has apparently ceased or by starting with hydrogen and iodine and allowing them to combine until combination has apparently ceased. In either case, the same equilibrium mixture ultimately results; i.e., the *point of equilibrium* is reached.

When chemical equilibrium is attained, the forward and reverse reactions proceed at the same rate. Deuterium has been used to prove that the forward and reverse reactions occur even after the point of equilibrium has been reached. If a small quantity of deuterium gas, D_2, is added to an equilibrium mixture of hydrogen, iodine, and hydrogen iodide, we find almost immediately that the mixture now contains all of the following chemical species: H_2, I_2, HI, D_2, DI, and HD.

The relative amounts of the various equilibrants in an equilibrium mixture depend on the reaction and the experimental conditions. For example, hydrogen iodide is 20 percent dissociated at the point of equilibrium when the temperature is 600°C. If the temperature is raised, the system will adjust to the change by establishing a new point of equilibrium at which the relative amounts of reactants and products will be different. Another reaction, of course, would have its own characteristic point of equilibrium under any given set of conditions. Some reactions are virtually complete, in one direction or the other, at the point of equilibrium, but other reactions attain equilibrium when they have reached some intermediate degree of completion.

15-3 The Law of Chemical Equilibrium

Generally, the point of equilibrium of a mixture cannot satisfactorily be described in terms of the amounts of the equilibrants but must be stated in terms of their concentrations. The rates at which chemical reactions take place frequently depend upon the concentrations of the reacting substances; for example, substances burn more rapidly in pure oxygen than in air. Since the forward and reverse reactions occur at the same rate at the point of equilibrium, it is not surprising that an equilibrium can best be described in terms of the concentrations of the components of the equilibrium mixture. In describing the composition of equilibrium mixtures, it is common practice to express the concentration of each equilibrant in moles per liter.

The point of equilibrium depends on the tendency of the forward reaction to take place rapidly or slowly in comparison with the reverse reaction. If the forward reaction is by nature a faster reaction than its reverse, it will proceed beyond the halfway point of completion before the concentrations of the products become sufficient to make the reverse reaction proceed with equal speed. During the approach to equilibrium, the rate of the forward reaction steadily decreases as the concentrations of the reactants diminish, and by the same token the rate of the reverse reaction steadily increases as the concentrations of the products continue to build up. As soon as the rate of the reverse reaction equals that of the forward reaction, equilibrium exists. After the equilibrium state has been reached, there will be no further changes in the concentrations of the equilibrants unless the equilibrium is disturbed by some change in conditions such as warming, cooling, the application of pressure, or the introduction of an additional quantity of one of the equilibrants from outside.

The relationship that prevails among the concentrations of the equilibrants in an equilibrium mixture was formulated in 1864 by Cato Guldberg, a Norwegian mathematician, and Peter Waage, a Norwegian chemist. We shall call this relationship the *Law of Chemical Equilibrium*, although Guldberg and Waage called it the Principle of Mass Action (they used the term "active mass" to mean what we mean by "concentration").

All reversible chemical reactions may be represented at once by the generalized equation

$$aA + bB \rightleftharpoons cC + dD$$

where a is the coefficient for reactant A, and so on. The Law of Chemical Equilibrium states that

$$\frac{[C]^c[D]^d}{[A]^a[B]^b} = K$$

where K represents an *equilibrium constant* that is characteristic of a particular reaction at a fixed temperature. Each capital letter in brackets represents the equilibrium concentration of the particular equilibrant.

The expression defining the equilibrium constant contains the concentrations of the products in the numerator and the concentrations of the reactants in the denominator. The small letters which were used to represent coefficients in the chemical equation appear as exponents in the equilibrium constant expression; the equilibrium concentration of each equilibrant must be raised to the power indicated by its exponent to satisfy the definition for the equilibrium constant.

The meaning of the Law of Chemical Equilibrium is that addition or removal of a quantity of one of the equilibrants (i.e., a change in its concentration) will be followed by a change in the concentration of each of the others such that the ratio of numerator to denominator in the equilibrium constant expression will be restored to the numerical value of K.

The size of the equilibrium constant of a reaction is large if a large fraction of the reactants is converted into products during the approach to equilibrium. A very small equilibrium constant characterizes a reaction which reaches equilibrium as soon as a small fraction of the starting materials has been converted into reaction products. Whether the equilibrium constant of a reaction is large or small, the reaction can be made more complete than otherwise by adding more of one of the starting materials.

Suppose that the starting materials for a reversible reaction are two substances, A (very scarce and expensive) and B (cheap and easily obtained). To make sure that A is not wasted, it is desirable to convert a large fraction of A into products. Consideration of the Law of Chemical Equilibrium indicates that establishment of a high concentration of B will result in a low concentration of A in the equilibrium mixture. Although we cannot force a reaction any closer to completion when it has already reached equilibrium, we can convert a larger fraction of one reactant by using an excess of the other reactant. Thus, when a reversible reaction is made more complete, it is often with respect to only one specified reactant, sometimes at the expense of the other reactant.

If we know the numerical value of the equilibrium constant for a reaction, we can calculate the equilibrium concentration of any one of the equilibrants when the concentrations of the other components of an equilibrium mixture are given.

15-4 Numerical Value of the Equilibrium Constant

The numerical value of K is calculated by substituting the equilibrium concentration of each equilibrant (obtained by chemical analysis) into the equilibrium constant expression.

EXAMPLE At 458°C an equilibrium mixture contains 5.617×10^{-3} mole per liter of hydrogen, 5.936×10^{-4} mole per liter of iodine, and 1.270×10^{-2} mole per liter of hydrogen iodide. Calculate the equilibrium constant for the reaction

$$H_2 + I_2 \rightleftharpoons 2HI$$

(Note the statement that the mixture is an *equilibrium* mixture.)

$$K = \frac{[HI]^2}{[H_2][I_2]}$$

$$= \frac{(1.270 \times 10^{-2} \text{ mole/liter})^2}{(5.617 \times 10^{-3} \text{ mole/liter})(5.936 \times 10^{-4} \text{ mole/liter})}$$

$$= 48.4$$

In Table 15-1 is presented a set of data listing the equilibrium concentration of each component in five different equilibrium mixtures of hydrogen, iodine, and hydrogen iodide at 458°C. A value for the equilibrium constant, calculated from each set of data, is listed in the last column of the table. Although this set of data was obtained under carefully controlled conditions, there is some variation in the calculated K values. This is due partly to experimental error and partly to inherent limitations in the simplified expression which states the Law of Chemical Equilibrium. The expression which we use to calculate K is for an ideal case, much as Boyle's Law and Charles' Law are for ideal gases.

15-5 Units of Equilibrium Constants

For the reversible reaction indicated by the chemical equation

$$CO_2 + H_2 \rightleftharpoons H_2O + CO$$

the equilibrium constant expression is

Table 15-1

Data on equilibrium mixtures of H_2, I_2, and HI at 458°C

Mixture	$[H_2]$ (moles/liter)	$[I_2]$ (moles/liter)	[HI] (moles/liter)	$K = \frac{[HI]^2}{[H_2][I_2]}$
1	5.617×10^{-3}	5.936×10^{-4}	1.270×10^{-2}	48.4
2	4.580×10^{-3}	9.733×10^{-4}	1.486×10^{-2}	49.5
3	4.567×10^{-3}	1.058×10^{-3}	1.544×10^{-2}	49.3
4	3.842×10^{-3}	1.524×10^{-3}	1.687×10^{-2}	48.6
5	1.696×10^{-3}	1.696×10^{-3}	1.181×10^{-2}	48.5
			AVERAGE =	48.9

$$K = \frac{[H_2O][CO]}{[CO_2][H_2]}$$

When the appropriate values are substituted into the above expression, the calculated value for K is independent of the units in which the concentrations are expressed.

EXAMPLE At 986°C an equilibrium mixture was found to have the following composition: H_2, 0.400 mole per liter; CO_2, 0.400 mole per liter; H_2O, 0.506 mole per liter; CO, 0.506 mole per liter. Calculate the equilibrium constant for the reaction.

$$K = \frac{(0.506\ \text{mole/liter})(0.506\ \text{mole/liter})}{(0.400\ \text{mole/liter})(0.400\ \text{mole/liter})}$$

$$= 1.60$$

If this same reaction is represented by the equation

$$H_2O + CO \rightleftharpoons CO_2 + H_2$$

then the correct expression for the equilibrium constant is

$$K = \frac{[CO_2][H_2]}{[H_2O][CO]}$$

which is, of course, the reciprocal of the previous equilibrium constant expression. Substitution of the same set of equilibrium concentrations into this expression gives, therefore, a calculated value for K (again, independent of concentration units) which is the reciprocal of 1.60:

$$K = \frac{(0.400\ M)(0.400\ M)}{(0.506\ M)(0.506\ M)}$$

$$= 0.625$$

For the reversible reaction

$$PCl_5(g) \rightleftharpoons PCl_3(g) + Cl_2(g)$$

the equilibrium constant expression is

$$K = \frac{[PCl_3][Cl_2]}{[PCl_5]}$$

Since this expression contains two concentrations in the numerator and only one in the denominator, the calculated value for K depends on the concentration units employed. It is customary to express equilibrium concentrations in moles per liter.

EXAMPLE At a certain temperature, 1 liter of an equilibrium mixture contains 0.035 mole of PCl_3, 0.147 mole of Cl_2, and 0.0083 mole of PCl_5, all gases. Calculate the equilibrium constant for the reaction

$$PCl_5(g) \rightleftharpoons PCl_3(g) + Cl_2(g)$$

$$K = \frac{(0.035 \text{ mole/liter})(0.147 \text{ mole/liter})}{(0.0083 \text{ mole/liter})}$$
$$= 0.62 \text{ mole/liter}$$

The practice of expressing equilibrium concentrations in moles per liter is so well established that K values such as the one just calculated are usually written without indicating the units. Nevertheless, it is well to be aware of the part played by the choice of concentration units in determining the numerical value of the calculated K.

If this reaction is represented by the equation

$$PCl_3(g) + Cl_2(g) \rightleftharpoons PCl_5(g)$$

the equilibrium constant expression is

$$K = \frac{[PCl_5]}{[PCl_3][Cl_2]}$$

Both the numerical value for K and the units which apply to it are the reciprocals of those which apply above (assuming the same concentration units are used):

$$K = \frac{(0.0083 \text{ mole/liter})}{(0.035 \text{ mole/liter})(0.147 \text{ mole/liter})}$$
$$= 1.6 \text{ liters/mole}$$

Note that one of the "mole/liter" expressions in the denominator cancels the "mole/liter" in the numerator. The other "mole/liter" in the denominator becomes "liters/mole" in the numerator. This is in accord with the nature of complex fractions. For example:

$$\frac{\frac{1}{2}}{\frac{1}{2} \times \frac{4}{3}} = \frac{3}{4}$$

To divide by a fraction you invert and multiply.

For the reaction

$$N_2(g) + 3H_2(g) \rightleftharpoons 2NH_3(g)$$

the equilibrium constant expression is

$$K = \frac{[NH_3]^2}{[N_2][H_2]^3}$$

An equilibrium mixture at 400°C contains 0.60 mole of N_2, 0.40 mole of H_2, and 0.14 mole of NH_3 in a volume of 1 liter. Therefore,

$$K = \frac{(0.14 \text{ mole/liter})^2}{(0.60 \text{ mole/liter})(0.40 \text{ mole/liter})^3}$$

$$= \frac{(0.0196 \text{ mole}^2/\text{liter}^2)}{(0.60 \text{ mole/liter})(0.064 \text{ mole}^3/\text{liter}^3)}$$

$$= \frac{(0.0196 \text{ mole}^2/\text{liter}^2)}{(0.0384 \text{ mole}^4/\text{liter}^4)}$$

$$= 0.51 \text{ liter}^2/\text{mole}^2$$

If the chemical equation is reversed, the equilibrium constant expression is inverted, and the calculated K is the reciprocal, i.e., 1.96 $\text{moles}^2/\text{liter}^2$.

15-6 Effect of Temperature on an Equilibrium

At a particular temperature the value of the equilibrium constant for a reaction is characteristic of the reaction at that temperature. Virtually all reactions have different equilibrium constants at different temperatures. In 1884 the Dutch chemist Jacobus van't Hoff concluded that the equilibrium constant for an endothermic reaction increases with rising temperature and for an exothermic reaction it decreases with rising temperature. This relationship is known as *van't Hoff's Law*. Bearing in mind that the reverse reaction is endothermic if the forward reaction is exothermic, we can predict with confidence whether a change in temperature will shift an equilibrium "toward the right" or "toward the left," i.e., whether it will be the forward reaction or the reverse reaction that is made more complete by the change in temperature.

The only kind of reaction whose equilibrium constant would not be affected by a change of temperature would be one for which the heat of reaction is zero. Strictly speaking, no such reaction is known. Thus, all equilibria are shifted somewhat by any change in temperature. The extent to which an equilibrium is shifted, and the equilibrium constant changed, by a specified change in temperature is related to the magnitude of the heat of reaction.

15-7 Shifting the Point of Equilibrium

If a reversible reaction has reached a state of chemical equilibrium and the conditions are then altered in some way, an automatic adjustment in the equilibrium will occur and a new point of equilibrium will be established. The concentration of each equilibrant will change to a new value which will remain constant as long as the conditions are not altered again. The change in the composition of the equilibrium mixture which occurs each time one of the conditions is changed is referred to as a *shift* in the equilibrium. If the adjustment in the equilibrium occurs so as to produce additional amounts of products, i.e., substances appearing on the right side of the chemical equation as it is written, we say that the equilibrium has shifted to the right. If the adjustment results in the formation of additional amounts of reactants, i.e., substances appearing on the left side of the chemical equation as it is written, we say the equilibrium has shifted to the left.

The following changes in conditions will result in a shift of equilibrium:

1. The concentration of one of the equilibrants can be changed by addition or removal.
2. The volume of the system can be reduced or increased; i.e., additional pressure can be applied or the applied pressure can be reduced.
3. The temperature of the equilibrium mixture can be raised or lowered.

The effect of any one of these changes upon the equilibrium can be predicted by applying the principles which have already been presented in this chapter.

Suppose we consider again the reversible reaction

$$H_2(g) + CO_2(g) \rightleftharpoons H_2O(g) + CO(g)$$

in a state of equilibrium. If to the equilibrium mixture we add more H_2 without changing the volume of the container which the mixture occupies, we are establishing a higher concentration of H_2. Considering the expression for the equilibrium constant for this reaction, we see that the numerical value of the denominator has been increased. Automatically, the other concentrations change so that the ratio of numerator to denominator is restored to its previous value, i.e., the value of K. This reestablishment of equilibrium entails shifting the equilibrium to the right with a resultant reduction of CO_2 concentration and a corresponding increase in the concentrations of H_2O and CO. In general, an increase in the concentration of one of the *reactants* will shift an equilibrium to the *right*; an increase in the concentration of one of the *products* will shift an equilibrium to the *left*.

If we were to bring into contact with this equilibrium mixture a hygroscopic material, the concentration of the gaseous H_2O in the mixture would be reduced. Considering again the expression for the equilibrium constant for this reaction, we see that this time we have *reduced* the numerical value of the numerator. Equilibrium cannot be maintained unless the concentrations of the other equilibrants change so as to preserve the ratio of numerator to denominator. In order to maintain equilibrium in the above situation, the equilibrium must shift to the right (resulting in the increase of CO concentration and a corresponding decrease in the concentrations of H_2 and CO_2). In general, a decrease in the concentration of one of the products will shift an equilibrium to the right; a decrease in the concentration of one of the reactants will shift an equilibrium to the left.

If the volume of the container of the equilibrium mixture is reduced, subjecting the mixture to increased pressure, the concentration of each equilibrant is increased in the same proportion. Whether this has any effect on the point of equilibrium depends on the type of equilibrium constant expression that applies to the reaction. An increase in pressure has no effect on an equilibrium if the equilibrium constant is a unitless number. More simply stated, the sum of the coefficients of the gaseous products in the chemical equation must be equal to the sum of the coefficients of the gaseous reactants if the application of pressure is to be without effect on the equilibrium.

Suppose the reaction in question is the one represented by the equation

$$PCl_5(g) \rightleftharpoons PCl_3(g) + Cl_2(g)$$

Here the sum of the coefficients of the products is 2, but for the reactant it is only 1. Doubling the pressure doubles the concentration of each equilibrant. Consider the equilibrium constant expression

$$K = \frac{[PCl_3][Cl_2]}{[PCl_5]}$$

If each concentration were doubled, the ratio of numerator to denominator would also double, and the value of the expression would no longer be that characteristic of this particular reaction. Instead, the equilibrium shifts to the left so that the concentrations of the products decrease and the concentration of the reactant correspondingly increases. In general, the effect of reducing the volume of an equilibrium system involving gases, i.e., increasing the applied pressure, is to shift the equilibrium *in the direction of fewer molecules* of gaseous substances. Conversely, reducing the pressure shifts the equilibrium in the direction of more molecules.

If the temperature of an equilibrium mixture is changed, the equilibrium will shift either toward the right or toward the left. If the forward reaction is endothermic, an increase in temperature will shift the equilibrium to the right. This was discussed in Section 15-6.

15-8 The Principle of Le Chatelier

The effects of changes in concentration, pressure, and temperature upon equilibrium were summarized in a single principle of broad application by the French chemist Henri Le Chatelier in 1888. The Principle of Le Chatelier (also called the Principle of Mobile Equilibrium) can be applied to all types of dynamic equilibria, chemical and physical.

Once a system has reached an equilibrium state, it *never* changes from that state spontaneously. The only changes which any system undergoes spontaneously are those in the direction toward equilibrium, and such a reaction will proceed only far enough to attain the equilibrium state. As we have already seen, this equilibrium state may be approached from either direction.

After a system has reached an equilibrium state, it will remain indefinitely in this state until conditions are changed. *Le Chatelier's Principle* states that the change which takes place in an equilibrium system when conditions are altered is a change that tends to restore the original conditions.

Van't Hoff's Law is only a special case of the Principle of Le Chatelier. An increase in temperature favors the endothermic reaction—the reaction whose occurrence tends to prevent an increase in temperature.

An increase in the pressure applied to an equilibrium mixture favors the reaction that results in a decrease in pressure. For example, in an equilibrium mixture of gases, high pressure favors the change which reduces the number of molecules present in the system.

Again, in accord with Le Chatelier's Principle, the effect of increasing the concentration of one of the equilibrants is to shift the equilibrium in the direction that reduces the concentration of that particular substance.

The Principle of Le Chatelier can be applied only to systems that are already in a state of equilibrium when the change in conditions occurs. A mixture of H_2 and O_2 gases can be preserved for an indefinite length of time with no apparent reaction at room temperature. If the temperature is increased enough, a sudden reaction of explosive violence occurs. This reaction is exothermic. It appears that here an increase in temperature has favored an exothermic reaction. However, this situation does not violate Le Chatelier's Principle, for a mixture containing only H_2 and O_2 is not in a state of equilibrium—true chemical equilibrium does not exist unless all reactants and products are present, with the opposing reactions occurring at equal rates. Le Chatelier's Principle affords no basis for predicting what will happen if conditions are changed in a system that is not already in equilibrium.

When an appropriate catalyst is added to a reaction mixture, the speed of the reaction is increased. If the reaction is reversible, the catalyst increases the rates of both the forward and reverse reactions to the same extent. Therefore, since a reversible reaction can proceed only as far as the point of equilibrium, a catalyst serves only to reduce the time required for equilibrium to be reached. This is not the same as shifting the equilibrium. If a catalyst is added to a system that has already attained equilibrium, there are no changes in the concentrations of any of the equilibrants.

The Principle of Le Chatelier also applies to equilibria that involve no chemical reactions. For example, if a liquid and its vapor are in equilibrium within a closed container, we can predict correctly that an increase in temperature will shift the equilibrium in the direction of more vapor and less liquid because the conversion of liquid to vapor is endothermic.

Since ice has a smaller density than water, the application of pressure to ice at 0°C will cause the ice to melt. The system responds to increased pressure by forming more of the material that has less volume; i.e., the higher pressure favors the process that tends to prevent an increase in pressure.

15-9 An Application of the Principle of Le Chatelier: The Haber Process

Ammonia is an economically important substance because of its many useful properties. Since elementary nitrogen is abundant in the atmosphere and elementary hydrogen can be manufactured at small cost, it has seemed for many decades that it might be economically feasible to synthesize ammonia from its elements. When we examine the equation

$$N_2 + 3H_2 \rightleftharpoons 2NH_3 \qquad \Delta H = -22 \text{ kcal}$$

we note that the reaction is reversible, that the forward reaction is the exothermic one, and that the product consists of half as many molecules as the reactants. One more fact required to complete the picture is that the reaction is by nature a slow one.

According to the Principle of Le Chatelier, the formation of ammonia, i.e., shifting the equilibrium toward the right, is favored by the use of high pressure. On the other hand, the use of high temperature shifts the equilibrium to the left. Finally, of course, the continuous removal of ammonia, as it is formed, favors the forward reaction.

In practice, low temperatures should be avoided because the rate of any reaction is reduced by lowering the temperature, and this reaction is normally slow. However, since high temperatures have an unfavorable effect on the equilibrium, a compromise is necessary, and the temperature used is about 500°C. Fortunately, the use of high pressure not only shifts the equilibrium toward the right but increases the concentration of each reactant, thus increasing the rate at which the reaction proceeds. Even with the use of the highest feasible pressures and the optimum temperature, the reaction is still too slow to be practical unless an effective catalyst is used. Fritz Haber, a German chemist, received the Nobel prize in chemistry in 1918 for developing this process for the synthesis of ammonia. The *Haber process* is one of the most successful methods for the "artificial fixation" of atmospheric nitrogen, i.e., the conversion of atmospheric nitrogen into useful compounds.

15-10 Chemical Equilibrium Involving Solids and Gases

The equilibrium involving copper oxide, copper metal, hydrogen gas, and gaseous water,

$$CuO(s) + H_2(g) \rightleftharpoons Cu(s) + H_2O(g)$$

involves two phases, the gas phase and the solid phase. As such, it is an example of *heterogeneous equilibrium*. In setting up equilibrium constant expressions for heterogeneous equilibria, we must remember that the concentrations apply to the phases specified in the chemical equation. For the above system, the relationship

$$\frac{[Cu(s)][H_2O(g)]}{[CuO(s)][H_2(g)]} = \text{constant}$$

correctly describes the equilibrium. Since, however, the concentration of a solid substance is proportional to its density, the concentration remains constant so long as the temperature is not changed. Recognizing, then, that [Cu(s)] and

$[CuO(s)]$ are constants in the above expression, we write the equilibrium constant for the reaction as follows:

$$K = \frac{[H_2O(g)]}{[H_2(g)]}$$

Therefore, the ratio between the concentration of gaseous H_2O and the concentration of H_2 gas is constant as long as chemical equilibrium prevails. Introducing more H_2 gas into the system that is already in equilibrium will shift the equilibrium toward the right, increasing the amount of gaseous H_2O present, until the equilibrium ratio of $[H_2O(g)]$ to $[H_2(g)]$ is restored. On the other hand, the introduction of more solid copper oxide into the system will have no effect on the equilibrium. For this reason, the concentrations of pure solids are usually omitted from equilibrium constant expressions for equilibria involving solids and gases.

The equilibrium constant expression for the reaction

$$CaCO_3(s) \rightleftharpoons CaO(s) + CO_2(g)$$

involves the concentration of only one substance:

$$K = [CO_2(g)]$$

Since the concentration of a gas is proportional to its pressure, this expression means that there is only one pressure at which CO_2 gas can be in equilibrium with solid $CaCO_3$ and solid CaO. At 900°C this pressure is 790 torrs. If solid $CaCO_3$ and solid CaO are present in the same container where the temperature is 900°C, three possibilities exist: (1) if CO_2 is present at a pressure of 790 torrs, equilibrium prevails and there will be no change in the amounts of $CaCO_3$ and CaO in the system; (2) if CO_2 is present at a pressure in excess of 790 torrs, the reaction will proceed to the left, producing more $CaCO_3$ and consuming CaO and CO_2 until the quantity of CO_2 consumed is sufficient to reduce the pressure of the remaining CO_2 to 790 torrs; and (3) if CO_2 is present at a pressure less than 790 torrs, or if no CO_2 is present, the reaction will proceed to the right, consuming $CaCO_3$ and producing CaO and CO_2 until a sufficient quantity of CO_2 is present to exert a pressure of 790 torrs.

15-11 Saturated Solutions of Solids

A solid can be dissolved in a liquid, a small quantity at a time, to give a series of solutions of differing concentrations; however, this cannot continue indefinitely. Eventually a point is reached beyond which no more solid dissolves; addition of more solid to the system simply results in the solid falling to the bottom of the container without dissolving. If the solid remains there indefinitely, in undiminished quantity, in spite of intermittent stirring of the solution or shaking of the vessel, the solution has reached the highest concentration possible at that particular temperature. When this situation exists, we say that the solution is *saturated.*

In a saturated solution, in the presence of excess solid, there is a state of equilibrium. The equilibrium is dynamic; i.e., the solid continues to dissolve, but the number of molecules going into solution, say, each minute, is the same as the number of molecules leaving the solution and attaching themselves to the solid each minute. After this equilibrium condition has been established, the concentration of solute in the solution undergoes no further change, and neither does the amount of undissolved solid in contact with the solution. If more solid is placed in the container, none of it dissolves. The concentration of the saturated solution is not affected by the amount of undissolved solid that happens to be present.

If all the undissolved solid is removed from contact with the saturated solution, as by filtration, we still have a solution of the same concentration. A saturated solution, therefore, may be defined as a solution that is or would be in equilibrium with the undissolved solid.

The *solubility* of a substance describes the concentration of its saturated solution. The factors determining the solubility of a solid are (1) the "nature" of the solid, (2) the "nature" of the solvent, and (3) the temperature. The act of dissolving is usually accompanied by the absorption or liberation of heat. On the basis of Le Chatelier's Principle, we can predict whether the solubility will increase or decrease with rising temperature if we know whether the solution process is endothermic or exothermic.

If we wish, we can describe the equilibrium between undissolved solid and saturated solution by means of a "chemical equation," even though the process is nonchemical. For example, iodine dissolves in water without reacting with it to any appreciable extent. The equilibrium between the saturated solution and excess undissolved solid may be expressed in the following manner:

$$I_2(s) \rightleftharpoons I_2 \text{ (saturated solution)}$$

From this point of view, the concentration of the saturated solution is the "equilibrium constant" for the process:

$$K = [I_2 \text{ (saturated solution)}]$$

If the solid is an ionic one, K takes on a special form, which is discussed in Section 16-12.

New Terms

Equilibrant: a component of an equilibrium system. Each reactant and each product in an equilibrium mixture is an equilibrant.

Equilibrium constant: a constant characteristic of a particular reaction that is in a state of chemical equilibrium at a particular temperature, found by calculating the ratio of the equilibrium concentrations of products to reactants, each raised to the appropriate power.

Equilibrium mixture: a system, consisting of a mixture of substances, in which a state of chemical equilibrium prevails.

Forward reaction: the reaction in which the reactants are consumed; the left-to-right reaction. (The right-to-left reaction is the *reverse reaction.*)

Heterogeneous equilibrium: equilibrium involving more than one phase, e.g., the equilibrium between a liquid and its vapor or between a solid and its vapor, or a chemical equilibrium involving both solid and gaseous substances.

Le Chatelier's Principle: the change that takes place in an equilibrium system when conditions are altered is a change that tends to restore the original conditions.

Reactants: in an equation representing a reversible reaction, the substances whose formulas are written on the left. (The substances represented on the right are *products.*)

Saturated solution: a solution of such a concentration that if excess solute is present, equilibrium exists between the dissolved and undissolved phases.

Solubility: the concentration of the saturated solution of a substance.

Van't Hoff's Law: the equilibrium constant for an endothermic reaction increases with rising temperature, and for an exothermic reaction it decreases with rising temperature.

Exercises

15-1 For each of the following substances, find the weight (in grams) required to prepare 1 liter of a 1 *M* solution, 3 liters of a 4 *M* solution, 5 liters of a 0.03 *M* solution, 250 ml of a 3 *M* solution, and 125 ml of a 0.01 *M* solution:

(a) sucrose, $C_{12}H_{22}O_{11}$
(b) sodium chloride, NaCl
(c) magnesium sulfate 7-hydrate, $MgSO_4 \cdot 7H_2O$

15-2 Write the expression for the equilibrium constant for each of the following reactions:

(a) $4NH_3(g) + 5O_2(g) \rightleftharpoons 4NO(g) + 6H_2O(g)$
(b) $COCl_2(g) \rightleftharpoons CO(g) + Cl_2(g)$
(c) $2NO(g) + O_2(g) \rightleftharpoons 2NO_2(g)$

15-3 At 400°C 1 liter of an equilibrium mixture of nitrogen, hydrogen, and ammonia gases was found to contain 0.40 mole of H_2, 0.60 mole of N_2, and 0.14 mole of NH_3. Calculate the equilibrium constant.

15-4 In which direction will the equilibrium be shifted by an increase in (1) the concentration of O_2, (2) pressure, and (3) temperature, in each of the following reactions:

(a) $2SO_2(g) + O_2(g) \rightleftharpoons 2SO_3(g) + \text{heat}$
(b) $N_2(g) + O_2(g) + \text{heat} \rightleftharpoons 2NO(g)$
(c) $2H_2(g) + O_2(g) \rightleftharpoons 2H_2O(g) + \text{heat}$
(d) $2CO(g) + O_2(g) \rightleftharpoons 2CO_2(g) + \text{heat}$

15-5 Write the expressions for the equilibrium constants for the following reactions:

(a) $NH_4Cl(s) \rightleftharpoons NH_3(g) + HCl(g)$
(b) $C(s) + CO_2(g) \rightleftharpoons 2CO(g)$

References

Carnell, P. H., and Reusch, R. N., *Molecular Equilibrium*, Saunders, Philadelphia, 1963 (paperbound). Programmed instruction in the basic principles of equilibrium.

Hiller, L. A., and Herber, R. H., *Principles of Chemistry*, McGraw-Hill, New York, 1960, Chapter 10, "The Nature of Chemical Equilibrium."

Hutchinson, E., *Chemistry: The Elements and Their Reactions*, 2nd ed., Saunders, Philadelphia, 1964, Chapters 15 and 16. Extensive lists of problems.

Sienko, M. J., and Plane, R. A., *Chemistry*, 3rd ed., McGraw-Hill, New York, 1966, Chapter 13, "Chemical Equilibrium." Concisely written; excellent questions.

16

Ionic Equilibrium

16-1 Ionization of Weak Acids and Weak Bases

A weak electrolyte is a covalent compound that yields a small concentration of ions in solution by incomplete reaction with the solvent. Most acids and bases are weak electrolytes.

Pure acetic acid (hydrogen acetate) is a covalent compound composed of $HC_2H_3O_2$ molecules. When dissolved in water, this compound partially ionizes; i.e., a fraction of the $HC_2H_3O_2$ molecules react with the water in accordance with the equation

$$HC_2H_3O_2 + H_2O \rightleftharpoons H_3O^+ + C_2H_3O_2^-$$

In consequence of this incomplete reaction, an aqueous solution of acetic acid contains a small concentration of H_3O^+ ions and $C_2H_3O_2^-$ ions; but the concentration of $HC_2H_3O_2$ molecules is considerably greater than that of the ions.

There is a dynamic equilibrium between the molecular and ionized forms of the acid, as indicated by the double arrows in the equation. Thus, the proportion of molecules and ions remains constant.

Application of the Law of Chemical Equilibrium to this reaction gives the expression

$$\frac{[H_3O^+][C_2H_3O_2^-]}{[H_2O][HC_2H_3O_2]} = \text{constant}$$

Except at high concentrations of the solute, the concentration of water in aqueous solutions varies so slightly that the value of $[H_2O]$ in such expressions as this one may be regarded as constant without introducing any substantial error. Therefore, the equilibrium constant that is customarily written for this reaction is the one defined by the expression

$$K = \frac{[H_3O^+][C_2H_3O_2^-]}{[HC_2H_3O_2]}$$

This constant is called the *ionization constant* or *dissociation constant* of acetic acid.

It is permissible to define the ionization constant for acetic acid by the expression

$$K = \frac{[H^+][C_2H_3O_2^-]}{[HC_2H_3O_2]}$$

where the hydrogen ion is symbolized as H^+ instead of H_3O^+; i.e., its hydration is ignored. If hydration is ignored, the ionization constant expression is the same as the one obtained directly from the simplified equation for the ionization of acetic acid:

$$HC_2H_3O_2 \rightleftharpoons H^+ + C_2H_3O_2^-$$

The numerical value for the ionization constant of acetic acid may be calculated by substituting into the ionization constant expression the values for the individual concentrations of a particular solution. For example, by measurements of electrical conductance, it is known that in a 0.1000 *M* solution of acetic acid at room temperature, 1.35 percent of the molecules form ions, while the remaining 98.65 percent retain their identity as molecules. Thus,

$$[H^+] = [C_2H_3O_2^-] = 0.0135 \times 0.1000\ M = 0.00135\ M$$

and

$$[HC_2H_3O_2] = 0.9865 \times 0.1000\ M = 0.09865\ M$$

Therefore,

$$K = \frac{(0.00135\ \text{mole/liter})(0.00135\ \text{mole/liter})}{(0.09865\ \text{mole/liter})}$$

$$= 1.85 \times 10^{-5}\ \text{mole/liter}$$

In Table 16-1 are presented the degrees of ionization of acetic acid at various concentrations and the corresponding values for the ionization constants. The last solution is almost a hundred times as dilute as the first, and the degree of ionization is more than nine times as great, but the value of the ionization constant is essentially the same in each case. This type of result is usually obtained for weak electrolytes and is indicative of a true equilibrium between molecules and ions.

Ammonia is one of the best-known weak bases. The equilibrium between molecules and ions in an aqueous solution of ammonia can be represented by the equation

$$NH_3 + H_2O \rightleftharpoons NH_4^+ + OH^-$$

The expression for the ionization constant is

$$K = \frac{[NH_4^+][OH^-]}{[NH_3]}$$

The conductance of a 0.100 *M* solution of ammonia indicates that 1.34 percent of the solute is ionized. Thus,

Table 16-1

Ionization of acetic acid at different concentrations

Concentration (moles/liter)	*Percent ionized*	*Ionization constant* (moles/liter)
0.10000	1.350	1.85×10^{-5}
0.05000	1.905	1.85×10^{-5}
0.02000	2.988	1.84×10^{-5}
0.009842	4.222	1.83×10^{-5}
0.005912	5.401	1.82×10^{-5}
0.001028	12.38	1.80×10^{-5}
	AVERAGE =	1.83×10^{-5} mole/liter

$$[NH_4^+] = [OH^-] = 0.0134 \times 0.100\ M = 0.00134 \text{ mole/liter}$$

and $$[NH_3] = 0.9866 \times 0.100\ M = 0.09866 \text{ mole/liter}$$

Therefore, $$K = \frac{(0.00134 \text{ mole/liter})(0.00134 \text{ mole/liter})}{(0.09866 \text{ mole/liter})}$$

$$= 1.82 \times 10^{-5} \text{ mole/liter}$$

The fact that the ionization constants of acetic acid and ammonia are substantially identical is an intriguing coincidence.

16-2 Pseudosalts

Weak electrolytes include not only the weak acids and weak bases but also the *pseudosalts*, which are covalent compounds containing a metallic element in combination with a nonmetallic element. They have comparatively low melting points and are poor conductors of electricity in the molten state, but their aqueous solutions are good conductors. An example of a pseudosalt is aluminum chloride, Al_2Cl_6. The anhydrous compound is composed of neutral molecules in which the atoms are linked together by covalent bonds. The molecules react with water, incompletely, to form ions:

$$Al_2Cl_6 + 12H_2O \rightleftharpoons 2Al(H_2O)_6^{+++} + 6Cl^-$$

The equilibrium constant for this reaction can be considered to be the ionization constant of aluminum chloride; i.e.,

$$K = \frac{[Al^{+++}]^2[Cl^-]^6}{[Al_2Cl_6]}$$

16-3 Use of Ionization Constants

Knowledge of the numerical value for the ionization constant of a weak electrolyte makes it possible to calculate the degree of ionization of the electrolyte, and the concentrations of its ions, in any solution containing it.

EXAMPLE Given that the ionization constant for acetic acid is 1.8×10^{-5}, calculate the hydrogen-ion concentration and the degree (percent) of ionization in a 0.010 *M* solution of the acid.

Since we are interested in hydrogen-ion concentration, we may begin as follows:

Let $$x = [H^+]$$

When we inspect the chemical equation

$$HC_2H_3O_2 \rightleftharpoons H^+ + C_2H_3O_2^-$$

we are reminded that acetate ions and hydrogen ions are formed in equal numbers. Therefore,

$$x = [C_2H_3O_2^-]$$

The chemical equation also shows that the number of molecules of acetic acid undergoing ionization is the same as the number of hydrogen ions formed. The equilibrium concentration of acetic acid molecules is the concentration which they would have in the absence of any ionization, minus the reduction in their concentration that results from the ionization; i.e.,

$$0.010 - x = [HC_2H_3O_2]$$

Substituting these values into the expression for the ionization constant, we have

$$\frac{[H^+][C_2H_3O_2^-]}{[HC_2H_3O_2]} = K$$

$$\frac{x^2}{0.010 - x} = 1.8 \times 10^{-5}$$

$$x^2 = 1.8 \times 10^{-7} - (1.8 \times 10^{-5})x$$

Written in the more familiar form of a quadratic equation,

$$x^2 + (1.8 \times 10^{-5})x - 1.8 \times 10^{-7} = 0$$

Remember that the solution of the generalized quadratic equation

$$ax^2 + bx + c = 0$$

is given by the formula

$$x = \frac{-b \pm \sqrt{b^2 - 4ac}}{2a}$$

Therefore, substituting $a = 1$, $b = 1.8 \times 10^{-5}$, and $c = -1.8 \times 10^{-7}$, we obtain the result

$$x = 4.16 \times 10^{-4}\ M$$

Since there are only two significant digits in the numbers for the ionization constant and for the concentration of the acid, we are justified in retaining only two digits of our answer. Thus, the final answer for the hydrogen-ion concentration is

$$x = 4.2 \times 10^{-4}\ M$$

The solution to a quadratic equation is time-consuming and frequently can be avoided. Reviewing our solution to this problem, we recognize that the small value for the ionization constant, 1.8×10^{-5}, means that the ratio

$$\frac{x^2}{0.010 - x}$$

is small. Therefore, x must be small compared to the number 0.010, from which it is to be subtracted. Therefore, $0.010 - x$ is substantially equal to 0.010. This enables us to write the simplified equation

$$\frac{x^2}{0.010} = 1.8 \times 10^{-5}$$

Solution of this equation does not require the quadratic formula.

$$x^2 = 1.8 \times 10^{-7} = 18 \times 10^{-8}$$

$$x = \sqrt{18 \times 10^{-8}} = \sqrt{18} \times \sqrt{10^{-8}}$$

$$= 4.24 \times 10^{-4}\ M$$

Again, we retain only two digits of the answer:

$$x = 4.2 \times 10^{-4}\ M$$

Thus, we see that we were justified in neglecting x in the expression $0.010 - x$, for we have obtained the same answer as by solving the quadratic equation.

Use of the approximate solution is justified when the degree of ionization is less than about 5 percent. In case of doubt, we can calculate the approximate solution and then ascertain whether the value for x thus calculated is in excess of 5 percent of the molarity of the solution. (Fortunately, the error introduced by the simplified solution increases the value of x.) Thus, if the resulting x is greater than 5 percent of the acid concentration, the results of the simplified solution should be discarded and the answer obtained by solution of the quadratic equation.

The degree of ionization is calculated by dividing the concentration of the acid in ionic form by the total concentration:

$$\frac{4.2 \times 10^{-4}\ M}{0.010\ M} = 0.042 = 4.2\% \text{ ionized}$$

16-4 Reactions Between Ions

The Law of Chemical Equilibrium does not apply to aqueous solutions of strong acids, such as HCl and HNO_3, for these react completely with water to form ions in dilute solutions:

SIMPLIFIED EQUATION: $$HCl \rightarrow H^+ + Cl^-$$

COMPLETE EQUATION: $$HCl + H_2O \rightarrow H_3O^+ + Cl^-$$

Nor does the Law of Chemical Equilibrium apply to solutions of compounds which are ionic in the solid state, such as NaOH or NaCl, for in their solutions these compounds remain completely ionic. There are no NaOH or NaCl molecules in these solutions. There are, therefore, no equilibria involving the ionic and molecular forms of such substances.

If an aqueous solution of a strong electrolyte is mixed with an aqueous solution of another strong electrolyte, there is frequently no effect of either on the other. For example, addition of a solution of NaCl to a solution of KBr produces no effect; the addition of NaCl to a solution of HCl has little if any effect on the acidity of the latter; addition of NaCl to a solution of NaOH has no detectable effect upon the alkalinity of the solution.

A reaction between strong electrolytes will occur if one of the possible products is a weak electrolyte. For example, consider a mixture of hydrochloric acid and sodium acetate in dilute solution. The vinegarlike odor of acetic acid is easily detected in such a solution; this is understandable in terms of the reaction

$$H^+ + Cl^- + Na^+ + C_2H_3O_2^- \rightleftharpoons HC_2H_3O_2 + Na^+ + Cl^-$$

Since the Na^+ and Cl^- ions do not participate in the reaction, the equation can be accurately and concisely written in the form

$$H^+ + C_2H_3O_2^- \rightleftharpoons HC_2H_3O_2$$

The reaction is reversible and therefore incomplete, but most of the H^+ and $C_2H_3O_2^-$ ions combine to form molecules. Of course, if acetic acid were a strong acid, this would not happen.

One of the most important reactions involving the combination of ions to form molecules occurs when an acid and a base are brought together in aqueous solution. If both the acid and the base are strong electrolytes, the only chemical event that occurs is the one represented by the equation

$$H^+ + OH^- \rightleftharpoons H_2O$$

The same amount of heat, 13.7 kcal, is liberated per mole of water formed by the neutralization of any strong acid by any strong base.

If either the acid or the base participating in a neutralization reaction is a weak electrolyte, the reaction cannot be correctly represented as merely the combination of H^+ and OH^- ions. For example, the concentration of H^+ ions in a solution of acetic acid is much smaller than the concentration of OH^- ions in a solution of sodium hydroxide of equal concentration. Nevertheless, an acetic acid solution neutralizes sodium hydroxide with practically the same speed as does hydrochloric acid, both reactions being so rapid as to be immeasurable by any ordinary means. Furthermore, neutralization of 1 mole of acetic acid requires the same amount of base as does the neutralization of 1 mole of hydrochloric acid. The neutralization of the acetic acid is best represented by the equation

$$HC_2H_3O_2 + OH^- \rightarrow H_2O + C_2H_3O_2^-$$

Since this reaction is different from the reaction that occurs with a strong acid, the amount of heat liberated per mole of water formed is less than 13.7 kcal.

16-5 The Common-Ion Effect

A 0.1 *M* solution of acetic acid has the sour taste typical of acidic solutions and causes indicators to show their "acid colors." If sodium acetate is added to the solution, the sourness lessens and the indicator changes its hue. If enough sodium acetate is added, the sourness is almost completely lost, and the color of the indicator shows that the solution has become almost neutral. Other acetates, such as potassium acetate and calcium acetate, produce the same effect as sodium acetate. In general, the ionization of a weak acid is reduced, i.e., the concentration of H^+ ions is decreased, in the presence of one of its own salts.

A dilute solution of ammonia has the properties characteristic of weak bases; e.g., it causes the indicator, phenolphthalein, to turn pink. If a sufficient amount of an ammonium salt, such as ammonium chloride or ammonium nitrate, is dissolved in the solution, the phenolphthalein no longer shows its "alkaline color" because the solution becomes almost neutral. In general, the ionization of a weak base is reduced, i.e., the concentration of OH^- ions is decreased, in the presence of one of its own salts.

Since sodium acetate and the other acetates have the acetate ion in common with acetic acid, the influence of an acetate in decreasing the ionization of acetic acid is known as the *common-ion effect*. The effect of ammonium salts in reducing the ionization of ammonia is another example of the common-ion effect. In general, if we add to a weak electrolyte a substance providing an ion in common with it, ionization of the electrolyte is repressed. Thus, addition of HCl to a solution of acetic acid reduces the concentration of acetate ions, and the addition of a strong acid to a solution containing H_2S reduces the concentration of sulfide ions.

The magnitude of the common-ion effect can be calculated from the value for the ionization constant.

EXAMPLE If a liter of 0.10 *M* acetic acid solution contains 0.50 mole of sodium acetate, calculate the hydrogen-ion concentration.

We may begin by saying

let $$x = [H^+]$$

Acetate ions are derived from both sodium acetate and acetic acid. For each H^+ ion derived from acetic acid, there is one $C_2H_3O_2^-$ ion; all of the sodium acetate is ionized. Thus,

$$0.50 + x = [C_2H_3O_2^-]$$

and $$0.10 - x = [HC_2H_3O_2]$$

Since x is small in comparison with 0.10, it is also small in comparison with 0.50. Therefore, we may say

$$0.50 = [C_2H_3O_2^-]$$

$$0.10 = [HC_2H_3O_2]$$

$$K = 1.8 \times 10^{-5} = \frac{[H^+][C_2H_3O_2^-]}{[HC_2H_3O_2]} = \frac{0.50x}{0.10}$$

$$x = \frac{1.8 \times 10^{-5}}{5}$$

$$= 3.6 \times 10^{-6}\ M$$

In Section 16-1 we saw that in a 0.10 *M* solution of acetic acid, in the absence of other solutes, the concentration of hydrogen ions is 0.00135 *M*. Thus, the presence of the 0.50 *M* sodium acetate in the 0.10 *M* acetic acid solution has reduced the hydrogen-ion concentration from $135 \times 10^{-5}\ M$ to $0.36 \times 10^{-5}\ M$.

In qualitative terms, the common-ion effect is merely another special case of the Principle of Le Chatelier. The acetate ion is one of the products in the ionization of acetic acid. Increasing the concentration of one of the products of a reaction always shifts the equilibrium to the left. If we increase the concentration of acetate ions, we thereby decrease the concentration of hydrogen ions and increase the concentration of hydrogen acetate molecules.

16-6 Ionization of Water

Although pure water is not a good conductor of electricity, its slight conductance is measurable. Measurements indicate that a very small fraction of the molecules, one in 555 million, are electrolytically dissociated, i.e., ionized, in accordance with the equation

$$H_2O + H_2O \rightleftharpoons H_3O^+ + OH^-$$

Water thus dissociates to give equal numbers of the hydronium ions characteristic of aqueous solutions of acids and the hydroxide ions which characterize aqueous solutions of bases. Historically, acidic and basic properties have been attributed to the H_3O^+ and OH^- ions, respectively, because acids and bases have most often been encountered in aqueous solutions where these ions are present.

The hydronium ion may be represented as H^+ (remember that this is merely a convenient simplification). The actual ion is a hydrated hydrogen ion, i.e., a hydrated proton, $[H(H_2O)]^+$. The H^+ ion is attached to a water molecule by a coordinate bond like the coordinate bonds present in any other hydrated cation:

$$\mathrm{H^+ + H:\underset{\displaystyle H}{\overset{\cdot\cdot}{\underset{\cdot\cdot}{O}}}: \;\rightarrow\; \left[\begin{array}{ccc} H: & \overset{\cdot\cdot}{\underset{\cdot\cdot}{O}} & :H \\ & H & \end{array}\right]^+}$$

All ions are probably hydrated in aqueous solution, but unless the hydration is being discussed, it is not ordinarily represented in formulas and equations. Thus, the ionization of water can be represented by the simplified equation

$$H_2O \rightleftharpoons H^+ + OH^-$$

The secondary ionization,

$$OH^- \rightleftharpoons O^{--} + H^+$$

is so small that it need not be considered.

Water itself can be regarded as simultaneously an acid and a base because its dissociation produces both H^+ and OH^- ions. Since the two kinds of ions are present in equal number, the properties of neither one predominate. Thus, water does not have the sour taste often present in aqueous solutions of acids nor the bitter taste characteristic of aqueous solutions of bases, and it does not affect the color of either red or blue litmus. Water is *neutral*, and any aqueous solution is neutral if the concentrations of H^+ and OH^- ions are equal. For example, a solution of sodium chloride is neutral because neither the Na^+ nor the Cl^- ions have any effect on the ionization of water or on either of the ions produced from water.

16-7 The Ion-Product Constant of Water

Since the ionization of a molecule of water produces one H^+ ion and one OH^- ion,

$$HOH \rightleftharpoons H^+ + OH^-$$

pure water always contains equal concentrations of these ions. If an acid is dissolved in water, it forms more H^+ ions. The increase in the hydrogen-ion concentration causes the equilibrium to shift to the left, thereby reducing the concentration of OH^- ions. It is impossible, however, to reduce the hydroxide-ion concentration to zero in any aqueous solution. If a base is dissolved in water, the increase in hydroxide-ion concentration shifts the equilibrium to the left, reducing the hydrogen-ion concentration; but the hydrogen-ion concentration cannot be reduced to zero in any aqueous solution.

In pure water at 25°C, the concentration of H^+ ions has been measured to be 1×10^{-7} mole per liter; the concentration of OH^- ions is also 1×10^{-7} mole per liter. In pure water, and in all aqueous solutions, the concentrations of H^+, OH^-, and H_2O are such as to satisfy the equilibrium relationship:

$$\frac{[H^+][OH^-]}{[H_2O]} = K$$

Except at high concentrations of solute, $[H_2O]$ varies so little that it may be regarded as a constant. Therefore, it is combined with K to give a constant, K_w:

$$[H^+][OH^-] = K_w$$

The constant K_w is the *ion-product constant of water*, and its value at 25°C is obtained by substituting into the K_w expression the concentrations of H^+ and OH^- in pure water at 25°C:

$$(1 \times 10^{-7}\text{ mole/liter})(1 \times 10^{-7}\text{ mole/liter}) = 1 \times 10^{-14}\text{ mole}^2/\text{liter}^2$$

Thus, the $[OH^-]$ of any aqueous solution can be calculated if its $[H^+]$ is known:

$$[OH^-] = \frac{1 \times 10^{-14}}{[H+]}$$

Likewise, the $[H^+]$ of any aqueous solution can be calculated if its $[OH^-]$ is known.

EXAMPLE If the hydroxide-ion concentration of a solution is $1 \times 10^{-3}\ M$, what is its hydrogen-ion concentration?

$$[H^+] = \frac{1 \times 10^{-14}}{1 \times 10^{-3}} = 1 \times 10^{-11}\ M$$

Since either the $[H^+]$ or the $[OH^-]$ can be calculated if the other is known, we do not have to state both $[H^+]$ and $[OH^-]$ to characterize a solution. Usually, the acidity, alkalinity, or neutrality of a solution is expressed in terms of $[H^+]$. To understand such characterization of solutions, it is necessary to remember these definitions:

1. In a neutral solution, $[H^+] = 1 \times 10^{-7}\ M$.
2. In an acidic solution, $[H^+] > 1 \times 10^{-7}\ M$.
3. In an alkaline solution, $[H^+] < 1 \times 10^{-7}\ M$.

16-8 The pH Scale

The hydrogen-ion concentration of a solution is conveniently expressed in terms of a function called *pH* (suggested in 1909 by the Danish biochemist

S. P. Sørensen). The pH is sometimes called the "hydrogen-ion exponent" since it refers to the negative power of 10 which represents the hydrogen-ion concentration. Thus, if $[H^+] = 10^{-3}$, pH = 3.

Expressed mathematically, pH is the negative logarithm of the hydrogen-ion concentration; in other words, it is the logarithm of the reciprocal of the hydrogen-ion concentration:

$$\text{pH} = -\log [H^+] = \log \frac{1}{[H^+]}$$

EXAMPLE What is the pH of 0.01 *M* HCl?

Since HCl is completely ionized in dilute solution, the hydrogen-ion concentration is 0.01 *M*, i.e., 10^{-2} *M*.

METHOD 1: $\text{pH} = -\log 10^{-2} = -(-2) = 2$

METHOD 2: $\text{pH} = \log \frac{1}{[H^+]} = \log \frac{1}{10^{-2}} = \log 10^2 = 2$

EXAMPLE If the pH of a solution is 8, what is the hydrogen-ion concentration?

$$\begin{aligned} \text{pH} = -\log [H^+] &= 8 \\ \log [H^+] &= -8 \\ [H^+] &= 10^{-8} \end{aligned}$$

We may go beyond the suggestion of Sørensen and define pOH as the negative logarithm of the hydroxide-ion concentration:

$$\text{pOH} = -\log [OH^-]$$

EXAMPLE What is the pOH of 0.001 *M* NaOH?

Since NaOH is a completely ionized compound, the hydroxide-ion concentration is 0.001 $M = 10^{-3}$ *M*. Therefore,

$$\text{pOH} = -\log 10^{-3} = 3$$

We may go a step further and define pK_w as the negative logarithm of the ion-product constant of water:

$$pK_w = -\log K_w = -\log 10^{-14} = 14$$

Table 16-2

Methods for expressing acidity or alkalinity of aqueous solutions

Solution	$[H^+]$	$[OH^-]$	pH	pOH	*Nature of solution*
0.1 *M* HCl	10^{-1}	10^{-13}	1	13	Strongly acidic
10^{-5} *M* HCl	10^{-5}	10^{-9}	5	9	Acidic
Pure water	10^{-7}	10^{-7}	7	7	Neutral
NaCl solution	10^{-7}	10^{-7}	7	7	Neutral
10^{-5} *M* NaOH	10^{-9}	10^{-5}	9	5	Basic
0.1 *M* NaOH	10^{-13}	10^{-1}	13	1	Strongly basic

This leads to the convenient relationship

$$pH + pOH = 14$$

so that the pH of a solution is readily calculated if its pOH is known, and the pOH can likewise be determined from the pH:

$$pH = 14 - pOH$$
$$pOH = 14 - pH$$

The relationships among the various methods for expressing the acidity or alkalinity of a solution are indicated in Table 16-2. A solution can be characterized, with respect to its acidity or alkalinity, by any one of the four quantities: $[H^+]$, $[OH^-]$, pH, or pOH. If one of these is known, the other three can be calculated from it.

In most instances the concentration of H^+ or OH^- ions in a solution is not exactly a negative integral power of 10 but must be expressed as a negative power of 10 multiplied by some factor. For example, in 0.002 *M* HCl the concentration of H^+ ions is 2×10^{-3} *M*. In such cases, a logarithm table must be consulted in order to find the pH.

EXAMPLE Calculate the pH of a 0.002 *M* HCl solution.

$$\begin{aligned} pH &= -\log(2 \times 10^{-3}) = -(\log 2 + \log 10^{-3}) \\ &= -(0.30 - 3) = -(-2.70) \\ &= 2.70 \end{aligned}$$

In Table 16-3 the two-digit logarithms of the numbers 2 through 9 are listed. This is all the information needed for calculating pH when the hydrogen-ion concentration is expressed as a power of 10 multiplied by an integral factor.

Table 16-3

Logarithms of the integers 2-9

Number	*Logarithm*	*Number*	*Logarithm*
2	0.30	6	0.78
3	0.48	7	0.85
4	0.60	8	0.90
5	0.70	9	0.95

16-9 Buffer Solutions

Certain solutions, called *buffer solutions*, maintain their pH values at virtually constant levels despite the addition of small quantities of strong acid or strong base. There are many substances and mixtures which can serve as buffers, but the ones that can be most simply explained in terms of chemical equilibrium are prepared by dissolving a weak acid and one of its salts, or a weak base and one of its salts, in the same solution.

If a solution contains both acetic acid (a weak acid) and sodium acetate (a completely ionized salt), the common-ion effect suppresses the ionization of the acetic acid so that the following equilibrium is shifted far to the left:

$$HC_2H_3O_2 \rightleftharpoons H^+ + C_2H_3O_2^-$$

If a small amount of NaOH (a strong base) is added to this acid buffer solution, H^+ ions join with the added OH^- ions to form water molecules, but the hydrogen-ion concentration is maintained at almost the same value through further ionization of $HC_2H_3O_2$ molecules. If, on the other hand, a small amount of HCl (a strong acid) is added to the solution, $C_2H_3O_2^-$ ions combine with most of the added H^+ ions to form more $HC_2H_3O_2$ molecules, leaving the hydrogen-ion concentration virtually unaffected.

Solutions that contain a weak base and one of its salts are basic buffer solutions which maintain virtually constant hydroxide-ion concentrations against the addition of small quantities of strong acid or strong base. The mechanism of their buffer action is analogous to that of a solution containing a weak acid and one of its salts.

Most body fluids are protected by buffer action against appreciable change in pH. The pH of the blood is normally between 7.35 and 7.45; gastric juice has a pH of 1.6 to 1.8; and saliva is nearly neutral (pH 7). Since many of the reactions that take place in body tissues form acidic products, the blood must have a mechanism for preventing appreciable increases in its acidity. If illness causes the pH of the blood to go below 7.0 or above 7.8, death occurs.

16-10 pH Indicators

Certain organic substances occurring in nature, and others of synthetic origin, possess different colors in solutions having different pH values. These substances, called acid-base indicators or *pH indicators*, are often employed to determine the pH values of solutions. Such substances are themselves weak acids or weak bases.

In a solution containing the indicator methyl orange (a weak acid), there is an equilibrium between the ionized and the nonionized forms. If we symbolize this weak-acid indicator by the formula HInd, we may represent the equilibrium by the equation

$$\underset{\text{red}}{\text{HInd}} \rightleftharpoons \text{H}^+ + \underset{\text{yellow}}{\text{Ind}^-}$$

The feature that makes this particular weak acid useful as an indicator is the difference between the red color of the molecule and the yellow color of the anion.

An increase in hydrogen-ion concentration of the solution, brought about by addition of an acid, shifts the equilibrium to the left toward the red form of the indicator. In solutions of lower hydrogen-ion concentration, the equilibrium is shifted to the right toward the yellow form. The color which the indicator shows in a solution of a particular pH is determined by the relative amounts of the yellow form and the red form, i.e., by the ratio of their concentrations. From the expression for the ionization constant of the indicator,

$$\frac{[\text{H}^+][\text{Ind}^-]}{[\text{HInd}]} = K$$

we can see that the ratio in question is given by the following expression:

$$\frac{[\text{Ind}^-]}{[\text{HInd}]} = \frac{K}{[\text{H}^+]}$$

When the hydrogen-ion concentration of the solution has the same numerical value as the ionization constant of the indicator, i.e., when the ratio is equal to 1, the red form and yellow form are present in equal amounts. This causes the solution to have an orange color, i.e., a color intermediate between yellow and red. When the pH is such that 90 percent of the indicator is in the red form, the eye cannot detect any change in color arising from further increase in hydrogen-ion concentration. If the pH of the solution is such that 90 percent of the indicator is in the yellow form, the eye cannot detect any change in color from further decrease in hydrogen-ion concentration. The limits between which this indicator undergoes color changes perceptible to the eye are pH 3.1 and pH 4.4. Within these limits the pH of a solution can be determined by observing the color which it imparts to methyl orange. The determination is made by comparison with color standards obtained by adding the indicator to a series of buffer solutions of known pH values.

A great number of pH indicators are known, each with its characteristic color-change interval. Determination of the pH of a particular solution requires the use of an indicator with the proper color-change interval; i.e., the pH of the solution must be between the limits within which the indicator undergoes perceptible color changes. The measurement and control of pH is important in scientific work, both pure and applied. This is done most conveniently and most accurately with electrical instruments known as pH meters.

16-11 Hydrolysis of Salts

Some hydrated cations, such as $[Cu(H_2O)_4]^{++}$, produce H_3O^+ ions in aqueous solutions; therefore, solutions of salts containing such ions are acidic. When copper(II) sulfate (either anhydrous or hydrated) is dissolved in water, the solution contains the hydrated copper(II) ion, which is an acid because it produces H_3O^+ ions:

$$[Cu(H_2O)_4]^{++} + H_2O \rightleftharpoons [Cu(H_2O)_3(OH)]^+ + H_3O^+$$

Only a low concentration of H_3O^+ ions is produced by this reaction, so the $[Cu(H_2O)_4]^{++}$ ion is classified as a weak acid. Some other hydrated metal ions that act as weak acids when they are present in aqueous solutions are $[Al(H_2O)_6]^{+++}$ and $[Zn(H_2O)_4]^{++}$.

The reactions between water and the ions from salts to produce either H_3O^+ ions or OH^- ions are classified together as *hydrolysis of salts*. Aqueous solutions of salts, such as NaCl, whose ions do not react with water are neutral. But aqueous solutions of such salts as $CuSO_4$, $Al_2(SO_4)_3$, or $ZnSO_4$ are acidic because of the reaction between water and the hydrated cations. Practically all metal ions other than those of the metals of Groups IA and IIA hydrolyze to give acidic solutions. The slight tendency of water molecules to yield protons is apparently greatly increased when the water molecules are bonded to a highly charged positive ion, presumably because of a strong repulsion between this ion and the positive proton.

Aqueous solutions of such salts as NaCN are alkaline because the hydrolysis of the anion produces OH^- ions:

$$CN^- + H_2O \rightleftharpoons HCN + OH^-$$

The reaction occurs because HCN is a weak acid.

16-12 The Solubility Product Principle

The equilibrium that exists in a saturated solution of a slightly soluble ionic solid in the presence of excess solid (compare Section 15-11) may be represented by a chemical equation. For example,

$$AgCl(s) \rightleftharpoons Ag^+ + Cl^-$$

The formation of a precipitate upon mixing solutions containing the necessary ions is a chemical reaction from any point of view; the above equation is simply the reverse of the net ionic equation representing the formation of a silver chloride precipitate when solutions of, say, silver nitrate and sodium chloride are mixed.

Application of the Law of Chemical Equilibrium to the reversible process represented by the above equation leads to the following expression for the equilibrium constant:

$$K_{sp} = [Ag^+][Cl^-]$$

In this expression the equilibrium constant, K_{sp}, is the *solubility product constant* for silver chloride.

If we add a concentrated solution of NaCl, or any other soluble ionic chloride, to a saturated solution of AgCl, we increase the concentration of Cl^- ion. The product, $[Ag^+][Cl^-]$, would now be greater than the numerical value of K_{sp} for AgCl if it were not for the fact that solid AgCl separates from the solution. This is an example of the common-ion effect (Section 16-5) as it applies in the case of the heterogeneous equilibrium existing between a strong electrolyte and its ions in solution.

If we add a concentrated solution of $AgNO_3$, or any other soluble silver salt, to a saturated solution of AgCl, we increase $[Ag^+]$ so that the product, $[Ag^+][Cl^-]$, would again exceed the numerical value of K_{sp} for AgCl if solid AgCl did not separate from the solution. This example also illustrates the common-ion effect.

The fact that in a saturated solution of AgCl the product of the ion concentrations, $[Ag^+][Cl^-]$, is a constant, is an example of the *solubility product principle*. Before attempting to state this principle, it would be well for us to inspect the expressions for the solubility product constants listed in Table 16-4. Note that if an ionic solid has more than one ion of one kind in its formula, the concentration of that ion must be raised to the corresponding power when the solubility product constant for the compound is calculated. This leads to the

Table 16-4

Solubility product expressions for selected compounds

Chemical formula	*Solubility product expression*
$BaSO_4$	$K_{sp} = [Ba^{++}][SO_4^{--}]$
CaF_2	$[Ca^{++}][F^-]^2$
Ag_2S	$[Ag^+]^2[S^{--}]$
$MgNH_4PO_4$	$[Mg^{++}][NH_4^+][PO_4^{---}]$
$Fe(OH)_3$	$[Fe^{+++}][OH^-]^3$
$Ca_3(PO_4)_2$	$[Ca^{++}]^3[PO_4^{---}]^2$
$FePO_4$	$[Fe^{+++}][PO_4^{---}]$
Bi_2S_3	$[Bi^{+++}]^2[S^{--}]^3$

following statement of the solubility product principle: the product of the concentrations of the ions in a saturated solution of a slightly soluble strong electrolyte is a constant for a given temperature, provided each concentration is raised to the power equal to the subscript of the corresponding ion in the chemical formula.

The solubility product principle applies only to strong electrolytes that are slightly soluble, i.e., compounds that form saturated solutions in which the ion concentrations are of the order of 0.01 *M* or less. The principle is useful because it applies to any solution that is in equilibrium with a slightly soluble ionic solid, regardless of how the ions got into the solution and regardless of the proportions in which they are present.

By analysis of a saturated solution of AgCl containing no other electrolytes, it has been found that the concentration of Ag^+ ions in the solution equals 1.28×10^{-5} *M* at 25°C. With no other electrolytes present, the concentration of Cl^- ions must also equal 1.28×10^{-5} *M*. For the solubility product constant we then have

$$\begin{aligned} K_{sp} = [Ag^+][Cl^-] &= (1.28 \times 10^{-5} \text{ mole/liter})(1.28 \times 10^{-5} \text{ mole/liter}) \\ &= 1.64 \times 10^{-10} \text{ mole}^2/\text{liter}^2 \text{ at } 25°\text{C} \end{aligned}$$

A simple experimental method for determining K_{sp} values is to determine the volumes of properly chosen solutions that must be mixed in order to produce the first detectable precipitation.

EXAMPLE It was found, at 25°C, that only 0.16 ml of 0.001 *M* $AgNO_3$ needed to be added to 1.000 liter of 0.001 *M* NaCl to produce the first signs of cloudiness (indicating precipitation) in the stirred solution. What is the solubility product constant for AgCl?

$$\begin{aligned} [Cl^-] &= 0.001\ M \\ &= 1 \times 10^{-3}\ M \\ [Ag^+] &= 0.001\ M \times \frac{0.16}{1000} \\ &= 1.6 \times 10^{-7}\ M \\ K_{sp} &= [Ag^+][Cl^-] \\ &= (1 \times 10^{-3} \text{ mole/liter})(1.6 \times 10^{-7} \text{ mole/liter}) \\ &= 1.6 \times 10^{-10} \text{ mole}^2/\text{liter}^2 \text{ at } 25°\text{C} \end{aligned}$$

In principle, the most direct procedure is to shake the slightly soluble solid with a large volume of water until a saturated solution is obtained, filter it to separate the solution from the excess solid, measure exactly 1 liter of the saturated solution, evaporate the water from the solution, and weigh the solid residue. It is sometimes difficult, though, to be sure when saturation has been reached by this procedure.

EXAMPLE A saturated solution of $PbCl_2$ at 20°C was found to contain 4.50 g of $PbCl_2$ per liter. Calculate the solubility product constant of lead chloride.

$$4.50 \text{ g} \div 278 \text{ g/mole} = 1.62 \times 10^{-2} \text{ mole } PbCl_2$$

Therefore,

$$\begin{aligned}
[Pb^{++}] &= 1.62 \times 10^{-2} \text{ mole/liter} \\
[Cl^-] &= 3.24 \times 10^{-2} \text{ mole/liter} \\
K_{sp} &= [Pb^{++}][Cl^-]^2 \\
&= (1.62 \times 10^{-2} \text{ mole/liter})(3.24 \times 10^{-2} \text{ mole/liter})^2 \\
&= (1.62 \times 10^{-2})(10.5 \times 10^{-4}) \text{ mole}^3\text{/liter}^3 \\
&= 1.70 \times 10^{-5} \text{ mole}^3\text{/liter}^3
\end{aligned}$$

If the numerical value for the solubility product constant for a compound is known, it can be used to determine whether a precipitate will form when a stated volume of an electrolyte solution of given concentration is added to a specified volume of a solution of another electrolyte of known concentration.

EXAMPLE Will a precipitate form if 5.0 ml of 0.00001 *M* NaCl is added to 95.0 ml of 0.0001 *M* $AgNO_3$?

Each solution is diluted by the other when they are mixed. If no precipitation occurs to reduce the ion concentrations, the concentrations of the ions after mixing can be calculated from the dilution ratios:

$$\begin{aligned}
[Ag^+] &= \frac{95.0}{100.0} \times 0.0001\ M \\
&= 9.5 \times 10^{-5}\ M \\
[Cl^-] &= \frac{5.0}{100.0} \times 0.00001\ M \\
&= 5.0 \times 10^{-7} M \\
[Ag^+][Cl^-] &= (9.5 \times 10^{-5})(5.0 \times 10^{-7}) \\
&= 4.75 \times 10^{-11}
\end{aligned}$$

Since the product of the ion concentrations, 4.75×10^{-11}, is substantially less than K_{sp} for AgCl, 1.6×10^{-10}, precipitation is impossible in this instance. Precipitation cannot occur from a solution that is less than saturated.

The solubility product principle can likewise be employed to calculate the concentration of an ion remaining in solution after precipitation has occurred.

Precipitation cannot proceed beyond the point where the ion concentrations have been reduced sufficiently so that the solution in contact with the precipitate is a saturated solution.

EXAMPLE If 10 ml of 6.0 M HCl is added to 90 ml of 0.20 M $AgNO_3$, what will be the silver-ion concentration in the solution after precipitation of AgCl has ceased?

If no precipitation were to occur, the $[Ag^+]$ after mixing would be $\frac{90}{100} \times 0.20\ M = 0.18\ M$, and the $[Cl^-]$ would be $\frac{10}{100} \times 6.0\ M = 0.60\ M$. If *all* the Ag^+ ions were to precipitate out as AgCl, the $[Cl^-]$ in the solution would only be reduced to $0.60 - 0.18 = 0.42\ M$. Actually, of course, not all of the Ag^+ ions will precipitate out, but we may be sure that enough will do so to make the remaining $[Cl^-]$ approximately 0.42 M. At equilibrium, $[Ag^+][Cl^-] = 1.6 \times 10^{-10}$. If $[Cl^-]$ is approximately 0.42 M, then the approximate value for $[Ag^+]$ is

$$[Ag^+] = \frac{1.6 \times 10^{-10}}{0.42} \cong 3.8 \times 10^{-10}\ M$$

New Terms

Buffer solution: a solution capable of maintaining its pH value at a virtually constant level despite the addition of a small quantity of a strong acid or strong base.

Common-ion effect: the reduction in the degree of ionization of a weak electrolyte brought about by the presence of a strong electrolyte providing an ion in common.

Ionization constant (of a weak electrolyte): the equilibrium constant for the reaction in which molecules of a weak electrolyte dissociate into ions.

Ion-product constant (of water): the constant, K_w, which represents the product $[H^+][OH^-]$ for any aqueous solution.

pH: a number used to designate the degree of acidity or alkalinity of a solution, defined as the negative logarithm of the hydrogen-ion concentration.

Pseudosalt: a covalent compound containing a metallic element and a nonmetallic element, e.g., Al_2Cl_6, $SnCl_4$.

Solubility product constant: the equilibrium constant, K_{sp}, which applies to the equilibrium between crystals of an ionic solid and its saturated solution.

Solubility product principle: the product of the concentrations of the ions in a saturated solution of a slightly soluble strong electrolyte is a constant for a given temperature, provided each concentration is raised to the power equal to the subscript of the corresponding ion in the chemical formula.

Exercises

16-1 Hydrofluoric acid, HF, is 11 percent ionized in a 0.050 *M* solution. Calculate its ionization constant.

16-2 Calculate the hydrogen-ion concentration in a 0.050 *M* solution of HCN, for which the ionization constant is 4.0×10^{-10}.

16-3 Calculate the hydrogen-ion concentration of a solution containing 0.050 mole of HCN and 1.50 moles of NaCN per liter.

16-4 Calculate the hydroxide-ion concentration in a 0.05 *M* solution of hydrochloric acid, HCl.

16-5 Calculate the pH and pOH of the following solutions:
(a) A solution in which the hydrogen-ion concentration is 1×10^{-2} mole per liter
(b) A solution in which the hydroxide-ion concentration is 3×10^{-4} mole per liter
(c) A 1.00×10^{-3} *M* solution of acetic acid, 12.6 percent ionized
(d) A 0.05 *M* solution of formic acid, for which the ionization constant is 1.8×10^{-4}

16-6 In an aqueous solution containing 0.20 mole of acetic acid in a total volume of 350 liters, the acid is 16.4 percent ionized. Calculate the weight (in grams) of H^+ ions in the entire solution.

16-7 The solubility of silver chromate at room temperature is 0.00435 g of Ag_2CrO_4 per 100 ml of H_2O. Calculate K_{sp} for silver chromate.

16-8 If 0.01 mole of sodium chromate is dissolved in 100 ml of water, how many grams of silver chromate can then be dissolved?

References

Benson, S. W., *Chemical Calculations*, 2nd ed., Wiley, New York, 1963 (paperbound). Chapters 12, 13, and 14 deal with equilibria in ionic solutions.

Sienko, M. J., *Chemistry Problems*, Benjamin, New York, 1967 (paperbound). Each of the following topics has a chapter devoted to it: pH, buffer solutions, solubility products, hydrolysis.

Nebergall, W. H., Schmidt, F. C., and Holtzclaw, H. F., *General Chemistry*, 2nd ed., Heath, Boston, 1963, Chapter 45, "The Solubility Product Principle." A discussion of the formation and dissolution of precipitates.

17

Energy and Chemical Change

17-1 Heats of Formation and Bond Energies

The heat of formation of a compound (Section 8-2) is related to the attraction between the atoms within the molecules of the compound, but it is not a direct measure of bond strength. The situation is complicated because both bond-forming and bond-breaking are involved in a chemical reaction. For example, in the formation of water from elementary hydrogen and elementary oxygen, H—H bonds and O—O bonds are broken and H—O bonds are formed. Energy is required to break the H—H bonds and the O—O bonds; the formation of the H—O bonds releases energy. The reaction is exothermic, i.e., the heat of formation of water is negative, because *more* energy is released in the formation of the H—O bonds than is required to break the H—H and O—O bonds.

The reaction of elementary hydrogen with elementary iodine is endothermic, i.e., the heat of formation of hydrogen iodide is positive, because the energy released in the formation of H—I bonds in the hydrogen iodide molecules is *less* than the energy required to break the H—H bonds in the hydrogen molecules and the I—I bonds in the iodine molecules.

These examples show that the heat of formation is not a simple, direct measure of bond strength because it involves *differences* between bond-breaking and bond-forming energies.

17-2 The Laws of Thermochemistry

Thermochemistry is the branch of chemistry that deals with the thermal changes accompanying both chemical and physical transformations. One of the first laws of thermochemistry (1780) resulted from the work of Lavoisier and Laplace who observed that the quantity of heat required for the decomposition of a compound into its constituent elements is equal to the quantity of heat released when the same quantity of the compound is formed from its elements. We now accept this principle as a necessary corollary to the Law of Conservation of Energy, which has been generally accepted since about 1850. The significant implication of the Law of Lavoisier and Laplace is that we can reverse any thermochemical equation, provided we also reverse the sign of the heat of reaction.

As stated in Section 7-4, the heat of a reaction is simply the difference between the heat content of the products and the heat content of the reactants. If we reverse a thermochemical equation and adhere to the definition (Section 15-1) that the reactants are the substances written on the left in a chemical equation, the reactants in the one equation are the products in the other equation. Therefore, the heat of reaction is the same in magnitude but opposite in sign. For example,

$$2H_2(g) + O_2(g) \rightarrow 2H_2O(g) \qquad \Delta H = -116 \text{ kcal}$$
$$2H_2O(g) \rightarrow 2H_2(g) + O_2(g) \qquad \Delta H = +116 \text{ kcal}$$

One of the most useful principles of thermochemistry was stated in 1840 by the Swiss-Russian chemist G. H. Hess. This law, another necessary corollary to the Law of Conservation of Energy, is known as Hess's Law of Constant Heat Summation. It states that the quantity of heat evolved or absorbed in a reaction is the same whether it occurs in a single step or in a series of steps. For example, the heat of the reaction

$$CS_2(l) + 3O_2(g) \rightarrow CO_2(g) + 2SO_2(g) \qquad \Delta H = -257 \text{ kcal}$$

can be calculated from the heats of the following reactions:

$$C(s) + 2S(s) \rightarrow CS_2(l) \qquad \Delta H = +21 \text{ kcal}$$
$$C(s) + O_2(g) \rightarrow CO_2(g) \qquad \Delta H = -94 \text{ kcal}$$
$$S(s) + O_2(g) \rightarrow SO_2(g) \qquad \Delta H = -71 \text{ kcal}$$

If we reverse the first equation and double the third one, we have three equations which can be added to give the equation for the combustion of carbon disulfide:

$$\begin{array}{ll} CS_2(l) \rightarrow C(s) + 2S(s) & \Delta H = -21 \text{ kcal} \\ C(s) + O_2(g) \rightarrow CO_2(g) & \Delta H = -94 \text{ kcal} \\ 2S(s) + 2O_2(g) \rightarrow 2SO_2(g) & \Delta H = -142 \text{ kcal} \\ \hline CS_2(l) + 3O_2(g) \rightarrow CO_2(g) + 2SO_2(g) & \end{array}$$

$$\Delta H = (-21 - 94 - 142) \text{ kcal} = -257 \text{ kcal}$$

The heat of the reaction thus calculated is in agreement with the experimental value.

17-3 Predicting Heats of Reaction

The laws of thermochemistry enable us to predict heats of reaction, as illustrated above for the heat of combustion of carbon disulfide. The usefulness of these laws becomes most apparent when we use them to calculate heats of reaction that would be difficult or even impossible to determine experimentally.

EXAMPLE Calculate the heat of formation of *n*-pentane. This means we are to calculate the heat of the reaction

$$5C(s) + 6H_2(g) \rightarrow n\text{-}C_5H_{12}(l)$$

This reaction cannot be carried out in the laboratory. However, the heats of the following reactions have been determined:

$$\begin{aligned} n\text{-}C_5H_{12}(l) + 8O_2(g) &\rightarrow 5CO_2(g) + 6H_2O(l) & \Delta H &= -837 \text{ kcal} \\ C(s) + O_2(g) &\rightarrow CO_2(g) & \Delta H &= -94 \text{ kcal} \\ 2H_2(g) + O_2(g) &\rightarrow 2H_2O(l) & \Delta H &= -136 \text{ kcal} \end{aligned}$$

If we reverse the first equation, multiply the second one by 5, and multiply the third one by 3, we have three equations which add to give the desired one:

$$\begin{aligned} 5CO_2(g) + 6H_2O(l) &\rightarrow n\text{-}C_5H_{12}(l) + 8O_2(g) & \Delta H &= +837 \text{ kcal} \\ 5C(s) + 5O_2(g) &\rightarrow 5CO_2(g) & \Delta H &= -470 \text{ kcal} \\ 6H_2(g) + 3O_2(g) &\rightarrow 6H_2O(l) & \Delta H &= -408 \text{ kcal} \\ \hline 5C(s) + 6H_2(g) &\rightarrow n\text{-}C_5H_{12}(l) \end{aligned}$$

$$\Delta H = (837 - 470 - 408) \text{ kcal} = -41 \text{ kcal}$$

Thus, the heat of formation of liquid *n*-pentane is -41 kcal per mole.

EXAMPLE Calculate the heat of formation of carbon monoxide. This means we are to calculate the heat of the reaction

$$2C(s) + O_2(g) \rightarrow 2CO(g)$$

The heat of the reaction would be very difficult to determine experimentally because carbon monoxide burns more easily than carbon. If we attempt to burn a sample of carbon in the theoretical amount of oxygen required to form carbon monoxide by the above equation, we actually get a mixture of carbon monoxide, carbon dioxide, and unburned carbon. However, the heats of combustion of carbon (to form carbon dioxide) and of carbon monoxide have both been determined:

$$\begin{aligned} C(s) + O_2(g) &\rightarrow CO_2(g) & \Delta H &= -94.05 \text{ kcal} \\ 2CO(g) + O_2(g) &\rightarrow 2CO_2(g) & \Delta H &= -135.26 \text{ kcal} \end{aligned}$$

If we double the first equation and reverse the second one, we have two equations which can be added to give the desired one:

$$
\begin{array}{rlr}
2C(s) + 2O_2(g) \rightarrow 2CO_2(g) & & \Delta H = -188.10 \text{ kcal} \\
2CO_2(g) \rightarrow 2CO(g) + O_2(g) & & \Delta H = +135.26 \text{ kcal} \\
\hline
2C(s) + O_2(g) \rightarrow 2CO(g) & &
\end{array}
$$

$$\Delta H = (-188.10 + 135.26) \text{ kcal} = -52.84 \text{ kcal}$$

Thus, the heat of formation of carbon monoxide is -26.42 kcal per mole.

EXAMPLE How does the heat content (enthalpy) of diamond compare with that of graphite?

The actual value of H cannot be determined for any substance, but the difference between the H values of diamond and graphite is simply the heat (ΔH) of the reaction

$$\text{C(diamond)} \rightarrow \text{C(graphite)}$$

It is impossible to convert diamond to graphite or graphite to diamond under conditions that enable us to measure the heat of reaction, which is small. However, the heats of combustion have been determined for both diamond and graphite and have been found to be slightly different:

$$\text{C(diamond)} + O_2(g) \rightarrow CO_2(g) \qquad \Delta H = -94.50 \text{ kcal}$$

$$\text{C(graphite)} + O_2(g) \rightarrow CO_2(g) \qquad \Delta H = -94.05 \text{ kcal}$$

If we reverse the second equation and add it to the first one, we obtain

$$\text{C(diamond)} \rightarrow \text{C(graphite)}$$

$$\Delta H = (-94.50 + 94.05) \text{ kcal} = -0.45 \text{ kcal}$$

Thus, the heat content, or enthalpy, of diamond exceeds that of graphite by 0.45 kcal per mole.

17-4 Thermochemical Calculations with Heats of Formation

The heat of any reaction may readily be calculated by subtracting the sum of the heats of formation of the reactants from the sum of the heats of formation of the products. All elementary substances taking part in the reaction, either as reactants or products, are omitted because the "heat of formation" is zero, by definition, for an elementary substance. The heat of formation per mole of each substance, of course, must be multiplied by the number of moles involved in the reaction.

EXAMPLE Calculate the heat of the reaction

$$CS_2(l) + 3O_2(g) \rightarrow CO_2(g) + 2SO_2(g)$$

From Table 17-1 we obtain the following heats of formation:

$$CS_2(l),\ +21.0 \text{ kcal/mole}$$
$$CO_2(g),\ -94.0 \text{ kcal/mole}$$
$$SO_2(g),\ -71.0 \text{ kcal/mole}$$

For the products of the reaction the sum is

$$-94.0 \text{ kcal} + 2(-71.0) \text{ kcal} = -94.0 \text{ kcal} - 142.0 \text{ kcal} = -236.0 \text{ kcal}$$

For the reactants the sum is +21.0 kcal. Subtracting the sum for the reactants from the sum for the products, we find for the heat of reaction that

$$\Delta H = -236.0 \text{ kcal} - 21.0 \text{ kcal} = -257.0 \text{ kcal}$$

A converse application is to calculate the heat of formation of a compound from the heats of formation of the other substances taking part in a reaction. This requires, of course, that the heat of the reaction be determined experimentally.

EXAMPLE The heat of combustion of benzene has been found to be −757.7 kcal per mole. Calculate its heat of formation.

Writing the equation for the combustion of benzene,

$$2C_6H_6(l) + 15O_2(g) \rightarrow 12CO_2(g) + 6H_2O(l) \qquad \Delta H = -1514.4 \text{ kcal}$$

we see that we need to obtain the heats of formation of carbon dioxide and water from Table 17-1:

$$CO_2(g),\ -94.0 \text{ kcal/mole}$$
$$H_2O(l),\ -68.3 \text{ kcal/mole}$$

For the products of our reaction the sum is

$$12(-94.0) \text{ kcal} + 6(-68.3) \text{ kcal} = -1128.0 \text{ kcal} - 409.8 \text{ kcal} = -1537.8 \text{ kcal}$$

For the reactants the sum is $2x$, where x is the heat of formation per mole of benzene. Subtracting the sum for the reactants from the sum for the products, we obtain the expression for the heat of reaction:

Table 17-1

Standard heats of formation

Compound	ΔH_f° (kcal/mole at 25°C and 1 atm)
$H_2O(l)$	−68.32
$H_2O(g)$	−57.80
$H_2S(g)$	−4.82
$NH_3(g)$	−11.04
$CH_4(g)$	−17.89
$CO(g)$	−26.42
$CO_2(g)$	−94.05
$CS_2(l)$	+21.0
$CCl_4(l)$	−33.34
$C_2H_6(g)$	−20.24
$C_2H_4(g)$	+12.50
$C_2H_2(g)$	+54.19
$C_6H_6(l)$	+11.72
$HF(g)$	−64.20
$HCl(g)$	−22.06
$HBr(g)$	−8.66
$HI(g)$	+6.20
$SO_2(g)$	−70.96

$$\Delta H = -1537.8 \text{ kcal} - 2x = -1514.4 \text{ kcal}$$
$$\therefore 2x = -1514.4 \text{ kcal} + 1537.8 \text{ kcal}$$
$$= +23.4 \text{ kcal}$$
$$x = +11.7 \text{ kcal}$$

17-5 Use of Bond Energies

It is sometimes desired to obtain an estimate for the heat of a reaction in which the heat of formation for one of the reactants or products is not known. In this event it is necessary to make an estimate of the unknown heat of formation. Such an estimate can be arrived at by use of bond energies.

As implied in Section 17-1, the heat of a reaction is the difference between the bond energies of the product molecules and the bond energies of the reactant molecules. The definition of bond energy is easiest for diatomic molecules. The bond energy for H_2, for example, is the energy required to break the bond between the H atoms in a mole of H_2 molecules. This energy can be measured by spectroscopic methods. It was alluded to in Section 8-2, where it was stated that energy in the amount of 104 kcal is absorbed in the dissociation of 1 mole

Table 17-2

Average bond energies

Bond	*Energy* (kcal/mole)
H—H	104
F—F	36
Cl—Cl	57
Br—Br	46
I—I	36
H—F	135
H—Cl	102
H—Br	86
H—I	71
O—H	111
S—H	88
N—H	93
C—H	99
C—C	83
C=C	148
C≡C	198
O—O	33
N—N	38

of H_2 molecules. Thus, bond energy and dissociation energy are identical for diatomic molecules.

Bond energies for fluorine, chlorine, bromine, iodine, and the corresponding hydrogen halides are listed in Table 17-2. From this information we can predict the heats of formation of the hydrogen halides from their elements. For example, the reaction

$$H_2(g) + Cl_2(g) \longrightarrow 2HCl(g)$$

requires breaking an H—H bond and a Cl—Cl bond, but two H—Cl bonds are formed. Therefore, the heat evolved in the formation of 2 moles of HCl from its elements is (2×102) kcal $-$ $(104 + 57)$ kcal $=$ 43 kcal. In other words, we can predict from the bond energies that the heat of formation of HCl from its elements is approximately -22 kcal per mole, which agrees with the experimental value.

Some of the bond energies listed in Table 17-2 are only average values. For example, the heat absorbed in the process

$$CH_4(g) \longrightarrow C(g) + 4H(g)$$

is 397 kcal per mole of CH_4. Since there are four C—H bonds in each CH_4 molecule, the average C—H bond energy is one fourth of this value or 99 kcal per mole of C—H bonds. This value is the one to be used in estimating the heats of reactions in which all four C—H bonds are broken. However, spectroscopic determination of the bond-dissociation energies of the individual bonds shows that the values for the different steps in the atomization of a CH_4 molecule are not identical.

$$CH_4(g) \rightarrow CH_3(g) + H(g) \qquad \Delta H = 103 \text{ kcal}$$

$$CH_3(g) \rightarrow CH_2(g) + H(g) \qquad \Delta H = 87 \text{ kcal}$$

$$CH_2(g) \rightarrow CH(g) + H(g) \qquad \Delta H = 125 \text{ kcal}$$

$$CH(g) \rightarrow C(g) + H(g) \qquad \Delta H = 81 \text{ kcal}$$

The sum of the four bond-dissociation energies, 396 kcal, is in good agreement with the thermochemically calculated value of 397 kcal. The bond-dissociation energy of a bond is the energy required to break that bond alone. For diatomic molecules the bond energy and the bond-dissociation energy are the same, but for polyatomic molecules they are different. The bond energy should be called the *average* bond energy when reference is made to polyatomic molecules. The average bond energy is the sum of the successive bond-dissociation energies divided by the number of bonds.

The Pauling method for establishing a scale of electronegativities involves the use of bond energies. The relation between the bond energy concept and the electronegativity concept can be seen by comparing the dissociation energies of H_2, Cl_2, and HCl. The H—H bond energy is 104 kcal per mole. In the H_2 molecule the sharing of the electron pair by the two H atoms is equal, so we can reasonably assume that each H atom contributes half of the bond energy, or 52 kcal per mole of H—H bonds. Similarly, from the dissociation energy of Cl_2, 57 kcal per mole, we can assume that each Cl atom contributes 28.5 kcal per mole of Cl—Cl bonds. The bond in HCl is polar, but suppose for a moment that it is nonpolar, i.e., that the bonding electron pair is shared equally by the H and Cl atoms. Each H atom would contribute 52 kcal per mole of H—Cl bonds, and each Cl atom would contribute 28.5 kcal per mole of H—Cl bonds, making the total bond energy 52 kcal + 28.5 kcal = 80.5 kcal. Actually, the H—Cl bond energy is 102 kcal per mole. The electrons are *not* equally shared—the bond is stronger than if the sharing were equal. The additional binding energy is assumed to depend on the difference between the electron-attracting abilities of the bonded atoms. In other words, there is an *ionic* contribution to the bond energy that is in addition to the covalant contribution. The heat of formation of a hydrogen halide is a measure of how much stronger the H—X bond is (because of its partial ionic character) than the average of H—H and X—X.

When we make similar calculations for the other hydrogen halides, we find that the differences between the calculated values and the observed values decrease regularly as follows:

	Obs.		*Calc.*	
HF	135	−	70	= 65
HCl	102	−	80	= 22
HBr	86	−	75	= 11
HI	71	−	70	= 1

The fact that the discrepancy is greatest for HF and least for HI implies that the sharing of electrons is most unequal in the HF molecule and most nearly equal in the HI molecule. This thought is often expressed by saying that in HF the partial ionic character of the bond is greatest whereas in HI it is least. Pauling's electronegativity values were selected by a complex empirical procedure to account for the actual bond energies.

17-6 The Direction of Chemical Change

One of the most important questions with which a chemist is faced is whether or not a particular reaction can take place. Such a question, of course, can be answered by experiment, but the chemist would prefer not to waste time and effort on impossible reactions. Furthermore, a negative experimental result may by misleading, for it does not prove that a reaction cannot occur. It may only mean that the reaction has taken place too slowly to be detected—in which case a good catalyst might lead to success. The ideal is to establish criteria for *predicting in advance* whether a reaction can occur spontaneously, i.e., whether it *can* take place, *regardless of rate*, without the use of some kind of external driving force.

Spontaneous changes are changes that can occur within systems when the systems are left to themselves. We are familiar with the fact that water runs downhill, that heat flows from hotter bodies to colder bodies, that gases expand into regions of lower pressure, and that chemical reactions proceed toward equilibrium. Our experience tells us that spontaneous processes do not reverse themselves. We use the term *nonspontaneous change* in referring to the reverse of a spontaneous change. Nonspontaneous changes do not occur when systems are left to themselves. Water does not run uphill, heat does not flow from a cold body into a hotter one, gases do not compress themselves, and chemical reactions that have reached equilibrium do not proceed further.

It is possible to "force" a nonspontaneous change to take place within a system by supplying energy from outside the system. In such a case we say we are doing work upon the system. We can use outside energy to pump water uphill, to transfer heat from a colder region to a hotter region (as in a refrigerator), to compress a gas (as when we inflate a tire), or to make chemical reactions "run backward" (as in electrolytic cells). Conversely, it is always possible—in principle at least—to devise a means for obtaining useful work from a system in which a spontaneous change is taking place. Falling water can operate a turbine, hot and cold reservoirs may be used to operate a heat engine, an expanding gas can push a piston, and a chemical reaction may be harnessed to produce an electric current (in a voltaic cell). Since a spontaneous process can be utilized for the performance of useful work, it follows that a system loses capacity for doing useful work when a spontaneous change occurs within it—whether or not *any* useful work is actually done.

Intuitively, we tend to feel that the ability of substances to react with each other should be related in some way to the energy change that accompanies the reaction. The thermochemical studies of Julius Thomsen (1854) in Denmark and Marcellin Berthelot (1867) in France led them to assume that heat of reaction, ΔH, is a measure of "chemical affinity." In other words, it was accepted that only exothermic reactions could occur spontaneously. However, the "Thomsen-Berthelot Principle" was wrong. It has been definitely established that there are many spontaneous endothermic reactions. The relation between chemical change and energy change is a more subtle one than was at first realized.

We can usually predict whether a reaction is spontaneous at room temperature and atmospheric pressure by noting the sign of ΔH as calculated from tables of heats of formation, but this cannot be used as a general criterion at other temperatures and pressures. Although ΔH changes only slowly with temperature, the spontaneity of a reaction is highly sensitive to variations in temperature and pressure.

The notable American mathematician J. Willard Gibbs, who founded the science of chemical thermodynamics, was the first to hold the modern view that a spontaneous reaction is one that is capable of doing useful work.

17-7 The Concept of Free Energy

Most chemical reactions are carried out in open containers. If the products occupy a greater volume than the reactants, work is required to push back the atmosphere. Since this work of expansion cannot be utilized for any other purpose, it is excluded from consideration when we use the term *useful work* in connection with a reaction taking place at constant temperature and pressure. Useful work would be, for example, the electrical work done by a voltaic cell. How nearly the useful work actually performed approaches the maximum of which the reaction is capable depends on the operating efficiency of the device by means of which the work is done. Our concern here is with the maximum amount of useful work of which the reaction is capable. If the reaction is permitted to occur without doing any useful work, the system nevertheless has lost its capacity for doing useful work in the same amount as if the work had been done. By analogy, water running downhill loses its capacity for doing useful work by an amount that depends on how far it falls, and this in no way is related to whether or not it impinges against the blades of a turbine as it falls.

Every substance is assumed to have a certain amount of internal energy of which only a portion is extractable as useful work. This portion of the internal energy that is potentially extractable as useful work is referred to as *Gibbs free energy*. The maximum amount of useful work that can be obtained from a process taking place at constant temperature and pressure is the difference in Gibbs free energy between the products and the reactants:

$$\Delta G_{\text{reaction}} = G_{\text{products}} - G_{\text{reactants}}$$

The quantity ΔG is called the Gibbs free energy change (or simply free energy change) because it refers to energy that is "free" to do useful work. The name implies that the remainder of the energy contained within the system is unavailable for doing useful work in a constant-temperature, constant-pressure process.

For reactions taking place at constant temperature and constant pressure, the quantity ΔG is the ultimate criterion for judging whether they will occur spontaneously. For a process in which useful work is done or could be done, ΔG is negative and equal in amount to the *maximum* useful work that could be done. Therefore, the criteria for spontaneity are as follows:

1. A reaction is spontaneous if ΔG is negative.
2. A reaction is nonspontaneous, i.e., the reverse reaction is spontaneous, if ΔG is positive.
3. A system is in a state of equilibrium if $\Delta G = 0$.

It must be emphasized that this criterion of spontaneity says nothing about how long it may take for a spontaneous process to occur.

The ΔG value for a reaction, like ΔH, is independent of the number of steps taken to convert the original reactants to the final products. The ΔG value for a reaction can be calculated by adding the ΔG values for the individual steps, but it is important that each ΔG should pertain to the same temperature and pressure. The ΔG value for a reaction at 25°C (298°K) and 1 atmosphere is labeled ΔG°_{298}.

EXAMPLE Calculate the standard free energy of formation for sulfur dioxide from the following data:

$$SO_2(g) + Cl_2(g) \rightarrow SO_2Cl_2(g) \qquad \Delta G^\circ_{298} = -2.27 \text{ kcal}$$

$$\text{S(rhombic)} + O_2(g) + Cl_2(g) \rightarrow SO_2Cl_2(g) \qquad \Delta G^\circ_{298} = -74.06 \text{ kcal}$$

Reversing the first equation and adding it to the second one, we obtain the desired result:

$$\text{S(rhombic)} + O_2(g) \rightarrow SO_2(g)$$

$$\Delta G^\circ_{298} = -74.06 \text{ kcal} + 2.27 \text{ kcal} = -71.79 \text{ kcal}$$

17-8 Free Energies of Formation

Free energies of reactions are obtainable by spectroscopic, thermochemical, and electrochemical methods. Tables of free energies of formation that are comparable to tables of heats of formation, have been set up. From these tables the free energy change of a proposed reaction can be calculated by subtracting the sum of the free energies of formation of the reactants from the sum of the free energies of formation of the products.

EXAMPLE Calculate the free energy change for the reaction

$$CH_4(g) + 4Cl_2(g) \rightarrow CCl_4(l) + 4HCl(g)$$

From Table 17-3 we obtain the following free energies of formation:

$$CH_4(g), \ -12.1 \text{ kcal/mole}$$
$$CCl_4(l), \ -16.4 \text{ kcal/mole}$$
$$HCl(g), \ -22.8 \text{ kcal/mole}$$

Table 17-3

Standard free energies of formation

Compound	ΔG_f° (kcal/mole at 25°C and 1 atm)
$H_2O(l)$	−56.7
$H_2O(g)$	−54.6
$H_2S(g)$	−7.9
$NH_3(g)$	−4.0
$CH_4(g)$	−12.1
$CO(g)$	−32.8
$CO_2(g)$	−94.3
$CS_2(l)$	+15.2
$CCl_4(l)$	−16.4
$C_2H_6(g)$	−7.9
$C_2H_4(g)$	+16.3
$C_2H_2(g)$	+50.0
$C_6H_6(l)$	+29.8
$HF(g)$	−64.7
$HCl(g)$	−22.8
$HBr(g)$	−12.7
$HI(g)$	+0.3
$SO_2(g)$	−71.8

For the products the sum is

$$-16.4 \text{ kcal} + 4(-22.8 \text{ kcal}) = -16.4 \text{ kcal} - 91.2 \text{ kcal}$$
$$= -107.6 \text{ kcal}$$

Subtracting the free energy of formation of methane from this sum, we find for the reaction that

$$\Delta G^\circ = -107.6 \text{ kcal} - (-12.1 \text{ kcal}) = -107.6 \text{ kcal} + 12.1 \text{ kcal}$$
$$= -95.5 \text{ kcal}$$

Since the calculated ΔG° value is a large negative quantity, the proposed reaction is a feasible one. (Chlorine *does* react with methane.)

Most of the compounds in Table 17-3 are shown to have negative free energies of formation. This means that most of them are stable with respect to decomposition into their elements at 25°C and 1 atmosphere. For example, the value for the free energy of formation of ammonia (−4.0 kcal per mole) means that the decomposition of ammonia into its elements is *not* a spontaneous reaction at room temperature and atmospheric pressure. Quite the contrary, the formation of ammonia from its elements is a spontaneous reaction, and it ought to be possible to produce ammonia by reacting hydrogen with nitrogen. In practice, the reaction is so slow at 25°C and 1 atmosphere that no detectable quantities of ammonia are formed even after a mixture of hydrogen and nitrogen gases has stood for many years. Higher temperatures, higher pressures, and a good catalyst are necessary for the success of the synthesis (Section 15-9). However, Fritz Haber never would have attempted to make the synthesis a success if the value calculated for ΔG under the proposed conditions had been a large positive number.

17-9 Free Energy Change and the Equilibrium Constant

It is an oversimplification to ask, simply, whether a reaction is spontaneous or nonspontaneous. The proper question to ask is not "Is this reaction possible?" but rather "How far will this reaction proceed before equilibrium is reached?" The value of the equilibrium constant tells how much reaction will occur before equilibrium is reached—or, what is tantamount to the same thing, the proportions in which the different reactants and products must be present initially in order to constitute an equilibrium mixture.

The relationship between the equilibrium constant and the standard free energy change of a reaction is given by the equation

$$\Delta G^\circ = -2.303RT \log K$$

where R is the molar gas constant (Section 9-9). The value of R to be used with this equation is 1.987 cal per mole-degree when ΔG° is given in cal per mole.

Thus, we can calculate the value of the equilibrium constant for any proposed reaction by consulting a table of standard free energies of formation.

EXAMPLE Calculate the value of the equilibrium constant for the reaction

$$CO(g) + H_2O(l) \rightleftharpoons CO_2(g) + H_2(g)$$

at 25°C.

From Table 17-3 we obtain the following standard free energies of formation:

$$\begin{aligned} &CO(g), \ -32.8 \text{ kcal/mole} \\ &H_2O(l), \ -56.7 \text{ kcal/mole} \\ &CO_2(g), \ -94.3 \text{ kcal/mole} \end{aligned}$$

Subtracting the sum of the standard free energies of formation of the reactants from the standard free energy of formation of carbon dioxide, we find

$$\begin{aligned} \Delta G^\circ_{\text{reaction}} &= -94.3 \text{ kcal} - (-32.8 \text{ kcal} - 56.7 \text{ kcal}) \\ &= -94.3 \text{ kcal} + 89.5 \text{ kcal} = -4.8 \text{ kcal} \\ &\qquad\qquad\qquad\qquad\qquad\quad = -4800 \text{ cal} \\ \log K &= -\frac{\Delta G^\circ}{2.303RT} \\ &= -\frac{-4800 \text{ cal/mole}}{2.303 \times 1.987 \text{ cal/mole-deg} \times 298^\circ} \\ &= 3.5 \\ \therefore K &= 3 \times 10^3 \end{aligned}$$

The magnitude of this equilibrium constant is large enough to indicate clearly that if only carbon monoxide and water are initially present, considerable reaction will take place before equilibrium is reached. In practice, however, this reaction is so slow at 25°C and 1 atmosphere that equilibrium would never be reached within the foreseeable future. At higher temperatures, with the H_2O in the gaseous state, the reaction proceeds rapidly.

17-10 The Concept of Entropy

Our experience tells us that some natural processes occur spontaneously with no energy change involved. For example, suppose we have a bulb filled with helium connected, through a stopcock, to another bulb filled with argon. When the stopcock is opened, each gas diffuses into the other bulb until a uniform mixture is present in both bulbs. It is conceivable that the chance motion of the molecules might, at some later time, result in all the helium molecules being in one bulb again and all the argon molecules in the other bulb. Our experience leads us not to expect this to happen soon, for the probability of the occurrence is vanishingly small when a macroscopic sample is employed.

With all the helium molecules in one bulb and all the argon molecules in the other bulb, a kind of order prevails that no longer exists after each bulb contains a mixture of the same composition as the mixture in the other bulb. We may generalize and say that a system, left to itself, tends to seek a state of maximum disorder.

Disorder can be described quantitatively in terms of a thermodynamic property called *entropy*. When a system undergoes a change that includes an increase in disorder, we say that the entropy of the system has increased. Using S to represent entropy and ΔS to represent change in entropy, we can say that ΔS is positive when there is an increase in disorder.

The tendency toward increase in entropy, i.e., the tendency toward maximum disorder, can be observed when two similar liquids are brought together. For example, heptane and octane molecules are so much alike that the attraction of heptane molecules for octane molecules is approximately the same as the attraction of heptane molecules for other heptane molecules or of octane molecules for other octane molecules. When the two liquids are brought together there is no operative factor to prevent them from mixing to form a homogeneous mixture, i.e., a solution. By mixing, they achieve maximum disorder—but no significant energy change occurs.

The "driving force" behind many chemical reactions is the tendency toward increased entropy. Consider, for example, the following reaction:

$$C(s) + H_2O(g) \rightarrow CO(g) + H_2(g) \qquad \Delta H = +31 \text{ kcal}$$

This reaction is endothermic, but it occurs spontaneously if the necessary energy is available. There is a high degree of order in the crystalline structure of carbon. Since both products of the reaction are gases, it is apparent that the reaction involves an increase in the disorder of the system.

The change in entropy for a chemical reaction is given by the equation

$$\Delta S_{\text{reaction}} = S_{\text{products}} - S_{\text{reactants}}$$

We can generalize and say that there is a *tendency* for those reactions to occur which are accompanied by an increase in entropy within the system, just as there is a *tendency* for those reactions to occur which are exothermic. However, just as some spontaneous reactions are endothermic, other spontaneous reactions are accompanied by a decrease in entropy *within the system*. An endothermic reaction cannot be spontaneous unless it is accompanied by an increase in entropy within the system. Neither can a spontaneous reaction be accompanied by a decrease in entropy within the system unless it is exothermic.

In order for a reaction to be spontaneous, ΔG must be negative—and in order for ΔG to be negative, either ΔH must be negative or ΔS must be positive.

17-11 The Relation Between ΔG and ΔH

One of the most important relationships in chemical thermodynamics is given by the equation

$$\Delta G = \Delta H - T\,\Delta S$$

This equation explains why the Thomsen-Berthelot principle was wrong. The sign of the free energy change, ΔG, tells whether or not a constant-temperature, constant-pressure reaction is spontaneous at a particular temperature and pressure; but at ordinary temperatures the sign of ΔG is not the same as the sign of ΔH unless ΔS is small. On the other hand, the equation explains why the change in entropy within the system cannot, by itself, determine whether a reaction is spontaneous. If ΔH is positive, ΔS must be highly positive in order for ΔG to be negative. On the other hand, a reaction may be spontaneous in spite of ΔS being negative if the reaction is sufficiently exothermic.

The meaning of the above equation can be summed up by saying that under conditions of constant temperature and constant pressure any system tends to go spontaneously toward a condition of minimum energy and maximum disorder.

For most (but not all) reactions near room temperature, $T\,\Delta S$ is small compared to ΔH. Therefore, in most cases, exothermic reactions are probable (i.e., ΔG is usually negative) and endothermic reactions are improbable. At high temperatures the large value of T makes $T\,\Delta S$ so large that the sign of ΔS determines the sign of ΔG. At low temperatures $T\,\Delta S$ becomes so small that the sign of ΔH determines the sign of ΔG.

In general, ΔS will be positive if large particles are decomposing into smaller ones. At very high temperatures all complex molecules cease to exist. Above 100,000°K even the atoms "decompose" into ions and electrons. Above 1,000,000°K the atoms are stripped down to their nuclei. The nuclei disintegrate at still higher temperatures.

17-12 Temperature and the Equilibrium Constant

The fact that $\Delta G°$ is temperature dependent is indicated by the equation $\Delta G° = \Delta H° - T\,\Delta S°$. The equilibrium constant is also temperature dependent for $\Delta G° = -RT \ln K$. (The symbolism "ln K" means "the natural logarithm of K" and is the same as 2.303 log K where "log K" is the logarithm of K to the base 10.)

In chemical thermodynamics it can be shown that in order for $\Delta H°$ to be constant over a range of temperatures, $\Delta S°$ must also be constant. It is an experimental fact that $\Delta H°$ and $\Delta S°$ do not change rapidly with temperature. Let us consider, then, a range of temperature over which $\Delta H°$ and $\Delta S°$ are constant. At the lower temperature,

$$\ln K_1 = \frac{-\Delta G_1^\circ}{RT_1}$$

and at the higher temperature

$$\ln K_2 = \frac{-\Delta G_2^\circ}{RT_2}$$

But $$-\Delta G_1^\circ = T_1\,\Delta S^\circ - \Delta H^\circ \text{ and } -\Delta G_2^\circ = T_2\,\Delta S^\circ - \Delta H^\circ$$

Thus, $$\ln K_1 = \frac{\Delta S^\circ}{R} - \frac{\Delta H^\circ}{RT_1}$$

and $$\ln K_2 = \frac{\Delta S^\circ}{R} - \frac{\Delta H^\circ}{RT_2}$$

Subtracting the next-to-last equation from the last equation, we obtain

$$\ln K_2 - \ln K_1 = \ln\frac{K_2}{K_1} = \frac{\Delta H^\circ}{R}\left(\frac{1}{T_1} - \frac{1}{T_2}\right)$$

This equation states that K_2 is greater than K_1 if the reaction is endothermic, in agreement with van't Hoff's Law (Section 15-6).

17-13 Chemical Kinetics

The ΔG value for a reaction tells us how nearly a reaction will need to approach completion to reach equilibrium, and the ΔH value tells us whether a reaction is exothermic or endothermic, but neither ΔG nor ΔH enables us to predict whether a reaction will take place rapidly or slowly. It is common knowledge that some reactions occur much more rapidly than others. The branch of chemistry which studies the rates at which reactions occur, i.e., the rate at which a chemical system approaches equilibrium, is called *chemical kinetics*.

It is found by experiment that reaction rates depend on the concentrations of the reactants. The usual procedure for determining how the rate of reaction changes with the concentrations of the different reactants is to keep all factors constant except the concentration of one of the reactants. Then the rate of reaction is determined in a series of experiments in which the concentration of the one reactant is changed in a systematic fashion (doubling, tripling, etc.). The rate is arrived at by observing how fast one reactant disappears or how fast one product forms. In reactions involving gases, the rate of the reaction can sometimes be followed by noting the change of pressure with time. For example, the reaction between nitrogen(II) oxide and hydrogen,

$$2NO(g) + 2H_2(g) \rightarrow N_2(g) + 2H_2O(g)$$

has been studied kinetically by measuring the change in pressure with time. The pressure drops as the reaction proceeds because only 3 moles of gas are formed when 4 moles of gas react.

17-14 Concentration and Reaction Rate

In studies of the reaction between nitrogen(II) oxide and hydrogen it has been established that doubling the initial concentration of H_2 doubles the reaction rate, tripling the concentration of H_2 triples the reaction rate, etc. In other words, the rate of reaction is found to be proportional to the concentration of H_2. On the other hand, doubling the initial concentration of NO quadruples the reaction rate, tripling the concentration of NO makes the reaction take place nine times as fast, etc. Thus, the reaction rate is found to be proportional to the *square* of the NO concentration. These findings can be expressed in the form of an equation known as the *rate law* for the reaction

$$\text{Rate} = k[H_2][NO]^2$$

The proportionality constant k is called the *specific rate constant*. It is characteristic of a particular reaction but it varies with temperature; the higher the temperature, the greater the value of k.

All rate laws are of the form

$$\text{Rate} = k[A]^m[B]^n \ldots$$

where m is the power to which the concentration of A must be raised, n is the power to which the concentration of B must be raised, and so on (as indicated by the dots) for each of the other reactants.

It is important to note that the exponents (m, n, etc.) in the rate law *must be determined by experiment*. In general, the exponents are *not* the same as the coefficients in the balanced net equation for the reaction. As we have just seen, the exponents for the reaction between H_2 and NO are 1 and 2, respectively, although the coefficients in the equation for the reaction are 2 and 2. This will be explained in the next section.

17-15 Reaction Mechanism

One of the challenging aspects of chemical kinetics is the problem of accounting for the exponents in the rate law. The *collision theory* assumes that a reaction takes place in steps and that each step requires a collision between the reacting particles. The slowest step in the sequence is the one that determines the measured reaction rate for it controls the rate at which any subsequent step can occur.

The rate law found for the reaction between nitrogen(II) oxide and hydrogen can be accounted for by assuming the following steps:

$$\begin{array}{rr} \text{STEP 1:} & H_2(g) + NO(g) + NO(g) \rightarrow N_2O(g) + H_2O(g) \\ \text{STEP 2:} & H_2(g) + N_2O(g) \rightarrow N_2(g) + H_2O(g) \\ \hline \text{NET EQUATION:} & 2H_2(g) + 2NO(g) \rightarrow N_2(g) + 2H_2O(g) \end{array}$$

If Step 1 is assumed to be the slow step, then we can explain why the reaction rate is dependent on the first power of the concentration of H_2 and on the second power of the concentration of NO. Surely Step 1 would be the slow one for it involves a three-body collision, whereas Step 2 involves a two-body collision. The simultaneous collision of three molecules is much less probable than a collision involving only two molecules.

Three-body collisions are so highly improbable that many chemical kineticists object to the mechanism proposed above in favor of an alternative mechanism that involves only two-body collisions:

$$\text{STEP 1:} \qquad 2NO(g) \rightarrow N_2O_2(g)$$
$$\text{STEP 2:} \qquad N_2O_2(g) + H_2(g) \rightarrow N_2O(g) + H_2O(g)$$
$$\text{STEP 3:} \qquad N_2O(g) + H_2(g) \rightarrow N_2(g) + H_2O(g)$$

The second step is assumed to be the rate-determining step. Since the collision of two NO molecules is required to form the N_2O_2 for the second step, the concentration of N_2O_2 is proportional to the second power of the NO concentration; i.e., $[N_2O_2] = k_1[NO]^2$ where k_1 is the specific rate constant for the first step. Therefore, the rate of the second step is proportional to the square of the concentration of NO multiplied by the concentration of H_2; i.e., $k_2[N_2O_2][H_2] = k_2k_1[NO]^2[H_2]$. This mechanism accounts for the observed rate law as well as the first one and it involves no three-body collisions. There is no way of knowing for certain whether this is really the mechanism for the reaction. All that we can do is to find a mechanism consistent with the data. Even if we can find no other mechanism at the moment this does not mean that our proposed mechanism is the only possible one.

17-16 Energy of Activation

The collision theory assumes that molecules do not undergo reaction upon colliding unless they collide with sufficient energy to cause bond breaking. The kinetic theory of gases (Section 9-12) assumes that most molecules of a gas have velocities near the average value but that some molecules have velocities far in excess of the average. Only the exceptional molecules with the highest velocities have sufficient kinetic energy to break bonds upon impact.

The extra energy colliding molecules must have to produce chemical reaction is called *energy of activation*. The magnitude of the energy of activation depends on the nature of the reactants, for some bonds are more easily broken than others. A reaction that has a large energy of activation is a slow reaction because only a small fraction of the colliding particles have as much kinetic energy as is required to bring about the necessary bond breaking. Reactions with small energies of activation take place more rapidly because a greater fraction of the collisions are effective.

An increase in temperature makes the average kinetic energy greater and causes a greater fraction of the colliding molecules to have the necessary energy to break bonds upon impact. Increasing the temperature has no effect on the energy of activation, but it causes a greater fraction of the molecules to have as much kinetic energy as they need to furnish the required energy of activation.

All reactions take place more rapidly with increasing temperature, but the magnitude of the effect is different for different reactions. Many reactions double or even triple their rates for each 10°C rise in temperature.

It seems logical to assume that the energy of activation of the slowest step in the reaction mechanism is the one that is important in determining the rate at which the reaction products are formed. The effect of a catalyst is somehow to cause a change of mechanism. The activation energy of the rate-determining step in the new mechanism must be less than for the old mechanism.

The details of how a catalyst operates have to be investigated separately in each instance. A substance that is a catalyst for one reaction usually is not a catalyst for another reaction. The highly specific nature of catalysis is especially remarkable in the reactions that take place in living organisms. Most of the reactions essential to life are catalyzed by organic substances of high molecular weight called *enzymes*. Each enzyme catalyzes a particular reaction. A better understanding of how enzymes operate might enhance significantly our control over biochemical processes and our hope for victory over such enemies as malignancy.

New Terms

Bond-dissociation energy: the energy required to break a particular bond. It is the same as (average) bond energy if a molecule is diatomic.

Bond energy: the energy required to break the bond between two atoms. Average bond energy is the total energy required to break a set of similar bonds divided by the number of bonds. Bond energy always means average bond energy.

Chemical kinetics: the branch of chemistry that deals with reaction rates and mechanisms.

Chemical thermodynamics: the general study of the relations between energy and chemical change. (Thermochemistry is included, but chemical kinetics is not. Thermodynamics is concerned with what is *possible*, kinetics is concerned with *how*, and *how fast*, a process actually takes place.)

Collision theory: the theory that assumes that a reaction takes place in one or more steps involving collisions between reacting particles.

Electrolytic cell: an electrochemical cell in which an oxidation-reduction reaction is brought about by an electric current from an outside source.

Energy of activation: the extra energy colliding particles must have in order to produce chemical reaction.

Entropy: the thermodynamic measure of the disorder within a system.

Enzymes: organic substances of high molecular weight which catalyze reactions taking place within living organisms.

Gibbs free energy: internal energy that is extractable as useful work.

Law of Hess (Hess's Law of Constant Heat Summation): the quantity of heat evolved or absorbed in a reaction is the same when it occurs in a series of steps as when it occurs in a single step.

Law of Lavoisier and Laplace: the quantity of heat required for the decomposition of a compound into its constituent elements is the same as the quantity of heat released when the same quantity of the compound is formed from its elements.

Rate law: an equation that states the relation between the rate of a reaction and the concentrations of the reactants.

Spontaneous change: a change that can occur within a system, regardless of rate, without the use of an external driving force.

Thermochemistry: the branch of chemistry that deals with the changes in heat energy that accompany chemical reactions and such physical transformations as melting and evaporation.

Thomsen-Berthelot principle: a fallacious idea, no longer accepted, according to which only exothermic processes are spontaneous.

Voltaic cell (galvanic cell): a device that produces electric current by means of an oxidation-reduction reaction.

Exercises

17-1 Using the data in Table 17-1, calculate the heats of the following reactions:

(a) $CH_4(g) + 4Cl_2(g) \rightarrow CCl_4(l) + 4HCl(g)$
(b) $2HI(g) + Cl_2(g) \rightarrow 2HCl(g) + I_2(s)$
(c) $3C_2H_2(g) \rightarrow C_6H_6(l)$
(d) $C_2H_2(g) + H_2(g) \rightarrow C_2H_4(g)$
(e) $C_2H_2(g) + 2H_2(g) \rightarrow C_2H_6(g)$
(f) $2C_2H_2(g) + 5O_2(g) \rightarrow 4CO_2(g) + 2H_2O(g)$

17-2 Given: $2SO_2(g) + O_2(g) \rightarrow 2SO_3(g) \quad \Delta H = -43$ kcal
Calculate the heat of formation of sulfur trioxide from its elements. (Consult Table 17-1.)

17-3 Using Table 17-1, calculate the molar heats of combustion of ethane, ethene, and ethyne, and then calculate the amounts of heat released upon burning 1 g of each.

17-4 Tungsten is obtained by the reaction $3H_2(g) + WO_3(s) \rightarrow W(s) + 3H_2O(g)$. Calculate ΔH for the reaction, given that the heat of formation of $WO_3(s)$ is -200.84 kcal per mole. (See Table 17-1.)

17-5 Given the following thermochemical equations:

$$H_2(g) + I_2(s) \rightarrow 2HI(g) \qquad \Delta H_{298} = +12.4 \text{ kcal}$$
$$H_2(g) + I_2(g) \rightarrow 2HI(g) \qquad \Delta H_{298} = -2.7 \text{ kcal}$$

Calculate the heat of sublimation of iodine.

17-6 From information given in Table 17-3, determine whether the following reactions are spontaneous:

(a) $2NH_3(g) \rightarrow N_2(g) + 3H_2(g)$
(b) $C_2H_2(g) \rightarrow 2C(s) + H_2(g)$

(c) $C_2H_6(g) \rightarrow 2C(s) + 3H_2(g)$
(d) $Cl_2(g) + 2HI(g) \rightarrow 2HCl(g) + I_2(s)$

17-7 At its boiling point a liquid is in equilibrium with its vapor when the vapor is at a pressure of 1 atmosphere. Therefore, $\Delta G = 0$ and $\Delta H = T\Delta S$. Trouton's Rule (Section 10-9) states, in effect, that the entropy of vaporization is about 22 cal per mole-deg. In terms of the entropy concept, explain the failure of water to conform to Trouton's Rule (see Sections 12-2 and 12-5).

17-8 For the reaction $2NO_2(g) \rightleftharpoons N_2O_4(g)$, the equilibrium constant is 172 at 25°C and 79.4 at 35°C. Is the dimerization of NO_2 exothermic or endothermic? Explain.

17-9 "Heating a reaction mixture will increase the rate of reaction if the reaction is endothermic, but not if it is exothermic." Do you believe this? Defend your answer.

17-10 How would you prepare a "three-minute egg" on a high mountain? Explain.

References

Allen, J. A., *Energy Changes in Chemistry*, Allyn & Bacon, Boston, 1966 (paperbound). An elementary presentation of basic ideas of thermodynamics and kinetics.

Campbell, J. A., *Why Do Chemical Reactions Occur?* Prentice-Hall, Englewood Cliffs, N.J., 1965 (paperbound). Stimulating discussion of thermodynamic topics.

King, E. L., *How Chemical Reactions Occur*, Benjamin, New York, 1963 (paperbound). An introduction to kinetics and reaction mechanisms.

Mahan, B. H., *Elementary Chemical Thermodynamics*, Benjamin, New York, 1963 (paperbound). Thermodynamic principles and some applications.

Nash, L. K., *Chemical Thermodynamics*, Addison-Wesley, Reading, Mass., 1962 (paperbound). The first and second laws of thermodynamics and their consequences.

Strong, L. E., and Stratton, W. J., *Chemical Energy*, Reinhold, New York, 1965 (paperbound). Presents thermodynamics without calculus.

18

Hydrocarbons

18-1 Methane, the Simplest Hydrocarbon

Of all the chemical elements, carbon is the one which forms the greatest number of different compounds. There are thousands of compounds consisting of carbon and hydrogen alone. These compounds are called *hydrocarbons*.

Methane is the only hydrocarbon that contains a single carbon atom per molecule. Its molecular formula is the same as its empirical formula, CH_4. Methane constitutes the major portion of the natural gas obtained from gas wells and used as a fuel. Quantities of it occur in solution in the complex hydrocarbon mixture known as petroleum, obtained from oil wells. Methane is formed in nature by the action of anaerobic bacteria on cellulose and other organic matter. In marshes and swamps the gas bubbles which are often seen rising to the surface of the water contain methane, carbon dioxide, and nitrogen. An old and popular name for methane is "marsh gas." It is also present in coal mines, where it is one of the causes of explosions and is known as "fire damp."

Methane, a nonpolar substance, does not dissolve in water because its molecules are unable to form hydrogen bonds with water molecules. Condensation of the gas to liquid requires cooling to $-160°C$ at standard pressure. The gas is colorless and odorless.

18-2 Methane and the Octet Rule

The atoms of the noble gases, except helium, have eight electrons in the outermost shell, and these electrons, occupying four orbitals, are called an octet. Atoms of other elements frequently achieve a noble-gas structure by acquiring enough additional electrons to complete the octet (Section 5-1). The acquisition may be made by sharing or by outright transfer. In a stable molecule each atom frequently has a noble-gas structure.

Fluorine has seven electrons in its outer shell, three pairs in three of the orbitals and a lone electron in the fourth orbital. With the use of this odd electron, the fluorine atom can form a single covalent bond with another atom and thus achieve the neon configuration. The structure of the hydrogen fluoride molecule can be represented by the electronic formula H : F̤̈ : . A line is often used to symbolize the covalent bond, H—F̤̈ : , and frequently the unshared pairs are omitted from the formula. All the halogens form compounds of similar structure: H—F, H—Cl, H—Br, and H—I.

Oxygen has six electrons in its outer shell, two pairs in two of the orbitals and a lone electron in each of the other two orbitals. The two odd electrons form the covalent bonds with the two hydrogen atoms in the water molecule:

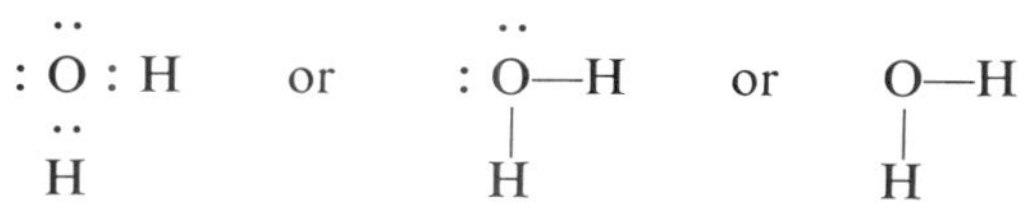

The relatives of oxygen form similar compounds with hydrogen:

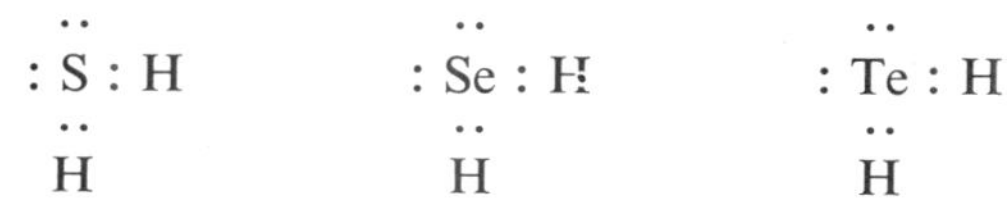

Nitrogen, in Group VA of the periodic table, has five electrons in its outermost shell and achieves the neon configuration by forming three covalent bonds with hydrogen atoms. The relatives of nitrogen form similar compounds:

ammonia phosphine arsine stibine

Carbon, with four electrons in its outermost shell, can form covalent bonds with four hydrogen atoms, forming methane:

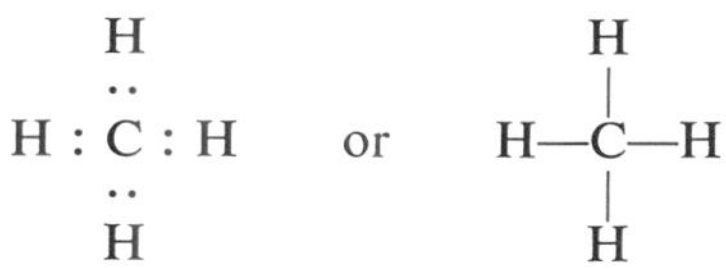

We might expect carbon to form only two covalent bonds, since a carbon atom in its ground state has its 2*s* orbital filled and has one electron in each of two 2*p* orbitals. However, if this were to happen, one *p* orbital would remain empty, and the octet rule would not be satisfied. Actually, the octet rule is satisfied; and when the four atoms bonded to the carbon are identical, as in CH_4, the four bonds are also identical.

18-3 The Four Bonds in Methane

The fact that the four bonds in methane are identical can be inferred from certain types of chemical evidence. By reaction with bromine, methane can be converted to bromomethane (methyl bromide):

$$CH_4 + Br_2 \rightarrow CH_3Br + HBr$$

In this type of reaction, known as *substitution*, a hydrogen atom is removed from a hydrocarbon, and another atom or group is inserted in its place. In the above reaction, a bromine atom has been substituted for a hydrogen atom. If bromomethane is treated with zinc in the presence of acid, methane is regenerated. No matter how many times the cycle of substitution and regeneration is repeated, the same bromomethane is obtained; i.e., the substitution product each time has properties that are identical. Unless a particular one of the four hydrogens is substituted preferentially each time (a possibility that we shall keep in mind), we must conclude that it makes no difference which of the four hydrogen atoms is replaced and that all four hydrogen atoms are equivalent with all bond angles equal.

There are three possible arrangements of the atoms that would make all four hydrogen atoms equivalent and also make all bond angles equal. In the first arrangement, Figure 18-1a, the molecule is planar, i.e., flat, with a hydrogen atom at each corner of a square and the carbon atom at the center. In the second arrangement, Figure 18-1b, the molecule has the shape of a pyramid with a square base; the carbon atom is at the apex, and a hydrogen atom is at each corner of the base. In the third arrangement, Figure 18-1c, the molecule is in the shape of a regular tetrahedron, with the carbon atom at the center and a hydrogen atom at each of the four corners.

Figure 18-1

Three spatial arrangements that would make all four hydrogen atoms in the methane molecule equivalent

(a) Planar. (b) Pyramidal. (c) Tetrahedral.

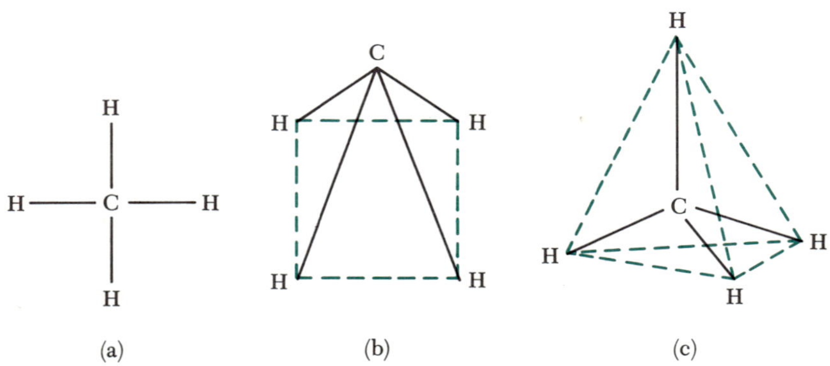

Figure 18-2

The two kinds of CH_2Br_2 molecules that would exist if the molecules were flat

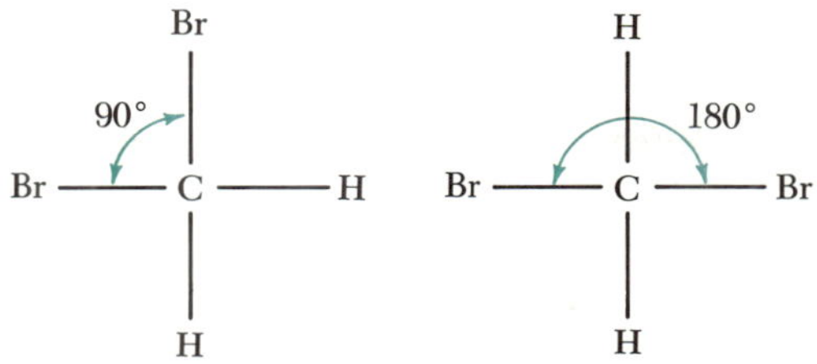

18-4 Shape of the Methane Molecule

The shape of the methane molecule, i.e., planar, pyramidal, or tetrahedral, can be inferred from additional chemical evidence. Further reaction of bromine with bromomethane forms dibromomethane (methylene bromide):

$$CH_3Br + Br_2 \rightarrow CH_2Br_2 + HBr$$

There is only one compound with the molecular formula CH_3Br, and there is only one CH_2Br_2. If the CH_4 molecule (and the CH_3Br molecule) were flat, then presumably a CH_2Br_2 molecule would also be flat. But if this were true, then two kinds of CH_2Br_2 molecule could exist, one with a 90° angle between C—Br bonds and the other with a 180° angle between C—Br bonds (Figure 18-2). If two kinds of CH_2Br_2 molecule existed, it would presumably be possible to separate them by some means. However, all of the experimental evidence indicates that there is only one CH_2Br_2, and we take this to mean that the CH_4 molecule is not flat. We also infer that it is not pyramidal, for if it were, there could be two kinds of CH_2Br_2 molecule, one with the two bromine atoms at neighboring corners of the base and the other with the bromine atoms at diagonally opposite corners of the base (Figure 18-3). However, the third possibility, a tetrahedral CH_4 molecule, would permit only one CH_2Br_2 to

Figure 18-3

The two kinds of CH_2Br_2 molecules that would exist if the molecules were pyramidal

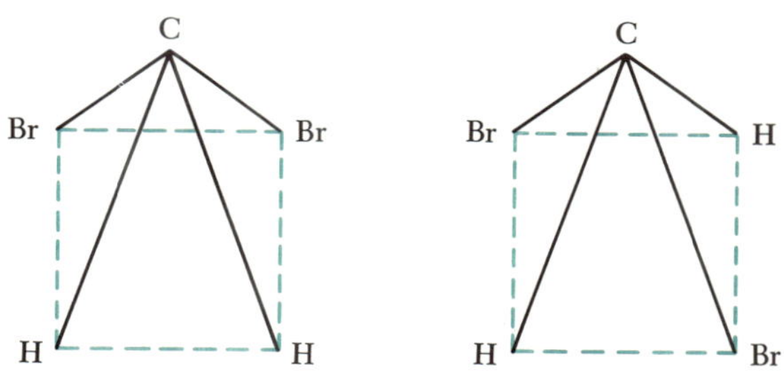

Figure 18-4

The three identical arrangements of atoms in the tetrahedral CH_2Br_2 molecule

The arrows indicate the spatial relationship between the two bromines and carbon, which is the same in each case.

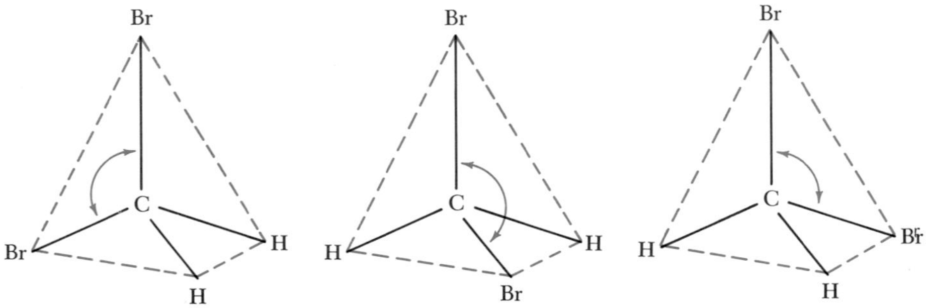

exist. The tetrahedral arrangement of CH_2Br_2 is illustrated in Figure 18-4; we can see that the spatial relationship of the second bromine to the first is the same in each case. This evidence that the CH_4 molecule is tetrahedral reinforces the earlier conclusion that the four C—H bonds in methane are identical and removes our suspicion that a particular hydrogen atom in methane might be substituted preferentially.

Further reaction of bromine with dibromomethane forms tribromomethane (bromoform):

$$CH_2Br_2 + Br_2 \rightarrow CHBr_3 + HBr$$

The fact that only one compound is known with the formula $CHBr_3$ does not, in itself, constitute evidence for or against the tetrahedral shape of the CH_4 molecule. However, the nonpolar character of tetrabromomethane (carbon tetrabromide), which is formed when tribromomethane undergoes further reaction with bromine, indicates a symmetrical molecule:

$$CHBr_3 + Br_2 \rightarrow CBr_4 + HBr$$

Since bromine is more electronegative than carbon, each C—Br bond is polar. If the CBr_4 molecule were pyramidal, it would also be polar. However, since CBr_4 is nonpolar, the four C—Br bonds must be symmetrically arranged in space so that their polarities cancel. Methane itself is nonpolar, and from this fact alone we could have concluded that the molecule is either planar or tetrahedral.

There is, of course, much evidence in addition to the fact that there is only one CH_2Br_2. In 1874 the Dutch chemist Jacobus van't Hoff and a French chemist, Jules Le Bel, concluded, independently of each other and on purely logical grounds, that the methane molecule is tetrahedral. Their deduction has been verified by modern physical methods of investigating molecular structure, such as X-ray diffraction, electron diffraction, and molecular spectroscopy.

When we speak of "the tetrahedral carbon atom," we mean merely that when a carbon atom is bonded to four other atoms, the spatial distribution of the four bonds is symmetrical in three dimensions. In the CH_4 molecule the angles between the bonds are 109.5°. There are six such bond angles, and they are all the same. Each C—H bond length (distance between nuclei) is 1.09 Å.

On a flat surface, such as a sheet of paper or a blackboard, the bonds connecting a carbon atom to four other atoms are usually written

as when we write the formula for methane,

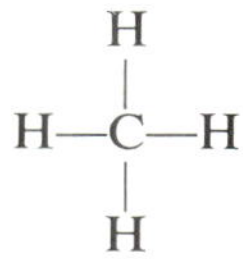

You must remember that the atoms are *not* all in the same plane.

18-5 The Theory of Hybrid Bond Orbitals

We need an explanation for the fact that a carbon atom with the $1s^2 2s^2 2p^2$ configuration forms four covalent bonds instead of only two. Likewise, the fact that all four of the C—H bonds in the CH_4 molecule are alike requires explanation. The most successful explanation of the tetrahedral carbon atom has been the *theory of hybrid bond orbitals*, proposed by Linus Pauling in 1931.

The *ground state* of an atom is the most stable state of the uncombined atom, and the *bonding state* is an excited state in which the atom is ready to form bonds with other atoms. Ground states and bonding states can be identical for atoms which have 1, 5, 6, or 7 electrons in the outermost shell, but they are *not* the same for atoms with 2, 3, or 4 outermost electrons. Thus, a chlorine atom has seven electrons and one vacancy in its outermost shell; therefore, it unites with one hydrogen atom, which has one electron and one vacancy, to form the molecule H—$\ddot{\underset{..}{Cl}}$: . An atom of oxygen has six outermost electrons which, according to Hund's Rule, are distributed most stably with two electrons in the *s* orbital, two electrons in one of the *p* orbitals, and one electron in each of the remaining *p* orbitals. The oxygen atom has vacancies in two orbitals and forms two covalent bonds, as in

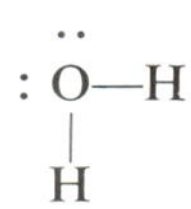

A nitrogen atom has five outermost electrons, with vacancies in three orbitals, so that it can form three bonds, as in

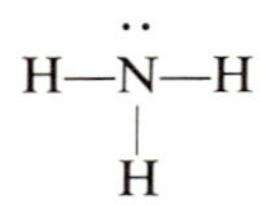

The ground state of a carbon atom has two electrons in the 2*s* orbital, two unpaired *p* electrons, and one empty *p* orbital; in the bonding state, one of the *s* electrons is promoted to the empty *p* orbital. The energy required to promote one electron of the *s* pair to the empty *p* orbital is much less than the energy released by the formation of the two additional bonds which the promotion makes possible. In general, atoms tend to form all of the bonds of which they are capable; thus, the simplest hydrocarbon is CH_4, not CH_2.

Although *s* orbitals do not have the same shape as *p* orbitals, all the bonds in CH_4 are alike. An essential feature of the theory of hybrid bond orbitals is that new orbitals are formed which are all alike. These new orbitals partake of the nature of the original orbitals, but they become equalized. The term *hybridization* indicates in general the mixing of a number of orbitals, not all of the same type, to form the same number of hybrid orbitals that are all alike. The name given to a particular set of hybrid orbitals indicates how many orbitals of each type have mixed to form the new hybrid orbitals. Thus, in methane we have sp^3 hybridization, and each orbital is an sp^3 hybrid orbital, because one *s* orbital and three *p* orbitals have mixed to form four hybrid orbitals.

In the ground state a beryllium atom has only two electrons in the outer shell, and both are in the 2*s* orbital. Without promotion to a higher state, it could form no covalent bonds. Actually, it forms compounds such as $BeCl_2$ which, in the gaseous state, have linear molecules. The linear structure can be explained if we consider the hybridization involved. One of the *s* electrons is promoted into a *p* orbital so that in the bonding state there are two equivalent *sp* hybrid orbitals. These two *sp* hybrid orbitals, because of electronic repulsion, are situated as far apart as possible, i.e., on opposite sides of the beryllium atom, thus accounting for the linear nature of the molecule formed.

Boron, with three electrons in its outermost shell, has the s^2p configuration in the ground state, but it can form three bonds that are all alike. These bonds are designated as sp^2 hybrid bonds since they result from the promotion of one of the *s* electrons to a *p* orbital, making available one *s* orbital and two *p* orbitals for hybridization.

18-6 Explaining the Shapes of Molecules

The shapes of molecules can be understood by considering both hybridization and the mutual repulsion of electron pairs. If there are only two electron pairs, they are located at opposite sides of the atom, forming a linear molecule, e.g., $BeCl_2$, in which the bond angle is 180°. If there are three pairs, they are located at the corners of an equilateral triangle with the atomic nucleus at the center, as in BF_3, which has all four atoms in the same plane, all bond angles being 120°.

With four electron pairs, maximum separation is achieved by their location at the corners of a regular tetrahedron with the nucleus at the center, as we have found for CH_4. Here the bond angle is 109.5°.

Thus, *sp* ("digonal") hybridization gives a bond angle of 180°; sp^2 ("trigonal") hybridization makes a bond angle of 120°; and sp^3 ("tetrahedral") hybridization gives a bond angle of 109.5°.

The tetrahedral configuration is very common in pentatomic molecules and ions, such as NH_4^+ and SO_4^{--}. A molecule or ion consisting of four atoms, with three of them bonded to the fourth (as in NH_3 and ClO_3^-), is pyramidal when there is an unshared pair of electrons on the central atom. Thus, the NH_3 molecule is not flat like BF_3 because the fourth pair of electrons (unshared) repels the other three pairs more than they repel one another, making each bond angle 108°. In the H_2O molecule there are two unshared pairs, which repel each other even more strongly than they repel the shared pairs, and this distorts the bond angle somewhat more than in NH_3, making it only 105°.

18-7 The Alkane Series: Methane and Ethane

Methane is the first member of a series of hydrocarbons with the general formula C_nH_{2n+2}, where *n* is the number of carbon atoms in the molecule. The second member of the series is another gas, ethane, C_2H_6, which boils at −89°C. In the ethane molecule the two carbon atoms are bonded to each other in the same manner as each hydrogen is bonded to a carbon:

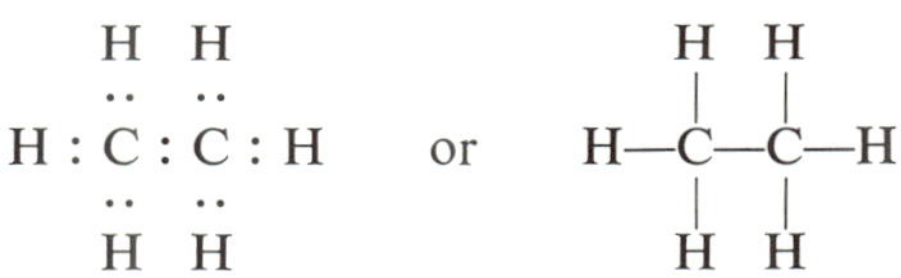

The four bonds which unite each carbon to three hydrogen atoms and to the other carbon are tetrahedrally arranged. The molecule may be said to contain two overlapping tetrahedra. Each carbon atom is at the center of one tetrahedron and at one of the corners of the other tetrahedron.

Carbon atoms are unique in their ability to become bonded to one another, as in the ethane molecule. This is the primary reason for the existence of such a large number of hydrocarbons.

As in methane, each carbon atom in ethane has the neon structure, and each hydrogen atom has the helium structure. This is true for all the hydrocarbons in this series. There are no unshared electron pairs and no vacancies in any of the molecules of the series. Furthermore, each bond is a single bond; i.e., only one pair of electrons is shared. These facts account for the relative lack of chemical reactivity of the members of the series.

The C_nH_{2n+2} series is officially called the *alkane series*. It is also known as the *methane series* or the *paraffin series* of hydrocarbons. The name *paraffin*, which means "having little affinity," was originally applied to the white, waxy substance, a mixture of C_nH_{2n+2} hydrocarbons of high molecular weights, that is now commonly known as paraffin wax.

18-8 Types of Formulas

The chemical analysis of ethane leads to CH_3 as its empirical formula. Determination of its molecular weight shows that its molecular formula is C_2H_6. Its *extended structural formula* is

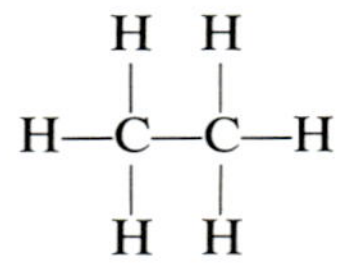

For convenience in writing, ethane is often represented by the *condensed structural* or *semistructural formula*, $CH_3—CH_3$. In the semistructural formula, the line refers to the bond between the two carbons, just as if it were written $H_3C—CH_3$, as is sometimes done.

In ethane the bond angles (109.5°) and C—H bond lengths (1.09 Å) are the same as in methane. The C—C bond length is 1.54 Å, the same as in diamond. Virtually the same values hold for all alkanes. Therefore, it is usually not necessary to call attention to the actual shape of a particular alkane molecule.

There is a possible variation in the shape of an ethane molecule which we did not encounter with methane. The molecule could have an arrangement like that indicated in Figure 18-5a, or one like that in Figure 18-5b, or any intermediate arrangement. Formulas of the type given in Figure 18-5 are called *projection formulas* and represent the molecule as viewed from one end, with one carbon behind the other. The nearer carbon atom and the three atoms attached to it are represented by the inverted Y; the more distant carbon and the atoms attached to it are indicated by the circle and the broken lines.

Figure 18-5

Two conformations of ethane

(a) Eclipsed conformation. (b) Staggered conformation.

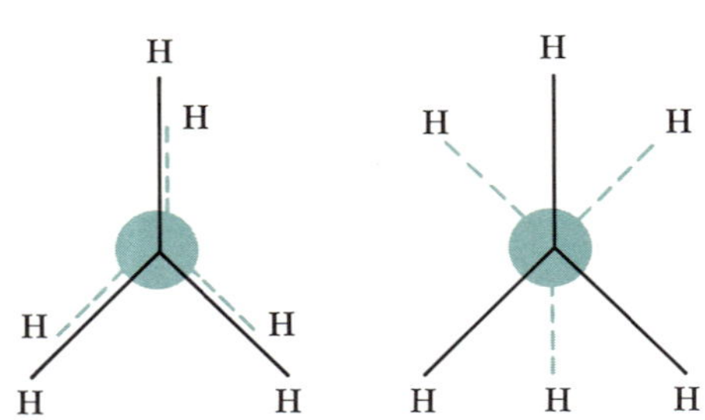

(a) (b)

18-9 Conformations

Any of the possible arrangements of the ethane molecule (two of which are shown in Figure 18-5) could be changed to another simply by rotating one end of the molecule about the C—C bond. Since there is only one compound with the formula CH_2X—CH_2X (where X is Cl, Br, etc.), we must conclude that the molecule has considerable freedom to rotate about this bond.

These different arrangements that can be converted into one another by rotating a part of a molecule without breaking any bonds are called *conformations*. In Figure 18-5 the arrangement in (a) is called the *eclipsed* conformation of ethane; the arrangement in (b) is called the *staggered* conformation.

There is some evidence that when the molecule is in the eclipsed conformation the electron pairs by which hydrogen atoms are bonded to carbon atoms repel one another more strongly than in any other conformation. In the staggered conformation the repulsion is least because the distances between electron pairs are greatest. Consequently, the eclipsed conformation of ethane tends to convert spontaneously into the staggered conformation.

The energy of the eclipsed conformation of ethane is about 3 kcal per mole greater than the energy of the staggered conformation. Since 3 kcal per mole is not a very high energy barrier, the molecule is relatively free to rotate from one staggered conformation to another. A substantial fraction of the collisions of ethane molecules with one another would be sufficiently energetic to ensure a continuous and rapid transformation between staggered conformations at room temperature. None of the molecules could be expected to remain for long in the eclipsed conformation.

18-10 The Alkane Series: Propane and the Butanes

The third member of the alkane series is propane, C_3H_8, which boils at $-42°C$. Although its condensed structural formula is written as CH_3—CH_2—CH_3, the three carbon atoms do not actually lie on a straight line. The tetrahedral distribution of all the atoms bonded to a given carbon demands that the carbon atoms have a zigzag arrangement in all conformations. Nevertheless, propane is referred to as a *straight-chain* hydrocarbon because its carbon atoms are combined in a chain. All members of the alkane series have an *open-chain* structure and are therefore *aliphatic* hydrocarbons. Since their hydrogen content is at a maximum, they are also said to be *saturated* hydrocarbons.

There are two ways of writing the structural formula for a four-carbon alkane, C_4H_{10}:

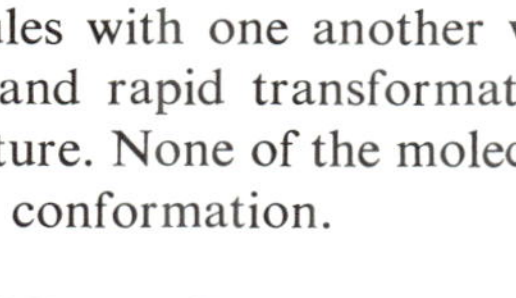

(a)

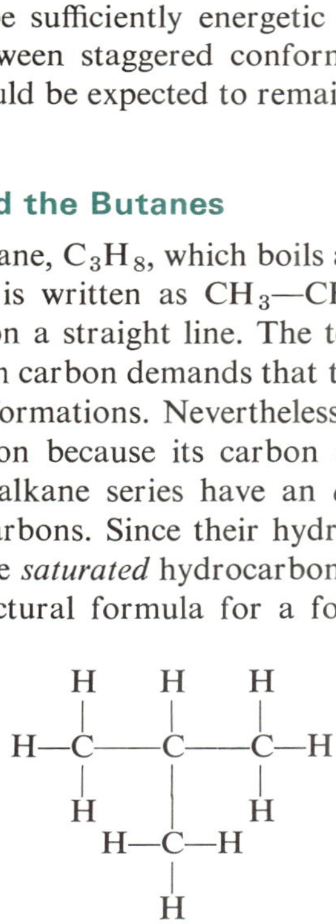

(b)

No amount of rotating or twisting would enable a molecule having one of these structures to become transformed into a molecule having the other structure without the breaking of bonds. In (a), no carbon atom is bonded to more than two other carbon atoms, but one of the carbon atoms in (b) is bonded to all of the other three carbon atoms. This is not a mere difference in conformation; it is a structural difference known as *isomerism*. Two or more different compounds having the same number and kinds of atoms per molecule are called *isomers*. The two compounds represented above have the same percentage composition, molecular weight, and molecular formula (C_4H_{10}), but they have different physical and chemical properties. The substance with the *branched chain*, (b), boils at $-12°C$ and is known as isobutane; the compound with the so-called straight (i.e., unbranched) chain, (a), boils at 0°C and is known as *n*-butane ("normal butane"). The two isomeric butanes have different melting points, densities, solubilities, and reactivities toward various chemical reagents. Thus, isobutane and *n*-butane are different compounds, even though they have the same molecular formula.

18-11 The Alkanes: A Homologous Series

Comparison of the molecular formulas of methane, ethane, propane, and the butanes reveals that each succeeding member of the alkane series has one more carbon atom and two more hydrogen atoms. A *homologous series* is defined as a series of compounds in which each member differs from the preceding member by the increment CH_2. The systematic study of the thousands of carbon compounds is simplified greatly by grouping them into homologous series.

There is a general formula for all members of a particular homologous series, for example, C_nH_{2n+2} for the alkanes. The members of a homologous series share certain general properties as a result of their similar molecular structures. Thus, all alkanes are held together entirely by single covalent bonds which are either nonpolar, because they join two atoms of the same kind, C—C, or are only slightly polar because they join two atoms that differ only slightly in electronegativity, C—H. Furthermore, the four bonds of each carbon are directed tetrahedrally, i.e., symmetrically, so that alkane molecules are nonpolar.

The boiling point increases regularly with increasing molecular weight if we compare the *n*-alkanes. However, branched-chain isomers have lower boiling points than the straight-chain isomers.

In accordance with the rule "like dissolves like," the alkanes are soluble in nonpolar solvents, such as benzene and carbon tetrachloride, and insoluble in polar solvents, such as water. The liquid alkanes serve as solvents for compounds of low polarity, but they do not dissolve highly polar compounds.

The chemical reactions of the alkanes are comparatively few. They include substitution and combustion. Combustion of hydrocarbon fuels gives water and carbon dioxide or carbon monoxide, depending on the adequacy of the oxygen supply:

$$CH_4 + 2O_2 \rightarrow CO_2 + 2H_2O$$
$$2CH_4 + 3O_2 \rightarrow 2CO + 4H_2O$$
$$2C_8H_{18} + 25O_2 \rightarrow 16CO_2 + 18H_2O$$
$$2C_8H_{18} + 17O_2 \rightarrow 16CO + 18H_2O$$

All members of the alkane series have names ending with *-ane*. Except for the first four members, the names are derived from prefixes indicating the number of carbon atoms. Thus, C_5H_{12} is pentane; C_6H_{14} is hexane; C_7H_{16} is heptane; C_8H_{18} is octane; C_9H_{20} is nonane; etc.

As the number of atoms increases, the number of possible arrangements of those atoms increases rapidly. There are 3 isomeric pentanes, 5 hexanes, 9 heptanes, 18 octanes, 35 nonanes, etc. It has been calculated that the number of possible eicosanes, $C_{20}H_{42}$, is approximately 366,000. A similar situation prevails in other homologous series.

18-12 Names for Higher Hydrocarbons

We have seen that the two butanes can be distinguished by assigning the names *n*-butane to the isomer with the unbranched chain and isobutane to the one with the branched chain. Since there are three isomeric pentanes, there is need for a third name. The name *n*-pentane has been assigned to the isomer with the unbranched chain, isopentane to the one with the least branching, and neopentane to the one with the most branching.

$$CH_3{-}CH_2{-}CH_2{-}CH_2{-}CH_3$$

n-pentane
bp 36°C

$$CH_3{-}CH_2{-}\underset{\displaystyle CH_3}{\underset{|}{CH}}{-}CH_3$$

isopentane
bp 28°C

$$CH_3{-}\overset{\displaystyle CH_3}{\overset{|}{\underset{\displaystyle CH_3}{\underset{|}{C}}}}{-}CH_3$$

neopentane
bp 10°C

With five isomeric hexanes the problem of assigning names to individual hydrocarbons becomes more complicated. Several systems have been used for naming hydrocarbons. All of them require that we have names for the various groups of atoms ($-CH_3$, $-C_2H_5$, etc.) that are present in the hydrocarbon molecules. Each group is itself a hydrocarbon less one hydrogen atom. These C_nH_{2n+1} groups are called *alkyl* groups. Each alkyl group is named by dropping *-ane* from the name of the hydrocarbon and adding *-yl*. Thus, $-CH_3$ is the *methyl* group and $-C_2H_5$ is *ethyl*. When there is need to indicate which carbon atom lacks a hydrogen atom, this is done by attaching a line to it in the formula for the group and by giving the group a name that is designed to convey the same information. Thus, $CH_3CH_2CH_2-$ is *n*-propyl ("normal propyl") and

$$CH_3\underset{|}{C}HCH_3$$

is isopropyl.

The following groups are related to the butanes:

$CH_3CH_2CH_2CH_2—$

n-butyl

secondary butyl

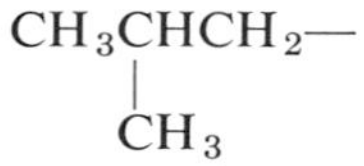

isobutyl

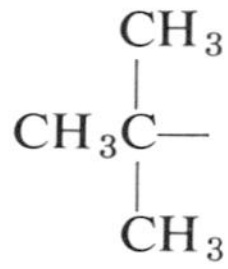

tertiary butyl

The *secondary* butyl group gets its name from the fact that the carbon atom which lacks a hydrogen is united to *two* other carbon atoms. Such a carbon atom is called a *secondary carbon atom*. The $(CH_3)_3C—$ group is known as the *tertiary* butyl group because the carbon atom that lacks a hydrogen is a *tertiary carbon atom*; i.e., it is united directly to three other carbon atoms.

The most widely used system for naming hydrocarbons and their derivatives was adopted in 1892 at the International Congress held at Geneva, Switzerland, and is often referred to as the *Geneva system*. It was extended at the meeting of the International Union of Chemistry (IUC) in 1930 and further developed at the meeting of the International Union of Pure and Applied Chemistry (IUPAC) in 1949. Therefore, it has been called, first, the IUC system and, later, the IUPAC system of nomenclature. The following is a summary of the IUPAC rules for naming alkanes:

1. Each alkane has a name ending in *-ane*.
2. The normal or "straight-chain" (i.e., continuous chain) alkanes are referred to by their common names: methane, ethane, propane, butane, pentane, hexane, heptane, octane, etc.
3. Branched-chain hydrocarbons are named as derivatives of normal hydrocarbons, with the longest "normal chain" in the molecule regarded as the parent hydrocarbon. The names of the branches are attached as prefixes to the parent name. Examples:

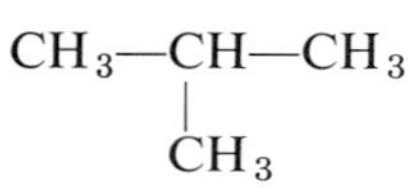

methylpropane

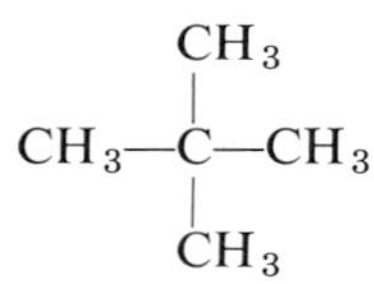

dimethylpropane

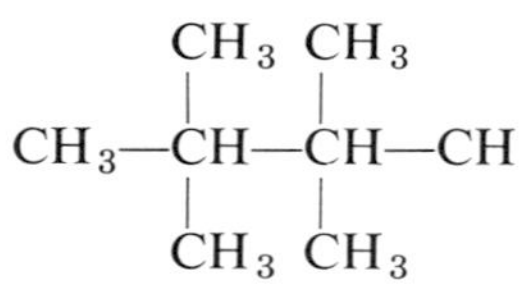

tetramethylbutane

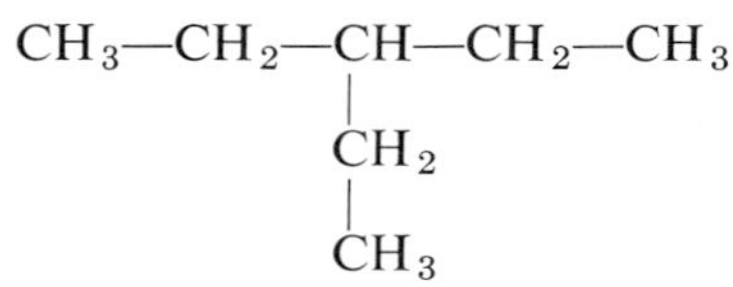

ethylpentane

4. When it is necessary to indicate the positions of the branches, this is done by numbering the carbon atoms of the parent alkane and using

the number that indicates the position of a branch as part of the name. The numbering of the parent alkane is done from the end that gives the branches the smaller numbers. Examples:

2-methylpentane
(*not* 4-methylpentane)

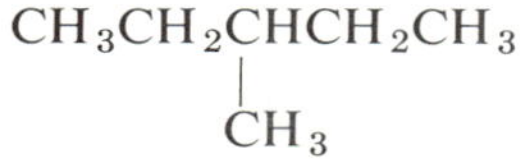
3-methylpentane

5. Consecutive numbers in a name are separated by commas. If two branches are on the same carbon atom, the number is repeated. Examples:

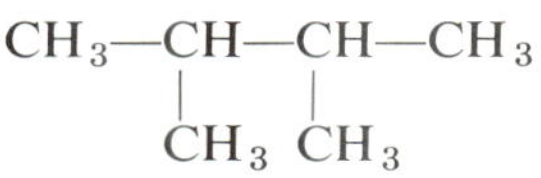
2,3-dimethylbutane

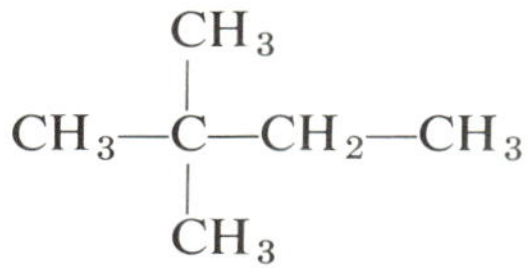
2,2-dimethylbutane

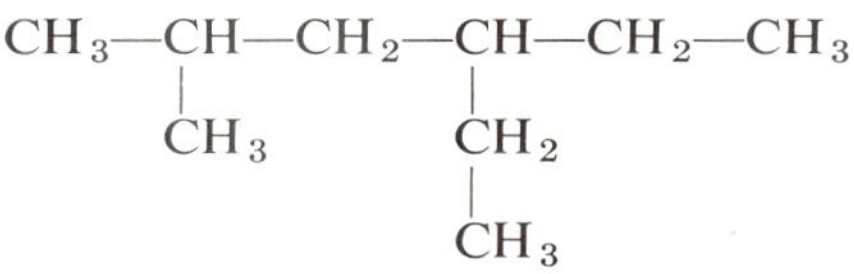
2-methyl-4-ethylhexane

All alkanes having a total of, say, five carbon atoms are pentanes, regardless of the fact that they are named as derivatives of simpler alkanes. Thus, dimethylpropane is a pentane because it has five carbon atoms (one in each of two methyl groups and three in the "normal chain" indicated by the parent name). Similarly, tetramethylbutane is an octane because it has a total of eight carbon atoms.

18-13 Cycloalkanes

Petroleum from certain areas, for example, California, is rich in hydrocarbons known to the petroleum industry as naphthenes. In the molecules of these hydrocarbons, the carbon atoms are arranged to form rings instead of open chains. All the C—C bonds are single bonds, as in the alkanes, so that much of the chemistry of these compounds is essentially the same as that of the alkanes. To the chemist the members of this series are known as *cycloparaffins* or *cycloalkanes*. Any hydrocarbon that has one or more rings is a *cyclic* hydrocarbon.

The general formula for the cycloalkanes is C_nH_{2n}. They are named by adding the prefix *cyclo-* to the name of the corresponding alkane, i.e., the alkane having the same number of carbon atoms. Cyclopropane, the simplest possible cycloalkane, is a planar molecule with the three carbon atoms forming a triangle. Thus, the bond angle is 60°, and this differs from the tetrahedral angle enough to make the molecule unstable compared with others in which the bond angles are tetrahedral. For example, cyclopropane undergoes a ring-opening reaction with bromine, or with hydrogen in the presence of a catalyst:

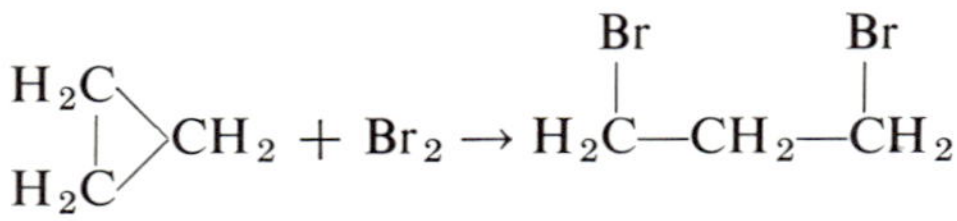

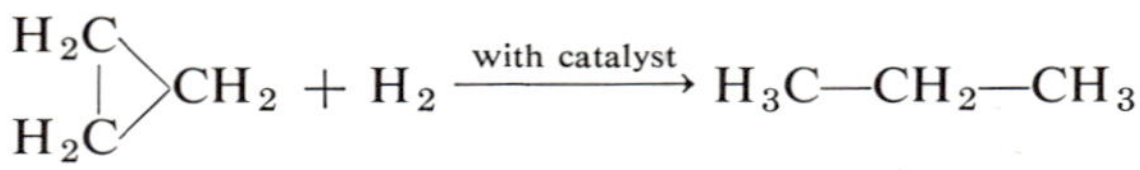

The second member of the series, cyclobutane, also has a flat molecule, with the four carbon atoms forming a square; i.e., the bond angles are 90°. Here the deviation of the bond angle from the tetrahedral angle is not as great as in cyclopropane. Cyclobutane does not undergo the ring-opening reaction with bromine. Addition of hydrogen occurs in the presence of a catalyst, but it requires a higher temperature (200°C) than that required for the hydrogenation of cyclopropane (80°C):

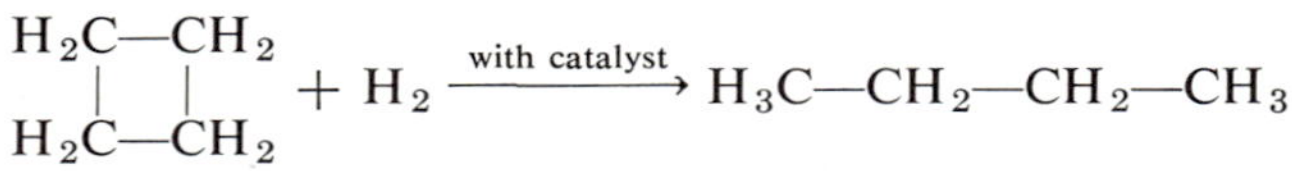

In cyclopentane a nearly planar arrangement of the carbon atoms permits the bond angles to have the tetrahedral value. All rings larger than this are *puckered* so that each bond angle can be 109.5°. The puckered ring of cyclohexane is shown in Figure 18-6 in the "chair" conformation and in the "boat" conformation. If the cyclohexane ring were flat, the bond angles would be 120°. For larger rings a planar arrangement would make the bond angle differ even more from the tetrahedral angle.

Regardless of whether the rings are flat or puckered, cycloalkanes are often represented by simple geometric figures. Thus, the following are accepted "formulas":

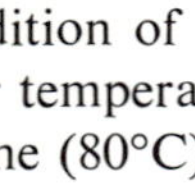

cyclopentane, C_5H_{10}

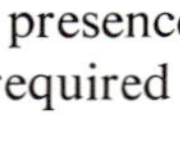

cyclohexane, C_6H_{12}

Figure 18-6

Two conformations of cyclohexane

(a) Chair conformation. (b) Boat conformation.

(a) (b)

The pentagon represents a ring of five carbon atoms, each singly bonded to two other carbons and also to two hydrogen atoms; the hexagon represents a molecule of similar structure, with six carbon atoms in the ring. Likewise, a heptagon represents cycloheptane, C_7H_{14}, and an octagon represents cyclooctane, C_8H_{16}.

Rings containing as many as thirty carbon atoms have been prepared. Except for the first two members, the chemical behavior of the cycloalkanes is very similar to that of the alkanes.

18-14 The Alkenes

The *alkenes*, with the general formula C_nH_{2n}, are the hydrocarbons with a double bond. Alkenes, also known as *olefins*, can be formed by removal of hydrogen from alkanes. Indeed, dehydrogenation is a widely used method for the industrial preparation of alkenes, two hydrogen atoms from adjacent carbon atoms of an alkane molecule being removed.

The simplest member of this homologous series is ethene, C_2H_4, more commonly known as ethylene. Ethene is manufactured from ethane by dehydrogenation, with chromium(III) oxide serving as a catalyst, at 500°C:

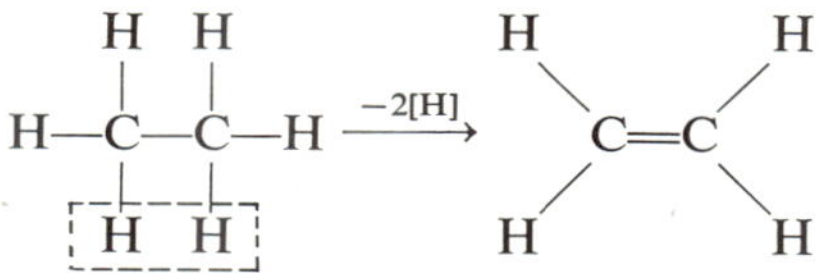

The removal of the two hydrogen atoms is represented by the broken lines. In the ethene molecule the two carbon atoms are bonded by two shared pairs of electrons, i.e., a double bond. The electron-dot formula, the extended structural formula, and the condensed structural formula are as follows:

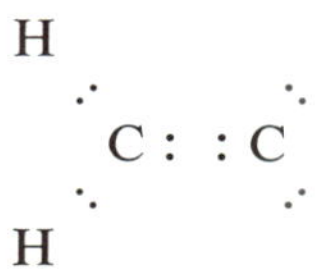

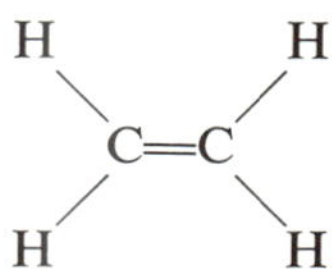

$$H_2C{=}CH_2$$

The ethene molecule has all six atoms in the same plane, and all bond angles are 120°. The carbon-to-carbon distance is 1.34 Å, as compared to 1.54 Å in ethane.

Since alkenes contain less hydrogen than alkanes, they are referred to as *unsaturated* aliphatic hydrocarbons; i.e., they have the same open-chain structure as the alkanes but contain less than the maximum quantity of hydrogen. Their unsaturation is responsible for the type of reaction known as *addition*. In the addition of hydrogen to ethene one hydrogen atom is added to each doubly bonded carbon, converting the double bond to a single bond. The product of the reaction is ethane:

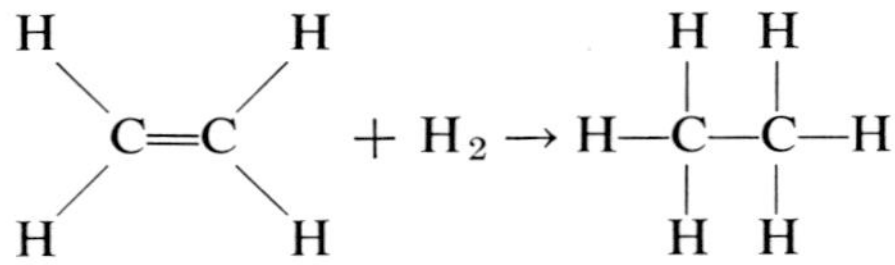

Hydrogenation of ethene is, of course, the reverse of the dehydrogenation of ethane. Addition reactions are characteristic of compounds containing the carbon-carbon double bond. Many compounds can be readily obtained by addition of properly chosen reagents to an alkene. For example:

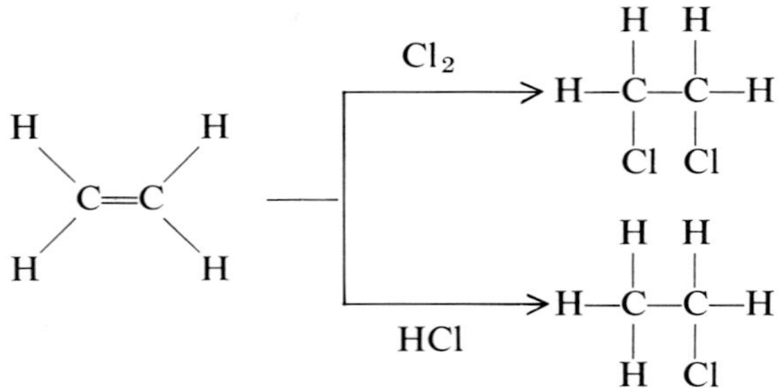

When dehydrogenation of a higher alkane occurs, two hydrogen atoms can be removed from adjacent carbon atoms in more than one way. Dehydrogenation of propane gives the same product, propene (commonly called propylene), regardless of which two carbon atoms lose hydrogen atoms:

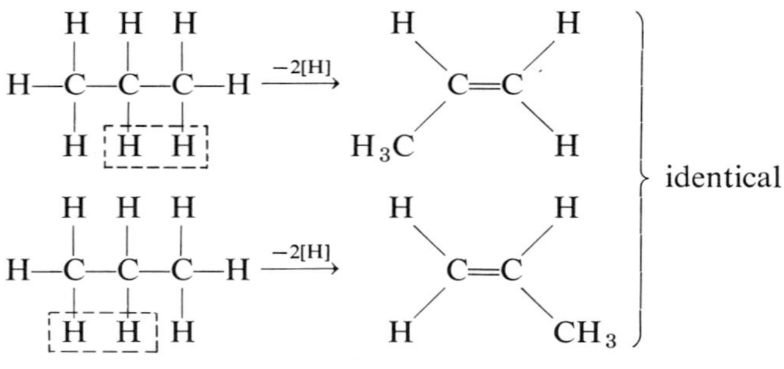

Thus, there is only one propene, just as there is only one propane.

When we consider the structural formula for a four-carbon alkene, C_4H_8, we find that there are several ways of writing it. First of all, the molecule may have the straight-chain structure of *n*-butane or the branched-chain structure of isobutane. Next we note that in an unbranched chain the double bond can be between the first and second carbons or between the second and third carbons. So far we have three structures:

$H_3C—CH_2—CH=CH_2$	$H_3C—CH=CH—CH_3$	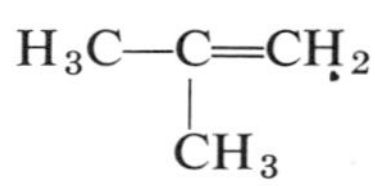
1-butene (α-butylene)	2-butene (β-butylene)	methylpropene (isobutylene)

Actually, there are four alkenes having the molecular formula C_4H_8, for there are two 2-butenes (this is discussed in Section 18-16).

In the petroleum industry, alkenes are obtained by the "cracking" of alkanes, i.e., decomposition brought about by heat. The products of cracking are alkenes, hydrogen, and alkanes smaller than the starting alkanes. For example, *n*-butane yields the mixtures shown below:

$$CH_3{-}CH_2{-}CH_2{-}CH_3 \xrightarrow{400\text{–}600°C} \begin{cases} H_2 + C_4H_8 \\ CH_4 + C_3H_6 \\ C_2H_6 + C_2H_4 \end{cases}$$

The alkenes of low molecular weight are important raw materials for the chemical industry.

18-15 The Nomenclature of Alkenes

The IUPAC rules for naming alkenes are as follows:

1. The ending *-ane* in the name of the corresponding alkane is replaced by *-ene*: ethene, propene, butene, pentene, etc.
2. The parent compound is taken to be the longest normal chain *containing the double bond.*
3. The carbon atoms of the parent compound are numbered from the end *nearest* the double bond.
4. The position of the double bond is indicated by the smaller of the two numbers that refer to the carbon atoms to which it is attached. Examples:

$CH_3CH_2CH{=}CH_2$
1-butene
(*not* 2-butene)

$CH_3CH_2CH{=}CHCH_3$
2-pentene
(*not* 3-pentene)

5. The names and positions of side chains are indicated in the same manner as in naming alkanes. Examples:

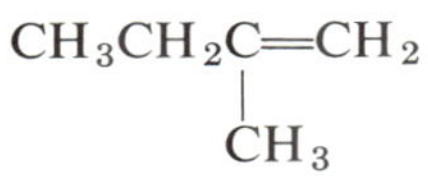
2-methyl-1-butene

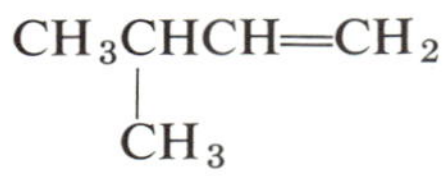
3-methyl-1-butene

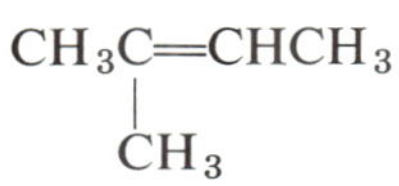
2-methyl-2-butene

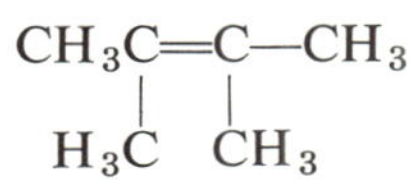
dimethyl-2-butene

18-16 Geometric Isomerism

To understand why there are two 2-butenes, we must recall that all six atoms of the ethene molecule lie in the same plane. The one end of the molecule is not free to rotate about the carbon-carbon bond as is the case with ethane. *A*

Figure 18-7

The spatial configurations of the 2-butenes

(a) *cis*-2-butene. (b) *trans*-2-butene.

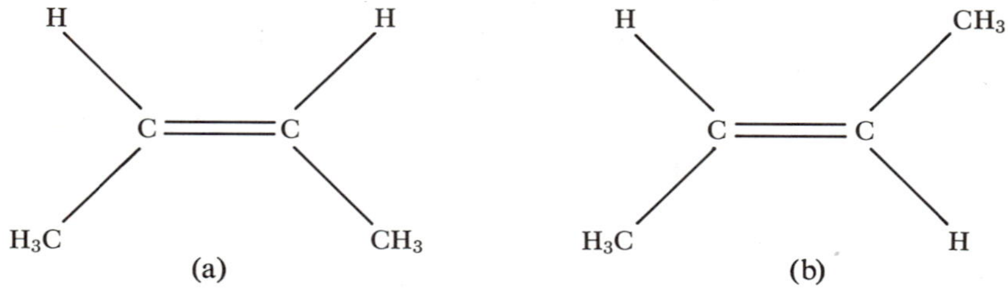

double bond does not permit freedom of rotation. Therefore, a portion of any alkene molecule is flat; i.e., the two doubly bonded carbon atoms and the four atoms bonded to them all lie in the same place. This is the case in 2-butene, but there are two different ways in which the atoms attached to the doubly bonded carbon atoms can be arranged in space. In one arrangement, Figure 18-7a, the CH_3 groups lie on the same side of the molecule; in the other arrangement, Figure 18-7b, the CH_3 groups lie on opposite sides. The one compound is designated by the prefix *cis*- (Latin: on this side) and the other by *trans*- (Latin: on the other side).

The type of isomerism that distinguishes methylpropene (isobutylene) from the other butenes is *skeletal* isomerism, and the type that distinguishes 1-butene from the 2-butenes is *positional* isomerism. The type distinguishing *cis*-2-butene from *trans*-2-butene is *geometric* isomerism. Geometric isomers have the same structural constitution, i.e., they are alike with respect to which atoms are joined to which, but they differ from one another in the way the atoms are oriented in space.

The increased possibilities of isomerism among the alkenes result in the existence of more pentenes than pentanes, more hexenes than hexanes, etc.

18-17 The Alkynes and the Alkadienes

Alkynes are aliphatic hydrocarbons with the general formula C_nH_{2n-2} and have a carbon-carbon triple bond as their distinguishing structural feature. The simplest alkyne is ethyne, C_2H_2, well known by the older name acetylene. All four atoms of the acetylene molecule lie along the same straight line; i.e., all bond angles are 180°: H—C≡C—H. The length of the carbon-carbon triple bond is 1.20 Å and that of the C—H bonds is 1.06 Å. Like the alkenes, and quite unlike the alkanes, the alkynes are highly reactive because of their unsaturation.

Acetylene, the only alkyne of industrial importance, is used as a fuel for the oxyacetylene torch. Large quantities of acetylene are used each year as the starting material for the synthesis of many important compounds, including acetic acid. Acetylene can be prepared by the action of water on calcium carbide, CaC_2:

$$CaC_2 + 2HOH \rightarrow Ca(OH)_2 + HC{\equiv}CH$$

Calcium carbide is prepared by the reaction between calcium oxide and coke at electric furnace temperatures:

$$CaO + 3C \xrightarrow{2000^\circ} CaC_2 + CO$$

Calcium oxide (quicklime) is manufactured from limestone in lime kilns:

$$CaCO_3 \rightarrow CaO + CO_2$$

Coke is obtained from coal. Thus, acetylene is obtained in a few steps from three abundant raw materials: coal, limestone, and water.

There are many hydrocarbons that contain more than one point of unsaturation. The *alkadienes* (or *diolefins*) are hydrocarbons that contain two carbon-carbon double bonds. They are named in the same way as alkenes, except that the suffix *-diene* is used to indicate two double bonds, and each name has two numbers which indicate the positions of the two double bonds (except for propadiene, in which there can be no variation in the positions of the bonds). One of the most important alkadienes, used in the manufacture of synthetic rubber, has the formula

$$CH_2{=}CH{-}CH{=}CH_2$$

and is called 1,3-butadiene (the numbers indicate that the double bonds follow the first and third carbon atoms). When both double bonds involve the same carbon atom, the compounds are usually unstable.

The alkadienes are isomeric with the alkynes; i.e., they have the same general formula, C_nH_{2n-2}. Unsaturated hydrocarbons, of course, may contain three or more double bonds, or both double and triple bonds. For example,

$$CH_2{=}CH{-}CH{=}CH{-}CH{=}CH_2 \quad \text{1,3,5-hexatriene}$$

$$HC{\equiv}C{-}CH{=}CH_2 \quad \text{1-buten-3-yne } \textit{or} \text{ vinyl acetylene}$$

All the hydrocarbons that have an open-chain structure (i.e., no rings) are classified together as aliphatic hydrocarbons. They include both saturated and unsaturated species (alkanes, alkenes, alkadienes, alkynes, etc.).

18-18 Unsaturated Cyclic Hydrocarbons

As we would expect, it is possible for one or more double bonds to be present in a ring of carbon atoms, as in cyclopentene:

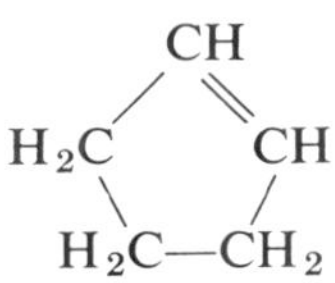

or

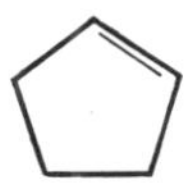

An example of a cyclic hydrocarbon with two double bonds is 1,3-cyclopentadiene:

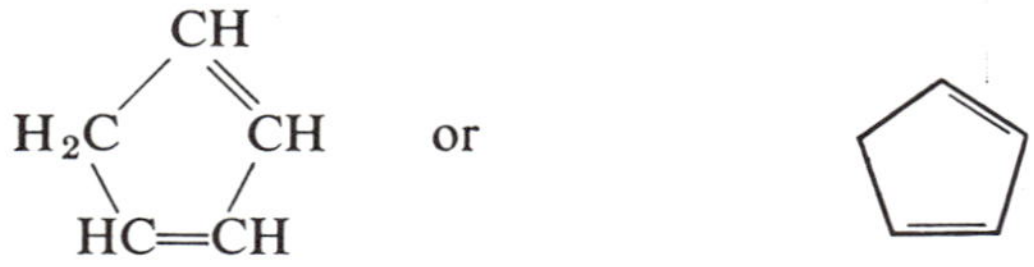

In general, the properties of cyclic compounds resemble those of open-chain compounds. Addition reactions occur at double bonds in unsaturated cyclic hydrocarbons. Cycloalkanes have no double bonds, so addition does not readily occur, except with cyclopropane and cyclobutane, which undergo ring-opening addition reactions. Cycloalkenes show the addition reactions typical of alkenes, and cyclic dienes resemble open-chain dienes in their ability to add twice as much H_2 or Br_2 per mole as an alkene or a cycloalkene.

Three different unsaturated cyclic compounds containing six carbon atoms per molecule are conceivable:

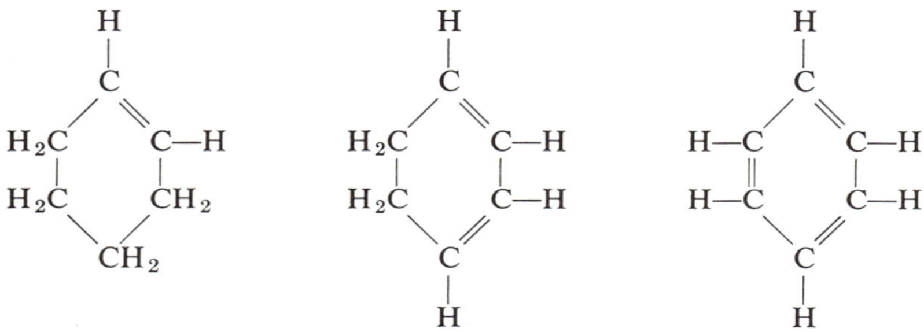

cyclohexene   cyclohexadiene   cyclohexatriene

Cyclohexene undergoes the reactions typical of alkenes and cycloalkenes, and cyclohexadiene acts like most other dienes, responding readily to tests for unsaturation. If cyclohexadiene is passed over a palladium catalyst at 100°C, it loses a molecule of H_2. The resulting "cyclohexatriene" proves to be the well-known compound commonly called benzene, which does not behave as if it really has three double bonds.

The cycloalkanes, cycloalkenes, and cycloalkadienes are classified together as *alicyclic* hydrocarbons, but hydrocarbons containing one or more benzene rings are classified separately as *aromatic* hydrocarbons.

18-19 Benzene

Benzene, C_6H_6, was discovered in 1825 by Michael Faraday. Its low hydrogen content suggests a high degree of unsaturation, but benzene exhibits considerable inertness toward those chemical reagents that ordinarily add to carbon-carbon double bonds. The properties of benzene have been investigated with unusual thoroughness because of its paradoxical nature and the importance of the many compounds derived from it.

In 1865 a German chemist, August Kekulé, proposed the ring structure for benzene that we have referred to above by the name *cyclohexatriene*. Although this was a great step forward, it left the stability of the ring unaccounted for.

Modern methods for investigating structure, such as X-ray determination, have shown that the benzene molecule is flat and that the six carbon-carbon bonds are of equal length, 1.39 Å. If there were three single and three double bonds, we would expect their lengths to be 1.54 Å and 1.34 Å, respectively, since these are the bond lengths in alkanes and alkenes. The fact that the bonds in benzene seem to be intermediate between single and double bonds has received its most satisfactory explanation by the application of the theory of resonance.

Two equally reasonable structures which differ only in the location of electrons can be written for the benzene molecule:

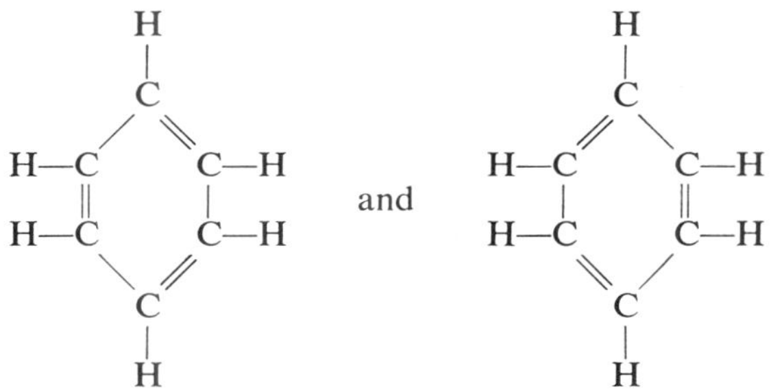

The "real" structure is neither of these but is a resonance hybrid that is intermediate in character between the two. All six carbon-carbon bonds are equivalent. Each is more than single and less than double, i.e., a hybrid bond that cannot satisfactorily be represented on paper or with models.

One method for attempting to represent the structure of benzene consists in writing the two contributing structures with braces enclosing them and a two-headed arrow separating them:

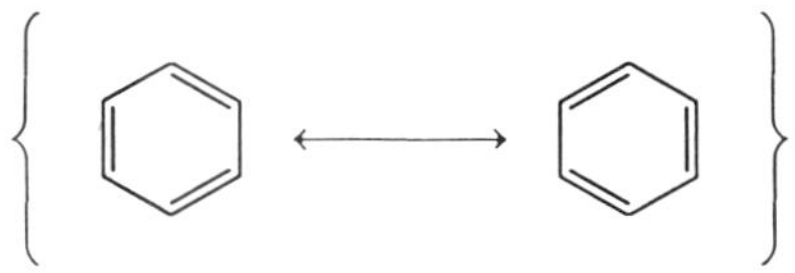

This symbolism must not be taken to mean that molecules of two structures exist in equilibrium, but that only one kind of molecule exists: a hybrid of the two contributing structures.

A simplified symbolism frequently employed for the benzene ring consists of a hexagon with a circle inside:

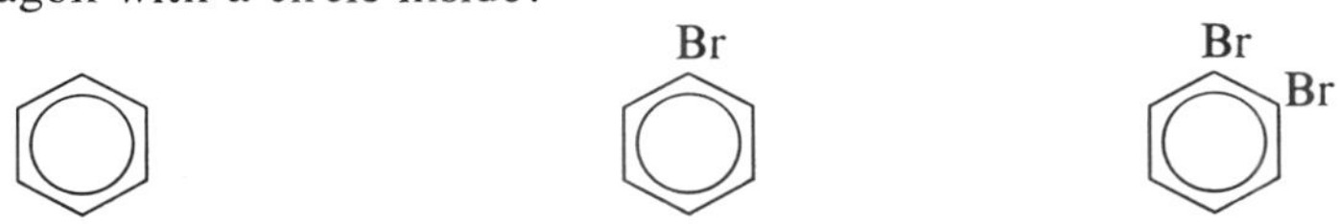

benzene bromobenzene 1,2-dibromobenzene

Each corner of the hexagon represents a carbon atom with a hydrogen atom bonded to it; if some other atom or group has replaced the hydrogen atom, it is shown in the formula. The purpose of the circle is to distinguish the benzene ring from the cyclohexane ring.

18-20 Aromatic Hydrocarbons

Benzene is the simplest hydrocarbon containing a benzene ring. When benzene is heated to a high temperature, it loses hydrogen and forms biphenyl:

$$2C_6H_6 \xrightarrow{800°C} C_6H_5—C_6H_5 + H_2$$

The biphenyl molecule consists of two benzene rings connected by a single bond. It is also possible for two benzene rings to share two carbon atoms; this is the situation in naphthalene (which constitutes "mothballs").

biphenyl, $C_{12}H_{10}$ naphthalene, $C_{10}H_8$

There are, of course, compounds containing more than two benzene rings. Any compound containing one or more benzene rings is referred to as an aromatic hydrocarbon.

Many of the aromatic hydrocarbons contain both aliphatic and aromatic units. The simplest of these is toluene, $C_6H_5—CH_3$. Ethylbenzene and styrene are hydrocarbons of this type.

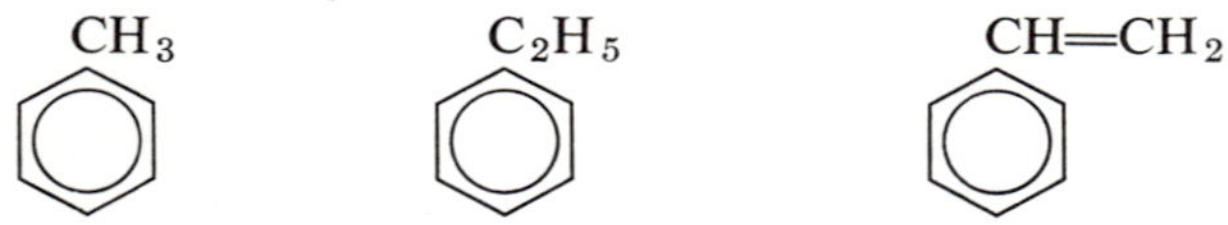

toluene ethylbenzene styrene

The group $—C_6H_5$, obtained by dropping a hydrogen atom from the benzene ring, is called the *phenyl* group. It is frequently represented by the Greek letter Φ (phi). Thus, ΦBr is phenyl bromide, i.e., bromobenzene. The group $C_6H_5CH_2—$, obtained by dropping a hydrogen from the methyl group of toluene, is called the *benzyl* group. Thus, $ΦCH_2Br$ is benzyl bromide.

All groups obtained by dropping a hydrogen atom from an aromatic nucleus (benzene ring) are called *aryl* groups. Examples of aryl groups are the phenyl group and the naphthyl group, $C_{10}H_7—$. The benzyl group is not an aryl group.

Coal is the major source of aromatic compounds. The coking of bituminous coal yields as a by-product a viscous material of unpleasant odor known as coal tar. Distillation of coal tar yields aromatic compounds which include hydrocarbons such as benzene, toluene, and naphthalene.

18-21 Polymers

Polymers are compounds of high molecular weight which have been obtained by a large number of simple molecules reacting with one another. Polyethylene is a well-known polymer that is used in the manufacture of such everyday items as plastic bottles and vegetable bags. It is flexible, tough, resistant to water, and an excellent electrical insulator. It is synthesized by the polymerization of ethylene:

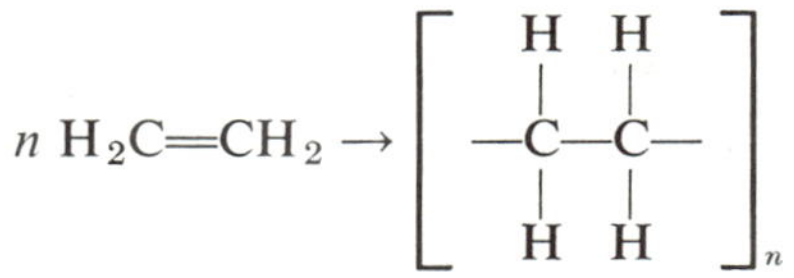

n molecules of ethylene — one molecule of polyethylene

The molecular weight of the polyethylene depends on *n*, the number of ethylene units that have reacted. This number is difficult to control with precision, and *n* might even have a different numerical value for each polyethylene molecule in a sample of the polymer. The average value of the molecular weight of commercial polyethylene is of the order of a few million; thus, polyethylene is eminently entitled to be classified as a *high polymer*. The lack of crystallinity typical of most high polymers is due in part to the varying sizes and shapes of the molecules.

Polypropylene, a polymer that is in increasing use, is made by polymerization of propylene:

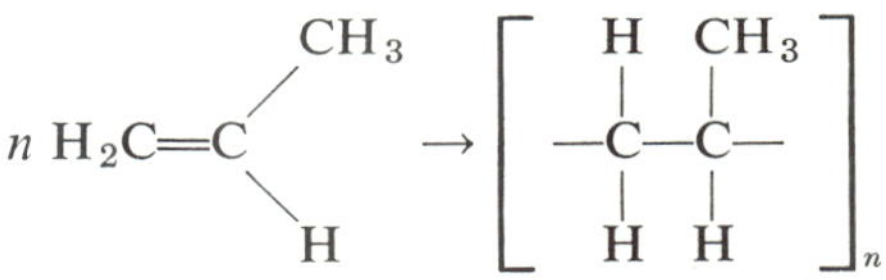

n molecules of propylene — one molecule of polypropylene

Various other compounds containing the ethylenic structure are used in the manufacture of high polymers. Examples are vinyl chloride, $H_2C{=}CHCl$, used in the synthesis of polyvinyl chloride for garden hoses, plastic tile, etc.; tetrafluoroethylene, $F_2C{=}CF_2$, used to make Teflon, a polymer possessing remarkable resistance to solvents and acids; and styrene, from which is made the polystyrene used in Styrofoam.

Not all polymers are synthetic. Among the many polymeric substances that occur in nature are rubber, cellulose, and the proteins. Rubber consists of long, coiled molecules produced by the polymerization of a diolefin called isoprene:

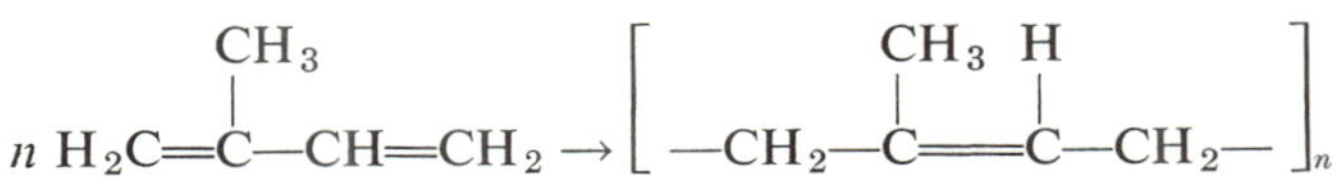

isoprene — polyisoprene

The first synthetic rubber was not an exact duplicate of natural rubber, but in 1958 the development of synthetic polyisoprene rubber was announced by several American manufacturers.

Synthetic polymers are frequently classified according to their uses. If they are stretchable, like rubber, they are called *elastomers*. Some can be extruded through spinnerets to produce synthetic *fibers*. The term *plastics* is applied to the synthetic polymers from which a variety of commerical products are formed, including plastic wrappings and molded objects of various shapes and sizes.

The plastics industry uses the term *resin* for a synthetic polymeric substance as it is initially obtained from the reaction mixture, and it uses the term *plastic* for the material developed from the resin by addition of other substances such as fillers, plasticizers, and coloring materials.

New Terms

Addition: a type of reaction characteristic of compounds containing double (or triple) bonds, in which two H atoms (or two Br atoms, etc.) are added to the double bond, i.e., one H atom to each doubly bonded carbon, converting the double bond to a single bond.

Alicyclic hydrocarbons: hydrocarbons that contain rings but resemble aliphatic hydrocarbons in chemical properties. The name is a combination of *aliphatic* and *cyclic*. Alicyclic hydrocarbons include cycloalkanes, cycloalkenes, cycloalkadienes, etc. (but not the aromatic hydrocarbons).

Aliphatic hydrocarbons: hydrocarbons that have an open-chain structure (i.e., no rings). They include the alkanes, alkenes, alkadienes, alkynes, etc.

Alkadiene: a hydrocarbon containing two double bonds.

Alkane: any hydrocarbon that is a member of the homologous series with the general formula C_nH_{2n+2}. The alkane series is also known as the methane series and as the paraffin series.

Alkene: a hydrocarbon containing a double bond. The general formula for alkenes is C_nH_{2n}. The alkene series is also known as the ethylene series and as the olefin series.

Alkyne: a hydrocarbon containing a triple bond. The general formula for alkynes is C_nH_{2n-2}. The alkyne series is also known as the acetylene series.

Aromatic hydrocarbon: a hydrocarbon containing one or more benzene rings. (The three general classes of hydrocarbons are the aliphatic hydrocarbons, the alicyclic hydrocarbons, and the aromatic hydrocarbons.)

Conformations: the different arrangements that can be converted into one another by rotating a part of a molecule without breaking any bonds.

Cyclic hydrocarbon: a hydrocarbon in which the carbon atoms are arranged in one or more rings.

Homologous series: a series of compounds in which each member differs from the preceding member in having one more carbon and two more hydrogen atoms. Examples of homologous series include the alkanes, the alkenes, the cycloalkanes, the cycloalkenes, etc.

Hybridization of orbitals: the mixing of a number of orbitals, not all of the same type, to form the same number of hybrid orbitals that are all alike.

Hydrocarbons: compounds consisting of hydrogen and carbon alone.

Isomers: two or more different compounds having the same number and kinds of atoms per molecule.

Saturated hydrocarbons: hydrocarbons that contain no double or triple bond. They are said to be saturated because their hydrogen content is at a maximum.

Substitution: the removal of a hydrogen atom from a hydrocarbon and the insertion of another atom or group in its place. Elementary hydrogen is not a product of the reaction, as it would be if the reaction were a displacement reaction.

EXAMPLE OF SUBSTITUTION: $CH_4 + Br_2 \rightarrow CH_3Br + HBr$
EXAMPLE OF DISLPACEMENT: $2HCl + Zn \rightarrow ZnCl_2 + H_2$

Exercises

18-1 Review the chemical evidence that the four bonds in methane are equivalent.

18-2 What reason is there for concluding that the methane molecule is tetrahedral?

18-3 Suppose two isomeric compounds with the molecular formula CH_2Br_2 were discovered. Would the discovery be considered significant? Why or why not?

18-4 What is the general formula for each of the following hydrocarbons: alkanes, alkenes, alkynes, alkadienes, cycloalkanes, cycloalkenes?

18-5 Compare the following types of formulas: empirical, molecular, extended structural, condensed structural, projection.

18-6 Define these terms: conformations, eclipsed conformation, staggered conformation, isomerism.

18-7 Write the structural formulas for all the isomeric pentanes, pentenes, and hexanes. Assign the correct IUPAC name to each compound.

18-8 Do you believe the existence of cyclobutyne is possible? Why or why not?

18-9 Define the following terms so as to distinguish among them: aliphatic, alicyclic, and aromatic hydrocarbons.

18-10 Illustrate three types of isomerism.

18-11 Why are all six carbon-carbon bonds in benzene considered equivalent?

References

Brown, T. L., *General Chemistry*, Merrill, Columbus, Ohio, 1963. Chapter 8, "The Covalent Bond," includes consideration of hybrid orbitals and resonance. Hydrocarbons are considered in the first seven sections of Chapter 23.

Herz, W., *The Shape of Carbon Compounds*, Benjamin, New York, 1963 (paperbound). One of the volumes in the "General Chemistry Monograph Series." An introduction to organic chemistry with emphasis on structure.

Mark, H. F., and the Editors of Life, *Giant Molecules*, Time, Inc., New York, 1966. An explanation of natural and synthetic polymers for interested persons who have not had formal training in science. Supplemented by elegant pictorial essays.

Rossini, F. D., "Hydrocarbons in Petroleum," *J. Chem. Educ.*, **37** (11), 554 (1960). An account of present knowledge of what petroleum is chemically—the relative amounts of different classes of hydrocarbons and of individual compounds within given classes.

19

Oxygen Derivatives of Hydrocarbons

19-1 The Nature of Organic Chemistry

The number of compounds containing carbon is so great that the study of their preparation and properties constitutes a special division of chemistry. For historical reasons this branch of chemistry is called organic chemistry, but it might better be called the chemistry of the compounds of carbon.

Man's basic needs for food, fuel, and clothing are all related to organic chemistry. The three important classes of organic compounds in foods are fats, carbohydrates, and proteins, and their digestion involves organic chemical reactions. These foodstuffs are of biological origin. The important fuels include coal, wood, natural gas, and petroleum, all of which are of biological origin. Cotton, wool, and silk are natural organic substances used for clothing, and they also are of biological origin.

Originally no distinction was made between the chemistry of substances of mineral origin and the chemistry of substances of biological origin. As more and more compounds were isolated from natural sources, the differences became so apparent that chemists began to differentiate between organic compounds (of biological origin) and inorganic compounds (of mineral origin). The term *organic chemistry* was first used in 1808 by the Swedish chemist J. J. Berzelius. The essential difference between inorganic and organic compounds—that organic compounds contain carbon—was first recognized by the German chemist Leopold Gmelin in 1848.

The number of carbon compounds obtained by chemical synthesis in laboratories and chemical factories now far exceeds the number that have been isolated from natural products. It is no longer correct, therefore, to regard organic chemistry as the branch of chemistry concerned with substances of biological origin; although concerned with these substances, organic chemistry also includes the synthetic compounds containing carbon. The term organic chemistry will probably continue to be used, but its orginal meaning no longer applies.

Any organic compound can be derived, in one or more real or imaginary steps, from the corresponding hydrocarbon. This fact is emphasized in an alternative definition which describes organic chemistry as the chemistry of the hydrocarbons and their derivatives.

The study of the chemical reactions occurring in living organisms is now called *biochemistry*.

19-2 Methyl Alcohol

Organic compounds are classified into groups having similar properties and similar structures. The *alcohols* comprise one of the most important of these groups. Several of the alcohols of low molecular weight are liquids which can dissolve molecular compounds readily and ionic compounds somewhat less readily. Any alcohol is a potential starting material for the preparation of other organic compounds.

The simplest of the alcohols is methyl alcohol, also known as wood alcohol because it can be obtained by heating dried hardwood (oak, hickory, etc.) in the absence of air. By chemical analysis and molecular weight determinations, its molecular formula has been established to be CH_4O. As its formula suggests, methyl alcohol can be obtained by controlled oxidation of methane from natural gas, but a far more important synthesis, used since 1920, consists in the catalytic hydrogenation of carbon monoxide:

$$CO + 2H_2 \rightarrow CH_3OH$$

In the above equation we have represented methyl alcohol by its condensed structural formula. The extended structural formula and the electronic formula are as follows:

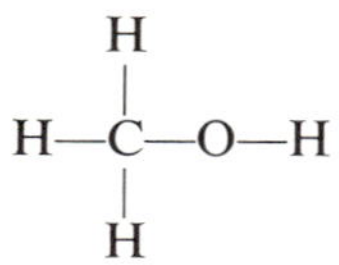

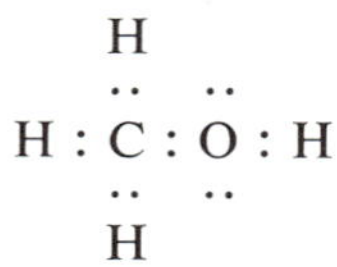

The important feature of the structure of this or any other alcohol molecule is that one of the hydrogen atoms is bonded to the oxygen instead of to the carbon. We may think of a methyl alcohol molecule as consisting of an —OH (hydroxyl) group and a $—CH_3$ (methyl) group.

Since the one structural feature common to all alcohols is the —OH group, this is called the *functional group* characteristic of alcohols. An alcohol that contains only one —OH group per molecule is known as a *monohydric* alchohol. Any alcohol derived from an alkane is called a *saturated* alchohol. The general formula for the saturated monohydric alcohols is $C_nH_{2n+1}OH$, which may be written more briefly as *R*OH, where *R* is an *alkyl* group, C_nH_{2n+1}.

An alternative method (IUPAC) for naming alcohols consists in the use of one-word names based on the names of the corresponding hydrocarbons. By this method, the name for CH_3OH is *methanol.* The suffix *-ol* indicates an alcohol.

All alcohols contain the —OH group, but they are nonelectrolytes; their aqueous solutions are neutral and do not conduct electric current. Alcohols react with sodium to give hydrogen and sodium alkoxide, just as water reacts with it to give hydrogen and sodium hydroxide:

$$2HOH + 2Na \rightarrow \underset{\text{sodium hydroxide}}{2Na^{+}OH^{-}} + H_2$$

$$2CH_3OH + 2Na \rightarrow \underset{\text{sodium methoxide}}{2Na^{+}OCH_3^{-}} + H_2$$

$$2R\mathrm{OH} + 2Na \rightarrow \underset{\text{sodium alkoxide}}{2Na^{+}\mathrm{O}R^{-}} + H_2$$

A good method for disposing of sodium metal wastes is to dissolve them in alcohol, with which the sodium reacts more slowly than with water.

Anhydrous hydrogen chloride and other strong acids react with alcohols to give ions comparable to the hydronium ion:

$$HCl + HOH \rightarrow \left[\begin{array}{c} \mathrm{H} \\ \cdot\cdot \\ \mathrm{H : O : H} \\ \cdot\cdot \end{array}\right]^{+} + Cl^{-}$$

$$HCl + R\mathrm{OH} \rightarrow \left[\begin{array}{c} \mathrm{H} \\ \cdot\cdot \\ R\mathrm{ : O : H} \\ \cdot\cdot \end{array}\right]^{+} + Cl^{-}$$

Methyl alcohol is a highly toxic liquid; even the inhalation of its vapor is hazardous. As in the case of water, the fact that methyl alcohol exists as a liquid despite its low molecular weight is explained in terms of hydrogen bonding, which can occur between the H of an —OH group and the O of another —OH group:

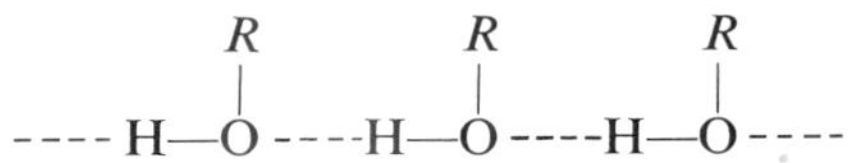

Here we have used R to represent the methyl group. The H atoms of the methyl group, like those in methane itself, are unable to form hydrogen bonds. Thus, there is no way that a methyl alcohol molecule could be hydrogen-bonded to *three* other molecules at the same time, as is the case with a water molecule.

Although methane is insoluble in water, methyl alcohol is miscible with water in all proportions. This, again, is due to hydrogen bonding. A hydrogen bond can form between a molecule of methyl alcohol and a molecule of water just as one can form between two molecules of the alcohol or between two molecules of water.

19-3 Homologs of Methyl Alcohol

The second member of the saturated monohydric series of alcohols is ethyl alcohol, or ethanol, also known as grain alcohol since it can be obtained by the fermentation of grain (e.g., rye or corn). It is the alcohol contained in alcoholic beverages, where it is the product of the fermentation of the low molecular weight sugar called glucose. The fermentation process is catalyzed by the enzyme known as zymase; carbon dioxide is also a product of the reaction:

$$C_6H_{12}O_6 \xrightarrow{\text{zymase}} 2CO_2 + 2C_2H_5OH$$

Ethyl alcohol bears the same formal relationship to ethane that methyl alcohol bears to methane; i.e., one of the H atoms of ethane has been replaced by an —OH group:

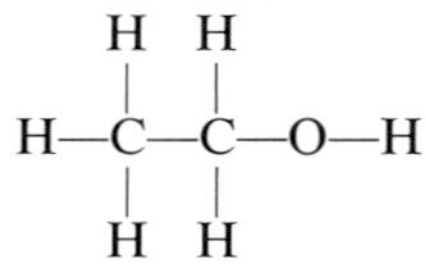

The structure is clearly implied by the condensed structural formula CH_3CH_2OH, or even by C_2H_5OH. We may think of an ethyl alcohol molecule as consisting of a hydroxyl group and an ethyl group.

Because of hydrogen bonding, ethyl alcohol has a higher boiling point than any hydrocarbon with a comparable molecular weight and is miscible with water in all proportions. As we consider successive members of this homologous series, the alkyl group C_nH_{2n+1} becomes larger in increments of CH_2 with the result that the higher alcohols are less soluble in water because they are more like hydrocarbons.

The third member of the series should be an alcohol with the formula C_3H_7OH. Actually, there are two such alcohols, one with the —OH group attached to an end carbon and the other with the —OH group attached to the middle carbon of the chain:

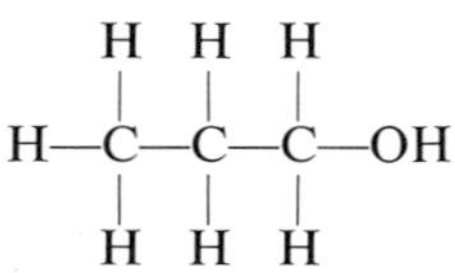

1-propanol
(*n*-propyl alcohol)
boiling point, 97°C

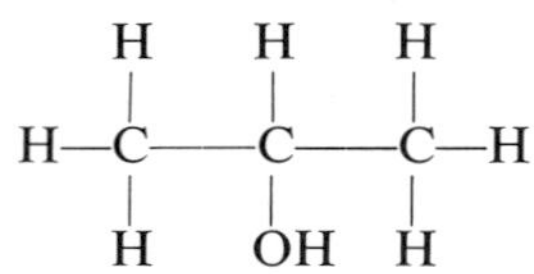

2-propanol
(isopropyl alcohol)
boiling point, 81°C

These two alcohols provide an illustration of position isomerism. Normal propyl alcohol, as it is commonly called, can be represented by the condensed structural formula $CH_3CH_2CH_2OH$, and isopropyl alcohol can be represented as $CH_3CHOHCH_3$, but to represent either of them as C_3H_7OH is ambiguous.

There are increasing numbers of isomeric alcohols as the number of carbon atoms increases. For example, there are 4 four-carbon (butyl) alcohols and 8 five-carbon (amyl or pentyl) alcohols.

Many of the alcohols are useful compounds, but methyl and ethyl alcohols are produced in the largest quantities. Methyl alcohol is used for the manufacture of formaldehyde and other chemicals, as a radiator antifreeze, as a solvent, and as a denaturant for ethyl alcohol. Denaturing is the addition of substances which make the alcohol unfit to drink. Ethyl alcohol for industrial use must be denatured if payment of the alcoholic beverage tax is to be avoided.

Ethyl alcohol is used in the manufacture of a great variety of dyes, medicines, perfumes, insecticides, etc. Drugs that are insoluble in water are often used in ethyl alcohol solutions called *tinctures*. Much of the industrial ethyl alcohol is now obtained synthetically by the catalyzed combination of water with ethylene resulting from the "cracking" of petroleum:

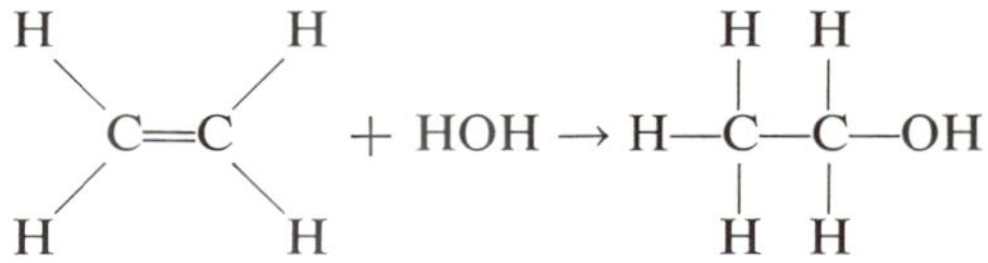

Isopropyl alcohol is well known as a component of "rubbing alcohol." Higher alcohols are used where a definite number of carbon atoms per molecule is required for the purpose at hand; e.g., the manufacture of butyl acetate requires butyl alcohol.

19-4 Classes of Alcohols

The number of hydrogen atoms attached to the same carbon atom as the —OH group determines some of the important chemical properties and uses of alcohols. By this method of classification, alcohols are termed primary, secondary, or tertiary. A *primary* alcohol has two hydrogen atoms attached to the alcohol carbon; i.e., it contains the —CH_2OH group. Examples are ethyl alcohol, CH_3CH_2OH, and *n*-propyl alcohol, $CH_3CH_2CH_2OH$. A *secondary* alcohol has only one hydrogen atom attached to the alcohol carbon because two carbon atoms are attached to it. An example is isopropyl alcohol, $CH_3CHOHCH_3$. A *tertiary* alcohol has three carbon atoms attached to the alcohol carbon and therefore has *no* hydrogen atom attached to it. An example is tertiary butyl alcohol:

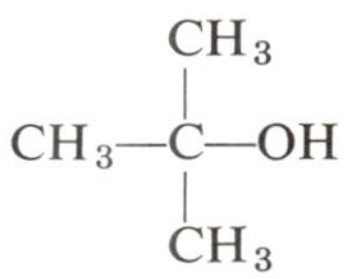

The general formula for tertiary alcohols is R_3COH, where the three alkyl groups may all be different or identical; for secondary alcohols, the general formula is R_2CHOH; for primary alcohols, RCH_2OH.

The differences in the three classes of alcohols show up in the presence of mild oxidizing agents in aqueous solution. Let us consider, for example, the oxidation of isopropyl alcohol by permanganate ions in aqueous solution. The oxidation of the isopropyl alcohol produces an important organic compound known as acetone, and the reduction of the purple permanganate produces a brown precipitate, MnO_2.

$$3CH_3CHOHCH_3 + 2MnO_4^- \rightarrow 3CH_3COCH_3 + 2MnO_2\downarrow + 2H_2O + 2OH^-$$

Acetone is the simplest member of a class of compounds known as *ketones*, which have the general formula

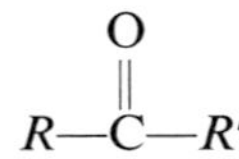

where R and R' are alkyl groups which may or may not be identical. Mild oxidation of a secondary alcohol always produces a ketone.

The oxidation of ethyl alcohol with aqueous permanganate produces acetic acid,

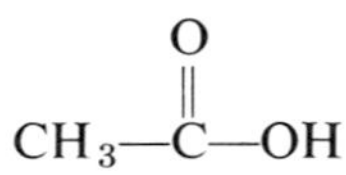

Acetic acid is a member of the class of compounds known as *carboxylic acids*, which have the general formula

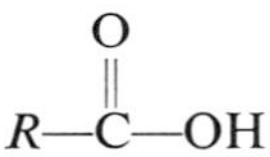

Oxidation of a primary alcohol produces a carboxylic acid because an oxygen atom takes the place of the two hydrogen atoms attached to the alcohol carbon. However, if the oxidation of a primary alcohol is carefully regulated, an intermediate product known as an *aldehyde* can be obtained. Thus, careful oxidation of ethyl alcohol produces acetaldehyde,

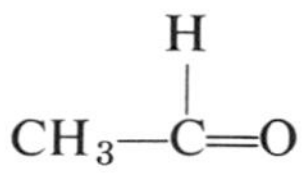

Aldehydes have the general formula

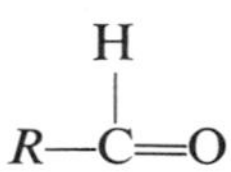

The differences among primary, secondary, and tertiary alcohols with respect to mild oxidation may be summarized as follows:

PRIMARY ALCOHOL:

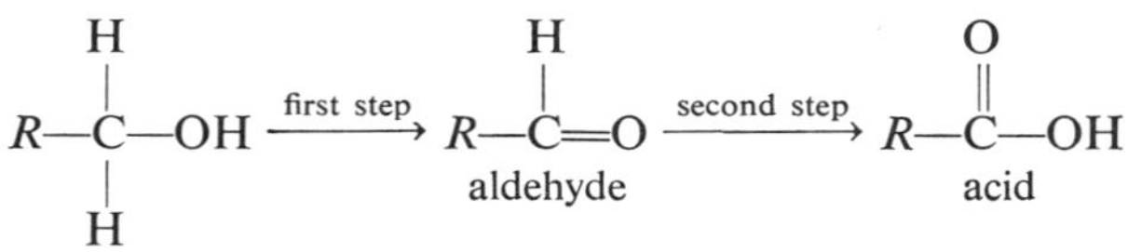

SECONDARY ALCOHOL:

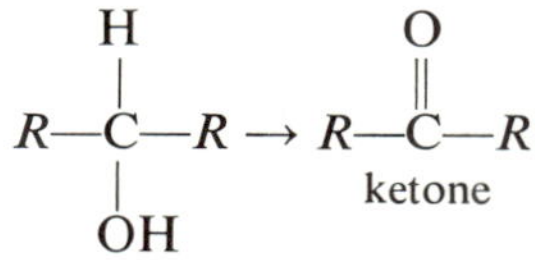

TERTIARY ALCOHOL: Oxidation impossible without breaking carbon chain.

Strong oxidizing agents can oxidize a tertiary alcohol, but since this involves breaking the carbon chain, the products have fewer carbon atoms than the original alcohol.

Although methyl alcohol is a primary alcohol, it is in a sense in a class of its own. The general formula for primary alcohols, RCH_2OH, applies to methyl alcohol only if we let R represent a hydrogen atom instead of an alkyl group. The presence of *three* hydrogen atoms on the alcohol carbon makes methyl alcohol unique and results in there being three possible steps in its oxidation:

STEP 1: STEP 2: STEP 3:

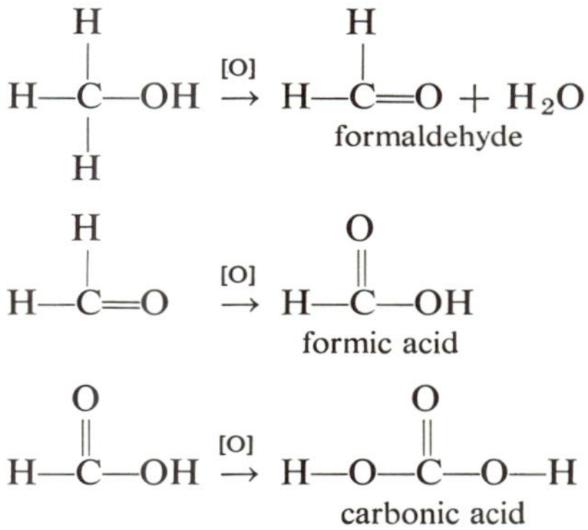

Carbonic acid is a weak and unstable acid which decomposes to give CO_2 and H_2O. It is present in so-called carbonated beverages, where it is formed (in small concentrations) by carbon dioxide dissolving in water.

19-5 Varieties of Structures in Alcohols

So far we have considered only alcohols that are derivatives of alkanes. Actually, an alcohol, like a hydrocarbon, may be either open-chain or cyclic; it may contain a double bond, a halogen atom, an aromatic ring, etc. Some idea of the variety of alcohols may be obtained by considering the following examples:

$H_2C{=}CH{-}CH_2OH$

allyl alcohol

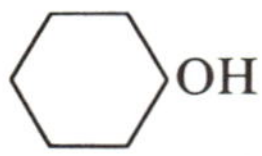

cyclohexanol

benzyl alcohol

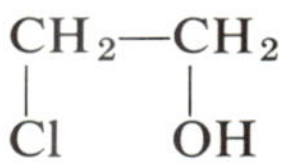

ethylene chlorohydrin

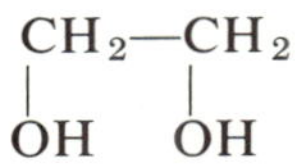

ethylene glycol

19-6 Phenols

Phenols are compounds which have an —OH group attached directly to an aromatic ring. They have the general formula *Ar*OH, where *Ar* is an *aryl* group: a phenyl group, a substituted phenyl group, a naphthyl group, etc. When it is necessary to make the distinction, the hydroxyl group in phenols is called a *phenolic hydroxyl* while that in alcohols is called an *alcoholic hydroxyl.*

The simplest member of the phenol family is *phenol,* ΦOH. This compound is a powerful antiseptic, popularly known as carbolic acid. The hydroxy derivatives of toluene are known as *cresols.*

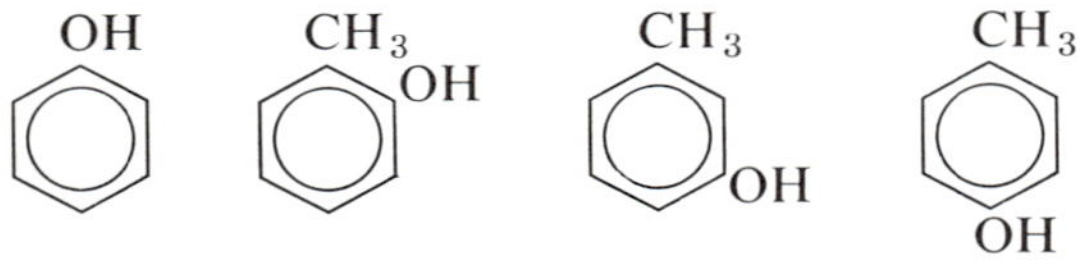

phenol *o*-cresol *m*-cresol *p*-cresol

In naming a compound that has more than one group attached to a benzene ring, we must indicate their relative positions. A disubstituted benzene has three possible isomers which are differentiated by the prefixes *ortho-* (*o*-), *meta-* (*m*-), and *para-* (*p*-). The ortho- isomer is the compound in which two *adjacent* hydrogen atoms of the benzene ring have been replaced. The meta-isomer is obtained by replacement of two *alternate* hydrogen atoms; and the para- isomer, by replacement of two *opposite* hydrogen atoms.

Phenol is somewhat soluble in water in consequence of hydrogen bonding between phenol and water molecules. Higher phenols are not water-soluble. Although colorless when pure, the phenols are easily oxidized to give colored oxidation products.

Aryl groups have more attraction for electrons than does hydrogen, whereas alkyl groups have slightly less attraction than hydrogen. Consequently, phenols are considerably more acidic than water, although alcohols are even less strongly acidic than water. The ionization constant for phenol is 1.1×10^{-10}. The water-insoluble phenols dissolve in dilute solutions of strong bases to form salts of phenols:

$$ArOH(s) + OH^- \rightleftharpoons ArO^- + H_2O$$

Despite their significant differences, alcohols and phenols resemble each other to a limited extent since both contain the —OH group. Both alcohols and phenols form ethers and esters.

19-7 Carboxylic Acids

The first three members of the homologous series of compounds known as the aliphatic carboxylic acids are

HCOOH	CH_3COOH	CH_3CH_2COOH
formic acid	acetic acid	propionic acid

The functional group characteristic of these acids is the —COOH group, which is called the *carboxyl* group. It has the structural formula

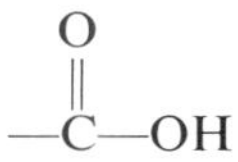

We may write *R*COOH as the general formula for aliphatic carboxylic acids if we allow *R* to be a hydrogen atom instead of an alkyl group in the first member of the series, formic acid. The general formula for aromatic carboxylic acids is *Ar*COOH.

The compounds of this series undergo partial ionization when dissolved in water; in this respect they differ from most organic compounds:

$$\underset{\text{}}{H_2O} + \underset{\text{acetic acid}}{CH_3COOH} \rightleftharpoons \underset{\text{acetate ion}}{CH_3COO^-} + \underset{\text{hydronium ion}}{H_3O^+}$$

Since the ionization produces the H_3O^+ ion characteristic of acidic aqueous solutions, these compounds are acids. The hydrogen atom that is attached to the oxygen atom in the carboxyl group is responsible for the acidity. Carboxylic acids are considerably stronger acids than phenols.

The aliphatic monocarboxylic acids are known as *fatty acids* because some of them, both saturated and unsaturated, occur in the form of esters of glycerol in animal fats and in animal and vegetable oils.

The solubility of the carboxylic acids in water is due to hydrogen bonding between acid and water. Long-chain acids are not very soluble, for the acids become more like hydrocarbons as the carbon chain lengthens. In the pure state, and in solution in nonpolar solvents, carboxylic acids are dimerized, two molecules being held together by two hydrogen bonds:

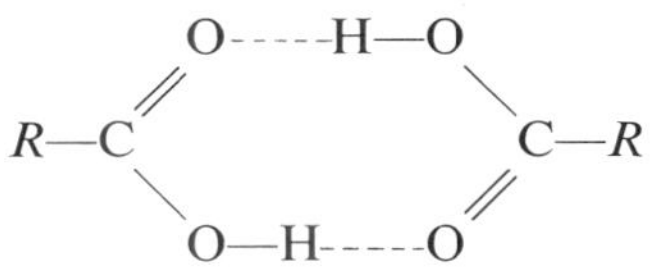

Since the carboxylic acids were first isolated from natural sources, especially fats, they were given common names indicating what these sources were. For example:

Formic acid (1 carbon atom)	Latin: *formica*, ant
Acetic acid (2 carbons)	Latin: *acetum*, vinegar
Butyric acid (4 carbons)	Latin: *butyrum*, butter
Caproic acid (6 carbons)	Latin: *caper*, goat
Palmitic acid (16 carbons)	palm oil
Stearic acid (18 carbons)	Greek: *stear*, tallow
Arachidic acid (20 carbons)	Latin: *arachis*, peanut

A mixture of stearic and palmitic acids is used in the manufacture of candles. Magnesium and zinc stearates are used in face powders.

One of the most important aromatic carboxylic acids is the first member of the series, ΦCOOH, benzoic acid. Sodium benzoate is used extensively as a food preservative.

Examples of unsaturated carboxylic acids are oleic acid, $C_{17}H_{33}COOH$, and cinnamic acid, ΦCH═CHCOOH.

19-8 Salts and Esters

Carboxylic acids react with metal hydroxides, carbonates, and bicarbonates to form salts. For example,

$$\underset{\text{formic acid}}{HCOOH} + Na^+OH^- \rightarrow \underset{\text{sodium formate}}{HCOO^-Na^+} + H_2O$$

$$\underset{\text{acetic acid}}{2CH_3COOH} + (Na^+)_2CO_3^{--} \rightarrow \underset{\text{sodium acetate}}{2CH_3COO^-Na^+} + CO_2\uparrow + H_2O$$

$$\underset{\text{propionic acid}}{CH_3CH_2COOH} + Na^+HCO_3^- \rightarrow \underset{\text{sodium propionate}}{CH_3CH_2COO^-Na^+} + CO_2\uparrow + H_2O$$

Because of their ionic nature, the salts of the carboxylic acids are nonvolatile. The alkali metal salts are much more soluble than the acids themselves. *Soaps* are alkali metal salts of acids having from ten to eighteen carbon atoms. An example is sodium stearate, $C_{17}H_{35}COO^-Na^+$.

Carboxylic acids react with alcohols to form *esters*; therefore, the reaction is called *esterification*. For example,

$$\underset{\text{formic acid}}{H\overset{\overset{O}{\|}}{C}{-}OH} + \underset{\text{methyl alcohol}}{HOCH_3} \rightarrow \underset{\text{methyl formate}}{H\overset{\overset{O}{\|}}{C}{-}OCH_3} + H_2O$$

$$\underset{\text{acetic acid}}{CH_3\overset{\overset{O}{\|}}{C}{-}OH} + \underset{\text{ethyl alcohol}}{HOC_2H_5} \rightarrow \underset{\text{ethyl acetate}}{CH_3\overset{\overset{O}{\|}}{C}{-}OC_2H_5} + H_2O$$

The general formula for esters, *RCOOR′*, indicates that an ester differs from an acid only in having an alkyl group in the place of the carboxylic hydrogen. (The two alkyl groups, *R* and *R′*, may be, but usually are not, identical.) It can also be said that a salt differs from an acid only in having a metal or other positive ion in the place of the carboxylic hydrogen. Esters are named as if they were alkyl salts of the carboxylic acids.

The reaction of an acid with an alcohol may superficially resemble neutralization of an acid by a base. However, esters are covalent, not ionic; neutralization is almost instantaneous, esterification is slow; neutralization goes to completion, esterification does not go to completion. It has been established experimentally, e.g., by means of studies with isotopes, that in esterification the —OH from the acid and the —H from the alcohol unite to form water; this is, of course, the exact opposite of neutralization.

Esters have some of the most pleasant fragrances known. Many of them are responsible for distinctive fruit flavors. For example,

$CH_3\overset{\overset{\large O}{\parallel}}{C}—O(CH_2)_4CH_3$	amyl acetate	bananas
$CH_3\overset{\overset{\large O}{\parallel}}{C}—O(CH_2)_7CH_3$	octyl acetate	oranges
$CH_3CH_2CH_2\overset{\overset{\large O}{\parallel}}{C}—OC_2H_5$	ethyl butyrate	pineapples

Examples of esters containing aromatic groups are $CH_3COO\Phi$, phenyl acetate; $\Phi COOCH_3$, methyl benzoate; and $\Phi COO\Phi$, phenyl benzoate.

One of the most important reactions of esters is *saponification*, a reaction with a strong base in aqueous solution. For example,

$$\underset{\text{ethyl acetate}}{CH_3\overset{\overset{\large O}{\parallel}}{C}—OC_2H_5} + Na^+OH^- \rightarrow \underset{\text{sodium acetate}}{CH_3\overset{\overset{\large O}{\parallel}}{C}—O^-Na^+} + \underset{\text{ethyl alcohol}}{HOC_2H_5}$$

Saponification of an ester, $RCOOR'$, with sodium hydroxide produces the sodium salt of the parent acid, $RCOO^-Na^+$, and liberates the parent alcohol, $R'OH$. The word *saponification* (Latin: *saponis*, soap) means "making soap." Ordinary soap consists of salts obtained by reaction of a strong base with fats, which are esters of the trihydric alcohol called glycerol.

The reduction of esters is an important method for preparing higher alcohols from natural fats. The catalytic reduction is accomplished with hydrogen at 200°C and 200 atmospheres:

$$RCOOR' + 2H_2 \rightarrow RCH_2OH + R'OH$$

19-9 Aldehydes and Ketones

It will be recalled that aldehydes can be obtained by mild oxidation of primary alcohols. Actually, aldehydes are more easily oxidized than the alcohols, so that the preparation of aldehydes by this method requires that they be distilled out of the reaction mixture as fast as they form. Fortunately, there are other methods for preparing aldehydes. One such procedure involves the catalytic dehydrogenation of primary alcohols:

$$RCH_2OH \xrightarrow{\text{Cu, 250°}} \underset{\text{an aldehyde}}{R—\overset{\overset{\large H}{|}}{C}{=}O} + H_2$$

The name *aldehyde* is derived from *Al. dehyd.*, which is an abbreviation of *Alcohol dehydrogenatum* (alcohol that has been dehydrogenated).

Ketones can also be obtained by catalytic dehydrogenation if secondary alcohols are used:

$$R_2CHOH \xrightarrow{Cu,\ 250°} \underset{\text{a ketone}}{R_2CO} + H_2$$

The common name of an aldehyde is determined by the name of the acid into which it is readily converted by oxidation. For example,

$$\begin{array}{c} \text{H} \\ | \\ \text{H—C=O} \end{array}$$

is formaldehyde because its oxidation yields formic acid;

$$\begin{array}{c} \quad\ \ \text{H} \\ \quad\ \ | \\ CH_3\text{—C=O} \end{array}$$

is acetaldehyde because it is oxidizable to acetic acid; and so on.

The common name of a ketone usually consists of the names of its two alkyl groups followed by the word *ketone*. Thus,

$$\begin{array}{c} \text{O} \\ \| \\ CH_3\text{—C—}C_2H_5 \end{array}$$

is methyl ethyl ketone. If the two alkyl groups are identical, the group is identified only once. For example,

$$\begin{array}{c} \text{O} \\ \| \\ C_2H_5\text{—C—}C_2H_5 \end{array}$$

is ethyl ketone. The simplest of the ketones, CH_3COCH_3, has long been known as acetone.

Formaldehyde is a gas, used as a germicide and fungicide, and commonly handled as an aqueous solution. Acetaldehyde is also a gas, now available synthetically by the catalyzed hydration of acetylene:

$$H\text{—C}\equiv\text{C—}H + H_2O \xrightarrow{HgSO_4} CH_3CHO$$

Acetaldehyde is used chiefly in making acetic acid. Most aldehydes are used for making alcohols and acids. The aliphatic aldehydes with 7 to 16 carbon atoms are used as ingredients of perfumes. Aromatic aldehydes include benzaldehyde, ΦCHO, long known as "oil of bitter almonds," and cinnamaldehyde, ΦCH=CHCHO, chief ingredient of oil of cinnamon.

Acetone is the most important ketone. It is used widely as a solvent and as the starting material for the manufacture of other chemicals. Water and nonpolar solvents are miscible with acetone. Methyl ethyl ketone, known industrially by the initials MEK, is used as a solvent for removing wax from lubricating oil. It is also familiar as fingernail-polish remover. Examples of aromatic ketones are acetophenone, which is methyl phenyl ketone, $\Phi COCH_3$; and benzophenone, which is phenyl ketone, $\Phi CO\Phi$. Both are used in perfumery. Ketones of more complicated structure include camphor and the sex hormones.

19-10 Dehydration of Alcohols

The removal of the elements of water from an alcohol may occur in either of two ways. *Intramolecular dehydration*, i.e., removal of an —H and an —OH from adjacent carbon atoms of the same alcohol molecule, produces an *alkene*. For example,

$$CH_3CH_2OH \rightarrow CH_2{=}CH_2 + H_2O$$

Intermolecular dehydration of an alcohol, i.e., removal of —H and —OH from different alcohol molecules, produces an *ether*. For example,

$$2C_2H_5OH \rightarrow (C_2H_5)_2O + H_2O$$

Either intramolecular or intermolecular dehydration of ethanol may be carried out by reaction with concentrated sulfuric acid, which is an excellent dehydrating agent.

The reaction of ethanol with sulfuric acid is an excellent example of the way a chemical reaction may be modified by controlling the conditions. When ethanol and sulfuric acid are brought together at room temperature, the only significant reaction is the formation of an oxonium salt:

$$C_2H_5OH + H_2SO_4 \rightleftharpoons C_2H_5OH_2^+ + HSO_4^-$$

This reaction is analogous to the reaction of a strong acid with water (e.g., $H_2O + HCl \rightleftharpoons H_3O^+ + Cl^-$), and it may be said to occur because the alcohols are stronger bases than the anions of strong acids. When ethanol is warmed with excess sulfuric acid, ethyl hydrogen sulfate is formed:

$$C_2H_5OH + H_2SO_4 \rightleftharpoons C_2H_5HSO_4 + H_2O$$

Heating the mixture to 180°C causes the ethyl hydrogen sulfate to decompose, forming ethylene.

$$C_2H_5HSO_4 \rightleftharpoons C_2H_4 + H_2SO_4$$

If, on the other hand, sulfuric acid is mixed with excess ethanol and the mixture heated under reduced pressure, ethyl sulfate distills:

$$C_2H_5HSO_4 + C_2H_5OH \rightleftharpoons (C_2H_5)_2SO_4 + H_2O$$

However, if a mixture of ethanol and sulfuric acid is heated to 140°C and maintained at this temperature while ethanol is continuously added below the surface to keep it in excess, ether distills:

$$C_2H_5HSO_4 + C_2H_5OH \rightleftharpoons (C_2H_5)_2O + H_2SO_4$$

All of these reactions are reversible; therefore, to obtain a particular compound, it is necessary to choose the optimum conditions for obtaining it rather than the other compounds.

19-11 Ethers

Ethers have the general formula *ROR′*. The simplest ether is CH_3OCH_3, methyl ether. The next ether in order of molecular weight is $CH_3OC_2H_5$, methyl ethyl ether; the next is $C_2H_5OC_2H_5$, ethyl ether; etc. If the two alkyl groups are different, the ether is named by identifying the two alkyl groups; if the alkyl groups are identical, it is only necessary to name the group once. The name *diethyl ether*, for example, is correct, but the name *ethyl ether* is adequate.

Since one molecule of an ether cannot form hydrogen bonds with another molecule, the ethers are much more volatile than alcohols of corresponding molecular weight. For example, the boiling point of methyl ether is −24°C, but the boiling point of ethyl alcohol is 78°C. These two compounds have the same molecular weight and the same molecular formula, C_2H_6O; i.e., methyl ether and ethyl alcohol are isomers. Since they contain different functional groups, they are called *functional isomers*. The group of atoms that is characteristic of all ethers is present in methyl ether,

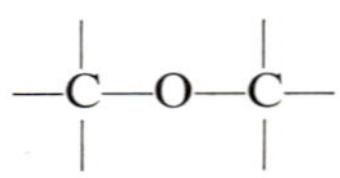

while in methyl alcohol there is the group of atoms always present in any alcohol,

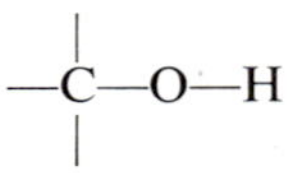

Ethyl ether, the best-known of the ethers, is the anesthetic usually referred to simply as "ether."

Important cyclic ethers include ethylene oxide and tetrahydrofuran:

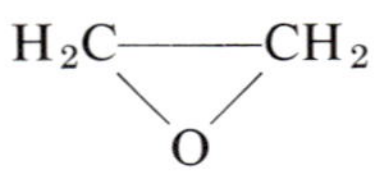

ethylene oxide

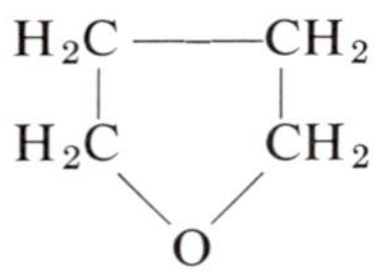

tetrahydrofuran

Important aromatic ethers include phenyl ether, Φ_2O; methyl phenyl ether, ΦOCH_3, which is called *anisole*; and ethyl phenyl ether, ΦOC_2H_5, called *phenetole*.

19-12 Polyfunctional Compounds

Many organic compounds contain more than one functional group. Thus, an alcohol may contain more than one —OH group, although two —OH groups are seldom attached to the same carbon atom. Alcohols containing two —OH groups per molecule are called *dihydric alcohols* or *glycols*. The simplest is the radiator antifreeze ethylene glycol, $HOCH_2—CH_2OH$. The simplest trihydric alcohol is glycerol, commonly known as glycerin, $CH_2OH—CHOH—CH_2OH$.

A molecule of a dihydric alcohol can react with two molecules of a carboxylic acid to form an ester:

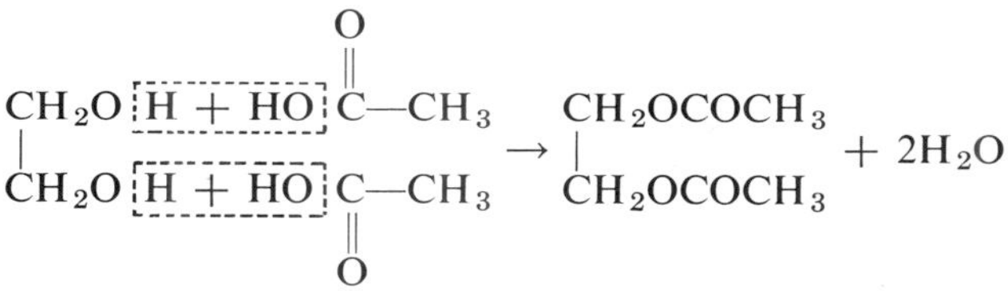

glycol — acetic acid — glycol diacetate

Similarly, a molecule of a trihydric alcohol can react with three molecules of an acid:

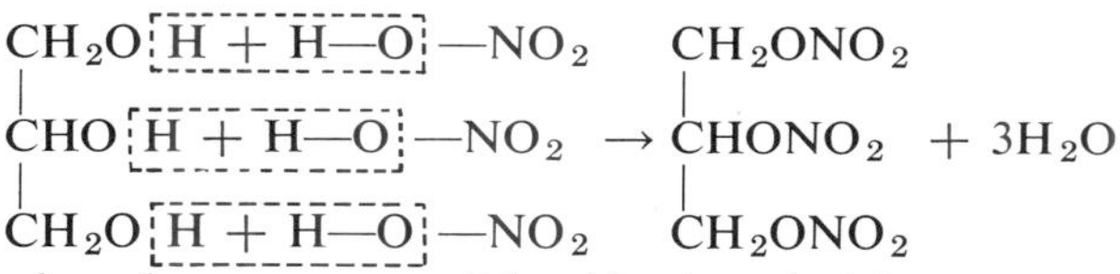

glycerol — nitric acid — glyceryl trinitrate (nitroglycerin)

An organic acid may have two or more carboxylic groups in the same molecule. The most important of the dicarboxylic acids commercially is adipic acid, $HOOC(CH_2)_4COOH$, an intermediate in the manufacture of nylon. The simplest dicarboxylic acid is oxalic acid, HOOC—COOH, used for the removal of stains and ink spots and responsible for the sour taste of rhubarb. The sourness of citrus fruits is due to citric acid, a tricarboxylic acid which also contains an —OH group:

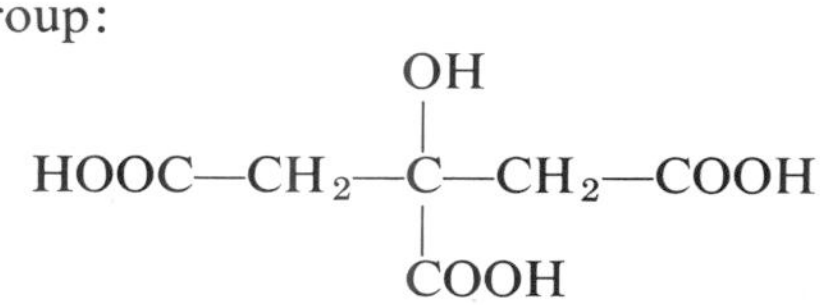

The simplest hydroxy acid is lactic acid,

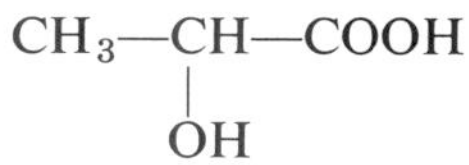

which occurs in sour milk and is produced in the body by the metabolism of glucose. During hard exercise lactic acid accumulates and causes soreness of muscles.

One of the most important bifunctional aromatic compounds is *o*-hydroxybenzoic acid, known as *salicylic acid*. The importance of this compound and its derivatives lies in their antipyretic (fever-reducing) and analgesic (pain-relieving) action. Sodium salicylate was used for the treatment of rheumatic fever as long ago as 1876. Its irritating action on the lining of the stomach led to the investigation of the effectiveness of other derivatives. The most widely used salicylic acid derivative now in use is acetylsalicylic acid, commonly called aspirin. The chief component of oil of wintergreen is methyl salicylate. It is used as a flavoring agent and, in rubbing liniments, as a counter-irritant for sore muscles.

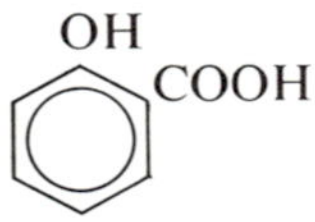

salicylic acid

acetylsalicylic acid

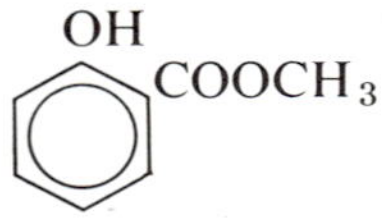

methyl salicylate

19-13 Optical Activity

The number of possible organic compounds is considerably greater than it would otherwise be because of an additional type of isomerism known as *optical isomerism.* This is the type of isomerism shown by two or more compounds that have the same structural formula but differ in *optical activity.* Optical activity is the ability to rotate the plane of polarized light.

When light is passing through any medium, according to the theory that light waves are electromagnetic waves, there is present an electric field perpendicular to the direction in which the light is traveling. There is, of course, an infinite number of perpendiculars to the line of propagation and the electric field exists in all directions. But certain media permit the passage of only that light that has an electric field aligned along a single perpendicular. The light transmitted by such media is said to vibrate in a single plane and to be *plane-polarized light.* Among the materials which polarize light are the mineral *calcite* and the synthetic material bearing the trade name Polaroid.

A calcite crystal, properly oriented, will divide a single ray of ordinary light into two rays. This phenomenon is known as *double refraction.* A single line viewed through a properly oriented crystal of doubly refracting material is seen as two lines. Each of the rays formed by double refraction is plane-polarized. The plane of vibration of one ray is at right angles to the plane of vibration of the other ray. Furthermore, one ray travels more slowly through the calcite than does the other ray.

In 1828, the Scottish physicist William Nicol invented a device, now known as a *Nicol prism,* for separating one of the plane-polarized rays from the other. A Nicol prism is constructed by cutting a piece of calcite diagonally and then cementing the pieces back together with Canada balsam (a fragrant resin obtained from the balsam fir). Canada balsam has an index of refraction such that one of the rays in the calcite is totally reflected by it while the other is transmitted as a single ray of polarized light.

When polarized light from a Nicol prism enters a second Nicol that is oriented like the first, the ray passes through without being affected. However, if the second prism is rotated through 90 degrees the light is totally reflected and does not pass through. Two prisms so oriented that polarized light transmitted by one is not transmitted by the other are referred to as *crossed Nicols.*

When a quartz plate, obtained by cutting a quartz crystal perpendicular to the crystal axis, is placed between crossed Nicols, some light passes through the second prism. In order for the second prism to become dark again it must be rotated through a definite angle. The ability to rotate the plane of polarization of plane-polarized light is called *optical activity.* The number of degrees, α, through which the second Nicol must be rotated to restore the original condition that existed with the crossed Nicols is called the *optical rotation.*

Optical rotation is measured by a *polarimeter.* This instrument consists of two Nicol prisms, one of which is in a fixed position while the other is attached to a disc, graduated in degrees, that can be rotated. The sample of the optically active substance is placed between the two Nicols.

René Just Haüy (Section 11-4) discovered that there are two kinds of quartz crystals which differ only in the location of two crystal facets which cause the crystals to be nonidentical mirror images. Because of the mirror-image relationship they are called *enantiomorphs* (Greek: opposite forms). The French physicist Jean Baptiste Biot discovered in 1815 that plates of equal thickness cut from the two kinds of quartz crystals rotate plane-polarized light the same amount but in opposite directions. The form of quartz crystal that rotates the plane of polarized light clockwise (i.e., to the right) is said to be *dextrorotatory* (Latin: *dexter*, right) and the form that rotates it counterclockwise (to the left) is said to be *levorotatory* (Latin: *laevus*, left).

Biot also discovered that turpentine is optically active, in both the liquid and vapor phases. The optical activity of liquids and gases cannot be attributed to the dextral or sinistral structure of crystals. It can only be attributed to the right-handedness or left-handedness of the molecules themselves. This conclusion is supported by Biot's further discovery that sugar solutions are optically active.

Chemists are usually concerned with measuring the optical activity of substances in solution. The amount of rotation, α, is proportional to the length, l, of the path through the optically active material. It is also proportional to the concentration of the optically active substance in solution, i.e.,

$$\alpha = [\alpha]cl$$

where c represents concentration and $[\alpha]$ is the *specific rotation*, a constant characteristic of the particular substance. The specific rotation is different for different wavelengths of light and undergoes some change with temperature. Usually, optical rotation is measured with light from a sodium lamp. For accurate work the temperature is held constant, usually at 25°C.

19-14 The Concept of the Asymmetric Carbon Atom

When van't Hoff and Le Bel proposed their theory of "the tetrahedral carbon atom" in 1874 (Section 18-4), they pointed out independently that the optically

Figure 19-1

Nonidentical mirror images

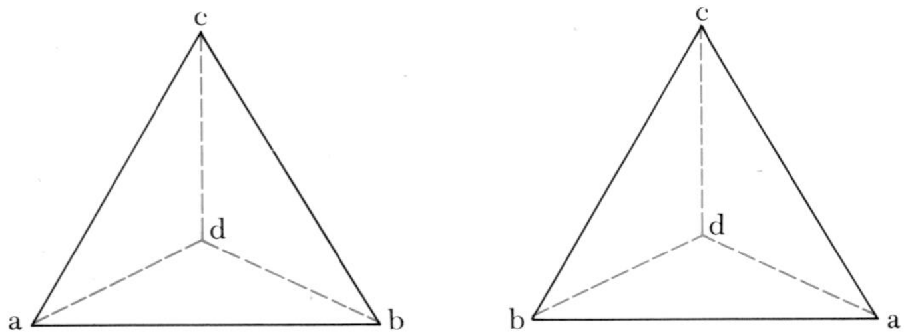

active carbon compounds then known contained at least one carbon atom that was combined with four *different* groups. Such a carbon atom is spoken of as an *asymmetric* carbon atom because a molecule containing such an atom is not symmetrical. This means that the four different groups situated at the corners of the tetrahedron can be arranged in two different configurations (Figure 19-1). One structure is the mirror image of the other, but it is not identical with it. An analogy exists in your own two hands. Certainly your left hand does not have the same configuration as your right hand, but if you look at the image of your left hand in a mirror you will see that the image has the same configuration as your right hand. Conversely, your left hand has the same configuration as the mirror image of your right hand. Optical isomers may be thought of as mirror-image isomers. A sample of a compound containing only "right-handed" molecules will rotate the plane of polarized light a certain amount in one direction, and a sample containing only "left-handed" molecules will rotate the light the same amount in the opposite direction.

A simple example of optical isomerism is the case of fluorochlorobromomethane, CHFClBr. Here four different atoms are bonded to the same carbon atom. Consequently, there are both dextrorotatory and levorotatory varieties of fluorochlorobromomethane molecules.

In Section 18-16 we learned that geometric (*cis*- and *trans*-) isomers have the same structure in the sense that they are alike with respect to which atoms are joined to which. In this sense optical isomers also have the same structure, for they too are alike with respect to which atoms are joined to which. In both cases the difference is in the way the atoms are distributed in space. Isomerism depending on the arrangement of atoms in space is called *stereoisomerism*. The two types of stereoisomerism are geometric isomerism and optical isomerism.

When chemists synthesize compounds containing asymmetric carbon atoms in the laboratory, their products are optically inactive because they contain equal numbers of right-handed and left-handed molecules. Such a mixture of dextrorotatory and levorotatory molecules is called a *racemic mixture*. The racemic mixtures synthesized by chemists can often be *resolved* (i.e., separated) by various methods, including biochemical ones. For example, many microorganisms when grown on a racemic mixture of appropriate compounds will destroy one isomer and leave the other unchanged.

The physiological actions of optical isomers in the body are often quite different. For example, levorotatory nicotine is far more poisonous than dextrorotatory nicotine.

Compounds of biochemical origin are often observed to be optically active, and many biochemical reactions are stereospecific, i.e., only one optical isomer is produced or consumed.

When we inspect the configurations of two optical isomers (Figure 19-1) we realize that we cannot tell by inspection alone which configuration will rotate polarized light to the right and which configuration will rotate it to the left. We can be confident, however, that all molecules that have exactly the same composition and spatial configuration will have exactly the same rotatory power. We can be equally confident that another set of molecules with the same composition but the opposite spatial configuration will show the same rotatory power in the opposite direction.

New Terms

Alcohol: any compound, *R*OH, that contains an —OH group and a hydrocarbon group, e.g., methyl alcohol (methanol), CH_3OH.

Aldehyde: any compound that contains a $\overset{\displaystyle H}{\overset{|}{—C}}{=}O$ group.

Alkyl group: any group, C_nH_{2n+1}, in which the number of hydrogen atoms is one more than twice the number of carbon atoms, e.g., the methyl group, $—CH_3$. An alkyl group is named according to its corresponding hydrocarbon.

Aryl group: any group obtained by dropping a hydrogen atom from an aromatic nucleus; e.g., the phenyl group, $—C_6H_5$; and the naphthyl group, $—C_{10}H_7$.

Asymmetric carbon atom: a carbon atom that is bonded to four different groups.

Biochemistry: the branch of chemistry that deals with chemical reactions occurring in living organisms.

Carboxylic acid: any compound, *R*COOH, that contains the carboxyl group, $\overset{\displaystyle O}{\overset{\|}{—C}}—OH$.

Ester: any compound conforming to the general formula $R—\overset{\displaystyle O}{\overset{\|}{C}}—O—R'$.

Esterification: the reaction of a carboxylic acid with an alcohol (so called because an ester is formed).

Ether: any compound conforming to the general formula $R—O—R'$.

Fatty acids: aliphatic monocarboxylic acids.

Functional group: a group of atoms occurring in the molecules of all compounds of a given class, e.g., the —OH group in alcohols.

Glycols (dihydric alcohols): organic compounds containing two —OH groups per molecule.

Ketone: any compound that conforms to the general formula $R\text{—}\overset{\displaystyle O}{\overset{\|}{C}}\text{—}R'$, where R and R' are both alkyl groups (i.e., neither can be a hydrogen atom).

Optical activity: the ability to rotate the plane of polarized light.

Organic chemistry: the chemistry of the hydrocarbons and their derivatives.

Phenol: any compound, *Ar*OH, that contains an —OH group attached to a benzene ring.

Polarimeter: an instrument with which optical rotation is measured.

Primary alcohol: an alcohol that contains a $—CH_2OH$ group, i.e., has two hydrogen atoms bonded to the same carbon as the —OH group.

Racemic mixture: a mixture of equal numbers of right-handed and left-handed molecules.

Saponification: the reaction of a strong base with an ester (so called because soap is formed by this type of reaction).

Secondary alcohol: an alcohol that has two carbon atoms (and only one hydrogen atom) bonded to the same carbon as the —OH group.

Tertiary alcohol: an alcohol that has three carbon atoms (and therefore no hydrogen) bonded to the same carbon as the —OH group.

Exercises

19-1 Organic chemistry is sometimes defined as the chemistry of carbon compounds and sometimes as the chemistry of hydrocarbons and their derivatives. What, if any, real difference is there between these definitions?

19-2 Of what importance is hydrogen bonding in determining the properties of alcohols? Ethers? Carboxylic acids? Ketones?

19-3 Explain the use of the terms *primary*, *secondary*, and *tertiary* in the classification of alcohols.

19-4 Write the structural formulas for all of the possible compounds with the molecular formula C_3H_8O.

19-5 How does esterification resemble neutralization? How do the two reactions differ?

19-6 Write the structural formulas for three compounds that share the molecular formula $C_2H_4O_2$.

19-7 The boiling point of methyl acetate (57°C) is less than that of either its parent alcohol (CH_3OH, 65°C) or its parent acid (CH_3COOH, 118°C); but the molecular weight of the methyl acetate is, of course, considerably greater than that of either of its parents. Explain the low boiling point of methyl acetate.

19-8 Write equations for the complete esterification of:

(a) ethylene glycol by acetic acid
(b) oxalic acid by ethyl alcohol
(c) ethyl alcohol by formic acid

19-9 The word *butane* is derived from the Latin *butyrum* ("butter"). Why should this particular hydrocarbon have this name?

19-10 Explain why lactic acid has optical isomers but citric acid does not.

References

Benfey, O. T., *From Vital Force to Structural Formulas*, Houghton Mifflin, Boston, 1964 (paperbound). Important developments in the early history of organic chemistry.

Fieser, L. F., *Chemistry in Three Dimensions*, Macalester Scientific Corp., Cambridge, Mass., 1964 (paperbound). An introduction to organic chemistry, written in a style to provide enjoyable reading.

Hart, H., and Schuetz, R. D., *Organic Chemistry*, 3rd ed., Houghton Mifflin, Boston, 1966. A brief text for students to acquire a knowledge of organic chemistry with a minimum of detail.

Noller, C. R., *Structure and Properties of Organic Compounds*, Saunders, Philadelphia, 1962. A modernized introduction to organic chemistry.

Pauling, L., and Hayward, R., *The Architecture of Molecules*, Freeman, San Francisco, 1964. Color drawings that show left-handed and right-handed molecules and many other aspects of molecular structure.

Schellman, J. A., and Schellman, C., "Optical Rotation and the Shape of Molecules," *Chemistry*, **40** (5), 12 (1967). An explanation of polarized light and its rotation by optically active molecules.

20

Oxidation-Reduction and Electrochemistry

20-1 Loss and Gain of Electrons

If a strip of metallic zinc is immersed in a solution containing a copper salt, such as copper sulfate or copper nitrate, the reaction which occurs can be represented by the equation

$$Zn + Cu^{++} \rightarrow Zn^{++} + Cu$$

In this reaction the zinc is said to undergo oxidation because its oxidation state changes upward, from 0 to +2; the copper is said to undergo reduction because its oxidation state changes downward, from +2 to 0.

The oxidation of the zinc consists in the conversion of zinc atoms to zinc ions through the loss of electrons:

$$Zn \rightarrow Zn^{++} + 2e^-$$

The reduction of the copper involves the conversion of copper ions to copper atoms by gaining electrons:

$$Cu^{++} + 2e^- \rightarrow Cu$$

Neither the oxidation of the zinc atoms nor the reduction of the copper ions could occur as an isolated event. The copper ions could not be reduced without a source of electrons, and zinc, when oxidized, needs another substance (in this case, copper ions) to take on the electrons that are given up. Thus, we say that the copper ions have been reduced *by* the zinc atoms and that the zinc atoms have been oxidized *by* the copper ions. In the above reaction, then, the oxidizing agent is the Cu^{++} ion and the reducing agent is elementary zinc.

The equations representing the oxidation of the zinc atoms (with loss of electrons) and the reduction of copper ions (with gain of electrons) are called *half-equations*, and the processes represented by them are called *half-reactions*. No half-reaction involving oxidation ever occurs unless accompanied by another half-reaction involving reduction.

20-2 Addition of Half-Equations

The total reaction that is made up of the two half-reactions, one involving oxidation and the other involving reduction, is called an *oxidation-reduction reaction* (or a "*redox*" *reaction*). The equation for an oxidation-reduction reaction can be obtained by adding the two appropriate half-equations. In the equation for the total reaction, no electrons should appear; i.e., the number of electrons lost must equal the number of electrons gained. This sometimes requires an adjustment, as in the case of aluminum metal displacing silver from a solution of a silver salt:

$$Al \rightarrow Al^{+++} + 3e^-$$
$$Ag^+ + e^- \rightarrow Ag$$

Before adding these two half-equations, we must adjust the second one by multiplying all items by 3 so that the number of electrons involved in the reduction of silver ions equals the number lost by the aluminum atom. Then the addition can be carried out with all electrons accounted for:

$$Al + 3Ag^+ \rightarrow Al^{+++} + 3Ag$$

The balanced equation for an oxidation-reduction reaction can be obtained, then, by writing the half-equations and adding them; but one or both equations may need to be multiplied by appropriate integers so that the number of electrons gained by the oxidizing agent is the same as the number lost by the reducing agent.

EXAMPLE Balance the oxidation-reduction equation

$$Al + Cu^{++} \rightarrow Al^{+++} + Cu$$

The equation is already balanced atomically, for there is one atom of each kind on each side of the equation. However, in the balanced equation the charges must also balance.

In balancing the equation, you may follow this procedure: (1) write the half-equations for the conversion of aluminum atoms to Al^{+++} ions and for the conversion of Cu^{++} ions to copper atoms; (2) note the least common multiple of the numbers of electrons in the two half-equations; (3) multiply all items in each half-equation by the number required to make the electrons lost equal to the electrons gained; and (4) add the half-equations:

$$Al \rightarrow Al^{+++} + 3e^-$$
$$Cu^{++} + 2e^- \rightarrow Cu$$

The least common multiple of 2 and 3 is 6. Therefore, multiply the first half-equation by 2 and the second by 3:

$$2Al \rightarrow 2Al^{+++} + 6e^-$$
$$3Cu^{++} + 6e^- \rightarrow 3Cu$$
$$\overline{2Al + 3Cu^{++} \rightarrow 2Al^{+++} + 3Cu}$$

20-3 The Chemical Nature of Voltaic Cells

The reduction of Cu^{++} ions by zinc atoms can proceed in the ingenious device known as a *voltaic cell* (or *galvanic cell*). In this type of cell electrons travel through an external circuit (when going from the zinc atoms to the Cu^{++} ions) rather than transfer directly from reducing agent to oxidizing agent. If a zinc strip is immersed directly in a solution containing Cu^{++} ions, the spontaneous reaction which occurs is accompanied by the release of energy in the form of heat. However, in the voltaic cell the energy is released in the form of electrical energy. Voltaic cells are an important means for producing electric current for operating cordless radios, electric toothbrushes, hearing aids, and many other devices requiring low-voltage direct current.

One of the simplest types of voltaic cell is diagramed in Figure 20-1. The cell consists of two *half-cells.* In the zinc half-cell the half-reaction that occurs is represented by the half-equation

$$Zn \rightarrow Zn^{++} + 2e^-$$

In the copper half-cell the half-reaction consists in the reduction of Cu^{++} ions:

$$Cu^{++} + 2e^- \rightarrow Cu$$

Thus, Zn^{++} ions enter the solution in the zinc half-cell, and Cu^{++} ions leave the solution in the copper half-cell. The zinc electrode loses weight and the

Figure 20-1

Voltaic cell made from zinc and copper electrodes immersed in solutions containing Zn^{++} and Cu^{++} ions

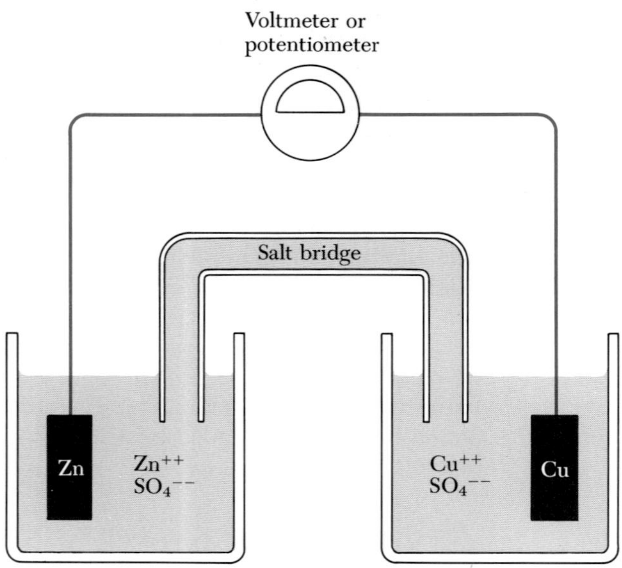

copper electrode gains weight as the cell reaction continues. Electrons left behind on the zinc electrode by the departing Zn^{++} ions flow through the external circuit toward the copper electrode from which electrons are removed by Cu^{++} ions.

The external circuit may include an ammeter to measure the rate at which the current is flowing, or a voltmeter or potentiometer (as shown in Figure 20-1) to measure the electromotive force developed by the cell, or some other device which would be operated by the electric current. The "salt bridge" connecting the two half-cells contains a solution of an electrolyte, e.g., KCl. From the solution contained in the salt bridge, Cl^- ions migrate into the zinc half-cell to balance the charge of the Zn^{++} ions produced there; K^+ ions migrate into the copper half-cell to replace Cu^{++} ions.

In all voltaic cells electric current is produced by oxidation-reduction reactions occurring spontaneously within them. In each half-cell of a voltaic cell, an element is present in two oxidation states. The two electrodes of a cell are distinguished by the names *anode* and *cathode*. The anode is the electrode at which oxidation occurs. The cathode is the electrode at which reduction occurs.

20-4 Oxidation Potentials

Every voltaic cell is characterized by the electromotive force (usually expressed in volts) which it develops. The voltage (or potential) is of a definite magnitude, depending upon the components of the cell. The voltage of a cell which employs two metals and their ions, such as the zinc-copper cell, depends primarily on the relation of the metals to each other in the activity series, but it also depends on the temperature and the concentrations of the two kinds of ions in the respective half-cells. The comparison of voltages of different components is made with solutions containing the ions at the standard concentration, 1 *M*.

The potential developed by the zinc-copper cell, with Zn^{++} ions and Cu^{++} ions at 1 *M* concentration, is 1.10 volts at 25°C. This voltage is a measure of the tendency of the cell reaction to occur, i.e., of the "drive" behind the reaction. Since the cell reaction is a composite of an oxidation half-reaction and a reduction half-reaction, we might hope to measure the voltage due to each separate half-reaction. However, a single half-reaction never occurs as an isolated event; therefore, it is impossible to measure the voltage of an individual electrode except on a relative basis.

A nickel-copper cell may be constructed like a zinc-copper cell, with one half-cell consisting of a nickel electrode immersed in a solution containing Ni^{++} ions. With both Ni^{++} and Cu^{++} ions at 1 *M* concentration, the potential of a nickel-copper cell at 25°C is 0.59 volt. Since this is 0.51 volt less than the potential of a zinc-copper cell, we can interpret this to mean that the reducing power of nickel atoms is 0.51 volt less than that of zinc atoms; i.e., the tendency of zinc atoms to become oxidized to Zn^{++} ions is 0.51 volt greater than the tendency of nickel atoms to become oxidized to Ni^{++} ions. Therefore, we would predict, accurately, the possibility of constructing a zinc-nickel cell in which the zinc would act as anode and the nickel as cathode and that the standard potential of this cell at 25°C would be 0.51 volt.

By measuring the potentials of properly chosen voltaic cells, we can therefore arrange the metals in order of their activity. For such an activity or electromotive series to be useful, there should be associated with each metal in the series a number such that the difference between the numbers for any two metals would be the potential developed by a cell constructed with these two metals as components. Since one concern of chemists is to classify metals on the basis of their ability or inability to replace hydrogen from acids, a voltage of 0 at 25°C has been assigned arbitrarily to the hydrogen-hydrogen ion half-reaction:

$$H_2 \rightleftharpoons 2H^+ + 2e^-$$

An oxidation half-reaction that has a *greater* tendency to occur than the hydrogen half-reaction is assigned a positive potential, and an oxidation half-reaction that has a lesser tendency to occur is assigned a negative potential. Examples of the entries in a table of standard *oxidation potentials* at 25°C are the following:

$$\begin{aligned} Zn &\rightleftharpoons Zn^{++} + 2e^- & E^\circ &= +0.76 \text{ volt} \\ Ni &\rightleftharpoons Ni^{++} + 2e^- & E^\circ &= +0.25 \text{ volt} \\ H_2 &\rightleftharpoons 2H^+ + 2e^- & E^\circ &= 0 \\ Cu &\rightleftharpoons Cu^{++} + 2e^- & E^\circ &= -0.34 \text{ volt} \end{aligned}$$

These and other oxidation potentials are listed in Table 20-1. The symbol E° refers to the voltage when all ions are at the standard concentration, 1 *M*. The voltage of a zinc-hydrogen cell (where zinc is the anode) is 0.76 volt; the voltage of a hydrogen-copper cell (where hydrogen is the anode) is 0.34 volt; and the voltage of a nickel-hydrogen cell (where nickel is the anode) is 0.25 volt. The voltage of a zinc-nickel cell is 0.76 volt − 0.25 volt = 0.51 volt; the voltage of a zinc-copper cell is 0.76 volt − (−0.34 volt) = 1.10 volts; and the voltage of a nickel-copper cell is 0.25 volt − (−0.34 volt) = 0.59 volt.

20-5 Types of Voltaic Cell Reactions

A voltaic cell can be designed to utilize the driving force behind the reaction of hydrogen or a metal with a nonmetal, such as chlorine. For example, electric current can be obtained from the reaction of hydrogen with chlorine to form HCl, i.e., H^+ and Cl^- ions, in aqueous solution:

$$H_2(g) + Cl_2(g) \rightarrow 2H^+ + 2Cl^-$$

One half-cell contains a hydrogen electrode, and the other contains a chlorine electrode. Gas electrodes consist of a gas bubbling about an "inert" metal strip immersed in a solution containing the appropriate ions. The metal strip (usually platinum) provides electrical contact for the half-cell and at the same time catalyzes the establishment of equilibrium between the gas and its ions. The platinum strip is "platinized," i.e., coated with a layer of finely divided platinum,

Table 20-1

Standard oxidation potentials at 25°C

Half-reaction	$E°$ (volts)
$K \rightleftharpoons K^+ + e^-$	2.93
$Ca \rightleftharpoons Ca^{++} + 2e^-$	2.87
$Na \rightleftharpoons Na^+ + e^-$	2.71
$Mg \rightleftharpoons Mg^{++} + 2e^-$	2.37
$Al \rightleftharpoons Al^{+++} + 3e^-$	1.66
$H_2 + 2OH^- \rightleftharpoons 2H_2O + 2e^-$	0.83
$Zn \rightleftharpoons Zn^{++} + 2e^-$	0.76
$Cr \rightleftharpoons Cr^{+++} + 3e^-$	0.74
$Fe \rightleftharpoons Fe^{++} + 2e^-$	0.44
$Cd \rightleftharpoons Cd^{++} + 2e^-$	0.40
$Pb + SO_4^{--} \rightleftharpoons PbSO_4 + 2e^-$	0.36
$Ni \rightleftharpoons Ni^{++} + 2e^-$	0.25
$Sn \rightleftharpoons Sn^{++} + 2e^-$	0.14
$Pb \rightleftharpoons Pb^{++} + 2e^-$	0.13
$H_2 \rightleftharpoons 2H^+ + 2e^-$	0.00
$Sn^{++} \rightleftharpoons Sn^{4+} + 2e^-$	−0.15
$Cu \rightleftharpoons Cu^{++} + 2e^-$	−0.34
$2I^- \rightleftharpoons I_2 + 2e^-$	−0.54
$Fe^{++} \rightleftharpoons Fe^{+++} + e^-$	−0.77
$Ag \rightleftharpoons Ag^+ + e^-$	−0.80
$NO + 2H_2O \rightleftharpoons NO_3^- + 4H^+ + 3e^-$	−0.96
$Au + 4Cl^- \rightleftharpoons AuCl_4^- + 3e^-$	−1.00
$2Br^- \rightleftharpoons Br_2 + 2e^-$	−1.07
$2H_2O \rightleftharpoons O_2 + 4H^+ + 4e^-$	−1.23
$2Cl^- \rightleftharpoons Cl_2 + 2e^-$	−1.36
$Au \rightleftharpoons Au^{+++} + 3e^-$	−1.50
$Mn^{++} + 4H_2O \rightleftharpoons MnO_4^- + 8H^+ + 5e^-$	−1.51
$2F^- \rightleftharpoons F_2 + 2e^-$	−2.87

called "platinum black" because of its color. The finely divided platinum provides a large surface area for the adsorption of H_2 molecules.

A standard hydrogen electrode or half-cell consists, then, of a strip of platinized platinum immersed in an aqueous solution in which the concentration of H^+ ions is 1 *M*, with H_2 gas at a pressure of 1 atmosphere passing continuously over the platinum strip. The voltage of a cell which consists of a standard hydrogen half-cell and a standard chlorine half-cell is found to be 1.36 volts at

Figure 20-2

Voltaic cell made from platinum electrodes immersed in solutions containing Sn^{++} and Fe^{+++} ions

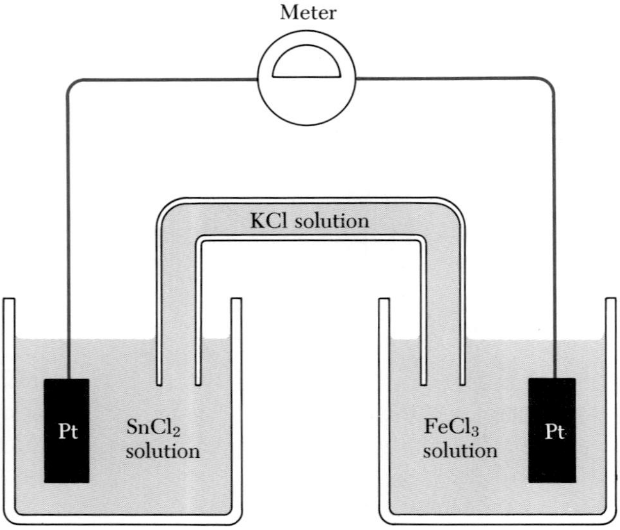

25°C. In this cell the hydrogen electrode is the anode and the chlorine electrode is the cathode:

$$H_2(g) \longrightarrow 2H^+ + 2e^-$$
$$Cl_2(g) + 2e^- \longrightarrow 2Cl^-$$

The oxidation potential of chlorine is therefore given in tables of oxidation potentials as -1.36 volts. The voltage of the cell is the difference between the oxidation potentials of the two half-cells:

$$\begin{aligned} E^\circ_{\text{cell}} &= E^\circ_{H_2} - E^\circ_{Cl_2} \\ &= 0.00 \text{ volts} - (-1.36 \text{ volts}) \\ &= 1.36 \text{ volts} \end{aligned}$$

A voltaic cell can derive its current from a reaction in which neither electrode is directly involved. An example is diagramed in Figure 20-2. Here the "inert" electrodes could be either platinum or graphite. No reaction occurs until the two electrodes are connected externally. The reaction that then occurs consists in the oxidation of stannous ions and the reduction of ferric ions. The equations for the half-reactions are as follows:

$$\text{ANODE:} \quad Sn^{++} \longrightarrow Sn^{4+} + 2e^-$$
$$\text{CATHODE:} \quad Fe^{+++} + e^- \longrightarrow Fe^{++}$$

In the stannous chloride-ferric chloride cell, stannic ions and ferrous ions are produced by the cell reaction. The voltage developed by the cell is affected somewhat by the concentration of each of the four kinds of ions. If all are present at 1 *M* concentration, the cell voltage is 0.62 volt. This can be calculated from the following entries in Table 20-1:

$$Sn^{++} \rightleftharpoons Sn^{4+} + 2e^- \qquad E^\circ = -0.15 \text{ volt}$$
$$Fe^{++} \rightleftharpoons Fe^{+++} + e^- \qquad E^\circ = -0.77 \text{ volt}$$

The standard cell voltage is always the difference between the standard oxidation potentials of the two half-cells:

$$\begin{aligned} E^\circ_{cell} &= -0.15 \text{ volt} - (-0.77 \text{ volt}) \\ &= 0.62 \text{ volt} \end{aligned}$$

The subtraction must always be done so as to give a positive answer. Algebraically, we change the sign of one potential, then add it to the other. The half-cell whose potential sign has to be changed is the one in which reduction is occurring. Thus, the half-equation representing what actually happens in the ferrous-ferric half-cell must show the reduction of ferric ions:

$$Fe^{+++} + e^- \rightarrow Fe^{++}$$

This half-equation, of course, must be doubled before it is added to the other half-equation to give the equation for the cell reaction (but this has no effect on the half-cell potential):

$$\begin{array}{r} Sn^{++} \rightarrow Sn^{4+} + 2e^- \\ 2Fe^{+++} + 2e^- \rightarrow 2Fe^{++} \\ \hline Sn^{++} + 2Fe^{+++} \rightarrow Sn^{4+} + 2Fe^{++} \end{array}$$

The coefficients in the equation have no effect on the cell potential.

20-6 Writing Equations for Half-Cell Reactions

Since the balanced equation for an oxidation-reduction reaction can be obtained by adding the equations for the two half-reactions, you should learn how to write the half-equations without resorting to memory or to the oxidation potential table. Each half-equation must be balanced atomically and electrically.

The first step in writing the equation for any chemical reaction, or for any half-reaction, is to know what the reactants and products are. For a half-reaction, this means knowing what is oxidized and what is formed by its oxidation or what is reduced and what is formed by its reduction. To write the half-equation for the oxidation of ferrous ion to ferric ion, one must know the formulas for the ions. The advantage of naming them iron(II) ion and iron(III) ion is that the formulas are implied by the names to be Fe^{++} and Fe^{+++}, respectively. The half-equation, $Fe^{++} \rightarrow Fe^{+++}$, is already balanced atomically; to balance it electrically, we must add an electron to the right-hand side:

$$Fe^{++} \rightarrow Fe^{+++} + e^-$$

Thus, we are forced to recognize that the oxidation involves the *loss* of an electron.

In order to write half-equations involving the important oxidizing ions containing oxygen, such as MnO_4^-, NO_3^-, or $Cr_2O_7^{--}$, we must realize that reactions occurring in aqueous solution may involve H_2O molecules, H^+ ions, or OH^- ions as reactants or as products. If the reaction is occurring in acidic solution, H^+ ions or H_2O molecules may appear on either side of the equation. If the reaction is occurring in alkaline solution, OH^- ions or H_2O molecules may appear on either side of the equation. The procedure for balancing complicated half-equations will be illustrated by three examples.

EXAMPLE Balance the half-equation for the reduction of permanganate ion in acidic solution.

First, you must know that the formula for permanganate ion is MnO_4^- and that its reduction in acidic solution produces manganese(II) ion. The half-equation, $MnO_4^- \rightarrow Mn^{++}$, must be balanced both atomically and electrically. There are two alternative procedures, depending on whether the atomic balancing or the insertion of the electrons is done first.

METHOD 1: If the atomic balancing is done first, you proceed as follows. Since the solution is acidic, you can feel free to add H^+ ions to the left; they, with the oxygen of the permanganate ion, form water:

$$MnO_4^- + H^+ \rightarrow Mn^{++} + H_2O \text{ (not balanced)}$$

Since there are four oxygen atoms in the MnO_4^- ion, you can balance the half-equation atomically by having eight H^+ ions on the left and four H_2O molecules on the right:

$$MnO_4^- + 8H^+ \rightarrow Mn^{++} + 4H_2O$$
(not balanced electrically)

Finally, to balance the half-equation electrically, add the necessary number of electrons where they are needed. Since the sum of the charges in the half-equation that is atomically (but not electrically) balanced is 7+ on the left and 2+ on the right, five electrons are needed on the left:

$$MnO_4^- + 8H^+ + 5e^- \rightarrow Mn^{++} + 4H_2O \text{ (balanced)}$$

Thus, the balanced half-equation can be obtained without any consideration of the change in oxidation state of the manganese.

METHOD 2: The other procedure for balancing the half-equation consists in noting *first* that the oxidation state of the manganese changes downward from +7 to +2, requiring a gain of five electrons:

$$MnO_4^- + 5e^- \rightarrow Mn^{++} \text{ (not balanced)}$$

Next, since the solution is acidic, add the necessary number of H^+ ions where they are needed to balance the half-equation electrically.

Since the sum of the charges in the unbalanced half-equation is 6− on the left and 2+ on the right, eight H^+ ions are needed on the left:

$$MnO_4^- + 8H^+ + 5e^- \rightarrow Mn^{++} \text{ (not balanced atomically)}$$

Finally, balance the half-equation atomically by adding four H_2O molecules on the right:

$$MnO_4^- + 8H^+ + 5e^- \rightarrow Mn^{++} + 4H_2O \text{ (balanced)}$$

Method 2 can be called the oxidation-number method, since it begins with consideration of the change in oxidation number (which is completely ignored in Method 1).

EXAMPLE Write the balanced half-equation for the oxidation of hydrogen peroxide in acidic solution.

It is not difficult to predict that the oxidation will yield elementary oxygen. In H_2O_2 the hydrogen is already in the +1 state, the highest oxidation state in which hydrogen is ever found; but the oxygen in H_2O_2 is in the −1 state, which is *not* the highest possible oxidation state for oxygen. Therefore, the oxidation of H_2O_2 must involve the oxidation of oxygen from the −1 state to the 0 state.

METHOD 1: First, the half-equation, $H_2O_2 \rightarrow O_2$, must be balanced atomically. In order to do this, add hydrogen to the right side; so as not to have the hydrogen changing its oxidation state, add H^+ *ions:*

$$H_2O_2 \rightarrow O_2\uparrow + 2H^+ \text{ (not balanced electrically)}$$

Then, to balance the half-equation electrically, add two electrons on the right:

$$H_2O_2 \rightarrow O_2\uparrow + 2H^+ + 2e^- \text{ (balanced)}$$

METHOD 2: Alternatively, you could note first that the change of two oxygen atoms from the −1 state to the 0 state requires a loss of two electrons:

$$H_2O_2 \rightarrow O_2\uparrow + 2e^- \text{ (not balanced)}$$

The next step is to balance the equation electrically by adding two H^+ ions on the right:

$$H_2O_2 \rightarrow O_2\uparrow + 2H^+ + 2e^- \text{ (balanced)}$$

The final step is to rejoice that the half-equation is already balanced atomically.

EXAMPLE Write the balanced half-equation for the oxidation of iodide ion to iodate ion in alkaline solution.

You must know the formulas for iodide ion and iodate ion to write the unbalanced half-equation, $I^- \rightarrow IO_3^-$.

METHOD 1: Since the solution is alkaline, you can feel free to add OH^- ions to the left to supply oxygen for the iodate ion. Since the removal of O^{--} from OH^- leaves H^+, which combines with another OH^- to form H_2O, the correct number of OH^- ions to add to the left is *twice* the number of oxygen atoms needed for the iodate (i.e., six instead of three):

$$I^- + 6OH^- \rightarrow IO_3^- + 3H_2O \text{ (not balanced electrically)}$$

To balance the half-equation electrically, add electrons where they are needed. Since the sum of the charges in the half-equation that is atomically (but not electrically) balanced is 7− on the left and 1− on the right, six electrons are needed on the right:

$$I^- + 6OH^- \rightarrow IO_3^- + 3H_2O + 6e^- \text{ (balanced)}$$

METHOD 2: Alternatively, you could note first that the change of oxidation state of the iodine atom from −1 to +5 requires a loss of six electrons:

$$I^- \rightarrow IO_3^- + 6e^- \text{ (not balanced)}$$

Next, since the solution is alkaline, add the necessary number of OH^- ions where they are needed to balance the half-equation electrically. Since the sum of the charges in the unbalanced half-equation is 1− on the left and 7− on the right, six OH^- ions are needed on the left:

$$I^- + 6OH^- \rightarrow IO_3^- + 6e^- \text{ (not balanced atomically)}$$

Finally, balance the half-equation atomically by adding three H_2O molecules on the right:

$$I^- + 6OH^- \rightarrow IO_3^- + 3H_2O + 6e^- \text{ (balanced)}$$

20-7 Writing Balanced Oxidation-Reduction Equations

The balancing of any chemical equation consists in finding coefficients for the different reactants and products such that the same number of each kind of atom appears on both sides of the equation. If the equation is in ionic form, the charges must also balance. An equation for a reaction in which there is no oxidation or reduction can be balanced by inspection (as can equations for the simpler oxidation-reduction reactions). In complicated oxidation-reduction reactions, balancing by inspection can be very difficult. A systematic method is helpful.

The balanced equation for an oxidation-reduction reaction can be obtained by adding the balanced half-equation for oxidation and the balanced half-equation for reduction. This procedure will now be illustrated with more complicated examples than the one in Section 20-2.

EXAMPLE Write the balanced equation for the oxidation of iron(II) ions by permanganate ions in acidic solution.

Earlier, we wrote the balanced half-equations for the reduction of permanganate ions in acidic medium and for the oxidation of iron(II) ions. All that remains is to add the two half-equations after multiplying each by the number required to make the electrons lost in the one equal to the electrons gained in the other:

$$\begin{array}{r} MnO_4^- + 8H^+ + 5e^- \rightarrow Mn^{++} + 4H_2O \\ 5 \times [Fe^{++} \rightarrow Fe^{+++} + e^-] \\ \hline 5Fe^{++} + MnO_4^- + 8H^+ \rightarrow 5Fe^{+++} + Mn^{++} + 4H_2O \end{array}$$

EXAMPLE Write the balanced equation for the reaction of copper with dilute nitric acid.

Here it must be known that the oxidation of the copper atoms produces bipositive copper ions and that the reduction of NO_3^- ions produces NO gas. (Under other circumstances, reduction of NO_3^- ions may produce NO_2 gas or elementary N_2 or NH_4^+ ions.) The half-equation for the oxidation of copper is easily constructed:

$$Cu \rightarrow Cu^{++} + 2e^-$$

but the half-equation for the reduction of NO_3^- ions must be developed as described in Section 20-6:

$$NO_3^- + 4H^+ + 3e^- \rightarrow NO\uparrow + 2H_2O$$

We can now obtain the balanced equation for the reaction by adding the balanced equations for the half-reactions (after multiplying each half-equation by the appropriate number):

$$\begin{array}{r} 3 \times [Cu \rightarrow Cu^{++} + 2e^-] \\ 2 \times [NO_3^- + 4H^+ + 3e^- \rightarrow NO\uparrow + 2H_2O] \\ \hline 3Cu + 2NO_3^- + 8H^+ \rightarrow 3Cu^{++} + 2NO\uparrow + 4H_2O \end{array}$$

20-8 Balancing Skeleton Equations

The balancing of a skeleton equation for an oxidation-reduction reaction, already complete except for coefficients, is most readily done by consideration of the changes in oxidation states.

EXAMPLE Hydrogen iodide reduces concentrated sulfuric acid to hydrogen sulfide, and the sulfuric acid oxidizes the hydrogen iodide to elementary iodine. Write the equation.

From the information given, it is possible to write most of the skeleton equation:

$$HI + H_2SO_4 \rightarrow H_2S + I_2 \text{ (incomplete)}$$

We can readily see that the equation is incomplete, for there is oxygen on the left but no oxygen on the right. If you have guessed that H_2O is also a product of the reaction, you have guessed right.

To proceed with the balancing of the equation, we note which element is reduced and write an "equation" for its change in oxidation state:

$$\overset{+6}{S} \rightarrow \overset{-2}{S} \text{ (change} = 8)$$

Then we note which element is oxidized and do the same for it:

$$\overset{-1}{I} \rightarrow \overset{0}{I} \text{ (change} = 1)$$

In the last "equation" we multiply through by 8 in order to make the total changes in the two equations equal; however, we write $4I_2$ rather than 8I because we know that the formula for elementary iodine is I_2. Now we combine the one "equation" with the other:

$$\overset{+6}{S} + 8\overset{-1}{I} \rightarrow \overset{-2}{S} + 4\overset{0}{I}_2$$

This gives us most of the coefficients we need to use to balance the skeleton equation:

$$8HI + H_2SO_4 \rightarrow 4I_2 + H_2S + H_2O \text{ (not balanced)}$$

All that remains is to find, by inspection, the coefficient needed for H_2O. This must be 4, for there are four oxygen atoms on the left:

$$8HI + H_2SO_4 \rightarrow 4I_2 + H_2S + 4H_2O$$

EXAMPLE Balance the equation

$$Fe^{++} + MnO_4^{-} + H^{+} \rightarrow Fe^{+++} + Mn^{++} + H_2O$$

First, we note which element is oxidized and how much it is oxidized. It is the iron that is oxidized, from the +2 state to the +3 state, a change of one unit. Second, we note which element is reduced and how much it is reduced. The manganese is reduced from the +7 state to the +2 state, a change of five units. Third, we note how many atoms of each element must undergo these changes simultaneously to make the total changes equal. Five iron atoms each changing one unit undergo the same total change as does one manganese atom changing five units:

$$5Fe^{++} + MnO_4^- + H^+ \rightarrow 5Fe^{+++} + Mn^{++} + H_2O$$
(not balanced)

Fourth, we note whether the equation is balanced; if it is not, we must complete the balancing by inspection. This one is not balanced. With four oxygen atoms on the left, we clearly need four H_2O on the right and therefore eight H^+ on the left:

$$5Fe^{++} + MnO_4^- + 8H^+ \rightarrow 5Fe^{+++} + Mn^{++} + 4H_2O$$

20-9 Predicting Oxidation-Reduction Reactions

A table of oxidation potentials, such as Table 20-1, may be consulted to determine whether certain proposed reactions are possible. Thus, when we find in the table the two entries

$$Cd \rightleftharpoons Cd^{++} + 2e^- \qquad E^\circ = +0.40 \text{ volt}$$
$$Cu \rightleftharpoons Cu^{++} + 2e^- \qquad E^\circ = -0.34 \text{ volt}$$

we can predict that the reaction

$$Cd + Cu^{++} \rightarrow Cd^{++} + Cu$$

will occur spontaneously and that the reaction

$$Cu + Cd^{++} \rightarrow Cu^{++} + Cd$$

will not occur spontaneously. Immersion of a strip of metallic copper into a solution containing Cd^{++} ions results in no observable reaction, but immersion of a strip of metallic cadmium in a solution containing blue Cu^{++} ions results in the formation of metallic copper accompanied by the decolorization of the solution.

The ability of chlorine to displace bromine from bromides is implied by the relative positions of these two halogens in the table of oxidation potentials:

$$2Br^- \rightleftharpoons Br_2 + 2e^- \qquad E^\circ = -1.07 \text{ volts}$$
$$2Cl^- \rightleftharpoons Cl_2 + 2e^- \qquad E^\circ = -1.36 \text{ volts}$$

The E° values indicate that bromide ions are more easily oxidized than chloride ions and that the reaction

$$Cl_2 + 2Br^- \rightarrow Br_2 + 2Cl^-$$

is possible, while the reverse reaction,

$$Br_2 + 2Cl^- \rightarrow Cl_2 + 2Br^-$$

is impossible. The table also includes the entry

$$Mn^{++} + 4H_2O \rightleftharpoons MnO_4^- + 8H^+ + 5e^- \qquad E^\circ = -1.51 \text{ volts}$$

From this, we can also predict that permanganate ion will oxidize either bromide or chloride.

In general, any oxidizing agent in the table will oxidize any reducing agent appearing above it: MnO_4^- will oxidize Fe^{++}; Fe^{+++} will oxidize Cu; and so on. You must remember, however, that $E°$ values refer to 1 M concentrations of ions in all cases and that most chemical reactions are reversible. The greater the spread between two $E°$ values in the table, the more complete will be the reaction between the corresponding substances.

20-10 Free Energy and Oxidation Potentials

The electrochemical method for obtaining free energies of reactions is based on the equation relating $\Delta G°$ to the $E°$ value for a voltaic cell in which the reaction occurs:

$$\Delta G° = -nFE°$$

Here n is the number of moles of electrons involved in the reaction and F is the value of the faraday (96,490 coulombs). With $E°$ in volts and F in coulombs, the calculated value for $\Delta G°$ will be in volt-coulombs, i.e., joules. To convert to kilocalories, we divide by 4184, the number of joules in 1 kcal.

EXAMPLE Calculate the standard free energy of the reaction

$$Cu(s) + 2Ag^+ \rightleftharpoons Cu^{++} + 2Ag(s)$$

From Table 20-1 we find $E° = -0.34$ volt $- (-0.80$ volt$) = 0.46$ volt at 25°C. For this reaction $n = 2$ moles.

$$\Delta G°_{\text{kcal}} = -\frac{2 \text{ moles} \times 96{,}490 \text{ coulombs/mole} \times 0.46 \text{ volt}}{4184 \text{ volt-coulombs/kcal}}$$

$$= -21.2 \text{ kcal at } 25°\text{C}$$

The large negative value for $\Delta G°$ means that the reaction is spontaneous and will proceed well toward completion before reaching equilibrium.

Once $\Delta G°$ has been determined, the equilibrium constant for a reaction can be calculated from the equation relating $\Delta G°$ and K:

$$\Delta G° = -2.303RT \log K$$

This type of calculation was illustrated in Section 17-9 and will now be illustrated by a further example.

EXAMPLE Calculate the value of the equilibrium constant for the reaction

$$Cu(s) + 2Ag^+ \rightleftharpoons Cu^{++} + 2Ag(s)$$

given that $\Delta G°$ for the reaction is -21.2 kcal.

The equilibrium constant can be calculated from the equation

$$\log K = -\frac{\Delta G^\circ}{2.303RT}$$

Here R is the molar gas constant. The value of R to be used with this equation is 1.987 cal/mole-deg.

$$\begin{aligned}\log K &= -\frac{-21.2 \text{ kcal/mole} \times 1000 \text{ cal/kcal}}{2.303 \times 1.987 \text{ cal/mole-deg} \times 298^\circ}\\ &\doteq 16\\ \therefore K &= 10^{16}\end{aligned}$$

20-11 Oxidation Potentials and Equilibrium Constants

The equilibrium constant can be calculated directly from the E° value for the voltaic cell in which the reaction occurs. In the equation relating these two quantities,

$$\log K = \frac{nFE^\circ}{2.303RT}$$

n is the number of moles of electrons involved in the reaction, F is the value of the faraday (96,490 coulombs), and R is the molar gas constant. The value of R to be used with this equation is 8.314 joules/mole-deg. (1 joule = 1 volt-coulomb.)

EXAMPLE Calculate the value of the equilibrium constant for the reaction

$$Cu(s) + 2Ag^+ \rightleftharpoons Cu^{++} + 2Ag(s)$$

at 25°C from E° values given in Table 20-1.

For this reaction $n = 2$ moles and (from Table 20-1) we find $E^\circ = -0.34 \text{ volt} - (-0.80 \text{ volt}) = 0.46 \text{ volt}$

$$\begin{aligned}\log K &= \frac{2 \text{ moles} \times 96{,}490 \text{ coulombs/mole} \times 0.46 \text{ volt}}{2.303 \times 8.314 \text{ volt-coulombs/mole-deg} \times 298^\circ}\\ &= 16\\ \therefore K &= 10^{16}\end{aligned}$$

If the E° values differ by 0.18 volt, the equilibrium constant for the reaction will be in excess of 1000 for $n = 1$ and in excess of 1,000,000 for $n = 2$. If the E° values differ by about 0.02 volt, the equilibrium constant will be about 2 for $n = 1$ and about 5 for $n = 2$. In general, then, as stated in Section 17-9, the proper question to ask is not "Is this reaction possible?" but rather "How far will this reaction proceed before equilibrium is reached?"

If we regard a so-called impossible reaction simply as one for which the equilibrium lies almost completely to the left, we are in the most favorable position for understanding certain chemical phenomena. This enables us, for example, to explain how a mixture of hydrochloric and nitric acids is able to dissolve gold, although neither acid alone is able to do so. By analogy with the reaction of nitric acid on other metals only slightly more active than gold, we can assume a tendency toward the following reaction:

$$Au + NO_3^- + 4H^+ \rightarrow Au^{+++} + NO + 2H_2O$$

This reaction is shifted far to the right in the presence of hydrochloric acid because of the union of Cl^- ions with Au^{+++} ions to form complex ions with the formula $AuCl_4^-$. The equation for the total reaction is

$$Au + NO_3^- + 4H^+ + 4Cl^- \rightarrow AuCl_4^- + NO + 2H_2O$$

The $E°$ value for the oxidation half-reaction,

$$Au \rightleftharpoons Au^{+++} + 3e^-$$

is only -1.50 volts. However, for the half-reaction

$$Au + 4Cl^- \rightleftharpoons AuCl_4^- + 3e^-$$

the $E°$ value is -1.00 volt, i.e., 0.50 volt greater.

A mixture consisting of one part concentrated nitric acid and three parts concentrated hydrochloric acid has long been known as *aqua regia*, meaning "royal water," because of its ability to dissolve the "noble metals," i.e., the metals of low activity, such as gold and platinum, that do not dissolve in either hydrochloric or nitric acid alone.

20-12 Practical Voltaic Cells

A practical voltaic cell must be compact and not too fragile. In principle, however, any reaction involving gain and loss of electrons can be utilized for the production of electric current.

A cell of historical interest is the gravity cell (Figure 20-3), which formerly supplied current for telegraph lines. The cathode was a copper grid placed in the bottom of a glass jar half-filled with saturated copper sulfate solution. Over the dense copper sulfate solution was a layer of a dilute solution of zinc sulfate. The anode was a zinc grid or "crowfoot" which was supported at the top of the cell where it would dip into the zinc sulfate solution.

The particular mechanical design of the gravity cell is only one of many which have been employed to utilize the reaction

$$Cu^{++} + Zn \rightarrow Zn^{++} + Cu$$

All such Cu-Zn cells are called *Daniell cells* because the English physicist J. F. Daniell invented the first one in 1841.

One type of battery that has been developed for specialized purposes uses a cell that is similar to the Daniell cell in principle but employs cadmium and nickel instead of zinc and copper.

Figure 20-3

The gravity cell

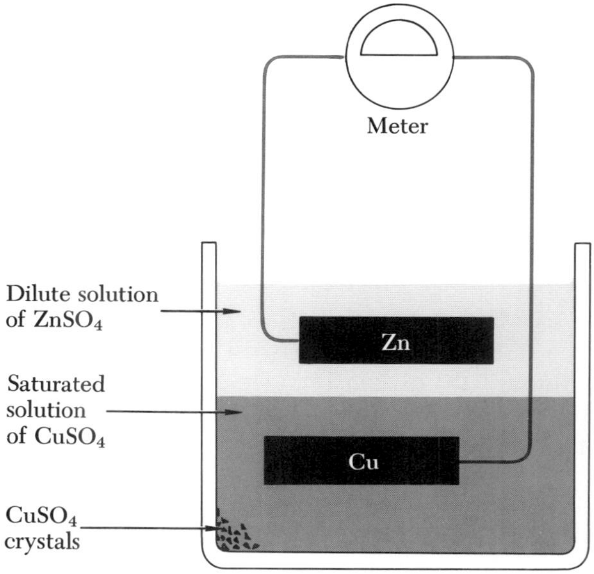

For producing small currents at intervals, the so-called dry cell is one of the most important of all voltaic cells. (Small versions of it are used as flashlight batteries.) The potential of a new dry cell is about 1.5 volts. This cell (Figure 20-4) is contained within a cylindrical zinc can which serves as the anode. The cathode, a carbon rod, is surrounded by an electrolyte paste (a moistened

Figure 20-4

The dry cell

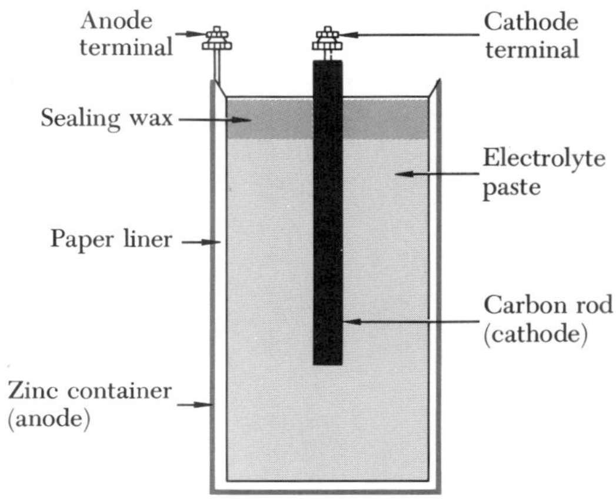

mixture of ammonium chloride, zinc chloride, and manganese dioxide with a porous inert material such as paper pulp or cotton waste). At the anode, the Zn atoms are oxidized:

$$Zn \rightarrow Zn^{++} + 2e^-$$

The cathode reaction is more complex; the following equation probably represents the net reaction:

$$2MnO_2 + 2NH_4^+ + 2e^- \rightarrow 2NH_3 + H_2O + Mn_2O_3$$

The NH_3 that is formed combines with Zn^{++} ions from the anode reaction to form complex ions:

$$Zn^{++} + 4NH_3 \rightarrow Zn(NH_3)_4^{++}$$

A paper liner separates the zinc can from the paste; therefore, the electrons required for the cathode reaction must pass through the external circuit and enter the cell through the carbon rod. The cell is sealed with sealing wax to prevent loss of water from the paste by evaporation. Water must be present in the paste to permit movement of ions. The only reason for calling the device a "dry cell" is that the sealed zinc can prevents the moist electrolyte from leaking out.

The mercury cell that is used in hearing aids is another type of dry cell.

The lead storage cell, used in automobile batteries, is a voltaic cell which operates reversibly. After it has served as a source of electric current, it can be "charged," i.e., restored to its original condition, by having electric energy supplied to it from an outside source. The charging is brought about by passing current through the cell in the direction opposite to the current produced by the spontaneous cell reaction. The unique feature of a storage cell is that the reactions taking place at the electrodes when current is passed through it are the reverse of those that occur when the cell acts as a source of current.

The electrodes in a lead storage cell are gratings made of a lead alloy. The openings in one electrode are packed with lead dioxide, and those in the other are filled with spongy lead. The electrolyte is dilute sulfuric acid. When the battery is delivering current, the spongy lead electrode acts as the anode; i.e., the atoms of the spongy lead are oxidized to Pb^{++} ions, and they combine with SO_4^{--} ions from the sulfuric acid to coat the electrode with lead sulfate. The lead dioxide electrode acts as cathode, for here the lead dioxide is reduced to Pb^{++} ions, the H^+ ions from the acid combining with the oxygen to form water, and here also the Pb^{++} ions combine with SO_4^{--} ions to coat the electrode with lead sulfate.

ANODE: $$Pb + SO_4^{--} \rightarrow PbSO_4(s) + 2e^-$$

CATHODE: $$PbO_2 + 4H^+ + SO_4^{--} + 2e^- \rightarrow PbSO_4(s) + 2H_2O$$

TOTAL CELL REACTION: $$Pb + PbO_2 + 2H_2SO_4 \underset{\text{charging}}{\overset{\text{discharging}}{\rightleftharpoons}} 2PbSO_4 + 2H_2O$$

As indicated by the equation, the concentration of sulfuric acid decreases during the discharge of the cell. This results in a decrease of the density of the electrolyte. Charging the cell regenerates the acid and restores the electrolyte to its higher density. Thus, the condition of charge of a lead storage cell can be determined by the use of a hydrometer to measure its density.

The potential of a fully charged lead storage cell is a little over 2 volts. A 6-volt automobile battery contains three cells; a 12-volt battery contains six cells.

20-13 Fuel Cells

Although voltaic cells are used for many practical purposes, most electrical energy is obtained from generators operated by steam turbines. The steam is obtained by heating water in a boiler with thermal energy released by burning coal, oil, or gas. The series of changes, chemical energy to thermal energy to mechanical energy to electrical energy, is relatively inefficient. A steam power plant using coal as fuel converts little more than a third of the chemical energy of the coal into electrical energy.

For several decades there has been discussion of the possibility of converting *directly* the chemical energy of fuels into electrical energy in a so-called "fuel cell." Since voltaic cells operate on the basis of spontaneous oxidation-reduction reactions, there is no theoretical reason why a cell should not employ the type of spontaneous oxidation-reduction reaction that occurs in the combustion of a fuel. However, the practical problems encountered in the actual design of fuel cells and in finding satisfactory catalysts for the electrode half-reactions have proved to be quite complex.

A fuel cell with promising possibilities, already employed in spacecraft, utilizes the reaction between hydrogen and oxygen gases under a pressure of 50 atmospheres at 250°C. The electrolyte is aqueous NaOH or KOH. With graphite electrodes, the following reactions occur *in the presence of certain catalysts:*

$$\begin{array}{ll} \text{ANODE:} & 2H_2(g) + 4OH^- \rightleftharpoons 4H_2O + 4e^- \\ \text{CATHODE:} & \underline{O_2(g) + 2H_2O + 4e^- \rightleftharpoons 4OH^-} \\ & 2H_2(g) + O_2(g) \rightleftharpoons 2H_2O \end{array}$$

The voltage developed by such a cell is about 1 volt.

20-14 Electrolytic Cells

Voltaic cells and electrolytic cells are types of electrochemical cells; i.e., electric current and chemical reactions are involved in the operation of both of them. In voltaic cells, oxidation-reduction reactions take place spontaneously, producing electric current; but in electrolytic cells, electric current from an outside source causes nonspontaneous oxidation-reduction reactions to take place. In either type of electrochemical cell, oxidation occurs at the anode, and reduction occurs at the cathode.

In the electrolysis of a molten salt, there is only one imaginable half-reaction at either electrode, provided the electrodes are made of materials that do not react when the cell is in operation. For example, the passage of direct current through molten sodium chloride in a cell provided with graphite electrodes can only result in the reduction of Na^+ ions at the cathode and the oxidation of Cl^- ions at the anode (Section 14-2).

If an electrolytic cell contains an aqueous solution of a salt, alternative half-reactions come to mind. Passage of direct current through a cell containing aqueous $CuBr_2$ solution, for example, might conceivably produce either of the following oxidation half-reactions at the anode:

$$2H_2O \rightleftharpoons O_2 + 4H^+ + 4e^- \qquad E^\circ = -1.23 \text{ volts}$$
$$2Br^- \rightleftharpoons Br_2 + 2e^- \qquad E^\circ = -1.07 \text{ volts}$$

Experience shows that bromine, not oxygen, is the anode product. We could predict this result by comparing the oxidation potentials of the two half-reactions. The half-reaction for the oxidation of Br^- ions has a higher (i.e., less negative) E° value than the half-reaction for the oxidation of water. Even though the oxidation potential for the one varies with the Br^- ion concentration and that for the other varies with the pH, our prediction that the electrolysis should produce bromine rather than oxygen is fulfilled.

Electrolysis of the $CuBr_2$ solution could conceivably produce either of the following reduction half-reactions at the cathode:

$$2H_2O + 2e^- \rightarrow H_2 + 2OH^-$$
$$Cu^{++} + 2e^- \rightarrow Cu$$

Experience shows that copper metal, not hydrogen gas, is formed at the cathode. This, too, we could predict on the basis of the E° values found in the oxidation potential table for oxidation half-reactions that are the reverse of the above reduction half-reactions:

$$H_2 + 2OH^- \rightleftharpoons 2H_2O + 2e^- \qquad E^\circ = +0.83 \text{ volt}$$
$$Cu \rightleftharpoons Cu^{++} + 2e^- \qquad E^\circ = -0.34 \text{ volt}$$

Since the tendency of elementary hydrogen to undergo oxidation spontaneously is greater than that of elementary copper, we would predict that the reduction of copper(II) ions would be more easily brought about than that of H_2O molecules.

Since H_2O molecules are more readily reduced than such ions as Na^+ or Ba^{++}, electrolysis of solutions of salts of metals of Group IA or Group IIA always produces H_2 gas. The product obtained at the anode depends on the identity of the anion. Chlorides yield chlorine; bromides yield bromine; and iodides yield iodine. But electrolysis of aqueous solutions of sulfates, nitrates, etc. yields oxygen at the anode because H_2O molecules are more easily oxidized than any of the oxyanions.

Industrially, electrolysis is used for the manufacture of aluminum (from bauxite dissolved in molten cryolite); magnesium (from molten $MgCl_2$); sodium (from molten NaCl); fluorine (from molten KHF_2); sodium hydroxide, hydrogen, and chlorine (from aqueous NaCl); and numerous other elements and compounds.

Electrolytic refining is a process used for the production of copper and other metals in their highest state of purity. In the electrolytic refining of copper, the anode is a bar of impure copper, and the cathode is a thin sheet of pure copper. These electrodes are immersed in an aqueous solution of $CuSO_4$. The applied voltage is adjusted to about 0.35 so that the copper will go into solution from the anode and plate out upon the cathode.

$$\text{ANODE:} \quad Cu \rightarrow Cu^{++} + 2e^-$$

$$\text{CATHODE:} \quad Cu^{++} + 2e^- \rightarrow Cu$$

Active metals present as impurities are also oxidized at the anode, but their positive ions remain in solution and are not reduced at the cathode. Metals less active than copper are not oxidized at the anode but drop to the bottom of the cell as an undissolved sludge. Precious metals, such as silver, gold, and platinum, are recovered from this "anode sludge."

New Terms

Anode (in any electrochemical cell, whether voltaic or electrolytic): the electrode at which oxidation occurs.

Cathode: the electrode at which reduction occurs.

Half-cell: that part of an electrochemical cell in which either the oxidation half-reaction or the reduction half-reaction takes place.

Half-equation: an equation representing a half-reaction.

Half-reaction: that part of an oxidation-reduction reaction involving either the oxidation of the reducing agent (with loss of electrons) or the reduction of the oxidizing agent (with gain of electrons).

Oxidation potential (of a particular oxidation half-reaction): a measure of the tendency of a specified oxidation half-reaction to occur, expressed in terms of the voltage of a cell in which the other half-cell is a hydrogen electrode.

Oxidation-reduction reaction ("redox" reaction): a reaction in which one element is oxidized and another element is reduced. (In some instances the atoms that are oxidized and those that are reduced are atoms of the same element.)

Exercises

20-1 Diagram a voltaic cell in which the reaction consists of the displacement of silver from $AgNO_3$ by metallic copper. Write the equation for the half-reaction that takes place in each half-cell, identifying each as oxidation or reduction; then write the equation for the total cell reaction.

20-2 What is the approximate voltage of the cell in the preceding question if the solutions used are 1 *M*? Indicate the direction of migration of the ions during the operation of the cell, and show how the electrons flow through the external circuit. Which electrode is the anode and which is the cathode?

20-3 Explain how a hydrometer can be used to determine the state of charge of an automobile battery.

20-4 What would be obtained at each electrode during the electrolysis of 1 *M* aqueous solutions of the following compounds, using "inert" electrodes:

(a) sodium fluoride
(b) silver nitrate
(c) copper chloride
(d) sodium hydroxide
(e) sulfuric acid

20-5 Predict whether the following reactions are possible:

(a) displacement of iodine from an iodide by bromine
(b) oxidation of bromide ion by chlorine
(c) reduction of iodine by fluoride ion
(d) oxidation of ferrous ion by permanganate ion
(e) oxidation of iron atoms by silver ions

20-6 Balance the following equations:

(a) $P + HNO_3 + H_2O \rightarrow H_3PO_4 + NO\uparrow$
(b) $Cu + H^+ + NO_3^- \rightarrow Cu^{++} + NO_2\uparrow + H_2O$
(c) $H_2SO_4 + HBr \rightarrow SO_2\uparrow + Br_2 + H_2O$
(d) $H_2SO_4 + HI \rightarrow H_2S\uparrow + I_2 + H_2O$
(e) $CuS + HNO_3 \rightarrow Cu(NO_3)_2 + S + NO\uparrow + H_2O$
(f) $MnO_4^- + H_2S + H^+ \rightarrow Mn^{++} + S + H_2O$
(g) $MnO_4^- + H_2O_2 + H^+ \rightarrow Mn^{++} + O_2\uparrow + H_2O$

References

Eblin, L. P., "Oxidation Number in Auto-Redox Reactions," *J. Chem. Educ.*, **28** (4), 221 (1951). A discussion of reactions in which one atom of an element oxidizes another atom of the same element.

Laubengayer, A. W., *General Chemistry*, 2nd ed., Rinehart, New York, 1957, Chapter 19, "Electrolysis." An account of the reactions that occur in electrolytic cells, as affected by the nature of the electrodes and the nature of the electrolyte.

Lockwood, K. L., "Redox Revisited," *J. Chem. Educ.*, **38** (6), 326 (1961). Emphasizes the distinction between stoichiometry and mechanism.

Mysels, K., "The Reactions of the Leclanché Dry Cell," *J. Chem. Educ.*, **32** (12), 638 (1955). A discussion of what is actually known about the chemistry of the dry cell.

Yalman, R. G., "Writing Oxidation-Reduction Equations," *J. Chem. Educ.*, **36** (5), 215 (1959). A review of textbook methods.

21

Concepts of Acids and Bases

21-1 Classical Concepts of Acids and Bases

Water has always been the most widely used solvent. Naturally, the early concepts of acids and bases emphasized behavior in water. Use of the English word *acid*, in consequence of the sourness of certain aqueous solutions, dates from the seventeenth century. Some of the properties which characterize acids were listed in 1680 by Robert Boyle. Bases were regarded essentially as antiacids, i.e., substances that neutralize acids.

Classically (Section 7-8), an acid is regarded as a substance whose aqueous solution turns blue litmus red, tastes sour, reacts with active metals to liberate hydrogen, and loses these properties on contact with alkalies. A base is regarded as a substance whose aqueous solution turns red litmus blue, tastes bitter, feels soapy, and loses these properties on contact with acids (Section 7-9).

Oxides such as CO_2 and SO_3 were at one time called acids, for their aqueous solutions possess the properties characteristic of solutions of acids. For example, CO_2 was once called "carbonic acid gas." There was logic in this, for it would be inconsistent to define acids in terms of the characteristic properties of their aqueous solutions and not classify as acids all substances whose aqueous solutions possess these properties.

Lavoisier attempted to shift the emphasis from properties to chemical composition. He regarded oxygen as the necessary and characteristic constituent of all acids. This idea had to be abandoned after Davy showed that some acids, e.g., hydrochloric acid, do not contain oxygen. Davy contended that acidity did not depend upon any particular element. Nevertheless, in the thinking of nineteenth-century chemists, the one-element concept persisted. Hydrogen replaced oxygen as the "acidifying principle" present in all acids.

The Arrhenius theory of ionization attributed the characteristic properties of aqueous solutions of acids to the hydrogen ion, H^+, and the properties of basic solutions to the hydroxide ion, OH^-. Thus, an acid was defined as a compound containing hydrogen atoms which can become hydrogen ions when the acid is dissolved in water. A base was defined as a compound containing OH groups which can become hydroxide ions when the base is dissolved in water.

The Arrhenius definitions do not include such compounds as NH_3, CaO, CO_2, and SO_3. It was assumed, for example, that when ammonia dissolves in water, it reacts with the solvent to form NH_4OH, which *can* be regarded as a base because it contains an OH group that becomes a hydroxide ion in the solution; in terms of the Arrhenius concept it is the NH_4OH, not the NH_3, that is the base. Similarly, CaO reacts with water to give the base $Ca(OH)_2$, which furnishes hydroxide ions to the solution; the oxide CaO is termed a basic oxide but not a base. The oxides CO_2 and SO_3 are classified as acidic oxides, but not as acids, because they react with water to produce H_2CO_3 and H_2SO_4, which *are* acids by the Arrhenius definition.

Arrhenius assumed that the excess hydrogen ions present in an aqueous solution of an acid are formed by the simple dissociation of some of the acid molecules into ions. Thus, in the case of hydrogen chloride, it was assumed that some of the HCl molecules dissociate into positive hydrogen ions and negative chloride ions and that these ions exist in the solution in equilibrium with undissociated HCl molecules.

$$HCl \rightleftharpoons H^+ + Cl^-$$

The only role credited to the solvent was that of providing the dielectric medium which reduces the mutual attraction of oppositely charged ions so that they can exist as separate particles in the solution.

We now know that a hydrogen ion, H^+, is unique among ions—it contains no electrons at all. In fact, it is nothing more than a proton, and its radius is only about 10^{-13} cm compared with 10^{-8} cm for other simple ions. This means that the H^+ ion has an extraordinarily high ratio of charge to radius. Certainly such a positive ion could not exist unhydrated in aqueous solution, where it would be surrounded by polar molecules of H_2O that have unshared electron pairs. It is therefore appropriate to represent a hydrogen ion in aqueous solution as H_3O^+, i.e., $[H(H_2O)]^+$, rather than simply as H^+.

Since water molecules are associated with one another by hydrogen bonding, each proton is actually hydrated with a variable number of molecules of water. In addition to H_3O^+ ions, aqueous solutions of acids contain $H_5O_2{}^+$ ions, $H_7O_3{}^+$ ions, etc., and their relative numbers vary with concentration and temperature. The formula H_3O^+ is used as a convenience to denote that the hydrogen ion *is* hydrated.

The Arrhenius concept of acids and bases is essentially correct for aqueous solutions if we attribute the characteristic properties of acids to the hydronium ion, H_3O^+, rather than the unhydrated proton, H^+. We must keep in mind that the "electrolytic dissociation" of an acid actually involves reaction with the solvent water:

$$HCl + H_2O \rightleftharpoons H_3O^+ + Cl^-$$

Neutralization reactions between strong acids and strong bases in aqueous solution are more adequately represented by

$$H_3O^+ + OH^- \rightleftharpoons 2H_2O$$

than by

$$H^+ + OH^- \rightleftharpoons H_2O$$

despite the fact that we so often use the latter equation for convenience.

21-2 Hydrolysis of Salts and the Arrhenius Concepts

An aqueous solution of sodium cyanide, NaCN, is basic, but NaCN does not meet the Arrhenius definition for a base. The solution is basic because it contains more OH^- ions than H_3O^+ ions. This situation arises as a consequence of *hydrolysis.*

In terms of classical Arrhenius concepts the hydrolysis of a salt is the reversible reaction of the salt with water, producing the parent acid and base. For example, NaCN is a salt that can be obtained from the acid HCN and the base NaOH. No other acid and base would react to produce this particular salt, and it therefore seems appropriate to refer to NaOH and HCN as the *parent base* and the *parent acid* of the salt NaCN. Conversely, NaOH is the base and HCN is the acid formed in an aqueous solution of NaCN by hydrolysis.

The Arrhenius theory is adequate for making correct predictions regarding the hydrolysis of salts as follows:

1. *A salt of a strong acid and a weak base will give an acidic solution.* Such a salt contains a cation that combines with OH^- ions, produced by the ionization of water, to form molecules of the weak base. The reaction takes place to an extent that depends on the weakness of the base. An example is the hydrolysis of ammonium chloride:

$$\begin{array}{r} HOH \rightleftharpoons H^+ + OH^- \\ NH_4{}^+ + OH^- \rightleftharpoons NH_4OH \\ \hline NH_4{}^+ + HOH \rightleftharpoons NH_4OH + H^+ \end{array}$$

The water ionization equilibrium is shifted to the right to an extent that depends upon the number of OH^- ions combined with by the $NH_4{}^+$ ions. The solution becomes acidic because of the excess hydrogen ions produced by this shift in the equilibrium. The net result is indicated by the third equation.

2. *A salt of a strong base and a weak acid will give an alkaline solution.* Such a salt contains an anion that combines with protons to form molecules of the weak acid. The reaction takes place to an extent that depends on the weakness of the acid. An example is the hydrolysis of sodium cyanide:

$$\begin{array}{r} HOH \rightleftharpoons H^+ + OH^- \\ H^+ + CN^- \rightleftharpoons HCN \\ \hline HOH + CN^- \rightleftharpoons HCN + OH^- \end{array}$$

The water ionization equilibrium is shifted to the right to an extent depending upon the number of H^+ ions combined with by the CN^- ions. The solution becomes alkaline because of the excess OH^- ions which accumulate in the solution.

3. *A salt of a strong acid and a strong base will give a neutral solution.* Such a salt does not contain an ion that combines with either of the ions of water. There is no hydrolysis, and the solution remains neutral like water itself.

4. *A salt derived from a weak acid and a weak base may be acidic, neutral, or alkaline, depending on the relative weakness of the parent acid and the parent base.* A salt derived from a weak acid and a weaker base will give an acidic solution. A salt whose parents are a weak base and a weaker acid will give an alkaline solution. If the parent acid and the parent base are equally weak, the solution will be neutral in spite of hydrolysis. Thus, a solution of ammonium formate is acidic, a solution of ammonium cyanide is alkaline, and a solution of ammonium acetate is neutral. We have already alluded (Section 16-11) to the fact that the ionization constants of acetic acid and ammonia are substantially identical, viz., 1.8×10^{-5}. Formic acid, $HCHO_2$, is a somewhat stronger acid than acetic acid; its ionization constant is 1.8×10^{-4}. The ionization constant of the weaker hydrocyanic acid, HCN, is only 4.0×10^{-10}.

21-3 Extension of Arrhenius Concepts to Nonaqueous Solvents

Liquid ammonia possesses a slight conductivity, which is attributed to the presence of a small concentration of positive ammonium ions, NH_4^+, and negative amide ions, NH_2^-, produced by the self-ionization of the ammonia. This process is analogous to the self-ionization of water, which produces hydronium and hydroxide ions in small concentration:

$$2NH_3 \rightleftharpoons NH_4^+ + NH_2^-$$
$$2H_2O \rightleftharpoons H_3O^+ + OH^-$$

Since the Arrhenius definition for an acid in aqueous solution is a substance that produces hydronium ion (the cation formed by the self-ionization of water), the analogous definition for an acid in liquid ammonia solution would be a substance that produces ammonium ion (the cation formed by the self-ionization of ammonia). Likewise, since the Arrhenius definition for a base in aqueous solution is a substance that produces hydroxide ion (the anion formed by the self-ionization of water), a base in liquid ammonia solution would be a substance that produces amide ion (the anion formed by the self-ionization of ammonia).

Since the Arrhenius concept of neutralization in aqueous solution involves the formation of water (solvent) molecules by the reaction of hydronium with hydroxide ions, neutralization in liquid ammonia solution can be defined as the formation of ammonia (solvent) molecules by the reaction of ammonium with amide ions. In both cases the reaction of a strong acid with a strong base involves nothing more than a process that is the exact opposite of the self-ionization of the solvent.

NEUTRALIZATION IN WATER SOLUTION: $H_3O^+ + OH^- \rightarrow 2H_2O$
NEUTRALIZATION IN AMMONIA SOLUTION: $NH_4^+ + NH_2^- \rightarrow 2NH_3$

The self-ionization of anhydrous acetic acid occurs in a manner analogous to the self-ionization of water and liquid ammonia and can be represented by the following equation:

$$2CH_3COOH \rightleftharpoons CH_3C(OH)_2^+ + CH_3COO^-$$

Thus, an acid in this solvent is a substance that produces $CH_3C(OH)_2^+$ ions, and a base is a substance that produces acetate, CH_3COO^-, ions. The $CH_3C(OH)_2^+$ is a solvated proton, and in this respect it is analogous to the H_3O^+ and NH_4^+ ions.

Neutralization in anhydrous acetic acid solution involves the formation of acetic acid (solvent) molecules. The reaction of a strong acid with a strong base involves nothing more than the reaction that is the reverse of the self-ionization of the solvent.

NEUTRALIZATION IN ACETIC ACID SOLUTION:

$$CH_3C(OH)_2^+ + CH_3COO^- \rightarrow 2CH_3COOH$$

The analogies that exist between the ammonia system, the water system, and the acetic acid system suggest that it might be possible to define acids and bases in a general manner that would be applicable to all three systems and, hopefully, to other systems also.

21-4 The Brønsted-Lowry Concept of Acids and Bases

The Arrhenius concept was greatly broadened to give a more general concept of acids and bases in the protonic definitions formulated independently in 1923 by the Danish chemist J. N. Brønsted and the English chemist T. Martin Lowry. Brønsted and Lowry defined an acid as any molecular or ionic species that gives up a proton to another species. They defined a base as any molecular or ionic species that combines with a proton. In the simplest terms, *an acid is a proton donor* and *a base is a proton acceptor*.

According to the Brønsted-Lowry definitions there must be a base corresponding to every acid and an acid corresponding to every base.

base	+ *proton*	⇌	*acid*
NH_3	+ H^+	⇌	NH_4^+
H_2O	+ H^+	⇌	H_3O^+
OH^-	+ H^+	⇌	H_2O
CN^-	+ H^+	⇌	HCN
Cl^-	+ H^+	⇌	HCl
SO_4^{--}	+ H^+	⇌	HSO_4^-
HSO_4^-	+ H^+	⇌	H_2SO_4

An acid and a base related in this manner are called a *conjugate acid-base pair.* Thus, NH_3 is the base conjugate to the acid NH_4^+; NH_4^+ is the acid conjugate to the base NH_3; NH_4^+ and NH_3 are a conjugate pair, for they differ only in the presence or absence of a proton.

Since protons under ordinary conditions are incapable of independent existence, the reaction of an acid with a base is a proton transfer reaction which produces a new acid and a new base:

$$\begin{array}{lllll} acid_1 & + \; base_2 & \rightleftharpoons base_1 & + \; acid_2 \\ HCl & + \; NH_3 & \rightleftharpoons Cl^- & + \; NH_4^+ \\ HCl & + \; H_2O & \rightleftharpoons Cl^- & + \; H_3O^+ \\ NH_4^+ & + \; H_2O & \rightleftharpoons NH_3 & + \; H_3O^+ \\ H_2O & + \; NH_3 & \rightleftharpoons OH^- & + \; NH_4^+ \\ H_3O^+ & + \; OH^- & \rightleftharpoons H_2O & + \; H_2O \\ HCN & + \; OH^- & \rightleftharpoons CN^- & + \; H_2O \end{array}$$

In these examples, $base_1$ is derived from $acid_1$ by loss of a proton; i.e., $base_1$ and $acid_1$ are a conjugate pair. Similarly, $acid_2$ and $base_2$ are a conjugate pair, for $acid_2$ is derived from $base_2$ by addition of a proton.

A substance capable of acting as a proton donor (i.e., an acid) can actually do so only in the presence of a proton acceptor (a base). Some substances are capable of acting as proton donors in the presence of bases and as proton acceptors in the presence of acids. Such substances are said to be *amphiprotic.* Examples are H_2O and the HSO_4^- ion. Water acts as a base when HCl dissolves in it but as an acid when the solute is NH_3, as indicated in the above equations. The HSO_4^- ion can accept a proton from a very strong acid, and it can donate a proton to a strong base:

$$H_3O^+ + HSO_4^- \rightleftharpoons H_2O + H_2SO_4$$
$$HSO_4^- + OH^- \rightleftharpoons SO_4^{--} + H_2O$$

To identify an amphoteric hydroxide as an amphiprotic substance requires that we recognize that it is really a hydrated compound. For example, writing zinc hydroxide as $Zn(H_2O)_2(OH)_2$ enables us to represent it as amphiprotic by indicating it to be a proton donor in the presence of hydroxide ions and a proton acceptor in the presence of hydronium ions:

$$\begin{array}{lllll} acid_1 & + \; base_2 & \rightleftharpoons base_1 & + \; acid_2 \\ Zn(H_2O)_2(OH)_2 & + \; 2OH^- & \rightleftharpoons Zn(OH)_4^{--} & + \; 2H_2O \\ 2H_3O^+ & + \; Zn(H_2O)_2(OH)_2 & \rightleftharpoons 2H_2O & + \; Zn(H_2O)_4^{++} \end{array}$$

All proton transfer reactions, i.e., all Brønsted-Lowry acid-base reactions, are included in the term *protolysis.* The Brønsted-Lowry concept of acids and bases applies to protolysis in any solvent. Indeed, it applies to protolysis reactions which occur when no solvent is present. The reaction between ammonia gas and hydrogen chloride gas is a Brønsted-Lowry acid-base reaction, for a proton is transferred from the HCl molecule (the acid) to the NH_3 molecule (the base):

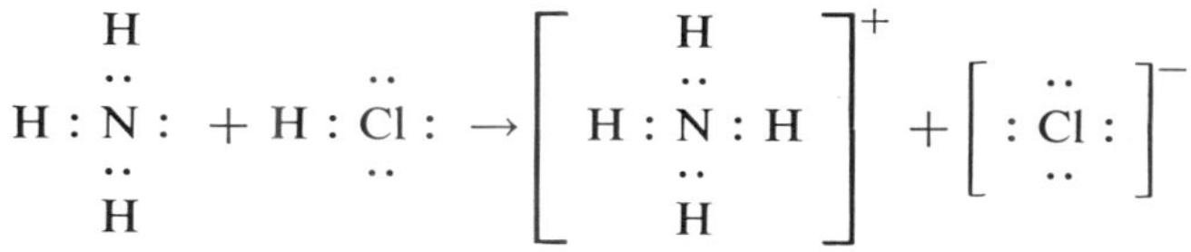

Exactly the same protolytic reaction occurs when hydrogen chloride dissolves in liquid ammonia or in an aqueous solution of ammonia.

21-5 Corollaries of Brønsted-Lowry Definitions

In terms of Brønsted-Lowry concepts the strength of an acid is measured by its tendency to donate a proton, and the strength of a base is measured by its tendency to accept a proton. The strengths of different acids vary between wide limits.

Hydrogen chloride is a strong acid; i.e., the HCl molecule gives up its proton readily. From this it follows that the chloride ion is necessarily a weak base; i.e., its tendency to acquire a proton, forming an HCl molecule, is very slight. Hydroxide ion, OH^-, is a strong base; i.e., it readily acquires a proton, forming an H_2O molecule. From this it follows that H_2O is a weak acid. In general, *the stronger an acid, the weaker its conjugate base; the stronger a base, the weaker its conjugate acid.*

In an aqueous solution of hydrogen chloride the protolytic reaction proceeds almost to completion; in Arrhenius terminology, hydrochloric acid is almost completely ionized:

$$\underset{acid_1}{HCl} + \underset{base_2}{H_2O} \rightleftharpoons \underset{acid_2}{H_3O^+} + \underset{base_1}{Cl^-}$$

In this equation we have used arrows of unequal length to emphasize that in the equilibrium solution the concentration of H_3O^+ ions is much greater than the concentration of HCl molecules. The two acids are competing to donate protons; the fact that the left-to-right reaction is virtually complete means that HCl is a stronger acid than H_3O^+. This is equivalent to saying that H_2O and Cl^- are competing to accept protons and that the left-to-right reaction is nearly complete because H_2O is a stronger base than Cl^-.

In an aqueous solution of HCN only a small fraction of the solute is ionized:

$$\underset{acid_1}{HCN} + \underset{base_2}{H_2O} \rightleftharpoons \underset{acid_2}{H_3O^+} + \underset{base_1}{CN^-}$$

As indicated again by arrows of unequal length, equilibrium is attained when one of the reactions is far more complete than its reverse. Here the competition for protons is between the bases H_2O and CN^-. The fact that the right-to-left reaction is more nearly complete than the left-to-right reaction means that CN^- ion is a stronger base than H_2O. This is equivalent to saying that the acids HCN and H_3O^+ are competing to donate protons to a base and that the right-to-left reaction is more nearly complete than the forward reaction because H_3O^+ is a stronger acid than HCN.

Every protolysis reaction favors the formation of the weaker acid and the weaker base. When the stronger acid donates a proton, the conjugate base that is formed is the weaker of the two bases. Likewise, when the stronger base accepts a proton, the conjugate acid that is formed is the weaker of the two acids. At the conclusion of a protolytic reaction there is established an equilibrium characterized by relatively greater amounts of the weaker acid and the weaker base. Hydrogen chloride ionizes almost completely in water because this reaction forms an acid, H_3O^+, that is weaker than HCl and a base, Cl^-, that is weaker than H_2O. The ionization of aqueous hydrogen cyanide reaches equilibrium when only a small fraction of the HCN molecules have ionized because this reaction produces an acid, H_3O^+, that is stronger than HCN and a base, CN^-, that is stronger than H_2O.

In general, whenever a proton donor and a proton acceptor are brought together, protolysis occurs. But the products of the protolysis are themselves a proton donor and a proton acceptor. If the starting acid and base are much stronger than the products, as in the reaction of hydrogen chloride with water, equilibrium will not be attained until the reaction has proceeded nearly to completion.

$$\text{STRONGER ACID} + \text{STRONGER BASE} \xrightleftharpoons[\leftarrow]{\longrightarrow} \text{weaker acid} + \text{weaker base}$$

If the starting acid and base are much weaker than the products, as in the reaction of hydrogen cyanide with water, equilibrium will be reached as soon as the reaction has proceeded only slightly to the right:

$$\text{weaker acid} + \text{weaker base} \rightleftharpoons \text{STRONGER ACID} + \text{STRONGER BASE}$$

21-6 Comparative Strengths of Acids and Bases

The proton-donating tendencies of different acids toward the same base may be compared as a method of arriving at their relative strengths or acidities. Thus, we may compare the acidities of acids of moderate strengths by comparing their ionization constants (Section 16-1). The ionization constants are equilibrium constants for the reactions of the acids with water. The smaller the ionization constant, the weaker is the acid.

$$\underset{\text{formic acid}}{HCOOH} + H_2O \rightleftharpoons H_3O^+ + \underset{\text{formate ion}}{HCOO^-} \qquad K = 1.8 \times 10^{-4}$$

$$\underset{\text{acetic acid}}{CH_3COOH} + H_2O \rightleftharpoons H_3O^+ + \underset{\text{acetate ion}}{CH_3COO^-} \qquad K = 1.8 \times 10^{-5}$$

$$\underset{\text{hydrocyanic acid}}{HCN} + H_2O \rightleftharpoons H_3O^+ + \underset{\text{cyanide ion}}{CN^-} \qquad K = 4.0 \times 10^{-10}$$

By comparing the ionization constants of the above acids we perceive immediately that formic acid is the strongest and hydrocyanic the weakest. Conversely, in Brønsted-Lowry terminology cyanide ion is a stronger base than formate ion or acetate ion, and formate ion is the weakest base of the three.

Table 21-1

Brønsted-Lowry acid-base chart

Conjugate acid		*Conjugate base*	
Name	*Formula*	*Formula*	*Name*
Perchloric acid	$HClO_4$	ClO_4^-	Perchlorate ion
Hydrogen iodide	HI	I^-	Iodide ion
Hydrogen bromide	HBr	Br^-	Bromide ion
Sulfuric acid	H_2SO_4	HSO_4^-	Hydrogen sulfate ion
Hydrogen chloride	HCl	Cl^-	Chloride ion
Nitric acid	HNO_3	NO_3^-	Nitrate ion
Hydronium ion	H_3O^+	H_2O	Water
Hydrogen sulfate ion	HSO_4^-	SO_4^{--}	Sulfate ion
Formic acid	$HCOOH$	$HCOO^-$	Formate ion
Acetic acid	CH_3COOH	CH_3COO^-	Acetate ion
Ammonium ion	NH_4^+	NH_3	Ammonia
Hydrogen cyanide	HCN	CN^-	Cyanide ion
Water	H_2O	OH^-	Hydroxide ion
Ethyl alcohol	C_2H_5OH	$C_2H_5O^-$	Ethoxide ion
Ammonia	NH_3	NH_2^-	Amide ion
Methylamine	CH_3NH_2	CH_3NH^-	Methylamide ion
Hydrogen	H_2	H^-	Hydride ion
Methane	CH_4	CH_3^-	Methide ion

In the Brønsted-Lowry acid-base chart (Table 21-1), certain selected acids and their conjugate bases have been listed side by side. The strongest acid, $HClO_4$, and the weakest base, ClO_4^-, are listed at the top of the chart; the weakest acid, CH_4, and the strongest base, CH_3^-, at the bottom. As stated in Section 21-5, every protolysis reaction favors the formation of the weaker acid and the weaker base. Any acid in the chart will react with any base that is lower in the chart to the extent of more than 50 percent. How nearly the reaction approaches 100 percent completion depends qualitatively on how far the acid and base are separated in the chart.

21-7 The Leveling Effect of the Solvent

When hydrogen chloride dissolves in liquid ammonia or in water the protolysis is virtually complete:

$$HCl + NH_3 \rightarrow NH_4^+ + Cl^-$$
$$HCl + H_2O \rightarrow H_3O^+ + Cl^-$$

Water is not as strong a base as ammonia, but HCl is an acid with such strong proton-donating tendencies that both the above reactions are virtually complete.

However, some acids, weaker than HCl, which react virtually completely with liquid ammonia do not react to the same extent with water because of the lesser proton-attracting strength of water.

The acids listed above hydronium ion in the acid-base chart (Table 21-1) have traditionally been classified as the strong acids. In aqueous solution all of them ionize virtually completely, and it is impossible to distinguish between their acid strengths without using nonaqueous solvents.

Some of the acids that react completely with water do not react completely with solvents that are less basic than water. For example, anhydrous acetic acid is less basic than water and only the strongest of the Arrhenius strong acids react completely to form ions when they are dissolved in this solvent. By using anhydrous acetic acid as a solvent it was found that $HClO_4 > HI > HBr > H_2SO_4 > HCl > HNO_3$ gives the relative proton-donating strengths of these strong acids. In acetic acid the reaction

$$HClO_4 + CH_3COOH \rightleftharpoons CH_3C(OH)_2^+ + ClO_4^-$$

takes place to a substantially greater extent than the reaction

$$HCl + CH_3COOH \rightleftharpoons CH_3C(OH)_2^+ + Cl^-$$

The acids $HClO_4$, HI, HBr, H_2SO_4, HCl, and HNO_3 appear to have the same acid strength in aqueous solution because the same acid, H_3O^+ ion, is responsible for the acidity in each instance. This phenomenon, which occurs because these acids react completely with water, is known as the *leveling effect*. Acids stronger than H_3O^+ are leveled to the strength of H_3O^+ when placed in aqueous solution. Acids stronger than NH_4^+ are leveled to the strength of NH_4^+ when placed in liquid ammonia solution. Acids stronger than $CH_3C(OH)_2^+$ are leveled to the strength of $CH_3C(OH)_2^+$ when placed in anhydrous acetic acid solution. In summary, *the strongest acid that can exist in substantial concentration in a solvent that can act as a base is the conjugate acid of that solvent base.*

The leveling effect is also exhibited when bases dissolve in acidic solvents. By *acidic solvent* is meant any solvent that can donate protons to a solute capable of accepting them. All the very strong bases, such as amide ion and hydride ion, react completely when they are added to water, forming hydroxide ion:

$$NH_2^- + H_2O \rightarrow NH_3\uparrow + OH^-$$
$$H^- + H_2O \rightarrow H_2\uparrow + OH^-$$

Thus, addition of sodium amide to water results in a vigorous protolytic reaction with liberation of ammonia. Similarly, sodium hydride reacts vigorously with water, liberating hydrogen gas, because H^- ion is a much stronger base than OH^- ion. *The strongest base that can exist in significant concentration in a solvent that can act as an acid is the conjugate base of that solvent acid.* Thus, the strongest base that can exist in water is OH^- ion, the strongest base in liquid ammonia is NH_2^- ion, and the strongest base in anhydrous acetic acid is CH_3COO^- ion.

The intrinsic differences in the proton-donating tendencies of various acids will be evident only in a solvent that has relatively weak proton-accepting tendencies, such as anhydrous acetic acid. This solvent is said to act as a *differentiating solvent* for acids. Conversely, liquid ammonia serves as a differentiating solvent for bases. Anhydrous acetic acid is a leveling solvent for bases, and liquid ammonia is a leveling solvent for acids. Water is a differentiating solvent for acids and bases sufficiently weak to be distinguishable as to strength by their ionization constants. Water is a leveling solvent for those acids classified as strong or very strong and for the very strong bases.

Solvents with neither strong proton-donating nor strong proton-accepting tendencies are called *aprotic solvents*. They can act as differentiating solvents for both acids and bases. In many important reactions of organic chemistry strong acids or strong bases are used as catalysts. In these reactions aprotic solvents are employed so that the strength of the catalyst will not be reduced by the leveling effect of a solvent with proton-accepting or proton-donating tendencies, such as water. Among the aprotic solvents extensively used are hexane, benzene, and carbon tetrachloride.

21-8 The Protonic Concept of Neutralization

The self-ionization of an amphiprotic solvent is a protolytic or acid-base reaction in which a proton is transferred from one solvent molecule to another:

$$\begin{array}{ccccccccc} acid_1 & + & base_2 & \rightleftarrows & acid_2 & + & base_1 \\ H_2O & + & H_2O & \rightleftarrows & H_3O^+ & + & OH^- \\ NH_3 & + & NH_3 & \rightleftarrows & NH_4^+ & + & NH_2^- \\ CH_3COOH & + & CH_3COOH & \rightleftarrows & CH_3C(OH)_2^+ & + & CH_3COO^- \end{array}$$

We have seen that the strongest acid that can exist in water is the H_3O^+ ion; the strongest acid that can exist in liquid ammonia is the NH_4^+ ion; the strongest acid that can exist in anhydrous acetic acid is the $CH_3C(OH)_2^+$ ion. In general, the strongest acid that can exist in a solvent is the cation that is formed by the self-ionization of that solvent. Conversely, the strongest base that can exist in a solvent is the anion that is formed by the self-ionization of the solvent: OH^- ion in water, NH_2^- ion in liquid ammonia, CH_3COO^- ion in anhydrous acetic acid.

Addition of any acid (proton donor) to water increases the concentration of H_3O^+ ion more or less, depending on whether the acid is strong or weak. Addition of an "ammono acid" to liquid ammonia increases the concentration of NH_4^+ ion. Thus, an acid in a particular solvent system may be defined as any substance that increases the concentration of the cation formed by the self-ionization of the solvent. Similarly, within a particular solvent system, a base is a substance that increases the concentration of the anion formed by the self-ionization of the solvent.

Neutralization of strong acids with strong bases in aqueous solutions involves the formation of water molecules by the reaction

$$H_3O^+ + OH^- \rightleftharpoons H_2O + H_2O$$

This reaction is the exact opposite of the self-ionization of water. Since H_3O^+ ion is the conjugate acid of H_2O and OH^- ion is the conjugate base of H_2O, we can say that the neutralization reaction occurring when aqueous solutions of strong acids and strong bases are mixed is the reaction of the conjugate acid of water with the conjugate base of water to form water. Similarly, the neutralization reaction typical of liquid ammonia solutions of strong acids and strong bases is the reaction of the conjugate acid of ammonia with the conjugate base of ammonia to form ammonia.

$$NH_4^+ + NH_2^- \rightleftharpoons NH_3 + NH_3$$

In general, the typical neutralization reaction between strong acids and strong bases within a given solvent involves formation of molecules of solvent by a proton transfer reaction between the conjugate acid and the conjugate base of the solvent.

It is desirable to define neutralization to include the reactions of strong bases with *weak* acids and the reactions of strong acids with *weak* bases. We can do this by defining neutralization as any protolytic reaction in which the solvent is a product (water in aqueous solutions, ammonia in liquid ammonia solutions, etc.). In aqueous solution the following reactions are neutralization reactions because they are proton transfer reactions in which water is formed:

$$CH_3COOH + OH^- \rightleftharpoons CH_3COO^- + H_2O$$
$$H_3O^+ + NH_3 \rightleftharpoons NH_4^+ + H_2O$$

A classical method for the preparation of a weak acid is to add a strong acid to a salt of the weak acid. The procedure works because the strong acid donates a proton to the conjugate base of the weak acid. As examples, acetic acid or hydrogen cyanide can be obtained by adding a strong acid, such as hydrochloric acid, to a solution containing an acetate or a cyanide:

$$H_3O^+ + CH_3COO^- \rightarrow CH_3COOH + H_2O$$
$$H_3O^+ + CN^- \rightarrow HCN + H_2O$$

Each reaction proceeds far to the right because H_3O^+ ion is a stronger acid than CH_3COOH or HCN. These reactions are protolysis reactions in which water is a product and are, therefore, examples of neutralization as defined in the previous paragraph.

The preparation of a weak base by adding a strong base to a salt of the weak base likewise involves Brønsted-Lowry neutralization:

$$NH_4^+ + OH^- \rightarrow NH_3 + H_2O$$
$$CH_3NH_3^+ + OH^- \rightarrow CH_3NH_2 + H_2O$$

These reactions proceed far to the right because OH^- ion is a stronger base than NH_3 or CH_3NH_2.

21-9 The Protonic Concept of Hydrolysis

Hydrolysis is essentially the opposite of neutralization in aqueous solution and is defined as a protolytic reaction in which water is a reactant. If the substance undergoing hydrolysis is a base, e.g., cyanide ion, the water acts as an acid:

$$CN^- + H_2O \rightleftharpoons HCN + OH^-$$

If the substance undergoing hydrolysis is an acid, e.g., ammonium ion, the water acts as a base:

$$NH_4^+ + H_2O \rightleftharpoons NH_3 + H_3O^+$$

When a salt derived from a strong base and a weak acid (Section 21-2) is dissolved in water, the anion (being a strong Brønsted-Lowry base) reacts with water to form OH^- ions which cause the solution to be alkaline. An example is sodium cyanide. The reaction does not proceed far because CN^- ion is a weaker base than OH^- ion and H_2O is a weaker acid than HCN. The fact that it occurs at all is significant, however, because it makes the aqueous solution of the salt, NaCN, alkaline rather than neutral.

The aqueous solution of a salt is acidic if the salt is derived from a weak base and a strong acid. This is because the cation (being a strong Brønsted-Lowry acid) reacts with water to form H_3O^+ ions which cause the solution to be acidic. An example is ammonium chloride. The hydrolysis of the ammonium ion does not proceed far because NH_4^+ ion is a weaker acid than H_3O^+ ion and H_2O is a weaker base than NH_3. But it is a significant reaction, even though far from complete, because it makes the aqueous solution of the salt, NH_4Cl, acidic.

The aqueous solution of a salt is neutral if the parents of the salt are a strong acid and a strong base. In such a salt the cation is a very weak Brønsted-Lowry acid and has very little tendency to react with water to form H_3O^+ ions; the anion is a very weak Brønsted-Lowry base and has very little tendency to react with water to form OH^- ions.

If a salt is derived from a weak acid and a weak base, both its cation and its anion tend to react with water. The result may be an acidic, alkaline, or neutral solution, depending on which competing factor is the winner—the tendency of the cation to make the solution acidic by its hydrolysis or the tendency of the anion to hydrolyze and make the solution alkaline. Furthermore, since the cation of such a salt is a Brønsted-Lowry acid and the anion is a Brønsted-Lowry base, the protolytic reaction between the two ions may be more important than the hydrolysis of either of them. In an aqueous solution of ammonium acetate the simultaneous presence of ammonia and acetic acid may be attributed to the proton transfer reaction between ammonium ions and acetate ions:

$$NH_4^+ + CH_3COO^- \rightleftharpoons NH_3 + CH_3COOH$$

By the protonic concept, hydrolysis of a cation is not possible unless it contains hydrogen, for otherwise it could not possibly act as a proton donor. If we are to explain, for example, the acidity of a solution of aluminum nitrate in Brønsted-Lowry terms, we cannot ignore the fact that the aluminum ion in aqueous solution is actually an aquo complex. The acidity of a solution of aluminum nitrate may be attributed to the proton transfer reaction occurring between the hexaquoaluminum ion and water:

$$[Al(H_2O)_6]^{+++} + H_2O \rightleftharpoons [Al(H_2O)_5OH]^{++} + H_3O^+$$

This protolytic reaction is quite properly referred to as the hydrolysis of the *hydrated* aluminum ion. (See Section 16-11.)

The hydrolysis of a salt is essentially complete only if its anion is a stronger base than OH^- ion. Thus, sodium amide, sodium hydride, and calcium acetylide undergo complete hydrolysis when placed in contact with water because NH_2^- ion, H^- ion, and C_2^{--} ion are stronger bases than OH^- ion.

21-10 The Lewis Concept of Acids and Bases

The Brønsted-Lowry concept of acids and bases is more general than the Arrhenius concept because it applies to both aqueous and nonaqueous solutions and also to reactions that occur in the absence of a solvent. In spite of its breadth, the Brønsted-Lowry concept is limited by its emphasis on the proton. Only compounds containing hydrogen can qualify as acids by the Brønsted-Lowry definition.

Let us reconsider the reaction between hydrogen chloride and water:

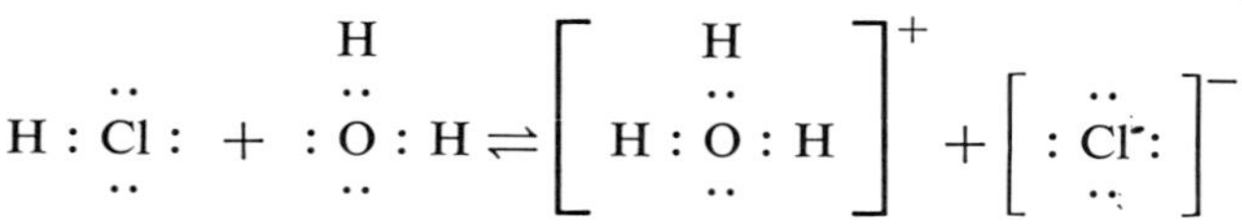

Instead of emphasizing that a proton is transferred to a water molecule, we can point to the fact that the oxygen atom of the water molecule donates a pair of electrons to the proton and thereby forms a coordinate covalent bond. In each of the following examples the Brønsted-Lowry base donates a pair of electrons to a proton:

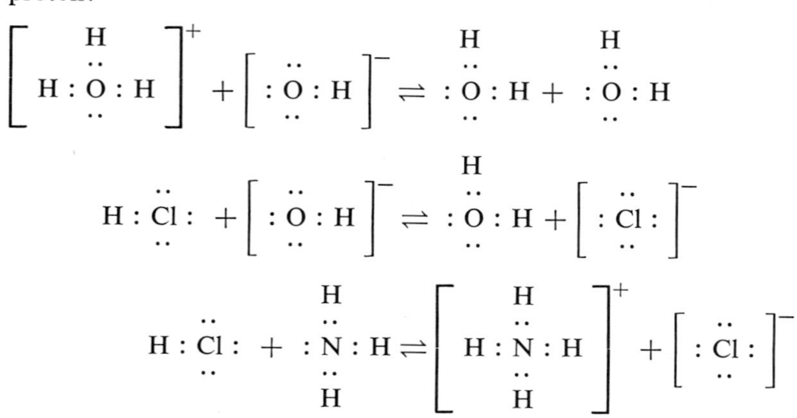

In every Brønsted-Lowry acid-base reaction the base donates a pair of electrons to a proton. In many other reactions one substance donates a pair of electrons to another substance, but these reactions are not included in the Brønsted-Lowry definition because a proton is not involved.

In 1923 the American chemist Gilbert Newton Lewis proposed generalized definitions for acids and bases that do not restrict them to compounds containing hydrogen. Lewis defined an acid as any molecular or ionic species that contains an atom capable of sharing a pair of electrons furnished by a base. He defined a base as any molecular or ionic species that contains an atom having a pair of electrons available for sharing by an acid. In simplest terms, a Lewis acid is an electron acceptor and a Lewis base is an electron donor.

In every instance of protolysis the Brønsted-Lowry base is also a Lewis base; but whereas the Brønsted-Lowry acid is the substance that donates the proton, the Lewis acid is the proton itself. The following examples are representative of the many Lewis acid-base reactions that do *not* conform to the Brønsted-Lowry definition for acid-base reactions:

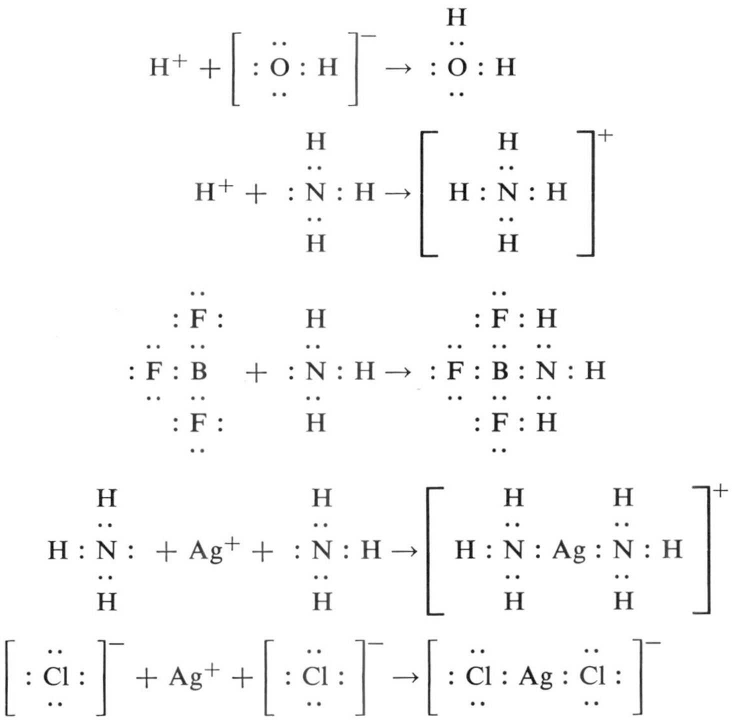

Sodium oxide and sulfur trioxide react to form the same salt that is formed when sodium hydroxide reacts with sulfuric acid. Classically, sodium oxide has been classified as a basic oxide because it reacts with water to form sodium hydroxide and with acids to form salts. Similarly, sulfur trioxide has been classified as an acidic oxide because it reacts with water to form sulfuric acid.

Nevertheless, the Brønsted-Lowry definition of acid-base reactions is not broad enough to include the reaction of sodium oxide with sulfur trioxide. The Lewis definition of an acid-base reaction, however, is sufficiently general to include both the reaction of sodium hydroxide with sulfuric acid and the reaction of sodium oxide with sulfur trioxide:

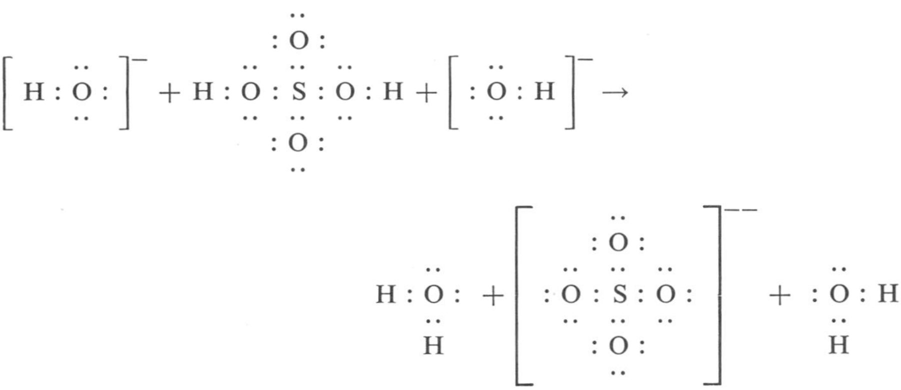

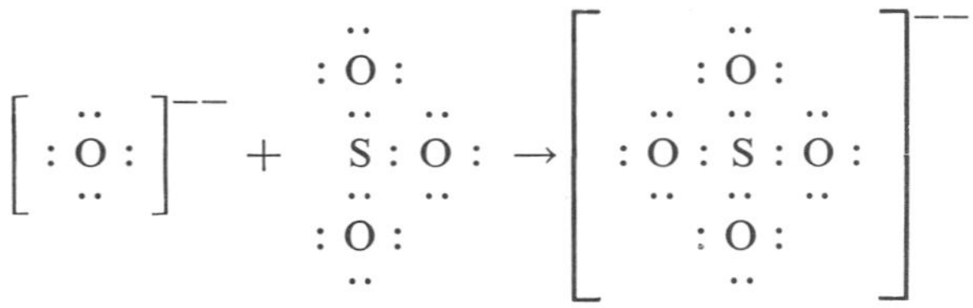

All positive ions are potential Lewis acids and all negative ions are potential Lewis bases. Any molecule in which the central atom has an incomplete octet, e.g., BF_3 or $AlCl_3$, is a potential Lewis acid. Any molecule in which an atom has one or two unshared pairs of electrons, e.g., NH_3 or H_2O, is a potential Lewis base.

The compounds formed by the reaction of Lewis acids with Lewis bases are coordination compounds. Indeed, the fundamental acid-base reaction in the Lewis sense is the formation of a coordinate bond.

Those substances that are bases in the Lewis system are also bases in the Brønsted-Lowry system, for electron donors can share electrons with any Lewis acid, including the proton. The Brønsted-Lowry definition limits acids to those substances which contain one or more protons available for sharing electrons made available by any base. The Lewis definition of acids includes all substances which contain any kind of atom that can share electrons with any base.

21-11 The Classification of Chemical Reactions

In the reactions of covalent compounds, bonds between atoms or ions are either formed or broken. If we use the symbol A to represent a particular atom or group of atoms and B to represent another atom or group of atoms, we can say that there are two ways in which a covalent bond between A and B can be formed or broken. In the first of these the bond can be formed or broken so that A supplies or retains one electron of the shared pair while B supplies or retains the other electron of the shared pair:

$$A\cdot + \cdot B \rightleftharpoons A:B$$

Thus, in this way of forming or breaking a covalent bond there is a *pairing of electrons* or an *unpairing of electrons*. Any atom, group of atoms, or ion that contains an unpaired electron is called a *free radical*. Therefore, this type of reaction, in which electrons are paired or unpaired, is called a *free radical reaction*.

The second way of forming or breaking a covalent bond between A and B may be represented as follows:

$$A^{+} + :B^{-} \rightleftharpoons AB$$

or

$$A + :B \rightleftharpoons AB \text{ (for a coordinate bond)}$$

In this way of forming or breaking a covalent bond there is a *sharing of electrons* or an *unsharing of electrons*. This type of reaction, in which an electron-pair donor and an electron-pair acceptor are involved, is called a *polar reaction*. Polar reactions include all the typical acid-base reactions (ionization, neutralization, solvolysis) and a great many organic chemical reactions that take place in solution.

All polar reactions are included in the broadened concept of acid-base reactions proposed by Lewis. This generalized acid-base concept has provided a useful and logical basis for the classification and interpretation of many reactions that involve the sharing or unsharing of electron pairs. The Lewis concept has provided a common basis for understanding a variety of reactions in organic and inorganic chemistry.

21-12 Complex Ions

If a solution contains silver ions, addition of another solution containing chloride ions gives an immediate white precipitate of silver chloride, AgCl. Upon addition of a sufficient quantity of aqueous ammonia, the precipitate completely dissolves, forming a clear and colorless solution. However, addition of excess nitric acid to the colorless solution causes the white precipitate to reappear. This set of facts can be understood in terms of the two equilibria represented by the following equations:

$$Ag^{+} + Cl^{-} \rightleftharpoons \underset{\text{white}}{AgCl\downarrow}$$

$$AgCl(s) + 2NH_3 \rightleftharpoons \underset{\text{colorless solution}}{Ag(NH_3)_2^{+} + Cl^{-}}$$

The white precipitate forms because AgCl is not soluble in water. It dissolves in excess ammonia because of the formation of the stable complex ion,

$Ag(NH_3)_2{}^+$; in other words, the complex salt $Ag(NH_3)_2Cl$ *is* soluble in water. Addition of excess nitric acid causes the second equilibrium to shift to the left, and this results in the reappearance of the white precipitate. The equilibrium shifts to the left because of the decrease in concentration of NH_3 molecules as a result of their reaction with H^+ to form $NH_4{}^+$.

In terms of the solubility product principle, we can think of the Cl^- ions and NH_3 molecules competing for the Ag^+ ions:

$$\mathrm{Ag^+ + Cl^- \rightleftharpoons AgCl}(s)$$
$$\mathrm{Ag^+ + 2NH_3 \rightleftharpoons Ag(NH_3)_2{}^+}$$

With a high concentration of NH_3 molecules, the Ag^+ ion concentration is reduced to such a small value that the product of the two ion concentrations, $[Ag^+] \times [Cl^-]$, is less than the solubility product constant of AgCl, and this enables the precipitate to dissolve completely. On the other hand, addition of acid (i.e., H^+ ions) reduces the NH_3 concentration to such a small value that the second equilibrium is shifted far to the left, and this increases the Ag^+ ion concentration so much that the product of the two ion concentrations, $[Ag^+] \times [Cl^-]$, reaches the value of the solubility product constant for AgCl so that the precipitate reappears.

The reaction of ammonia with H^+ to form $NH_4{}^+$ and with Ag^+ to form $Ag(NH_3)_2{}^+$ produces a coordinate covalent bond like that in NH_3BF_3 (Section 5-4). These are, therefore, Lewis acid-base reactions:

$$\mathrm{H^+} + \begin{array}{ccc} & \mathrm{H} & \\ & \cdot\cdot & \\ : & \mathrm{N} & :\mathrm{H} \\ & \cdot\cdot & \\ & \mathrm{H} & \end{array} \rightarrow \left[\begin{array}{ccc} & \mathrm{H} & \\ & \cdot\cdot & \\ \mathrm{H}: & \mathrm{N} & :\mathrm{H} \\ & \cdot\cdot & \\ & \mathrm{H} & \end{array}\right]^+$$

$$\begin{array}{ccc} & \mathrm{H} & \\ & \cdot\cdot & \\ \mathrm{H}: & \mathrm{N} & : \\ & \cdot\cdot & \\ & \mathrm{H} & \end{array} + \mathrm{Ag^+} + \begin{array}{ccc} & \mathrm{H} & \\ & \cdot\cdot & \\ : & \mathrm{N} & :\mathrm{H} \\ & \cdot\cdot & \\ & \mathrm{H} & \end{array} \rightarrow \left[\begin{array}{ccccc} & \mathrm{H} & & \mathrm{H} & \\ & \cdot\cdot & & \cdot\cdot & \\ \mathrm{H}: & \mathrm{N} & :\mathrm{Ag}: & \mathrm{N} & :\mathrm{H} \\ & \cdot\cdot & & \cdot\cdot & \\ & \mathrm{H} & & \mathrm{H} & \end{array}\right]^+$$

In the first reaction, H^+ is the Lewis acid; in the second reaction, Ag^+ is the Lewis acid. In both reactions, NH_3 is the Lewis base. It is characteristic of complex ions that their formation involves a Lewis acid-base reaction.

New Terms

Amphiprotic substance: a substance capable of acting as a proton donor in the presence of a proton acceptor and as a proton acceptor in the presence of a proton donor.

Aprotic solvent: a solvent that has neither strong proton-donating nor strong proton-accepting tendencies, e.g., benzene.

Arrhenius acid: a hydrogen compound which forms hydrogen ions (hydronium ions) in aqueous solution. [A modernized version of the definition states that acids are substances that increase the hydrogen-ion concentration when they are added to water. This definition includes such species as CO_2, NH_4^+, and $Al(H_2O)_6^{+++}$.]

Arrhenius base: a compound that dissociates to give hydroxide ions in aqueous solution. The compound may be ionic (e.g., NaOH) or covalent (e.g., NH_4OH). [A modernized version of the definition states that bases are substances that increase the hydroxide-ion concentration when they are added to water. This definition includes such species as CaO, NH_3, CN^-, and $Zn(OH)_4^{--}$.]

Brønsted-Lowry acid: a proton donor.

Brønsted-Lowry base: a proton acceptor.

Complex ion: an ion formed by the union of two species of simple ions or by the union of simple ions with molecules. Usually, one of the species is a metal ion. Examples: $Ag(CN)_2^-$, $Ag(NH_3)_2^+$. (In the broadest sense, a complex ion is a charged particle consisting of more than one atom. It is not common practice, however, to speak of such ions as SO_4^{--}, NO_3^-, or OH^- as complex ions.)

Conjugate acid-base pair: two species so related that one is formed from the other by transfer of a proton. For example, NH_3 is the base conjugate to the acid NH_4^+; NH_4^+ is the acid conjugate to the base NH_3; NH_4^+ and NH_3 are a conjugate acid-base pair.

Differentiating solvent: a solvent that does not exert a significant leveling effect upon the acids or bases referred to in the designation. Thus, each acid (or base) is able to display its intrinsic acid (or basic) strength in that particular solvent.

Free radical: any atom, group of atoms, or ion that contains an unpaired electron.

Free radical reaction: a reaction in which electrons become paired or unpaired, $A\cdot + \cdot B \rightleftharpoons A:B$.

Hydrolysis (Arrhenius definition): the reaction of a salt with water, producing the parent acid and the parent base.

Hydrolysis (Brønsted-Lowry definition): any protolytic reaction in which water is a reactant.

Leveling effect: the effect of a solvent in reducing the acid strength of a solute to the level of the acid strength of the solvent cation, or the basic strength of a solute to the level of the basic strength of the solvent anion. Thus, H_3O^+ is the strongest acid that can exist in aqueous solution, and OH^- is the strongest base in aqueous solution.

Lewis acid: an electron-pair acceptor.

Lewis base: an electron-pair donor.

Neutralization (Arrhenius definition): the formation of water by the reaction of hydrogen ions with hydroxide ions.

Neutralization (Brønsted-Lowry definition): any protolytic reaction in which the solvent is a product.

Neutralization (Lewis definition): the reaction of a Lewis acid with a Lewis base, i.e., the formation of a coordinate bond.

Polar reaction: a reaction in which there is a sharing or an unsharing of an electron pair.

Protolysis: any proton-transfer reaction (Brønsted-Lowry acid-base reaction).

Exercises

21-1 By means of equations, compare the self-ionizations of water, liquid ammonia, and anhydrous acetic acid.

21-2 What is an amphiprotic solvent? An aprotic solvent?

21-3 Define neutralization according to (a) the Arrhenius concept, (b) the Brønsted-Lowry concept, and (c) the Lewis concept.

21-4 State and illustrate the Brønsted-Lowry concept of hydrolysis.

21-5 List several strong acids and their conjugate bases. What can be said of the strength of these bases?

21-6 List several strong bases and their conjugate acids. Discuss the strength of these conjugate acids.

21-7 Write equations for the reaction of the following acids and bases with water: HCl, $HClO_4$, NH_4^+, NH_3, CN^-. Identify each of these as a Brønsted-Lowry acid, a Brønsted-Lowry base, a Lewis acid, or a Lewis base. Identify the role played by the water in each reaction.

21-8 Compare ammono acids with aquo acids and ammono bases with aquo bases.

21-9 Compare neutralization in water with neutralization in liquid ammonia and in anhydrous acetic acid.

21-10 Construct definitions for the following terms in the light of the information presented in this chapter: salt, ammonolysis, solvolysis.

21-11 Discuss the leveling effect of water on strong acids and strong bases.

21-12 How can the relative strengths of a given list of acids be determined if all of them are completely ionized in water?

References

Lewis, G. N., *Valence and the Structure of Atoms and Molecules*, Dover, New York, 1966 (paperbound). Reprint of the 1923 classic explaining the Lewis concepts of the covalent bond and acid-base reactions.

Luder, W. F., and Zuffanti, S., *The Electronic Theory of Acids and Bases*, 2nd ed., Dover, New York, 1961 (paperbound). A discussion of the many ramifications of the Lewis concept of acids and bases.

Sisler, H. H., *Chemistry in Non-Aqueous Solvents*, Reinhold, New York, 1961 (paperbound). An account of the relation between the properties of solvents and the processes which occur in them, including acid-base reactions.

VanderWerf, C. A., *Acids, Bases, and the Chemistry of the Covalent Bond*, Reinhold, New York, 1961 (paperbound). An organic chemist's interpretation of the Brønsted-Lowry and Lewis concepts.

22

The Alkali Metals

22-1 Properties of the Alkali Metals

Group IA of the periodic table consists of six elements that have only one *s* electron in the outermost shell of each atom: lithium, sodium, potassium, rubidium, cesium, and francium (Figure 22-1). The ease with which this one electron can be removed, as indicated by the first ionization energies and the standard oxidation potentials (Table 22-1), makes these six elements *the most active group of metals.* Most compounds of these metals are ionic. The atoms have noble-gas kernels and each element is in the +1 oxidation state in all its compounds. The ions are isoelectronic with the noble-gas atoms.

Because of their chemical activity these metals are never found free in nature. They react vigorously with water, liberating hydrogen gas and forming hydroxides that are highly soluble, strong bases:

$$2M + 2HOH \rightarrow 2(M^+ + OH^-) + H_2\uparrow$$

Sodium hydroxide, NaOH, and potassium hydroxide, KOH, have long been known as *alkalies.* For this reason the metals of this family are known as the *alkali metals.* They react in similar fashion with alcohols to give alkoxides; they react violently with acids.

The atoms of the alkali metals are the largest atoms in their respective rows of the periodic table. This means that the single electron in the outermost shell is relatively far from the nucleus. The ease with which this electron is lost increases with increasing atomic number because the atomic size increases with increasing atomic number. Thus, cesium is the most active of the members of the family that have been thoroughly investigated.

Though the amount of energy required to remove the single *s* electron from the outermost shell of an alkali-metal atom is relatively small, the removal of an electron from the next level requires many times this amount of energy.

Figure 22-1

Electronic configurations of the alkali-metal atoms

	1	2	3	4	5	6	7
$_{3}$Li	s^2	s^1					
$_{11}$Na	s^2	s^2p^6	s^1				
$_{19}$K	s^2	s^2p^6	s^2p^6	s^1			
$_{37}$Rb	s^2	s^2p^6	$s^2p^6d^{10}$	s^2p^6	s^1		
$_{55}$Cs	s^2	s^2p^6	$s^2p^6d^{10}$	$s^2p^6d^{10}$	s^2p^6	s^1	
$_{87}$Fr	s^2	s^2p^6	$s^2p^6d^{10}$	$s^2p^6d^{10}f^{14}$	$s^2p^6d^{10}$	s^2p^6	s^1

(Compare the first and second ionization energies in Table 22-1.) Thus, the formation of 1+ ions occurs with ease, but the formation of 2+ ions never occurs in ordinary reactions. As a result, the chemistry of these elements is about as simple as possible.

When the alkali-metal atoms come together physically to form the solid or liquid metal, 1+ ions form in a cloud of loose electrons that are free to move about among the positive ions. The freedom of movement of these loose electrons makes the metals good conductors of heat. The easy displacement of the electrons, when the ends of the metal are at different electrical potentials, accounts for the high electrical conductivity of these elements.

Table 22-1

Properties of the alkali metals

Element	*Density at 20°C* (g/ml)	*Melting point* (°C)	*Boiling point* (°C)	*First ionization energy* (kcal/mole)	*Second ionization energy* (kcal/mole)	*Standard oxidation potential* (volts)
$_{3}$Li	0.53	180	1326	124	1744	3.02
$_{11}$Na	0.97	97.5	889	118	1091	2.71
$_{19}$K	0.86	63.4	757	100	734	2.93
$_{37}$Rb	1.53	38.8	679	96	634	2.99
$_{55}$Cs	1.90	28.7	690	89	579	3.02

The force holding a crystal of an alkali metal together is due to the attraction between the metal ions and the electron cloud. Since this is not a strong force, the structure is not a tightly packed one. Consequently, all the alkali metals except lithium are so soft that they easily can be cut with a knife, and they are the least dense of all metals. Lithium, sodium, and potassium are less dense than water; lithium is the lightest metal known. The melting points of the members of the group are quite low (Table 22-1) and they decrease with increasing atomic numbers. Cesium (melting point, 28.7°C) is a liquid on warm days.

Since their outermost electrons are loosely held, the alkali metals are strong reducing agents. All but lithium combine readily with oxygen on contact with air, the dull oxide scale obscuring the silvery-white appearance of a freshly cut surface. The "normal" oxides are composed of metal (M^+) and oxide (O^{--}) ions and react vigorously with water to form the corresponding hydroxides:

$$M_2O + H_2O \rightarrow 2(M^+ + OH^-)$$

The normal oxide of lithium is formed when the metal is heated with oxygen, but the preparation of the normal oxides of the other alkali metals requires controlled oxidation. With the exception of lithium, heating the metal with oxygen easily gives the peroxide, containing the O_2^{--} ion:

$$2M + O_2 \rightarrow M_2O_2$$

Lithium peroxide can be obtained as a precipitate by adding hydrogen peroxide to a solution of lithium hydroxide. The peroxides react with water to give hydrogen peroxide. Sodium, potassium, rubidium, and cesium also form superoxides, MO_2, which contain the O_2^- ion (Section 7-15).

At high temperatures the alkali metals combine with hydrogen to form ionic hydrides:

$$2M + H_2 \rightarrow 2M^+H^-$$

They likewise combine with the halogens to form alkali-metal halides, which are superb examples of ionic compounds:

$$2M + X_2 \rightarrow 2M^+X^-$$

All alkali-metal salts are colorless except those that contain colored anions, such as yellow Na_2CrO_4 and purple $KMnO_4$.

22-2 Occurrence

Sodium and potassium compounds, especially salt (sodium chloride), soda (sodium carbonate), and potash (potassium carbonate), have been known and used since very early times. Sodium compounds are so widely distributed that they can be detected in almost any specimen of matter that has not been highly purified.

Sodium and potassium are present in many complex silicate rocks such as the feldspars; examples are albite, $NaAlSi_3O_8$, and orthoclase, $KAlSi_3O_8$, both of which are quite abundant. As such minerals disintegrate by weathering to form clays, the sodium and potassium are converted to water-soluble compounds, which are leached out of the soil by water. As these compounds pass through the soil, the potassium compounds are absorbed by the colloidal particles of soil to a greater extent than are the sodium compounds. The potassium compounds are needed by growing plants.

The sodium compounds that are carried into the sea account for most of its salt content. Geological changes that have resulted in the gradual evaporation of old seas and lakes have produced salt beds and underground brines that serve as excellent sources of sodium salts. In some cases they are good sources of potassium salts.

Sodium chloride is one of our most abundant minerals and is the chief source of sodium and chlorine compounds. It is usually contaminated with magnesium chloride and calcium chloride. Both of these absorb water from a humid atmosphere, causing salt to "cake" in the salt shaker. Sodium carbonate occurs on almost every continent. Sodium nitrate is the only nitrate that is found in nature in large amounts.

The potassium minerals that are economically important contain the chloride, KCl, and the sulfate, K_2SO_4. These occur in dry deposits and in brines. Their greatest importance lies in their use as agricultural fertilizers.

Lithium occurs in relatively small amounts in the form of complex silicate minerals, such as spodumene, $LiAlSi_2O_6$. Rubidium and cesium occur to a small extent in similar minerals. Francium is highly radioactive and occurs only in traces.

Since they are soluble in water and occur in abundance, sodium and potassium salts are often used in preference to salts of other metals in industry and in the laboratory. Because in most cases they are less expensive, sodium salts are used more extensively than potassium salts.

22-3 The Carbonates

Since the formula for carbonic acid is H_2CO_3, there are two carbonates of each alkali metal. Sodium hydrogen carbonate, $NaHCO_3$, is well known as "baking soda" or "bicarbonate of soda." Sodium carbonate, Na_2CO_3, is an important commercial chemical known as "soda ash." Although both these substances occur in nature, the quantities are insufficient to meet the demand. They are produced from sodium chloride by a process developed in 1863 by Ernest Solvay in Belgium.

The great success of the Solvay process is due to its ingenious utilization of raw materials and by-products. Essentially, the process consists of passing NH_3 and CO_2 gases into a cold, saturated solution of NaCl. The two gases react slightly with water, the one forming a weak base and the other a very weak acid; together they form a salt, and the salt participates in a metathetic reaction with NaCl to produce the desired product:

$$NH_3 + H_2O \rightleftharpoons NH_4OH$$
$$CO_2 + H_2O \rightleftharpoons H_2CO_3$$
$$NH_3 + H_2O + CO_2 \rightarrow (NH_4^+ + HCO_3^-)$$
$$(NH_4^+ + HCO_3^-) + (Na^+ + Cl^-) \rightarrow NaHCO_3 \downarrow + (NH_4^+ + Cl^-)$$

The product separates from the solution as a white solid and is removed by filtration, since it is much less soluble than ammonium chloride.

The CO_2 required for the Solvay process is obtained by heating limestone:

$$CaCO_3 \rightarrow CaO + CO_2$$

The calcium oxide ("quicklime") obtained in this reaction is treated with water and the resulting calcium hydroxide is used to recover NH_3 from the ammonium chloride:

$$CaO + H_2O \rightarrow Ca(OH)_2$$
$$Ca(OH)_2 + 2NH_4Cl \rightarrow CaCl_2 + 2NH_3 + 2H_2O$$

Since the NH_3 can be reused, the only true by-product of the Solvay process is $CaCl_2$.

Although the $NaHCO_3$ has some important uses of its own, e.g., in baking, most of it is converted to sodium carbonate by heating:

$$2NaHCO_3 \rightarrow Na_2CO_3 + H_2O + CO_2$$

Commercially, sodium carbonate is the most important sodium compound. Large quantities are used in the manufacture of soap and glass and a variety of industrial chemicals.

Potassium carbonate cannot be produced by the Solvay process because $KHCO_3$ is too highly soluble in aqueous solutions of ammonia to be precipitated. The ash from the waste matter of the beet-sugar industry is a source of K_2CO_3, which is used in the manufacture of soft soap and hard glass.

22-4 The Hydroxides

Sodium hydroxide is obtained by electrolysis of aqueous sodium chloride:

$$2(Na^+ + Cl^-) + 2H_2O \rightarrow 2(Na^+ + OH^-) + H_2\uparrow + Cl_2\uparrow$$

It is also produced by the reaction of sodium carbonate with calcium hydroxide:

$$(2Na^+ + CO_3^{--}) + (Ca^{++} + 2OH^-) \rightarrow CaCO_3 + 2(Na^+ + OH^-)$$

After the insoluble $CaCO_3$ ("precipitated chalk") settles to the bottom, the NaOH solution is removed; if solid NaOH is desired, the water can be removed from the solution by evaporation.

Sodium hydroxide ("caustic soda") is an extremely hygroscopic white solid that is very soluble in water. Its aqueous solutions are sometimes called "lye";

they have the slippery feel characteristic of solutions of strong bases and are very corrosive (caustic) to the skin. NaOH finds extensive use as a very soluble, strong base since it is cheaper than KOH and stronger than the less expensive $Ca(OH)_2$.

NaOH and KOH melt and boil without decomposition. Lithium hydroxide is a weaker base, much less soluble in water. When lithium hydroxide is heated, it loses water and forms the oxide.

$$2LiOH \rightarrow Li_2O + H_2O\uparrow$$

Growing plants remove potassium ions from the soil and form potassium salts of organic acids. When plant materials are burned these salts are changed into potassium carbonate, which therefore is present in wood ashes. In former times it was the practice to leach wood ashes with water and to evaporate the resulting solution in iron pots. The residue, which was mostly K_2CO_3, was called potash. It was converted to "potash lye" or "caustic potash" by allowing an aqueous solution of it to flow through a bed of "slaked lime" (calcium hydroxide):

$$(2K^+ + CO_3^{--}) + Ca(OH)_2 \rightarrow CaCO_3 + 2(K^+ + OH^-)$$

The caustic potash (potassium hydroxide) was then heated with fat to make soap. Sodium carbonate was similarly converted to "soda lye" or "caustic soda" by treatment with lime, and the caustic soda (sodium hydroxide) was also used in making soap.

The making of soap is still an important use for sodium hydroxide and potassium hydroxide. Lithium hydroxide is used in making lithium soaps for use in certain lubricating greases.

22-5 The Nitrates

Since it is highly soluble in water, sodium nitrate is found only in very arid regions, notably the high, arid regions along the coast of Chile. It is used as a nitrogenous fertilizer and for the production of potassium nitrate. Prior to the development of the Haber and Ostwald processes, it was the chief source of nitric acid.

Potassium nitrate, commonly known as saltpeter, is found in a few arid regions of the Orient. It finds extensive use in making ammunition for small firearms. Because of its deliquescent nature, sodium nitrate is unsuited for this purpose. Most potassium nitrate is obtained by mixing hot solutions of sodium nitrate and potassium chloride and then increasing the concentration of the solution by evaporation. Of the four salts that might conceivably crystallize from a solution containing Na^+, K^+, Cl^-, and NO_3^- ions, NaCl is the least soluble in hot water. Since KNO_3 is very soluble in hot water, the two salts are separated by filtration. Potassium nitrate crystallizes out when the filtrate is cooled.

Table 22-2

Melting points and boiling points of the alkali-metal halides

	Fluoride		*Chloride*		*Bromide*		*Iodide*	
	mp (°C)	bp (°C)	mp (°C)	bp (°C)	mp (°C)	bp (°C)	mp (°C)	bp (°C)
Lithium	870	1670	613	1382	547	1310	450	1190
Sodium	992	1705	801	1430	755	1390	653	1300
Potassium	880	1500	776	1500	748	1376	693	1330
Rubidium	775	1410	717	1390	682	1350	642	1305
Cesium	684	1250	646	1300	636	1300	521	1280

22-6 The Halides

Reactions of the alkali metals with the halogens yield typical ionic compounds which are remarkably similar. The reaction is often so violent that it is dangerous. It is much safer to prepare an alkali-metal halide by reaction of the hydroxide of the metal with the appropriate hydrohalic acid:

$$(M^+ + OH^-) + (H^+ + X^-) \rightarrow (M^+ + X^-) + HOH$$

Since the alkali metals and the halogens are groups of elements that are strongly similar among themselves, it is to be expected that the alkali-metal halides should be very similar. Furthermore, since the alkali metals are the least electronegative and the halogens are the most electronegative elements, it is to be expected that the alkali-metal halides should be typical ionic compounds with high melting points and high boiling points (Table 22-2). All of them dissolve in water to give electrically conducting solutions, and all are electrically conducting in the pure liquid state. The electrolysis of a fused alkali-metal halide produces the halogen at the anode and liberates the free alkali metal at the cathode:

$$2X^- \rightarrow X_2 + 2e^-$$

$$2M^+ + 2e^- \rightarrow 2M$$

One of the alkali-metal halides, sodium chloride, is familiar as common salt for preserving and flavoring food, but this use accounts for only a small portion of the millions of tons of this substance mined every year. It is used extensively for the manufacture of sodium metal, chlorine, sodium hydroxide, sodium carbonate, and hydrochloric acid.

22-7 The Born-Haber Cycle

The reaction of an alkali metal with a halogen liberates considerable amounts of energy, and it is a temptation to attribute this solely to the electron affinity of the halogen and to the ease with which an alkali-metal atom gives up its outer-shell electron. The incompleteness of this explanation becomes evident, however, when we compare the electron affinity of chlorine, the ionization energy of sodium, and the heat of formation of sodium chloride. The ionization energy of sodium is 118 kcal per mole; i.e., the removal of the outer-shell electrons from 6.023×10^{23} sodium atoms requires an input of 118 kcal of energy. The electron affinity of chlorine is 87 kcal per mole; i.e., the conversion of 6.023×10^{23} chlorine atoms to chloride ions (by a gain of one electron by each atom) is accompanied by the liberation of only 87 kcal of energy. These two facts alone cannot account for the liberation of 98 kcal of energy when one mole of sodium chloride is formed from its elements!

The formation of crystalline sodium chloride from sodium and chlorine involves considerably more than the mere transfer of an electron from each sodium atom to a chlorine atom. We imply as much when we write the equation

$$Na(s) + \tfrac{1}{2}Cl_2(g) \rightarrow Na^+Cl^-(s)$$

This equation states that *solid* sodium is reacting with *diatomic* chlorine to form *crystalline* sodium chloride. The very existence of the solid state implies the existence within the solid of attractive forces that hold the constituent entities of the solid together, for otherwise the kinetic energies of these particles would cause them to disperse as a gas and to become distributed throughout all the space available to them. Therefore, the reaction in which sodium chloride is produced from its elements involves overcoming the forces of attraction that hold the sodium atoms together within the solid sodium. This requires an input of energy. However, an interionic attraction exists between the Na^+ ions and the Cl^- ions within the sodium chloride crystals formed in the reaction. Just as an input of energy is required to overcome the interatomic attraction within solid sodium, so is energy released because of the attraction between the positive and negative ions when the crystal is formed as a result of the reaction. The energy released when gaseous ions form a crystal is called *lattice energy*. When the arrangement of the ions in a crystalline solid is known, including the interionic distances, it is possible to calculate a value for the lattice energy of that substance. The theoretical value for the lattice energy of sodium chloride thus calculated from electrostatic theory is -184 kcal per mole. Since this is substantially greater in numerical magnitude than the heat of formation, -98 kcal per mole, it is the key to the explanation for the exothermic character of the reaction between sodium and chlorine.

To account for the difference between the lattice energy and the heat of formation of sodium chloride, we must have numerical values for the energy required to convert solid sodium to gaseous sodium and to convert diatomic chlorine to monatomic chlorine. The conversion of a solid directly to the gaseous state is called sublimation (Section 11-3), and the energy input required to effect this conversion is called the *heat of sublimation*. The heat of sublimation is the heat of fusion plus the heat of vaporization. For sodium the value is 26 kcal per mole. The value for the heat of dissociation of diatomic chlorine is 58 kcal per mole, i.e., 29 kcal for $\frac{1}{2}$ mole.

We now have the necessary numerical information for calculating the heat of formation of crystalline sodium chloride from solid sodium and diatomic chlorine gas by means of a cycle devised in 1919 by Max Born and Fritz Haber in Germany. The heat of formation must be the sum of the energy changes accompanying the five imaginary steps in the following sequence:

1. Conversion of 1 mole of solid sodium to gaseous sodium: heat of sublimation = +26 kcal.
2. Conversion of 1 mole of gaseous sodium atoms to gaseous sodium ions: energy of ionization = +118 kcal.
3. Conversion of $\frac{1}{2}$ mole of gaseous diatomic chlorine to 1 mole of gaseous monatomic chlorine: heat of dissociation = +29 kcal.
4. Conversion of 1 mole of gaseous chlorine atoms to gaseous chloride ions: electron affinity = −87 kcal.
5. Conversion of 1 mole of gaseous Na^+ ions and 1 mole of gaseous Cl^- ions to 1 mole of crystalline $Na^+Cl^-(s)$: lattice energy = −184 kcal.

The heat of formation is therefore $26 + 118 + 29 - 87 - 184 = -98$ kcal per mole.

The Born-Haber cycle makes it possible to calculate electron affinities of elements. This operation requires that it be possible to compute reliable theoretical values for the lattice energies of ionic compounds.

Heats of formation and lattice energies for alkali-metal halides are tabulated in Table 22-3. Note that there is complete regularity in the variation of the lattice energies; smaller ions give rise to greater lattice energies without exception. A lithium halide always has a higher numerical value for its lattice energy than the corresponding halides of the other alkali metals, the order being Li > Na > K > Rb > Cs. The fluoride of an alkali metal has the largest lattice energy of the halides of that metal, the order being F > Cl > Br > I. With the heats

Table 22-3

Lattice energies and heats of formation of alkali-metal halides (kcal/mole)

	Fluoride		*Chloride*		*Bromide*		*Iodide*	
	Lattice energy	*Heat of formation*	*Lattice energy*	*Heat of formation*	*Lattice energy*	*Heat of formation*	*Lattice energy*	*Heat of formation*
Lithium	−243	−146	−202	−98	−191	−84	−178	−65
Sodium	−217	−136	−184	−98	−176	−86	−165	−69
Potassium	−192	−134	−168	−104	−161	−94	−152	−78
Rubidium	−185	−131	−163	−103	−157	−93	−148	−79
Cesium	−176	−127	−157	−104	−151	−94	−144	−81

of formation, there is not the same predictable variation from metal to metal. Although the lattice energy is an important factor governing the heat of formation, the other quantities in the Born-Haber cycle combine to place the heats of formation in different sequences than the lattice energies.

The tendency for reaction between a given alkali metal and a given halogen is correctly expressed in terms of the Gibbs free energy change, $\Delta G°$, rather than $\Delta H°$. However, the value for $\Delta G°$ is largely determined by the value for $\Delta H°$ in these reactions. Entropy effects are small, as can be seen by comparing the free energy of formation of an alkali halide with its heat (enthalpy) of formation. As an example, ΔG_f° for LiF is -140 kcal per mole, which differs by only 6 kcal per mole from its ΔH_f° value (Table 22-3). For CsI, ΔG_f° is -80 kcal per mole, or only 1 kcal per mole different from ΔH_f°.

22-8 Production and Uses of the Alkali Metals

Since the elements of Group IA are active metals that always occur in nature in the $+1$ oxidation state, production of the free metals requires reduction. Because oxidation of metals to their ions is easy, the ions do not readily accept electrons. Electrolysis is therefore the most commonly used method for the production of the metals. Sodium is manufactured by the electrolysis of fused sodium chloride to which calcium chloride has been added to lower the melting point; chlorine is a valuable coproduct. Sodium is also obtained by electrolysis of fused sodium hydroxide, which has a much lower melting point than the chloride. Potassium is produced by electrolysis of fused potassium hydroxide, and lithium is obtain by electrolysis of a fused mixture of lithium chloride and potassium chloride.

The most important use for sodium is to make a sodium-lead alloy required for the manufacture of tetraethyllead, $(C_2H_5)_4Pb$, which is used in large quantities in the "ethyl fluid" added to gasoline to serve as an "antiknock" agent. Sodium metal serves as a powerful reducing agent with both organic and inorganic applications, and it is used in the manufacture of a variety of sodium compounds, such as the peroxide and the cyanide.

Sodium vapor lamps are used for lighting highways because their yellow light penetrates fog well. The excellent thermal conductivity of sodium finds application in the manufacture of exhaust valves for engines used in airplanes and heavy trucks. The hollow steel valves are filled with sodium, which becomes liquid at the operating temperatures of the engines and helps to prevent overheating by transferring heat from the heads of the valves to the stems. Circulating liquid sodium also serves to transfer heat from the nuclear reactor to the steam boiler in nuclear-powered submarines and electrical generators.

Potassium and rubidium have very few uses; their potential uses can usually be served equally well by sodium, which is cheaper. Cesium is used in photoelectric cells because it loses electrons easily when struck by light. A potential future use for cesium is in the proposed *cesium ion engine* for space travel. In this device, atoms of cesium vapor are converted to ions, which are accelerated to very high speeds in an electric field. As the high velocity ions are ejected through a rocket nozzle (incidentally picking up electrons to prevent the spacecraft from accumulating a high negative charge), they exert a thrust on the spaceship. Cesium is the element under consideration for this use because of the ease with which it can be ionized and because of the heavy mass of its atoms.

Because of the small mass of lithium atoms, lithium hydride became important during World War II as an efficient portable source of hydrogen for filling balloons on location for military and meteorological observations. Hydrogen is generated when water comes into contact with the hydride:

$$LiH + HOH \rightarrow (Li^+ + OH^-) + H_2\uparrow$$

Lithium has become of considerable economic importance in recent years. Among the alkali metals it now ranks next to sodium in quantity of production. It is used as a minor component of various alloys; e.g., small amounts added to magnesium alloys make them harder and increase their tensile strength. Lithium aluminum hydride, $LiAlH_4$, is an important industrial chemical and is used in organic chemistry as a reducing agent; e.g., it is used in the production of alcohols by reduction of the corresponding organic acids and aldehydes. It is made by reaction of lithium hydride with aluminum chloride:

$$4LiH + AlCl_3 \rightarrow 3LiCl + LiAlH_4$$

A significant part of the lithium metal produced is used in the production of energy by nuclear fusion. The hydrogen in a so-called hydrogen bomb may be present as solid lithium deuteride.

22-9 Organometallic Compounds

In recent years rapid progress has been made in developing knowledge and uses of organometallic compounds. These compounds are defined as compounds possessing carbon-metal bonds. This definition excludes such substances as sodium acetate and sodium ethoxide, which contain both carbon and metal but do not contain carbon-to-metal bonds.

Organometallic compounds of lithium, magnesium, and aluminum are now commercially available and are used on a large scale in synthetic processes in the chemical industries. The reaction of the metal with an organic halide is a method used for the preparation of organometallics derived from active metals. For example, *n*-butyllithium is made by the reaction of lithium metal with *n*-butyl bromide:

$$n\text{-}C_4H_9Br + 2Li \rightarrow n\text{-}C_4H_9Li + LiBr$$

Butyllithium is used in large quantities as a polymerization initiator or catalyst in the production of synthetic rubber from isoprene (Section 18-21) and from butadiene (Section 18-17).

Commercially, the most important organometallic is tetraethyllead (Section 22-8), the gasoline additive that is made by the reaction of a sodium-lead alloy with ethyl chloride:

$$4C_2H_5Cl + Pb + 4Na \rightarrow (C_2H_5)_4Pb + 4NaCl$$

22-10 Differences Between Lithium and the Other Alkali Metals

The first member of an A group of the periodic table usually resembles the other members of the group less than the other members resemble one another. In accordance with this generalization, lithium has certain unique properties, many of which can be attributed to the small size of the lithium atom.

One of the illustrative "anomalies" of lithium is the high value of its oxidation potential (Table 22-1). There is a progressive increase in the oxidation potentials of the alkali metals from 2.71 volts for sodium to 3.02 volts for cesium; but lithium, instead of having a potential less than that of sodium, has a value equal to that of cesium. The small size of the lithium ion apparently enhances its tendency to become hydrated in aqueous solution. The amount of energy released in the hydration of the lithium ion is so much greater than the amount released in the hydration of the other alkali-metal ions that lithium, *in the presence of water*, is as powerful a reducing agent as cesium. Although ΔG°, rather than ΔH°, is directly related to the reducing power of a metal, ΔG° for the half-reaction involving conversion of a metal atom to the hydrated metal ion is largely determined by ΔH°.

Another generalization states that the first member of an A group bears considerable resemblance to the members of the next group and particularly to the second member of that group. In accordance with this, certain salts of lithium are sparingly soluble like the corresponding salts of most of the Group IIA metals. Lithium fluoride, lithium carbonate, and lithium phosphate are much less soluble than the corresponding salts of the other alkali metals. Lithium chloride is a deliquescent salt like magnesium chloride and calcium chloride. Lithium is the only alkali metal that reacts with nitrogen gas to form a nitride, Li_3N; the formation of nitrides, such as Mg_3N_2, is characteristic of metals of Group IIA. The Li^+ ion and the Mg^{++} ion are of very similar size.

The partially covalent character of LiCl is indicated by its solubility in organic solvents such as acetone and ethanol. The organometallic compounds containing lithium are predominantly covalent and soluble in organic solvents, but the corresponding compounds containing the other alkali metals are nonvolatile ionic solids.

22-11 Resemblance of Ammonium Ion to Potassium Ion

The reaction of ammonia with acids produces ammonium salts containing the ammonium ion:

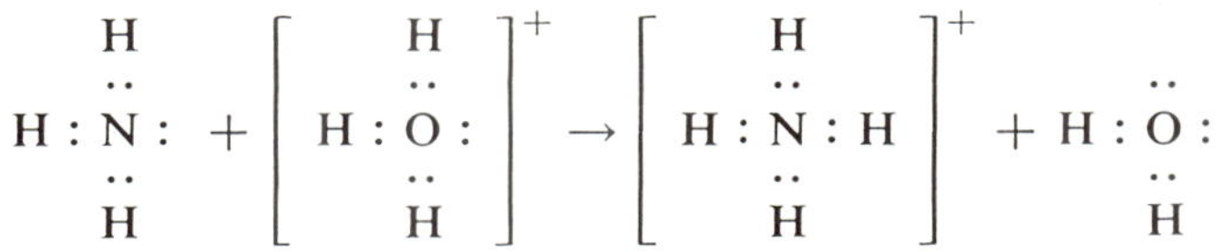

The ammonium ion is tetrahedral, like the methane molecule, and has the same charge and approximately the same size as the potassium ion. Chemically, the ammonium ion behaves very much like the potassium ion. The solubilities of ammonium salts closely parallel those of potassium salts. Like the K^+ ion, the NH_4^+ ion is precipitated from solution by the complex anions $Co(NO_2)_6^{---}$ and $PtCl_6^{--}$. In the absence of ammonium ions, these complex anions can be used to test for the presence of potassium ions. If ammonium ions are present they must be removed from a solution before the tests for the potassium ion are carried out. Such removal can be accomplished by adding a strong base and heating:

$$NH_4^+ + OH^- \rightarrow NH_3\uparrow + H_2O$$

New Terms

Alkalies: highly soluble, strong bases. The best known alkalies are NaOH and KOH.

Alkali metals: the elements of Group IA of the periodic table, the most active group of metals.

Born-Haber cycle: a hypothetical stepwise cycle for calculating heats of formation of ionic compounds.

Heat of sublimation: the quantity of heat absorbed during sublimation of a stated quantity (e.g., 1 g or 1 mole) of a solid without change of temperature.

Lattice energy: the energy released when gaseous ions form a crystal.

Organometallic compound: a compound possessing a metal-to-carbon bond.

Solvay process: a process for manufacturing sodium carbonate from sodium chloride, carbon dioxide, and ammonia.

Exercises

22-1 Why are the elements of Group IA called alkali metals?

22-2 If new and important uses should be developed for calcium chloride, what would you suggest as a cheap source?

22-3 Which alkali-metal halide has the largest lattice energy? Which has the smallest? How do these values correlate with the sizes of the ions?

22-4 What is a Born-Haber cycle?

22-5 Identify: soda ash, caustic potash, saltpeter, lithium aluminum hydride, and ethyllithium.

22-6 Describe the Solvay process. Why can $KHCO_3$ not be made by this process?

22-7 Compare potassium and ammonium salts.

22-8 A piece of sodium metal was placed in a beaker containing 200 ml of water. Titration of the resulting solution required 47.5 ml of 0.50 *M* HCl solution. What was the weight of the sodium metal?

References

Flood, W. E., *Dictionary of Chemical Names*, Philosophical Library, New York, 1963. Articles on potash, potassium, soda, sodium, alkali, etc.

Fulmer, J. Z., "Discovery of Potassium," *Chemistry*, **37** (8), 18 (1964). An account of the experimental work of Sir Humphry Davy and the background provided by Galvani, Volta, and others.

Rochow, E. G., *Organometallic Chemistry*, Reinhold, New York, 1964 (paperbound). A brief presentation of the chemistry of compounds containing carbon-metal bonds.

Verhoek, F. H., "What is a Metal?" *Chemistry*, **37** (11), 6 (1964). A discussion of metallic properties and the nature of the binding holding the atoms together within a metal.

Weeks, M. E., and Leicester, H. M., *Discovery of the Elements*, 6th ed., Chemical Education Publishing Co., Easton, Pa., 1956. Consult index for individual elements.

23

The Alkaline Earth Metals

23-1 Introduction

The word "earth" was formerly used in chemistry to denote oxides such as alumina, Al_2O_3, that are not easily reduced to give the free metal and are not easily dissolved in water. The oxides CaO, SrO, and BaO resemble typical earths in that they are hard to reduce, but they react with water to give alkaline solutions. For this reason these oxides became known as *alkaline earths*, and the elements of Group IIA are called the alkaline earth metals: beryllium, magnesium, calcium, strontium, barium, and radium.

Figure 23-1

Electronic configurations of the atoms of the alkaline earth metals

	1	2	3	4	5	6	7
$_4$Be	s^2	s^2					
$_{12}$Mg	s^2	s^2p^6	s^2				
$_{20}$Ca	s^2	s^2p^6	s^2p^6	s^2			
$_{38}$Sr	s^2	s^2p^6	$s^2p^6d^{10}$	s^2p^6	s^2		
$_{56}$Ba	s^2	s^2p^6	$s^2p^6d^{10}$	$s^2p^6d^{10}$	s^2p^6	s^2	
$_{88}$Ra	s^2	s^2p^6	$s^2p^6d^{10}$	$s^2p^6d^{10}f^{14}$	$s^2p^6d^{10}$	s^2p^6	s^2

The first alkaline earth known was lime or "lime earth," which was used by the Romans in the making of mortar. The Latin word for lime (CaO) is *calx*. Since lime is a crumbly powder, the alchemists came to refer to any friable material obtained by heating something as a calx (Section 1-2).

23-2 Properties of the Alkaline Earth Metals

All the elements of this group have only two electrons in the outer shells of their atoms (Figure 23-1) and all show the oxidation number +2 in their compounds. Consequently, the only oxidation-reduction reactions involving these elements are reactions in which the metals are oxidized to form compounds (as when they react with a nonmetal or an acid) and reactions in which the metals are liberated from their compounds by reduction.

Beryllium and magnesium have a tendency to form covalent as well as ionic compounds, but the other members of the group form ionic compounds almost exclusively. Since the atoms of these elements have noble-gas kernels, their ions are isoelectronic with noble-gas atoms.

Beryllium does not react with water, even at high temperatures; magnesium reacts only with boiling water and steam; the other metals of the group are very reactive with cold water:

$$Ca + 2HOH \rightarrow (Ca^{++} + 2OH^-) + H_2\uparrow$$

Although on the whole the alkaline earth metals are less reactive than the alkali metals, they are too reactive ever to be found free in nature. All but beryllium have relatively large atoms from which the outer-shell electrons are easily removed. Since the atomic size increases with increasing atomic number, the ease with which the atoms are converted to ions increases in the same order. This is reflected in the ionization energies and oxidation potentials of the elements (Table 23-1).

Table 23-1

Properties of the alkaline earth metals

Element	*Density at 20°C* (g/ml)	*Melting point* (°C)	*Boiling point* (°C)	*Ionization energies* (kcal/mole) 1st	2nd	3rd	*Standard oxidation potential* (volts)
$_4Be$	1.86	1283	1500	215	420	3548	1.70
$_{12}Mg$	1.74	650	1120	176	346	1848	2.37
$_{20}Ca$	1.55	850	1490	140	274	1181	2.87
$_{38}Sr$	2.54	770	1384	131	254	?	2.89
$_{56}Ba$	3.5	704	1638	120	200	?	2.90

Since the second ionization energy is considerably greater than the first, the nonexistence of Group IIA ions in the +1 oxidation state in aqueous solution requires explanation. The increase in the energy of hydration with ionic charge is responsible for this phenomenon. The exclusive occurrence of the +2 state in ionic crystals of Group IIA compounds is explained in an analogous manner. The gain in lattice energy obtained by forming a dipositive ion is more than sufficient to compensate for the additional energy required to remove the second electron from a Group IIA atom.

The fact that alkaline earth metals have oxidation potentials comparable to those of alkali metals in spite of much higher ionization energies is attributed to the greater energy of hydration of dipositive ions. Even though Na^+ ion and Ca^{++} ion have nearly the same size, the energy of hydration of calcium ion is nearly four times that of sodium ion.

The metals of this group have low densities, although on the average they are more dense than the alkali metals. Their melting points are much higher than those of the alkali metals.

The metals have a typically metallic appearance when their surfaces are clean, but beryllium and magnesium are the only ones that retain this appearance after exposure to air. The other metals become covered with white oxide. Beryllium and magnesium form a thin oxide film which protects them against further oxidation.

Barium is the only member of this group that forms a peroxide when heated in oxygen:

$$Ba + O_2 \rightarrow BaO_2$$

The other metals form "normal" oxides when heated (although beryllium must be finely powdered to do so).

$$2Ca + O_2 \rightarrow 2CaO$$

When heated in an atmosphere of nitrogen, the metals form nitrides (lithium is the only alkali metal that does this):

$$3Mg + N_2 \rightarrow Mg_3N_2$$

When heated in air, the metals form a mixture of oxide and nitride.

The metals react with sulfur to form sulfides and with halogens to form halides:

$$Mg + S \rightarrow MgS$$
$$Mg + Cl_2 \rightarrow MgCl_2$$

Calcium, strontium, and barium, when heated with hydrogen, form ionic hydrides, MH_2, which readily react with water, liberating hydrogen gas:

$$Ca + H_2 \rightarrow CaH_2$$
$$CaH_2 + 2HOH \rightarrow Ca(OH)_2 + 2H_2\uparrow$$

Magnesium combines with hydrogen only when heated at high pressures; beryllium hydride must be prepared by indirect means.

23-3 Occurrence

Calcium is the fifth most abundant element (Figure 1-3). Calcium carbonate is one of the most plentiful compounds and is next in abundance to silicate rocks in the earth's crust. Limestone is impure calcium carbonate that has been separated out from seawater by the action of minute marine organisms and deposited on the ocean bottom. Fossil remains of these organisms are often present in limestone, as are the remains of other marine life, e.g., shells of clams and stems of sea lilies. Marble was once limestone but has been converted by heat and pressure into a crystalline rock. Both marble and limestone are used extensively as building stone.

Calcium sulfate occurs in large amounts as the dihydrate, $CaSO_4 \cdot 2H_2O$, called gypsum. Other natural calcium compounds of importance are the fluoride, CaF_2, and the phosphate, $Ca_3(PO_4)_2$. Complex silicates of calcium are present in glass, cement, slag, and certain rocks. Most of the calcium in the human body is found in the bones, which are mainly calcium phosphate. Dairy products and many seafoods are sources of calcium in our diet.

The chloride and the sulfate of magnesium are very soluble in water; when present in groundwater they cause "permanent hardness," i.e., hardness that remains after the water has been heated. These salts are also present in seawater, which is now an important commercial source of magnesium. The salt beds at Stassfurt, Germany, contain carnallite, which is the hydrated double salt $KCl \cdot MgCl_2 \cdot 6H_2O$. Magnesium sulfate occurs as the hydrate, $MgSO_4 \cdot 7H_2O$. Geologically, one of the most important natural substances is dolomite, $MgCO_3 \cdot CaCO_3$. Dolomite and dolomitic limestones form whole mountain ranges. Complex silicates of magnesium include asbestos, soapstone, and meerschaum. In green plants magnesium occurs as the central atom in the molecule of the complex pigment known as chlorophyll, which acts as the light receptor for photosynthesis.

Barium occurs as $BaSO_4$ in barite (barytes, heavy spar), and strontium occurs as $SrSO_4$ in celestite. Carbonates of barium and strontium also occur in nature. Fallout from nuclear explosions in the atmosphere contains radioactive strontium-90 which is extremely dangerous because of its tendency to replace calcium in the bones.

Beryllium occurs in small amounts, chiefly in the form of silicates, of which beryl, $Be_3Al_2(SiO_3)_6$, found chiefly in Brazil, is the most important. Emeralds are particularly good crystals of beryl, prized as precious stones; their color is due to traces of chromium.

Radium occurs in extremely small amounts in all uranium minerals. Most of the world's radium comes from the minerals pitchblende and carnotite, found in the Congo, in the Great Bear Lake region of Canada, and in the Colorado-Utah area in the United States.

23-4 The Carbonates

Calcium carbonate is only slightly soluble in pure water but dissolves readily in solutions of acids:

$$CaCO_3 + 2H^+ \rightarrow Ca^{++} + H_2O + CO_2\uparrow$$

Thus, treatment of the solid with hydrochloric acid gives a solution containing calcium chloride, $CaCl_2$, and nitric acid gives calcium nitrate, $Ca(NO_3)_2$.

Calcium carbonate even dissolves in water containing dissolved carbon dioxide (i.e., carbonic acid, H_2CO_3) to produce the soluble calcium hydrogen carbonate, $Ca(HCO_3)_2$, otherwise referred to as calcium bicarbonate:

$$CaCO_3 + H_2O + CO_2 \rightarrow Ca^{++} + 2HCO_3^-$$

This reaction accounts for the formation of underground caverns in limestone regions. Water containing CO_2 dissolves the limestone with which it comes in contact; when this happens underground, a cavern forms. When water containing dissolved $Ca(HCO_3)_2$ seeps through cracks in the ceiling of such a cavern, CO_2 escapes from the solution, thus redepositing calcium carbonate as a stalactite growing downward:

$$Ca^{++} + 2HCO_3^- \rightarrow CaCO_3\downarrow + H_2O + CO_2\uparrow$$

Where water containing $Ca(HCO_3)_2$ drips onto the floor of the cave, the same reaction deposits calcium carbonate as a stalagmite growing upward. The water tends to drip at the greatest rate from the tips of the stalactites so that the stalagmites form directly beneath the stalactites; often they grow together and form a column extending from floor to ceiling.

Water containing calcium hydrogen carbonate readily precipitates calcium carbonate on the walls when it is heated in kettles and boilers. The precipitated $CaCO_3$ is a poor conductor of heat. This not only wastes fuel, but since the boiler metal must be heated to a high temperature to bring the water to the boiling point, the formation of a crack in the boiler scale will permit the water to make contact with the hot metal and form vast quantities of steam with dangerous rapidity. The danger is enhanced by the possibility of a reaction between the water and the hot metal, producing hydrogen gas.

Water containing $Ca(HCO_3)_2$ reacts with soluble soaps (Section 19-8) to form a precipitate of insoluble calcium soap, commonly referred to as soap curd:

$$\underset{\text{calcium ion}}{Ca^{++}} + \underset{\text{stearate ion}}{2C_{17}H_{35}COO^-} \rightarrow \underset{\text{calcium stearate}}{Ca(C_{17}H_{35}COO)_2\downarrow}$$

The insoluble calcium soap not only has no cleansing power but tends to adhere to fabrics, giving them an undesirable appearance. Water that behaves in this manner when soap is added is known as hard water (Section 12-10). The hardness may be due to any salt of calcium, magnesium, or iron(II), all of which are likely to be present in underground water. Hardness due to $Ca(HCO_3)_2$, $Mg(HCO_3)_2$, or $Fe(HCO_3)_2$ is known as "temporary hardness" because the water is no longer hard after it has been heated and the carbonates have precipitated.

The *limewater test* for carbon dioxide consists in the formation of a calcium carbonate precipitate when the gas is bubbled through limewater (calcium hydroxide solution):

$$(Ca^{++} + 2OH^-) + CO_2 \rightarrow CaCO_3 \downarrow + H_2O$$

Continued admittance of CO_2 causes the precipitate to disappear:

$$CaCO_3(s) + H_2O + CO_2 \rightarrow Ca^{++} + 2HCO_3^-$$

Similar results are obtained by bubbling CO_2 through solutions of $Sr(OH)_2$ or $Ba(OH)_2$; the initial carbonate precipitate eventually dissolves through the formation of soluble $Sr(HCO_3)_2$ or $Ba(HCO_3)_2$.

Although $MgCO_3$ occurs in nature, the precipitate obtained by adding, say, sodium carbonate to a solution of a magnesium salt is a basic carbonate having approximately the composition $3MgCO_3 \cdot Mg(OH)_2 \cdot 3H_2O$. This compound is used commercially in toothpaste and silver polish. Addition of sodium carbonate to a solution of a beryllium salt precipitates basic carbonates of variable composition, i.e., mixtures of $Be(OH)_2$ and $BeCO_3$.

Like most carbonates (except those of Group IA metals), the carbonates of the alkaline earth metals lose CO_2 when heated.

$$MgCO_3 \rightarrow MgO + CO_2 \uparrow$$

The temperatures at which the dissociation pressure of CO_2 reaches a value of 1 atmosphere vary as follows: $MgCO_3$, 540°C; $CaCO_3$, 900°C; $SrCO_3$, 1290°C; $BaCO_3$, 1360°C. Thus, as the basicities of the hydroxides increase (see Section 23-5), the stabilities of the carbonates also increase. It is in keeping with this trend that beryllium carbonate is very unstable and only can be made and kept in an atmosphere of carbon dioxide.

23-5 The Oxides and Hydroxides

The oxides of the alkaline earth metals have unusually high melting points (Table 23-2), and some of their uses depend upon this fact. Magnesia, MgO, mixed with clay is used extensively to make a refractory brick for furnace linings. Lime, CaO, emits an intense light when it is heated to a high temperature. (The part of a theater stage where the chief action occurred was once illuminated with the "limelight" produced by heating lime.)

Calcium oxide is manufactured by heating limestone in tall furnaces known as lime kilns.

$$CaCO_3 \rightarrow CaO + CO_2$$

Because of this method of manufacture it is sometimes called "burnt lime"; more often it is called quicklime (or caustic lime, since it produces blisters similar to those caused by burning). The term "quick" in quicklime means

Table 23-2

Melting points of oxides and chlorides of the alkaline earth metals

	Melting points of oxides (°C)	*Melting points of chlorides* (°C)
Beryllium	2580	440
Magnesium	2800	710
Calcium	2570	770
Strontium	2430	870
Barium	1920	960

"living" and refers to its being a "lively" substance that reacts vigorously with water:

$$CaO + H_2O \rightarrow Ca(OH)_2$$

This reaction is referred to as the slaking of lime because the quicklime is quenched or slaked by the water. The common name for calcium hydroxide is slaked lime or hydrated lime. The slaking of quicklime generates so much heat that it is likely to ignite paper or wood. (Dry slaked lime is usually sold in preference to unslaked lime so that it can be handled in paper bags.)

Calcium hydroxide is a strong base but not highly soluble. Because it is both active and cheap, it is used more extensively than any other base. Calcium oxide and calcium hydroxide are used in the manufacture of many important industrial chemicals; because of this fact limestone ranks as a very important chemical raw material along with air, water, coal, salt, sulfur, natural gas, and petroleum.

A saturated solution of calcium hydroxide is known as limewater; a suspension of the solid hydroxide in the saturated solution is known as milk of lime or whitewash. Mortar is a mixture of slaked lime, sand, and water. The first step in the "setting" of mortar is the loss of water by evaporation and by absorption into the bricks. The ultimate hardening is due to the action of the atmospheric carbon dioxide, producing insoluble calcium carbonate:

$$Ca(OH)_2 + CO_2 \rightarrow CaCO_3 + H_2O$$

The same reaction occurs in the limewater test for CO_2 (Section 23-4).

The oxides of strontium and barium resemble lime in reacting vigorously with water to produce the hydroxides. Barium hydroxide, $Ba(OH)_2$, is the most soluble of the Group IIA hydroxides and is the strongest base of the group.

Magnesium oxide reacts with water at high temperatures to form the slightly soluble hydroxide, $Mg(OH)_2$. Magnesium hydroxide is easily obtained as a precipitate by adding a base to a solution of a magnesium salt.

$$Mg^{++} + 2OH^- \rightarrow Mg(OH)_2 \downarrow$$

A suspension of magnesium hydroxide in water is familiar as milk of magnesia.

Unlike the ionic oxides formed by the other members of Group IIA, beryllium oxide is a covalent or network solid. It is very hard and refractory and does not react with water. Beryllium hydroxide is the least soluble of the Group IIA hydroxides; it is precipitated when solutions of beryllium salts, such as $BeSO_4$ or $Be(NO_3)_2$, are made alkaline:

$$Be^{++} + 2OH^- \rightarrow Be(OH)_2 \downarrow$$

It may be dissolved by treatment with either acid or alkali, i.e., it is an amphoteric hydroxide:

$$Be(OH)_2 + 2H^+ \rightleftharpoons Be^{++} + 2H_2O$$
$$Be(OH)_2 + 2OH^- \rightleftharpoons Be(OH)_4^{--} \rightleftharpoons BeO_2^{--} + 2H_2O$$

The $Be(OH)_4^{--}$ ion is the beryllate ion, and BeO_2^{--} is its "dehydrated" form. None of the other hydroxides of Group IIA metals are amphoteric; in this respect, as in many others, beryllium shows a resemblance to aluminum (the second member of Group IIIA).

When beryllium hydroxide is heated, it loses water and forms the oxide. The stability of Group IIA hydroxides with respect to this type of reaction increases with the atomic number of the metals. Barium hydroxide is the only one that can be heated until it melts without decomposing.

Barium oxide is the only oxide of Group IIA metals (except possibly radium oxide) that combines with additional oxygen to form a peroxide when it is heated in air:

$$2BaO + O_2 \rightleftharpoons 2BaO_2$$

However, peroxides of magnesium, calcium, and strontium may be obtained as precipitates by adding sodium peroxide, Na_2O_2, to solutions of their salts:

$$Mg^{++} + O_2^{--} \rightarrow MgO_2 \downarrow$$

All Group IIA peroxides decompose into the normal oxide and oxygen on being heated; barium peroxide decomposes least readily and magnesium peroxide most readily. Beryllium does not form a peroxide.

Before the development of liquid-air machines, pure oxygen was obtained by alternately heating barium oxide to 500°C in air and then heating the resulting peroxide to 800°C to decompose it into oxygen and the normal oxide. The normal oxide could then be brought into contact with air at 500°C, and the cycle could be repeated.

Barium peroxide was once an important source of hydrogen peroxide, which can be obtained from it by treatment with sulfuric acid:

$$BaO_2 + 2H^+ + SO_4^{--} \rightarrow BaSO_4 \downarrow + H_2O_2$$

Some hydrogen peroxide is still made by this method. Barium peroxide is also used to some extent as a bleach.

23-6 The Sulfates

Beryllium sulfate and magnesium sulfate are quite soluble in water, but the sulfates of the other Group IIA metals decrease in solubility from calcium to barium. The insolubility of barium sulfate, $BaSO_4$, is the basis of qualitative tests for both Ba^{++} and SO_4^{--} ions, as well as for their quantitative determination. Mixing a solution of a sulfate and a solution containing a barium salt precipitates both Ba^{++} and SO_4^{--} so completely that the amount of either ion originally present can be accurately calculated from the weight of the $BaSO_4$ precipitate. The barium sulfate remains insoluble in the presence of acids.

Calcium sulfate is found in nature as gypsum, $CaSO_4 \cdot 2H_2O$. Selenite is a transparent form of gypsum; satin spar is a fibrous, lustrous form; and alabaster, used in sculpture, is an opaque form that has an appearance similar to marble, although it is much softer. Even though calcium sulfate is not highly soluble, it is the cause of most of the noncarbonate hardness (permanent hardness) of natural waters.

When gypsum is mildly heated, loss of water occurs:

$$2(CaSO_4 \cdot 2H_2O) \rightleftharpoons (CaSO_4)_2 \cdot H_2O + 3H_2O$$

The product, calcium sulfate hemihydrate, sometimes represented by the formula $CaSO_4 \cdot \frac{1}{2}H_2O$, is called plaster of Paris because it was once prepared from gypsum obtained from Montmartre, Paris. When finely ground plaster of Paris is mixed with water the reaction reverses and a solid mass of interlocking gypsum crystals is formed within a few minutes. Plaster of Paris is used in medicine for keeping parts of the body strictly rigid for extended periods of time.

Calcium sulfate is the chief component of so-called writing chalk. (True chalk, such as occurs in the white cliffs of Dover, is a soft limestone, made up of shells of tiny marine orgainsms.)

Barium sulfate is used in paper-making to impart the desired weight and opacity. In medicine it is given to patients prior to X-ray examination of the gastrointestinal tract. The high atomic number of barium causes strong absorption of X rays, thus giving a clear outline to the stomach and intestines in the picture. All soluble barium salts are highly toxic, but the insolubility of $BaSO_4$ makes it nonpoisonous.

A mixture of barium sulfate and zinc sulfide is used as a white pigment, called lithopone, in paints.

Magnesium sulfate has long been used in medicine as a powerful purgative under the name bitter salts. It is also known as Epsom salts because the source of supply was once the spring water at the Epsom spa in England.

23-7 Production and Uses of the Metals

Seawater is now the most important source of magnesium metal. Bromine and magnesium are both extracted in a large plant operating at Freeport, Texas. Seawater from which bromine has already been extracted is treated with slaked lime to precipitate magnesium hydroxide, which is filtered off and dissolved in hydrochloric acid:

$$Mg^{++} + Ca(OH)_2 \rightarrow Mg(OH)_2 \downarrow + Ca^{++}$$
$$Mg(OH)_2 + 2(H^+Cl^-) \rightarrow Mg^{++}(Cl^-)_2 + 2H_2O$$

The $MgCl_2$ is crystallized from the resulting solution and electrolyzed in the molten state, producing magnesium metal at the cathode and chlorine gas at the anode:

$$\text{CATHODE:} \quad Mg^{++} + 2e^- \rightarrow Mg$$
$$\text{ANODE:} \quad 2Cl^- \rightarrow Cl_2 + 2e^-$$

The most important use for magnesium metal is in making lightweight alloys for the construction of airplanes, speedboats, and automobile parts. In wartime it is used in tracer bullets. Large magnesium plates are attached to the hulls of steel ships to reduce corrosion in seawater. The combination of magnesium and steel acts as a short-circuited galvanic cell so that the magnesium suffers corrosion and thus protects the steel. Underground pipelines can be protected in a similar manner by connecting them to pieces of magnesium buried every mile or so.

Calcium metal is prepared by electrolysis of fused calcium chloride obtained as a by-product in the Solvay soda process. It serves as a hardening agent for the lead used in storage batteries and cable coverings. It is also used in metallurgy as a scavenger; i.e., it is used to remove gases from liquid metals.

Although several methods are used for the production of strontium and barium metals, the most important method consists in the electrolysis of their molten chlorides. Barium is used as a thin film on the glass inside of vacuum tubes to serve as a "getter"; i.e., it is used to pick up any gases present in the tubes. Strontium is far less plentiful than either calcium or barium, and the metal has no commercial importance.

Very little metallic radium has ever been made because its use as a radioactive material, e.g., in the treatment of malignant growths, offers no advantage over its compounds, which are cheaper.

Beryllium is obtained by electrolysis of a fused mixture of $BeCl_2$ and NaCl. It has become an important metal mainly because it forms a hard, flexible alloy with copper. Heavy springs made of this alloy can endure repeated flexing over long periods of time. Small quantities of beryllium metal are used as windows in X-ray tubes. Because of its low atomic number beryllium absorbs X rays the least of any solid that is stable in air. Windows made of a special glass, rather than metallic beryllium, have the advantage of making the inside of the X-ray tube visible for inspection. The special glass is made by replacing the sodium and calcium in ordinary glass with lithium and beryllium.

Special health precautions are required in the beryllium industry to avoid the dust of the metal or any of its compounds. All beryllium compounds are poisonous.

23-8 Organometallic Compounds

Organoberyllium compounds are covalent, but the organometallic compounds of the other metals of Group IIA are highly ionic. The only ones of importance are organomagnesium compounds.

In synthetic organic chemical laboratories and industrial plants the most widely used organometallic compounds are the Grignard reagents, discovered by the French chemist Victor Grignard in 1900. These reagents are prepared by reacting an alkyl halide (RX, where X is a halogen) with magnesium turnings:

$$RX + Mg \rightarrow RMgX$$

They are used in the synthesis of various organic compounds, including other organometallics. For example, a tertiary alcohol of a prescribed structure can be obtained by hydrolysis of the addition compound obtained by reacting the appropriate Grignard reagent with the properly chosen ketone:

$$R_2CO + R'MgX \rightarrow \underset{\displaystyle \text{OMgX}}{R_2\overset{}{C}R'} \xrightarrow{H_2O} \underset{\displaystyle \text{OH}}{R_2CR'} + Mg(OH)X$$

Many organometallics can be obtained most readily by reacting the appropriate Grignard reagents with the halides of the metals; e.g., triethylaluminum can be made by reacting aluminum chloride with ethylmagnesium chloride:

$$AlCl_3 + 3C_2H_5MgCl \rightarrow (C_2H_5)_3Al + 3MgCl_2$$

23-9 Resemblances Between Beryllium and Aluminum

Most simple beryllium salts are mainly covalent, but it is possible to estimate the ionic radius for Be^{++} at about 0.31 Å. Thus, the charge/radius ratio for beryllium ion is about 6.5, which is greater than that for any other cation. The charge/radius ratio for the aluminum ion, Al^{+++}, is 6.0, or nearly as great. Accordingly, there are strong chemical similarities between beryllium and aluminum. These two elements provide an excellent example of the "diagonal relationship" which results in elements of the first row of the periodic table having certain resemblances to the second members of the next groups.

Beryllium resembles aluminum more than lithium resembles magnesium. For example, the metal is rather resistant to acids but dissolves in bases; and the oxide is a hard solid with a high melting point.

The tendency of beryllium to form covalent bonds far exceeds that of the other Group IIA metals. This is a direct consequence of the fact that the beryllium atom is small and has only two electrons in its penultimate shell.

New Terms

Alkaline earths: oxides of Group IIA metals: BeO, beryllia (beryl earth); MgO, magnesia (bitter earth); CaO, quicklime (lime earth); SrO, strontia; BaO, baryta.

Alkaline earth metals: the elements of Group IIA of the periodic table (beryllium, magnesium, calcium, strontium, barium, and radium).

Earths: metallic oxides that are difficult to reduce, once thought to be elements. Examples: Al_2O_3, alumina; Y_2O_3, yttria.

Grignard reagents: compounds of the general formula RMgX, where R is an alkyl group and X is a halogen atom, used as synthetic reagents for preparing a great variety of organic compounds.

Exercises

23-1 Why are the elements of Group IIA called alkaline earth metals?

23-2 Identify the following: plaster of Paris, whitewash, limewater, milk of magnesia, Epsom salts, quicklime, slaked lime, gypsum.

23-3 List the eight most important chemical raw materials.

23-4 How many pounds of limestone, 82% pure, would have to be heated to produce 450 lb of quicklime?

23-5 What is the difference between temporary hardness and permanent hardness of water? What causes each?

23-6 Discuss applications of the high absorption of X rays by elements of high atomic number and their low absorption by elements of low atomic number.

23-7 Compare beryllium with aluminum and magnesium with lithium.

23-8 The recommended emergency antidote for barium compounds that have been swallowed accidentally is to drink sodium sulfate solution. Explain.

References

Moody, B. J., *Comparative Inorganic Chemistry*, American Elsevier, New York and London, 1965. Chapter 16 is devoted to alkaline earth metals. Industrial processes are included.

Weeks, M. E., and Leicester, H. M., *Discovery of the Elements*, 6th ed., Chemical Education Publishing Co., Easton, Pa., 1956. Calcium, barium, strontium, and magnesium are discussed in Chapter 20, beryllium is discussed in Chapter 22, and radium in Chapter 29.

24

The Halogens and the Halides

24-1 Introduction

Group VIIA of the periodic table consists of five elements that have the s^2p^5 configuration in the outermost shell of each atom: fluorine, chlorine, bromine, iodine, and astatine (Figure 24-1). The atoms of these elements easily acquire the s^2p^6 configuration of the noble-gas atoms. They do this by gaining an additional electron and forming a mononegative ion or by establishing a covalent bond with another atom.

$$:\underset{\cdot\cdot}{\overset{\cdot\cdot}{\mathrm{Cl}}}\cdot + e^- \rightarrow \left[:\underset{\cdot\cdot}{\overset{\cdot\cdot}{\mathrm{Cl}}}:\right]^-$$

$$:\underset{\cdot\cdot}{\overset{\cdot\cdot}{\mathrm{Cl}}}\cdot + \cdot\underset{\cdot\cdot}{\overset{\cdot\cdot}{\mathrm{Cl}}}: \rightarrow \ :\underset{\cdot\cdot}{\overset{\cdot\cdot}{\mathrm{Cl}}}:\underset{\cdot\cdot}{\overset{\cdot\cdot}{\mathrm{Cl}}}:$$

$$\mathrm{H}\cdot + \cdot\underset{\cdot\cdot}{\overset{\cdot\cdot}{\mathrm{Cl}}}: \rightarrow \ \mathrm{H}:\underset{\cdot\cdot}{\overset{\cdot\cdot}{\mathrm{Cl}}}:$$

This is the most active family of nonmetals, and fluorine is the most active of all nonmetallic elements. None of the elements are found free in nature, but all occur as salts containing the mononegative ions. The elements of the family are known as *halogens* (Greek, sea-salt formers) in reference to the fact that chlorine, bromine, and iodine occur in the sea as salts.

24-2 Physical Properties

In the elementary state the halogens exist as diatomic molecules, $:\underset{\cdot\cdot}{\overset{\cdot\cdot}{\mathrm{X}}}:\underset{\cdot\cdot}{\overset{\cdot\cdot}{\mathrm{X}}}:$, in which the two atoms are bonded together by a shared pair of electrons. The distance between the centers of the atoms increases with increasing atomic

Figure 24-1

Electronic configurations of the halogen atoms

	1	2	3	4	5	6
$_{9}F$	s^2	s^2p^5				
$_{17}Cl$	s^2	s^2p^6	s^2p^5			
$_{35}Br$	s^2	s^2p^6	$s^2p^6d^{10}$	s^2p^5		
$_{53}I$	s^2	s^2p^6	$s^2p^6d^{10}$	$s^2p^6d^{10}$	s^2p^5	
$_{85}At$	s^2	s^2p^6	$s^2p^6d^{10}$	$s^2p^6d^{10}f^{14}$	$s^2p^6d^{10}$	s^2p^5

number. The halogens show great similarity in their physical and chemical properties, but all their properties vary in regular order according to atomic number. The color deepens, the density increases, and the melting points and boiling points rise in the order of increasing atomic numbers (Table 24-1). Fluorine, F_2, is a pale-yellow gas that is about 1.3 times as dense as air. Chlorine, Cl_2, is a greenish-yellow gas that has a density about 2.5 times that of air. Bromine, Br_2, has the distinction of being the only nonmetal that is liquid under ordinary conditions of temperature and pressure. It is a reddish-brown liquid with a high vapor pressure; the red vapor can easily be seen in a bottle that is only partly filled with liquid bromine. Iodine, I_2, is a shiny purple solid so dark that it looks almost black; the crystals have a high vapor pressure, and if they are carefully heated they change into a purple vapor without melting. This makes it convenient to purify iodine by sublimation, i.e., it is vaporized directly from the solid state by heating, and the pure vapor is condensed by cooling.

Table 24-1

Properties of the halogens

Element	*Melting point* (°C)	*Boiling point* (°C)	*Ionization energy* (kcal/mole)	*Electro-negativity* (Pauling scale)	*Standard oxidation potential* *Half-reaction*	*E*° (volts)
$_{9}F$	−218	−188	402	4.0	$2F^- \rightleftharpoons F_2 + 2e^-$	−2.87
$_{17}Cl$	−101	−34	300	3.0	$2Cl^- \rightleftharpoons Cl_2 + 2e^-$	−1.36
$_{35}Br$	−7	59	273	2.8	$2Br^- \rightleftharpoons Br_2 + 2e^-$	−1.07
$_{53}I$	114	184	241	2.5	$2I^- \rightleftharpoons I_2 + 2e^-$	−0.54

Bromine is soluble in such common solvents as carbon tetrachloride, chloroform, alcohol, and ether. If the solutions are dilute they are yellow; if they are concentrated they are brown. Iodine is soluble in the same solvents, but the colors of its solutions vary with the solvent. With nonpolar solvents, such as carbon tetrachloride, iodine forms violet solutions; with polar solvents, such as alcohol, it forms brown solutions. The solutions formed with nonpolar solvents have the same color as iodine vapor because the iodine is present as I_2 molecules in both cases. The brown color of the solutions formed with polar solvents is due to chemical association of the I_2 molecules with molecules of solvent.

24-3 Chemical Properties

The solubility of chlorine in water at 20°C and 1 atmosphere pressure is 0.09 mole per liter; bromine, 0.21 mole per liter; iodine, 0.0013 mole per liter. Fluorine reacts completely with water, liberating oxygen:

$$2F_2 + 2H_2O \rightarrow 4HF + O_2$$

The yellowish-green solution of chlorine in water is known as "chlorine water." Some of the dissolved chlorine reacts with water to form an equilibrium mixture containing hydrochloric acid, a strong acid, and hypochlorous acid, a weak acid:

$$Cl_2 + H_2O \rightleftharpoons (H^+ + Cl^-) + HClO$$

The hypochlorous acid is a strong oxidizing agent and is responsible for the bleaching and germicidal action of chlorine in the presence of water. Bromine reacts only slightly with water. Iodine hardly reacts at all with water; on the contrary, atmospheric oxygen slowly liberates iodine from solutions of iodides.

Large quantities of iodine will dissolve in aqueous solutions of iodides because of the reversible combination of I_2 molecules and iodide ions to form complex I_3^- ions:

$$I^- + I_2 \rightleftharpoons I_3^-$$

All of the halogens react with hydrogen to form hydrogen halides, and they combine directly with most metals to form metal halides. Fluorine and chlorine are much more reactive than bromine or iodine toward both metals and nonmetals. Fluorine combines directly with all nonmetals except nitrogen and the lightest noble gases; fluorides of nitrogen can be obtained by indirect means. Chlorine combines directly with all nonmetals except carbon, nitrogen, oxygen, and the noble gases; oxides of chlorine and chlorides of carbon and nitrogen can be obtained by indirect means. Bromine and iodine combine directly with only certain nonmetals.

The halogens combine directly with certain covalent compounds, as in the following examples:

$$SO_2 + Cl_2 \rightarrow SO_2Cl_2$$
sulfur dioxide → sulfuryl chloride

$$C_2H_4 + Cl_2 \rightarrow C_2H_4Cl_2$$
ethylene → ethylene chloride

$$CO + Cl_2 \rightarrow COCl_2$$
carbon monoxide → carbonyl chloride ("phosgene")

In general, the electron-attracting powers of the halogen atoms decrease with increasing atomic number. Chlorine can be demonstrated to be more active than bromine or iodine by the liberation of bromine upon adding chlorine water to a solution of a bromide and the formation of elementary iodine upon adding chlorine water to a solution of an iodide. Likewise, bromine is more active than iodine, as can be seen by the liberation of iodine upon adding bromine water to a solution of an iodide:

$$Cl_2 + 2Br^- \rightarrow 2Cl^- + Br_2$$
$$Cl_2 + 2I^- \rightarrow 2Cl^- + I_2$$
$$Br_2 + 2I^- \rightarrow 2Br^- + I_2$$

These are oxidation-reduction reactions. The fact that the iodide ion is a stronger reducing agent (i.e., more easily oxidized) than the bromide ion is indicated by the fact that excess hydrogen iodide reduces sulfuric acid to hydrogen sulfide while excess hydrogen bromide reduces it only to sulfur dioxide:

$$8HI + H_2SO_4 \rightarrow H_2S + 4I_2 + 4H_2O$$
$$2HBr + H_2SO_4 \rightarrow SO_2 + Br_2 + 2H_2O$$

Neither HCl nor HF is able to reduce sulfuric acid.

24-4 Reaction of the Halogens with Hydrogen

All the halogens react with hydrogen to form hydrogen halides. The formation of hydrogen fluoride from its elements is accompanied by the release of 64 kcal of energy per mole. The reaction of fluorine with hydrogen occurs with explosive violence, even at $-253°C$, the boiling point of hydrogen, where fluorine is solid. A jet of fluorine can be burned in an atmosphere of hydrogen:

$$H_2 + F_2 \rightarrow 2HF$$

The reaction of hydrogen with chlorine is highly exothermic, although substantially less so than the hydrogen-fluorine reaction, and a jet of either gas will burn in an atmosphere of the other. Since hydrogen and chlorine are available as products of the electrolysis of aqueous sodium chloride, one commercial method for producing hydrogen chloride is the combustion of hydrogen with chlorine in special burners designed for this purpose.

Exposure of a mixture of hydrogen and chlorine to sunlight results in an explosion. Chemical reactions brought about by the effect of light are called *photochemical reactions.* The absorption of a photon (a quantum of light energy) by a Cl_2 molecule is the first step in a succession of rapid events that constitute a chain reaction. The absorption of the photon causes the Cl_2 molecule to dissociate into two *activated,* i.e., energy-rich, chlorine atoms:

$$Cl_2 + \text{energy} \rightarrow 2Cl$$

Either of the chlorine atoms can then react with an H_2 molecule to form a molecule of hydrogen chloride and a hydrogen atom:

$$Cl + H_2 \rightarrow HCl + H$$

The hydrogen atom can then react with a Cl_2 molecule to form another molecule of hydrogen chloride and a new chlorine atom:

$$H + Cl_2 \rightarrow HCl + Cl$$

The new chlorine atom, in turn, can react with another H_2 molecule, and thus a *chain reaction* that can result in the formation of a very large number of HCl molecules is established. The chain will not be terminated until an atom combines with another atom. For example,

$$H + H \rightarrow H_2$$

or

$$Cl + Cl \rightarrow Cl_2$$

or

$$H + Cl \rightarrow HCl$$

The reactions that terminate the chain do not occur with great frequency because fewer atoms are present at any moment than H_2 or Cl_2 molecules. The probability of a collision between an atom and an H_2 or Cl_2 molecule is much greater than the chance that two atoms will collide.

The energy released when hydrogen combines with bromine is less than when it combines with chlorine, and the formation of hydrogen iodide from the elements is an endothermic reaction. Conversely, hydrogen iodide is most easily decomposed into its elements by heating, and hydrogen fluoride is the most stable hydrogen halide with respect to thermal decomposition.

24-5 The Hydrogen Halides

The hydrogen halides are colorless gases and have sharp, penetrating odors. Hydrogen fluoride becomes liquid at 20°C; the other hydrogen halides must be cooled to subzero temperatures to liquefy (Table 24-2). All are polar covalent compounds, the polarity being greatest for HF and least for HI. Because of the polarity of the molecules, all the hydrogen halides are highly soluble in water; and because of their solubility, the gases fume in moist air. In dilute aqueous solutions, the chloride, bromide, and iodide ionize completely so that the solutions are excellent conductors of electric current:

Table 24-2

Properties of hydrogen halides

	Melting point (°C)	*Boiling point* (°C)	*Heat of fusion* (kcal/mole)	*Heat of vaporization* (kcal/mole)	*Heat of formation* (kcal/mole)	*Solubility in water* (g/100 g solution at 20°C and 1 atm)
HF	−83	20	1.09	7.24	−64.0	35
HCl	−115	−85	0.476	3.85	−22.0	42
HBr	−87	−67	0.575	4.21	−13.5	49
HI	−51	−35	0.686	4.72	0.8	57

$$HX + H_2O \rightarrow H_3O^+ + X^-$$

The ionization of hydrogen fluoride is very incomplete.

The aqueous solutions of the hydrogen halides are indicated by the names hydrofluoric acid, hydrochloric acid, hydrobromic acid, and hydriodic acid. The salts of the hydrohalic acids are called fluorides, chlorides, bromides, and iodides. (An old name for hydrochloric acid is muriatic acid; its salts are sometimes called muriates in commerce.)

The hydrohalic acids neutralize bases to form salts and water; the same salts are formed when the acids react with metals that are above hydrogen in the activity series.

$$Mg(OH)_2 + 2(H^+ + Cl^-) \rightarrow (Mg^{++} + 2Cl^-) + 2H_2O$$
$$Mg + 2(H^+ + Cl^-) \rightarrow (Mg^{++} + 2Cl^-) + H_2\uparrow$$

In the latter reaction, hydrogen ion is the oxidizing agent. The anion of a hydrohalic acid often acts as a reducing agent, as in the following examples:

$$2(H^+ + Br^-) + Cl_2 \rightarrow Br_2 + 2(H^+ + Cl^-)$$
$$14(H^+ + Cl^-) + 2(H^+ + MnO_4^-) \rightarrow 2(Mn^{++} + 2Cl^-) + 5Cl_2\uparrow + 8H_2O$$

In another common reaction, the anion of a hydrohalic acid acts as a precipitating agent:

$$(Ag^+ + NO_3^-) + (H^+ + Cl^-) \rightarrow AgCl\downarrow + (H^+ + NO_3^-)$$

24-6 Hydrogen Fluoride

Hydrogen fluoride is unique among the hydrogen halides: its molecules are associated by hydrogen bonding, it attacks sand and glass, and in its aqueous solution it behaves as a weak acid (ionization constant, 6.7×10^{-4}).

In the solid state the structure of hydrogen fluoride differs from that of the other hydrogen halides. It consists of zigzag chains of HF molecules held together by hydrogen bonding. As a result, its melting point is higher than that of HCl or HBr (in spite of its smaller "molecular weight") because hydrogen bonds must be broken to convert the solid to a liquid. A substantial amount of hydrogen bonding is still present in the liquid state; this accounts for the boiling point of HF being higher than the boiling points of the other hydrogen halides. Even in the vapor state there is still some hydrogen bonding, so that the density of the vapor does not correspond to the formula HF until the temperature has reached 88°C.

The weakness of hydrofluoric acid is related to the fact that the bond between hydrogen and fluorine is more difficult to break than is any other H—X bond (X represents a halogen atom). This can be attributed to the short interatomic distance (center to center), which is only 1.0 Å in HF, and to the high electronegativity of fluorine. The ionization of HF in dilute solutions is complicated by the formation of the HF_2^- ion by hydrogen bonding of fluoride ion to HF:

$$F^- + HF \rightarrow HF_2^-$$

Acid fluorides of the alkali metals can be prepared by mixing the constituents in appropriate proportions; an example is $KF \cdot HF$ (melting point 239°C), often written KHF_2.

The reaction of HF with sand is exothermic and one of the products, silicon tetrafluoride, escapes as a gas:

$$SiO_2 + 4HF \rightarrow SiF_4\uparrow + 2H_2O$$

Electric light bulbs and various other glass objects are "frosted" or "etched" by the action of hydrofluoric acid. Glass is a mixture of silicates, chiefly sodium and calcium. The reaction of HF with calcium silicate is given by the equation

$$CaSiO_3 + 6HF \rightarrow CaF_2 + SiF_4 + 3H_2O$$

Thermometers, burets, and other items of laboratory glassware are marked by etching with HF. The glass is first covered with paraffin, and then a sharp instrument is used to remove the paraffin from the glass at those places where the markings are to be made by the HF. Since hydrofluoric acid cannot be stored in glass containers, special bottles are made for this purpose from wax and certain plastics such as polyethylene.

24-7 Preparation and Uses of Hydrogen Halides

Hydrogen fluoride is prepared by heating the mineral fluorspar, CaF_2, with concentrated sulfuric acid:

$$CaF_2 + H_2SO_4 \rightarrow CaSO_4 + 2HF\uparrow$$

Hydrogen chloride is likewise prepared by heating sodium chloride with concentrated sulfuric acid:

$$NaCl + H_2SO_4 \rightarrow NaHSO_4 + HCl\uparrow$$

The preparation of a hydrogen halide by the action of concentrated sulfuric acid upon a metallic halide is possible because the hydrogen halides are gases at temperatures at which sulfuric acid is still liquid. Since HBr and HI react with sulfuric acid to give the elementary halogens, their preparation requires substitution of another acid that lacks the oxidizing power of sulfuric acid. Phosphoric acid is such an acid, and it has the required high boiling point:

$$NaBr + H_3PO_4 \rightarrow NaH_2PO_4 + HBr \uparrow$$
$$NaI + H_3PO_4 \rightarrow NaH_2PO_4 + HI \uparrow$$

Hydrogen chloride is the only hydrogen halide prepared commercially by direct union of the elements. It is obtained in large quantities as a coproduct of the chlorination of hydrocarbons, e.g., in the manufacture of ethyl chloride from ethane:

$$C_2H_6 + Cl_2 \rightarrow C_2H_5Cl + HCl$$

Next to sulfuric acid, hydrochloric acid is the most important acid in commerce. It is used for cleaning metals and for making many chemicals. Hydrogen fluoride is used for etching glass, as an important catalyst in the synthesis of gasoline hydrocarbons, and in making fluorides of metals which are finding such important uses as in the production of organic fluorine compounds.

24-8 Halides of Metals and Nonmetals

Halides of the most metallic elements, e.g., the alkali metals and the heavier alkaline earth metals, are ionic; those formed by nonmetallic elements and the less active metals are polar covalent. Thus, NaCl is an excellent example of an ionic compound with its high melting point, high boiling point, and good electrical conductivity both in the fused state and in aqueous solution. On the other hand, PCl_3 is a good example of a polar covalent compound: a volatile liquid unstable in the presence of water. (The C—Cl bonds in carbon tetrachloride are polar covalent, but the molecule is nonpolar because of its tetrahedral symmetry.)

If a metal forms ions that have a large charge/radius ratio, it tends to form halides that are predominantly covalent because of the strong attraction of such ions for the electrons of halide ions. Examples are $BeCl_2$, Al_2Cl_6, and $SnCl_4$. Such halides are volatile, soluble in nonpolar solvents, and conduct electric current poorly if at all in the liquid state; they undergo extensive hydrolysis in water:

$$SnCl_4 + 2H_2O \rightarrow SnO_2 + 4(H^+ + Cl^-)$$

In the case of aluminum chloride, the hydrated salt, $AlCl_3 \cdot 6H_2O$, may be obtained by evaporation of the aqueous solution. The solid hydrate is apparently ionic, consisting of chloride anions and hydrated aluminum cations, $Al(H_2O)_6^{+++} + 3Cl^-$. When the hydrate is heated, the anhydrous aluminum chloride is not obtained; instead, the oxide and hydrogen chloride are formed:

$$2[Al(H_2O)_6{}^{+++} + 3Cl^-] \rightarrow Al_2O_3(s) + 6HCl\uparrow + 9H_2O\uparrow$$

Iron(III) chloride, $FeCl_3$, is an example of a metal halide intermediate between typically ionic and typically covalent halides in its properties; it is volatile and easily hydrolyzable, but in the liquid state it is a good conductor.

Many nonmetal halides are so easily hydrolyzed that they must be kept out of contact with moist air. As an example, a good laboratory method for preparing HBr consists in hydrolyzing phosphorus(III) bromide:

$$PBr_3 + 3H_2O \rightarrow 3HBr\uparrow + H_3PO_3$$

In their fluorides, a number of elements exist in a higher oxidation state than in their other halides. Examples include AgF_2, AsF_5, SF_6, and IF_7. Fluorides of metals are usually more ionic than chlorides; e.g., AlF_3 is ionic, Al_2Cl_6 is covalent, SnF_4 is ionic, and $SnCl_4$ is covalent.

The compounds formed by the union of two different halogens are listed in Table 24-3. The diatomic interhalogen compounds have many properties intermediate between the properties of the constituent halogens. This can be illustrated by comparing the boiling points of F_2 (−188°C), ClF (−100°C), and Cl_2 (−34°C).

The composition of many molecules is determined by the relative sizes of the component atoms. The halogen atoms increase in size from fluorine to iodine. A small central atom cannot be bonded simultaneously to three large atoms. The chlorine atom is large enough to form a trifluoride, but you will note in Table 24-3 that there is no $ClBr_3$ or ClI_3.

Table 24-3

Interhalogen compounds

Type	*Formula*	*Description*	*Melting point (°C)*	*Boiling point (°C)*
AB	IBr	Gray-violet solid	42	116
	ICl	Red solid	27	?
	BrCl	Unstable		
	BrF	Unstable		
	ClF	Colorless gas	−156	−100
AB_3	ICl_3	Orange solid	Sublimes	
	BrF_3	Greenish-yellow liquid	8.8	127
	ClF_3	Colorless gas	−83	12
AB_5	IF_5	Colorless liquid	9.6	98
	BrF_5	Colorless liquid	−61	40.5
AB_7	IF_7	Colorless gas	4.5	5.5

Numerous metal halides are important in commerce. Sodium chloride (table salt) is an essential constituent of food. (The word salary is derived from *sal*, the Latin word for salt, as a result of the fact that the Roman soldiers were sometimes paid with salt.) Sodium chloride is the industrial source of most other sodium and chlorine compounds. Crude potassium chloride is used as a fertilizer and is the source of all other potassium compounds. Calcium fluoride has extensive use as a metallurgical flux. Various bromides, such as NaBr and KBr, are used in medicine as sedatives. Silver bromide is the essential light-sensitive compound in photographic emulsions.

Important nonmetal halides include the solvents CCl_4 and S_2Cl_2; BF_3, used as a catalyst for many different types of organic reactions, including the synthesis of gasoline hydrocarbons; and sulfur hexafluoride, SF_6, used as an insulating gas for high-voltage equipment. Organic compounds containing fluorine and chlorine are considered in Chapter 25.

24-9 Production and Uses of Elementary Chlorine

Chlorine is the most abundant element of the halogen family. The three commonest chlorides are NaCl, KCl, and $MgCl_2$. Sodium chloride is by far the most abundant of these; it occurs in seawater and in salt beds formed by the evaporation of ancient seas and lakes. The most important commercial source of elementary chlorine is the electrolysis of aqueous sodium chloride. Sodium hydroxide and hydrogen gas are coproducts of the process. On the other hand, chlorine is a coproduct of the manufacture of sodium, magnesium, and other active metals by electrolysis of their fused chlorides. In the laboratory it is more convenient to prepare chlorine by oxidation of the chloride ion with chemical oxidizing agents; e.g., HCl is oxidized by manganese dioxide:

$$MnO_2 + 4H^+ + 2Cl^- \rightarrow Cl_2\uparrow + Mn^{++} + 2H_2O$$

Chlorine is now used on such an extensive scale that it is shipped, as a liquid, in tank cars and barges. One of its most familiar uses is in the treatment of drinking water and swimming pools for the purpose of killing microorganisms. This practice has helped greatly to eliminate typhoid fever as a major cause of death in the United States. Considerable quantities of chlorine are used for bleaching wood pulp in the paper industry and for manufacturing bleaching powders and household bleaching solutions. Much larger quantities are used in the manufacture of organic chlorine compounds.

Chlorine is a poisonous gas that was first used in chemical warfare by the Germans in World War I. The first attack in April, 1915, was much more effective than subsequent ones because of the element of surprise and the attendant lack of defensive preparations. Some of the most important chlorine compounds that have been used or considered for use in chemical warfare are phosgene, mustard gas, and chloropicrin. Phosgene, $COCl_2$, made by reaction of chlorine with carbon monoxide, is more poisonous than chlorine. So-called mustard gas, $(ClC_2H_4)_2S$, is really an oily liquid that produces serious burns and attacks the lungs; it is made by the reaction of ethylene with S_2Cl_2:

$$2H_2C{=}CH_2 + S_2Cl_2 \rightarrow ClCH_2CH_2SCH_2CH_2Cl + S$$

Chloropicrin, or nitrochloroform, Cl_3CNO_2, is also a liquid, made by chlorinating nitromethane.

24-10 The Bromine Industry

Bromine occurs as bromide ion in small amounts in seawater. The same bromides ($NaBr$, KBr, $MgBr_2$, and $CaBr_2$) are found in salt deposits, but seawater is now the most important source of bromine. In a large plant at Freeport, Texas, bromine is first extracted by displacement with chlorine, and then magnesium is recovered.

The bromine industry developed as a consequence of the discovery that knocking in internal combustion engines could be prevented by the addition of tetraethyllead to the gasoline. If ethylene bromide, $C_2H_4Br_2$, is also added, combustion of the gasoline is accompanied by the formation of $PbBr_2$, which is a gas at engine temperatures and is vented with the exhaust. In the absence of some such substance, the tetraethyllead would burn to PbO, a solid which would clog the valves. The ethylene bromide is produced by reaction of bromine with ethylene:

$$H_2C{=}CH_2 + Br_2 \rightarrow BrCH_2CH_2Br$$

Nine tenths of the bromine produced in the United States is used in making ethylene bromide.

All compounds of bromine, organic and inorganic, are prepared, directly or indirectly, from the element. These include KBr, used for the production of AgBr in the manufacture of photographic film; CH_3Br, used as a grain fumigant; and CBr_2Cl_2, an expensive but effective and nontoxic fire extinguisher used in aircraft.

24-11 Sources and Uses of Iodine

Iodine occurs in very small quantities as iodide ion in seawater. Certain marine organisms, e.g., seaweed and oysters, concentrate the iodine in organic form. Seaweed ash is still an important local source of iodine in France and Japan. However, the world's most important source of iodine is crude sodium nitrate (Chile saltpeter), which contains small quantities of sodium iodate, $NaIO_3$, a salt that is isomorphic with $NaNO_3$. The sodium iodate is allowed to concentrate in the mother liquors from crystallization of $NaNO_3$ and is then treated with the exact quantity of $NaHSO_3$ solution required to reduce all the iodate to iodide:

$$IO_3^- + 3HSO_3^- \rightarrow I^- + 3HSO_4^-$$

The mixture is next treated with the exact amount of fresh mother liquor to liberate all the iodine:

$$5I^- + IO_3^- + 3HSO_4^- + 3H^+ \rightarrow 3I_2 + 3SO_4^{--} + 3H_2O$$

In the United States iodine is obtained by chlorination of certain California oil-well brines that have been found to contain small quantities of sodium iodide.

Iodine is used in alcoholic solution ("tincture of iodine") as an antiseptic, and it is the starting material for the production of all compounds of iodine, organic and inorganic. The most important iodine compound in commerce is KI. Small quantities of KI are now commonly added to table salt; since this practice was instituted, there has been a marked decrease in the incidence of goiter in the United States. Goiter is often the result of a lack of iodine in the diet. This type of dietary deficiency is associated with the consumption of food grown in soil which has a low iodine content. Those who eat plenty of seafood are not likely to have this type of dietary deficiency. Since iodine concentrates in the thyroid gland, where it is used in the biosynthesis of thyroxin (a hormone that is an iodine compound), radioactive iodine, prepared in an atomic pile, has been used for the treatment of thyroid malignancy.

24-12 The New Importance of Fluorine

Fluorine is far more abundant than bromine or iodine. It occurs chiefly as calcium fluoride, CaF_2, which is found in large deposits in Kentucky and Illinois. It also occurs as cryolite, Na_3AlF_6; and as fluorapatite, $CaF_2 \cdot Ca_3(PO_4)_2$, in phosphate rock. Fluorapatite is also an essential ingredient of bones and teeth. Drinking water containing about one part per million of fluoride ion prevents the development of dental caries in the majority of children. In some localities the natural water supply contains fluoride ion at the proper concentration for good teeth. In a few communities children develop teeth which are badly discolored, although free from caries, because of an excessively high concentration of fluoride ions in the natural water supply. More commonly the water supply contains insufficient fluoride for the prevention of caries. Communities which are unfortunate in this respect can remedy the situation very simply by adding the necessary amount of sodium fluoride to the water supply. This procedure is called *fluoridation.*

Since elementary fluorine is the most powerful oxidizing agent known, the element occurs only in the -1 oxidation state. The element can be obtained in the free state only by electrolytic oxidation of a fluoride in a nonaqueous medium, employing procedures based on the pioneering work of the French chemist Henri Moissan. Elementary fluorine remained a chemical curiosity until a demand for large quantities of it developed suddenly during World War II for the production of uranium hexafluoride in connection with the atomic energy project. For the production of elementary fluorine, a solution of KF in anhydrous liquid HF is electrolyzed in a carbon steel cell, with copper-impregnated graphite as the anode and carbon steel as the cathode.

When KF dissolves in HF, the F^- ions combine with HF molecules to form HF_2^- ions. During electrolysis, reduction of HF_2^- ions produces H_2 gas at the cathode, and oxidation of HF_2^- ions produces F_2 gas at the anode:

$$4F^- + 4HF \rightarrow 4HF_2^-$$

$$\text{CATHODE:} \quad 2HF_2^- + 2e^- \rightarrow H_2\uparrow + 4F^-$$

$$\text{ANODE:} \quad 2HF_2^- \rightarrow 2HF + F_2\uparrow + 2e^-$$

$$\text{NET REACTION:} \quad 2HF \rightarrow H_2\uparrow + F_2\uparrow$$

As electrolysis continues, the electrolyte is replenished by addition of anhydrous HF.

The usefulness of fluorine is limited by its extreme reactivity. Its power as an oxidizing agent is indicated by the fact that it ignites many substances immediately upon contact with them. Asbestos will burn when held in a stream of fluorine, and water will burn if a stream of fluorine gas is allowed to flow onto its surface. In spite of the difficulties in handling it, there are certain uses for which elementary fluorine is indispensable. One of the first important uses of fluorine gas was in the wartime production of uranium hexafluoride. Treatment of uranium(IV) oxide with hydrogen fluoride produces UF_4, but elementary fluorine is required for the conversion of UF_4 to UF_6:

$$UO_2 + 4HF \rightarrow UF_4 + 2H_2O$$

$$UF_4 + F_2 \rightarrow UF_6$$

Much of the present-day importance of fluorine is associated with its use, directly or indirectly, in the production of a variety of highly useful organic fluorine compounds.

24-13 Discovery of the Halogens

Chlorine was first identified in 1774 when the Swedish chemist Karl Wilhelm Scheele noted that a suffocating odor was obtained by allowing muriatic acid (which we now call hydrochloric acid) to stand in contact with pyrolusite (crude manganese dioxide). He also noted that the gas was slightly soluble in water, that it bleached green leaves and colored flowers, and that it attacked metals. Since the gas was obtained by oxidation of muriatic acid, it was called oxymuriatic acid.

Antoine Lavoisier's teaching that all acids contain oxygen led to the assumption that muriatic acid contains oxygen and that oxymuriatic acid contains still more. Sir Humphry Davy performed experiments in which he attempted to remove the oxygen from oxymuriatic acid (e.g., by the use of phosphorus). In no case did he succeed in obtaining from oxymuriatic acid any substance that was known to contain oxygen. Davy concluded (1810) that the so-called oxymuriatic acid is an element and gave it the name chlorine (Greek: *chloros*, greenish-yellow).

Elementary iodine was first observed in 1811 by Bernard Courtois, a French soap maker, upon adding excess sulfuric acid to the ash obtained by burning seaweed. It was given the name iodine (Greek: *iodos*, violet) because of its color.

In 1771 Scheele obtained a new acid which he prepared by heating the mineral *fluorspar* (CaF_2) with sulfuric acid. It was assumed, of course, that the acid contains oxygen; but Davy, in 1813, concluded that the acid is similar to hydrochloric acid and is a compound of hydrogen with an element resembling chlorine, for which he proposed the name fluorine. Numerous attempts to isolate elementary fluorine failed because of its great chemical reactivity. It was isolated for the first time in 1886 when Moissan succeeded in preparing it by electrolyzing a solution of KF in anhydrous liquid HF, using electrodes and an electrolytic cell that were made of fluorine-resistant materials.

Bromine was discovered in 1826 by the young French chemist Antoine Balard while he was investigating the waste liquor remaining after sodium chloride had crystallized out from concentrated seawater. It was given the name bromine (Greek: *bromos*, stench) because of its irritating vapor.

The search for element 85 was pursued for many years in the expectation that it would prove to be the missing halogen provided for by an empty space in the periodic table. Definite proof of its existence was first obtained in 1940 at the University of California, where it was prepared on a miniature scale in the cyclotron by Emilio Segré, D. R. Corson, and K. R. MacKenzie. It was named astatine (Greek: *astatos*, unstable) because it is the only halogen without a stable isotope.

New Terms

Chain reaction: a reaction which yields products that cause further reactions of the same kind and thus becomes self-sustaining.

Fluoridation (of drinking water): addition of NaF to drinking water for the purpose of preventing dental caries.

Halogens: the elements of Group VIIA of the periodic table (fluorine, chlorine, bromine, iodine, and astatine).

Interhalogen compound: a compound composed of two different elements of Group VIIA; e.g., ClF_3.

Photochemical reaction: a chemical reaction brought about by the effect of light.

Exercises

24-1 Why are the elements of Group VIIA called halogens?

24-2 Discuss the relationship of fluorides, chlorides, bromides, and iodides to health and medicine.

24-3 What is a chain reaction?

24-4 Compare fluorine with the other halogens, and compare hydrogen fluoride with the other hydrogen halides.

24-5 List the formulas of several chemical warfare gases.

24-6 What is the relationship of the automotive industry to the extraction of bromine and magnesium from seawater?

24-7 Calculate the weights of sodium hydroxide, hydrogen, and chlorine that would be obtained from 50 tons of sodium chloride by electrolysis, assuming the process is 94% efficient.

References

Jolly, W. L., *The Chemistry of the Non-Metals*, Prentice-Hall, Englewood Cliffs, N.J., 1966. Chapter 3 discusses fluorine; Chapter 4, chlorine, bromine, iodine, and astatine.

Kleinberg, J., Argersinger, Jr., W. J., and Griswold, E., *Inorganic Chemistry*, Heath, Boston, 1960. Chapter 16 is devoted to the halogens.

25

Organic Compounds and Oxyacids of the Halogens

25-1 Organic Compounds of Chlorine

Almost two thirds of the chlorine manufactured in the United States is used for the production of organic compounds. Several of these are simple derivatives of methane.

Methyl chloride, CH_3Cl, is a gas, used chiefly in the production of silicone resins and rubbers (Section 29-16). Most of it is made from methanol and hydrogen chloride:

$$CH_3OH + HCl \rightarrow CH_3Cl + H_2O$$

Methylene chloride, CH_2Cl_2, is a liquid, used in paint and varnish removers and as a solvent for aerosol paints. It is made by chlorination of methyl chloride or methane:

$$CH_3Cl + Cl_2 \rightarrow CH_2Cl_2 + HCl$$
$$CH_4 + 2Cl_2 \rightarrow CH_2Cl_2 + 2HCl$$

Chloroform, $CHCl_3$, is also made by chlorination of methyl chloride or methane:

$$CH_3Cl + 2Cl_2 \rightarrow CHCl_3 + 2HCl$$
$$CH_4 + 3Cl_2 \rightarrow CHCl_3 + 3HCl$$

The chief use of chloroform is for the production of chlorodifluoromethane (Freon-22).

The most important of the chlorinated methanes is carbon tetrachloride. The method most used for its manufacture involves the reaction of chlorine with carbon disulfide:

$$CS_2 + 3Cl_2 \rightarrow CCl_4 + S_2Cl_2$$

The industrial usefulness of carbon tetrachloride is based largely on its conversion to dichlorodifluoromethane (Freon-12), trichlorofluoromethane (Freon-11), and tetrachloroethylene.

Ethyl chloride, C_2H_5Cl, is made by an addition reaction between ethylene and hydrogen chloride:

$$H_2C{=}CH_2 + HCl \rightarrow H_3C{-}CH_2Cl$$

Most of the ethyl chloride made in the United States is used for the manufacture of tetraethyllead. Physicians use it as a general anesthetic, and also as a local anesthetic in minor surgery.

Ethylene chloride, $C_2H_4Cl_2$, is manufactured in greater quantity than any other organic chlorine compound. It is made by an addition reaction between ethylene and chlorine. The most important use of ethylene chloride is in the production of vinyl chloride:

$$ClCH_2CH_2Cl \xrightarrow{400°C} H_2C{=}CHCl + HCl$$

Vinyl chloride, by polymerization, is converted to resins and plastics that are used in numerous consumer goods such as phonograph records, raincoats, shoe soles, and floor tile.

Tetrachloroethylene, also known as perchloroethylene to indicate that all of the hydrogen in ethylene has been replaced by chlorine, is made by pyrolysis of carbon tetrachloride:

$$2CCl_4 \xrightarrow{900°C} Cl_2C{=}CCl_2 + 2Cl_2$$

Tetrachloroethylene is now used extensively as a nonflammable dry-cleaning solvent.

Benzene reacts rapidly with chlorine in the absence of sunlight and in the presence of a catalyst (such as $FeCl_3$) to give chlorobenzene, a substitution product:

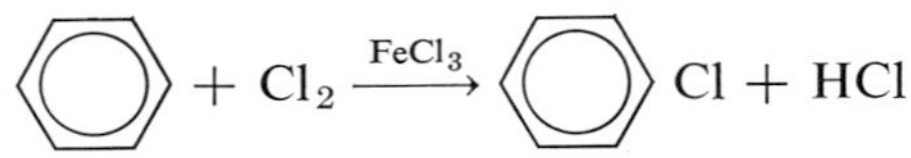

In the presence of light of short wavelength to provide the necessary activation energy, benzene *adds* three moles of chlorine to form benzene hexachloride:

$$C_6H_6 + 3Cl_2 \xrightarrow{light} C_6H_6Cl_6$$

The product is commercially important because of its insecticidal properties. It is called BHC or 666, abbreviations of the name or the formula.

Nine stereoisomers of benzene hexachloride are possible. The insecticidal properties of the mixture of isomers obtained in the above reaction are due entirely to the so-called gamma isomer, which makes up only 18 percent of the mixture. This isomer is known by the names "gammexane" and "lindane" and has the structure indicated by the *perspective formula* in Figure 25-1.

Structurally, $C_6H_6Cl_6$ is a substituted cyclohexane, i.e., the ring is alicyclic and puckered, not aromatic and flat. In the perspective formula no attempt is made to show that the ring is puckered. The ring is shown as a hexagon slightly tilted out of the plane of the paper, with thick lines representing the bonds nearest the viewer and thin lines representing those that are farthest away. Ring

Figure 25-1

Perspective formula for γ-$C_6H_6Cl_6$

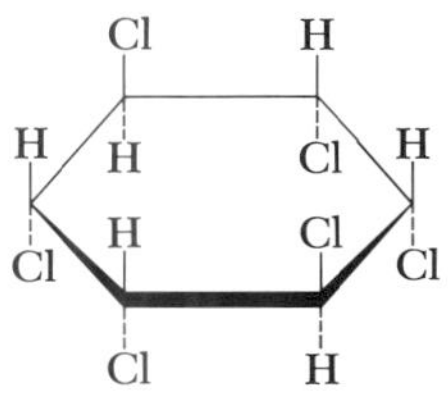

carbon atoms are omitted. Solid lines indicate positions above the ring, and broken lines indicate positions below the ring.

25-2 Organic Compounds of Fluorine

Organic fluorine compounds have attained great importance in recent years as a result of the commercial production of the fluorine and hydrogen fluoride required in the preparation of many of them. The properties of fluorine compounds vary in the extreme. Some are relatively inert and others are highly reactive. Some are highly toxic, whereas others are completely nontoxic. Completely fluorinated hydrocarbons, called *fluorocarbons*, are almost as inert as the noble gases.

The reaction of fluorine with hydrocarbons is so violent that methods had to be devised for moderating the reaction. One method consists in substituting for fluorine a less reactive fluorinating agent. Cobalt(III) fluoride is often used for this purpose. When so used it is reduced to cobalt(II) fluoride; elementary fluorine is used to convert CoF_2 to CoF_3 again. Many of the organic fluorine compounds are prepared by exchange reactions between organic chlorine compounds and inorganic fluorides. Thus, carbon tetrachloride is converted to Freon-11 by reaction with antimony pentafluoride:

$$CCl_4 + SbF_5 \rightarrow CCl_3F + SbF_4Cl$$

HF converts SbF_4Cl to SbF_5 again. If the replacement of chlorine with fluorine is carried one step further, Freon-12 is produced:

$$CCl_3F + SbF_5 \rightarrow CCl_2F_2 + SbF_4Cl$$

Freon-11 and Freon-12 are the two most important Freons; they are used extensively as refrigerants and in aerosol sprays.

The Freons have been given numerical designations because of the possibility of confusion if the chemical names were used in commerce. Actually, the numbers indicate the chemical formula. The digit furthest to the right is the number of fluorine atoms, the digit next to the right is one more than the number of hydrogen atoms, and if there is a third digit at the left it is one less than the number of carbon atoms. Thus, dichlorodifluoromethane is designated Freon-12 because there are two fluorine atoms in the molecule, no hydrogen atoms, and only one carbon atom (the third digit from the right would be zero in this case and is not written). The common refrigerant for household refrigerators with rotary-type compressors is Freon-114; from this numerical designation one can tell that its formula is $CClF_2CClF_2$ (it being understood that the chlorine and fluorine atoms are symmetrically distributed in the two-carbon Freons).

Although the Freon-22 made from chloroform is important as a refrigerant, it is even more important as an intermediate in the manufacture of tetrafluoroethylene:

$$2CHClF_2 \xrightarrow{700°C} F_2C{=}CF_2 + 2HCl$$

Polymerization of tetrafluoroethylene gives Teflon, a remarkable polymer which is chemically so inert that it is not attacked by acids, alkalies, oxidizing agents, or organic solvents.

25-3 Oxyacids of the Halogens

All the halogens exhibit the oxidation number -1 in the halides, and this is the only oxidation number shown by fluorine in any of its compounds. However, chlorine, bromine, and iodine exhibit positive oxidation numbers in compounds in which their atoms are covalently bonded to oxygen, i.e., in oxides and oxyacids. The oxyacids of the halogens are listed in Table 25-1 and in Figure 25-2.

25-4 Hypochlorous Acid and Hypochlorites

Hypochlorous acid, HClO, is a weak, unstable acid, for which the ionization constant is 3×10^{-8}. It is formed in aqueous solution, together with hydrochloric acid, when chlorine reacts with water:

Table 25-1

Oxyacids and oxyanions of the halogens

Oxidation state	*Acid*	*Name*	*Ionization constant*	*Anion*	*Name*
+1	$HClO$	Hypochlorous acid	3×10^{-8}	ClO^-	Hypochlorite ion
+3	$HClO_2$	Chlorous acid	1×10^{-2}	ClO_2^-	Chlorite ion
+5	$HClO_3$	Chloric acid	Strong acid	ClO_3^-	Chlorate ion
+7	$HClO_4$	Perchloric acid	Strong acid	ClO_4^-	Perchlorate ion
+1	$HBrO$	Hypobromous acid	2×10^{-9}	BrO^-	Hypobromite ion
+5	$HBrO_3$	Bromic acid	Strong acid	BrO_3^-	Bromate ion
+1	HIO	Hypoiodous acid	3×10^{-11}	IO^-	Hypoiodite ion
+5	HIO_3	Iodic acid	2×10^{-1}	IO_3^-	Iodate ion
+7	HIO_4	Metaperiodic acid	Strong acid	IO_4^-	Metaperiodate ion
+7	H_5IO_6	Paraperiodic acid	5×10^{-4}	$H_4IO_6^-$	Paraperiodate ion

Figure 25-2

Oxyacids of the halogens

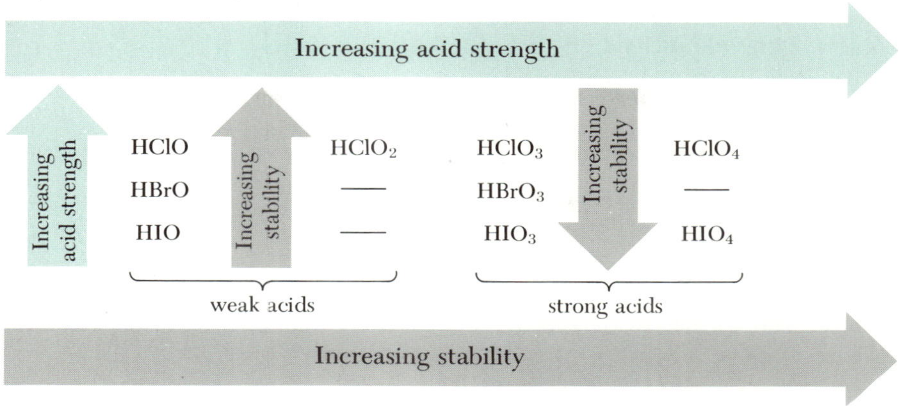

$$Cl_2 + H_2O \rightleftharpoons H^+ + Cl^- + HClO$$

It has not been isolated and is known only in solution. It is a powerful oxidizing agent and readily decomposes:

$$2HClO \rightarrow 2(H^+ + Cl^-) + O_2$$

Its salts, the hypochlorites, are obtained by reaction between chlorine and cold solutions of bases:

$$Cl_2 + 2OH^- \rightarrow Cl^- + ClO^- + H_2O$$

Commercially, sodium hypochlorite is obtained by electrolysis of concentrated sodium chloride solutions, with stirring to bring the Cl_2 formed at the anode into contact with the OH^- ions formed at the cathode. The net result is represented by the following equation:

$$Cl^- + H_2O \rightarrow ClO^- + H_2\uparrow$$

Solutions containing sodium hypochlorite at a concentration of 5 percent are sold for use as laundry bleach. Such solutions are much more stable than hypochlorous acid itself. However, only a few hypochlorites have been isolated as solids; they are salts of some of the metals of Groups IA and IIA.

Acidification of solutions containing both hypochlorite and chloride ions produces chlorine by interaction between hypochlorous and hydrochloric acids:

$$H^+ + Cl^- + HClO \rightarrow H_2O + Cl_2\uparrow$$

This reaction is, of course, the reverse of the one that occurs between chlorine and water; it shows that hypochlorous acid is a powerful oxidizing agent.

A solution or a solid containing hypochlorite ions is useful as a bleaching or disinfecting agent. Commercial bleaching powder, a mixture often improperly referred to as "chloride of lime," is prepared by passing chlorine over dry calcium hydroxide; one of its components is calcium hypochlorite. The material known commercially as HTH ("high-test hypochlorite") is essentially pure calcium hypochlorite. The first step in its preparation consists of passing chlorine into milk of lime:

$$(Ca^{++} + 2OH^-) + Cl_2 \rightarrow Ca^{++} + ClO^- + Cl^- + H_2O$$

Calcium hypochlorite crystallizes out from the solution as the tetrahydrate, $Ca(ClO)_2 \cdot 4H_2O$.

25-5 Chloric Acid and Chlorates

Chlorates are formed by the reaction of chlorine with hot concentrated solutions of bases:

$$3Cl_2 + 6OH^- \rightarrow ClO_3^- + 5Cl^- + 3H_2O$$

The most important chlorates in commerce are the potassium and sodium salts, obtained by the action of chlorine on hot KOH or NaOH solution. When the reaction mixture cools, the solid $KClO_3$ or $NaClO_3$ crystallizes from solution. Potassium chlorate is less deliquescent than sodium chlorate and therefore is more easily handled. Because it is much more soluble in water, sodium chlorate has partially displaced potassium chlorate as a weed killer in recent years. Potassium chlorate is used in the manufacture of matches and fireworks. The inclusion of $KClO_3$ with the combustible material in the head of a match makes it possible for the match to start burning immediately when a sufficiently high temperature has been produced by friction. One type of red flare is made from a mixture of potassium chlorate, sugar, and strontium nitrate. The heat from the oxidation of the sugar causes the strontium salt to emit red light.

The ease with which potassium chlorate is decomposed by heat when mixed with a suitable catalyst (such as MnO_2) accounts for its use as a laboratory source of oxygen:

$$2KClO_3 \rightarrow 2KCl + 3O_2\uparrow$$

Potassium chlorate should never be mixed with combustible materials. Samples used in laboratory work should be completely decomposed by heating before they are discarded.

Chloric acid, $HClO_3$, is obtained in solution by the reaction of barium chlorate with dilute sulfuric acid; advantage is taken of the insolubility of barium sulfate to effect separation of the products:

$$(Ba^{++} + 2ClO_3^-) + (2H^+ + SO_4^{--}) \rightarrow 2(H^+ + ClO_3^-) + BaSO_4\downarrow$$

Chloric acid is a fairly strong acid. It is a very strong oxidizing agent and is stable only in dilute solution. Unless they are kept cold, concentrated solutions of chloric acid are likely to undergo violent decomposition.

25-6 Perchloric Acid and Perchlorates

Metal perchlorates are obtained by carefully heating chlorates without a catalyst:

$$4KClO_3 \rightarrow 3KClO_4 + KCl$$

At higher temperatures perchlorates are decomposed, liberating oxygen:

$$KClO_4 \rightarrow KCl + 2O_2\uparrow$$

Potassium perchlorate is much safer than potassium chlorate for use in flares and fireworks.

Perchloric acid, $HClO_4$, is the only oxyacid of chlorine that has been obtained in pure condition. It is a colorless substance that melts at −112°C. Although the aqueous solution is fairly stable, the pure compound decomposes with explosive violence when it is heated. It reacts violently with reducing agents. The anhydride (Cl_2O_7, called chlorine heptoxide) can be obtained by treating the anhydrous acid with a strong dehydrating agent (e.g., P_4O_{10}). It is a colorless, oily liquid that is even more dangerously unstable than the anhydrous acid.

Perchloric acid is one of the strongest acids known. In dilute aqueous solution it is completely ionized:

$$HClO_4 + H_2O \rightarrow H_3O^+ + ClO_4^-$$

It becomes a powerful oxidizing agent in hot concentrated solutions. However, it is a poor oxidizing agent in cold dilute solutions; e.g., the reaction of cold dilute perchloric acid with zinc liberates hydrogen gas with no reduction of the perchlorate ion:

$$Zn + 2(H_3O^+ + ClO_4^-) \rightarrow (Zn^{++} + 2ClO_4^-) + H_2\uparrow + 2H_2O$$

Under no conditions is perchloric acid as strong an oxidizing agent as the other oxyacids of chlorine.

Perchloric acid can be obtained by treating potassium perchlorate with sulfuric acid and removing the perchloric acid from the reaction mixture by distillation under reduced pressure. This is a hazardous operation.

All metal perchlorates are highly soluble in water except $KClO_4$, $RbClO_4$, and $CsClO_4$. Ammonium perchlorate is also only slightly soluble. Some of the perchlorates, e.g., $Mg(ClO_4)_2$, are valuable drying agents.

Perchloric acid monohydrate, $HClO_4 \cdot H_2O$, is a colorless solid which melts at 50°C. X-ray analysis has shown this solid to be an ionic compound, $(H_3O^+)(ClO_4^-)$, analogous to ammonium perchlorate, $(NH_4^+)(ClO_4^-)$, which has the same crystal structure.

25-7 Chlorous Acid and Chlorites

Chlorous acid, $HClO_2$, can be obtained in aqueous solution by adding sulfuric acid to a suspension of barium chlorite:

$$Ba(ClO_2)_2 + (2H^+ + SO_4^{--}) \rightarrow BaSO_4\downarrow + 2HClO_2$$

Chlorous acid is a moderately weak acid; its ionization constant is 1×10^{-2}. The acid is a powerful oxidizing agent. It is so unstable that it decomposes whenever its isolation is attempted.

Chlorites result from the reaction of chlorine dioxide with solutions of bases:

$$2ClO_2 + 2OH^- \rightarrow ClO_2^- + ClO_3^- + H_2O$$

Commercially, sodium chlorite is prepared by the reaction of chlorine dioxide with sodium peroxide:

$$Na_2O_2 + 2ClO_2 \rightarrow 2NaClO_2 + O_2$$

In recent years sodium chlorite has become important as a mild bleaching agent for cellulose fibers. The chlorine dioxide used in its preparation is a highly explosive orange-yellow gas that is obtained commercially by the action of nitrogen dioxide on sodium chlorate solution:

$$NO_2 + ClO_3^- \rightarrow NO_3^- + ClO_2$$

Chlorine dioxide is also used for bleaching flour and paper pulp.

25-8 Geometry of Oxyanions of Chlorine

The perchlorate ion has the same tetrahedral structure as the methane molecule. It is frequently true that a tetrahedral arrangement results when four identical atoms or groups are attached to a central atom. The structure of the chlorate ion is that of a trigonal pyramid; as such, it has three oxygen atoms located much as they are in the perchlorate ion, with one unshared electron pair in the place of the fourth oxygen (Figure 25-3). The chlorite ion is angular rather than

Figure 25-3

Structures of oxyanions of chlorine

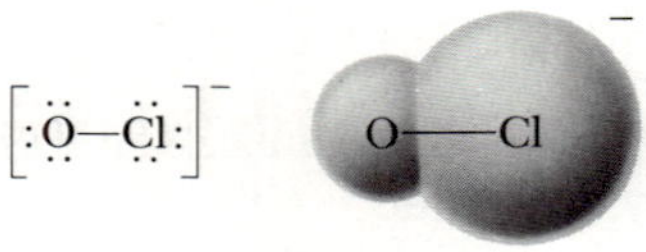

Hypochlorite ion: linear

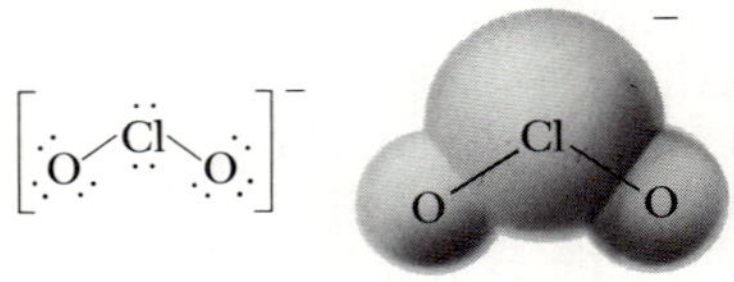

Chlorite ion: bent chain

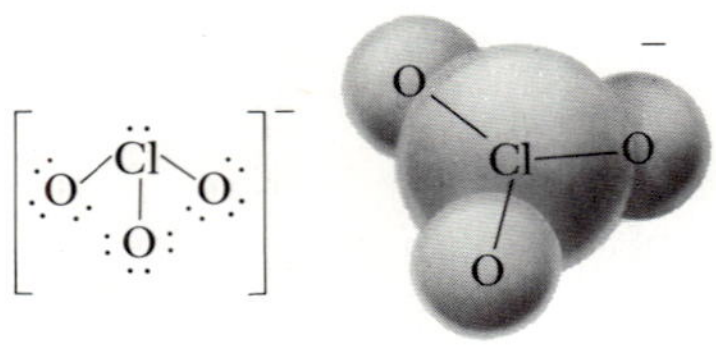

Chlorate ion: trigonal pyramidal

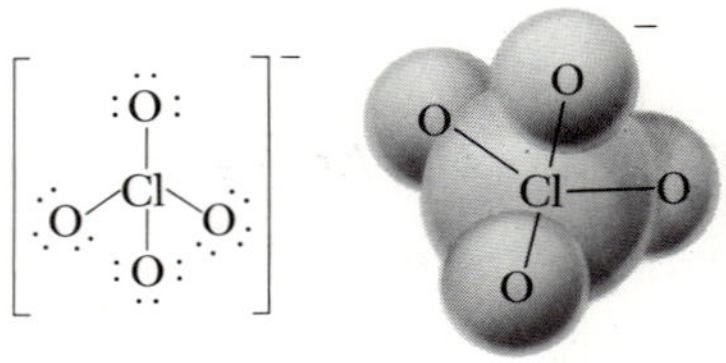

Perchlorate ion: tetrahedral

linear, and the angle between the bonds is once again approximately the same as in the chlorate and perchlorate ions; in this ion the two unshared electron pairs may be thought of as occupying the positions of two of the oxygen atoms in the perchlorate ion. The hypochlorite ion is, of necessity, linear; it may be thought of as having the structure of a perchlorate ion, with three unshared electron pairs occupying the positions of three of the oxygen atoms.

25-9 Oxyacids of Bromine

Hypobromous acid, HBrO, is a much weaker acid than HClO and is also less stable. No solid hypobromites have been isolated.

No proof has been obtained for the existence of bromous acid or the bromite ion.

Bromates are formed when bromine reacts with hot concentrated solutions of bases:

$$3Br_2 + 6OH^- \rightarrow BrO_3^- + 5Br^- + 3H_2O$$

Potassium bromate, $KBrO_3$, is a useful oxidizing agent. Like $KClO_3$, it can be obtained as a pure solid by recrystallization from solution. Bromic acid, $HBrO_3$, is a strong acid and a strong oxidizing agent, known only in solution.

All attempts to prepare perbromic acid and perbromates have failed. The reasons for their apparent nonexistence are not known.

25-10 Oxyacids of Iodine

Hypoiodous acid, HIO, is even weaker than HBrO and is still more unstable. No hypoiodites have been isolated.

Iodous acid and the iodite ion seem not to exist.

Iodic acid is more stable than chloric acid or bromic acid; it can readily be obtained in crystalline form as a white solid. It is a fairly strong acid and a good oxidizing agent. Iodates are much more stable than chlorates or bromates, but explosions result from heating them with combustible substances. One method for producing iodic acid consists in the reaction of iodine with concentrated nitric acid:

$$10HNO_3 + 3I_2 \rightarrow 6HIO_3 + 10NO + 2H_2O$$

Iodates result from gently warming powdered iodine with solutions of chlorates:

$$I_2 + 2ClO_3^- \rightarrow 2IO_3^- + Cl_2$$

When iodic acid is heated to about 200°C, iodine pentoxide remains in the form of a white powder:

$$2HIO_3 \rightarrow I_2O_5 + H_2O$$

A series of periodic acids exists; the most important are HIO_4, metaperiodic acid, and H_5IO_6, paraperiodic acid (sometimes called orthoperiodic acid). The

Figure 25-4

Octahedral structure of the paraperiodate ion, $H_4IO_6^-$

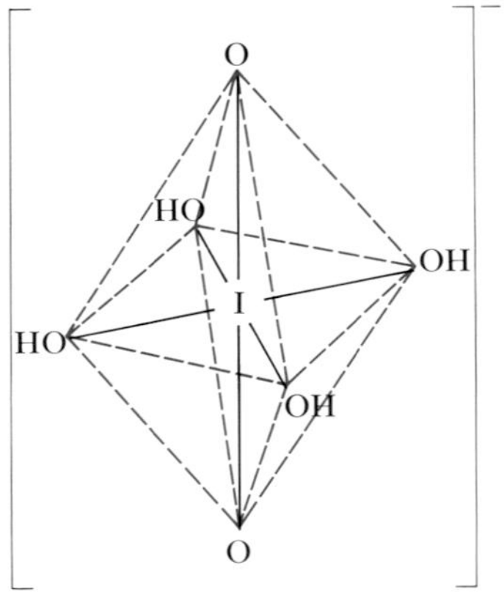

paraperiodates are the commonest of the periodates. Thermal decomposition of alkaline earth iodates produces periodates:

$$5Ba(IO_3)_2 \rightarrow Ba_5(IO_6)_2 + 4I_2\uparrow + 9O_2\uparrow$$

Paraperiodic acid is a good oxidizing agent. It is obtained as a white solid by evaporating the filtered solution resulting from reacting barium periodate with sulfuric acid:

$$Ba_5(IO_6)_2 + 5H_2SO_4 \rightarrow 5BaSO_4\downarrow + 2H_5IO_6$$

Under a vacuum it can be dehydrated in stages, giving first $H_4I_2O_9$ and then HIO_4. Further heating gives iodic acid and oxygen; the anhydride, I_2O_7, seems not to exist.

The metaperiodate ion, IO_4^-, has a tetrahedral structure, but the paraperiodate ion, $H_4IO_6^-$, has six oxygen atoms bonded to the central iodine atom in an octahedral arrangement (Figure 25-4). It is usually true that an octahedral structure results when six atoms or groups are attached to a central atom. The size of the central atom determines whether it is possible for more than four atoms or groups to be attached to it.

25-11 Predicting the Strengths of Oxyacids

Oxyacids, as illustrated by the oxyacids of the halogens, consist of oxygen atoms O and hydroxyl groups OH bonded to a central atom. We can represent oxyacids in general by the formula $XO_m(OH)_n$. From the value of m in the formula for a particular acid it has been found possible to make an approximate prediction of the strength of the acid. If $m = 0$, as in hypochlorous acid (HClO or ClOH), it can be predicted that the acid is *very weak*, with an ionization constant $K \leq 10^{-7}$. If $m = 1$, as in $HClO_2$, i.e., ClO(OH), the prediction is that the acid is *weak*, with an ionization constant $K \leq 10^{-2}$. Note that the acids

Table 25-2

Predicted and actual strengths of acids

Acid	*Predicted strength*	*Actual strength*
HClO or ClOH	$K \leq 10^{-7}$	$K = 3 \times 10^{-8}$
HBrO or BrOH	$K \leq 10^{-7}$	$K = 2 \times 10^{-9}$
HIO or IOH	$K \leq 10^{-7}$	$K = 3 \times 10^{-11}$
$HClO_2$ or ClO(OH)	$K \leq 10^{-2}$	$K = 1 \times 10^{-2}$
$HClO_3$ or $ClO_2(OH)$	Strong acid	Strong acid
$HBrO_3$ or $BrO_2(OH)$	Strong acid	Strong acid
HIO_3 or $IO_2(OH)$	Strong acid	$K = 2 \times 10^{-1}$
$HClO_4$ or $ClO_3(OH)$	Very strong acid	Very strong acid
HIO_4 or $IO_3(OH)$	Very strong acid	Apparently strong
H_5IO_6 or $IO(OH)_5$	$K_1 \leq 10^{-2}$	$K_1 = 5 \times 10^{-4}$

of the weak class have ionization constants about 10^5 times as great as acids of the very weak class. If $m = 2$, as in chloric acid, $ClO_2(OH)$, we would predict that the compound is a *strong* acid. If $m = 3$, as in perchloric acid, our prediction is a *very strong* acid.

In general, the "rule" works out very well with the oxyacids of the halogens, as shown in Table 25-2. We can understand that the attraction between a proton and a chlorite ion, $[OClO]^-$, is divided between two oxygen atoms and that the force between the proton and each one of the oxygen atoms is therefore less than the attraction between a proton and the single oxygen of a hypochlorite ion, $[ClO]^-$. Therefore, chlorous acid can be expected to be more highly dissociated than hypochlorous acid. An acid of the third type ($m = 2$) would be still more highly dissociated by the same reasoning.

New Terms

Fluorocarbons: completely fluorinated hydrocarbons, e.g., tetrafluoroethylene, C_2F_4.

Freons: halogenated hydrocarbons having one or more fluorine atoms in each molecule, used as refrigerants and in aerosol sprays. The two most important Freons are CCl_3F and CCl_2F_2.

Perspective formulas: formulas designed to give a three-dimensional perspective of the arrangement of the atoms in stereoisomers.

Exercises

25-1 In the electrolysis of aqueous sodium chloride, what are the products if the electrode products are kept separate? If they are mixed?

25-2 Describe the method used to designate individual Freons in commerce. Write the formulas for Freon-11, Freon-12, Freon-22, and Freon-114.

25-3 Identify: Freons, fluorocarbons, Teflon, and vinyl chloride.

25-4 Describe how the strengths of oxyacids can be predicted.

25-5 State the most important uses for chloroform, carbon tetrachloride, ethyl chloride, ethylene chloride, tetrachloroethylene, sodium hypochlorite, sodium chlorite, and chlorine dioxide.

25-6 What is a perspective formula?

25-7 Write the perspective formulas for the nine stereoisomers of benzene hexachloride.

References

Buttle, J. W., Daniels, D. J., and Beckett, P. J., *Chemistry: A Unified Approach*, Butterworths, Washington, D.C., 1966. Chapter 10 discusses the halogens and Chapter 20, organic halogen compounds.

Noller, C. R., *Chemistry of Organic Compounds*, 3rd ed., Saunders, Philadelphia, 1965. Chapter 35 is devoted to chlorinated and fluorinated aliphatic hydrocarbons.

26

Sulfur and Related Elements

26-1 Oxidation States of the Elements

Group VIA of the periodic table consists of five elements that have the s^2p^4 configuration in the outermost shells of their atoms: oxygen, sulfur, selenium, tellurium, and polonium (Figure 26-1). The atoms of these elements can acquire the s^2p^6 configuration characteristic of noble-gas atoms by gaining two electrons from atoms of metals to form binegative ions: O^{--}, oxide ion; S^{--}, sulfide ion; Se^{--}, selenide ion; Te^{--}, telluride ion; and Po^{--}, polonide ion. Oxygen is the most electronegative of all elements except fluorine. It has the greatest tendency of any of the members of its family to form negative ions. The other members of the family form negative ions only when reacting with the most active metals.

Figure 26-1

Electronic configurations of the oxygen-sulfur family

	1	2	3	4	5	6
$_{8}O$	s^2	s^2p^4				
$_{16}S$	s^2	s^2p^6	s^2p^4			
$_{34}Se$	s^2	s^2p^6	$s^2p^6d^{10}$	s^2p^4		
$_{52}Te$	s^2	s^2p^6	$s^2p^6d^{10}$	$s^2p^6d^{10}$	s^2p^4	
$_{84}Po$	s^2	s^2p^6	$s^2p^6d^{10}$	$s^2p^6d^{10}f^{14}$	$s^2p^6d^{10}$	s^2p^4

Noble-gas configurations can also be acquired by forming covalent bonds with other atoms. Such bonds are present in H_2O, H_2S, H_2Se, H_2Te, and H_2Po and in halides of the elements, e.g., OF_2, SCl_2, and $SeBr_2$. However, oxygen is the only member of the family limited to a maximum of eight electrons in its outer shell. The outer shell of oxygen is the second shell, and it has reached its maximum with the s^2p^6 configuration. The outermost shell of the sulfur atom is the third shell, and it can accommodate more than eight electrons, as in SF_6 where it has twelve. However, fluorine is the only halogen with atoms small enough for six of them to be bonded to the same sulfur atom. Selenium and tellurium form hexafluorides and various other compounds in which six atoms are covalently bonded to the central atom.

Positive oxidation states are shown in the molecules and ions in which the Group VIA element is covalently bonded to oxygen or a halogen. The only compounds in which oxygen shows a positive oxidation state are its fluorides. The other members of the family form numerous compounds in which they show positive oxidation states. Particularly important are the +4 and +6 states. The +4 state is shown, for example, in SO_2, $SeBr_4$, and TeI_4. The +6 state is observed in SO_3, H_2SO_4, SF_6, H_6TeO_6, etc.

There is some evidence that polonium forms a positive ion, Po^{++}. Our knowledge of the chemistry of polonium is seriously limited, however, because very little of it has been available for study. Polonium is a highly radioactive element, and its intense radiation necessitates special techniques for the protection of chemists working with its compounds.

26-2 Physical Properties: Allotropic Forms

Although oxygen is a gas, the other members of Group VIA are solids at ordinary temperatures. The melting points and boiling points of oxygen, sulfur, selenium, and tellurium rise with increasing atomic number (Table 26-1).

Table 26-1

Properties of the elements of Group VIA

Element	*Melting point* (°C)	*Boiling point* (°C)	*Ionization energy* (kcal/mole)	*Electronegativity* (Pauling scale)	*Specific resistivity* (microhm-cm, 0°C)
$_{8}O$	−219	−183	314	3.5	
$_{16}S$	119	445	239	2.5	2×10^{23}
$_{34}Se$	217	688	225	2.4	2×10^{11}
$_{52}Te$	450	1390	215	2.1	2×10^{5}
$_{84}Po$	254	965	194	2.0	43

In addition to the ordinary diatomic molecules, O_2, oxygen exists in the extremely unstable allotropic modification, O_3, known as ozone. A number of modifications of elementary sulfur are known in all three physical states. Although much work has been done on the allotropism of sulfur, many uncertainties await clarification. It can at least be said definitely that the crystalline modification of sulfur that is stable at room temperature is the rhombic form, and the monoclinic form is stable above 95.6°C. If rhombic sulfur is heated to this *transition temperature*, it undergoes transformation into monoclinic sulfur. The transition is extremely slow, however, and if rhombic sulfur is heated rapidly to 112.8°C it is converted to liquid sulfur without having been converted to monoclinic sulfur. The melting point of monoclinic sulfur is 119.0°C.

If liquid sulfur is allowed to cool slowly, it freezes at 119.0°C, forming monoclinic crystals. On standing at temperatures below 95.6°C, the monoclinic form is slowly transformed to the rhombic variety. If either rhombic or monoclinic sulfur is dissolved in toluene (boiling point, 110.6°C) and the solution is evaporated to dryness, the sulfur forms rhombic crystals if the temperature is below 95.6°C and monoclinic crystals if the temperature is above this.

Both rhombic and monoclinic sulfur are soluble in carbon disulfide, CS_2. Neither form is soluble in water. Both are yellow solids composed of octatomic molecules, S_8, in which the atoms are arranged in a puckered ring (Figure 26-2).

The facts regarding the allotropic varieties of selenium are awaiting clarification. The form that is stable at ordinary temperatures consists of gray crystals belonging to the hexagonal system. It is commonly referred to as "metallic" selenium. Rapid reduction of selenium compounds in aqueous solution gives a red precipitate that is designated amorphous selenium. Extraction of this red substance with CS_2, followed by evaporation of the solvent, gives red crystals of monoclinic selenium composed of Se_8 molecules. Both the amorphous and the monoclinic selenium change to the gray form when heated. Gray selenium is insoluble in CS_2 and melts at 217°C. Although it has a metallic luster, it is not a good conductor of electric current. Its conductivity is substantially greater in the light than in the dark.

Figure 26-2

The S_8 molecule: a puckered ring

Crystalline tellurium has a silvery-white metallic appearance. It is isomorphic with gray selenium and is not a good conductor in comparison with metals.

The trend toward more metallic character with increasing atomic number may be said to be more or less complete with polonium. This can be shown by comparing the specific resistivities of the elements. The *specific resistivity* of a substance is the resistance of a body of the substance 1 cm^2 in cross section and 1 cm in length. The fundamental unit of specific resistivity is the ohm-centimeter. In Table 26-1 the values are given in microhm-centimeters (1 microhm-cm $= 10^{-6}$ ohm-cm) at 0°C. The value for polonium is within the range for metals. Furthermore, the resistance of polonium increases with rising temperature, which is characteristic of metals. Sulfur is a true insulator with a high specific resistivity. Selenium and tellurium have intermediate values, and this is one reason for classifying them as metalloids. The resistances of sulfur, selenium, and tellurium decrease with rising temperature, which is characteristic of nonmetals. Their chemical properties identify all three of these elements as nonmetals. It is unfortunate that selenium and tellurium have names ending in *-ium* and that they are frequently referred to as metals in commerce.

26-3 Liquid and Vapor States of the Elements

At temperatures not far above the melting point, sulfur is a yellow, transparent liquid of low viscosity. If the temperature is raised to about 160°C, it turns brown and becomes increasingly viscous; at 200°C it will not flow from an inverted test tube. Above 200°C the viscosity decreases with rising temperature. This strange behavior of liquid sulfur can be attributed to the opening of the S_8 rings, forming open chains, when the sulfur is heated above its melting point. Near the melting point the rings roll over one another easily, and the sulfur is quite fluid. When higher temperatures are reached and the rings are broken into open chains which then join together to form long molecules, the viscosity increases greatly by entanglement of the long molecules with one another. Above 200°C the long chains break up more and more to form shorter chains, and the viscosity decreases as the temperature rises.

When boiling liquid sulfur is poured into cold water, the sulfur is cooled below its freezing point too rapidly for the atoms to orient themselves to reconstitute S_8 molecules and for the molecules to arrange themselves in crystalline form. As a result, the sulfur is amorphous and plastic. On standing, it crystallizes completely, losing its flexibility and becoming brittle.

Melted selenium is a black, viscous liquid, which yields, on rapid cooling, an amorphous material that is brittle and exhibits an irregular fracture like glass. Crystallization of the black vitreous selenium takes place slowly.

Rapid cooling of molten tellurium produces a mixture of amorphous and crystalline tellurium. A rather remarkable property of liquid tellurium is the fact that, like water, it has its maximum density at a temperature above its melting point.

Sulfur vapor appears to be a mixture of S_8, S_6, S_4, and S_2 molecules, the relative amounts changing with the temperature. Significant amounts of monatomic sulfur are present above 2000°C. Oxygen gas is about 1 percent monatomic and 99 percent diatomic at 2300°C. Selenium vapor is almost entirely monatomic above 2000°C. At lower temperatures Se_8, Se_6, and Se_2 molecules are present. Tellurium forms a golden yellow vapor consisting of Te_2 molecules. Above 2000°C there is considerable dissociation into atoms. The Te_2, Se_2, S_2, and O_2 molecules are all paramagnetic.

26-4 Discovery of the Elements

Sulfur was one of the nine elements known to the ancient world (along with gold, silver, copper, iron, lead, tin, mercury, and carbon). An old term for elementary sulfur is brimstone ("the stone that burns"). Lavoisier included sulfur and oxygen in his first list of elements. Tellurium was discovered in 1782 by Franz Müller (Baron von Reichenstein) in a Transylvanian gold ore, but the discovery was overlooked at the time. J. J. Berzelius discovered selenium in 1817 in the sludge from a sulfuric acid plant. Tellurium was named for the earth (Latin: *tellus*) and selenium for the moon (Greek: *selene*). Group VIA was completed with the discovery of polonium by the Curies in 1898.

26-5 Behavior Toward Other Elements

Elementary sulfur, selenium, and tellurium resemble elementary oxygen in their behavior toward other elements. They combine directly with numerous metals and nonmetals. Unlike oxygen, they combine directly with halogens. When the elements burn in air, they form the dioxides SO_2, SeO_2, and TeO_2.

Sulfur is much less reactive with hydrogen than is oxygen. The preparation of H_2S by direct union requires bubbling hydrogen through molten sulfur. The combination of selenium with hydrogen is endothermic and incomplete. Hydrogen telluride cannot be obtained by direct union.

26-6 The −2 Oxidation State

The elements of Group VIA show the -2 oxidation state in their binary compounds with metals and with nonmetals less electronegative than themselves. This oxidation state is also shown in organic compounds, e.g., CH_3OH and CH_3SH.

With the exception of water, the hydrogen compounds of these elements have melting points and boiling points that rise regularly with increasing molecular weight. Because of hydrogen bonding, water has the highest melting and boiling points of any of these compounds. (See Table 12-1.)

The gases H_2S, H_2Se, and H_2Te have highly offensive odors and are extremely poisonous. The odor of hydrogen sulfide can best be described as the odor of rotten eggs. When the shells of rotten eggs are broken, quantities of H_2S are released because this compound is a product of the putrefaction of sulfur-containing proteins when air is excluded. So-called "sulfur springs" contain dissolved H_2S, which is responsible for the unpleasant taste of the water. The odor of sewers is due in part to H_2S. It is also one of the gases issuing from volcanoes.

The heats of formation of gaseous H_2O, H_2S, H_2Se, and H_2Te are -58, -4.8, $+18.5$, and $+34$ kcal per mole, respectively. In view of these values it is not surprising that H_2Se and H_2Te are unstable and slowly decompose into the elements at room temperature, especially in the presence of oxygen, which converts the hydrogen to water. Because the oxygen atom is smallest and the tellurium atom is largest, the oxygen atom has the most attraction for the electron pairs bonding it to hydrogen atoms (in H_2O) and the tellurium atom has the least (in H_2Te). Thus, H_2O is the most thermally stable of these four hydrogen compounds, and H_2Te is the least stable.

Hydrogen sulfide burns to form water and sulfur dioxide gas if the oxygen supply is adequate. Water and sulfur are the products if the oxygen supply is limited:

$$2H_2S + 3O_2 \rightarrow 2H_2O + 2SO_2$$
$$2H_2S + O_2 \rightarrow 2H_2O + 2S$$

Burning hydrogen selenide with an adequate supply of oxygen produces water and a white smoke of selenium dioxide. Water and a deposit of red selenium are produced if the oxygen supply is limited. Hydrogen telluride similarly produces water and TeO_2 or water and Te, depending upon the oxygen supply, when it is burned. It reacts with atmospheric oxygen at ordinary temperatures, depositing elementary tellurium. If moisture is present, H_2S and H_2Se are also attacked by atmospheric oxygen. If an aqueous solution of H_2S has been standing on the shelf for some time, a precipitate can be seen consisting of elementary sulfur on the bottom of the bottle. The bottoms of "sulfur springs" are likewise covered with a sulfur sediment.

The three gases dissolve in water to the extent of about 0.1 mole per liter at ordinary temperatures and pressures. Their aqueous solutions are mildly acidic, the acid strengths of the three compounds increasing from H_2S to H_2Te, as indicated by their ionization constants.

HYDROSULFURIC ACID:	$H_2S \rightleftharpoons H^+ + HS^-$	$K = 1.1 \times 10^{-7}$
HYDROSELENIC ACID:	$H_2Se \rightleftharpoons H^+ + HSe^-$	$K = 1.7 \times 10^{-4}$
HYDROTELLURIC ACID:	$H_2Te \rightleftharpoons H^+ + HTe^-$	$K = 2.3 \times 10^{-3}$

The second ionization constants are very small, and the concentrations of S^{--}, Se^{--}, and Te^{--} are of a very low order except in alkaline solutions.

If we pass H_2S gas into NaOH solution until no more is absorbed, it is an indication that the OH^- ions have been completely replaced by HS^- ions:

$$(Na^+ + OH^-) + H_2S \rightarrow (Na^+ + HS^-) + H_2O$$

Thus, the solution now contains the "acid salt" sodium hydrogen sulfide, NaHS, which is also known as sodium hydrosulfide. If we wish to prepare a solution of the "normal salt" sodium sulfide, Na_2S, we divide the NaOH solution into two equal portions, saturate one portion with H_2S, and then mix the two portions:

$$(Na^+ + OH^-) + (Na^+ + HS^-) \rightarrow (2Na^+ + S^{--}) + H_2O$$

Solutions of soluble sulfides are used to remove hair from hides in the tanning industry. Almost all sulfides of metals except those of the elements of Groups IA and IIA are relatively insoluble, i.e., only very slightly soluble in water. The traditional scheme of systematic qualitative analysis for cations is based on the differences between the values of the solubility product constants for different sulfides. For example, when H_2S is passed into a solution that is 0.3 M in H^+ ions, 0.1 M in Cd^{++} ions, and 0.1 M in Zn^{++} ions, cadmium sulfide precipitates but zinc sulfide does not. Although both sulfides are relatively insoluble in water, there is a substantial difference between their K_{sp} values (Section 16-12). For ZnS, K_{sp} is 1.1×10^{-20}, but the K_{sp} value for CdS is only 3.6×10^{-29}. By establishing the required hydrogen-ion concentration (e.g., by adding the proper quantity of hydrochloric acid) before passing H_2S into the solution, the ionization of the H_2S may be sufficiently suppressed by the common-ion effect (Section 16-5) to keep the sulfide-ion concentration too small for the K_{sp} value for ZnS to be reached but large enough for the K_{sp} value for CdS to be attained.

The industrial method for preparing H_2S consists in bubbling hydrogen gas through molten sulfur, but one of the more convenient methods for obtaining it in the laboratory consists in treating certain sulfides of metals with a nonoxidizing acid, e.g., HCl. Iron(II) sulfide is suitable for this purpose and is available commercially. It is made by heating powdered iron with sulfur:

$$Fe + S \rightarrow FeS$$
$$FeS + 2H^+ \rightarrow Fe^{++} + H_2S\uparrow$$

The action of water alone is sufficient to obtain hydrogen sulfide from aluminum sulfide, or hydrogen selenide from aluminum selenide:

$$Al_2S_3 + 6H_2O \rightarrow 2Al(OH)_3 + 3H_2S\uparrow$$
$$Al_2Se_3 + 6H_2O \rightarrow 2Al(OH)_3 + 3H_2Se\uparrow$$

The readiness with which sulfide and selenide ions accept protons is the converse of the weakness of H_2S and H_2Se as acids.

Insoluble sulfides of metals occur extensively as minerals, and many of them are important ores. They often have associated with them small quantities of the selenides and tellurides of the metals.

The binary compounds of sulfur are regarded as sulfides only if the sulfur is the more electronegative of the two elements. Thus, P_4S_3 is regarded as a sulfide of phosphorus, but S_4N_4 is called sulfur nitride. One of the most important sulfides in commerce is carbon disulfide, CS_2. It is manufactured by the reaction of sulfur vapor with white-hot coke in an electric furnace. It is an excellent solvent for sulfur, fats, waxes, and rubber. Large quantities are used for making carbon tetrachloride and viscose rayon.

26-7 Polysulfides

Although elementary sulfur is very insoluble in water, it dissolves in aqueous solutions of soluble metal sulfides by forming *polysulfides*, which impart a yellow to dark red color to the solutions. The simplest of the polysulfide ions is the disulfide ion, S_2^{--}, which is analogous in formula and structure to the peroxide ion, O_2^{--}. Polysulfide ions in general can be formulated as S_n^{--}, where n is usually 2, 3, 4, or 5 but is in some instances larger. The greater the amount of sulfur dissolved in a sulfide solution, the greater are the relative amounts of S_4^{--} and S_5^{--} ions.

In its greater ability to undergo self-linkage through covalent bonds, sulfur differs markedly from oxygen. The electronic formulas for disulfide and pentasulfide ions are written as follows:

$$\left[:\ddot{\underset{..}{S}}:\ddot{\underset{..}{S}}:\right]^{--} \qquad \left[:\ddot{\underset{..}{S}}:\ddot{\underset{..}{S}}:\ddot{\underset{..}{S}}:\ddot{\underset{..}{S}}:\ddot{\underset{..}{S}}:\right]^{--}$$

disulfide ion — pentasulfide ion

This should not be taken to mean that the pentasulfide ion is linear. Although the atoms are in a continuous chain, the bond angle is more nearly 90°. The oxidation number of each sulfur atom in the disulfide ion is taken to be -1, the same as that of each oxygen atom in the peroxide ion. In the pentasulfide ion, the *average* oxidation number of the five sulfur atoms is -0.4.

Acidification of solutions containing polysulfide ions produces a white liquid, known as "milk of sulfur," which contains elementary sulfur in finely divided form:

$$S_2^{--} + 2H^+ \rightarrow H_2S\uparrow + S\downarrow$$

Polysulfide ions, like the peroxide ion, can act as oxidizing agents. This is illustrated by the reaction in which tin(II) sulfide is dissolved by treatment with sodium polysulfide solution, forming thiostannate ion:

$$SnS + S_2^{--} \rightarrow SnS_3^{--}$$

This equation shows disulfide ion oxidizing tin from the +2 state to the +4 state. When a solution containing SnS_3^{--} ion is acidified, SnS_2 precipitates.

Selenium dissolves in aqueous solutions of selenides to form polyselenide anions, Se_n^{--}, and tellurium, to a limited extent, dissolves in solutions containing telluride ion to form polyanions, Te_n^{--}. Polyselenides and polytellurides have a gray-black color.

When calcium hydroxide, sulfur, and water are heated together, a dark red liquid results that has been used extensively as a horticultural spray when diluted with water. The reactions of sulfur with base are complex and the "lime-sulfur" mixture contains calcium thiosulfate, CaS_2O_3, in addition to various calcium polysulfides, formed by such reactions as the following:

$$6OH^- + 4S \rightarrow 2S^{--} + S_2O_3^{--} + 3H_2O$$
$$S^{--} + 4S \rightarrow S_5^{--}$$

The use of lime-sulfur spray has greatly diminished because of the development of organic sulfur compounds as fungicides and insecticides for horticultural use.

Some disulfides occur as minerals. Pyrite ("fool's gold") is iron(II) disulfide, FeS_2. The crystalline structure of pyrite is similar to that of sodium chloride, with S_2^{--} ions occupying Cl^- positions and Fe^{++} ions occupying Na^+ positions.

26-8 Organic Compounds

Alcohols, ethers, aldehydes, ketones, carboxylic acids, and esters all contain oxygen (Chapter 19). In a formal sense these different classes of organic compounds can be regarded as derivatives of water. The formula of an alcohol is like that of water except for having an alkyl group in place of one of the hydrogen atoms. An ether can be regarded, in a strictly formal sense, as a derivative of water in which alkyl groups have taken the place of both hydrogen atoms.

Some organic compounds can be regarded as derivatives of hydrogen peroxide, and not merely in a formal sense. For example, the acid-catalyzed addition of hydrogen peroxide to isobutylene yields organic peroxides:

$$(CH_3)_2C{=}CH_2 + H_2O_2 \rightleftharpoons \underset{t\text{-butyl hydroperoxide}}{(CH_3)_3C{-}O{-}O{-}H}$$

$$(CH_3)_2C{=}CH_2 + (CH_3)_3C{-}O{-}O{-}H \rightleftharpoons \underset{t\text{-butyl peroxide}}{(CH_3)_3C{-}O{-}O{-}C(CH_3)_3}$$

Organic peroxides are frequently used as initiators or catalysts for important polymerizations.

Mercaptans, also known as thioalcohols or thiols, bear the same relationship to hydrogen sulfide that alcohols bear to water; and the organic sulfides (thioethers) are related to hydrogen sulfide exactly as ethers are related to water:

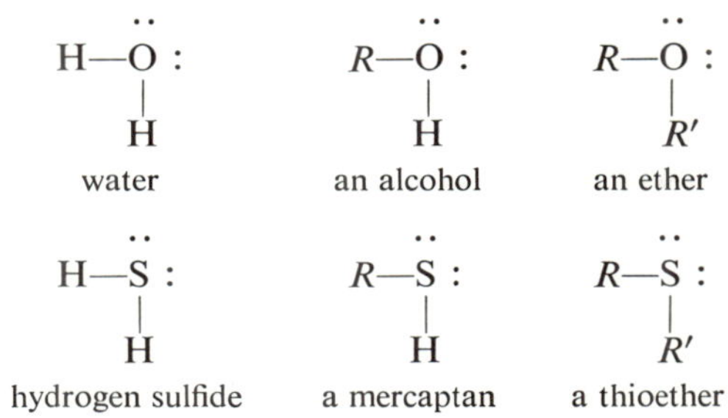

Even though alcohols and ethers are not made from water, and mercaptans and thioethers are not prepared from hydrogen sulfide, these formal relationships and analogies are useful.

Mercaptans are formed when an alkyl halide reacts with an alcoholic solution of sodium hydrogen sulfide:

$$RX + NaHS \rightarrow RSH + NaX$$

Like hydrogen sulfide, the volatile mercaptans are toxic. They have a disagreeable odor, which decreases with increasing molecular weight. The lachrymatory agent escaping from freshly cut onion is *n*-propyl mercaptan. Skunks use *n*-butyl mercaptan in their chemical warfare.

Mercaptans react readily with aqueous solutions of strong bases to form salts, called mercaptides:

$$(Na^+ + OH^-) + HSR \rightarrow (Na^+ + SR^-) + H_2O$$

Mercaptides of heavy metals are insoluble in water. The ease with which mercury can be precipitated as the insoluble salt gave rise to the name mercaptan (Latin: *mercurium captans*, mercury-catching).

Organic sulfides can be prepared by reaction of alkyl halides with sodium sulfide in alcoholic solution:

$$2RX + Na_2S \rightarrow R_2S + 2NaX$$

Mixed sulfides can be obtained by substituting sodium mercaptides for the sodium sulfide:

$$RX + NaSR' \rightarrow RSR' + NaX$$

Methyl sulfide, obtained commercially from lignin by heating the spent liquor from the manufacture of paper pulp with sodium sulfide, is used to impart an odor to fuel gas so that gas leaks may become known before they become dangerous. A variety of organic sulfides occur naturally in petroleum. Some crude oil contains as much as 5 percent sulfur in the form of sulfides.

Organic disulfides can be prepared by the reaction of alkyl halides with sodium disulfide:

$$2RX + Na_2S_2 \rightarrow R{-}S{-}S{-}R + 2NaX$$

Allyl disulfide, $(CH_2{=}CHCH_2)_2S_2$, is the chief component of oil of garlic. The sulfur-sulfur bond is easily broken, the disulfide being reduced to mercaptan. The chemical reactions that occur in the permanent waving of hair involve the reduction of disulfide linkages to permit shaping of the hair and the subsequent reconstitution of the linkages by oxidation to set the wave.

Other organic sulfur compounds in which sulfur atoms appear to have "replaced" oxygen atoms include the thioketones,

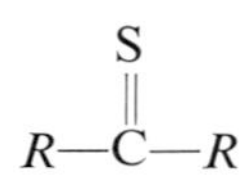

and the thiocarboxylic acids. Thioketones of low molecular weight are red oils with obnoxious odors. In theory there should be two isomeric types of thiocarboxylic acids,

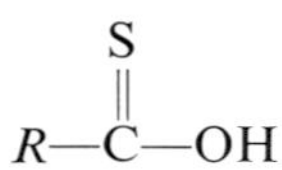 and 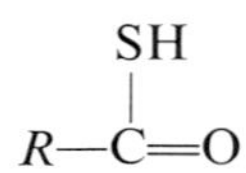

but only one substance can be isolated because of the rapid interconversion of the two forms. The dithio acids,

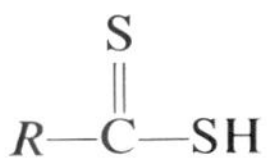

are colored oils with highly offensive odors.

26-9 Production and Uses of the Elements

Sulfur occurs free in nature, and until the present century its main sources were the volcanic regions of Italy and Japan. Since 1900 it has been obtained chiefly from the extensive underground deposits in Texas and Louisiana. More recently, important deposits have been discovered and developed on the Isthmus of Tehuantepec in Mexico.

Sulfur is obtained from the underground deposits by the hot-water mining process developed by the German-American chemical engineer Herman Frasch, in the decade 1890–1900. In the Frasch process, water is heated under pressure to a temperature of 170°C and pumped down to the sulfur deposits, which may be as deep as 2000 feet beneath the surface. The sulfur is melted out of the rocks in which it is embedded and forced to the surface by compressed air. The superheated water is forced down the outermost of three concentric pipes, and the compressed air is forced down the innermost pipe. Liquid sulfur mixed with air rises through the third pipe in the form of a froth. The sulfur is conveyed to large storage bins in which it settles and solidifies. The sulfur is at least 99.5 percent pure and can be used without further purification for most purposes.

Sulfur is also obtained as a by-product in the refining of natural gas and petroleum. Hydrogen sulfide from the gas or from the oil refinery is burned to sulfur dioxide, and the resulting sulfur dioxide is then allowed to react with more hydrogen sulfide to form sulfur:

$$2H_2S + 3O_2 \rightarrow 2H_2O + 2SO_2$$
$$SO_2 + 2H_2S \rightarrow 2H_2O + 3S\downarrow$$

These same reactions account for the formation of sulfur in volcanic craters.

Numerous sulfates and sulfides of metals are found in nature, and many of them are useful minerals. They are not exploited as sources of elementary sulfur, but the burning of pyrite and the roasting of sulfide ores are important sources of sulfur dioxide.

The most important use for sulfur is in making sulfuric acid. Large quantities are also used in vulcanizing rubber. Other important uses include the synthesis of carbon disulfide and the making of gunpowder. So-called sulfur monochloride, S_2Cl_2, is a good solvent for rubber and is used in the vulcanization of rubber. It is an orange liquid with an offensive odor, obtained by the direct action of chlorine on molten sulfur.

Selenium occurs in small amounts, mostly as metal selenides in sulfide ores of metals. Several thousand acres of semiarid soil in Wyoming and South

Dakota contain enough selenium to make certain plants toxic to animals. Loco disease, a nervous disease that can result in "blind staggers" or even death, is often caused by eating such plants, which are known as locoweed or crazyweed.

The chief commercial source of selenium is the anode mud from the electrolytic purification of copper. The green color in glass that is due to its iron content is neutralized by addition of small amounts of selenium. Larger quantities give the glass a clear red color suitable for signal lights.

Many of the uses of selenium are related to its unusual electrical properties. Selenium develops a great increase in electrical conductance when it is exposed to light. It regains its resistance almost immediately when the light is turned off. Selenium cells are used in burglar alarms, automatic door openers, and the portable exposure meters used by photographers. Selenium of high purity has been widely used to rectify alternating current in electronic equipment.

Tellurium occurs in small amounts, most often combined with gold, silver, or lead as a telluride. The town of Telluride, Colorado, was named for the type of gold ore that was mined there. Like selenium, tellurium is obtained mainly as a by-product of the purification of copper. Small quantities of tellurium are alloyed with lead to increase its resistance to corrosion. Extensive use of this element awaits further developments.

26-10 The +4 Oxidation State

The +4 oxidation state is shown in the dioxides SO_2, SeO_2, and TeO_2; in the sulfites, selenites, and tellurites; and in the tetrahalides.

The dioxides can be obtained by burning the elements in air. Sulfur dioxide is a colorless gas with a choking odor (the odor of burning sulfur), but selenium dioxide and tellurium dioxide are white solids. It has been established that in the gaseous state the dioxides of sulfur and selenium exist as simple SO_2 and SeO_2 molecules having bond angles of about 120°. If the octet rule is assumed to apply to all three atoms in the molecule, it is possible to write the electronic formula for SO_2 in two different ways:

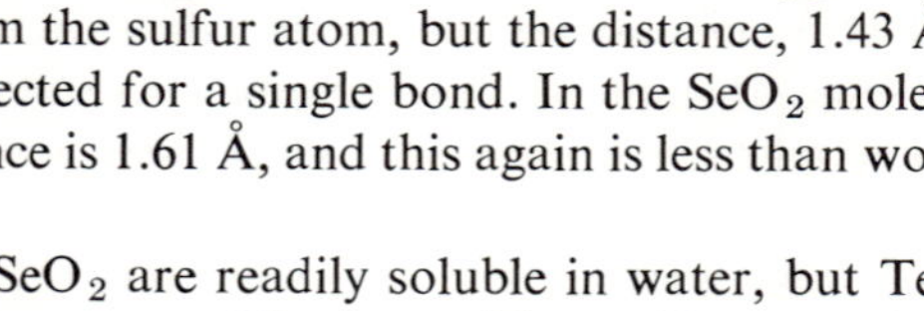

The structure of the SO_2 molecule is described as a resonance hybrid involving both of these contributing structures. The two oxygen atoms are situated equally distant from the sulfur atom, but the distance, 1.43 Å, is somewhat less than would be expected for a single bond. In the SeO_2 molecule, the selenium-oxygen bond distance is 1.61 Å, and this again is less than would be expected for a single bond.

Both SO_2 and SeO_2 are readily soluble in water, but TeO_2 is only slightly soluble. The solutions are acidic, but neither sulfurous acid nor tellurous acid has been isolated. Selenous acid, H_2SeO_3, is a white solid that is hygroscopic in a humid atmosphere but readily loses water, reverting to SeO_2, in a dry atmosphere. Addition of an acid to a solution of potassium tellurite, K_2TeO_3, gives a hydrous precipitate of TeO_2. Acidification of sulfite solutions liberates SO_2 gas.

$$TeO_3^{--} + 2H^+ \rightarrow TeO_2\downarrow + H_2O$$
$$SO_3^{--} + 2H^+ \rightarrow SO_2\uparrow + H_2O$$

Each of the three acids is diprotic:

SULFUROUS ACID:	$H_2SO_3 \rightleftharpoons H^+ + HSO_3^-$	$K_1 = 1.2 \times 10^{-2}$
	$HSO_3^- \rightleftharpoons H^+ + SO_3^{--}$	$K_2 = 5.6 \times 10^{-8}$
SELENOUS ACID:	$H_2SeO_3 \rightleftharpoons H^+ + HSeO_3^-$	$K_1 = 2.7 \times 10^{-3}$
	$HSeO_3^- \rightleftharpoons H^+ + SeO_3^{--}$	$K_2 = 2.5 \times 10^{-7}$
TELLUROUS ACID:	$H_2TeO_3 \rightleftharpoons H^+ + HTeO_3^-$	$K_1 = 2 \times 10^{-3}$
	$HTeO_3^- \rightleftharpoons H^+ + TeO_3^{--}$	$K_2 = 1 \times 10^{-8}$

Normal salts, e.g., Na_2SO_3, and acid salts, e.g., $NaHSO_3$, are known for all three acids. The acid salts are more soluble than the normal salts. The only normal sulfites that are soluble are the alkali-metal sulfites. Solutions of the normal salts are alkaline as a result of hydrolysis:

$$SO_3^{--} + H_2O \rightleftharpoons HSO_3^- + OH^-$$

Because sulfur is in an intermediate oxidation state in SO_2, sulfurous acid, and the sulfites, these compounds can serve either as oxidizing agents or as reducing agents. The reducing power of sulfites is indicated by their slow oxidation to sulfates on standing in contact with atmospheric oxygen. With halogens, conversion to sulfates is rapid and complete. Selenous and tellurous acids are much weaker reducing agents than sulfurous acid. Conversely, they are much stronger oxidizing agents; i.e., they are more readily reduced than sulfurous acid. The common procedure for obtaining red selenium is to precipitate it by reducing a selenite with SO_2 in acid solution.

$$SeO_3^{--} + 2SO_2 + H_2O \rightarrow Se\downarrow + 2SO_4^{--} + 2H^+$$

Sulfur dioxide occurs in volcanic gases, and it is sometimes encountered in the atmosphere near industrial plants where it is formed, for example, by burning low-grade coal containing sulfur compounds. Commercially, SO_2 is obtained by burning sulfur or pyrite and by roasting sulfide ores, i.e., heating them in air, which is the first step in their metallurgical treatment:

$$S + O_2 \rightarrow SO_2$$
$$4FeS_2 + 11O_2 \rightarrow 2Fe_2O_3 + 8SO_2$$
$$Cu_2S + 2O_2 \rightarrow 2CuO + SO_2$$

In the laboratory, sulfur dioxide may be conveniently obtained by treating sodium sulfite or sodium hydrogen sulfite (sodium "bisulfite") with sulfuric acid.

The reaction produces sulfurous acid, which decomposes to give sulfur dioxide and water:

$$2H^+ + SO_3^{--} \rightarrow H_2SO_3 \rightarrow H_2O + SO_2$$
$$H^+ + HSO_3^{--} \rightarrow H_2SO_3 \rightarrow H_2O + SO_2$$

The commercial uses of sulfur dioxide include the bleaching of straw, paper, and wool. It is also used as a preservative for dried fruits. In aqueous solution, as sulfurous acid, it acts as a bleaching agent, either by reducing colored compounds to colorless compounds or by combining with colored compounds to form colorless addition products. The colorless addition compounds are often decomposed by light. Thus, paper that has been bleached by SO_2 turns yellow in sunlight. The same is true of straw hats. A solution of calcium bisulfite, $Ca(HSO_3)_2$, made by reaction of sulfur dioxide with calcium hydroxide, is used for treating wood chips to dissolve lignin, the substance that cements the cellulose fibers together in wood. This treatment also bleaches the wood, so the result is white wood pulp suitable for making paper.

About three fourths of the sulfur dioxide produced each year in the United States is used for manufacturing sulfuric acid. Half of the remainder is used in the paper industry.

26-11 The +6 Oxidation State

Sulfur, selenium, and tellurium show the oxidation state +6 in the trioxides SO_3, SeO_3, and TeO_3; in the oxyacids H_2SO_4, H_2SeO_4, and H_6TeO_6; and in the hexafluorides SF_6, SeF_6, and TeF_6.

The hexafluorides are obtained by direct union of the elements. The chemical inertness of SF_6 is comparable to that of the noble gases. It is not attacked by oxygen, even at high temperatures, but it will react with sodium at 250°C. Its inertness and high dielectric strength make it useful as a gaseous insulator in high-voltage equipment. The hexafluorides of selenium and tellurium are considerably less inert than SF_6. On contact with water for 24 hours, TeF_6 undergoes complete hydrolysis.

Because the burning of the elements in air produces the dioxides, the preparation of the trioxides is not simple. Dehydration of telluric acid gives TeO_3, but dehydration of selenic acid gives selenium dioxide and oxygen. Pure selenium trioxide has never been prepared. Sulfur trioxide is manufactured by the reaction of sulfur dioxide with oxygen:

$$2SO_2 + O_2 \rightleftharpoons 2SO_3$$

In terms of economics there is no more important chemical reaction than this one, for the manufacture of sulfuric acid depends on it. The success of the reaction requires the use of moderate temperatures and effective catalysts.

Sulfur trioxide is a highly reactive compound. It is a powerful oxidizing agent that liberates all the halogens except fluorine from metallic and nonmetallic halides. It combines with water to give sulfuric acid and with hydrogen chloride to form chlorosulfonic acid:

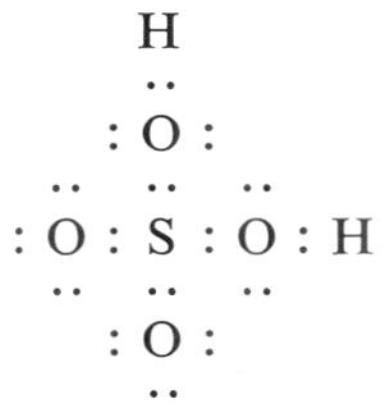

sulfuric acid

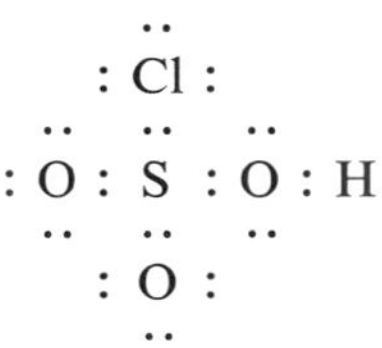

chlorosulfonic acid

The SO_3 molecule is planar and has the configuration of an equilateral triangle with the sulfur atom at the center. Its electronic formula can be written in three different ways with each atom having an octet:

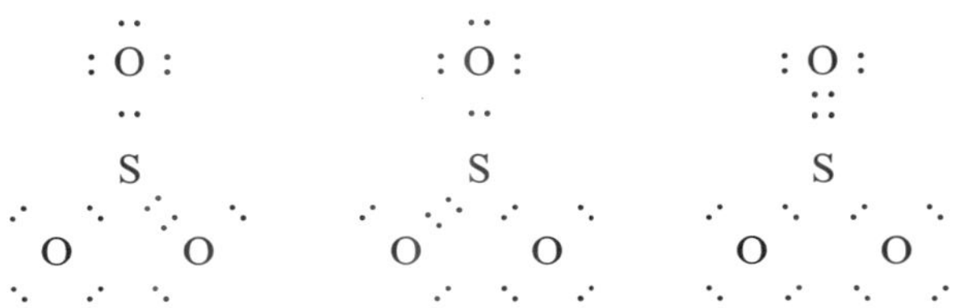

The structure of the SO_3 molecule is described as a hybrid of the various contributing structures.

Commercial SO_3 is a polymer, which is a mixture of two crystalline forms. It has a melting point that approximates 40°C but is dependent on the proportions of the two forms present. Liquid sulfur trioxide boils at 44.8°C.

Anhydrous H_2SO_4 is a viscous, colorless liquid that freezes at 10.4°C. It is sometimes called oil of vitriol. When heated to 290°C it begins to boil, but the boiling is accompanied by the formation of SO_3 vapor. One of the notable characteristics of sulfuric acid is its tremendous affinity for water. Its aqueous solutions are strongly acidic. In dilute solution the first stage of its dissociation is virtually complete; the second stage is incomplete:

$$H_2SO_4 \rightleftharpoons H^+ + HSO_4^- \qquad \text{virtually complete}$$
$$HSO_4^- \rightleftharpoons H^+ + SO_4^{--} \qquad K = 1.2 \times 10^{-2}$$

Selenic acid is very similar to sulfuric acid in its acidic properties:

$$H_2SeO_4 \rightleftharpoons H^+ + HSeO_4^- \qquad \text{virtually complete}$$
$$HSeO_4^- \rightleftharpoons H^+ + SeO_4^{--} \qquad K = 1.2 \times 10^{-2}$$

Anhydrous selenic acid is a solid that melts at 58°C. It is less stable than sulfuric acid and begins to evolve oxygen when heated above 210°C. Selenic acid is a more powerful oxidizing agent than sulfuric acid.

No compound of the formula H_2TeO_4 is known. The common form of telluric acid is H_6TeO_6, orthotelluric acid. It exists in two crystalline modifications. In each molecule of the acid the six —OH groups that are bonded to the central Te atom are arranged octahedrally. The acid is a weak, dibasic acid in aqueous solution, the first dissociation constant being 6×10^{-7}. Above 300°C telluric acid decomposes, giving TeO_3. The acid is a strong oxidizing agent.

The salts of sulfuric acid and selenic acid are similar, and corresponding salts are usually isomorphous. The sulfate and selenate ions are tetrahedral, the sulfur-oxygen bond distance being 1.51 Å and the selenium-oxygen bond distance being 1.61 Å. The acids form normal salts, e.g., Na_2SO_4 and Na_2SeO_4, and acid salts, e.g., $NaHSO_4$ and $NaHSeO_4$. Telluric acid forms salts of the types represented by NaH_5TeO_6 and $Na_2H_4TeO_6$.

26-12 Sulfuric Acid in Nature

The sulfide minerals, in the presence of moisture and air, are converted to sulfates by weathering. For example, pyrite is a common impurity in coal seams, and its weathering (hastened by the bacterium *Thiobacillus thio-oxidans*) produces dilute sulfuric acid, which drains continuously from the mine and causes a serious stream pollution problem:

$$2FeS_2 + 7O_2 + 2H_2O \rightarrow 2FeSO_4 + 2H_2SO_4$$

Iron(II) sulfate occurs native in small amounts, always in conjunction with an iron sulfide mineral. It crystallizes from solution in the form of greenish crystals of the heptahydrate, $FeSO_4 \cdot 7H_2O$, which is sometimes called green vitriol. Various salts of sulfuric acid have been known as vitriols for hundreds of years. They have been given names that indicate their colors. White vitriol is $ZnSO_4 \cdot 7H_2O$, and blue vitriol is $CuSO_4 \cdot 5H_2O$.

26-13 Manufacture of Sulfuric Acid

Sulfuric acid is used for industrial purposes in greater quantities than any other man-made compound. The amount of it produced each year in the United States can be used as a measure of the condition of the nation's economy. There are three steps in its synthesis: the production of SO_2 by burning sulfur or pyrite; the oxidation of SO_2 to SO_3; and the conversion of SO_3 to H_2SO_4. The acid is obtained by the *lead-chamber process* and the *contact process.*

The first lead-chamber plant was built in England in 1746. The reactions are carried out in the gaseous phase in large lead-lined rooms. These reactions are complicated and not completely understood. Essentially, the process consists of the catalyzed conversion of SO_2 to SO_3, with gaseous nitric oxide serving as a "carrier catalyst":

$$2NO + O_2 \rightarrow 2NO_2$$
$$SO_2 + NO_2 \rightarrow SO_3 + NO$$

In its mechanism the process is much more complicated than is indicated by the above equations. It is believed that nitrosyl sulfuric acid, $H(NO)SO_4$, is an intermediate, which then reacts with water to form sulfuric acid and regenerate the catalyst. The product is of low concentration (66 percent) and is impure.

For many purposes, pure concentrated sulfuric acid is needed. This is obtained directly in the modern contact process. Almost all sulfuric acid plants constructed in recent decades use this process, in which SO_2 is oxidized by means of air in the presence of a solid catalyst with which the reacting gases must make contact. The original contact catalyst for this process was platinized asbestos, but it has been replaced by the less expensive and more serviceable vanadium(v) oxide, V_2O_5.

The oxidation of SO_2 to SO_3 in the absence of a catalyst is a very slow reaction. Since the reaction is exothermic and reversible, the high temperatures needed to make the reaction a rapid one serve to prevent a high equilibrium yield. The rate of an endothermic reaction is always increased more than the rate of its exothermic opposite when a system in equilibrium is raised to a higher temperature (van't Hoff's Law). Thus, the manufacture of SO_3 requires a catalyst. In both processes the catalyst is so effective that the use of high pressure is not necessary.

For practical reasons, the gaseous SO_3 formed in the contact process is absorbed in concentrated sulfuric acid rather than in water. The absorption of the oxide by water from a gaseous mixture of air and SO_3 is complicated by the formation of a fine mist of H_2SO_4 in the gas stream. Consequently, the absorption of SO_3 in water is very incomplete. The gas is readily absorbed by concentrated H_2SO_4, with which it reacts to form disulfuric acid, $H_2S_2O_7$ (sometimes called "pyrosulfuric acid"). Commercial fuming sulfuric acid or oleum is an equilibrium mixture of $H_2S_2O_7$, SO_3, and H_2SO_4. If the desired product from the contact process is "concentrated" rather than "fuming" sulfuric acid, a regulated influx of dilute sulfuric acid is used to maintain the desired composition. The laboratory reagent called concentrated sulfuric acid usually contains about 2 percent water and 98 percent H_2SO_4.

26-14 Properties and Uses of Sulfuric Acid

The affinity of sulfuric acid for water is so tremendous that dilution with water releases large quantities of heat. The acid must be added slowly and cautiously to the water, with stirring and cooling. If the water is added to the acid, the less dense water floats on the surface of the acid, and the heat released at the interface causes boiling with a violent and hazardous spurting of the liquid. If the acid is added to the water, the dense acid sinks, thereby undergoing mixing while in contact with a large quantity of water that serves to distribute the heat.

Sulfuric acid is used as a drying agent for gases and for taking up the water formed in certain reversible reactions, e.g., esterification reactions. It will remove the elements of water from organic compounds that contain hydrogen and oxygen in the 2:1 ratio. For example, it causes sugar to char:

$$C_{12}H_{22}O_{11} \rightarrow 12C + 11H_2O$$

A convenient method for generating carbon monoxide consists in allowing formic acid to drop into hot sulfuric acid:

$$HCOOH \rightarrow H_2O + CO\uparrow$$

Concentrated sulfuric acid reacts with aromatic hydrocarbons to form sulfonic acids in which sulfur is linked to carbon.

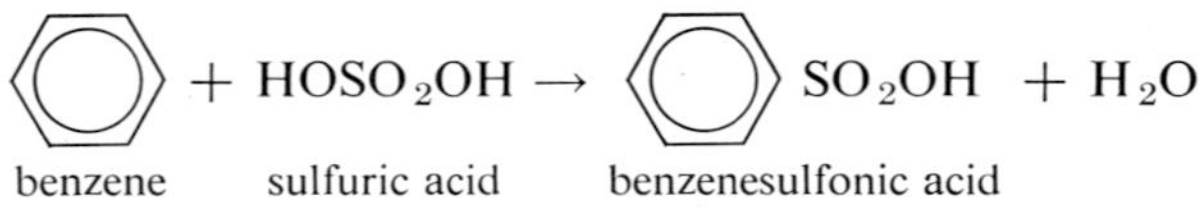

Sulfuric acid forms two series of esters in which oxygen is linked to carbon. These are illustrated by methyl sulfate, $(CH_3)_2SO_4$, and methyl hydrogen sulfate, CH_3HSO_4.

Hot, concentrated sulfuric acid acts as an oxidizing agent toward many elements and compounds. The product of its own reduction is SO_2, S, or H_2S, depending on the strength of the reducing agent and other circumstances. Such metals as zinc and aluminum liberate hydrogen from *dilute* sulfuric acid, but the concentrated acid is reduced to SO_2 by metals both above and below hydrogen in the activity series.

$$Zn + H_2SO_4\ (\text{dilute}) \rightarrow ZnSO_4 + H_2$$
$$Cu + H_2SO_4\ (\text{dilute}) \rightarrow \text{no reaction}$$
$$Zn + 2H_2SO_4\ (\text{concentrated}) \rightarrow ZnSO_4 + 2H_2O + SO_2$$
$$Cu + 2H_2SO_4\ (\text{concentrated}) \rightarrow CuSO_4 + 2H_2O + SO_2$$

Zinc is a sufficiently powerful reducing agent that quantities of sulfur are also formed by its reaction with concentrated sulfuric acid.

$$3Zn + 4H_2SO_4 \rightarrow 3ZnSO_4 + 4H_2O + S$$

Since nonmetals do not form sulfates, their reaction with concentrated sulfuric acid forms either an acidic oxide or an oxyacid:

$$C + 2H_2SO_4 \rightarrow CO_2 + 2SO_2 + 2H_2O$$
$$S + 2H_2SO_4 \rightarrow 3SO_2 + 2H_2O$$
$$2P + 5H_2SO_4 \rightarrow 2H_3PO_4 + 5SO_2 + 2H_2O$$

One of the most important uses for sulfuric acid is in the preparation of other more volatile acids by displacement from their salts. Thus, hydrogen chloride is obtained by heating chlorides with sulfuric acid:

$$NaCl + H_2SO_4 \rightarrow NaHSO_4 + HCl$$

One of the major uses of sulfuric acid is in the fertilizer industry, where it is used to convert gaseous ammonia to solid ammonium sulfate and to convert insoluble calcium phosphate rock to soluble calcium dihydrogen phosphate:

$$2NH_3 + H_2SO_4 \rightarrow (NH_4)_2SO_4$$
$$Ca_3(PO_4)_2 + 2H_2SO_4 \rightarrow Ca(H_2PO_4)_2 + 2CaSO_4$$

The economic importance of sulfuric acid is due to a combination of its cheapness, its acid strength, its high boiling point, and its affinity for water.

26-15 Hydrogen Peroxide and the Peroxysulfuric Acids

When sulfuric acid solutions which are about 50 percent H_2SO_4 are subjected to electrolysis at low temperature, the HSO_4^- ion is oxidized at the anode with the formation of peroxydisulfuric acid, $H_2S_2O_8$:

$$2HSO_4^- \rightarrow H_2S_2O_8 + 2e^-$$

Peroxydisulfuric acid is also obtained by electrolysis of acid sulfates, e.g., $KHSO_4$. Electrolytic oxidation of concentrated solutions of sulfates gives the corresponding peroxydisulfates:

$$2SO_4^{--} \rightarrow S_2O_8^{--} + 2e^-$$

The peroxydisulfate ion is a very powerful oxidizing agent.

Fractional distillation of a solution containing peroxydisulfuric acid yields a distillate that contains about 30 percent hydrogen peroxide resulting from the hydrolysis of the peroxydisulfuric acid:

$$H_2S_2O_8 + 2H_2O \rightarrow H_2O_2 + 2H_2SO_4$$

One of the most important processes for the manufacture of hydrogen peroxide involves fractional distillation of the 30 percent solution under reduced pressure. Solutions containing about 90 percent H_2O_2 can be obtained by this procedure. Fractional crystallization is required to give a product of still higher concentration.

At room temperature 30 percent hydrogen peroxide reacts with concentrated sulfuric acid to give peroxymonosulfuric acid, H_2SO_5:

$$H_2O_2 + H_2SO_4 \rightarrow H_2SO_5 + H_2O$$

Salts of peroxymonosulfuric acid are unstable and have not been obtained in pure form.

The peroxysulfuric acids can be obtained in the anhydrous state by the reaction of anhydrous hydrogen peroxide with chlorosulfonic acid. The product obtained is determined by the ratio in which the reactants are brought together:

$$HSO_3Cl + H_2O_2 \rightarrow H_2SO_5 + HCl$$
$$2HSO_3Cl + H_2O_2 \rightarrow H_2S_2O_8 + 2HCl$$

Both peroxysulfuric acids may be considered to be derivatives of hydrogen peroxide. Both contain the peroxy group characteristic of hydrogen peroxide. In peroxymonosulfuric acid one of the H atoms of H_2O_2 has been replaced by the HSO_3 group. In peroxydisulfuric acid both H atoms of H_2O_2 have been replaced by HSO_3 groups:

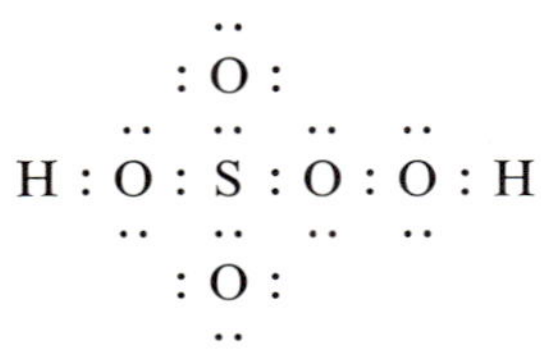

peroxymonosulfuric acid
(melting point 45°C)

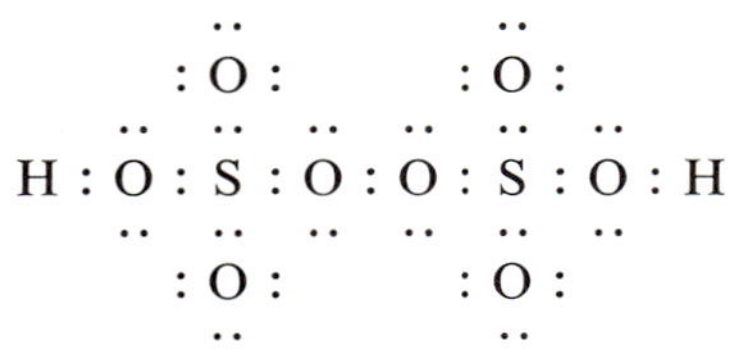

peroxydisulfuric acid
(melting point 65°C)

The oxidizing action of the peroxysulfuric acids involves the reduction of the peroxy oxygen atoms from the -1 oxidation state to the -2 state; there is no change in the oxidation state of the sulfur.

Peroxy acids containing selenium and tellurium are not known.

26-16 Thiosulfates

Elementary sulfur dissolves in aqueous solutions of sulfites to form thiosulfates:

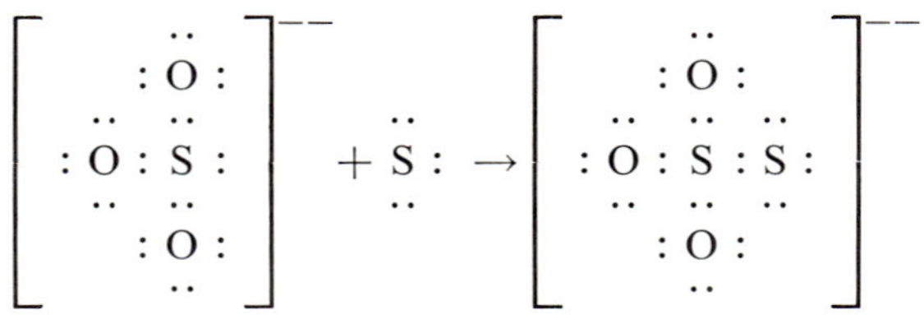

This reaction is analogous to the oxidation of sulfite by oxygen. Note that the two sulfur atoms in the $S_2O_3^{--}$ ion are not equivalent. The ion may be thought of as being formally derived from the sulfate ion, SO_4^{--}, through replacement of one oxygen atom by a sulfur atom (even though this is not the procedure by which it is prepared). The name thiosulfate is intended to imply this formal relationship.

In the thiosulfate ion it is reasonable to assign the oxidation number $+6$ to the central sulfur atom because that is the value assigned to it in the sulfate ion. It is likewise reasonable to assign the oxidation number -2 to the other sulfur atom because that is the value assigned to each oxygen atom in the sulfate ion (and also in the thiosulfate ion). Thus, the average of the oxidation numbers of the two sulfur atoms in the thiosulfate ion is $+2$, and in a formal sense sulfur may be said to be in the $+2$ state in thiosulfates. This is the value immediately arrived at by inspection of the formula $S_2O_3^{--}$, using the customary rules for determining oxidation numbers.

The thiosulfate ion forms rather stable complexes with several of the heavy metal cations. For example, silver halides, which are only very slightly soluble in water, are readily dissolved by sodium thiosulfate solution:

$$AgBr(s) + 2S_2O_3^{--} \rightleftharpoons [Ag(S_2O_3)_2]^{---} + Br^-$$

This reaction accounts for the usefulness of sodium thiosulfate ("hypo") as a fixing agent in photography.

Thiosulfuric acid is unstable in the presence of water and cannot be isolated. When a solution of a thiosulfate is acidified, there is rapid decomposition to give sulfur dioxide and a sulfur precipitate:

$$S_2O_3^{--} + 2H^+ \rightarrow H_2O + SO_2\uparrow + S\downarrow$$

When an aqueous solution of hypo is exposed to the air it becomes white because carbon dioxide dissolves in the solution and produces enough carbonic acid to bring about this reaction in which free sulfur is liberated. Thiosulfuric acid has been isolated from the reaction between sodium thiosulfate and hydrogen chloride in ether.

Mild oxidation of sodium thiosulfate by iodine gives sodium tetrathionate:

$$I_2 + 2S_2O_3^{--} \rightarrow 2I^- + S_4O_6^{--}$$

Sodium thiosulfate is therefore used in conjunction with potassium iodide for the quantitative determination of oxidizing agents by volumetric analysis. The iodine liberated by the oxidation of iodide ions is titrated with a standard solution of sodium thiosulfate.

Bromine and chlorine, being more powerful oxidizing agents than iodine, cause oxidation of thiosulfate to sulfate:

$$S_2O_3^{--} + 4Cl_2 + 5H_2O \rightarrow 2SO_4^{--} + 8Cl^- + 10H^+$$

In bleaching operations, sodium thiosulfate is used as an "antichlor" to remove residual chlorine from the bleached fabric.

26-17 Dithionites and Dithionates

Sodium dithionite, $Na_2S_2O_4$, can be obtained by reducing sodium hydrogen sulfite solution with zinc dust. Formally, the sulfur changes in this reaction from the +4 state to the +3 state. As dry solids, the dithionites are stable salts. In alkaline solution they are powerful reducing agents. Sodium dithionite can be used for the complete removal of oxygen from gas mixtures. It is an important reducing agent in the organic dye industry.

When a solution of a dithionite is acidified, there is rapid decomposition to give sulfur dioxide and a sulfur precipitate. Dithionous acid is, however, a fairly strong acid:

$$H_2S_2O_4 \rightleftharpoons H^+ + HS_2O_4^- \qquad K_1 = 4.5 \times 10^{-1}$$
$$HS_2O_4^- \rightleftharpoons H^+ + S_2O_4^{--} \qquad K_2 = 3.5 \times 10^{-3}$$

Dithionic acid, $H_2S_2O_6$, is a strong acid and is fairly stable in solution. Dithionates are obtained by oxidation of sulfites, the sulfur changing formally from the +4 state to the +5 state. For reasons not well understood, dithionic acid and its salts are not attacked by such strong oxidizing agents as hypochlorites, and neither are they attacked by any but the most powerful reducing agents.

In the dithionite and dithionate ions the sulfur atoms are bonded to each other. In the dithionite ion, two oxygen atoms are bonded to each sulfur atom. In the dithionate ion, each sulfur atom has three oxygen atoms bonded to it.

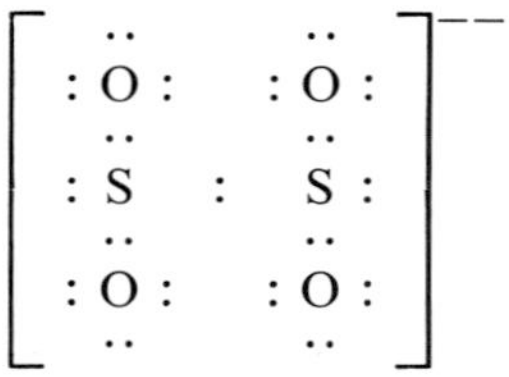
dithionite ion

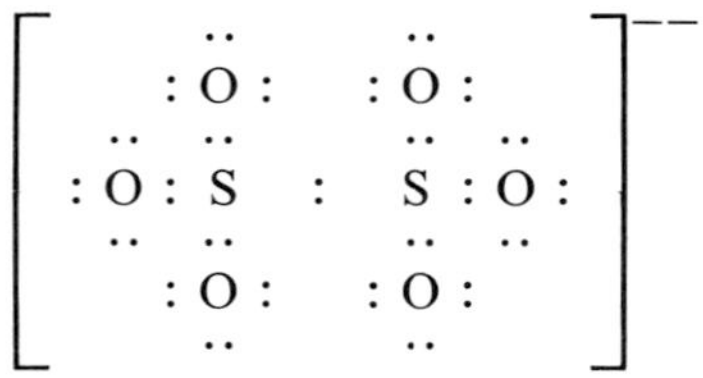
dithionate ion

26-18 Polythionates

The polythionate ions have the general formula $[O_3SS_nSO_3]^{--}$, where n is 1, 2, 3, or 4. Their individual names indicate the total number of sulfur atoms. Thus, $S_3O_6^{--}$ is the trithionate ion, $S_4O_6^{--}$ is the tetrathionate ion, $S_5O_6^{--}$ is the pentathionate ion, and $S_6O_6^{--}$ is the hexathionate ion. The free acids are apparently strong acids, but they are highly unstable.

Mixtures of polythionates are obtained by reaction between thiosulfate ions and sulfur dioxide; for unknown reasons this reaction requires the presence of arsenic(III) oxide. A mixture of polythionic acids is formed by passing H_2S into an aqueous solution of SO_2. Fortunately, some of the individual polythionates can be obtained by selective methods. For example, the action of hydrogen peroxide on cold sodium thiosulfate solution produces sodium trithionate:

$$2S_2O_3^{--} + 4H_2O_2 \rightarrow S_3O_6^{--} + SO_4^{--} + 4H_2O$$

Sodium tetrathionate is formed in the reaction between iodine and sodium thiosulfate.

Structurally, the distinguishing feature of the polythionates is the presence of one or more sulfur atoms that are bonded only to other sulfur atoms:

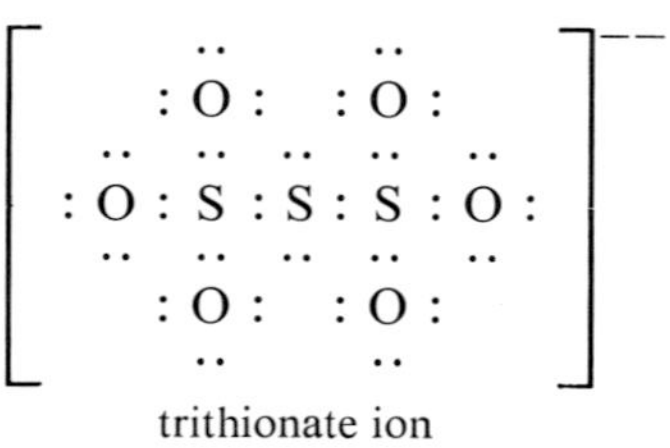
trithionate ion

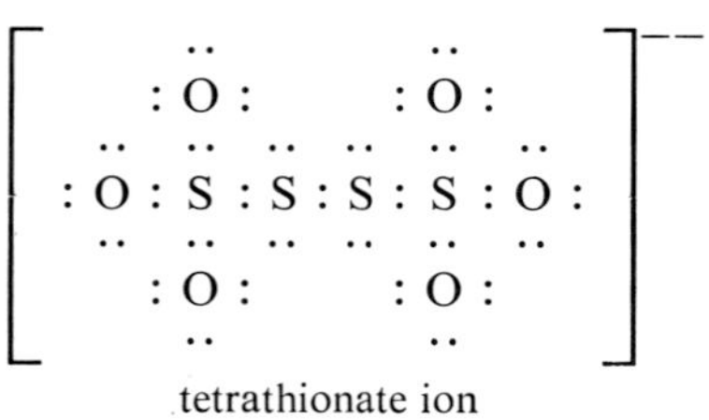
tetrathionate ion

The average oxidation numbers of the sulfur atoms in the polythionates vary from $+3\frac{1}{3}$ in the trithionates down to $+1\frac{2}{3}$ in the hexathionates. All of the polythionates contain sulfur in a lower oxidation state than do sulfites, whereas dithionates contain sulfur in a higher oxidation state than do sulfites. Polythionates are easily oxidized but dithionates are not.

New Terms

Mercaptan (thioalcohol): any compound, *R*SH, that contains an —SH group and a hydrocarbon group, e.g., methyl mercaptan, CH_3SH.

Polysulfide: a compound which contains a polysulfide ion, S_n^{--}, where n is usually 2, 3, 4, or 5 but is in some instances larger.

Specific resistivity: the electrical resistance offered by a piece of material 1 cm^2 in cross section and 1 cm in length.

Thioether: any compound conforming to the general formula *R*—S—*R*′.

Transition temperature: the temperature at which one crystalline form of a substance undergoes transformation into another crystalline form.

Exercises

26-1 List the important oxidation states of the elements of the oxygen-sulfur family, with examples of compounds in each of the states.

26-2 Show the analogy between peroxides and polysulfides.

26-3 Identify the following: "milk of sulfur," "lime-sulfur," fool's gold, "hypo," oil of vitriol, oleum.

26-4 Account for the fact that sulfuric acid is the most important of all industrial chemicals.

26-5 How could you easily distinguish a sulfite from a sulfate? Why would you expect to find sulfate present as an impurity in a sulfite?

26-6 Account for the apparent oxidation states of sulfur in thiosulfates, dithionates, dithionites, polythionates and polysulfides.

26-7 Why do helium, selenium, and tellurium have names ending in *-ium*?

26-8 Write the electronic formulas for peroxydisulfuric acid and tetrathionic acid.

26-9 What is a mercaptan? A mercaptide? A thioketone? A thiol? A dithiocarboxylic acid?

26-10 Account for the use of sodium thiosulfate (a) as a "fixer" in photography, (b) as an antichlor, and (c) in the volumetric determination of oxidizing agents.

References

Haynes, W., *Brimstone: The Stone that Burns*, Van Nostrand, Princeton, N.J., 1959. An account of the Frasch chemical industry in the United States and Mexico.

Weeks, M. E., and Leicester, H. M., *Discovery of the Elements*, 6th ed., Chemical Education Publishing Co., Easton, Pa., 1956. Sulfur is discussed (pp. 52–58) and the discoveries of tellurium and selenium are recounted (Chapter 11).

27

Nitrogen

27-1 Physical and Chemical Properties

Nitrogen is the first member of Group VA of the periodic table, which also includes phosphorus, arsenic, antimony, and bismuth. As is usually the case with the first members of the periodic groups, nitrogen exhibits some anomalous properties. Since the chemistry of this element involves several different types of compounds of unusual importance, we shall first study nitrogen apart from the other members of its family. Then, in the next chapter, we shall study the elements related to nitrogen.

Nitrogen is a colorless, odorless, and tasteless gas and is the major component of the earth's atmosphere. The discovery of nitrogen is usually credited to Daniel Rutherford of Scotland, who prepared it in 1772 by burning phosphorus in a volume of confined air. It has a density of 1.251 g per liter under standard conditions, corresponding to the molecular formula N_2. It is not quite as dense as air, for approximately one fifth of the molecules in air are the heavier O_2 molecules.

Under a pressure of 1 atmosphere, liquid nitrogen boils at $-196°C$ (13° below oxygen). It freezes at $-210°C$ (9° below oxygen).

The N_2 molecule is the most stable diatomic molecule known. The lack of chemical reactivity that is characteristic of elementary nitrogen under ordinary conditions is due to this molecular stability.

The nitrogen atom has two *s* and three *p* electrons in its outermost shell. Three pairs of shared electrons bond the two atoms together in the N_2 molecule:

$$:N:::N:$$

The great strength of this triple bond is responsible for the stability of the N_2 molecule and the consequent relative inertness of elementary nitrogen. The dissociation of 1 mole of N_2 molecules involves the absorption of 171 kcal of energy. The same quantity of energy is released when 2 moles of nitrogen atoms form 1 mole of N_2 molecules. This is the basic reason for the instability of many nitrogen compounds. Some of the compounds are so unstable that they decompose with explosive violence. Most chemical explosives are nitrogen compounds, and elementary nitrogen is a product of their decomposition.

In spite of the chemical sluggishness of elementary nitrogen and the instability of many nitrogen compounds, this is a highly electronegative element. With the same electronegativity value as chlorine (3.0), it is exceeded only by oxygen (3.5) and fluorine (4.0). This fact underscores the importance of realizing that the triple bond in the N_2 molecule provides the explanation for the reluctance of nitrogen to leave the zero oxidation state and for its readiness to return to it.

At room temperature nitrogen combines with metallic lithium to form lithium nitride, Li_3N, and undergoes "fixation" by the bacteria that occur on the root nodules of leguminous plants. When heated, nitrogen becomes more reactive, combining with magnesium and calcium to form nitrides. Under suitable conditions it combines with hydrogen to form ammonia and with oxygen to form nitric oxide, NO. It also reacts at high temperatures with certain compounds, e.g., with calcium carbide to form calcium cyanamide:

$$CaC_2 + N_2 \rightarrow CaCN_2 + C$$

Other compounds of nitrogen can be obtained by indirect methods even though they cannot be obtained by direct union of the elements, e.g., cyanogen (C_2N_2) and nitrogen trifluoride (NF_3).

Nitrogen exhibits every integral oxidation state from -3 to $+5$ in its compounds (Table 27-1).

Table 27-1

Oxidation states of nitrogen

Oxidation state	*Compounds*
-3	Ammonia, amines, amides, nitrides
-2	Hydrazine, N_2H_4
-1	Hydroxylamine, NH_2OH
0	Elementary nitrogen, N_2
$+1$	Nitrous oxide, N_2O; hyponitrous acid, $H_2N_2O_2$
$+2$	Nitric oxide, NO
$+3$	Dinitrogen trioxide; nitrous acid, HNO_2
$+4$	Nitrogen dioxide, dinitrogen tetroxide
$+5$	Dinitrogen pentoxide; nitric acid, HNO_3

27-2 Occurrence and Uses

Nitrogen exists chiefly in the elementary state, N_2 molecules comprising 78 percent of the earth's atmosphere. Industrially, the element is obtained by the fractional distillation of liquid air.

Pure nitrogen can be prepared by heating a solution of ammonium nitrite. Since this compound is so unstable that it cannot be stored, a mixture of sodium nitrite and ammonium chloride is used to furnish the ions of ammonium nitrite:

$$NH_4^+ + NO_2^- \rightarrow N_2\uparrow + 2H_2O$$

Some of the uses for elementary nitrogen are based on its lack of chemical activity. When a process requires an "inert atmosphere" as a protection against oxidation, it is frequently possible to use nitrogen for this purpose. Canned foods, e.g., coffee, retain their flavor better if the cans are filled in the presence of nitrogen in place of air.

By far the largest single use for nitrogen is for the manufacture of synthetic ammonia. Lesser quantities are used for making calcium cyanamide.

In mineral form, nitrogen occurs as sodium nitrate in great deposits located in various parts of the world, e.g., the desert region of Chile.

The importance of nitrogen to life is due to the fact that it is an essential element of all plant and animal proteins.

27-3 Nitrides

Nitrides are formed by a large number of metals. A small number of active metals, e.g., magnesium and lithium, react at moderate temperatures, and many others react at elevated temperatures. Thus, the burning of magnesium in air produces not only magnesium oxide but also another white solid, magnesium nitride. The odor of ammonia is easily detected upon addition of water to the ash:

$$Mg_3N_2 + 6HOH \rightarrow 3Mg(OH)_2 + 2NH_3\uparrow$$

The formula for magnesium nitride corresponds to what we would expect from the combination of bipositive magnesium ions with trinegative nitride ions.

Many nitrides of nonmetals also exist. An example is S_4N_4, a covalent compound obtained by the reaction of ammonia with sulfur chlorides. It forms orange-yellow crystals, which are subject to detonation by shock.

27-4 Ammonia

One of the most important industrial compounds of nitrogen is ammonia, NH_3, a colorless gas with a characteristic pungent odor. Because ammonia is easily liquefied (bp, $-33°C$) and has a higher heat of vaporization than any other liquid except water, it has had extensive use as a refrigerant.

Ammonia is very soluble in water, being comparable in this regard with the hydrogen halides. At 35°C a saturated aqueous solution contains 28 percent

Figure 27-1

The NH_3 molecule

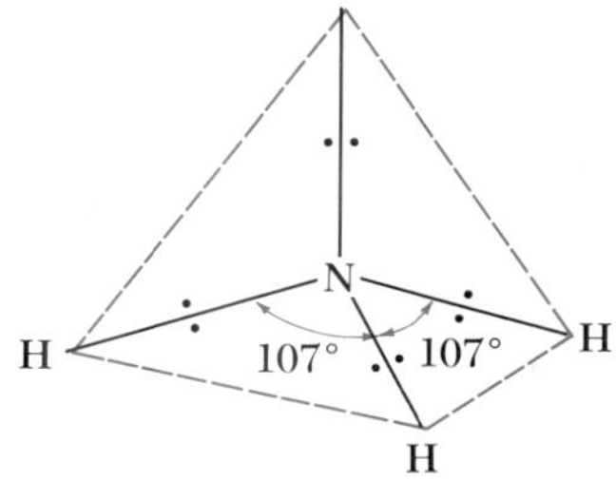

ammonia by weight. This is the reagent that is labeled "ammonium hydroxide" in the chemical trade.

In the ammonia molecule the three hydrogen atoms are covalently bonded to the nitrogen atom. The two *s* electrons of the nitrogen atom remain unshared. The molecule has the configuration of a pyramid with the nitrogen atom at the apex (Figure 27-1). The center of the nitrogen atom is 0.36 Å distant from the plane of the three hydrogen atoms. Each H—N—H bond angle is 107°, which is very nearly the tetrahedral angle found in methane. We can regard the ammonia molecule as essentially tetrahedral with the nitrogen atom using sp^3 hybrid orbitals, one of which accommodates an unshared pair of electrons.

The ammonia molecule is polar, because it is not symmetrical. Like water, ammonia is highly associated in the liquid state as a result of the polar character of the molecules and the strong hydrogen bonding.

At high temperatures ammonia undergoes rapid and almost complete decomposition. This is the reason that 500°C is about the highest feasible temperature to employ in the synthesis of ammonia from its elements. On the other hand, the combination of nitrogen with hydrogen is inefficiently slow if the synthesis is attempted at temperatures much lower than this, even with the use of high pressure and the best catalysts. Therefore, the temperature used in the Haber process (Section 15-9) is 500°C.

At high temperatures ammonia acts like a mixture of hydrogen and nitrogen when it reduces oxides of certain metals and when it burns in oxygen:

$$3CuO + 2NH_3 \rightarrow 3Cu + N_2 + 3H_2O$$
$$4NH_3 + 3O_2 \rightarrow 2N_2 + 6H_2O$$

However, a mixture of air and ammonia in the presence of platinum produces nitric oxide:

$$4NH_3 + 5O_2 \rightarrow 4NO + 6H_2O$$

This reaction is the first step in the Ostwald process for manufacturing nitric acid (Section 27-11).

When ammonia is passed through a tube containing heated magnesium, hydrogen is displaced and magnesium nitride is formed:

$$3Mg + 2NH_3 \rightarrow Mg_3N_2 + 3H_2$$

If ammonia is passed over hot sodium, sodium amide is formed:

$$2Na + 2NH_3 \rightarrow 2NaNH_2 + H_2$$

In both reactions ammonia is acting as an oxidizing agent, hydrogen being reduced from the +1 state to the 0 state. Magnesium nitride is formally related to ammonia in the same way that magnesium oxide is related to water, whereas sodium amide bears the same formal relationship to ammonia that sodium hydroxide bears to water. An intermediate type of ammonia derivative is represented by lead imide, PbNH, in which the metal atom has taken the place of two of the three hydrogen atoms of ammonia.

One of the most important properties of ammonia is its ability to react with acids to form ammonium salts:

$$NH_3 + HCl \rightarrow NH_4^+ + Cl^-$$
$$NH_3 + HNO_3 \rightarrow NH_4^+ + NO_3^-$$
$$2NH_3 + H_2SO_4 \rightarrow 2NH_4^+ + SO_4^{--}$$

In chemical laboratories gaseous ammonia and gaseous hydrogen chloride react to form a white cloud composed of tiny particles of ammonium chloride. The solid settles out to form a white deposit that is easily noted on windowpanes and laboratory glassware.

The transfer of a proton from an acid molecule to an ammonia molecule is possible because of the unshared electron pair on the nitrogen atom. In terms of the Brønsted-Lowry concept of acids and bases, the ammonia molecule is acting as a base when it accepts a proton, regardless of the source of the proton.

Aqueous solutions of ammonia are commonly referred to as solutions of "ammonium hydroxide." The alkalinity of the solutions is due to the presence of relatively small concentrations of OH^- ions:

$$NH_3 + HOH \rightleftharpoons NH_4^+ + OH^- \qquad K = 1.8 \times 10^{-5}$$

The equilibrium constant for this reaction is sometimes referred to as "the ionization constant of ammonium hydroxide." Its small value indicates that aqueous ammonia is only weakly basic. Heating an ammonia solution reverses the reaction completely and drives most of the ammonia out of solution before much water has evaporated.

If an acid is added to aqueous ammonia, protons are transferred to the NH_3 molecules (forming NH_4^+ ions) and to the OH^- ions (forming H_2O molecules). The "neutralized" solution contains an ammonium salt, which may be recovered as a crystalline solid upon evaporation of the solvent water.

When dry ammonium salts are heated, they decompose. Most of them give ammonia and the parent acid. If the acid is volatile, the salt is completely converted to gases:

$$NH_4Cl(s) \rightarrow NH_3(g) + HCl(g)$$

The essential ingredient of "smelling salts" is ammonium bicarbonate, which decomposes slowly, even without being heated:

$$NH_4HCO_3 \rightarrow NH_3\uparrow + H_2O + CO_2\uparrow$$

The gaseous mixture that is released serves as a stimulant.

Ammonium nitrite and ammonium nitrate differ from most ammonium salts in the way they decompose when heated. In these cases the anion oxidizes the ammonium ion:

$$NH_4NO_2 \rightarrow N_2\uparrow + 2H_2O$$
$$NH_4NO_3 \rightarrow N_2O\uparrow + 2H_2O$$

Ammonia molecules share electrons with certain metallic ions, forming *ammine complexes:*

$$Ag^+ + 2NH_3 \rightleftharpoons \underset{\text{diamminesilver ion}}{Ag(NH_3)_2{}^+}$$

$$Cu^{++} + 4NH_3 \rightleftharpoons \underset{\text{tetramminecopper(II) ion}}{Cu(NH_3)_4{}^{++}}$$

Reactions of this type are utilized frequently in systematic qualitative analysis. For example, addition of ammonia to a precipitate containing silver chloride will dissolve the silver chloride, thus separating it from other substances present in the precipitate. Such separations usually precede the performance of tests for specific ions in any systematic analytical scheme.

Under high pressure ammonia and carbon dioxide react at 175°C to form urea:

$$CO_2 + 2NH_3 \rightarrow CO(NH_2)_2 + H_2O$$

Urea is used as a fertilizer, as an additive for commercial cattle feeds, and as a raw material for the manufacture of an important group of plastics known as the urea-formaldehyde resins.

Sulfamic acid, NH_2SO_2OH, is a white solid that is obtained commercially by the reaction of urea with fuming sulfuric acid. Its salts are useful as weed killers and as flameproofing agents.

The most important uses for ammonia are for the manufacture of ammonium salts, nitric acid, and urea. A fairly new use for ammonia has resulted from the development of methods for the direct application of aqueous ammonia and anhydrous liquid ammonia to the soil to serve as a quick-acting fertilizer.

27-5 Liquid Ammonia as a Solvent

Ammonia and water at first thought might not seem to be very similar since one is a gas and the other a liquid at ordinary temperatures. But ammonia is easily liquefied and the liquid ammonia, like water, has a high heat of vaporization so that it can be kept in the liquid state for extended periods of time in Dewar flasks without the use of pressure. Liquid ammonia has been extensively studied as a solvent, especially since the extensive researches of Edward C. Franklin in the early decades of the present century.

Ammonia is not as good a solvent for electrolytes as is water because its dielectric constant is smaller than that of water. In liquid ammonia solutions, the oppositely charged ions of dissolved electrolytes are therefore not as well insulated as in aqueous solutions.

Like water, liquid ammonia is a poor conductor of electricity. Its slight conductivity is due to the presence of a very small concentration of positive ammonium ions and negative amide ions produced by its "self-ionization":

$$2NH_3 \rightleftharpoons NH_4^+ + NH_2^-$$

In liquid ammonia solutions the NH_4^+ ion plays the acidic role that the H_3O^+ ion plays in aqueous solutions, and the NH_2^- ion is analogous to the OH^- ion. Therefore, such compounds as NH_4Cl behave as acids in liquid ammonia. Indeed, a solution obtained by dissolving NH_4Cl in ammonia cannot be distinguished from a solution resulting from dissolving HCl in ammonia, for there is no difference between the two solutions. Sodium amide, $NaNH_2$, in liquid ammonia is analogous to NaOH in water. A liquid ammonia solution of NH_4Cl can be titrated with a liquid ammonia solution of $NaNH_2$ by using a suitable indicator.

Solubility relationships are not always the same in ammonia as in water. For example, silver chloride is not soluble in water, but it is soluble in liquid ammonia. On the other hand, barium chloride is not soluble in ammonia, although it is soluble in water. A reaction occurs between aqueous solutions of silver nitrate and barium chloride because of the insolubility of silver chloride in water:

$$2(Ag^+ + NO_3^-) + (Ba^{++} + 2Cl^-) \rightarrow 2AgCl\downarrow + (Ba^{++} + 2NO_3^-)$$

Ammonia solutions of silver chloride and barium nitrate react to form a precipitate consisting of barium chloride. The metathetic reaction occurring in this instance is the opposite of the first reaction:

$$2(Ag^+ + Cl^-) + (Ba^{++} + 2NO_3^-) \rightarrow 2(Ag^+ + NO_3^-) + BaCl_2\downarrow$$

Circumstances alter cases!

27-6 Amines

The *amines* are compounds that are related to ammonia as alcohols and ethers are related to water. The simplest of the amines is CH_3NH_2, methylamine. Any alkyl group may be present in the place of one of the hydrogen atoms of ammonia. The analogy of methylamine to methyl alcohol, of ethylamine to ethyl alcohol, etc. is apparent from comparison of their formulas:

CH_3OH methyl alcohol (two words)	CH_3NH_2 methylamine (one word)
CH_3CH_2OH ethyl alcohol	$CH_3CH_2NH_2$ ethylamine

$$\mathrm{CH_3CH_2CH_2OH}$$
n-propyl alcohol

$$\mathrm{CH_3CH_2CH_2NH_2}$$
n-propylamine

$$\mathrm{CH_3\underset{\displaystyle |\atop \displaystyle OH}{C}HCH_3}$$
isopropyl alcohol

$$\mathrm{CH_3\underset{\displaystyle |\atop \displaystyle NH_2}{C}HCH_3}$$
isopropylamine

The amines mentioned thus far are analogous to alcohols because only one of the hydrogen atoms of ammonia has been replaced by an alkyl group. These are called *primary* amines. Their general formula is $R\mathrm{NH_2}$.

Secondary amines contain two alkyl groups, and in this respect they are analogous to ethers. Since the term *amine* is retained in naming these compounds, it is mandatory that the prefix *di-* be used if the amine contains two identical alkyl groups:

$$\mathrm{CH_3{-}O{-}CH_3}$$
methyl ether (two words)

$$\mathrm{CH_3{-}\overset{\displaystyle H \atop \displaystyle |}{N}{-}CH_3}$$
dimethylamine (one word)

$$\mathrm{CH_3{-}O{-}C_2H_5}$$
methyl ethyl ether (three words)

$$\mathrm{CH_3{-}\overset{\displaystyle H \atop \displaystyle |}{N}{-}C_2H_5}$$
methylethylamine (one word)

$$\mathrm{C_2H_5{-}O{-}C_2H_5}$$
ethyl ether

$$\mathrm{C_2H_5{-}\overset{\displaystyle H \atop \displaystyle |}{N}{-}C_2H_5}$$
diethylamine

Compounds in which a nitrogen atom is bonded to three alkyl groups are called *tertiary* amines. These compounds are named in much the same manner as primary and secondary amines. Thus, $\mathrm{(CH_3)_3N}$ is trimethylamine; $\mathrm{(CH_3)_2NC_2H_5}$ is dimethylethylamine; etc.

The most important amine from the viewpoint of the chemical industry is aniline, $\mathrm{C_6H_5NH_2}$, the simplest of the aromatic amines. Millions of pounds are used each year in the manufacture of rubber accelerators and synthetic dyes. ("Accelerators" are vulcanization catalysts.)

Primary and secondary amines are associated through hydrogen bonding, but tertiary amines have no hydrogen atom attached to nitrogen, and hydrogen bonding between molecules is absent. Consequently, trimethylamine, for example, with a molecular weight of 59 boils at 4°C, which is 3° below the boiling point of dimethylamine (molecular weight, 45). Although a tertiary amine molecule cannot form a hydrogen bond with another molecule of its own kind, the unshared electron pair of a tertiary amine molecule can form a hydrogen bond with a water molecule. Like ammonia itself, all amines of low molecular weight are soluble in water and in most organic solvents.

The hydrogen bonding of ammonia and the amines is summarized below.

1. An NH_3 molecule can form a hydrogen bond with another NH_3 molecule or with an H_2O molecule:

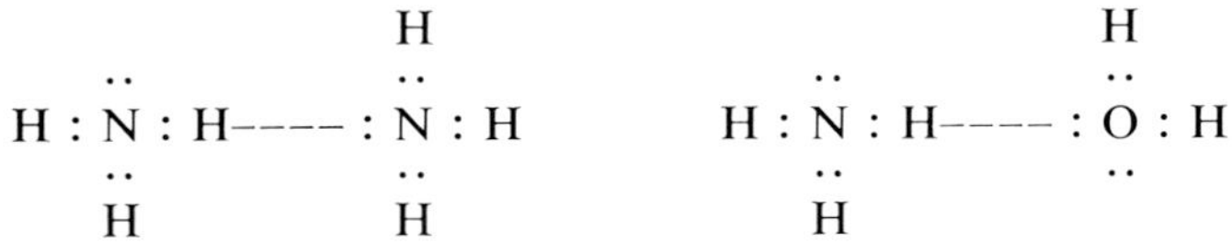

2. A molecule of a primary amine can form a hydrogen bond with another molecule like itself or with an H_2O molecule:

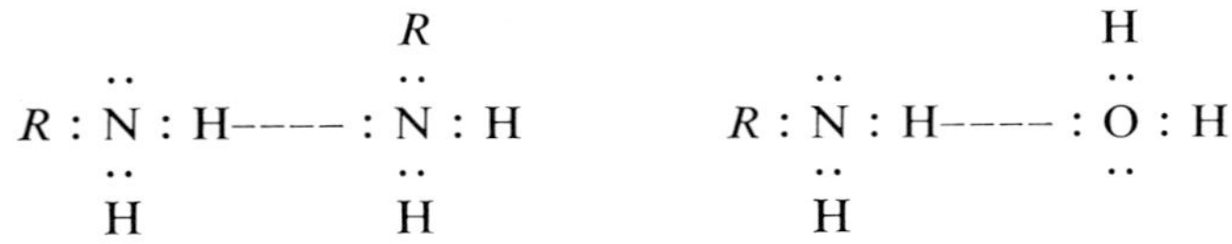

3. A molecule of a secondary amine can form a hydrogen bond with another molecule like itself or with an H_2O molecule:

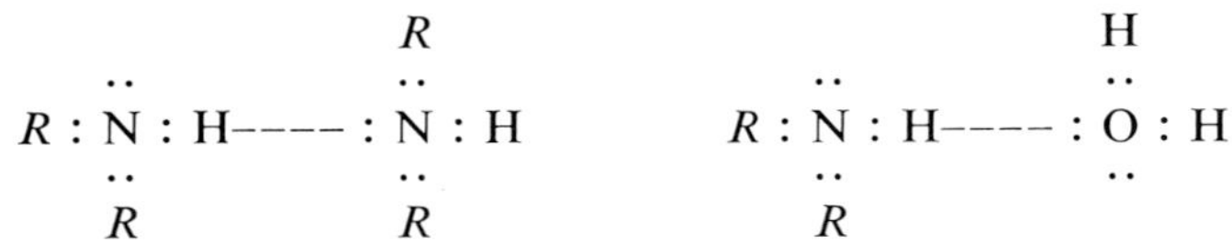

4. A molecule of a tertiary amine can form a hydrogen bond with an H_2O molecule but not with another molecule of tertiary amine:

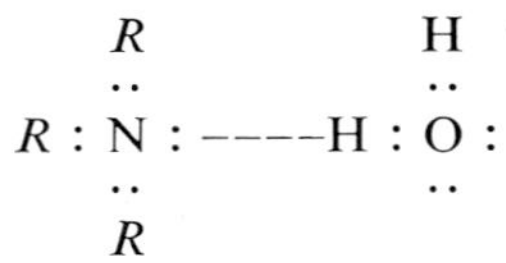

An aqueous solution of an aliphatic amine is somewhat more alkaline than an aqueous solution of ammonia. Neutralization of this solution produces a substituted ammonium salt. Just as $NH_4^+Cl^-$ is ammonium chloride, so $CH_3NH_3^+Cl^-$ is methylammonium chloride.

Heating an ammonium salt with NaOH regenerates ammonia; heating a methylammonium salt with NaOH regenerates methylamine; etc.:

$$NH_4^+Cl^- + Na^+OH^- \rightarrow Na^+Cl^- + NH_3\uparrow + H_2O$$
$$CH_3NH_3^+Cl^- + Na^+OH^- \rightarrow Na^+Cl^- + CH_3NH_2\uparrow + H_2O$$
$$RNH_3^+Cl^- + Na^+OH^- \rightarrow Na^+Cl^- + RNH_2\uparrow + H_2O$$
$$R_2NH_2^+Cl^- + Na^+OH^- \rightarrow Na^+Cl^- + R_2NH\uparrow + H_2O$$
$$R_3NH^+Cl^- + Na^+OH^- \rightarrow Na^+Cl^- + R_3N\uparrow + H_2O$$

The reaction of a tertiary amine with an alkyl halide gives a quaternary ammonium salt:

$$R_3N + RX \rightarrow [R_4N]^+X^-$$

Quaternary ammonium hydroxides may be obtained by the action of moist silver oxide ("silver hydroxide") on quaternary ammonium salts:

$$(C_2H_5)_4NI + AgOH \rightarrow AgI + (C_2H_5)_4NOH$$

The quaternary ammonium hydroxides are completely ionized; i.e., they have the same basic strength in water as NaOH or KOH.

If one of the alkyl groups attached to nitrogen in a quaternary ammonium salt is a long-chain group, the compound has properties similar to soaps. An example is cetyltrimethylammonium chloride, $[C_{16}H_{33}(CH_3)_3N]^+Cl^-$. Such compounds are known as *invert soaps* or *cationic detergents* because the detergent action is due to the positive ion rather than the negative ion as is the case with an ordinary soap. Many of the invert soaps have both detergent and germicidal properties.

27-7 Amino Acids

One of the most important classes of bifunctional organic compounds are the *amino acids*, compounds that contain both $—NH_2$ and $—COOH$ groups. The simplest is aminoacetic acid,

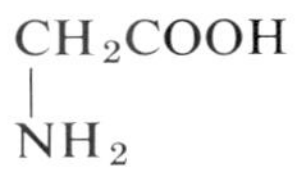

usually called glycine. Because it contains both a basic $—NH_2$ group and an acidic $—COOH$ group, glycine can act either as a base or as an acid. In the reaction

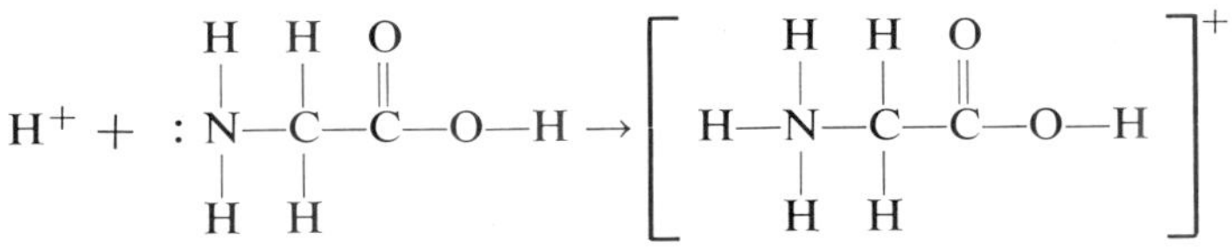

the H^+ ion joins the glycine molecule through the unshared electron pair on the nitrogen atom. In the presence of a base, the proton from the carboxyl group neutralizes OH^- ions:

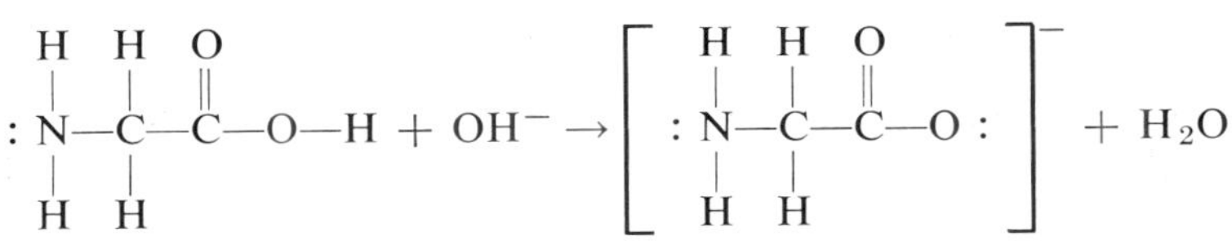

All amino acids can act as both acids and bases, i.e., they are amphoteric.

The most common source of amino acids is from the chemical breakdown of proteins. All of them obtained from this source are *alpha* (α) amino acids, i.e., the $—NH_2$ group is attached to the same carbon as the $—COOH$ group.

27-8 Hydrazine

Hydrazine, N_2H_4, contains nitrogen in the -2 state. Therefore, as might be expected, it can be obtained by the oxidation of ammonia. In the commercial preparation of hydrazine, the oxidizing agent is sodium hypochlorite in aqueous solution. The preparation is carried out in two steps. Chloramine, NH_2Cl, is made at low temperatures in the first step:

$$NH_3 + NaClO \rightarrow NH_2Cl + NaOH$$
$$NH_3 + NH_2Cl + NaOH \rightarrow N_2H_4 + NaCl + H_2O$$

Hydrazine was formerly considered a mere laboratory curiosity, but it became important during World War II as a fuel for guided missiles. Its reaction with hydrogen peroxide is highly exothermic and yields only gaseous products:

$$N_2H_4 + 2H_2O_2 \rightarrow N_2 + 4H_2O$$

It is apparent from the structural formulas of hydrazine and hydrogen peroxide that the former is the nitrogen analogue of the latter; i.e., hydrazine bears the same relation to ammonia that hydrogen peroxide bears to water:

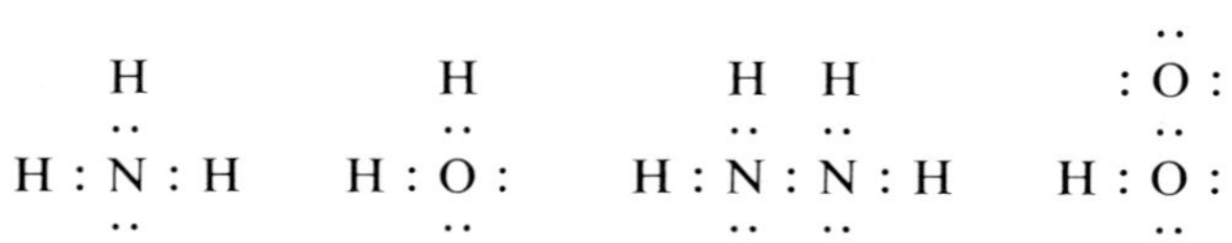

ammonia | water | hydrazine | hydrogen peroxide

Anhydrous hydrazine (mp, 2°C; bp, 114°C) is a colorless liquid which, like anhydrous liquid ammonia, is a good solvent for many salts. It is very soluble in water. In aqueous solution it is a somewhat weaker base than ammonia:

$$N_2H_4 + H_2O \rightleftharpoons N_2H_5^+ + OH^- \qquad K = 8.5 \times 10^{-7}$$

Organic derivatives of hydrazine can be synthesized readily. Alkyl and aryl derivatives were well known before hydrazine itself was isolated. Any or all of the hydrogen atoms of hydrazine may be replaced by organic groups. Thus, there are five types of substituted hydrazines: RNHNH$_2$, R_2NNH$_2$, RNHNHR, R_2NNHR, and R_2NNR_2.

Unsymmetrical dimethylhydrazine (UDMH) is a liquid that reacts spontaneously with nitrogen(IV) oxide to give only gaseous products:

$$(CH_3)_2NNH_2 + 2N_2O_4 \rightarrow 2CO_2 + 3N_2 + 4H_2O$$

The reaction is exothermic. In combination with hydrazine, UDMH is used as a rocket fuel. Both the fuel and the oxidizing agent can be stored without refrigeration.

27-9 Hydroxylamine

In hydroxylamine, NH_2OH, nitrogen is in the -1 oxidation state. The compound is obtained by electrolytic reduction of nitric acid.

$$HONO_2 + 6H^+ + 6e^- \rightarrow HONH_2 + 2H_2O$$

Pure hydroxylamine (mp, 33°C) is a white solid that must be kept on ice because it is unstable above 15°C and can undergo explosive decomposition at high temperatures. In aqueous solution it is still a weaker base than hydrazine:

$$HONH_2 + H_2O \rightleftharpoons HONH_3^+ + OH^- \qquad K = 6.6 \times 10^{-9}$$

Hydroxylamine is usually prepared and stored as a salt, such as hydroxylammonium chloride, $HONH_3Cl$; hydroxylammonium nitrate, $HONH_3NO_3$; or hydroxylammonium sulfate, $(HONH_3)_2SO_4$. These are stable white solids that are soluble in water. They are poisonous.

27-10 Oxides of Nitrogen

Nitrous Oxide. When ammonium nitrate is melted by gentle heating, the liquid undergoes a decomposition reaction that involves oxidation of NH_4^+ ions by NO_3^- ions, producing nitrogen(I) oxide, which is commonly called nitrous oxide or "laughing gas":

$$NH_4NO_3 \rightarrow N_2O + 2H_2O$$

The N_2O molecule is linear but not symmetrical. Its structure is described as a resonance hybrid:

$$:\ddot{N}::\ddot{N}::\ddot{O}: \longleftrightarrow :N:::N:\underset{\cdot\cdot}{\ddot{O}}:$$

Nitrous oxide is relatively unreactive at ordinary temperatures but decomposes to give nitrogen and oxygen at higher temperatures. This fact may explain why nitrous oxide causes a glowing splint to burst into flame, as oxygen does. It also explains why nitrogen and oxygen are produced by decomposition of ammonium nitrate when it is used in explosives:

$$2NH_4NO_3 \rightarrow 2N_2 + O_2 + 4H_2O$$

The name laughing gas arose from the fact that small doses of N_2O cause mild intoxication and a desire to laugh. General anesthesia is produced by large doses, and oxygen-nitrous oxide mixtures are frequently used for this purpose in dentistry. The chief commercial use for nitrous oxide is in self-whipping cream. This is an application of its solubility in fats. The cream is packaged with the gas dissolved in it under pressure. When the pressure is released the gas comes out of solution and forms tiny bubbles in the cream.

Nitric Oxide. Many reactions involving the reduction of dilute nitric acid give nitrogen(II) oxide, NO, commonly called nitric oxide. On a commercial scale, this oxide is obtained by the catalytic oxidation of ammonia (Section 27-4). In nature, it is formed from nitrogen and oxygen of the air by the action of lightning during a thunderstorm.

$$N_2 + O_2 \rightarrow 2NO$$

This reaction is highly endothermic and it can only occur when large surges of energy are available. The action of lightning bolts in making this reaction possible is one of the paths by which atmospheric nitrogen becomes available in the soil, even to nonleguminous plants.

The NO molecule is an "odd" molecule; i.e., it contains an odd number of electrons. This makes it impossible to write an electronic formula that does not violate the octet rule. The fact that NO is paramagnetic indicates that the molecule contains an unpaired electron. One way that has been proposed for representing its structure is to consider it a resonance hybrid.

$$\left\{ :\dot{\mathrm{N}}::\ddot{\mathrm{O}}: \longleftrightarrow :\ddot{\mathrm{N}}::\dot{\mathrm{O}}: \right\}$$

This representation implies that the unpaired electron is not solely on either the nitrogen atom or the oxygen atom.

The NO molecule is the most stable odd molecule known. In the gaseous state nitric oxide is colorless and does not dimerize. In the liquid state, however, NO is blue and is apparently dimerized.

Nitric oxide reacts instantly with oxygen to form brown fumes of NO_2.

$$2NO + O_2 \rightarrow 2NO_2$$

This property serves as a means of identifying NO in the absence of NO_2. It also explains why NO is an important industrial compound—it is the intermediate in the Ostwald process for the manufacture of nitric acid (Section 27-11).

Dinitrogen Trioxide. Cooling an equilibrium mixture of nitric oxide and nitrogen dioxide gives dinitrogen trioxide, a blue liquid which freezes to a blue solid. The compound is pure only in the solid state (mp, −103°C). When the blue liquid (bp, 3.5°C) is warmed, it becomes brown as the result of partial decomposition:

$$N_2O_3 \rightleftharpoons NO + NO_2$$

In the vapor state at 25°C and 1 atmosphere the equilibrium mixture contains only 10 percent N_2O_3.

The oxidation state of nitrogen is +3 in both dinitrogen trioxide and nitrous acid. Dinitrogen trioxide is therefore the anhydride of nitrous acid. Consequently, an equimolar mixture of NO and NO_2 is converted completely to nitrite ion when absorbed by an alkali solution:

$$NO + NO_2 + 2OH^- \rightarrow 2NO_2^- + H_2O$$

Nitrogen Dioxide and Dinitrogen Tetroxide. Reduction of concentrated nitric acid with copper gives a brown gas. The empirical formula of the gas is NO_2, but experimental determinations of its molecular weight yield values varying from 46 to 92, depending on the temperature. The higher the temperature, the deeper is the color of the gas. These facts are explained in terms of a chemical equilibrium involving nitrogen dioxide and its dimer, dinitrogen tetroxide:

$$\underset{\text{colorless}}{N_2O_4} \rightleftharpoons \underset{\text{brown}}{2NO_2}$$

Nitrogen is in the +4 state in both NO_2 and N_2O_4, and the mixture is commonly referred to simply as nitrogen dioxide. As a convenience in writing equations, $N_2O_4 \rightleftharpoons 2NO_2$ mixtures are usually represented by the formula NO_2. Thus, the reaction of copper with concentrated nitric acid is represented by the following equation:

$$Cu + 4H^+ + 2NO_3^- \rightarrow Cu^{++} + 2NO_2 + 2H_2O$$

This is done with the understanding that the monomer/dimer ratio varies with the temperature.

At the boiling point (21°C), the vapor is about 20 percent NO_2, and at 100°C and 1 atmosphere pressure it is about 90 percent NO_2. Dissociation of the dimer is complete at 140°C. Above 150°C the NO_2 begins to decompose:

$$2NO_2 \rightleftharpoons 2NO + O_2$$

At −11°C the +4 oxide freezes to a white solid, which is wholly N_2O_4. In the liquid state the oxide is an equilibrium mixture which may be thought of as a dilute solution of NO_2 in N_2O_4.

The NO_2 molecule is an "odd" molecule, and its color, paramagnetism, and ready dimerization can be attributed to an unpaired electron. It is regarded as a resonance hybrid, implying that the unpaired electron is not solely on one atom.

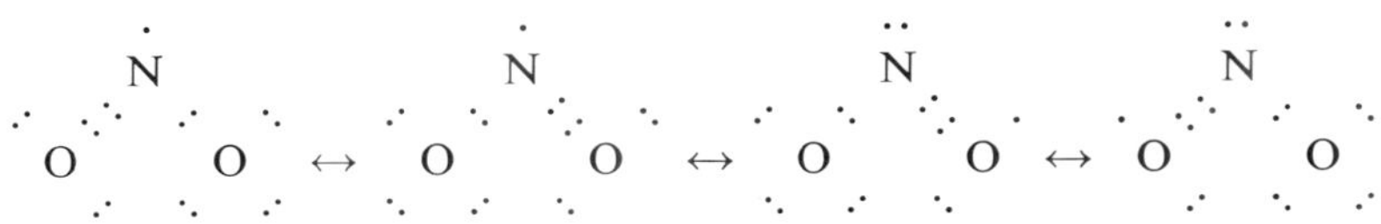

Two triangular NO_2 molecules combine to form a dimer molecule that is planar. Each NO_2 fragment in the dimer molecule has almost exactly the same dimensions as in the monomer (Figure 27-2). Because the dimer molecule contains no unpaired electrons, N_2O_4 is neither colored nor paramagnetic.

Commercially, nitrogen dioxide is obtained by bringing NO into contact with air. Its most important industrial use is its reaction with warm water to give nitric acid:

$$3NO_2 + H_2O \rightarrow 2HNO_3 + NO$$

Figure 27-2

The NO_2 and N_2O_4 molecules

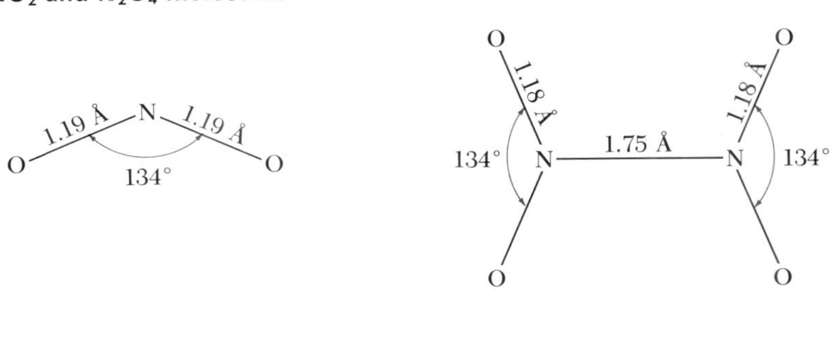

It is also useful in preparing chlorine dioxide (Section 25-7) and as the catalyst in the lead-chamber process for making sulfuric acid (Section 26-13).

Dinitrogen Pentoxide. A white solid, N_2O_5, is obtained by dehydration of pure nitric acid with phosphorus(V) oxide:

$$4HNO_3 + P_4O_{10} \rightarrow 2N_2O_5 + 4HPO_3$$

The same compound can be obtained more conveniently by the oxidation of nitrogen dioxide with ozone:

$$2NO_2 + O_3 \rightarrow N_2O_5 + O_2$$

The pentoxide is thermally unstable and sometimes decomposes explosively:

$$2N_2O_5 \rightarrow 4NO_2 + O_2$$

The oxidation state of nitrogen is +5 in both dinitrogen pentoxide and nitric acid. The oxide is therefore the anhydride of nitric acid. It is deliquescent and dissolves readily in water:

$$N_2O_5 + H_2O \rightarrow 2HNO_3$$

In the vapor state the compound consists of N_2O_5 molecules in which two NO_2 groups are bonded to the same oxygen atom. An electronic formula that conforms to the octet rule can be written for the molecule in four different ways. The N_2O_5 molecule therefore is regarded as a resonance hybrid. One of the contributing structures is:

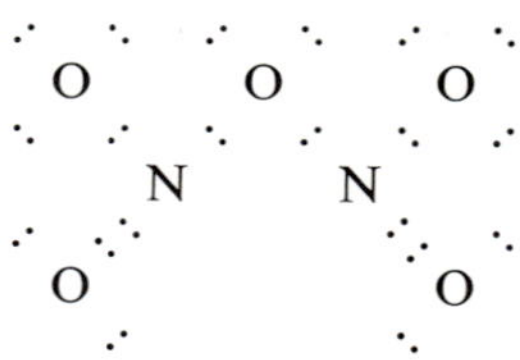

The molecule is bent (and therefore polar), the N—O—N angle being about 145°.

27-11 Nitric Acid and the Nitrates

Pure nitric acid is a colorless liquid, but it slowly decomposes on standing at room temperature unless it is kept in the dark:

$$4HNO_3 \rightleftharpoons 4NO_2 + O_2 + 2H_2O$$

The resulting NO_2 dissolves in the liquid and causes it to turn yellow. The concentrated nitric acid available commercially is 68 percent HNO_3. Even at this high concentration, there is extensive ionization of the acid:

$$HNO_3 + H_2O \rightleftharpoons H_3O^+ + NO_3^-$$

The nitric acid molecule has all its atoms in the same plane and is regarded as a resonance hybrid of two contributing structures:

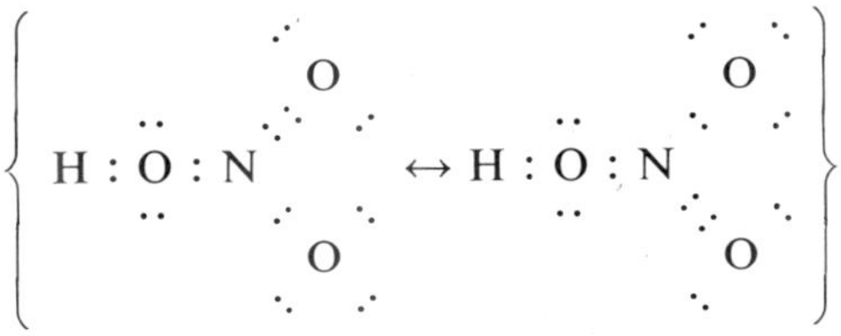

In the pure liquid there are also significant quantities of NO_2^+ and NO_3^- ions resulting from self-ionization:

$$2HNO_3 \rightleftharpoons NO_2^+ + NO_3^- + H_2O$$

The nitronium ion, NO_2^+, is also present in solutions of nitric acid in sulfuric acid:

$$HNO_3 + 2H_2SO_4 \rightleftharpoons NO_2^+ + H_3O^+ + 2HSO_4^-$$

Measurements of the lowering of the freezing point of anhydrous sulfuric acid by dissolved nitric acid indicate the formation of four ions from each molecule of HNO_3, as indicated by the above equation. Mixtures of nitric and sulfuric acids are used to bring about the *nitration* of benzene and other organic compounds. Nitration consists of the introduction of an $—NO_2$ group in the place of an H atom. The mechanism of the nitration of benzene and other aromatic compounds by the nitric-sulfuric acid mixture is believed to involve the nitronium ion, with the HSO_4^- ion acting as acceptor for the displaced proton:

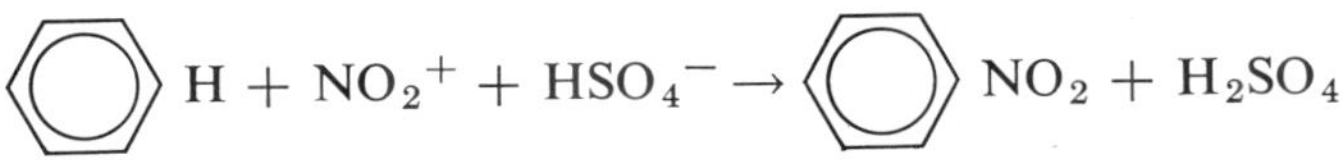

Pure nitric acid is a strong oxidizing agent that attacks many metals and organic compounds with great vigor, leading in some instances to explosive violence. The concentrated nitric acid (i.e., the 68 percent aqueous solution) is itself a vigorous oxidizing agent, but it is much safer to handle than pure nitric acid. It is obtained industrially by the Ostwald process, a three-stage operation that involves the catalyzed oxidation of ammonia (Section 27-4), the conversion of the resulting NO to NO_2 upon cooling, and absorption of the NO_2 in water:

$$4NH_3 + 5O_2 \xrightarrow[1600°C]{(Pt)} 4NO + 6H_2O$$

$$2NO + O_2 \longrightarrow 2NO_2$$

$$3NO_2 + H_2O \longrightarrow 2HNO_3 + NO$$

The NO is immediately oxidized by air to NO_2, and thus it also becomes converted to nitric acid.

When more NO_2 is dissolved in water than is required to give a 68 percent HNO_3 solution, the product is called "fuming nitric acid." This is a vigorous oxidizing agent that reacts violently with reducing agents such as hydrazine in one type of rocket engine fuel.

The salts of nitric acid, the nitrates, are produced when nitric acid reacts with bases, basic oxides, carbonates, and most metals:

$$NaOH + HNO_3 \rightarrow NaNO_3 + H_2O$$
$$NH_3 + HNO_3 \rightarrow NH_4NO_3$$
$$CaO + 2HNO_3 \rightarrow Ca(NO_3)_2 + H_2O$$
$$Na_2CO_3 + 2HNO_3 \rightarrow 2NaNO_3 + H_2O + CO_2$$
$$Cu + 4HNO_3\ (conc) \rightarrow Cu(NO_3)_2 + 2H_2O + 2NO_2$$

It would be exceedingly dangerous to allow an active metal, such as sodium, to come into contact with nitric acid because of the violence of the reaction.

The nitrate ion has the nitrogen atom at the center of an equilateral triangle with the oxygen atoms all at the same distance. The ion is regarded as a resonance hybrid of three contributing structures:

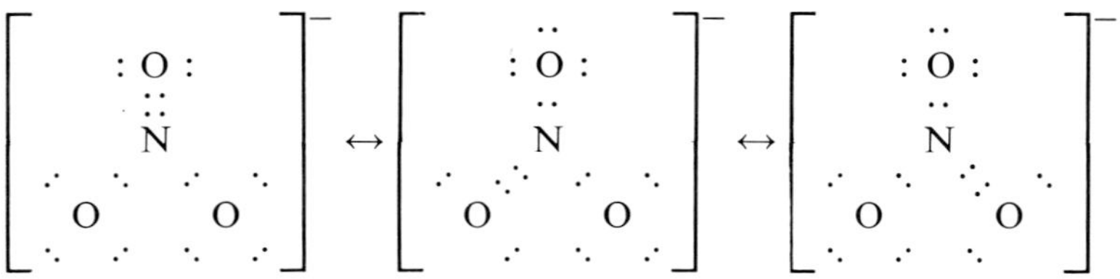

Practically all nitrates are soluble in water. The dry salts are rather easily decomposed by heat. Nitrates of very active metals, such as sodium, lose one third of their oxygen when they are heated, giving nitrites:

$$2NaNO_3 \rightarrow 2NaNO_2 + O_2$$

The nitrates of less active metals usually give the metal oxide when heated:

$$2Cu(NO_3)_2 \rightarrow 2CuO + 4NO_2 + O_2$$

It is hazardous to heat a nitrate in the presence of oxidizable material. Gunpowder is a mixture of potassium nitrate, charcoal, and sulfur.

The most important nitrates are KNO_3, used for making gunpowder; $NaNO_3$, used as a fertilizer and for making sodium nitrite; $Ca(NO_3)_2$, used as a fertilizer; NH_4NO_3, used as a fertilizer and as an explosive; and $AgNO_3$, used for making silver bromide for the photographic industry.

When nitric acid acts as an oxidizing agent, the product of its own reduction is determined by its concentration, the nature of the reducing agent, and the temperature. Pure nitric acid and concentrated nitric acid suffer minimum reduction, yielding NO_2, when they react with nonmetals or with the less active metals:

$$S + 6HNO_3 \rightarrow H_2SO_4 + 6NO_2\uparrow + 2H_2O$$
$$Ag + 2HNO_3 \rightarrow AgNO_3 + NO_2\uparrow + H_2O$$

The reduction of dilute nitric acid with similar reducing agents commonly gives NO:

$$3Cu + 8HNO_3\ (dilute) \rightarrow 3Cu(NO_3)_2 + 2NO\uparrow + 4H_2O$$

It is hardly to be expected that reduction of concentrated nitric acid should give NO when one realizes that NO is readily converted to NO_2 by the concentrated acid:

$$NO + 2HNO_3 \rightarrow 3NO_2 + H_2O$$

Strong reducing agents, such as active metals, reduce dilute nitric acid to give substances in which the nitrogen is in oxidation states even lower than +2. The products may be NO, N_2O, N_2, or the NH_4^+ ion:

$$\begin{aligned}
Zn + 4HNO_3 \text{ (very conc)} &\rightarrow Zn(NO_3)_2 + 2NO_2 + 2H_2O \\
3Zn + 8HNO_3 \text{ (moderate conc)} &\rightarrow 3Zn(NO_3)_2 + 2NO + 4H_2O \\
4Zn + 10HNO_3 \text{ (dilute)} &\rightarrow 4Zn(NO_3)_2 + N_2O + 5H_2O \\
5Zn + 12HNO_3 \text{ (dilute)} &\rightarrow 5Zn(NO_3)_2 + N_2 + 6H_2O \\
4Zn + 10HNO_3 \text{ (very dilute)} &\rightarrow 4Zn(NO_3)_2 + NH_4NO_3 + 3H_2O
\end{aligned}$$

The nitrate ion is so much stronger an oxidizing agent than the hydronium ion that hydrogen gas is seldom obtained by the reaction of nitric acid with a metal.

The oxidizing action of nitric acid on inorganic compounds is well illustrated by the reactions of concentrated and dilute nitric acid on copper(II) sulfide:

$$\begin{aligned}
3CuS + 8HNO_3 \text{ (dilute)} &\rightarrow 3Cu(NO_3)_2 + 3S + 2NO + 4H_2O \\
CuS + 8HNO_3 \text{ (conc)} &\rightarrow CuSO_4 + 8NO_2 + 4H_2O
\end{aligned}$$

Note that the sulfur is raised from the −2 state to the +6 state by the concentrated acid but only to the 0 state by the dilute acid.

Hydrochloric acid is oxidized by concentrated nitric acid to chlorine and nitrosyl chloride:

$$HNO_3 + 3HCl \rightarrow Cl_2 + NOCl + 2H_2O$$

A mixture of concentrated nitric acid and concentrated hydrochloric acid in a 1:3 ratio by volume is known as aqua regia. This reagent is useful for dissolving gold and platinum (Section 20-11).

One of the important uses of nitric acid is in the manufacture of organic explosives and other nitration products. The nitration of "glycerin" produces glyceryl trinitrate, an ester commonly called "nitroglycerin" (Section 19-12). Dynamite is nitroglycerin taken up in porous material, such as wood pulp. Guncotton is cellulose nitrate, obtained by the nitration of the —OH groups in cotton (cellulose). Another well-known explosive is trinitrotoluene (TNT), produced by the nitration of the benzene ring in toluene. Trinitrotoluene is an example of a true nitro compound, in which the nitrogen atoms of the nitro (—NO_2) groups are attached directly to carbon atoms. Amatol is a mixture of TNT and ammonium nitrate.

The reaction of nitric acid vapor with methane produces nitromethane, CH_3NO_2, the simplest of the *nitroparaffins*. The nitroparaffins are stable, nontoxic liquids which are widely used as solvents.

Concentrated nitric acid causes yellow stains on the skin by nitration of the aromatic rings present in proteins.

27-12 Nitrous Acid

Thermal decomposition of alkali metal nitrates yields nitrites:

$$2NaNO_3(s) \xrightarrow{\text{heat}} 2NaNO_2(s) + O_2\uparrow$$

Nitrites are salts of nitrous acid, which is a weak acid that cannot be obtained in the pure state:

$$HNO_2 \rightleftharpoons H^+ + NO_2^- \qquad K = 5.1 \times 10^{-4}$$

Nitrous acid is responsible for the pale blue color obtained when an acid is added to a cold solution of a nitrite. The acid is unstable, undergoing auto-oxidation-reduction slowly at room temperature and rapidly when heated:

$$3HNO_2 \rightarrow H^+ + NO_3^- + 2NO\uparrow + H_2O$$

Nitrites are much more stable than nitrous acid itself. The triangular nitrite ion is represented as a resonance hybrid:

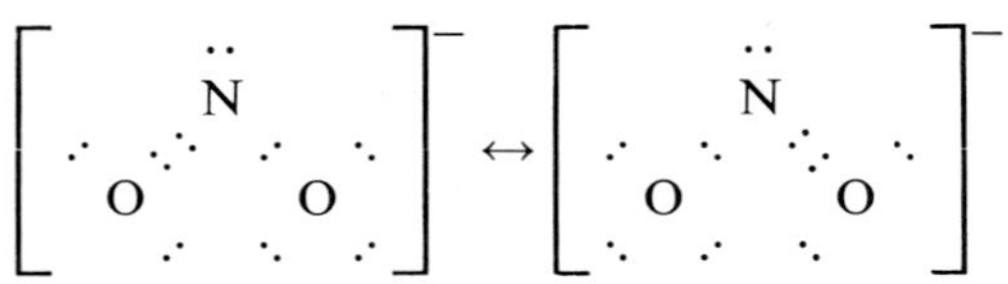

Since nitrogen is in the intermediate +3 oxidation state in nitrous acid, the compound acts as a good oxidizing agent toward strong reducing agents but can itself be oxidized by strong oxidizing agents:

$$2I^- + 2HNO_2 + 2H^+ \rightarrow I_2 + 2NO\uparrow + 2H_2O$$
$$2MnO_4^- + 5HNO_2 + H^+ \rightarrow 2Mn^{++} + 5NO_3^- + 3H_2O$$

Large quantities of nitrous acid, prepared as needed by mixing cold solutions of sulfuric acid and sodium nitrite, are used in the production of azo dyes containing the —N=N— group, called the azo group.

27-13 Hyponitrous Acid

Nitrogen is in the +1 state in hyponitrous acid, $H_2N_2O_2$. This acid is a white solid that is very unstable. It is extremely explosive when dry. Although it is very soluble in water, its solution is only weakly acidic.

$$H_2N_2O_2 \rightleftharpoons H^+ + HN_2O_2^- \qquad K = 9 \times 10^{-8}$$

Even in solution, as in the dry state, it breaks down on standing:

$$H_2N_2O_2 \rightarrow H_2O + N_2O$$

This reaction is irreversible. Although nitrous oxide is rather soluble in water, its solution is neutral. Nitrous oxide does not react with solutions of alkalies to give hyponitrites, as we would expect the anhydride of hyponitrous acid to do.

Infrared spectra of several hyponitrites indicate that the hyponitrite ion has the configuration

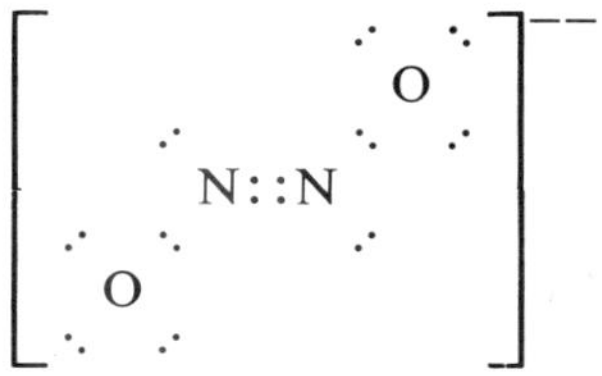

27-14 Hydrazoic Acid

Hydrazoic acid, HN_3, is a colorless liquid that boils at 36°C and freezes at −80°C. It gives off a vapor that is irritating and toxic. The acid is dangerously explosive. In aqueous solution it is of about the same strength as acetic acid.

$$HN_3 \rightleftharpoons H^+ + N_3^- \qquad K = 1.8 \times 10^{-5}$$

Salts of hydrazoic acid are called azides. Azides of alkali metals decompose smoothly when heated, but azides of heavy metals are easily detonated by heat or shock. The explosion of an azide readily causes detonation of other explosives. Lead azide, $Pb(N_3)_2$, is used as a detonator for military ammunition and for explosives used in blasting.

The azide ion is linear and symmetrical. The ion is a resonance hybrid with three contributing structures:

The hydrazoic acid molecule is a resonance hybrid with two contributing structures:

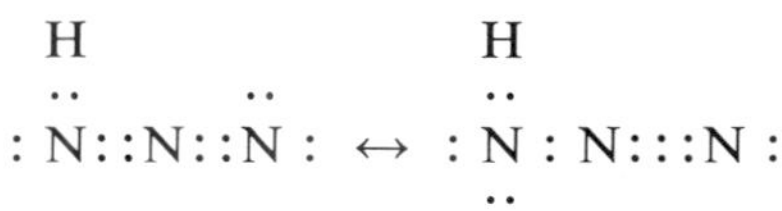

Here the three nitrogen atoms are linear, but the molecule does not have the symmetry of the azide ion (Figure 27-3).

Figure 27-3

The HN_3 molecule and the N_3^- ion

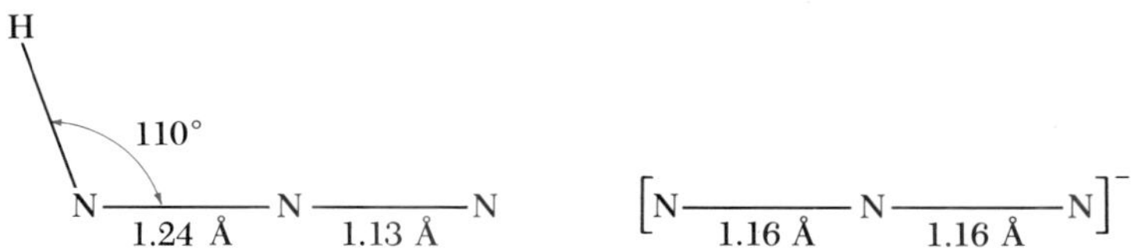

27-15 Cyanides

The cyanide ion, CN^-, is the anion of such salts as NaCN and KCN. Cyanides can be made in several different ways. A recent process for manufacturing sodium cyanide involves passing nitrogen over a heated mixture of sodium carbonate and coke:

$$Na_2CO_3 + 4C + N_2 \xrightarrow[\text{catalyst}]{\text{Fe}} 2NaCN + 3CO$$

The alkali-metal cyanides are soluble salts of very high toxicity. They are used in electroplating and in the metallurgy of gold. Chlorine reacts with metallic cyanides to form ClCN, cyanogen chloride (bp, 15.5°C), a chemical warfare gas.

The cyanides are salts of hydrocyanic acid, HCN, also known as prussic acid. It is formed when a cyanide is treated with an acid. It is a very volatile compound (bp, 25.6°C) with the odor of bitter almonds. Because of its exceptionally poisonous nature and the speed with which it acts, hydrogen cyanide is used in exterminating rats from ships and as a method of capital punishment in some states.

Hydrocyanic acid is such a weak acid ($K = 4 \times 10^{-10}$) that it does not turn litmus red. Consequently, solutions of its salts are alkaline by hydrolysis:

$$CN^- + HOH \rightleftharpoons HCN + OH^-$$

The HCN molecule is linear and highly polar. Consequently, hydrogen bonding occurs between molecules:

$$H{-}C{\equiv}N{-}{-}{-}{-}H{-}C{\equiv}N{-}{-}{-}{-}H{-}C{\equiv}N$$

This accounts for the fact that the substance is a liquid (although a very volatile one). Liquid HCN, as a polar solvent, is able to dissolve ionic compounds.

27-16 Cyanamide

When nitrogen is passed over hot barium carbide, barium cyanide is formed. Under the same conditions calcium carbide forms a dark gray mixture of calcium cyanamide and carbon.

$$BaC_2 + N_2 \rightarrow Ba(CN)_2$$
$$CaC_2 + N_2 \rightarrow CaCN_2 + C$$

The electronic structure of the cyanamide ion is $[:\ddot{N}::C::\ddot{N}:]^{--}$. Thus the CN_2^- ion is isoelectronic with the CO_2 molecule.

The cyanamide process was the first important industrial process for the fixation of atmospheric nitrogen. The product was treated with steam to produce ammonia by hydrolysis of the calcium cyanamide.

$$CaCN_2 + 3H_2O \rightarrow CaCO_3 + 2NH_3$$

As a source of ammonia, the cyanamide process was largely superseded by the

Haber process. The process is still in operation, however, since $CaCN_2$ is useful as a fertilizer.

Calcium cyanamide is the calcium salt of *cyanamide*, H_2NCN, which is a colorless crystalline compound that can be obtained by passing CO_2 into a suspension of calcium cyanamide in aqueous methanol.

27-17 Cyanates and Fulminates

Oxidation of cyanides by fusion with lead oxide gives cyanates.

$$KCN + PbO \rightarrow KOCN + Pb$$

In the dry state cyanates are stable, even at high temperatures. Acidification of a cyanate solution liberates cyanic acid but it undergoes immediate hydrolysis with effervescence.

$$HOCN + H_2O \rightarrow CO_2\uparrow + NH_3\uparrow$$

Aqueous solutions of cyanates gradually decompose on standing, with evolution of CO_2 and NH_3.

In 1828 the German chemist Friedrich Wöhler attempted to prepare ammonium cyanate by treating silver cyanate with ammonium chloride solution, but obtained urea instead. This was the first preparation of an animal product from inorganic compounds. Urea is the chief product of nitrogen metabolism in mammals. It was first isolated from urine in 1773. According to a widespread belief called vitalism, organic compounds were produced by a vital force that was present only in living organisms. Wöhler's synthesis of urea started the decline of this idea. The synthesis was probably the result of the dissociation of ammonium cyanate into ammonia and cyanic acid, followed by addition.

$$NH_4{}^+OCN^- \rightleftharpoons NH_3 + HOCN \rightleftharpoons H_2NCONH_2$$

Mercury(II) fulminate, $Hg(ONC)_2$, is among the most sensitive detonating agents for setting off high explosives. It is the product of a complex reaction involving mercury, nitric acid, and ethanol. It was first used by Alfred Nobel in 1864 for the explosion of nitroglycerin. Although it is still used to some extent in commercial blasting caps and military ammunition, it has been replaced largely by the less expensive lead azide (Section 27-14).

The German chemist, Justus Liebig showed in 1823 that silver fulminate, AgONC, has the same composition that had already been found for silver cyanate. This was the first recognition of the phenomenon that we now call isomerism. The probable electronic structure for fulminic acid is $H:\underset{..}{\overset{..}{O}}:N:::C:$ while the structure for cyanic acid may be either $H:\underset{..}{\overset{..}{O}}:C:::N:$ or $H:N:::C:\underset{..}{\overset{..}{O}}:$ or an equilibrium mixture of the two.

27-18 Thiocyanates

Thiocyanates are salts of thiocyanic acid, HSCN. Examples are KSCN, NaSCN, and NH_4SCN. The sodium and potassium salts can be made by melting the cyanide with sulfur.

$$KCN + S \rightarrow KSCN$$

Ammonium thiocyanate can be made by the action of carbon disulfide with aqueous ammonia at 110°C under pressure.

$$2NH_3 + CS_2 \rightarrow NH_4SCN + H_2S$$

The salts are colorless, very soluble in water, and resemble chlorides in many of their reactions. They are used in the laboratory for the analytical determination of silver, since AgSCN is insoluble in water and in dilute acids. The deep red color produced when a thiocyanate is added to a solution containing a ferric salt (or vice versa) constitutes a very sensitive test for Fe^{+++} ions (or SCN^- ions). Sodium thiocyanate has its most important use in the processing of color film. It is also used in medicine for the treatment of high blood pressure.

Thiocyanic acid melts at 5°C to give a colorless volatile liquid which rapidly decomposes, so that its boiling point is not known.

Ammonium thiocyanate, by a reaction analogous to Wöhler's synthesis of urea, is converted to thiourea when heated to 140°C.

$$NH_4SCN \rightarrow CS(NH_2)_2$$

27-19 The Nitrogen Cycle in Nature

Plant proteins are produced from combined nitrogen taken from the soil by growing plants. When animals consume plant material as food, their bodies convert the plant proteins into animal proteins, which are basic structural materials of animal organisms. When plants and animals die, the decay of their bodies partially replenishes the combined nitrogen content of the soil. Much of the nitrogen in their bodies is released to the atmosphere as elementary nitrogen.

The nitrogen cycle of nature is completed by the action of bacteria present in soils containing humus. Also important are the bacteria that live in nodules on the roots of leguminous plants, such as clover. The growth of legumes thus serves to replenish the soil nitrogen, especially if the plants are plowed under. Another natural process of importance in the nitrogen cycle is the action of lightning flashes in bringing about the chemical combination of nitrogen and oxygen to form compounds that dissolve in the falling rain.

Natural processes for replenishment of soil nitrogen are inadequate when plants are harvested and transported to distant points for their ultimate utilization. Even the conscientious use of animal manures and plant compost, plus the occasional plowing under of legume crops, leaves a soil nitrogen deficit which must be met with commercial fertilizers. Nitrogenous fertilizers include sodium nitrate, ammonium sulfate, ammonium nitrate, calcium nitrate, calcium cyanamide, urea, diammonium phosphate, aqueous ammonia, and anhydrous liquid ammonia.

New Terms

Amide (of a metal): a compound containing the amide ion, NH_2^-; e.g., sodium amide (sometimes called sodamide), $NaNH_2$.

Amine: any compound conforming to any of the following general formulas: R_3N (tertiary amines), R_2NH (secondary amines), or RNH_2 (primary amines).

Amino acids: compounds that contain both —NH_2 and —COOH groups.

Ammine complex: a complex ion in which ammonia molecules share electrons with a metallic ion; e.g., the diamminesilver ion, $Ag(NH_3)_2^+$.

Imide (of a metal): a compound containing the =NH group; e.g., lead imide, PbNH.

Invert soaps (cationic detergents): compounds that have detergent properties due to the positive ion.

Nitration: introduction of an —NO_2 group into an organic compound in the place of an H atom.

Vitalism: the doctrine that held that organic compounds were produced by a vital force that was present only in living organisms.

Exercises

27-1 Show the relationship between the inactivity of nitrogen and the explosiveness of many of its compounds.

27-2 How do you reconcile the inactivity of nitrogen with its high electronegativity?

27-3 Describe some of the natural and artificial methods for nitrogen fixation.

27-4 Discuss the chemistry and the economic importance of the Haber and Ostwald processes.

27-5 What are acids and bases (a) in water and (b) in liquid ammonia?

27-6 Identify the following: invert soaps, odd molecules, azides.

27-7 Show how nitrogen is important to life.

27-8 Show the structures of hydrazine and hydroxylammonium nitrate.

27-9 Identify the following: laughing gas, smelling salts, nitroglycerin, dynamite, guncotton, and TNT.

27-10 What is nitration? Dimerization?

27-11 Write the formulas for four contributing structures of the N_2O_5 molecule that conform to the octet rule.

27-12 Describe how the following are manufactured: NaCN, NaOCN, and NaSCN.

References

Fernelius, W. C., and Johnson, W. C., "Liquid Ammonia as a Solvent," *J. Chem. Educ.*, **5** (6), 664; (7), 828 (1928). A classic review by two distinguished chemists who were students of E. C. Franklin.

Fullmer, J. Z., "The Great Laughing Gas Experiment," *Chemistry*, **37** (4), 6 (1964). How the usefulness of nitrous oxide as an anesthetic was discovered.

Schaar, B. E., "Chance Favors the Prepared Mind. VI. Nobel's Dynamite," *Chemistry*, **39** (6), 19 (1966). An account of the background for the work of a great inventor who endowed the well-known prizes for peace, literature, medicine, physics, and chemistry.

"Sniffing Out the Enemy," *Time*, June 9, 1967. An explanation of the "E–63 Man-pack Personnel Detector," a chemical and electronic version of a nose, based on the reaction of ammonia with hydrogen chloride.

28

Phosphorus and Related Elements

28-1 Group Characteristics

Group VA is the only group of representative elements whose members were all known before the periodic table was devised. Lavoisier's first list of elements included all the members of this group: nitrogen, phosphorus, arsenic, antimony, and bismuth.

The elements of this group have s^2p^3 configurations in the outermost shells of their atoms (Figure 28-1). Because of this similarity in atomic structure, the elements of the group resemble one another in many ways. There is, however, a

Figure 28-1

Electronic configurations of Group VA elements

	1	2	3	4	5	6
$_7N$	s^2	s^2p^3				
$_{15}P$	s^2	s^2p^6	s^2p^3			
$_{33}As$	s^2	s^2p^6	$s^2p^6d^{10}$	s^2p^3		
$_{51}Sb$	s^2	s^2p^6	$s^2p^6d^{10}$	$s^2p^6d^{10}$	s^2p^3	
$_{83}Bi$	s^2	s^2p^6	$s^2p^6d^{10}$	$s^2p^6d^{10}f^{14}$	$s^2p^6d^{10}$	s^2p^3

Table 28-1

Properties of the elements of Group VA

Element	*Melting point* (°C)	*Boiling point* (°C)	*Ionization energy* (kcal/mole)	*Electro-negativity* (Pauling scale)	*Specific resistivity* (microhm-cm, 0°C)	*Density* (g/ml, 0°C)
$_{7}N$	−210	−196	335	3.0		
$_{15}P$	44 (white)	280 (white)	254	2.1	10^{17} (white)	1.83 (white)
$_{33}As$	814 (36 atm)	633 (sublimes)	226	2.0	35	5.73
$_{51}Sb$	630	1325	199	1.9	39	6.67
$_{83}Bi$	271	1560	168	1.9	107	9.80

substantial variation in properties as we go from the lightest to the heaviest member of the group. Nitrogen and phosphorus are nonmetals, arsenic and antimony are metalloids, and bismuth is a metal. This gradation in properties is a consequence of increasing size, smaller ionization energies, and diminished electronegativities (Table 28-1). The heavier elements are more inclined to form positive ions.

The atoms of these elements lack three electrons of having noble-gas configurations, but the energy required to convert the atoms to trinegative ions is extremely great so that only nitrogen and phosphorus form such ions. Loss of five electrons to form pentapositive ions is prevented by the high energy requirements. The tendency for the heavier atoms in a group to lose p electrons and form positive ions with an s^2 configuration in the outer shell is illustrated in this group by the formation of tripositive Sb^{+++} and Bi^{+++} ions, as in $Sb_2(SO_4)_3$ and $Bi(NO_3)_3$.

The elements form covalent compounds with a variety of other elements. Covalent compounds with a particular element, e.g., chlorine, become more polar as we go from nitrogen to bismuth. Nitrogen is the only member of the group that is limited to a maximum of eight electrons in its outer shell. In the nitrogen atom, as in the oxygen atom, the outer shell is the second shell, and its maximum is reached with the s^2p^6 configuration, as in NF_3, NH_3, and NH_4^+. The outermost shell of the phosphorus atom is the third shell and it, like sulfur, can accommodate more than eight electrons. For example, phosphorus forms the pentahalides PF_5, PCl_5, and PBr_5 (but not PI_5). The heavier elements of the family form the pentahalides AsF_5, SbF_5, $SbCl_5$, and BiF_5.

28-2 Allotropic Forms

Nitrogen is the only member of the group that is a gas under ordinary conditions and exists in the elementary state at ordinary temperatures as diatomic molecules. There are several allotropic forms of phosphorus, arsenic, and antimony. The most familiar phosphorus allotropes are *white phosphorus* and *red phosphorus*. White phosphorus, obtained by the condensation of phosphorus vapor, is soft like wax. It is transparent and colorless for a time, but it acquires a surface coating of red phosphorus after exposure to light. This gives it a yellow tint that explains why this form of the element is often called yellow phosphorus. The same P_4 molecules are present in white phosphorus as are present in liquid phosphorus. All forms of phosphorus vaporize as P_4, which undergoes partial dissociation into P_2 above 800°C.

The German alchemist Hennig Brand discovered phosphorus in 1669. Upon heating a mixture of evaporated urine and sand he distilled out a highly flammable waxy substance. Because it glows in the dark, a phenomenon known as *phosphorescence*, it became known by the name phosphorus. White phosphorus is the only allotropic variety of this element that exhibits phosphorescence. The phosphorescence of white phosphorus is due to slow oxidation.

In the P_4 molecule the four atoms have their centers located at the corners of a tetrahedron, each atom being joined to each of its three neighbors by a covalent bond (Figure 28-2).

White phosphorus is highly toxic. It has a low melting point (44°C) and a low boiling point (280°C). Inhalation of phosphorus vapor causes serious bodily damage, especially to the gums and jawbones ("phossy jaw"). The use of white phosphorus in the manufacture of matches is now forbidden by law.

White phosphorus is soluble in benzene, ether, carbon disulfide, etc. but not in water. It is so highly reactive that it must be kept under water, for it catches fire spontaneously on contact with air.

$$P_4 + 5O_2 \rightarrow P_4O_{10}$$

A lower oxide, P_4O_6, is formed if the oxygen supply is limited. Phosphorus combines energetically with the halogens to form trihalides and pentahalides,

Figure 28-2

The P_4 molecule

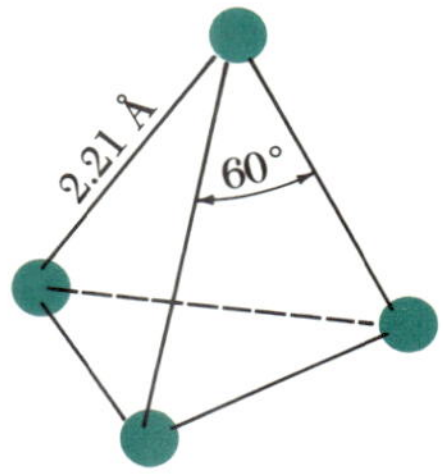

and it unites with sulfur to give sulfides. It reacts with many metals to form phosphides, e.g., Ca_3P_2, that are analogous to nitrides, such as Ca_3N_2.

When exposed to light, or when heated in the absence of air, white phosphorus is slowly transformed into red phosphorus. The true nature and exact structure of red phosphorus are still uncertain. Its properties are markedly different from those of white phosphorus. The density (2.2 g per ml) is greater than that of white phosphorus (1.82 g per ml), and it is insoluble in all known solvents. The latter property presumably explains why red phosphorus is nontoxic. It should be noted, however, that commercial red phosphorus usually contains small amounts of white phosphorus. When heated under atmospheric pressure red phosphorus sublimes before it melts; under 43 atmospheres pressure it melts at 590°C. Chemically, red phosphorus is much less active than the white variety; much higher temperatures are required for analogous reactions to occur.

A third variety of phosphorus is the so-called *black phosphorus* that results from subjecting white phosphorus to high pressure. It has an iron-gray color and a metallic luster. It conducts electric current 10^{11} times as well as white phosphorus. It has a layer structure, as does graphite, with the distance between layers (3.68 Å) substantially greater than the distance between nearest atoms in a layer (2.18 Å).

Both arsenic and antimony have allotropic modifications. The most stable form of each of these elements has a metallic type of crystal lattice. Bismuth occurs only in metallic form. "Metallic" arsenic is gray, brittle, and rather soft; it sublimes at 633°C under 1 atmosphere pressure. Metallic antimony is silver-white, brittle, and soft. Bismuth is reddish-white, brittle, and somewhat softer than either arsenic or antimony.

Arsenic and antimony vapors consist of tetratomic molecules. Bismuth vapor is a mixture of diatomic and monatomic molecules. *Yellow arsenic* and *yellow antimony* are obtained by rapid cooling of the vapors of the elements. These allotropic forms of the elements consist of discrete As_4 and Sb_4 molecules. They are even more reactive than white phosphorus.

28-3 The Metals of the Group

Antimony and bismuth have the unusual property of expanding upon solidifying. They are used in alloys to prevent them form shrinking when they solidify. For example, a lead-antimony-tin alloy is used for type metal.

As we go from arsenic to antimony to bismuth, the densities increase and the boiling points rise, but the melting points and electrical conductivities decrease (Table 28-1). Antimony and bismuth are the only metals whose electrical resistance is greater in the solid state than in the liquid state. The low melting point of bismuth is reflected in the low melting points of a number of its alloys. Such alloys are useful for electrical fuses, automatic sprinkler systems, and safety plugs for boilers.

Chemically, the increasingly metallic character of the successive members of the group is indicated by their behavior with nitric acid. Phosphorus gives phosphoric acid and bismuth gives bismuth nitrate:

$$3P_4 + 20HNO_3 + 8H_2O \rightarrow 12H_3PO_4 + 20NO\uparrow$$

$$Bi + 6HNO_3 \rightarrow Bi(NO_3)_3 + 3NO_2\uparrow + 3H_2O$$

With arsenic, dilute nitric acid gives arsenous acid, and concentrated nitric acid forms arsenic acid. Antimony forms antimony(III) oxide or antimony(V) oxide, depending on the concentration of the nitric acid.

28-4 Hydrogen Compounds

The elements of Group VA form binary compounds when heated with active metals, and these compounds hydrolyze to give hydrogen compounds with analogous formulas:

$$Mg_3N_2 + 6HOH \rightarrow 3Mg(OH)_2 + \underset{\text{ammonia}}{2NH_3\uparrow}$$

$$Ca_3P_2 + 6HOH \rightarrow 3Ca(OH)_2 + \underset{\text{phosphine}}{2PH_3\uparrow}$$

$$AlAs + 3HOH \rightarrow Al(OH)_3 + \underset{\text{arsine}}{AsH_3\uparrow}$$

$$Zn_3Sb_2 + 6HOH \rightarrow 3Zn(OH)_2 + \underset{\text{stibine}}{2SbH_3\uparrow}$$

$$Mg_3Bi_2 + 6HOH \rightarrow 3Mg(OH)_2 + \underset{\text{bismuthine}}{2BiH_3\uparrow}$$

The analogues of ammonia are colorless, evil-smelling, highly poisonous gases. Arsenic is present as an impurity in many acids and metals, and arsine therefore is present as a common poisonous impurity in hydrogen gas prepared in the laboratory.

Unlike ammonia, its analogues are only slightly soluble in water, and the solutions are not alkaline. This indicates that ammonia is the only one of these hydrogen compounds that reacts with water to form OH^- ion and an *onium* ion. Nevertheless, a few phosphonium compounds have been prepared; e.g., phosphine (PH_3) and hydrogen iodide react to produce solid phosphonium iodide:

$$PH_3(g) + HI(g) \rightarrow PH_4I(s)$$

Phosphonium compounds hydrolyze completely when they come into contact with water:

$$PH_4^+ + H_2O \rightarrow PH_3\uparrow + H_3O^+$$

Arsenic, antimony, and bismuth do not form compounds analogous with ammonium and phosphonium compounds. (But see Section 28-6 for quaternary antimonium and stibonium compounds.)

A gradation in the properties of the hydrogen compounds of the Group VA elements is to be expected since the nature of the elements varies from highly electronegative and completely nonmetallic (nitrogen) to mainly metallic (bismuth). The gradation is well illustrated by the radical decrease in thermal stability as we go from NH_3 to BiH_3. The bond between the central atom and the attached hydrogen atoms becomes weaker as the distance between the nuclei increases. Bismuthine decomposes even at room temperature. Only ammonia can successfully be prepared by direct synthesis from the elements. Only ammonia and phosphine are able to form a bond with an additional proton to form ammonium and phosphonium ions. The fact that each succeeding hydrogen compound among the Group VA elements is a weaker base is analogous to the increasing acidity as we go from H_2O to H_2Te in Group VIA or from HF to HI in Group VIIA.

28-5 Diphosphine and Diarsine

Phosphorus forms a compound, P_2H_4, called diphosphine, that is analogous to hydrazine. It is a volatile liquid obtained as a by-product in the preparation of phosphine by some methods. The spontaneous flammability of its vapor has given rise to the false impression that this is a property of phosphine.

Diarsine, As_2H_4, is so unstable that it decomposes even at temperatures as low as $-100°C$. Its existence was not known until 1948.

28-6 Compounds Analogous to Amines

Organic derivatives of phosphine and arsine are known in which one or more hydrogen atoms of the parent molecule are replaced by alkyl or aryl groups. For example,

Methylarsine,	CH_3AsH_2	(bp, 2°C)
Dimethylarsine,	$(CH_3)_2AsH$	(bp, 36°C)
Trimethylarsine,	$(CH_3)_3As$	(bp, 50°C)
Trimethylphosphine,	$(CH_3)_3P$	(bp, 38°C)
Triethylphosphine,	$(C_2H_5)_3P$	(bp, 128°C)
Triphenylphosphine,	$(C_6H_5)_3P$	(melts at 80°C)

The alkylphosphines and alkylarsines of low molecular weight are colorless liquids that tend to catch fire in air. In general, however, these compounds are thermally more stable than the parent hydrogen compounds.

Derivatives of stibine and bismuthine are known in which all the hydrogen atoms of the parent molecule are replaced by alkyl or aryl groups. The following are examples:

Trimethylstibine,	$(CH_3)_3Sb$	(bp, 78.5°C)
Triphenylstibine,	$(C_6H_5)_3Sb$	(melts at 48°C)
Trimethylbismuthine,	$(CH_3)_3Bi$	(bp, 110°C)
Triphenylbismuthine,	$(C_6H_5)_3Bi$	(melts at 75°C)

Compounds of the types $R\text{SbH}_2$, $R_2\text{SbH}$, $R\text{BiH}_2$, and $R_2\text{BiH}$ are unknown.

Trialkylphosphines, trialkylarsines, and trialkylstibines react with alkyl halides to give quaternary phosphonium, arsonium, and stibonium salts, which are crystalline and dissolve in water to give neutral solutions.

$$(CH_3)_3P + CH_3I \rightarrow \underset{\text{tetramethylphosphonium iodide}}{(CH_3)_4PI}$$

$$(CH_3)_3As + CH_3I \rightarrow \underset{\text{tetramethylarsonium iodide}}{(CH_3)_4AsI}$$

$$(CH_3)_3Sb + CH_3I \rightarrow \underset{\text{tetramethylstibonium iodide}}{(CH_3)_4SbI}$$

Reaction of moist silver oxide with these salts gives the corresponding hydroxides. Tetramethylphosphonium hydroxide, tetramethylarsonium hydroxide, and tetramethylstibonium hydroxide are strong bases, like tetramethylammonium hydroxide and the hydroxides of the alkali metals.

An organic derivative of diarsine having historical interest is tetramethyldiarsine, $(CH_3)_2As{-}As(CH_3)_2$, a very poisonous liquid with a disagreeable odor, traditionally called cacodyl (Greek, "ill-smelling"). The group $(CH_3)_2As{-}$ was the first group to be recognized as a unit (radical) that moves from one compound to another in many chemical reactions. It is present in such compounds as $(CH_3)_2AsCl$ and $[(CH_3)_2As)]_2S$.

28-7 The +3 Oxidation State

All the elements of Group VA show the +3 oxidation state in the trihalides and in the so-called sesquioxides, i.e., oxides in which the number of oxygen atoms is $1\frac{1}{2}$ times the number of atoms of the other element: N_2O_3, P_4O_6, As_4O_6, Sb_4O_6, and Bi_2O_3. The same oxidation state is shown, of course, in the corresponding oxyacids and hydroxides, e.g., HNO_2, H_3PO_3, and $Bi(OH)_3$, and in the related salts, such as $NaNO_2$, NaH_2PO_3, and $Bi_2(SO_4)_3$.

Phosphorus(III) *oxide* is a white crystalline solid (mp, 23.8°C) known, from measurements of the density of the vapor, to have the molecular formula P_4O_6. However, it has often been represented by the empirical formula P_2O_3 and called phosphorus trioxide. In the P_4O_6 molecule the four phosphorus atoms accupy the corners of a tetrahedron, and the oxygen atoms lie just outside the midpoints of the edges (Figure 28-3). It dissolves slowly in water, forming phosphorous acid, which functions as a diprotic acid. The hydrogen atom attached directly to the phosphorus atom is not removed, even with strong bases:

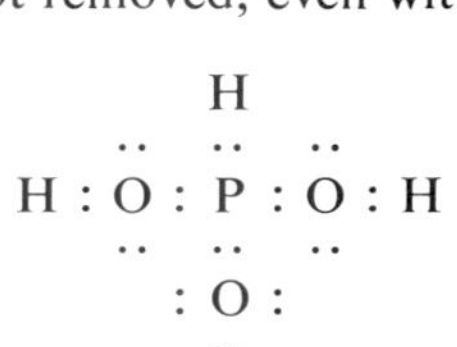

Figure 28-3

The P_4O_6 molecule

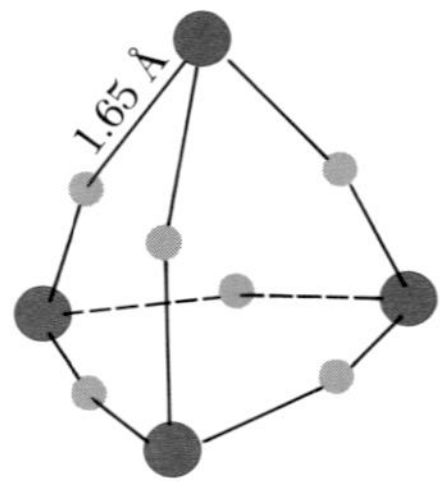

It is a moderately strong acid with respect to the first stage of its ionization:

$$H_2(HPO_3) \rightleftharpoons H^+ + H(HPO_3)^- \qquad K_1 = 1.6 \times 10^{-2}$$
$$H(HPO_3)^- \rightleftharpoons H^+ + HPO_3^{--} \qquad K_2 = 7.0 \times 10^{-7}$$

The salts of phosphorous acid, the phosphites, are of two types, illustrated by the sodium salts, NaH_2PO_3 and Na_2HPO_3.

Pure phosphorous acid is a deliquescent solid (mp, 73°C). The acid and its salts are strong reducing agents. The acid is most conveniently obtained by the hydrolysis of phosphorus trichloride:

$$PCl_3 + 3H_2O \rightarrow H_3PO_3 + 3HCl$$

Arsenic(III) *oxide*, As_4O_6, is usually referred to as arsenic trioxide by chemists and is often represented by its empirical formula, As_2O_3. It is commonly known as "white arsenic," or simply as "arsenic." It is the oxide resulting from the combustion of elementary arsenic. It is obtained commercially by roasting the mineral arsenopyrite, FeAsS, and as a by-product of roasting various ores. Like all arsenic compounds, the oxide is highly poisonous. It is so strongly toxic that even less than 0.1 g can be fatal if it is swallowed and not quickly removed from the stomach. It is used in the manufacture of glass as a decolorizing agent, and in taxidermy as a preservative for feathers and skins.

Arsenic(III) oxide is a white solid (mp, 310°C) that is somewhat soluble in water. The aqueous solution contains arsenous acid, which is commonly written as the ortho acid, H_3AsO_3, but it may also be present as the meta acid, $HAsO_2$. The acid is not known outside of solution. It is amphoteric, behaving as a very weak acid (of about the same strength as boric acid) and also dissociating as a very weak base.

$$HAsO_2 \rightleftharpoons H^+ + AsO_2^- \qquad K_a = 6 \times 10^{-10}$$
$$AsO(OH) \rightleftharpoons AsO^+ + OH^- \qquad K_b = 5 \times 10^{-15}$$

Most arsenites are derived from the meta acid. Examples are sodium arsenite, $NaAsO_2$, used as a weed killer; cupric arsenite, $Cu(AsO_2)_2$, a pigment (Scheele's green); and cupric acetate-arsenite, $Cu(C_2H_3O_2)_2 \cdot 3Cu(AsO_2)_2$, an insecticide

(Paris green). However, silver ions precipitate yellow silver orthoarsenite, Ag_3AsO_3.

Arsenic(III) oxide is easily reduced to elementary arsenic. Strong oxidizing agents convert it to arsenic acid, H_3AsO_4. The arsenite ion is a strong reducing agent in alkaline solution.

Antimony(III) *oxide*, Sb_4O_6, is a white solid (mp, 656°C) that is only slightly soluble in water. It is the oxide that results from the burning of elementary antimony. It is amphoteric; i.e., it dissolves in strong acids to form antimony salts, such as $SbCl_3$ or $Sb_2(SO_4)_3$, but also dissolves in strong bases to form antimonites, such as $NaSbO_2$. Antimony salts are extensively hydrolyzed, giving insoluble basic salts such as antimonyl chloride, SbOCl, also called antimony oxychloride. "Tartar emetic" is potassium antimonyl tartrate, $K(SbO)C_4H_4O_6$. The antimonyl salts may be regarded as salts of the partially dehydrated hydroxide SbO(OH), and the antimony salts may be looked upon as salts of $Sb(OH)_3$. This does not mean necessarily that either hydroxide actually exists. Attempts to prepare them as definite compounds have never succeeded. The precipitates obtained by addition of a limited quantity of base to an antimony salt or a limited quantity of acid to an antimonite are antimony(III) oxide gels of variable water content.

Bismuth(III) *oxide* is a yellow solid that is represented by its empirical formula Bi_2O_3, because it is an ionic compound (mp, 820°C). It is the oxide formed when elementary bismuth combines with oxygen. It is insoluble in water and in bases but dissolves readily in acids, forming bismuth salts. Addition of a base to a solution of a bismuth salt gives bismuth hydroxide, $Bi(OH)_3$, which has been shown to be a definite chemical compound. Both the oxide and the hydroxide are completely basic.

Bismuth salts dissolve in water to form clear solutions only if a high concentration of hydrogen ions is present to prevent hydrolysis. Basic salts are otherwise formed, and most of them are insoluble solids, e.g., BiOCl, bismuthyl chloride, also known as bismuth oxychloride. Under the name bismuth "subnitrate," $BiONO_3$ is used in medicine for the treatment of stomach disorders. The oxychloride is used in face powders.

28-8 The +5 Oxidation State

The elements of the group show the +5 oxidation state in pentahalides and in the so-called pentoxides, i.e., N_2O_5, P_4O_{10}, As_4O_{10}, Sb_2O_5, and Bi_2O_5. The last two oxides are represented by their empirical formulas because their structures are unknown. The structure of arsenic pentoxide has not been definitely established.

So-called phosphorus pentoxide is the main product obtained by burning phosphorus. It is a white powder that combines with water in violent fashion, forming phosphoric acid. Below 100°C it is an effective drying agent. It removes the elements of water from compounds ordinarily regarded as good dehydrating agents themselves, e.g., it converts H_2SO_4 to SO_3.

Figure 28-4

The P_4O_{10} molecule

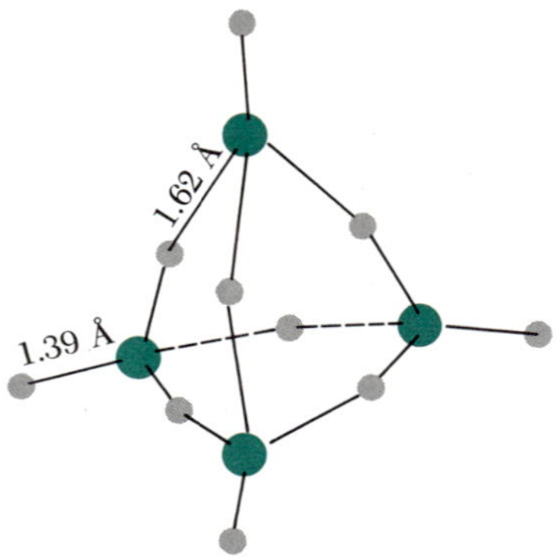

The P_4O_{10} molecule has the same structure as the P_4O_6 molecule with an additional oxygen atom on each phosphorus atom at a distance of only 1.39 Å (Figure 28-4).

Because the direct union of oxygen with arsenic, antimony, and bismuth produces only the trioxides, other procedures are required for the preparation of the pentoxides of these elements. Arsenic pentoxide is a white crystalline substance formed by heating arsenic acid:

$$4H_3AsO_4 \rightarrow As_4O_{10} + 6H_2O$$

At ordinary temperatures this reaction proceeds in the opposite direction. On the other hand, when heated to higher temperatures it loses part of its oxygen and is converted to the trioxide. Antimony pentoxide is a yellow substance obtained by carefully heating so-called antimonic acid; above 700°C it loses part of its oxygen. Bismuth pentoxide is a red solid that results from the oxidation of the trioxide by strong oxidizing agents. It is so unstable that it has never been isolated as a pure substance; even at 100°C it loses part of its oxygen and forms the trioxide. The pentoxide is only slightly soluble in water, but its acidic nature is shown by the fact that it dissolves in concentrated NaOH solution, forming sodium bismuthate, $NaBiO_3$, a powerful oxidizing agent.

Arsenic acid is obtained by oxidizing arsenic(III) oxide with concentrated nitric acid:

$$As_4O_6 + 8HNO_3 + 2H_2O \rightarrow 4H_3AsO_4 + 8NO_2\uparrow$$

It crystallizes from solution as a white deliquescent solid that is extremely soluble in water. It is a triprotic acid, somewhat weaker than phosphoric acid.

$$H_3AsO_4 \rightleftharpoons H^+ + H_2AsO_4^- \qquad K = 4.8 \times 10^{-3}$$

It is, however, a moderately strong oxidizing agent. Some of its salts are of technical importance as insecticides. Lead arsenate is used in the control of insects on fruit trees, and calcium arsenate is used against the cotton boll weevil. Sodium arsenate is used as a wood preservative.

Antimony is one of the few metals which do not dissolve in nitric acid. So-called antimonic acid is a white gelatinous precipitate obtained by oxidizing antimony with concentrated nitric acid. The precipitate is actually an antimony(v) oxide gel of variable water content. Potassium antimonate, $KSb(OH)_6$, is prepared by fusing antimony pentoxide with potassium hydroxide followed by dissolving the melt in a small quantity of water and allowing the salt to crystallize. Although $KSb(OH)_6$ is very soluble in water, the corresponding sodium compound is the least soluble of all sodium salts. Potassium antimonate therefore is used to test for sodium ions.

28-9 The Phosphoric Acids and the Phosphates

There are several phosphoric acids, all of which contain phosphorus in the +5 state. Each phosphoric acid represents a different degree of hydration of the oxide P_4O_{10}, which is the anhydride of all of them.

Pure *orthophosphoric acid*, H_3PO_4, forms colorless crystals (mp, 42.4°C) that are highly deliquescent. The acid is commonly sold as a viscous aqueous solution called "syrupy phosphoric acid" that contains 82 percent H_3PO_4. Orthophosphoric acid is a triprotic acid, since all three hydrogen atoms are bonded to oxygen atoms:

$$
\begin{array}{ccccccccc}
 & & & & \ddot{} & & & & \\
 & & & : & \mathrm{O} & : & & & \\
 & & \ddot{} & & \ddot{} & & \ddot{} & & \\
\mathrm{H} & : & \mathrm{O} & : & \mathrm{P} & : & \mathrm{O} & : & \mathrm{H} \\
 & & \ddot{} & & \ddot{} & & \ddot{} & & \\
 & & & : & \mathrm{O} & : & & & \\
 & & & & \ddot{} & & & & \\
 & & & & \mathrm{H} & & & &
\end{array}
$$

Only the first step in the ionization occurs to a significant extent, and it is far from complete:

$$
\begin{aligned}
H_3PO_4 &\rightleftharpoons H^+ + H_2PO_4^- & K_1 &= 7.5 \times 10^{-3} \\
H_2PO_4^- &\rightleftharpoons H^+ + HPO_4^{--} & K_2 &= 6.3 \times 10^{-8} \\
HPO_4^{--} &\rightleftharpoons H^+ + PO_4^{---} & K_3 &= 1.7 \times 10^{-12}
\end{aligned}
$$

The salts derived from orthophosphoric acid are of three types, depending on the number of hydrogen atoms replaced. Since orthophosphoric acid is the most important of the phosphoric acids, it is the custom to name the orthophosphates simply as *phosphates*. The following examples illustrate the types of formulas that are typical of orthophosphates and the two most commonly used systems for naming them.

NaH_2PO_4	sodium dihydrogen phosphate or primary sodium phosphate
Na_2HPO_4	disodium hydrogen phosphate or secondary sodium phosphate
Na_3PO_4	sodium (or trisodium) phosphate or tertiary sodium phosphate
$Ca(H_2PO_4)_2$	calcium dihydrogen phosphate or primary calcium phosphate

$CaHPO_4$	calcium monohydrogen phosphate or secondary calcium phosphate
$Ca_3(PO_4)_2$	calcium (or tricalcium) phosphate or tertiary calcium phosphate

Primary phosphates are most soluble, and tertiary phosphates are least so. All three types of alkali-metal phosphates are soluble. Solutions of trisodium phosphate are strongly alkaline by hydrolysis:

$$PO_4^{---} + HOH \rightleftharpoons HPO_4^{--} + OH^-$$

Solutions of NaH_2PO_4 are weakly acidic because the $H_2PO_4^-$ ion is an acid. Solutions of Na_2HPO_4 are slightly alkaline because the HPO_4^{--} is a stronger proton acceptor than donor:

$$HPO_4^{--} + HOH \rightleftharpoons H_2PO_4^- + OH^-$$

Calcium dihydrogen phosphate is used as a constituent of baking powders because it is a convenient solid acid that will react with sodium bicarbonate to liberate CO_2 gas when moisture is present:

$$H_2PO_4^- + HCO_3^- \rightleftharpoons HPO_4^{--} + H_2O + CO_2\uparrow$$

Trisodium phosphate (TSP) is an important commercial chemical because of its usefulness as a detergent, which is a consequence of the alkalinity of its solutions, and the dispersing action of phosphate ions. It also serves as a water softener by precipitating the calcium and magnesium ions from hard water.

Phosphate fertilizers are added to soils to replace the phosphorus removed by harvesting crops. The tricalcium phosphate present in finely ground phosphate rock is too slightly soluble to be of immediate value for growing plants. Acid phosphates produced from it by treatment with acids are usually employed for this purpose. The mixture called "superphosphate of lime" is obtained by the prolonged action of crude sulfuric acid over a period of about two weeks (this being a major use for sulfuric acid):

$$Ca_3(PO_4)_2 + 2H_2SO_4 \rightarrow Ca(H_2PO_4)_2 + 2CaSO_4$$

More than half the weight of the "superphosphate" is due to the calcium sulfate, an ingredient of doubtful value on which freight charges must nevertheless be paid. This problem is avoided by treating the calcium phosphate with phosphoric acid instead of sulfuric acid, giving a product known in the fertilizer trade as "triple superphosphate of lime":

$$Ca_3(PO_4)_2 + 4H_3PO_4 \rightarrow 3Ca(H_2PO_4)_2$$

A fertilizer containing both phosphorus and nitrogen, known commercially as "nitrophos," is obtained by treating the calcium phosphate with nitric acid:

$$Ca_3(PO_4)_2 + 4HNO_3 \rightarrow Ca(H_2PO_4)_2 + 2Ca(NO_3)_2$$

Impure phosphoric acid is made commercially by treatment of calcium phosphate with sulfuric acid, followed by dilution with water to bring about the precipitation of the calcium sulfate, which is removed by filtration:

$$Ca_3(PO_4)_2 + 3H_2SO_4 \rightarrow 2H_3PO_4 + 3CaSO_4$$

It is this impure phosphoric acid that is used in the manufacture of "triple superphosphate" fertilizer. It can also be used to recover ammonia from coke-oven gas, giving diammonium phosphate, $(NH_4)_2HPO_4$, a fertilizer with both nitrogen and phosphorus in one compound.

Pure phosphoric acid must be made from elementary phosphorus. This can be done by oxidation of the phosphorus with nitric acid or by combustion of the phosphorus with excess air and reaction of the resulting P_4O_{10} with water. Phosphoric acid is a stable compound and shows virtually no oxidizing properties below 400°C.

Heating orthophosphoric acid to 250°C produces *diphosphoric acid*, commonly called pyrophosphoric acid. It is a white solid (mp, 61°C), that gradually reverts to orthophosphoric acid when it is dissolved in water:

$$2H_3PO_4 \rightleftharpoons H_4P_2O_7 + H_2O$$

If orthophosphoric (or diphosphoric) acid is heated above 400°C it loses more water and forms *metaphosphoric acid*, which solidifies to form a glassy translucent material called "glacial" phosphoric acid:

$$nH_3PO_4 \rightleftharpoons (HPO_3)_n + nH_2O$$

Metaphosphoric acid has the empirical formula HPO_3, but this formula does not represent a molecule of the acid. By means of "oxygen bridges" between adjacent phosphorus atoms, chains and rings are built up in which four oxygen atoms are bonded to every phosphorus. As in the orthophosphate ion itself, each phosphorus atom is at the center of a tetrahedron with four oxygen atoms symmetrically disposed about it. Polymer molecules of assorted sizes are formed so that the best formula for representing metaphosphoric acid is $(HPO_3)_n$. The polymer slowly reverts to orthophosphoric acid when it is dissolved in water.

If sodium dihydrogen orthophosphate is heated to about 700°C and then rapidly cooled, it forms a water-soluble *polymetaphosphate:*

$$nNaH_2PO_4 \rightarrow (NaPO_3)_n + nH_2O$$

This is a long-chain polymer that forms a soluble complex with calcium ions and thereby prevents the formation of insoluble "scum" when soap is added to hard water. It is sold for this purpose under the trade name *Calgon*.

If sodium dihydrogen orthophosphate is heated only to 200°C, it yields disodium dihydrogen diphosphate:

$$2NaH_2PO_4 \rightarrow Na_2H_2P_2O_7 + H_2O\uparrow$$

Heating disodium hydrogen orthophosphate yields tetrasodium diphosphate:

$$2Na_2HPO_4 \rightarrow Na_4P_2O_7 + H_2O\uparrow$$

This compound is useful as a water softener in consequence of forming a soluble complex with the magnesium ions, and with part of the calcium ions, in hard water.

The P—O—P bonds present in polyphosphates are referred to as "energy-rich" bonds in biochemistry. By this is meant that the breaking of these bonds by hydrolysis is an exothermic process. The energy required for the contraction of muscles is derived from the hydrolysis of an organic compound called adenosine triphosphate (ATP). The hydrolysis is induced by enzyme action and releases 10 kcal per mole of ATP. If we let R represent the organic portion of the molecule, the reaction can be represented by the following equation:

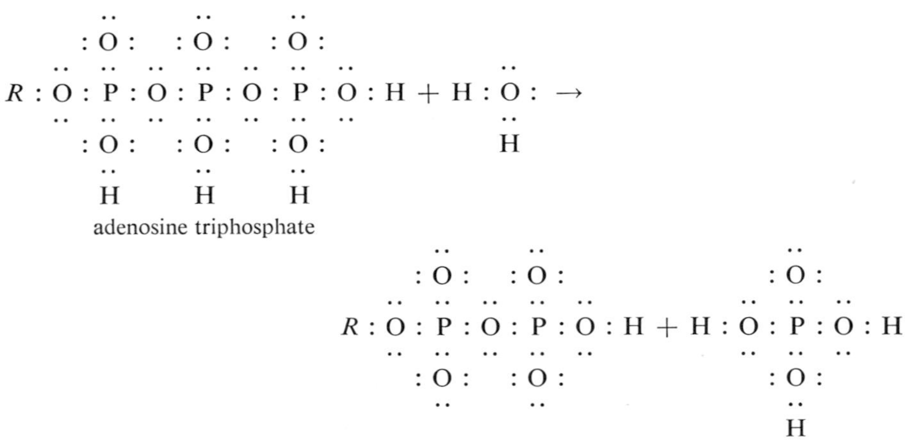

adenosine triphosphate

adenosine diphosphate   orthophosphoric acid

28-10 Hypophosphorous Acid

One of the most convenient methods for preparing phosphine consists of heating white phosphorus in a concentrated solution of sodium hydroxide:

$$P_4 + 3(Na^+ + OH^-) + 3H_2O \rightarrow 3(Na^+ + H_2PO_2^-) + PH_3\uparrow$$

The solution remaining contains sodium hypophosphite, NaH_2PO_2. Barium hypophosphite likewise may be obtained in solution by using barium hydroxide in the reaction. Addition of sulfuric acid to the barium hypophosphite solution causes barium sulfate to precipitate and leaves hypophosphorous acid in solution:

$$(Ba^{++} + 2H_2PO_2^-) + (2H^+ + SO_4^{--}) \rightarrow BaSO_4\downarrow + 2H_3PO_2$$

Hypophosphorous acid (mp, 27°C) is a monoprotic acid, somewhat stronger than phosphorous acid. The two nonacidic hydrogen atoms are bonded directly to the phosphorus atom:

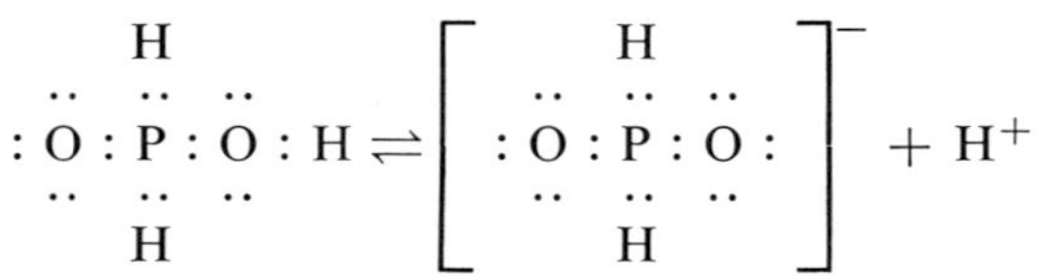

$K = 8.0 \times 10^{-2}$

The acid and its salts are strong reducing agents.

28-11 Sulfides of the Elements

Phosphorus forms three well-defined sulfides with the formulas P_4S_3, P_4S_7, and P_4S_{10}. Any one of these may be prepared by mixing the required amounts of the elements and heating the mixture out of contact with air. The head of a "strike anywhere" match is tipped with a mixture of P_4S_3, PbO_2, and powdered glass. Friction explodes the mixture of P_4S_3 and PbO_2, and the resulting heat ignites the head of the match, which is composed of a combustible mixture of potassium chlorate and sulfur.

Arsenic forms sulfides with the formulas As_4S_4, As_4S_6, and As_4S_{10}. They may be prepared by heating the elements together in the required proportions. The arsenic(III) sulfide is more commonly obtained by passing hydrogen sulfide into a hydrochloric acid solution of arsenic(III) chloride:

$$4AsCl_3 + 6H_2S \rightarrow As_4S_6\downarrow + 12(H^+ + Cl^-)$$

Arsenic(V) sulfide precipitates very slowly when hydrogen sulfide is passed into a hydrochloric acid solution of arsenic acid:

$$4H_3AsO_4 + 10H_2S \rightarrow As_4S_{10}\downarrow + 16H_2O$$

If the solution is cold and only weakly acidic, much of the arsenic is reduced to the +3 state, and the precipitate is mostly a mixture of arsenic(III) sulfide and sulfur. Iodide ions catalyze the reduction:

$$4H_3AsO_4 + 10H_2S \rightarrow As_4S_6\downarrow + 4S\downarrow + 16H_2O$$

Both As_4S_6 and As_4S_{10} dissolve readily in sodium sulfide solution, forming sodium thioarsenite and sodium thioarsenate, respectively:

$$As_4S_6 + 6S^{--} \rightarrow 4AsS_3^{---}$$
$$As_4S_{10} + 6S^{--} \rightarrow 4AsS_4^{---}$$

The acids corresponding to the thioarsenites and thioarsenates are not stable. Consequently, acidification reprecipitates the sulfides:

$$4AsS_3^{---} + 12H^+ \rightarrow As_4S_6\downarrow + 6H_2S\uparrow$$
$$4AsS_4^{---} + 12H^+ \rightarrow As_4S_{10}\downarrow + 6H_2S\uparrow$$

Both As_4S_6 and As_4S_{10} are yellow. The As_4S_6 occurs in nature as the golden-yellow mineral called orpiment. As_4S_4 occurs in nature as a bright red mineral called realgar. It ignites readily with oxidizing agents and is used in fireworks to produce "Bengal Fire" flares.

Antimony forms two sulfides, Sb_2S_3 and Sb_2S_5. The trisulfide occurs in nature in the form of the black mineral stibnite, the most important ore of antimony. The black form of the trisulfide also forms when antimony and sulfur are heated together. The trisulfide forms as a brilliant orange-red precipitate when hydrogen sulfide is passed into a weakly acidified solution of an antimony(III) salt. The precipitate dissolves in sodium sulfide solution to form sodium thioantimonite and in sodium polysulfide reagent to form the thioantimonate:

$$Sb_2S_3 + 3S^{--} \rightarrow 2SbS_3^{---}$$
$$Sb_2S_3 + 3S^{--} + 2S \rightarrow 2SbS_4^{---}$$

Acidification of the thioantimonite solution reprecipitates the orange-red Sb_2S_3. Acidification of the thioantimonate solution precipitates the orange-yellow Sb_2S_5:

$$2SbS_3^{---} + 6H^+ \rightarrow Sb_2S_3\downarrow + 3H_2S\uparrow$$
$$2SbS_4^{---} + 6H^+ \rightarrow Sb_2S_5\downarrow + 3H_2S\uparrow$$

Antimony trisulfide is responsible for the color in red rubber. It has been used in range-finding shells, which produce dense white smoke when they explode. The head of a safety match contains antimony sulfide as the combustible substance, mixed with an oxidizing agent such as potassium chlorate. The side of the box, or the striking surface on the outside of the "book," is coated with a mixture of powdered glass, red phosphorus, and glue. Friction of the match head against this prepared surface causes tiny explosions involving the phosphorus and potassium chlorate. The heat that thereby is liberated ignites the head of the match, which is not readily ignited by friction alone.

Bismuth forms only one sulfide, Bi_2S_3. It occurs in nature as the dark brown mineral bismuthinite, one of the important bismuth ores. It is obtained as a dark brown precipitate when hydrogen sulfide is passed into a solution of a bismuth salt. It does not dissolve in sodium sulfide solution, for it is not sufficiently acidic to form thiosalts. This property is the basis for separating bismuth from arsenic and antimony in qualitative analysis.

28-12 Occurrence and Uses of the Elements

Phosphorus is the only member of Group VA that, because of its activity toward oxygen, is never found free in nature. It occurs almost exclusively in the form of the PO_4^{---} ion. Calcium phosphate, $Ca_3(PO_4)_2$, is the characteristic constituent of the minerals called *apatites*, which may be represented as $Ca_5(PO_4)_3X$ or $3Ca_3(PO_4)_2 \cdot CaX_2$, where X may be OH^-, F^-, or Cl^-. The apatites are minerals of fossil origin and have a chemical composition similar to the structural materials of bones and teeth.

The salt sodium ammonium hydrogen phosphate, $NaNH_4HPO_4$, known for several centuries as "microcosmic salt," was originally obtained from human urine. The presence of phosphates in excreta is a reminder of their importance in the blood. The ratio between the concentrations of Na_2HPO_4 and NaH_2PO_4 serves to regulate the pH of the blood and tissue fluids. The Na_2HPO_4 is present in higher concentration in the blood, but the acid salt, NaH_2PO_4, is present to a greater extent in the urine. The phospholipids, or phosphatides, that are present in all living cells are esters of organic derivatives of phosphoric acid.

Industrially, elementary phosphorus is obtained from phosphate rock by heating it with sand and coke in an electric furnace:

$$2Ca_3(PO_4)_2 + 6SiO_2 \rightarrow P_4O_{10} + 6CaSiO_3$$
$$P_4O_{10} + 10C \rightarrow P_4 + 10CO$$

The phosphorus vapor that distills from the furnace is condensed under water. It is sold in the form of finger-sized sticks, stored under water. Most of it is converted into red phosphorus or P_4S_3, for use in the match industry, or into phosphoric acid and phosphates.

Arsenic is widely distributed in nature, especially as arsenides that form isomorphous mixtures with sulfides. The mineral *arsenopyrite* approximates the composition FeAsS but is actually an isomorphous mixture of FeS_2 and $FeAs_2$. Pure $FeAs_2$ is called *arsenical pyrite*. Arsenic is occasionally found in the "native" (i.e., elementary) state. Industrially, the principal sources of arsenic and its compounds are the by-products from smelting sulfide ores of copper and lead. Realgar and orpiment have been mined for use as pigments.

Small quantities of arsenic are added to lead shot to give more perfect spheres. Arsenic-copper alloys are used in making automobile radiators and other objects that are assembled with solder. Most arsenic is sold as arsenites and arsenates.

The principal ore of *antimony* is stibnite, which occurs in China, Mexico, and Bolivia. It is reduced by heating with scrap iron:

$$Sb_2S_3 + 3Fe \rightarrow 2Sb + 3FeS$$

Antimony is used chiefly in alloys, such as type metal and pewter. Lead that has been hardened by addition of antimony is used for shrapnel, antifriction metal, and storage battery plates.

Bismuth is most often found in the native state. Industrially, it is obtained as a by-product of the metallurgy of lead and copper. Most of it is used in making low-melting alloys.

New Terms

Native State: the free, uncombined, elementary state. (This term is used to refer to the occurrence of an elementary metal in an ore.)

Phosphorescence: the state or property of emitting light without sensible heat, either as the result of previous exposure to an exciting radiation or as the result of slow oxidation.

Sesquioxide: an oxide in which the number of oxygen atoms is $1\frac{1}{2}$ times the number of atoms of the other element; e.g., N_2O_3, P_4O_6.

Exercises

28-1 How do you explain the following facts: (a) nitrogen forms no pentahalides; (b) phosphorus does not form PI_5 although it forms PF_5, PCl_5, and PBr_5; (c) the only pentahalide formed by bismuth is BiF_5?

28-2 Identify the following: dimethylarsine, tetramethylphosphonium iodide, cacodyl, disodium hydrogen phosphate, and sodium thioantimonate.

28-3 Compare the structures of the following molecules: P_4, P_4O_6, and P_4O_{10}.

28-4 Explain why H_3PO_3 is diprotic, H_3PO_4 is triprotic, and H_3PO_2 is monoprotic.

28-5 Why is sodium bismuthate a powerful oxidizing agent?

28-6 What is "syrupy phosphoric acid"? "White arsenic"? "TSP"? "Superphosphate of lime"?

28-7 Describe the difference between safety matches and "strike anywhere" matches.

28-8 Show how potassium, nitrogen, and phosphorus are supplied to soils in commercial fertilizers.

28-9 Which elements of Group VA are found free in nature? How else do they occur?

28-10 Which elements of the group have uses in the free state? State some of these uses.

References

Bailar, Jr., J. C., Moeller, T., and Kleinberg, J., *University Chemistry*, Heath, Boston, 1965, Chapter 17, "The Nitrogen Family." A discussion of the chemistry of the elements of Group VA by three inorganic chemists.

Hamm, D. I., *Chemistry*, New York, Appleton-Century-Crofts, 1965, Chapter 31, "The Nitrogen Family." An extended consideration of the elements of Group VA.

29

The Elements of the Carbon Group

29-1 Group Characteristics

The elements of Group IVA (carbon, silicon, germanium, tin, and lead) have some points of similarity as a consequence of having the same electronic configuration, s^2p^2, in the outermost shells of their atoms (Figure 29-1). The differences between them are equally notable, however. In this group we see a striking example of the discontinuity in general properties between the first row and second row elements of the periodic table. We cannot infer much silicon chemistry from the chemistry of carbon, even though both these elements can be classified as nonmetals. As we go further down the group the general trend toward increasing metallic character continues. Germanium is a borderline metal, much as silicon is a borderline nonmetal; both are often described

Figure 29-1

Electronic configurations of Group IVA elements

	1	2	3	4	5	6
$_{6}C$	s^2	s^2p^2				
$_{14}Si$	s^2	s^2p^6	s^2p^2			
$_{32}Ge$	s^2	s^2p^6	$s^2p^6d^{10}$	s^2p^2		
$_{50}Sn$	s^2	s^2p^6	$s^2p^6d^{10}$	$s^2p^6d^{10}$	s^2p^2	
$_{82}Pb$	s^2	s^2p^6	$s^2p^6d^{10}$	$s^2p^6d^{10}f^{14}$	$s^2p^6d^{10}$	s^2p^2

Table 29-1

Properties of the elements of Group IVA

Element	*Ionization Energy* (kcal/mole)	*Electro-negativity* (Pauling scale)	*Melting point* (°C)	*Boiling point* (°C)	*Specific resistivity* (microhm-cm)	*Density* (g/ml, 20°C)
$_{6}C$	261	2.5	[a]D unknown	unknown	5×10^{20} at 15°C	3.5
			G 3570	4200	1400 at 20°C	2.2
$_{14}Si$	188	1.8	1414	2355	8.5×10^{4} at 20°C	2.4
$_{32}Ge$	182	1.8	958	2700	2×10^{5} at 20°C	5.4
$_{50}Sn$	169	1.8	232	2362	12 at 20°C	[b]W 7.3
						G 5.8
$_{82}Pb$	171	1.8	328	1755	21 at 20°C	11.3

[a] D, diamond; G, graphite.
[b] W, white; G, gray.

as semimetals or metalloids. Tin and lead are truly metallic, although they have amphoteric oxides.

Carbon is decidedly more electronegative than the other members of the group, and it has the highest ionization energy (Table 29-1). Because the atoms of this group have outer shells that are half-filled relative to the noble-gas configurations, it is conceivable that they might form tetranegative ions by gaining four electrons or tetrapositive ions by having four electrons removed. The energy requirements for doing either of these are very high. Only the most active metals form carbides that seem to be essentially ionic, e.g., Al_4C_3, containing C^{4-} ions, and CaC_2, containing C_2^{--} ions. It is doubtful whether any of the elements of the group form 4+ ions, but lead(II) fluoride, PbF_2, is definitely ionic. Much of the chemistry of tin(II) and lead(II) compounds in aqueous solution is interpreted, for convenience, in terms of Sn^{++} and Pb^{++} ions, but both ions are hydrolyzed in water, giving $Sn(OH)^{+}$ and $Pb(OH)^{+}$ ions, unless the solutions are sufficiently acidic to prevent this.

All of the elements of the group form covalent compounds in the +4 state, and this is the favored state for all except lead. Carbon and silicon are acid-forming elements, i.e., their dioxides are weakly acidic. The monoxides and dioxides of germanium, tin, and lead are amphoteric.

29-2 Physical Properties: Allotropic Forms

Two of the elements, carbon and tin, exist in allotropic forms. The two forms of carbon are called *diamond* and *graphite*. Diamond is the hardest natural substance known and is the only gemstone composed of a single element.

It is inert to all chemical action at ordinary temperatures but burns when it is heated in air. If it is heated to 1000°C in the absence of air, it changes to graphite.

In a diamond crystal each carbon atom is located at the center of a tetrahedron and is covalently bonded to four others located at the corners (Figure 11-5). In other words, all the atoms in the crystal are chemically bonded into a giant molecule. The hardness and the high melting point are consequences of this structure and the strength of the carbon-carbon bonds. Since all the outer-shell electrons are involved in the chemical bonding, there are no mobile electrons such as are present in a metal. Thus, diamond has a high electrical resistivity, i.e., it is a nonconductor.

Most diamonds are mined in Africa, where they occur in the conical pipes of extinct volcanoes. It is assumed that the diamonds were formed when very pure carbon was subjected to very great pressure at high temperature. This theory appears reasonable since the density of diamond is greater than that of graphite (Table 29-1). Attempts to synthesize diamond from graphite in pressure cells at high temperature met with success in the research laboratories of the General Electric Company in 1957. Although the synthetic diamonds are small, they are very useful in abrasive powders.

No man-made diamonds of gem quality have yet been synthesized. The brilliance of diamonds is due to the many internal reflections that result from the high index of refraction of the crystals and from the manner in which the facets are cut. The flashes of color are due to the high dispersive power of diamond; i.e., the index of refraction is different for the various wavelengths of white light.

Graphite is a soft, grayish-black solid with a metallic luster and an appreciable electrical conductivity. Graphite has often been confused with lead. Our use of the expression "lead pencil" is a relic of this confusion. The carbon atoms in graphite are arranged in layers (Figure 29-2), with each atom bonded to three other atoms that are its nearest neighbors. The electron count comes out correct if we represent half of the bonds as single and the other half as double. Resonance among the different equivalent arrangements makes all bonds alike and also makes all carbon-carbon distances equal, 1.42 Å. This distance is intermediate between the single-bond carbon-carbon distance found in alkanes, 1.54 Å, and the double-bond distance found in alkenes, 1.33 Å. It is a little greater than the carbon-carbon distance in benzene, 1.40 Å.

The distance between layers, 3.40 Å, indicates that they are held by forces of the van der Waals type and not by chemical bonds. Thus, the layers can be separated easily, as indicated by the soft, flaky character of graphite.

No other elementary substance has the graphite structure, although black phosphorus has a somewhat similar structure (Section 28-2). The diamond structure occurs in silicon, germanium, and gray tin.

Tin exists in three allotropic forms. The form we ordinarily see is called *white tin* or *malleable tin*. It forms tetragonal crystals and is silvery-white with a slight bluish tinge. When a rod of white tin is bent, it produces a sound known as "tin cry" as the crystals slip past one another. Above 161°C the stable form

Figure 29-2

Arrangement of carbon atoms in graphite

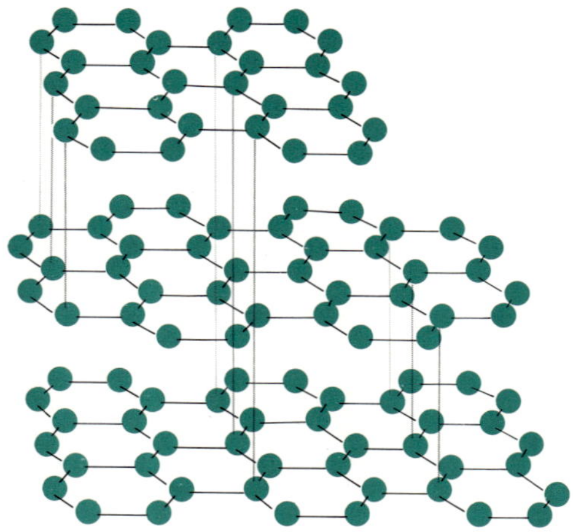

of tin is another metallic allotrope called *brittle tin*, which forms rhombic crystals. It shatters when it is hammered. Below 13.2°C white tin changes slowly into gray tin, which forms as a powder. Articles made of tin are therefore in danger of undergoing physical disintegration during long periods of cold weather. The temperature at which it is worst is about −48°C. The change progresses from a spot of origin, spreading like an infection. The phenomenon is called "tin disease." It was first noticed on organ pipes in unheated European cathedrals and was blamed on the Devil. In a later and more sophisticated time it was blamed on microorganisms. Both have now been absolved.

Only one crystalline form, the cubic metallic, is known for lead. It has a bluish-gray appearance but loses its luster when exposed to air because of the formation of a surface film of oxide and carbonate. Lead, like white tin, is soft and malleable. Neither metal is ductile. Both have low melting points and are used in low-melting alloys.

Like graphite, silicon and germanium become better electrical conductors with rising temperature. Tin and lead show an increase in electrical resistance as the temperature rises, as is characteristic of metallic conduction.

29-3 Chemical Trends Within the Group

Dilute nitric acid reacts with tin and lead, forming $Sn(NO_3)_2$ and $Pb(NO_3)_2$. Concentrated nitric acid reacts with germanium, tin, and lead, forming GeO_2, SnO_2, and $Pb(NO_3)_2$.

Silicon, tin, and lead dissolve slowly in strong bases, liberating hydrogen:

$$Si + 2NaOH + H_2O \rightarrow Na_2SiO_3 + 2H_2\uparrow$$
$$Sn + 2NaOH \rightarrow Na_2SnO_2 + H_2\uparrow$$
$$Pb + 2NaOH \rightarrow Na_2PbO_2 + H_2\uparrow$$

Germanium does not react with bases alone, but it dissolves readily if an oxidizing agent is also present:

$$Ge + 2NaOH + 2H_2O_2 \rightarrow Na_2GeO_3 + 3H_2O$$

Carbon is different from all other elements by its pronounced tendency to *catenation*, i.e., the formation of chains of identical atoms, as in the hydrocarbons (Chapter 18). Silicon has some tendency of this sort, and germanium has very little. Tin and lead are not known to form any compounds in which one of their atoms is bonded to more than one other atom of the same kind.

29-4 Oxides Obtained by Combustion

Carbon is relatively inert at ordinary temperatures, but when it is heated it combines with oxygen with the liberation of heat. The relative yields of CO and CO_2 depend on the temperature and on whether the carbon or the oxygen is present in excess. Burning silicon gives the dioxide, SiO_2; germanium gives GeO_2; and tin gives SnO_2. When lead is burned it ordinarily forms the monoxide, PbO.

29-5 Carbon Dioxide

Carbon dioxide is formed not only by combustion of elementary carbon but also when hydrocarbons, alcohols, or cellulose fuels (wood, paper, etc.) undergo complete combustion:

$$2C_8H_{18} + 25O_2 \rightarrow 16CO_2 + 18H_2O$$
$$C_2H_5OH + 3O_2 \rightarrow 2CO_2 + 3H_2O$$
$$(C_6H_{10}O_5)_n + 6nO_2 \rightarrow 6nCO_2 + 5nH_2O$$

It is also obtained by the action of acids on carbonates, by heating certain carbonates, and by the fermentation of glucose:

$$CaCO_3 + 2H^+ \rightarrow Ca^{++} + H_2O + CO_2\uparrow$$
$$CaCO_3 \xrightarrow{\text{heat}} CaO + CO_2\uparrow$$
$$C_6H_{12}O_6 \xrightarrow{\text{zymase}} 2C_2H_5OH + 2CO_2$$

Carbon dioxide is a colorless gas that is easily compressed to a liquid, provided the temperature is below 31°C (its critical temperature). Solid CO_2, well known as Dry Ice, has a sublimation pressure of 1 atmosphere at −79°C so that it ordinarily passes directly from the solid state to the gaseous state. The CO_2 molecule is linear and symmetrical, but the carbon-oxygen distance

(1.15 Å) is less than the theoretical double-bond distance (1.22 Å). It is regarded as a resonance hybrid:

$$O::C::O \leftrightarrow :O:C:::O: \leftrightarrow :O:::C:O:$$

Carbon dioxide is stable to about 1000°C and is only 1.8 percent dissociated into carbon monoxide and oxygen at 2000°C:

$$2CO_2 \rightleftharpoons 2CO + O_2$$

Therefore, it is only a weak oxidizing agent. However, it reacts readily with hot carbon or burning magnesium:

$$C + CO_2 \rightarrow 2CO$$
$$2Mg + CO_2 \rightarrow 2MgO + C$$

Carbon dioxide is slightly soluble in water. It has a mildly acid taste because of the formation of carbonic acid, H_2CO_3, when it reacts with water in the mouth. Carbonic acid is a weak acid ($K_1 = 4.4 \times 10^{-7}$ and $K_2 = 4.7 \times 10^{-11}$). Normal carbonates of the alkali metals (except lithium) are soluble in water, but carbonates of most other metals are insoluble. Bicarbonates (hydrogen carbonates) are usually more soluble than normal carbonates (Section 23–4). Solutions of both carbonates and bicarbonates are alkaline as a result of hydrolysis:

$$CO_3^{--} + H_2O \rightleftharpoons HCO_3^- + OH^-$$
$$HCO_3^- + H_2O \rightleftharpoons H_2CO_3 + OH^-$$

The carbonate ion is triangular, with 120° bond angles. Its structure can be represented as a resonance hybrid:

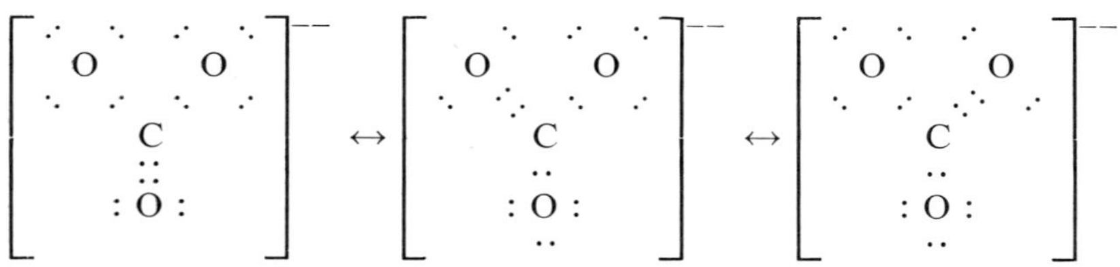

Carbamic acid, H_2NCOOH, can be regarded as derived from carbonic acid by substitution of an $—NH_2$ group for an —OH group. If the second —OH group is replaced by $—NH_2$, we have urea:

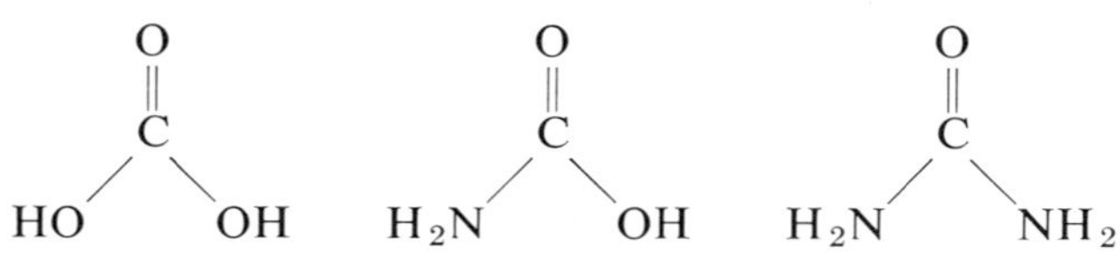

carbonic acid carbamic acid urea

Carbamic acid, like carbonic acid, is unstable and is known only in the form of its salts (carbamates) and esters (urethanes).

29-6 Carbon Monoxide

Carbon monoxide is formally the anhydride of formic acid, but it dissolves only slightly in water, and the solution is not acidic. Nevertheless, the most convenient method for obtaining the gas is to allow formic acid to drip into hot, concentrated sulfuric acid and thereby undergo instantaneous dehydration:

$$HCOOH \rightleftharpoons H_2O + CO\uparrow$$

Carbon monoxide is colorless, odorless, tasteless, and dangerously toxic. It burns in air with the formation of CO_2. The mixture of hydrogen and carbon monoxide gases obtained from the oxidation of coke by steam is a valuable fuel. At elevated temperatures carbon monoxide reduces metal oxides, and for this reason it plays an important role in the metallurgy of iron. It combines with chlorine in the presence of sunlight to form carbonyl chloride, $COCl_2$, commonly called phosgene. (The word *phosgene* is from the Greek, meaning "born of light," and has no reference to the element phosphorus.) Phosgene is a colorless gas with a suffocating odor. It is highly toxic because of its hydrolysis in the lungs, which liberates hydrochloric acid. Phosgene is now used in the manufacture of dyes and polyurethane plastics.

29-7 Carbon Suboxide

Carbon suboxide, C_3O_2, is a malodorous gas formed by dehydration of malonic acid with phosphorus(V) oxide:

$$HOOC\text{—}CH_2\text{—}COOH \rightarrow O{=}C{=}C{=}C{=}O + 2H_2O$$

Its molecule is linear. The gas dissolves readily in water, regenerating malonic acid.

29-8 Silicon Dioxide

Silicon dioxide is an acidic oxide, but no silicic acids have been isolated. The most common form of silica (SiO_2) is quartz; sand and sandstone consist chiefly of small quartz particles. Quartz crystals are used as semiprecious stones, in optical instruments, and in controlling the frequency of radio transmission signals.

No greater contrast can be found between two chemically analogous compounds than that which is shown by CO_2 and SiO_2. The unit of structure in solid carbon dioxide is the CO_2 molecule; weak intermolecular forces hold these units in position in the crystal lattice. In contrast, each silicon atom in silica is located at the center of a tetrahedron and is bonded to four oxygen atoms located at the corners of the tetrahedron. The SiO_4 tetrahedra are attached to one another by sharing the corner oxygen atoms, i.e., each oxygen is bonded to two silicon atoms. This structure extends in three dimensions to form a network solid (Section 11-6). In order to melt or break a quartz crystal it is necessary to break some silicon-oxygen bonds. Consequently, quartz is hard and has a high melting point (1600°C).

Because the silicon atom has a larger radius than the carbon atom, its nucleus attracts the outer electrons less strongly. In a silicon-oxygen linkage there is, therefore, a lesser attraction of the silicon for electrons than that of carbon for electrons in a carbon-oxygen linkage. This leads to the sharing of an oxygen atom between two different silicon atoms so that SiO_2 (empirical formula) forms a network structure whereas CO_2 (molecular formula) consists of individual molecules that do not polymerize.

Molten silica is highly viscous. It has a great tendency to supercool, i.e., the liquid reaches temperatures well below the original melting point without crystallizing. When cooled to 1500°C molten silica becomes too stiff to flow. In this state it is referred to as *silica glass*. It is also known as quartz glass or fused quartz. Silica glass differs from crystalline quartz in that the arrangement of the SiO_4 tetrahedra is irregular.

The coefficient of thermal expansion of silica glass is so small that vessels made of it do not break when they are subjected to sudden heating or cooling. This makes it useful for making chemical apparatus. It is used in mercury vapor lamps and in optical instruments because it is transparent to ultraviolet light. It has the additional quality of being chemically inert except to fluorine and hydrogen fluoride.

Addition of acid to a solution of a silicate yields a gel of hydrous silica. For convenience it is often represented as $Si(OH)_4$, i.e., H_4SiO_4; but it is more correctly represented as $SiO_2 \cdot nH_2O$, where n is variable. The gel is insoluble in acids and bases. Alkali-metal silicates, which are the only silicates soluble in water, must be obtained by fusing sand with alkali-metal hydroxides or carbonates. The composition of the product, which is called "waterglass," is determined by the relative amounts of the reactants used in the fusion. For convenience, waterglass can be represented as Na_2SiO_3 in equations representing its formation:

$$Na_2CO_3 + SiO_2 \rightarrow Na_2SiO_3 + CO_2\uparrow$$
$$2NaOH + SiO_2 \rightarrow Na_2SiO_3 + H_2O$$

The silicate ions actually present in solutions of silicates are of varying degrees of complexity, including some of colloidal size. Waterglass is used as an adhesive for sealing cardboard boxes.

29-9 Oxides of Germanium, Tin, and Lead

The dioxides of germanium and tin are white compounds that can be obtained by direct union of the elements. Lead dioxide is a dark brown compound that must be obtained by procedures involving strong oxidation, e.g., electrolytic oxidation of lead(II) salts. All three dioxides are amphoteric. They react with bases to give germanates, stannates, and plumbates. Both metagermanates, $M^{I}_2GeO_3$, and orthogermanates, $M^{I}_4GeO_4$, are known. (M^{I} denotes any metal in the +1 oxidation state.) Solutions of stannates and plumbates contain $Sn(OH)_6^{--}$ and $Pb(OH)_6^{--}$ ions. The dioxides dissolve in hydrohalic acids to give tetrahalides or, in some cases, hexahalo complexes, e.g., $SnCl_6^{--}$. Lead tetrachloride is unstable above 100°C:

$$PbCl_4 \rightarrow PbCl_2 + Cl_2$$

The monoxides of germanium, tin, and lead are also amphoteric. Germanium(II) oxide can be obtained by hydrolysis of $GeCl_2$ or by reduction of GeO_2. Hydrous tin(II) oxide is precipitated by addition of aqueous ammonia (i.e., a low concentration of OH^- ions) to $SnCl_2$ solution. Lead(II) oxide is obtained by heating metallic lead in air or by thermal decomposition of PbO_2 or Pb_3O_4.

In Pb_3O_4, commonly known as "red lead," the metal is in both the +2 and +4 states. It may be regarded as plumbous orthoplumbate, Pb_2PbO_4. It is formed by heating lead(II) oxide in air at 430°C:

$$6PbO + O_2 \rightarrow 2Pb_3O_4$$

No compounds of the formula $M(OH)_2$ or $M(OH)_4$ are definitely known for the members of Group IVA. The precipitates obtained by addition of OH^- ions to solutions of tin(II) or lead(II) salts are commonly called hydroxides and are represented by the formulas $Sn(OH)_2$ and $Pb(OH)_2$, but this is done merely as a convenience. The precipitates are really hydrous oxides, $SnO \cdot nH_2O$ and $PbO \cdot nH_2O$, where n is variable.

29-10 The Natural Silicates

The large number of important silicates found in nature include granite, mica, asbestos, kaolin, and many others. The almost bewildering confusion in attempting to classify the natural silicates is partly due to our simultaneous use of terms taken from mineralogical, chemical, and economic contexts. For example, granite is not the name of a compound, nor is it the name of a mineral; granite is a term used to refer to a common rock that is composed of different types of feldspar, quartz, and small amounts of mica. Feldspar, in turn, is a term used by mineralogists to refer to a group of closely related aluminosilicates; the feldspars are the most abundant of all minerals.

From the viewpoint of chemical structure, the silicates fall into the following categories:

1. The Simple Orthosilicates. These minerals contain discrete, individual SiO_4^{4-} anions, as in the *olivine* series $(Mg, Fe)_2SiO_4$. (This method of writing the formula implies that the composition may vary from Mg_2SiO_4 to Fe_2SiO_4, i.e., that there may be any relative amounts of Mg^{++} and Fe^{++} ions, but there must be two such ions for each SiO_4^{4-} anion to preserve the electroneutrality of the crystal.) The olivines are common in igneous rocks. Each SiO_4^{4-} ion is a tetrahedron with the silicon atom at the center and an oxygen atom at each corner. The metallic ions are situated between the silicate ions in the crystal.

2. The Disilicates. These minerals contain individual $Si_2O_7{}^{6-}$ anions. Each of these anions consists of two SiO_4 tetrahedra that are sharing a corner oxygen. The electron distribution is indicated as follows:

$$\left[\begin{array}{ccccccccc} & & & \cdot\cdot & & \cdot\cdot & & & \\ & & : & O & : \; : & O & : & & \\ & \cdot\cdot & & \cdot\cdot & \cdot\cdot & \cdot\cdot & & \cdot\cdot & \\ : & O & : & Si & : O : & Si & : & O & : \\ & \cdot\cdot & & \cdot\cdot & \cdot\cdot & \cdot\cdot & & \cdot\cdot & \\ & & : & O & : \; : & O & : & & \\ & & & \cdot\cdot & & \cdot\cdot & & & \end{array}\right]^{6-}$$

Only two or three minerals are known to have this structure.

3. Cyclic Silicates. A few silicate minerals contain cyclic silicate anions, but only two such ions are known: $Si_3O_9{}^{6-}$ and $Si_6O_{18}{}^{12-}$. In both these ions SiO_4 tetrahedra are sharing corner oxygens and forming closed rings. The charges of the ring ions are matched by the charges of the metallic ions present in the openings of the crystal structure. An example of a cyclic silicate is *wollastonite*, which has been traditionally represented by its empirical formula, $CaSiO_3$. Because it contains $Si_3O_9{}^{6-}$ anions, each matched by three Ca^{++} ions, we can best write the formula for wollastonite as $Ca_3Si_3O_9$. An example of a cyclic silicate containing the $Si_6O_{18}{}^{12-}$ ring is *beryl*, $Be_3Al_2Si_6O_{18}$.

4. Pyroxenes. Minerals of the *pyroxene* class contain *chain ions of indefinite length* formed from SiO_4 tetrahedra sharing two corner oxygens. The composition of the chain can be represented as $(SiO_3{}^{2-})_n$, where n is indefinite. The chains are parallel, with metal ions situated between them. An example of a pyroxene is the lithium mineral *spodumene*, $LiAl(SiO_3)_2$. This is, of course, an empirical formula, a simplification for $Li_nAl_n(SiO_3)_{2n}$.

5. Amphiboles. Minerals of the *amphibole* class contain *double-chain ions* of indefinite length, each chain formed from SiO_4 tetrahedra sharing two corner oxygens but with the additional feature that half the silicon atoms in each chain share another corner oxygen with a silicon atom in the other chain. The composition of the double chain can be represented as $(Si_4O_{11}{}^{6-})_n$, where n is indefinite. Most amphiboles contain at least two cations, and apparently all of them contain some hydroxyl groups which are not attached to the silicate chain.

The electrostatic forces between the metal ions and the silicate chains are not as strong as the silicon-oxygen bonds within the chains. Consequently, such minerals cleave most readily in directions parallel to the chains. Many of the amphiboles can easily be parted into fibers. The term *asbestos* includes all minerals having a fibrous structure. The principal commercial asbestos is *chrysotile*, $Mg_6Si_4O_{11}(OH)_6 \cdot H_2O$.

6. Layer Silicates. In some silicates the SiO_4 tetrahedra are linked into *two-dimensional networks*, forming sheets, which give the mineral a laminar or layered structure. Each silicon atom shares *three* oxygen atoms with its neighbors. The composition of each sheet can be represented as $(Si_2O_5{}^{2-})_n$, where n is indefinite. Such minerals tend to cleave into thin layers. An example is *talc*, $Mg_3(Si_2O_5)_2(OH)_2$, of which soapstone is composed. Better known are the *micas*, which readily separate into thin transparent sheets. The micas are aluminosilicates; i.e., some of the SiO_4 tetrahedra are replaced by AlO_4 tetrahedra. Since the charge on an aluminum nucleus is one less than that on a silicon nucleus, each such substitution increases the anion charge by one unit. This increase is balanced by the presence of additional metal ions. For example, *biotite* is a mica with a composition similar to talc except that every fourth silicon atom has been replaced by an aluminum atom and an equal number of potassium ions have been introduced to balance the additional charge. Consequently, the composition of biotite is $KMg_3AlSi_3O_{10}(OH)_2$.

7. *Framework Silicates*. The next logical step in our progression from simple orthosilicate ions to two-dimensional ions of indefinite extent would be to a three-dimensional structure in which each silicon atom shares all four of its oxygen atoms with its neighbors, so that *every* oxygen atom is shared by two silicon atoms. This would mean that the number of oxygen atoms present is twice the number of silicon atoms; i.e., the composition of the substance can be represented as $(SiO_2)_n$. In other words, the substance is silica, a network solid. However, if some silicon atoms are replaced by aluminum atoms, the three-dimensional network becomes negatively charged, and there must be metal ions distributed through it to preserve electroneutrality. Aluminosilicates of this type are known as *framework silicates*. The feldspars belong to this class. They are the major constituents of igneous rocks; examples are *orthoclase*, $KAlSi_3O_8$, and *albite*, $NaAlSi_3O_8$.

The slow processes of chemical weathering convert feldspars to clays. *Kaolinite*, one of the important clay minerals, is produced by the weathering of orthoclase:

$$\underset{\text{orthoclase}}{2KAlSi_3O_8} + 2CO_2 + 3H_2O \rightarrow \underset{\text{kaolinite}}{Al_2Si_2O_5(OH)_4} + 4SiO_2 + 2KHCO_3$$

As implied by its formula, kaolinite is a layer silicate. The layers are electrically neutral and slide past one another very readily, thus accounting for the softness and soapy feel of clay.

The *zeolites* are framework silicates characterized by corridors that run through the framework and have ample dimensions to permit ions to enter and leave freely. The zeolite minerals used for softening water by ion exchange belong to this class.

29-11 Ion Exchange

Two major types of porous solids are used as *ion exchangers:* (1) the zeolites, which are naturally occurring aluminosilicate minerals, e.g., $NaAl(SiO_3)_2$ or $NaAlSi_2O_6$, and (2) the ion-exchange resins, which are synthetic organic polymers.

In the process of ion exchange, ions previously held at electrically charged sites within the porous solid have been exchanged for ions in a solution which surrounds and invades the solid. For example, when water containing Ca^{++}

or Mg^{++} ions is allowed to filter through thick layers of zeolite, the Na^+ ions in the zeolite are replaced by Ca^{++} or Mg^{++} ions, and the Na^+ ions replace the Ca^{++} or Mg^{++} ions in the solution:

$$2NaAlSi_2O_6 + Ca^{++} \rightarrow Ca(AlSi_2O_6)_2 + 2Na^+$$

You will recall that water containing Ca^{++} or Mg^{++} ions is hard water. Thus, the effect of the ion exchange in the zeolite is to soften the water. This is the principle upon which domestic water softeners operate.

Ion-exchange reactions are reversible. The direction which the reaction takes is determined by the ion concentrations. If a concentrated NaCl solution is allowed to stand in contact with a zeolite which has been used for softening water, the calcium aluminosilicate is reconverted to sodium aluminosilicate:

$$Ca(AlSi_2O_6)_2 + 2Na^+ \rightarrow 2NaAlSi_2O_6 + Ca^{++}$$

This reaction, the reverse of the preceding one, leaves sodium zeolite ready to serve as a water softener again. When, through use, it is again converted to calcium zeolite, sodium zeolite can be regenerated again by the same procedure as before. Thus, the only material needed to keep an ion-exchange water softener operating over a period of many years is common salt.

Both cation exchangers and anion exchangers are available. Water can be completely demineralized by being passed through an exchanger which replaces all metal ions with H^+ ions, and then through another exchanger which exchanges OH^- ions for any other anions. The effluent water is as pure as carefully distilled water.

So-called *molecular sieves* are composed of synthetic zeolites. The cavities within them are of such size that small molecules can be accommodated within them and held by van der Waals forces. Large molecules cannot penetrate the cavities. Molecular sieves are used to separate straight-chain hydrocarbons from branched-chain and ring hydrocarbons in gasoline. Straight-chain hydrocarbons have poor antiknock properties.

29-12 Glass

The art of making glass is several thousand years old. The common glass used in windows and bottles is made by heating a mixture of sodium carbonate, limestone, and sand:

$$Na_2CO_3 + SiO_2 \rightarrow Na_2SiO_3 + CO_2\uparrow$$
$$CaCO_3 + SiO_2 \rightarrow CaSiO_3 + CO_2\uparrow$$

The product is a complex mixture of silicates and silica. The above equations do not mean that SiO_3^{--} ions are actually present in glass. Like the natural silicates, glass consists essentially of SiO_4 tetrahedra sharing corner oxygens.

When cooled, molten glass gradually becomes more and more viscous until finally it becomes rigid. It retains the amorphous structure characteristic of liquids, and when it is reheated it softens gradually. Because there is no particular temperature at which it observably melts on heating or freezes on cooling, glass is classified as a supercooled liquid of very high viscosity. Any material meeting these criteria may be termed a glassy solid (as opposed to a crystalline or true solid).

Structurally, the difference between a silicate glass and a crystalline silicate is that the SiO_4 tetrahedra in glass are not arranged in a regular pattern. If molten glass is cooled very slowly it will devitrify; i.e., it will form crystals. At ordinary temperatures devitrification requires hundreds or even thousands of years, but the process is speeded up if the glass is subjected to prolonged heating (below the melting point).

The actual composition of silicate glass varies a great deal. Inevitably there are impurities even in the best grade of sand and some of these impart color to the glass. Other impurities are deliberately introduced to offset undesirable colors or to develop colors when a colored glass is desired. If part of the sodium is replaced by potassium the glass is harder and has a higher melting point. If part of the SiO_2 is replaced by B_2O_3, the resulting borosilicate glass (Pyrex and Kimax) has a remarkably low coefficient of thermal expansion. Such glass can undergo rapid changes of temperature without breaking. Aluminosilicate glasses also have this property and can withstand higher temperatures. Borosilicate glass is used in oven baking ware, and aluminosilicate glass in glassware intended for cooking over an open flame.

One of the new types of glass consists essentially of silica and aluminum phosphate, $AlPO_4$. Since aluminum and phosphorus are the nearest neighbors of silicon in the periodic table, aluminum phosphate is isoelectronic with silica. Consequently, no metal ions are required in the composition of such a glass, for there are no negative ions. The structure of the glass is like that of silica glass with AlO_4 tetrahedra and PO_4 tetrahedra alternating with SiO_4 tetrahedra.

29-13 Sulfides

Carbon disulfide is obtained by reaction of sulfur vapor with white-hot carbon in an electric resistance furnace. It has wide use as a solvent and as a starting material for the manufacture of other chemicals, e.g., CCl_4. It is a colorless, highly volatile liquid (bp, 46°C) that is highly flammable, very poisonous, and (because of impurities) repulsively ill-smelling. It is a typical covalent compound, miscible with many organic compounds and an excellent solvent for iodine, sulfur, and white phosphorus.

Carbon disulfide and carbon oxysulfide are analogues of carbon dioxide:

O=C=O	O=C—S	S=C=S
carbon dioxide	carbon oxysulfide	carbon disulfide

Carbon oxysulfide is formed by the reaction of sulfur vapor with carbon monoxide. It is an odorless, toxic gas, sometimes called carbonyl sulfide.

Carbon disulfide reacts with sulfide ion to form thiocarbonate ion:

$$CS_2 + S^{--} \rightarrow CS_3^{--}$$

Thus, potassium thiocarbonate, an insecticide, is obtained by shaking a concentrated K_2S solution with CS_2. Free thiocarbonic acid, H_2CS_3, is a red oily liquid.

Reaction of carbon disulfide with alcoholic alkali gives ester salts commonly known as *xanthates:*

$$\underset{\text{}}{C_2H_5OH + CS_2 + NaOH \rightarrow \underset{\text{sodium ethyl xanthate}}{C_2H_5OCS_2^-Na^+} + H_2O}$$

The reaction is reversed by the addition of acid:

$$C_2H_5OCS_2^- + H^+ \rightarrow \underset{\substack{\text{xanthic acid}\\ \text{(unstable)}}}{\left[C_2H_5O{-}\overset{\overset{\displaystyle S}{\|}}{C}{-}SH\right]} \rightarrow C_2H_5OH + CS_2\uparrow$$

One of the most important uses for CS_2 is in the manufacture of viscose rayon from wood pulp or cotton linters. The —OH groups of the cellulose react with CS_2 in the presence of NaOH to give water-soluble xanthates:

$$ROH + CS_2 + NaOH \rightarrow ROCS_2^-Na^+ + H_2O$$

The viscous cellulose xanthate solution is called *viscose*. Extruding the solution through fine nozzles known as spinnerets into an acid bath reverses the reaction, forming filaments of regenerated cellulose fiber which are twisted into thread. Alternatively, the viscose can be converted into cellophane by extruding it through a narrow slot into the coagulating bath. Cellophane is an important transparent wrapping film, but polyethylene film is now available at less cost per square foot.

Silicon disulfide, SiS_2, is formed by melting a mixture of silicon and sulfur. When purified by sublimation it consists of colorless, silky crystals. Its fibrous nature is due to the fact that it consists of chains of indefinite length, each complete in itself:

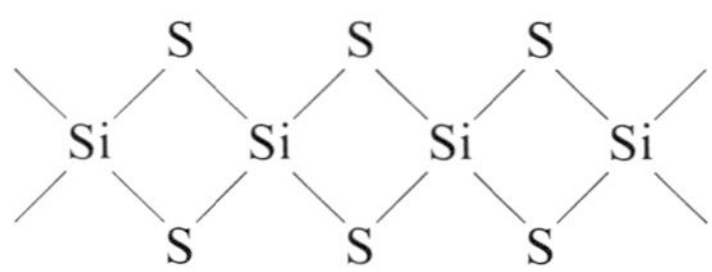

Each silicon is at the center of a tetrahedron with a sulfur atom at each corner, and each sulfur atom is joined to two silicon atoms. Two SiS_4 tetrahedra share two corner sulfur atoms simultaneously. In other words, the SiS_4 tetrahedra share edges. (SiO_4 tetrahedra can share only corners.) Silicon disulfide is stable in dry air, but in the presence of moisture it hydrolyzes:

$$SiS_2 + 2H_2O \rightarrow SiO_2 + 2H_2S\uparrow$$

Germanium disulfide, GeS_2, forms as a white precipitate when H_2S is passed into a strongly acidified solution of GeO_2. It dissolves in solutions of alkali sulfides, forming thiogermanates:

$$GeS_2 + S^{--} \rightarrow GeS_3^{--}$$

On the addition of acid, the GeS_2 reprecipitates.

Tin disulfide, SnS_2, forms as a yellow precipitate when H_2S is passed into weakly acidified solutions of tin(IV) salts. It dissolves in warm HCl of moderate concentration, forming $SnCl_4$ or $SnCl_6^{--}$. It also dissolves in solutions of alkali sulfides, forming thiostannates:

$$SnS_2 + S^{--} \rightarrow SnS_3^{--}$$
$$SnS_2 + 2S^{--} \rightarrow SnS_4^{----}$$

Acidification reprecipitates the SnS_2.

Germanium and tin disulfides (like CS_2 and SiS_2) can be obtained by direct union of the elements, but lead does not form a disulfide. Germanium, tin, and lead form +2 sulfides.

Germanium(II) *sulfide* forms as a red-brown precipitate when H_2S is passed into a hydrochloric acid solution of $GeCl_2$. It dissolves in hot, concentrated HCl. It also dissolves in solutions of alkali sulfides, forming thiogermanites. Upon addition of acid, the red-brown GeS reprecipitates. Ammonium polysulfide reagent dissolves GeS, forming thiogermanate; acidification of this solution precipitates the white GeS_2.

Tin(II) *sulfide* forms as a dark brown precipitate when H_2S is passed into a solution of a tin(II) salt. It is insoluble in dilute acids and is also insoluble in alkali sulfides. It dissolves in ammonium polysulfide reagent, forming thiostannate; acidification of this solution precipitates the yellow SnS_2.

Lead sulfide, PbS, forms as a black precipitate when H_2S is passed into a solution of a lead salt. It is not soluble in dilute HCl, alkali sulfides, or polysulfides.

29-14 Compounds with Hydrogen

Silicon forms hydrogen compounds, called *silanes* or *silicon hydrides*, of the general formula Si_nH_{2n+2}, analogous to the alkanes. However, the maximum number of silicon atoms known to occur in a continuous chain is six, indicating the drastically reduced catenation ability of silicon in comparison with carbon. The first two members of the series are gases; the others are liquids. All silanes are colorless and are spontaneously flammable in air.

With increasing atomic number the Group IVA elements show a diminishing ability for self-linkage. Only three *germanes* (germanium hydrides) are known: GeH_4, Ge_2H_6, and Ge_3H_8. There is only one *stannane*, SnH_4, and one *plumbane*, PbH_4.

The thermal stabilities of the hydrides decrease with increasing atomic number of the Group IVA element. Stannane decomposes slowly at room temperature and quite rapidly above 150°C. Plumbane is unstable above 0°C.

Like the hydrocarbons, the Group IVA hydrides must be prepared by indirect means rather than by direct union. When hydrochloric acid acts upon magnesium silicide, Mg_2Si, the main product is hydrogen gas, but the reaction also yields small quantities of various silanes, which can be separated by fractionation. Similar action with magnesium germanide, Mg_2Ge, gives a mixture of H_2 and the germanes. Pure SiH_4, Si_2H_6, GeH_4, or SnH_4 can be obtained by reduction of the corresponding chloride ($SiCl_4$, Si_2Cl_6, $GeCl_4$, or $SnCl_4$) with lithium aluminum hydride in ether solution. Plumbane is obtained with difficulty by procedures involving electrolytic reduction.

Silicon-carbon bonds are much more stable than silicon-hydrogen bonds, and the same is true of germanium, tin, and lead. The first compound containing a silicon-carbon bond was obtained in 1863 by the reaction between silicon tetrachloride and diethylzinc:

$$SiCl_4 + 2(C_2H_5)_2Zn \rightarrow (C_2H_5)_4Si + 2ZnCl_2$$

Tetraethylsilane (bp, 153°C) is a stable compound, as is generally true of the tetraalkyl and trialkyl silanes. Those containing only one or two alkyl groups, e.g., $C_2H_5SiH_3$ and $(C_2H_5)_2SiH_2$, are almost as active as silane itself.

The germanium tetraalkyls are colorless liquids of high stability. Even hot, concentrated nitric acid does not attack tetraethylgermane, $(C_2H_5)_4Ge$. The alkylstannanes are also stable, but those that have hydrogen bonded to tin are strong reducing agents similar to other metallic hydrides. Tetraethyllead, used as an antiknocking agent in motor fuels, is produced in far larger quantities than any other organometallic compound (Sections 22-8, 22-9, and 24-10). The tetraethyllead is a colorless liquid, but gasoline to which it is added is colored red by addition of a dye so that it may be recognized, for tetraethyllead, like all lead compounds, is toxic.

29-15 Tetrahalides

All of the possible tetrahalides of Group IVA elements are known except $PbBr_4$ and PbI_4. The nonexistence of these two can be attributed to the reducing powers of bromine and iodine in the -1 state and the oxidizing power of lead in the $+4$ state. Even $PbCl_4$ must be kept at low temperatures to prevent its decomposition into $PbCl_2$ and Cl_2. The low volatility of SnF_4 and PbF_4 in comparison with the other tetrahalides indicates that they have a considerably greater degree of ionic character.

Carbon tetrafluoride may be obtained by direct union of the elements, but the other carbon tetrahalides must be obtained by indirect means. The tetrahalides of silicon, germanium, and tin can usually be prepared by direct union, but SiF_4 is most often obtained by the action of hydrofluoric acid on silica, and $SiCl_4$ is made by passing chlorine over a hot mixture of carbon and silica:

$$SiO_2 + 4HF \rightarrow SiF_4 + 2H_2O$$
$$SiO_2 + 2C + 2Cl_2 \rightarrow SiCl_4 + 2CO$$

Tin(IV) chloride (stannic chloride, tin tetrachloride) is obtained by treating scrap tinplate with chlorine.

Carbon tetrahalides are not affected by water, but the tetrahalides of the other Group IVA elements undergo hydrolysis. Silicon tetrachloride is a pungent smelling liquid, which fumes strongly in air as the result of complete and irreversible hydrolysis:

$$SiCl_4 + 2H_2O \rightarrow SiO_2 \downarrow + 4HCl$$

The resulting smoke can be made still denser by allowing ammonia to escape at the same time. The mixture of solid silica and ammonium chloride constitutes an effective smoke screen used by warships.

Silicon tetrabromide and tetraiodide hydrolyze in the same fashion as the tetrachloride, but SiF_4 undergoes only partial hydrolysis because the HF produced by the hydrolysis unites with unhydrolyzed SiF_4 to form fluosilicic acid:

$$SiF_4 + 2H_2O \rightarrow SiO_2 + 4HF$$
$$SiF_4 + 2HF \rightarrow 2H^+ + SiF_6^{--}$$

Fluosilicic acid is a strong acid that can be isolated from solution as a hydrate. Its sodium and barium salts are useful insecticides. The fluosilicate anion is a stable anion, but chlorine, bromine, and iodine atoms are larger than fluorine atoms and it is not possible for six of them to get close enough to a silicon atom to form bonds.

Germanium tetrahalides hydrolyze to give GeO_2 and the halogen acid. The final product of hydrolysis of GeF_4 is fluogermanic acid, H_2GeF_6. The hydolysis of tin and lead tetrachlorides is incomplete and can be reversed by addition of acid; excess HCl results in the formation of chlorostannic and chloroplumbic acids (H_2SnCl_6 and H_2PbCl_6). Ammonium chlorostannate is industrially important because of its use as a mordant in dyeing: it precipitates SnO_2 upon the fibers of the cloth, where it takes up the dye. The $PbCl_6^{--}$ ion is distinctly more stable than $PbCl_4$ itself. Potassium bromoplumbate, K_2PbBr_6, and potassium iodoplumbate, K_2PbI_6, have been obtained by the reaction of chloroplumbates with KBr and KI even though $PbBr_4$ and PbI_4 are apparently nonexistent.

From one point of view the tetrahalides of the Group IVA elements are the completely halogenated derivatives of methane, silane, germane, stannane, and plumbane. We can say that CCl_4 is tetrachloromethane, $SiCl_4$ is tetrachlorosilane, etc. In addition, there are other compounds that can be regarded as partially halogenated derivatives of methane, silane, and germane. Examples are CH_3Cl, SiH_3Cl, GeH_3Cl; CH_2Cl_2, SiH_2Cl_2, GeH_2Cl_2; and $CHCl_3$, $SiHCl_3$, $GeHCl_3$.

The chlorosilanes react readily with water. It might be expected that the effect of hydrolysis would simply be replacement of a chlorine atom by an —OH group. We have already seen, however, that hydrolysis of $SiCl_4$ gives SiO_2 rather than $Si(OH)_4$. The SiO_2 is a high polymer characterized by Si—O—Si links. The tendency for the Si—O—Si link to form is so great that the product of the reaction of water with chlorosilane, H_3SiCl, is not H_3SiOH but $H_3Si—O—SiH_3$, called *disiloxane*, a colorless, odorless gas:

$$2H_3SiCl + H_2O \rightarrow H_3Si{-}O{-}SiH_3 + 2HCl$$

Hydrolysis of dichlorosilane, H_2SiCl_2, gives H_2SiO, which soon polymerizes to give a solid substance that closely resembles silica. Hydrolysis of trichlorosilane, $HSiCl_3$, also produces a silica-like polymer.

29-16 Silicones

A variety of compounds can be regarded as derivatives of silane, germane, stannane, and plumbane in which one or more hydrogen atoms have been replaced by halogen atoms whereas the others have been replaced by alkyl or aryl groups. An example is dimethyldichlorosilane, $(CH_3)_2SiCl_2$, a liquid (bp, 70°C), which is of great commercial importance as an intermediate from which millions of pounds of silicone polymers are made each year. Other examples are ethylgermanium trifluoride (or ethyltrifluorogermane), $C_2H_5GeF_3$; triphenyltin iodide, $(C_6H_5)_3SnI$; and trimethyllead bromide, $(CH_3)_3PbBr$.

Hydrolysis of dialkyldichlorosilanes gives dialkylsilanediols:

$$R_2SiCl_2 + 2H_2O \rightarrow R_2Si(OH)_2 + 2HCl$$

The diols lose water readily to form polymers that are most accurately described as *polyorganosiloxanes* but are more commonly referred to as *silicones:*

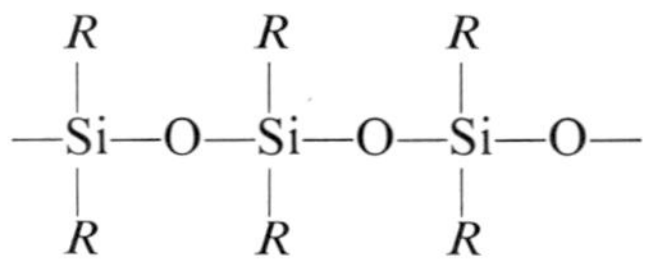

Two bonds of each silicon atom are occupied by organic groups, and the other two are linked to separate oxygen atoms. Note that the polymeric chain has the —Si—O—Si— links present in silica, not the —C—C— links present in organic polymers. Because the chain itself is a stable oxidized structure and the *R*—Si linkages are also resistant to oxidation, the polysiloxanes have a high degree of thermal stability and chemical inertness that makes them unique.

To illustrate the nature of silicone formation, suppose dimethyldichlorosilane hydrolyzes and that four of the resulting diol molecules condense to form a chain by splitting out water molecules between adjacent monomers:

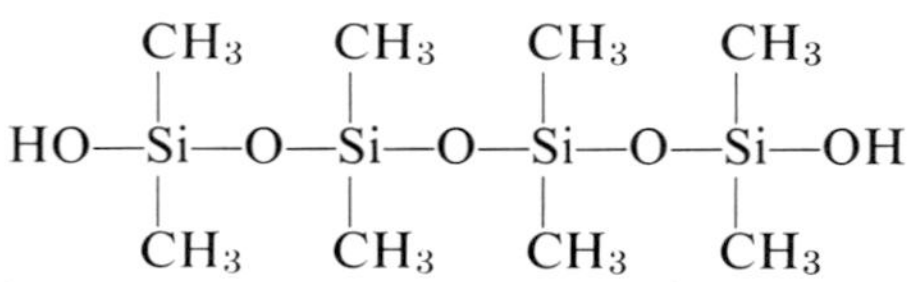

If a little trimethylchlorosilane is also present, it hydrolyzes to give $(CH_3)_3SiOH$, and one molecule of this product could condense with each end of the chain so that the final structure would be

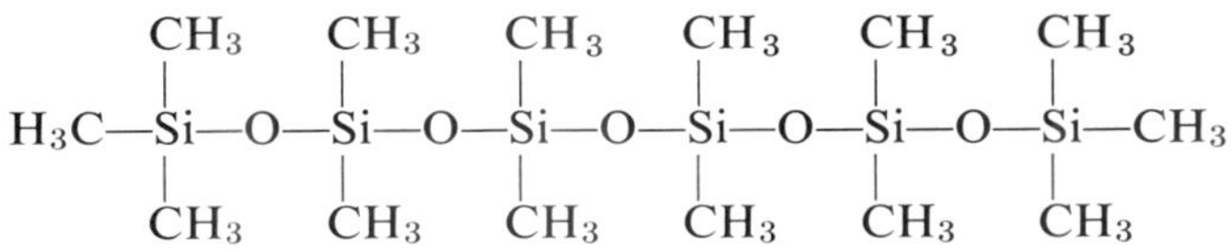

Chains of any desired length may be obtained by regulating conditions, such as the relative amounts of the two chlorosilanes in the starting mixture. The greater the proportion of dimethyldichlorosilane, the longer the chain that is formed. The longer the chain, the more nearly the composition of the polymer agrees with the empirical formula $(CH_3)_2SiO$. Note the similarity of this to the formula for acetone, $(CH_3)_2CO$. It is the similarity of the empirical formulas of these polymers to the formulas for ketones that is the basis for the name *silicones*.

Cross-links may be introduced into a silicone structure in any desired number by using a suitable proportion of alkyltrichlorosilane in the starting mixture.

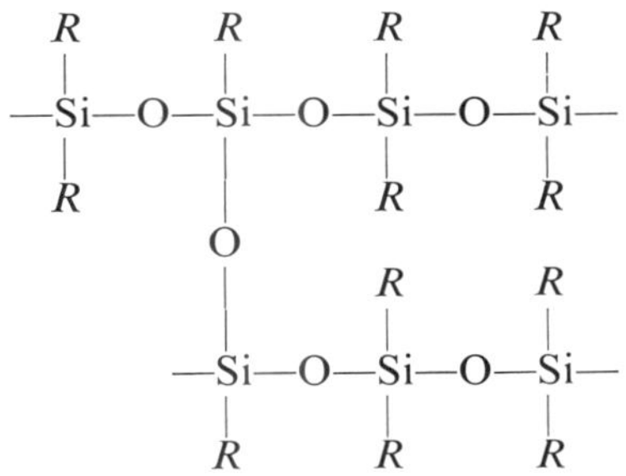

The usefulness of the silicones is due to their being heat resistant, electrically insulating, chemically inert, and water repellent. They are used extensively for insulating, fireproofing, and moistureproofing.

Because the nature of the organic radicals in silicones may be varied, as well as the relative numbers of different ones, many different polymers are possible. The most common starting material is dimethyldichlorosilane, obtained by passing methyl chloride over a mixture of silicon and copper heated to 350°C.

$$2CH_3Cl + Si \xrightarrow[\text{catalyst}]{\text{Cu}} (CH_3)_2SiCl_2$$

The diol obtained by hydrolysis is an oil, which is polymerized by churning with warm dilute acid or alkali for several hours. To obtain cross-linking, CH_3SiCl_3 is introduced into the original mixture, and $(CH_3)_3SiCl$ is added to serve as a "chain-stopper."

Silicone oils for use in lubrication or in hydraulic systems consist of molecules containing about 20 silicon atoms. They undergo very little change in viscosity with changing temperatures. At arctic temperatures they retain their fluidity, and at high temperatures they retain the viscosity necessary for effective lubrication.

Silicone elastomers are obtained by allowing polymerization to proceed until there are about 2000 silicon atoms per molecule. They retain their flexibility at low temperatures and are not attacked by ozone. By the proper choice of organic groups in the silicone, oil-resistant rubbers may be obtained. Silicone rubbers are used in such fascinating toys as "Super Ball" and "Silly Putty."

Silicone resins are cross-linked polymers used for insulating the wires in electric motor windings. Motors with such windings may be run at higher temperatures than those wound with wires insulated with hydrocarbon resins. Smaller motors therefore may be used for a given function.

Cellulosic materials, such as cotton, may be given a permanent water-repellent coating by brief exposure to the vapor of a methylchlorosilane. Because cellulosic materials have —OH groups in their molecules, dimethyldichlorosilane will react with the exposed surfaces of cellulosic fibers:

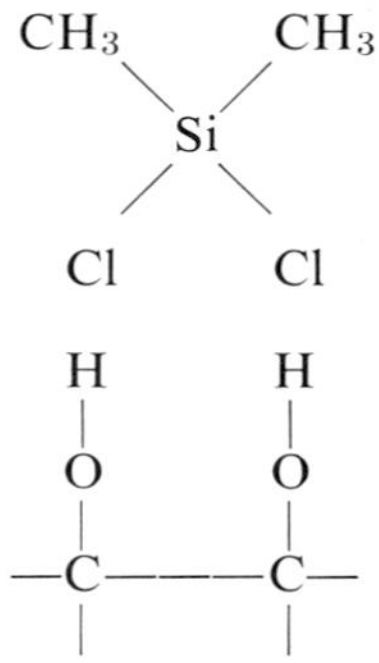

The surface of the fabric thus becomes covered with water-repellent methyl groups. The film is not affected by detergents or dry-cleaning fluids.

Water-repellent coatings may also be applied to glass, porcelain, and other ceramic surfaces. Water molecules adsorbed from the air onto the surfaces of ceramic materials will react with the vapors of methylchlorosilanes, forming hydrophobic silicone films, which completely coat the ceramic objects. Thereafter, when rain falls on such treated surfaces the water collects on them in large isolated globules which slide off under the influence of gravity. The discovery that insulators can be made water repellent by this treatment has proved useful, for they can now be kept free of hazardous aqueous films that might serve to carry electric current and destroy their effectiveness as insulators.

29-17 Production and Uses of the Elements

Carbon is only the eleventh element in order of abundance in the earth's crust, but it exceeds most of the others in the variety of forms and locations in which it occurs. More than 99.5 percent of the carbon occurs as limestone, $CaCO_3$. In addition, substantial quantities of hydrocarbons occur in petroleum and natural gas, and untold numbers of organic compounds of great variety occur in living organisms. Elementary carbon occurs in the form of diamond, graphite, and a variety of apparently amorphous or noncrystalline forms such as soot and lampblack. It is now known, from X-ray analysis, that the so-called amorphous forms of carbon are identical in structure with graphite. It is the extreme smallness of the particles that makes these forms of carbon appear superficially to be noncrystalline.

Africa is the greatest source of diamond, and South Korea is the largest producer of natural graphite. Only a small proportion of the diamonds that are mined possess the qualities required for use as jewels; the others are used for boring and grinding hard materials, for cutting glass, and as dies for drawing wires. Graphite is used for making "lead pencils," crucibles for casting metals, conductors for use in the electrical industry (e.g., generator brushes and electrodes), and in a finely divided form as a lubricant.

Artificial graphite is now made in large quantities by a process invented in 1896 by the American chemist Edward G. Acheson. The process involves passing a powerful alternating current through petroleum coke or granular anthracite (mixed with sand and iron oxide). The carbon is volatilized, and the vapor condenses as crystalline graphite. Synthetic diamonds are now made on a small scale (Section 29-1).

Peat, lignite, coal, and anthracite are minerals containing free carbon formed by the bacterial decomposition of primeval vegetable matter under pressure and in the absence of air. In the carbonization process most of the hydrogen, oxygen, and nitrogen originally combined with the carbon is eliminated. Geologically, anthracite is the oldest of these carbonization products, and it usually has a free-carbon content in excess of 85 percent.

Coke is the residue remaining after coal is heated in the absence of air. This process of artificial carbonization yields coal gas (mostly H_2 and CH_4), ammonia, coal tar, and coke. The coke is mostly free carbon and ash. It is used for the reduction of ores, as a fuel, and for making "water gas."

Charcoal is the residue remaining after wood has been heated in the absence of air. This "destructive distillation" yields methanol ("wood alcohol"), acetone, acetic acid, tar, and charcoal (which is impure carbon). Charcoal is used as fuel, as a reducing agent in metallurgy, and as an adsorbent for removing impurities from gases and solutions.

Bone black is obtained by the destructive distillation of bones and other refuse from the meat-packing industry and is used for decolorizing crude sugar solutions. Carbon black is made by burning natural gas in a limited supply of air and is used in the manufacture of automobile tires, printer's ink, and carbon paper.

Silicon is the second most abundant element in the earth's crust, about 95 percent of the rocky material at or near the earth's surface consisting of silica or silicates. Silicon never occurs free. A useful alloy of iron and silicon, known as ferrosilicon, is obtained by the simultaneous reduction of iron oxide and silica with carbon. Technically pure silicon is obtained by heating carbon with pure silica in an electric arc furnace:

$$2C + SiO_2 \rightarrow 2CO\uparrow + Si$$

The main use for this product is in the making of silicones. Small quantities are highly purified for use in transistors, rectifiers, and solar batteries. The "superpure" silicon is obtained by a process known as *zone-melting* or *zone-refining*. This is done by admitting the molten silicon into a silica-glass tube in which it is allowed to solidify. Then a narrow electrical heating element, encircling the tube like a ring, is moved along the tube, causing the silicon to melt in the narrow zone that lies in the center of the heating ring. As the heating element progresses along the length of the tube, the impurities advance along with it, for they remain in the liquid zone while silicon crystallizes out alone from the melt.

Germanium is a scarce and widely distributed element. It occurs as a trace element in many silicates. It has attained technical importance only since 1945 when its properties as a semiconductor provided the basis for a revolution in electronics. Commercially, it is obtained in limited amounts as a by-product of the metallurgy of American zinc sulfide ores. Crude GeO_2 from the zinc smelters is treated with HCl, and the volatile $GeCl_4$ is separated by distillation. The $GeCl_4$ is then hydrolyzed, and the pure GeO_2 is reduced with hydrogen. Its use in transistors and rectifiers requires that germanium be exceedingly pure. The preparation of superpure germanium requires use of the zone-melting technique.

The use of either germanium or silicon as a semiconductor requires the controlled addition of exceedingly small quantities of properly selected impurities to the superpure element. In some semiconductor applications silicon is preferred over germanium. Devices made with silicon will continue in operation at temperatures as high as 200°C, but the upper limit for germanium devices is about 70°C.

Germanium has the unusual property of expanding upon freezing. This property persists when the germanium is alloyed with as much as 92 percent gold. Such alloys may find use in precision castings, e.g., in dental inlays.

Tin is a relatively rare element, and the world output is not great. The price is comparatively high so that even the working of low-grade ores can be profitable. The most important tin ore is cassiterite, SnO_2.

After concentration, tin ore is reduced by carbon:

$$SnO_2 + 2C \rightarrow Sn + 2CO\uparrow$$

The liquid metal is drawn off from the furnace and cast into molds ("block tin"). The chief use for tin is in making *tinplate*, i.e., sheet iron coated with tin. This is used in making so-called tin cans and various other utensils. Formerly, tinplate was applied by the "hot-dip" method, which involved thoroughly cleaning the surface of the iron by "pickling" in dilute sulfuric acid and then bringing it into contact with molten tin. The present method involves electroplating, which gives a more uniform coating and uses only half as much tin.

Tinfoil was once used for wrapping food products, but it now has been replaced by aluminum foil and polyethylene. Among the important alloys containing tin are bronze, pewter, and solder.

Lead is not as scarce as tin, but it is one of the least abundant of the common metals. It occurs widespread in a variety of minerals, but the only important lead ore is galena, PbS. After being concentrated, the ore is roasted in a blast furnace with coke:

$$2PbS + 3O_2 \rightarrow 2PbO + 2SO_2\uparrow$$
$$PbO + C \rightarrow Pb + CO\uparrow$$

In terms of world production lead is the fifth most important metal; it is exceeded only by iron, copper, aluminum, and zinc.

The uses of lead depend on its low melting point, its resistance to corrosion, and the ease with which it can be worked. It is used in the manufacture of plumbing items and cable coverings. Important lead alloys include type metal and battery plate. Alloyed with 0.5 percent arsenic, lead is used in making shot and shrapnel. Significant quantities of lead are used in making white lead, $Pb(OH)_2 \cdot 2PbCO_3$, and red lead, Pb_3O_4, for use in paints.

29-18 Carbides

Compounds consisting of carbon and another element less electronegative than carbon are properly termed *carbides*. These compounds are of three types: ionic, covalent, and metallic or interstitial.

Ionic carbides are formed by metals of Groups IA, IIA, and IIIA. For example, Al_4C_3 is formed by direct union of aluminum with carbon at 2000°C, and CaC_2 is manufactured by heating quicklime with coke in an electric furnace. Ionic carbides, when pure, are colorless crystalline solids. On contact with water they undergo complete and irreversible hydrolysis, producing hydrocarbons. Usually the hydrocarbon is either methane or acetylene. Those that produce methane are sometimes called methanides:

$$Be_2C + 4H_2O \rightarrow 2Be(OH)_2 + CH_4\uparrow$$
$$Al_4C_3 + 12H_2O \rightarrow 4Al(OH)_3 + 3CH_4\uparrow$$

Most ionic carbides are acetylides:

$$CaC_2 + 2H_2O \rightarrow Ca(OH)_2 + C_2H_2\uparrow$$
$$BeC_2 + 2H_2O \rightarrow Be(OH)_2 + C_2H_2\uparrow$$
$$Al_2(C_2)_3 + 6H_2O \rightarrow 2Al(OH)_3 + 3C_2H_2\uparrow$$

Calcium acetylide (commonly called calcium carbide) is an important industrial chemical, used for making acetylene and calcium cyanamide.

The *covalent carbides*, formed by boron and silicon, have properties entirely different from all the other binary compounds of carbon with nonmetals. They are network solids characterized by extreme hardness, chemical inertness, and high melting points. Silicon carbide, SiC, is well known by the name carborundum (originally a trade name). It is widely used as an abrasive. It is made in electric resistance furnaces by the reaction of sand with coke, a process invented by Edward G. Acheson in 1891:

$$SiO_2 + 3C \rightarrow SiC + 2CO\uparrow$$

In carborundum the atoms are arranged as they are in diamond and in crystalline silicon, with silicon and carbon atoms alternating. The carbon-silicon distance (1.90 Å) is very nearly the arithmetic mean of the carbon-carbon distance in diamond (1.54 Å) and the silicon-silicon distance in crystalline silicon (2.35 Å).

Metallic carbides (*interstitial carbides*) are formed when carbon atoms are inserted into the spaces between the metal ions in the crystal lattices of transition metals. Their formulas depend to a considerable extent on the comparative sizes of the metallic ions and the carbon atoms. The metallic carbides are hard materials with high melting points. They have metallic luster, and they are good conductors of electricity. High-speed cutting tools are tipped with tantalum carbide, TaC, or tungsten carbide, W_2C. Iron carbide or *cementite*, Fe_3C, is an essential component of steel.

Germanium, tin, and lead do not form carbides.

29-19 Nitrides of Carbon and Silicon

Carbon and silicon are the only elements of Group IVA that form nitrides. Silicon nitride, Si_3N_4, is a chemically stable crystalline material that is obtained by heating silicon in a nitrogen atmosphere. Cyanogen, C_2N_2, is a very poisonous gas with the odor of bitter almonds. It cannot be obtained by direct union of the elements, but it is formed in a variety of other reactions. Addition of a cyanide to a solution of a cupric salt liberates cyanogen, paralleling the liberation of iodine upon addition of an iodide:

$$2Cu^{++} + 4I^- \rightarrow 2CuI + I_2$$
$$2Cu^{++} + 4CN^- \rightarrow 2CuCN + C_2N_2$$

Cyanogen is one of several substances called *pseudohalogens* because of this and other similarities to the halogen elements. It reacts with basic solutions in a manner analogous to the reaction of chlorine:

$$Cl_2 + 2OH^- \rightarrow Cl^- + ClO^- + H_2O$$
$$(CN)_2 + 2OH^- \rightarrow \underset{\text{cyanide ion}}{CN^-} + \underset{\text{cyanate ion}}{OCN^-} + H_2O$$

The cyanogen molecule is linear and symmetrical and therefore nonpolar:

$$:N:::C:C:::N:$$

New Terms

Catenation: the formation of chains of identical atoms, as in the hydrocarbons.

Ion exchange: the process by which ions held at electrically charged sites within a porous solid are exchanged for ions in a solution in contact with the solid.

Pseudohalogens: compounds that resemble the halogen elements in some of their properties (e.g., in their reactions with strong bases). Cyanogen is an example.

Zone-refining: a process for bringing a substance to a state of ultrapurity by melting it in a moving zone and allowing the pure material to crystallize out behind.

Exercises

29-1 How do the properties of the oxides of the Group IVA elements vary with the atomic numbers of the elements?

29-2 Compare the properties of the allotropes of carbon and explain their differences in terms of structure.

29-3 Discuss the effect of temperature on the electrical conductivity of the Group IVA elements.

29-4 Identify the following: carbamic acid, urea, phosgene, red lead, and white lead.

29-5 Outline the types of structure present in silicates.

29-6 Discuss weathering as the source of soil.

29-7 What are zeolites? Xanthates? Silanes? Germanes? Silicones? Pseudohalogens?

29-8 What is peat? Lignite? Coal? Anthracite? Charcoal? Coke? Tin plate?

29-9 Describe zone-refining and discuss its importance.

References

Day, F. H., *The Chemical Elements in Nature*, Reinhold, New York, 1964. A treatment of the principles of "chemical natural history" and the practical sources of chemical materials.

Johnson, R. C., *Introductory Descriptive Chemistry*, Benjamin, New York, 1966 (paperbound). The author classifies silicon, germanium, arsenic, antimony, tellurium, polonium, and astatine as semimetals.

Rochow, E. G., *An Introduction to the Chemistry of the Silicones*, 2nd ed., Wiley, New York, 1951. A compact but comprehensive survey for the nonspecialist, written by an eminent authority.

Speier, J. L., "The Si-C Bond and Silicones," *Chemistry*, **37** (7), 6 (1964). "The story of silicones is just beginning."

30

The Elements of the Boron Group

30-1 Group Characteristics

The elements of Group IIIA have s^2p configurations in the outer shells of their atoms (Figure 30–1). All the elements of the group show the +3 oxidation state. For boron and aluminum, the +3 state is the only important one. The +1 state is well defined for gallium, indium, and thallium. The stability of the +1 state increases with increasing atomic number; although rare for gallium and indium, +1 is the chief oxidation state for thallium. The +3 compounds of thallium are strong oxidizing agents.

Figure 30-1

Electronic configurations of Group IIIA elements

	1	2	3	4	5	6
$_5$B	s^2	s^2p				
$_{13}$Al	s^2	s^2p^6	s^2p			
$_{31}$Ga	s^2	s^2p^6	$s^2p^6d^{10}$	s^2p		
$_{49}$In	s^2	s^2p^6	$s^2p^6d^{10}$	$s^2p^6d^{10}$	s^2p	
$_{81}$Tl	s^2	s^2p^6	$s^2p^6d^{10}$	$s^2p^6d^{10}f^{14}$	$s^2p^6d^{10}$	s^2p

30-2 Comparison of Boron with the Metals of the Group

Boron is the only chemical element with fewer than four electrons in the outer shell that is not a metal. Like carbon and silicon, boron does not form a simple positive ion. Thus, it does not form a nitrate or a sulfate. This is a consequence of the small size of the boron atom. Despite its s^2p configuration, the boron atom always forms three bonds instead of one. This behavior is analogous to that of carbon, which has an s^2p^2 configuration but always forms four bonds instead of two. The explanation in both cases is based on hybrid bond orbitals (Section 18–5). The total energy released in the formation of three bonds in a BX_3 compound exceeds the amount that would be released in the formation of one bond in a BX compound by more than the amount required to promote one electron of the *s* pair to an empty *p* orbital. This promotion makes possible the formation of three bonds instead of only one. The three sp^2 hybrid bonds in a BX_3 compound are directed toward the corners of an equilateral triangle with the boron atom in the center. All molecules of the BX_3 type are flat and triangular.

Aluminum and the heavier members of Group IIIA are much more inclined to form positive ions (Table 30–1). Between boron and aluminum there is a remarkable discontinuity in properties that is unusual among the groups of the periodic table. This is related to the fact that the aluminum atom is considerably larger than the boron atom. Nevertheless, the radii of the tripositive ions Al^{+++}, Ga^{+++}, In^{+++}, and Tl^{+++} are small enough to inhibit the tendency for these ions to form. Not surprisingly, most of the binary compounds of these elements in the +3 oxidation state are covalent substances.

All the elements of the boron family are too reactive to exist in nature in the elementary state. It is very difficult to obtain elementary boron in a pure state because of its high melting point and its reactivity at high temperatures.

Table 30-1

Properties of the elements of Group IIIA

Element	*Ionization energy* (kcal/mole)	*Electronegativity* (Pauling scale)	*Melting point* (°C)	*Boiling point* (°C)	*Specific resistivity* (microhm-cm, 0°C)	*Density* (g/ml, 0°C)
$_{5}B$	191	2.0	2300	2550	1.8×10^{12}	2.32
$_{13}Al$	137	1.5	660	2327	2.6	2.70
$_{31}Ga$	138	1.6	30	1983	53	5.91
$_{49}In$	133	1.7	157	2000	8.4	7.36
$_{81}Tl$	140	1.8	304	1457	18	11.85

Amorphous boron is obtained as a dark powder by reduction of B_2O_3 with magnesium, but it is only 95–98 percent pure. Crystalline boron of high purity is obtained by reduction of BCl_3 with hydrogen at electric-arc temperatures. Except for diamond, crystalline boron has the greatest hardness of any of the chemical elements. Its electrical resistance is quite high, and it decreases with rising temperature (a characteristic of semiconductors). It is blackish-gray and resistant to chemical attack. Amorphous boron reacts at high temperatures with oxygen, sulfur, nitrogen, and the halogens. Boron is one of the few elements that combine directly with nitrogen. It combines directly with fluorine at ordinary temperatures but does not combine with hydrogen at any temperature. Boron hydrides are obtained by indirect methods. Boron and carbon combine at high temperatures to form boron carbide, B_4C, a black solid that is almost as hard as diamond and has replaced diamond dust for many industrial uses. (The carbide is manufactured by the reaction of carbon with B_2O_3 in an electric furnace.) Boron combines with many metals at high temperatures to form borides.

Although boron is definitely a nonmetal, the "metallic" luster of crystalline boron and the weakness of boric acid indicate that the element is on the borderline between metals and nonmetals. Aluminum and boron are the only elements in Group IIIA that form carbides. Aluminum combines with carbon at high temperatures to form Al_4C_3, which is hard but decomposes to give methane on contact with water. Aluminum acetylide, $Al_2(C_2)_3$ is made by passing acetylene over hot aluminum. At high temperatures aluminum also combines with oxygen, sulfur, nitrogen, and the halogens.

Since they do not form carbides, gallium, indium, and thallium can be obtained by reduction of their oxides with carbon. They can also be electroplated from aqueous solution.

30-3 Chemical Properties of the Metals

In most respects, aluminum bears greater resemblance to gallium, indium, and thallium than to boron. All four metals displace hydrogen from nonoxidizing acids; aluminum, gallium, and indium form tripositive ions (hydrated, of course), but thallium forms a monopositive ion.

$$2Al + 6H^+ \rightarrow 2Al^{+++} + 3H_2\uparrow$$
$$2Ga + 6H^+ \rightarrow 2Ga^{+++} + 3H_2\uparrow$$
$$2In + 6H^+ \rightarrow 2In^{+++} + 3H_2\uparrow$$
$$2Tl + 2H^+ \rightarrow 2Tl^+ + H_2\uparrow$$

All thallium salts are highly poisonous. Thallous sulfate is especially effective as a rat poison because, being odorless and tasteless, it gives rats no warning of its presence.

Aluminum and gallium dissolve readily in solutions of strong bases, liberating hydrogen gas:

$$2Al + 2NaOH + 2H_2O \rightarrow 2NaAlO_2 + 3H_2\uparrow$$
$$2Ga + 2NaOH + 2H_2O \rightarrow 2NaGaO_2 + 3H_2\uparrow$$

Indium and thallium do not react with strong bases. Boron does not react with solutions of strong bases, but it dissolves in fused sodium hydroxide, liberating hydrogen.

Dilute nitric acid converts boron to boric acid:

$$B + HNO_3 + H_2O \rightarrow H_3BO_3 + NO\uparrow$$

The reaction with concentrated nitric acid is so vigorous that the boron catches fire. Aluminum does not react with nitric acid, despite the fact that aluminum is a vigorous reducing agent and nitric acid a strong oxidizing agent. The resistance of aluminum to serious attack by atmospheric oxygen is attributed to the formation of a tightly adhering film of aluminum oxide on the surface of the metal. The failure of aluminum to displace hydrogen from water is also due to this protective coating. Nitric acid strengthens the coating and renders the metal passive. Nitric acid is often transported in aluminum tank cars.

Gallium dissolves in nitric acid to form $Ga(NO_3)_3$, indium forms $In(NO_3)_3$, and thallium forms $TlNO_3$.

30-4 Physical Properties of the Metals

The physical properties of aluminum, gallium, indium, and thallium are mostly the properties characteristic of metals. Aluminum has pronounced metallic character, including malleability, tensile strength, and low electrical resistance. The silvery appearance of uncorroded aluminum is well known. Gallium also has a shiny luster. Indium looks like tin and emits a "cry" when it is bent. The tinlike appearance of a fresh thallium surface gives way quickly to a dark tarnish due to the formation of black Tl_2O. Thallium has poor tensile strength and thus resembles lead, its neighbor in the periodic table. Both indium and thallium are softer than lead.

Gallium is harder than most low-melting metals. It is one of the few substances which, like water, expand upon freezing. It has the longest liquid range of any known substance. Its melting point is so low that it will melt in your hand. The molten gallium supercools easily and may remain liquid below its melting point over a long period of time unless it is seeded with pieces of the crystalline metal. The molten metal looks like mercury, but it has a greater tendency to wet other substances. In a glass tube a gallium meniscus is curved opposite to a mercury meniscus because gallium wets glass.

In comparison with the elements of Groups IA and IIA, the elements of Group IIIA show some irregularities. For example, the melting points of Group IA elements diminish in a regular fashion with increasing atomic number (Table 22–1); but in Group IIIA the melting points drop sharply from boron to aluminum and again from aluminum to gallium and then rise from gallium to thallium. Such irregularities are the result of the interplay of various factors, including the fact that gallium, indium, and thallium are separated from Groups IA and IIA by the elements of the B groups (the transition elements). The tripositive ions of gallium, indium, and thallium are unique among the positive ions formed by elements of the A groups of the periodic table, for the outermost shell of each of these three ions contains 18 instead of the usual 8 electrons.

30-5 Boric Acid and Borax

The principal oxide of boron is boron(III) oxide, or boric oxide, a white covalent solid with the empirical formula B_2O_3. Its polymeric nature is indicated by its high melting point (557°C) and boiling point (1500°C). Its heat of formation is quite large:

$$4B + 3O_2 \rightarrow 2B_2O_3 \qquad \Delta H = -698 \text{ kcal}$$

It is reduced to boron when heated with active metals, such as sodium.

Boric oxide dissolves in hot water. The aqueous solution, upon cooling, yields white crystals of orthoboric acid, H_3BO_3. This compound is commonly known simply as boric acid.

In aqueous solution orthoboric acid acts as a very weak acid. Only the first stage of its ionization is large enough to be measured; its ionization constant is about 6×10^{-10} at 25°C. The traditional method of representing its acidic ionization shows the transfer of a proton to a water molecule:

$$H_3BO_3 + H_2O \rightleftharpoons H_3O^+ + H_2BO_3^-$$

It is probably more nearly correct to represent the acid as an OH^- acceptor:

$$B(OH)_3 + H_2O \rightleftharpoons B(OH)_4^- + H^+$$

Neutralization of orthoboric acid with a strong base yields tetraborate ion:

$$4H_3BO_3 + 2(Na^+ + OH^-) \rightarrow (2Na^+ + B_4O_7^{--}) + 7H_2O$$

When sodium tetraborate is crystallized from solution it separates as the decahydrate, $Na_2B_4O_7 \cdot 10H_2O$, a substance well known by the name *borax*. Solutions of borax are alkaline by hydrolysis. This makes borax a useful cleansing agent, as OH^- ions have a strong dispersing effect on dirt particles. Borax is also used as a water softener, as a flux, and in making porcelain enamels and borosilicate glass (Pyrex and Kimax). Boric acid is obtained by heating an aqueous solution of borax with a strong acid.

30-6 The Chemistry of Aluminum Oxide

The reaction between aluminum and oxygen is highly exothermic:

$$4Al + 3O_2 \rightarrow 2Al_2O_3 \qquad \Delta H = -798 \text{ kcal}$$

The oxide occurs in nature in several different forms. Anhydrous alumina occurs as *corundum*, a very hard mineral that is used as an abrasive. Some varieties of corundum are prized as gemstones, which owe their colors to traces of impurities; examples include ruby and sapphire. Emery is a granular mixture of corundum and iron oxide. The present source of aluminum metal is *bauxite*, which is a mixture of $Al_2O_3 \cdot H_2O$ and $Al_2O_3 \cdot 3H_2O$.

Addition of ammonia to an aqueous solution of an aluminum salt produces a gelatinous precipitate of aluminum hydroxide, $Al(OH)_3$. Addition of excess strong base, e.g., NaOH, produces no precipitate because of the formation of soluble aluminate. The aluminate ion is commonly represented by the formula AlO_2^- as a convenience, but the ions actually present in aqueous solution are probably hydrated $Al(OH)_4^-$ ions:

$$Al^{+++} + 3OH^- \rightleftharpoons Al(OH)_3\downarrow$$
$$Al(OH)_3 + OH^- \rightleftharpoons Al(OH)_4^-$$
$$Al(OH)_4^- + H^+ \rightleftharpoons Al(OH)_3\downarrow + H_2O$$
$$Al(OH)_3 + 3H^+ \rightleftharpoons Al^{+++} + 3H_2O$$

Aluminum hydroxide is definitely amphoteric; the freshly precipitated hydroxide dissolves at once in either concentrated acid or concentrated base. The aluminum ions present in solutions of aluminum salts are commonly represented by the formula Al^{+++}, but the ions, of course, are hydrated. Solutions of aluminum salts are acidic because of hydrolysis; i.e., the hydrated aluminum ion acts as an acid (Section 16-11):

$$Al(H_2O)_6^{+++} + H_2O \rightleftharpoons Al(H_2O)_5OH^{++} + H_3O^+$$

Hydrolysis causes solutions of aluminates to be alkaline. The only soluble aluminates are those of the alkali metals.

The mineral known as *spinel*, with the empirical formula $MgAl_2O_4$ or $Mg(AlO_2)_2$, may be formally regarded as an aluminate. Structurally, it is a double oxide rather than a salt. The oxygen atoms are shared by the metal atoms. Each magnesium atom is surrounded tetrahedrally by four oxygen atoms, and each aluminum atom is surrounded octahedrally by six oxygen atoms. The term *spinel* is now used to refer to any mineral that has this same structure and fits the type formula $M^{II}(M^{III}O_2)_2$. Examples include zinc spinel, $Zn(AlO_2)_2$ or $ZnO \cdot Al_2O_3$; chromite, $Fe(CrO_2)_2$ or $FeO \cdot Cr_2O_3$, the most important ore of chromium; and magnetic oxide of iron, $Fe(FeO_2)_2$ or Fe_3O_4.

30-7 Hydroxides of Gallium, Indium, and Thallium

Gallium, indium, and thallium burn when heated in air, giving Ga_2O_3, In_2O_3, and a mixture of Tl_2O and Tl_2O_3. Gallium hydroxide is obtained as a white precipitate by making solutions of gallium salts slightly alkaline. It is amphoteric; i.e., it dissolves in acids to form gallium(III) salts, and it dissolves in bases to form gallates. Surprisingly, $Ga(OH)_3$ is more acidic than $Al(OH)_3$, for it will dissolve in strong ammonia solution.

Indium hydroxide is probably exclusively basic, and thallium(III) hydroxide is definitely so. Addition of a base to a solution of a thallium(I) salt does not produce a precipitate. Addition of barium hydroxide to thallium(I) sulfate precipitates barium sulfate and leaves a strongly alkaline solution:

$$(Ba^{++} + 2OH^-) + (2Tl^+ + SO_4^{--}) \rightarrow BaSO_4 \downarrow + 2(Tl^+ + OH^-)$$

Upon evaporation the solution yields yellow crystals of thallium(I) hydroxide, TlOH, a soluble strong base.

30-8 Halides of Boron

Boron forms trihalides with all halogens. The fluoride is made by the reaction of HF with the oxide:

$$B_2O_3 + 6HF \rightarrow 2BF_3 + 3H_2O$$

The other boron trihalides are prepared from the oxide by reaction with carbon and the halogen at high temperatures:

$$B_2O_3 + 3C + 3X_2 \rightarrow 2BX_3 + 3CO$$

The fluoride and chloride are gases at ordinary temperatures, BBr_3 is a viscous liquid, and BI_3 is a white crystalline solid.

In a boron trihalide molecule the boron still lacks two electrons of having a noble-gas structure. Consequently, these substances have the ability to act as Lewis acids and form *addition compounds* in which the boron atom establishes a fourth bond. Thus, BF_3 forms a hydrate, $F_3B : OH_2$, which melts at 10°C and decomposes, giving off BF_3, above 20°C:

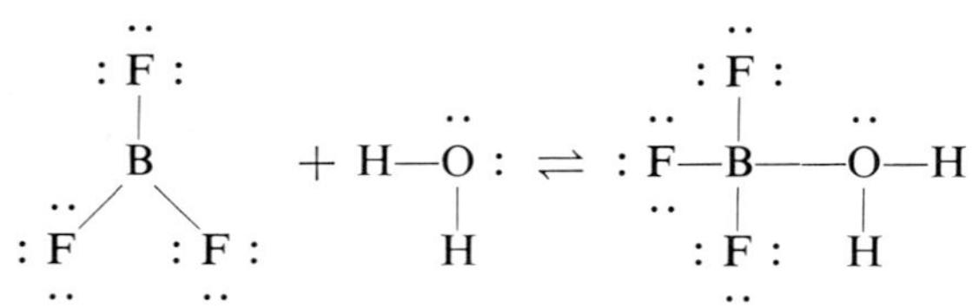

BF_3 is conveniently available as the ethyl etherate, $F_3B : O(C_2H_5)_2$, a liquid (bp, 126°C) that is stable below 50°C. Boron trifluoride unites with ammonia to form the solid addition compound $F_3B : NH_3$, which dissolves in water without decomposition. It also readily unites with alcohols, amines, phosphines, etc. in a 1 : 1 ratio. In these compounds the four bonds are tetrahedrally disposed about the boron atom.

Boron trifluoride is extensively used to promote various organic reactions for which it is loosely described as a catalyst. Its effectiveness in this regard is almost certainly due to its ability to form addition compounds with the reactants. It also has recently become of importance in the manufacture of boranes (Section 30-11).

With the exception of BF_3, the boron halides hydrolyze to give boric and hydrohalic acids:

$$BX_3 + 3H_2O \rightarrow H_3BO_3 + 3(H^+ + X^-)$$

Boron triiodide hydrolyzes explosively.

When a relatively small amount of BF_3 is passed into water, the HF produced by the hydrolysis combines with unhydrolyzed BF_3 to give fluoboric acid:

$$4BF_3 + 3H_2O \rightarrow H_3BO_3 + 3(H^+ + BF_4^-)$$

Fluoboric acid is a strong acid which has not been isolated. It may also be obtained, in solution, by dissolving boric acid in aqueous HF:

$$H_3BO_3 + 4HF \rightarrow (H^+ + BF_4^-) + 3H_2O$$

The fluoborate ion has a tetrahedral structure. Well-crystallized salts of fluoboric acid have been known for more than a century. They can be prepared by the usual methods for preparing salts, i.e., by reaction of metallic oxides, hydroxides, or carbonates with aqueous HBF_4.

30-9 Halides of the Metals

The trifluorides of aluminum, gallium, and indium have high melting points because they consist of extended arrays of metal ions and fluoride ions in which each metal ion is coordinated with six fluoride ions. The trichlorides, tribromides, and triiodides of aluminum, gallium, and indium are low-melting dimers in which the metal atoms are coordinated with four halogen atoms.

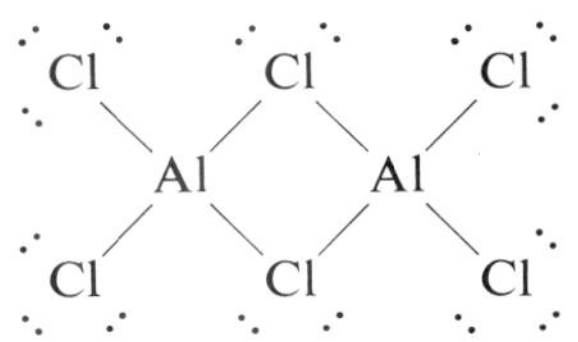

Aluminum fluoride, AlF_3 (empirical formula), is insoluble in water. Aluminum chloride, Al_2Cl_6, sublimes at 760 torrs at 180°C. The chloride, bromide, and iodide fume strongly in moist air as the result of hydrolysis by the atmospheric moisture:

$$Al_2X_6 + 6H_2O \rightarrow 2Al(OH)_3 + 6HX\uparrow$$

They are very soluble in water, and their aqueous solutions are strongly acidic because of hydrolysis.

Aluminum chloride forms addition compounds with many organic substances. It acts as a catalyst for attaching alkyl groups to aromatic rings and is used extensively for this purpose. The substance containing the aromatic ring may be benzene itself, certain derivatives of benzene, or more complicated substances such as naphthalene. The attachment of the alkyl group is brought about by the reaction of the aromatic compound with an alkyl halide in the presence of aluminum chloride. For example, addition of Al_2Cl_6 to a mixture of benzene and methyl chloride results in a vigorous reaction with the evolution of hydrogen chloride gas:

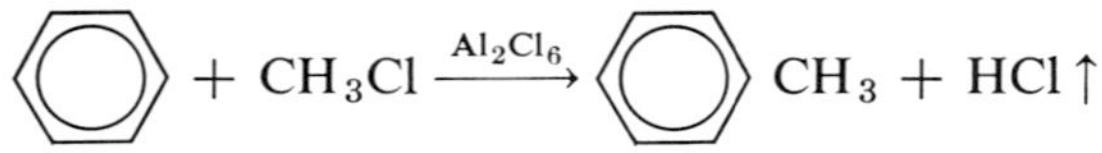

This type of reaction is called a Friedel-Crafts reaction, because the effectiveness of anhydrous aluminum chloride as the catalyst was discovered in 1877 at the University of Paris by the French chemist Charles Friedel and the American chemist James Mason Crafts. It is possible to use BF_3 or HF instead of Al_2Cl_6 as the catalyst. Note that these catalysts are Lewis acids.

Thallium(III) halides are unstable, decomposing to give thallium(I) halides and elementary halogen:

$$TlX_3 \rightarrow TlX + X_2$$

Thallium(I) halides resemble silver halides. Thus, TlCl, TlBr, and TlI are obtained as curdy precipitates similar to AgCl, AgBr, and AgI when the halide ions are added to a solution of a thallium(I) salt. On the other hand, TlF, like AgF, is soluble in water. TlCl, TlBr, and TlI are light sensitive and must be kept in the dark to avoid decomposition.

30-10 Nitrides

All elements of the boron family except thallium form giant-molecule nitrides of the type $(MN)_n$. Boron and aluminum are the only elements of the group that combine directly with nitrogen when heated in its presence.

Boron nitride, BN, is a white powder that looks and feels like talc. Its slipperiness is due to its layer structure, which is similar to that of graphite. In each layer the boron and nitrogen atoms occur alternately in hexagonal rings, the B—N distance being 1.45 Å. The distance between layers is 3.34 Å. The similarities of boron nitride to graphite have led to its being called "white graphite" or "inorganic graphite." The analogy must not be carried too far, however, for boron nitride is a nonconductor, unlike graphite. This indicates that the bonding electrons in boron nitride are not so loosely held as in graphite.

Boron nitride is such an excellent insulator, even at high temperatures, that it is used as such in high-temperature induction furnaces. Being an excellent high-temperature lubricant, it is also used as an antisticking agent for glassmaker's molds.

Although boron nitride can be obtained by heating boron in the presence of either nitrogen or ammonia, it is most conveniently prepared either by heating boron trifluoride ammoniate above 125°C or by heating borax with ammonium chloride:

$$4F_3B : NH_3 \rightarrow BN + 3NH_4BF_4$$

$$Na_2B_4O_7 + 4NH_4Cl \rightarrow 4BN + 2NaCl + 7H_2O + 2HCl$$

The analogy of boron nitride to graphite is heightened by the fact that under a pressure of 70,000 atmospheres at 1600°C it is converted to a diamondlike form called *borazon*. This material not only has the structure and hardness of diamond but also resists attack by air at high temperatures which would cause diamond to burn. It may become a valuable component in tools designed for high-temperature machining of refractory materials and may become an important refractory itself.

Boron nitride does not hydrolyze unless it is heated to a red heat with steam, but aluminum nitride hydrolyzes readily, even with cold water:

$$AlN + 3H_2O \rightarrow Al(OH)_3 + NH_3 \uparrow$$

Aluminum nitride is most successfully prepared by heating alumina with coke and nitrogen in an electric furnace:

$$Al_2O_3 + 3C + N_2 \rightarrow 2AlN + 3CO$$

It is apparent that this reaction, followed by the hydrolysis of the aluminum nitride, might be employed as a method for producing ammonia from atmospheric nitrogen. Attempts to develop the process to the point where it might compete with the Haber process have not been successful.

Gallium nitride, GaN, is formed as a gray powder by reaction of gallium with ammonia. An analogous phosphide, GaP, behaves as a semiconductor. Indium nitride, InN, is a gray powder which resembles GaN.

30-11 Hydrides of Boron

Boron forms a number of binary compounds with hydrogen, most of which are spontaneously flammable in air. We would expect the simplest of the boron hydrides to have the formula BH_3, but actually its formula is B_2H_6. The boron hydrides are called *boranes*; B_2H_6 is *diborane*. Diborane is the first member of a series with the general formula B_nH_{n+4}. The best-known members of this series are B_2H_6, B_5H_9, B_6H_{10}, and $B_{10}H_{14}$. Another series of boranes has the general formula B_nH_{n+6}. The members of this series that are well known are B_4H_{10}, B_5H_{11}, B_9H_{15}, and $B_{10}H_{16}$.

Until recently, interest in the boranes has been limited to amazement at their unexpected formulas and to curiosity concerning their structures. Present interest in these compounds arose originally because of their evident potential for use as high-energy rocket fuels. For example, the heat of combustion of diborane is -482 kcal per mole. The heat liberated in the combustion of a borane is about twice that produced by the combustion of an equal mass of hydrocarbon fuel. Unfortunately, the cost of producing boranes makes it difficult for them to compete with hydrocarbons.

Increased interest in the boranes has enhanced the importance of boron trifluoride, for the most convenient method of preparing diborane is by the reaction of BF_3 with sodium borohydride:

$$4BF_3 + 3NaBH_4 \rightarrow 3NaBF_4 + 2B_2H_6$$

Higher boranes are obtained from diborane by heating it with hydrogen at temperatures varying from 100 to 200°C, depending on which borane is sought.

30-12 Complex Hydrides

Sodium borohydride and lithium aluminum hydride are the most important of the complex hydrides that are formed by boron, aluminum, and gallium. $LiAlH_4$ is a nonvolatile, crystalline solid, white when pure, that is stable below 120°C. It is made by the reaction of lithium hydride with aluminum chloride and is available commercially:

$$4LiH + AlCl_3 \rightarrow LiAlH_4 + 3LiCl$$

It is used extensively as a reducing agent in both organic and inorganic reactions. Its importance is due in part to the fact that it is soluble in ethyl ether.

The importance of sodium borohydride is due to the fact that it dissolves in water without much decomposition. By contrast, $NaAlH_4$ and $NaGaH_4$ undergo rapid hydrolytic decomposition. Sodium borohydride is obtained by the reaction of sodium hydride with methyl borate (the methyl ester of boric acid) at 250°C:

$$4NaH + B(OCH_3)_3 \rightarrow NaBH_4 + 3CH_3ONa$$

It is a white, crystalline solid.

The BH_4^- ion, which is isoelectronic with the CH_4 molecule and the NH_4^+ ion, has a tetrahedral structure and may be formally regarded as a complex analogous to the BF_4^- ion. Presumably, the AlH_4^- and GaH_4^- ions have the same structure.

30-13 Structure of Diborane

It has now been established by infrared spectroscopy and electron diffraction studies that the diborane molecule has its two boron atoms and four of its hydrogen atoms in the same plane. The other two hydrogen atoms are located symmetrically outside this plane, one above and one below. The type of bonding that exists between the boron atoms and the two bridging hydrogen atoms has been the subject of much speculation. It may be supposed that the B_2H_6 molecule consists of two BH_2 groups united by two hydrogen bridges and that within each of these groups the B—H bonds are ordinary covalent bonds, which may be described as two-center, two-electron bonds. This leaves each boron with only one unpaired electron:

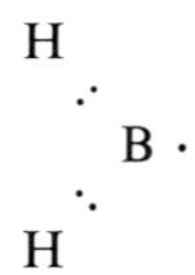

Because each bridging hydrogen also has only one electron, there are not enough electrons to account for the structure of the B_2H_6 molecule in terms of ordinary covalent bonds alone. Yet diborane is diamagnetic and presumably has equal numbers of electrons with opposite spins.

A theory that has gained considerable favor attempts to account for the known facts about diborane by broadening our concept of chemical bonding to include *three-center bonds*, i.e., the sharing of a pair of electrons by three atoms rather than two. In diborane each of two three-center bonds is assumed to hold the two boron atoms and one hydrogen atom together:

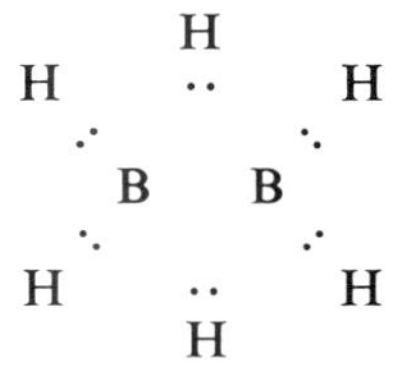

Recently, J. W. Linnett in England has proposed a structure for diborane that contains one-electron bonds:

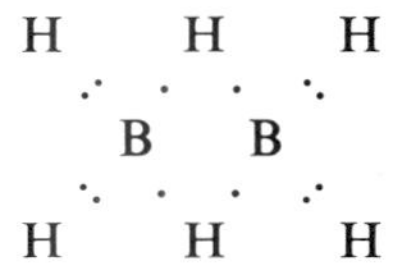

30-14 Boron Alkyls

In spite of the fact that there is no compound with the formula BH_3, the boron alkyls have the molecular formula BR_3. An example is trimethylboron, a spontaneously flammable gas, which can be made from a boron trihalide and a Grignard reagent:

$$BX_3 + 3CH_3MgX \rightarrow (CH_3)_3B + 3MgX_2$$

Diborane reacts with boron alkyls to form a series of alkyl-substituted diboranes, viz., $CH_3BH_2 \cdot BH_3$, $CH_3BH_2 \cdot BH_2CH_3$, $(CH_3)_2BH \cdot BH_3$, $(CH_3)_2BH \cdot BH_2CH_3$, and $(CH_3)_2BH \cdot BH(CH_3)_2$. It is noteworthy that only four of the hydrogen atoms in the diborane molecule can be replaced by alkyl groups.

Diborane reacts with several substances to give addition compounds of the hypothetical "borine," BH_3. Examples are $H_3B : N(CH_3)_3$ and $H_3B : O(CH_3)_2$.

30-15 Hydrides and Alkyls of the Metals

Aluminum hydride, $(AlH_3)_n$, and indium hydride, $(InH_3)_n$, are known only as highly polymeric solids. Aluminum hydride decomposes above 105°C, and indium hydride decomposes above 80°C. Gallium hydride, like diborane, exists as a dimer, Ga_2H_6. It is a liquid that decomposes above 130°C. Thallium hydride, $(TlH_3)_n$, decomposes to give $(TlH)_n$, which is also unstable when heated, decomposing into the elements.

Trialkylaluminum compounds can be made by a variety of procedures. Trimethylaluminum (bp, 130°C) is obtained by reaction of aluminum with methyl iodide:

$$4Al + 6CH_3I \rightarrow Al_2(CH_3)_6 + Al_2I_6$$

Trimethylaluminum is dimeric, even in the vapor state. Its structure, like that of diborane, may involve three-center bonding with the two methyl groups forming a bridge between the two aluminum atoms. The higher alkyls of aluminum are likewise dimeric, as is trimethylgallium, but the higher alkyls of gallium are monomeric. Even trimethylindium and trimethylthallium have been found to be monomeric in ether solution.

30-16 Uses of Alkylboron and Alkylaluminum Compounds

Alkylboron and alkylaluminum compounds have become articles of commerce. Trimethylboron, trimethylaluminum, and triethylaluminum, because they ignite spontaneously on contact with air, are used as jet fuel igniters. When added to jet fuel they help to prevent "flame-out" of jet engines. Triethylaluminum has also achieved importance as a catalyst for stereospecific polymerization, i.e., the formation of polymers in which the polymeric units have an ordered arrangement in space instead of a random or chance arrangement.

30-17 Occurrence and Uses of Boron

The most important deposits of borates are found in certain regions where salt lakes and inland seas have evaporated to dryness. The chief producer is the Mojave Desert of southern California, where the mineral named *kernite*, $Na_2B_4O_7 \cdot 4H_2O$, was discovered in Kern County in 1926. Kernite has replaced the natural borax, $Na_2B_4O_7 \cdot 10H_2O$, of Death Valley as the most important boron mineral. Kernite is extracted with water, from which it crystallizes as borax.

The chief use of elementary boron is in the metallurgy of steel. Because it is extremely reactive at high temperatures with both oxygen and nitrogen, boron is used to remove the remaining traces of these gases after their amount has been greatly reduced by a cheaper degasifier such as ferrosilicon. The presence of very small amounts of boron increases the "hardenability" of steel, i.e., the depth to which hardening occurs when the steel is quenched (suddenly cooled). The usual way of adding boron to steel is in the form of *ferroboron*, an alloy produced by reduction of a mixture of iron oxide and boric oxide by carbon in an electric furnace.

Borides are obtained by heating metals with boron to about 1600°C. Boron for this purpose is obtained by heating a mixture of boric oxide and magnesium powder. Typical borides are characterized by metallic conductivity, hardness approaching that of diamond, and a considerable degree of chemical inertness. The borides have formulas that are not predictable by simple oxidation state "rules," e.g., CaB_6 and WB_2. Many are used in tools and dies.

Ultrapure boron is used as a semiconductor. It has the advantage that it can function at higher temperatures than germanium or silicon.

The most important boron mineral in rocks is tourmaline, a very complex borosilicate of variable ionic content. It is common in granite deposits. The weathering of granite therefore provides soils with a small boron content. Although it is toxic to plant and animal life in certain concentrations, boron is one of the essential trace elements for the growth of land plants. An example of the effects of boron deficiency is "dry rot" of sugar beet. Most soils that have been under cultivation for long periods of time become deficient in available boron. Borax is now a standard component of many artificial fertilizers.

30-18 Occurrence and Uses of Aluminum

Aluminum, known as aluminium in European and British Commonwealth countries, is the most abundant of all metals. Ordinary soil consists primarily of kaolin and other hydrous aluminum silicates that have been produced by the weathering of such minerals as feldspar. Kaolin and other aluminum compounds occurring as clays are the raw materials of the ceramic industries. Common clay, which is used in making bricks and tile, contains impurities that are responsible for its color. China and porcelain are made from pure clay, which is iron-free and therefore white. Portland cement is obtained by heating a mixture of clay and limestone. It consists of complex silicates and aluminates of calcium. The "setting" of cement occurs as a result of hydration and hydrolysis reactions. These reactions continue for many years.

Procedures for obtaining alumina from clay have been developed to the pilot-plant stage, but the continued availability of high-grade bauxite has kept these processes from becoming competitive as sources of raw material for the extraction of aluminum metal. Bauxite is a mixture of different hydrates of alumina with an average composition approximating $Al_2O_3 \cdot 2H_2O$. Its name is based on the discovery of the first mineral of high aluminum content at Les Baux, France, in 1821.

Most aluminum made in the United States today is obtained from Jamaican bauxite by the process invented independently in 1886 by Charles Martin Hall (1863–1914) at Oberlin College and by Paul Heroult (1863–1914) in France. Until this process was invented, aluminum was an expensive metal in the jewelry class. The process involves the electrolytic decomposition of purified bauxite dissolved in molten cryolite and fluorspar. *Cryolite* is sodium fluoaluminate, Na_3AlF_6. Its name, from the Greek meaning "ice-stone," is based on its translucency and resemblance to ice. The only important source of cryolite is at Ivigtut, Greenland. A synthetic cryolite is obtained by reaction of sodium fluoride with aluminum sulfate:

$$12NaF + Al_2(SO_4)_3 \rightarrow 3Na_2SO_4 + 2Na_3AlF_6$$

Cryolite is used to provide opacity to milk glass and porcelain enamel.

The excellent thermal conductivity of aluminum, its resistance to corrosion, and its light weight are the properties that account for its use in the manufacture of cooking utensils. Its use as a structural material has now become of prime importance. Although not so light as magnesium, it is stronger. Aluminum and its alloys are used extensively in truck bodies and aircraft. Modern electrical transmission lines carry twice as much current as could be carried by copper lines of the same weight. Aluminum foil is familiar as a thermal insulator and moisture barrier in houses. Aluminum foil and aluminized paper have displaced tinfoil as a wrapping material.

Aluminum powder is used in paint, and it serves as a powerful reducing agent in the "thermite" welding process. Thermite is a mixture of aluminum and iron oxide. When the mixture is raised to a sufficiently high temperature, as by ignition with a mixture of powdered magnesium and barium peroxide, the components react with the evolution of so much heat that the temperature reaches 3000°C, which is considerably above the melting point of iron:

$$3Fe_3O_4 + 8Al \rightarrow 4Al_2O_3 + 9Fe \qquad \Delta H = -800 \text{ kcal}$$

The molten iron is allowed to run between the ends of broken steel rails or machine parts so that they become welded together when the iron solidifies.

The thermite process is a special case of a general reaction used for making rather pure specimens of metals like chromium and manganese, whose oxides are hard to reduce:

$$Cr_2O_3 + 2Al \rightarrow 2Cr + Al_2O_3 \qquad \Delta H = -109 \text{ kcal}$$

30-19 Sources and Uses of Gallium, Indium, and Thallium

Gallium, indium, and thallium are rare metals, which occur widely distributed as impurities in ores of other metals. Gallium is recovered from zinc ores and as a by-product of the purification of bauxite. Indium and thallium are by-products of smelting lead and zinc ores.

Gallium has some use as a thermometer fluid for use at high temperatures because of its wide liquid range (30 to 1983°C). It has been suggested for use in dental alloys. The chief use of indium is to provide protection against corrosion when electroplated on certain alloys. It is used in jewelry. Some of the alloys of thallium are useful as high-temperature solders. Various thallium compounds find use in infrared detecting and signaling systems. A mercury-thallium alloy (fp, −60°C) is used in place of mercury in thermometers, relays, and switches designed for arctic and stratospheric conditions.

30-20 The Alums

Aluminum sulfate is a white powder that is quite soluble, not only in hot water but also in cold. Therefore, it is difficult to purify by recrystallization. It has extensive industrial use in paper-making, tanning, and dyeing. For these applications it is important that it be free from iron salts.

If aluminum sulfate and potassium sulfate are dissolved together, the solution deposits transparent crystals of a *double salt*, $KAl(SO_4)_2 \cdot 12H_2O$, which has long been known as *alum*. Alum has often been used instead of aluminum sulfate in applications where impurities would be objectionable, for it is much easier to purify alum by recrystallization. The solubility of alum is considerably less than that of aluminum sulfate at low temperatures and increases greatly as the temperature rises.

Alum was originally extracted from *alum stone*, which is a basic double sulfate, $K(AlO)_3(SO_4)_2 \cdot 3H_2O$, that occurs in nature. The term alum is now used in reference to a variety of double salts with the general formula $M^{I}M^{III}(SO_4)_2 \cdot 12H_2O$. All the alums are *isomorphous*, i.e., they have the same crystal structure. The monopositive ion in an alum may be almost any 1+ ion except Li^+, which is too small to fit into the structure. The tripositive ion may be Al^{+++}, Ga^{+++}, In^{+++}, Cr^{+++}, Fe^{+++}, and many others. It is even possible for selenate (but not tellurate) to be present in place of sulfate.

The following are illustrative of the variety of alums and the names used in referring to them:

$NaAl(SO_4)_2 \cdot 12H_2O$	Sodium aluminum alum (sodium alum)
$KAl(SO_4)_2 \cdot 12H_2O$	Potassium aluminum alum ("potash alum")
$NH_4Al(SO_4)_2 \cdot 12H_2O$	Ammonium aluminum alum (ammonium alum)
$NH_4Fe(SO_4)_2 \cdot 12H_2O$	Ammonium iron(III) alum ("ferric alum")
$KCr(SO_4)_2 \cdot 12H_2O$	Potassium chromium(III) alum ("chrome alum")

"Alum," "common alum," and "ordinary alum" are expressions now used for both potassium alum and ammonium alum, because the identities of the products sold under these names depend on economic conditions. Industrially, aluminum sulfate has largely replaced them since it can be obtained in a satisfactorily pure state by dissolving iron-free aluminum hydroxide in sulfuric acid. This has made the confusion of names even worse, for now aluminum sulfate itself is frequently referred to as "alum" or "paper-maker's alum."

New Terms

Alum: a type of double salt with the general formula $M^{I}M^{III}(SO_4)_2 \cdot 12H_2O$.

Boranes: hydrides of boron. The simplest is diborane, B_2H_6.

Thermite process: a process for producing molten iron by the reaction of small particles of aluminum with iron oxide.

Exercises

30-1 Show how boron differs from other nonmetals in regard to atomic structure and how it differs from other Group IIIA elements in chemical properties and in electrical conductivity.

30-2 Which Group A elements are too reactive to occur in nature in the elementary state?

30-3 Which Group IIIA elements form carbides? Which ones form compounds in which their oxidation state is $+1$? Which ones react with nitric acid to form nitrates?

30-4 Identify the following: corundum, bauxite, fluoboric acid, borax, kernite, and "aluminium."

30-5 What are spinels? Boranes?

30-6 Discuss the properties of boron nitride and diborane.

30-7 Discuss the occurrence and the significance of aluminum and boron in soil.

30-8 Describe the thermite process.

30-9 Discuss the composition and the naming of alums.

References

Rochow, E. G., *The Metalloids*, Heath, Boston, 1966 (paperbound). The author classifies boron, silicon, germanium, arsenic, antimony, tellurium, polonium, and astatine as metalloids.

Scott, E. C., and Kanda, F. A., *The Nature of Atoms and Molecules*, Harper & Row, New York, 1962, Chapter 24, "Giant Molecules." Both organic and inorganic giant molecules are considered. This constitutes a refreshing review of much of the chemistry of Groups IIIA and IVA.

31

Metal Complexes

31-1 Definitions of Terms

Transition metal compounds are fascinating because of their colors, magnetic properties, and great tendency to form complexes. A *complex* is an ion or molecule composed of two or more species capable of existing independently of each other. It is characteristic of complexes that their formation involves a Lewis acid-base reaction (Section 21-12). Typical examples are the diammine-silver(I) ion and the dicyanoargentate(I) ion:

$$Ag^+ + 2\,:\underset{\underset{H}{\cdot\cdot}}{\overset{\overset{H}{\cdot\cdot}}{N}}:H \rightarrow \left[H:\underset{\underset{H}{\cdot\cdot}}{\overset{\overset{H}{\cdot\cdot}}{N}}:Ag:\underset{\underset{H}{\cdot\cdot}}{\overset{\overset{H}{\cdot\cdot}}{N}}:H \right]^+$$

$$Ag^+ + 2:C:::N^-: \rightarrow [:N:::C:Ag:C:::N:]^-$$

In these examples the Ag^+ ion is a Lewis acid. In the first example the Lewis base is the NH_3 molecule; in the second example the Lewis base is the CN^- ion.

Usually, a complex ion consists of a positive metal ion to which are coordinated a number of electron-rich molecules or anions. Any hydrated or ammoniated metal ion is a complex ion. Hydrated and ammoniated ions usually have similar formulas:

$$Zn^{++} + 4H_2O \rightarrow [Zn(H_2O)_4]^{++}$$
$$Zn^{++} + 4NH_3 \rightarrow [Zn(NH_3)_4]^{++}$$

The *coordination number* of an acceptor metal ion is the number of electron pairs coordinated to it in forming a complex. When each donor group furnishes only one electron pair, the coordination number is the number of such groups attached to the central ion in the complex. The coordination number for the silver ion is 2 in the $Ag(NH_3)_2^+$ and $Ag(CN)_2^-$ ions; the coordination number for the zinc ion is 4 in $Zn(H_2O)_4^{++}$ and $Zn(NH_3)_4^{++}$; the coordination number of platinum is 6 in $PtCl_6^{--}$; etc.

The groups attached to the central metal ion are called donors, coordinating groups, or *ligands*. The particular atom attached to the central ion by means of the coordinate bond is called the *donor atom*. Thus, in the $Ag(NH_3)_2^+$ complex the NH_3 molecules are the ligands, and the nitrogen atoms are the donor atoms.

31-2 The Werner Coordination Theory

Our present understanding of metal complexes is based on the foundations laid by the Swiss chemist Alfred Werner, whose research papers were published during the period from 1893 to 1918. Note that his work preceded the advent of the electronic theory of chemical combination.

A basic item in Werner's coordination theory is the postulate that the ligands arrange themselves in a definite geometrical pattern around the central ion. In this way he was able to account for many of the surprising facts he established in his work. For example, cobalt(III) chloride forms four ammoniates whose properties differ in a strikingly systematic fashion. The yellow crystalline compound with the formula $CoCl_3 \cdot 6NH_3$ is an excellent conductor in solution, and when silver nitrate is added to its solution all the chlorine is precipitated as AgCl, leaving $Co(NO_3)_3 \cdot 6NH_3$ in solution. The purple solid with the formula $CoCl_3 \cdot 5NH_3$ forms a solution that also conducts, but not so well, and addition of silver nitrate to its solution precipitates only two thirds of the chlorine. When silver nitrate is added to a solution of the green compound, $CoCl_3 \cdot 4NH_3$, only one third of the chlorine precipitates. The triammine, $CoCl_3 \cdot 3NH_3$, is not an electrolyte and contains no chloride ions.

These and other facts were explained by Werner's assumption that in each of these compounds six groups are attached to the central cobalt atom. Thus, the cation in the first compound is $[Co(NH_3)_6]^{+++}$, and 1 mole of the salt, $[Co(NH_3)_6]Cl_3$, furnishes 4 moles of ions, including 3 moles of Cl^- ions. The cation in the second compound is $[Co(NH_3)_5Cl]^{++}$, i.e., one chlorine in addition to the five ammonia molecules is coordinated to the central cobalt; thus, 1 mole of the salt, $[Co(NH_3)_5Cl]Cl_2$, furnishes 3 moles of ions, including 2 moles of Cl^- ions. The cation in the third compound is $[Co(NH_3)_4Cl_2]^+$, and 1 mole of the salt, $[Co(NH_3)_4Cl_2]Cl$, furnishes only 2 moles of ions, only one third of the chlorine being present as Cl^- ions. In the fourth compound all three chlorines as well as the three NH_3 molecules are coordinated with the central cobalt to give a neutral nonionic complex, $[Co(NH_3)_3Cl_3]^0$.

31-3 Naming Coordination Compounds

The system now used for naming coordination compounds is based on the system originated by Werner. By this system the compounds mentioned in Section 31-2 have been given the following names:

$[Co(NH_3)_6]Cl_3$	Hexamminecobalt(III) chloride
$[Co(NH_3)_5Cl]Cl_2$	Chloropentamminecobalt(III) chloride
$[Co(NH_3)_4Cl_2]Cl$	Dichlorotetramminecobalt(III) chloride
$[Co(NH_3)_3Cl_3]$	Trichlorotriamminecobalt(III)

The rules for naming coordination compounds are

1. If the compound is a salt, the cation is named first.
2. In the name of a complex, the ligands are named first and then the central atom, regardless of whether the complex is a cation, an anion, or a neutral molecule.
3. If more than one kind of ligand is coordinated, negative ligands are named first.
4. The names of negative ligands end in *-o*, e.g., Cl^-, chloro; CN^-, cyano; and OH^-, hydroxo.
5. Neutral ligands have no distinctive endings. Names in use include *ammine* for NH_3, *aquo* for H_2O, and *carbonyl* for CO.
6. If more than one ligand of a particular kind is coordinated to the central atom, the number is indicated by using the Greek prefixes *di-*, *tri-*, *tetra-*, *penta-*, *hexa-*, *hepta-*, and *octa-*.
7. If the complex is a cation or a neutral molecule, the name of the central metal is followed by a Roman numeral in parentheses to indicate the oxidation state of the metal. However, it is often superfluous or redundant to indicate the oxidation state if it is clearly indicated by the composition of the complex. Thus, trichlorotriamminecobalt is a fully sufficient name for the compound $[Co(NH_3)_3Cl_3]$.
8. If the complex is an anion the suffix *-ate* is added to the stem for the name for the central metal, regardless of which oxidation state the metal is showing. The oxidation state is then indicated by a Roman numeral in parentheses.

The following examples illustrate the naming of salts containing complex anions:

$K_2[PtCl_6]$	Potassium hexachloroplatinate(IV)
$K_2[PtCl_4]$	Potassium tetrachloroplatinate(II)
$K_3[Co(NO_2)_6]$	Potassium hexanitrocobaltate(III)
$K_3[Fe(CN)_6]$	Potassium hexacyanoferrate(III)
$K_4[Fe(CN)_6]$	Potassium hexacyanoferrate(II)

31-4 The Configurations of Complexes

The structures of complexes, i.e., their geometrical configurations, vary with the coordination numbers of the central atoms. The most common coordination numbers are 2, 4, and 6.

The coordination number 2 is characteristic of only a few metals. Most complexes with the coordination number 2 are *linear*. Examples of 2-coordinate complexes which X-ray investigations have shown to be linear are the silver(I)

Figure 31-1

Structure of the $[AgCl_2]^-$ ion

(a) Shape of the ion. (b) Electronic formula.

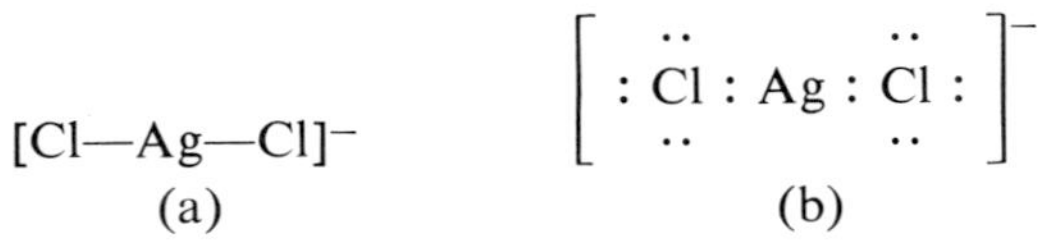

and gold(I) complexes $Ag(NH_3)_2^+$, $AgCl_2^-$, $Ag(CN)_2^-$, $Au(CN)_2^-$, and $AuCl_2^-$. The linearity of the $AgCl_2^-$ ion and its electronic formula are depicted in Figure 31-1. When we describe the $[AgCl_2]^-$ ion as linear, we mean that the silver nucleus is at the midpoint of a straight line connecting the two chlorine nuclei. In the $[Ag(NH_3)_2]^+$ complex the two nitrogen nuclei and the silver nucleus lie on the same line, so we say that this complex is also linear; within each NH_3 group the hydrogen nuclei have the same locations relative to one another and to the nitrogen nucleus as in a "free" ammonia molecule.

A number of complexes containing four ligands have been found to have a tetrahedral structure. Examples include $Zn(CN)_4^{--}$, $Zn(NH_3)_4^{++}$, BeF_4^{--}, $CoCl_4^{--}$, and $Ni(CO)_4$. The shape of the $Zn(NH_3)_4^{++}$ ion and its electronic formula are indicated in Figure 31-2.

In some 4-coordinate complexes the donor atoms of the ligands are located at the corners of a square. An example is the neutral complex $[Pt(NH_3)_2Cl_2]$, dichlorodiammineplatinum(II). Werner postulated that this complex has a square planar configuration to explain the existence of two compounds with this same formula. As with the two 2-butenes (Section 18-16), there are *cis* and *trans* isomers of dichlorodiammineplatinum(II). As indicated in Figure 31-3, the *cis* isomer has the two ammonia molecules coordinated in adjacent positions so that the two chloride ions, of necessity, are also coordinated in adjacent positions. The *trans* isomer has the two ammonia molecules attached on opposite sides of the central platinum; i.e., each ammonia molecule has two

Figure 31-2

Structure of the $[Zn(NH_3)_4]^{++}$ ion

(a) Shape of the ion. (b) Electronic formula.

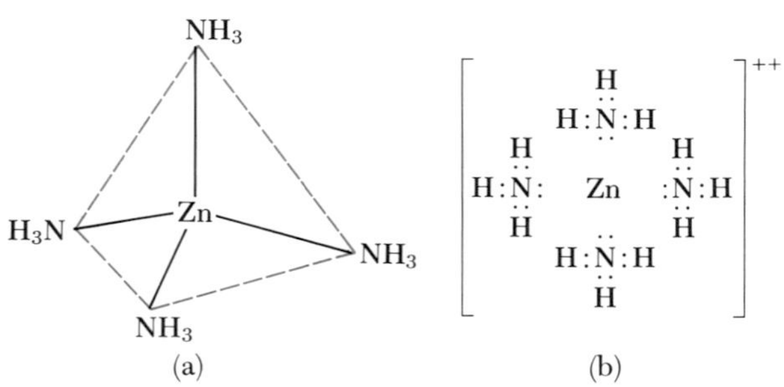

Figure 31-3

The spatial configurations of dichlorodiammineplatinum(II)

(a) The *cis* isomer. (b) The *trans* isomer.

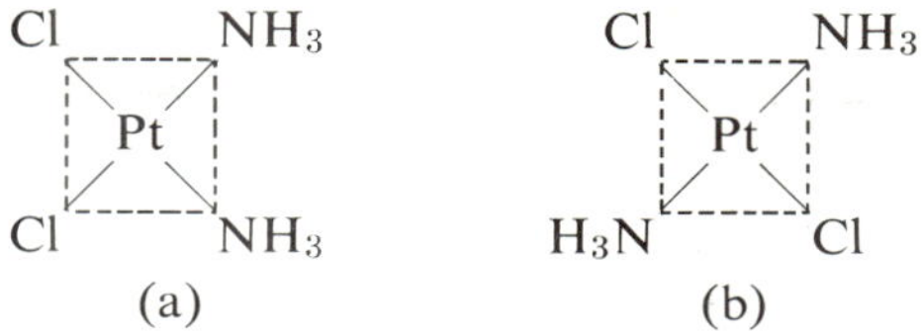

chloride ions for its nearest neighbors, each chloride ion has two ammonia molecules for its nearest neighbors, and one chloride is coordinated to the platinum at a position diagonally opposite to the location of the other. If these complexes had a tetrahedral configuration, isomerism would not be possible with two ligands of one kind and two ligands of another kind. The type of reasoning that leads us to conclude that the CH_2Br_2 molecule is tetrahedral (Section 18-4) led Werner to conclude that the $Pt(NH_3)_2Cl_2$ molecule *cannot* be tetrahedral. Direct evidence for the correctness of his postulate that this molecule is square coplanar has been obtained by X-ray crystallographic studies.

The 6-coordinate complexes are most numerous and have been investigated at length. In all complexes of this type, e.g., $Co(NH_3)_6^{+++}$, the six bonds appear to be equivalent. Such equivalency would result if the six ligands were located at the corners of an octahedron or at the corners of a trigonal prism or if the complex were a coplanar hexagon with the metal ion at the center and the ligands at the six apices. All lines of evidence, including the facts of isomerism and the direct evidence of X-ray crystallography, point to the octahedral structure as the preferred one (although several examples of the trigonal prism structure have been discovered since 1965). The structure of the $Co(NH_3)_6^{+++}$ ion and its electronic formula are shown in Figure 31-4.

Figure 31-4

Structure of the $[Co(NH_3)_6]^{+++}$ ion

(a) Shape of the ion. (b) Electronic formula.

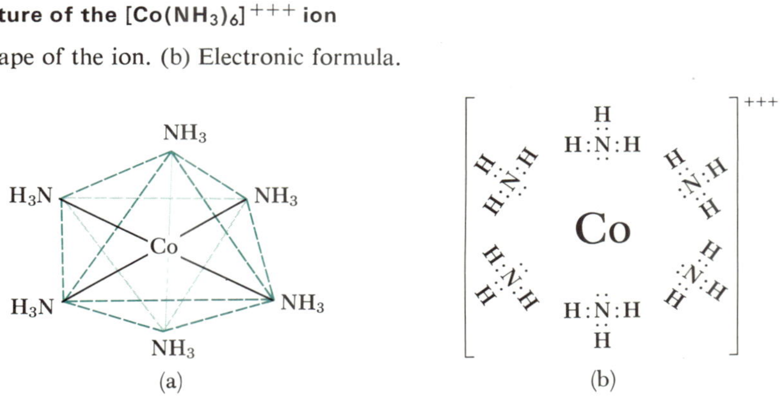

Figure 31-5

Monodentate and bidentate ligands

(a) The tetramminecopper(II) ion. (b) The bis(ethylenediamine)copper(II) ion.

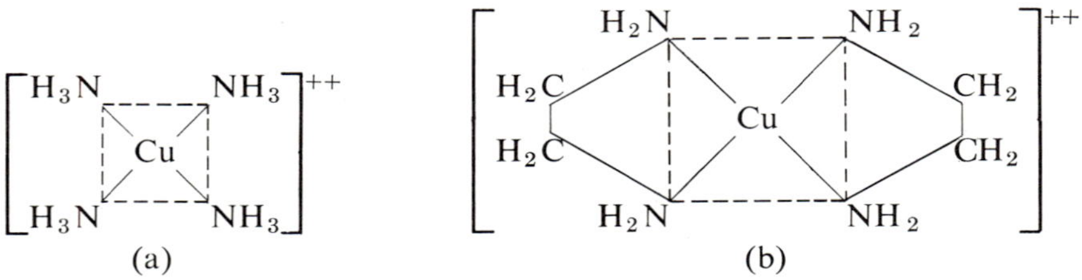

31-5 Chelates

Some ligands have more than one unshared pair of electrons located far enough apart that the ligands can occupy more than one coordination position about a given central atom. An example is ethylenediamine,

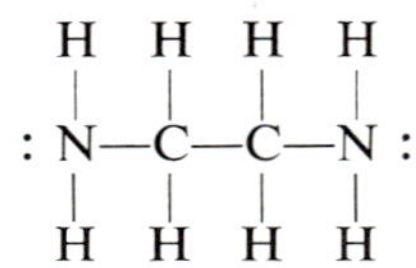

One molecule of ethylenediamine plays the same role as two molecules of ammonia in forming a metal complex, for each of its two nitrogen atoms acts as an electron-pair donor. Thus, two molecules of ethylenediamine coordinate with a copper(II) ion, taking the places of the four ammonia molecules in the coplanar tetramminecopper(II) ion (Figure 31-5).

A ligand that occupies two or more positions in the coordination sphere of a single metal ion is known as a *chelating group*, and the resulting complex is called a chelate (Greek: *chele*, claw). The bis(ethylenediamine)copper(II) ion is a typical chelate. Its existence is made possible by the fact that ethylenediamine is a *bidentate* (two-toothed) ligand. Ammonia, on the other hand, is a *monodentate* (one-toothed) ligand because it has only one unshared electron pair.

A chelate has a cyclic structure. Thus, the bis(ethylenediamine)copper(II) ion has two five-membered rings, the copper being a member of both rings. Five-membered rings are the most stable rings occurring in chelate complexes.

31-6 Stability of Complex Ions

Complex ions differ greatly in the extent to which they dissociate in solution into their component parts. The *stability constant* or *formation constant* for a complex is the equilibrium constant for its formation from its components. For example, the formation constant for tetramminezinc ion is given by the expression

$$K = \frac{[Zn(NH_3)_4{}^{++}]}{[Zn^{++}][NH_3]^4}$$

The expression for the formation constant is derived directly from the equation for the formation of the complex from its components,

$$Zn^{++} + 4NH_3 \rightleftharpoons Zn(NH_3)_4{}^{++}$$

The value for the formation constant for tetramminezinc ion is 3.2×10^9. Knowledge of the numerical value for this constant enables us to calculate, for example, the concentration of Zn^{++} ions remaining in a solution that was originally 0.1 M in $ZnSO_4$ but which has thereafter been made $3M$ in NH_3 by passing NH_3 gas into the solution (without changing the volume of the solution, let us suppose, to keep the calculation simple). Letting $[Zn^{++}] = x$, $[Zn(NH_3)_4{}^{++}] = 0.1 - x$, and $[NH_3] = 3M$,

$$\frac{0.1 - x}{3^4 x} = 3.2 \times 10^9$$

Solving, $x = 3.9 \times 10^{-13}$ M. Thus, we see to what a small value the ammonia has reduced the concentration of Zn^{++} ions.

31-7 Types of Metal Complexes

In solutions of salts the metal ions seldom, if ever, occur in the idealized simple form implied by the formulas we customarily employ to represent them, such as Ag^+ or Zn^{++}. In the absence of other coordinating groups the molecules of the solvent usually act as ligands in forming coordination complexes. Notably, the only liquids in which salts are soluble are those with molecules that have unshared electron pairs.

Anhydrous copper(II) sulfate is a white solid that dissolves in water to form a blue solution. The color of the solution is attributed to $[Cu(H_2O)_4]^{++}$ ions. When ammonia is added to this solution, the color deepens to an intensely dark blue. This color change is due to the conversion of an aquo complex to an ammine complex:

$$[Cu(H_2O)_4]^{++} + 4NH_3 \rightleftharpoons [Cu(NH_3)_4]^{++} + 4H_2O$$

It is such a well-established custom to ignore the hydration of ions in aqueous solution that the equation for this conversion is usually written in the following simplified manner:

$$Cu^{++} + 4NH_3 \rightleftharpoons [Cu(NH_3)_4]^{++}$$

Such simplified equations are written with the understanding that, in aqueous solution, the ions represented as having no coordinated ligands are actually hydrated.

The ligands in complexes may be (1) inorganic molecules, such as NH_3 in ammine complexes and H_2O in aquo complexes; (2) organic molecules, such as $H_2NCH_2CH_2NH_2$ in ethylenediamine complexes; (3) inorganic anions, such as Cl^- in chloro complexes; and (4) organic anions, such as $C_2O_4^{--}$ in the trioxalatoferrate(III) ion, $[Fe(C_2O_4)_3]^{---}$.

We shall now consider briefly some of the species of complexes frequently encountered in analytical and other applications of coordination chemistry.

Ammine Complexes. The high solubility of the complex salt $Ag(NH_3)_2Cl$ as compared to the low solubility of AgCl is the basis for dissolving AgCl with "ammonium hydroxide" reagent and thereby separating it from other compounds which do not dissolve under this treatment. For example, in the systematic qualitative analysis of aqueous solutions for metal ions, addition of HCl causes the precipitation of lead(II) chloride, mercury(I) chloride, and silver(I) chloride. From the mixed precipitate the lead chloride can be removed with hot water, mercury(I) chloride and silver(I) chloride being insoluble in both hot and cold water. Addition of "ammonium hydroxide" to the solid remaining at this point serves to dissolve the silver chloride by converting it into soluble diamminesilver(I) chloride:

$$AgCl + 2NH_3 \rightleftharpoons Ag(NH_3)_2^+ + Cl^-$$

Addition of nitric acid to the ammonia extract serves to reverse this reaction because H^+ is a better acid than Ag^+. This causes AgCl to reprecipitate and thus provides proof that silver ions were indeed present in the original solution:

$$Ag(NH_3)_2^+ + Cl^- + 2H^+ \rightarrow AgCl\downarrow + 2NH_4^+$$

Hydroxo Complexes. Certain hydroxides of metals are insoluble in water but will dissolve when treated with a solution of either a strong acid or a strong base. What is expected as typical behavior of hydroxides of metals is that they should behave as bases and for this reason dissolve in a solution of a strong acid. Those hydroxides of metals that dissolve when treated with strong acids *or* strong bases are called *amphoteric hydroxides.* Their reaction with bases involves the formation of hydroxo complexes. An application of this difference between the behavior of amphoteric hydroxides and that of other insoluble hydroxides of metals makes possible the separation of aluminum from iron in qualitative analysis. Unlike $Fe(OH)_3$, which dissolves when treated with acid but does not do so when treated with base, $Al(OH)_3$ dissolves when treated with either acid or base:

$$Al(OH)_3 + (Na^+ + OH^-) \rightarrow Na^+ + Al(OH)_4^-$$

The soluble $NaAl(OH)_4$ is commonly referred to as sodium aluminate. A more complete name is sodium tetrahydroxoaluminate.

Cyano Complexes. The stability of the dicyanoargentate ion, $Ag(CN)_2^-$, and the dicyanoaurate(I) ion, $Au(CN)_2^-$, is the basis for the extraction of silver and gold from low-grade ores. Silver occurs as the metal, as the sulfide, and as the chloride; but gold occurs chiefly as the metal. All these are readily dissolved by sodium cyanide in the presence of air:

$$4Ag + 8CN^- + O_2 + 2H_2O \rightarrow 4Ag(CN)_2^- + 4OH^-$$
$$2Ag_2S + 8CN^- + O_2 + 2H_2O \rightarrow 4Ag(CN)_2^- + 4OH^- + 2S\downarrow$$
$$AgCl + 2CN^- \rightarrow Ag(CN)_2^- + Cl^-$$
$$4Au + 8CN^- + O_2 + 2H_2O \rightarrow 4Au(CN)_2^- + 4OH^-$$

Atmospheric oxygen is not a sufficiently powerful oxidizing agent to act upon silver, silver sulfide, or gold in the absence of a complexing agent, but it readily does so in the presence of cyanide ions. From the solution containing sodium dicyanoargentate or sodium dicyanoaurate(I), silver and gold are obtained by replacement with zinc:

$$Zn + 2Ag(CN)_2^- \rightarrow Zn(CN)_4^{--} + 2Ag\downarrow$$
$$Zn + 2Au(CN)_2^- \rightarrow Zn(CN)_4^{--} + 2Au\downarrow$$

Halo Complexes. A striking illustration of halo complex formation can be obtained by adding potassium iodide, a small portion at a time, to a solution of mercury(II) nitrate. First, the iodide ions precipitate the mercury(II) ions as HgI_2, which is a brilliant orange-red compound. As soon as excess KI has been added, the precipitate dissolves in consequence of its conversion to the soluble complex salt, potassium tetraiodomercurate(II):

$$Hg^{++} + 2I^- \rightleftharpoons HgI_2\downarrow$$
$$HgI_2 + 2I^- \rightleftharpoons HgI_4^{--}$$

If the procedure is reversed and a small quantity of mercury(II) nitrate is added to a potassium iodide solution, no precipitate forms.

Among the halide ions the fluoride ion is the most effective complexing agent, as in the SiF_6^{--}, AlF_6^{---}, and BF_4^- complexes. Chloro complexes are also common, e.g., $SnCl_6^{--}$, $AuCl_4^-$, and $PtCl_6^{--}$.

Thiocyanato Complexes. One of the most colorful examples of the complexing action of thiocyanate ion can be observed by adding potassium thiocyanate to a solution containing an iron(III) salt. The red color formed in the solution, which serves as a delicate test for the iron(III) ion, is attributed to the complex cation, $Fe(SCN)^{++}$. Undoubtedly the coordination number for iron is really 6, and the composition of the complex can be varied from $[Fe(H_2O)_5SCN]^{++}$ to $[Fe(SCN)_6]^{---}$.

Silver and mercury form insoluble thiocyanates which dissolve in solutions containing excess thiocyanate ions in consequence of the formation of dithiocyanatoargentate and tetrathiocyanatomercurate(II) ions:

$$AgSCN + SCN^- \rightleftharpoons Ag(SCN)_2^-$$
$$Hg(SCN)_2 + 2SCN^- \rightleftharpoons Hg(SCN)_4^{--}$$

Thiosulfato Complexes. Several metals form thiosulfato complexes. The silver complex is one of the most important because of the role that it plays in photography. After a photographic film has been developed, it is "fixed" by immersion in a solution of sodium thiosulfate ("hypo"). The thiosulfate dissolves the silver bromide that has not been reduced by the developer, forming the soluble complex salt, sodium dithiosulfatoargentate:

$$AgBr + 2S_2O_3^{--} \rightleftharpoons Ag(S_2O_3)_2^{---} + Br^-$$

Carbonyl Complexes. When carbon monoxide is passed over powdered nickel metal at 50°C, the nickel is converted to volatile nickel tetracarbonyl, $Ni(CO)_4$. The nickel carbonyl vapor is carried along in the stream of carbon monoxide gas. If the gas is passed through heated tubes, the unstable carbonyl decomposes into carbon monoxide and nickel metal, which is recovered. This process has been used to extract nickel from its ores.

Several of the other transition metals also form coordination compounds in the *zero oxidation state* with CO as ligand, including chromium, manganese, iron, and cobalt. The nickel carbonyl is remarkable for the ease with which it forms, no pressure and only a little heating being required. The $Ni(CO)_4$ molecule is tetrahedral.

The highly poisonous nature of carbon monoxide is undoubtedly related to the tendency of CO molecules to coordinate with atoms of the transition metals. It is known that carbon monoxide readily combines with the hemoglobin of the blood and destroys its capacity for carrying oxygen. Iron is present in hemoglobin.

EDTA Complexes. Chelating agents are divided into two classes, depending on whether they form water-soluble chelates that remain in solution or water-insoluble chelates that form precipitates. Those that form water-soluble chelates are called *sequestering agents.* The suppression of a property or a reaction of a metal ion without its actual physical removal from solution is called *sequestration.* For all practical purposes a sequestered ion has been "removed" from the solution. It is actually still in the solution, but at the center of a coordination complex.

One of the most extensively used sequestering agents is ethylenediaminetetracetic acid (EDTA). About ten million pounds of EDTA and its tetrasodium salt (Na_4EDTA) are made and used each year. The most important use for this sequestering agent is in softening water. The ethylenediaminetetracetate ion (Figure 31-6) is a hexadentate ligand. The Ca^{++} ion, which causes hardness

Figure 31-6

Structural formula for the EDTA ion

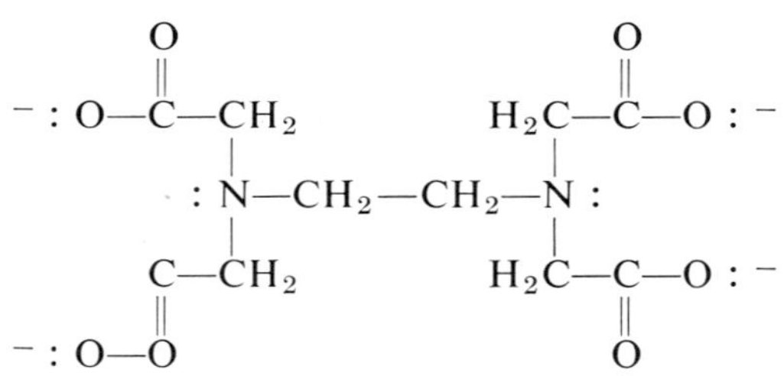

in water, is apparently just the right size for the six donor atoms of this ligand to attach themselves in octahedral positions to form a complex of great stability:

$$Ca^{++} + (EDTA)^{4-} \rightleftharpoons [Ca(EDTA)]^{--}$$

Thus, "free" Ca^{++} ions are not present to form a precipitate when soap is added.

31-8 Isomerism in Coordination Compounds

Different compounds having the same formula are referred to as isomers (Section 18-10). The types of isomerism among coordination compounds can be classified as ionization isomerism, hydrate isomerism, coordination isomerism, polymerization isomerism, linkage isomerism, ligand isomerism, geometric isomerism, and optical isomerism.

Ionization Isomerism. Compounds that have the same composition but consist of unlike ions are called ionization isomers. A striking example of ionization isomerism is provided by the compounds $[Pt(NH_3)_4SO_4](OH)_2$ and $[Pt(NH_3)_4(OH)_2]SO_4$. The first of these is a strong base because OH^- ions are present in its aqueous solution, but it gives no test for sulfate ions because they are coordinated to the central platinum. The second compound is not a base but a salt that responds in the normal manner to tests for sulfate ions; in this compound the OH^- ions are coordinated to the central platinum. Another example of ionization isomerism is provided by the pair of salts $[Pt(NH_3)_4Cl_2]Br_2$ and $[Pt(NH_3)_4Br_2]Cl_2$.

Hydrate Isomerism. Coordination compounds that have the same composition and include H_2O in their structure may differ as to the number of H_2O molecules coordinated with the central atom of the complex. As an example, $[Co(NH_3)_4(H_2O)Cl]Cl_2$ is a violet compound and $[Co(NH_3)_4Cl_2]Cl \cdot H_2O$ is green. It is apparent that hydrate isomerism is very similar to ionization isomerism. Some authors regard hydrate isomerism as a special case of ionization isomerism.

Coordination Isomerism. A salt that consists of a complex cation and a complex anion can show isomerism resulting from different distributions of ligands between the two coordination spheres. An instance of this is the isomerism of the salts $[Co(NH_3)_6][Cr(CN)_6]$ and $[Cr(NH_3)_6][Co(CN)_6]$. The first of these has cobalt in the complex cation $[Co(NH_3)_6]^{+++}$. In the second compound chromium is in the cation $[Cr(NH_3)_6]^{+++}$ and cobalt is in the anion $[Co(CN)_6]^{---}$.

Polymerization Isomerism. Compounds that have the same composition but different formula weights are not true isomers. Traditionally, however, coordination compounds having the same empirical formulas have been classified as polymerization isomers. For example, the yellow nonelectrolyte $[Pt(NH_3)_2Cl_2]$ and the green salt $[Pt(NH_3)_4][PtCl_4]$ share the same empirical formula and have been classically regarded as polymerization isomers.

Linkage Isomerism. Linkage isomerism results from the possibility that either of two different atoms in a ligand might serve as the donor atom. There are, for example, two compounds which, superficially regarded, have the same formula, $[Co(NH_3)_5(NO_2)]Cl_2$. There is evidence that in one of these, a red salt, an oxygen atom of the NO_2^- ion is coordinated to the central cobalt, as can be indicated by writing the formula $[Co(NH_3)_5(ONO)]Cl_2$ and naming the compound nitritopentamminecobalt(III) chloride. In the other compound, a yellow salt, the nitrogen atom of the NO_2^- ion is coordinated to the cobalt, so that it is appropriate to write the formula $[Co(NH_3)_5(NO_2)]Cl_2$ and to call the compound nitropentamminecobalt(III) chloride.

Ligand Isomerism. Isomerism in coordination compounds arises in some instances from isomerism within the ligands themselves. For example, 1,2-diaminopropane,

$$\begin{array}{lll} H_2C & —CH & —CH_3 \\ | & | & \\ H_2N & NH_2 & \end{array}$$

and 1,3-diaminopropane,

$$\begin{array}{lll} H_2C & —CH_2— & CH_2 \\ | & & | \\ H_2N & & NH_2 \end{array}$$

are position isomers. Each can coordinate with a central cobalt, serving as a bidentate ligand. It should come as no surprise that there are two complex salts of the same composition, $[Co(diaminopropane)_2Cl_2]Cl$. In one of these the organic ligand is 1,2-diaminopropane; in the other it is 1,3-diaminopropane. In one of these there is a five-membered metal-chelate ring; in the other, a six-membered ring.

Geometric Isomerism. We have already referred (Section 31-4) to Werner's postulation of a square coplanar configuration for the neutral complex $[Pt(NH_3)_2Cl_2]$ to explain the existence of its two isomers. The existence of *cis-trans* isomerism in a number of coordination compounds has now been firmly established. In 6-coordinated complexes, where there is an octahedral arrangement of the donor atoms of the ligands around the coordination center, complexes such as $[Co(NH_3)_4Cl_2]^+$ show geometric isomerism. By considering the structure of the $[Co(NH_3)_6]^{+++}$ ion (Figure 31-4), you can see that the substitution of chloride ions for two of the ammonia molecules could occur either in adjacent positions or in diagonally opposite positions, giving rise to *cis* and *trans* isomers, respectively. Salts of *cis*-dichlorotetramminecobalt(III) ion are violet, and salts containing the *trans* cation are green.

Optical Isomerism. Optical isomerism is the type of stereoisomerism that results when two compounds have configurations in space such that one structure is the mirror image of the other but is not identical with it (Sections 19-13 and 19-14).

In coordination chemistry the most important instances of optical isomerism occur among octahedral complexes. If all the ligands in an octahedral complex are monodentate, the existence of optical isomerism requires that there be at least three different kinds of ligands and that there be no more than two of any one kind. If these conditions are not met it is impossible to have two different steric arrangements of the ligands. Optical isomerism in tetrahedral complexes requires that all ligands be different if they are monodentate. Optical isomerism in square coplanar complexes is impossible for monodentate ligands. Most experimental work on optical isomerism among coordination compounds has been done with octahedral complexes containing chelate ligands (three bidentate ligands, or two bidentate and two monodentate ligands).

31-9 Bonding in Transition Metal Complexes

The theory of covalent bonding, which we have used to explain the structures of many compounds, has been extended to explain the structures of complexes in which various ligands are coordinated to a central atom or ion. The *valence bond theory* assumes that the electron pairs donated by the ligands enter hybrid orbitals associated with the metal atom. According to this theory, the linear structure of 2-coordinate complexes results from the use of *sp* hybrid orbitals, as in the $BeCl_2$ molecule (Section 18-5). For example, Figure 31-7 gives schematic representations of the electronic configurations present in the silver atom, in the Ag^+ ion, and in the dichloroargentate(I) complex ion, which is linear.

The valence bond theory attributes the tetrahedral structure of certain 4-coordinate complexes to sp^3 hybridization, which we have discussed in connec-

Figure 31-7

Schematic representation of *sp* hybridization

(a) The $_{47}Ag$ atom. (b) The Ag^+ ion. (c) Ag(I) in the $[AgCl_2]^-$ ion.

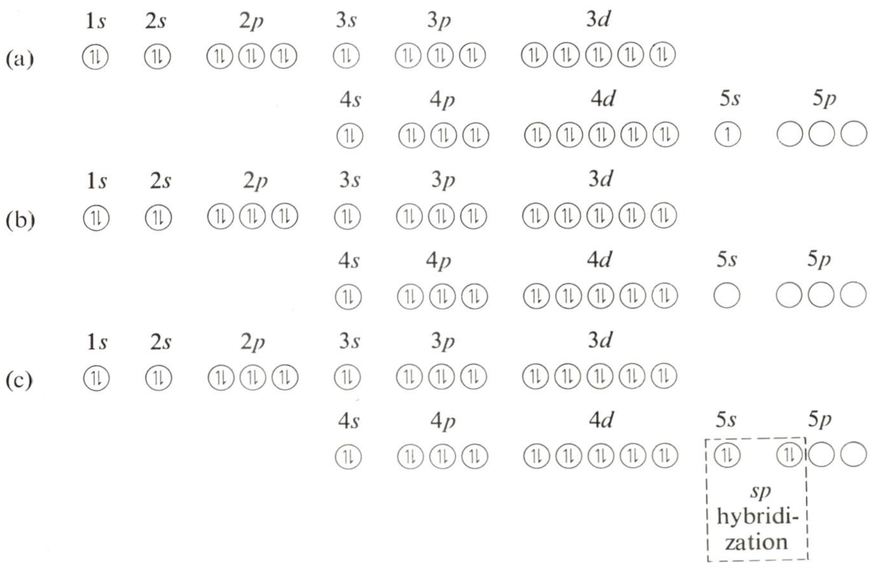

Figure 31-8

Schematic representation of sp^3 hybridization

(a) The $_{30}Zn$ atom. (b) The Zn^{++} ion. (c) Zn(II) in the $[Zn(CN)_4]^{--}$ ion.

	1s	2s	2p	3s	3p	3d	4s	4p
(a)	↑↓	↑↓	↑↓ ↑↓ ↑↓	↑↓	↑↓ ↑↓ ↑↓	↑↓ ↑↓ ↑↓ ↑↓ ↑↓	↑↓	○ ○ ○
(b)	↑↓	↑↓	↑↓ ↑↓ ↑↓	↑↓	↑↓ ↑↓ ↑↓	↑↓ ↑↓ ↑↓ ↑↓ ↑↓	○	○ ○ ○
(c)	↑↓	↑↓	↑↓ ↑↓ ↑↓	↑↓	↑↓ ↑↓ ↑↓	↑↓ ↑↓ ↑↓ ↑↓ ↑↓	↑↓	↑↓ ↑↓ ↑↓

(4s and 4p in (c) boxed: sp^3 hybridization)

tion with the structure of the CH_4 molecule (Section 18-5). Figure 31-8 shows schematic representations of the electronic configurations present in the zinc atom, in the Zn^{++} ion, and in the tetramminezinc(II) ion, which is tetrahedral. The existence of other 4-coordinate complexes with a square planar structure is attributed to dsp^2 hybridization, which is depicted in Figure 31-9, where schematic representations are given for the electronic configurations present in the nickel atom, in the Ni^{++} ion, in the "activated" Ni^{++} ion in which a lone electron has been "promoted" from one 3*d* orbital into another 3*d* orbital containing a lone electron, and in Ni(II) in the $Ni(CN)_4^{--}$ complex, which is square coplanar.

As an example of the application of valence bond theory in the explanation of the octahedral structure of 6-coordinate complexes, we may consider the $Co(NH_3)_6^{+++}$ ion. The d^2sp^3 hybridization to which its octahedral structure is attributed by the theory is indicated in Figure 31-10, where schematic representations are given for the electronic configurations present in the cobalt atom, in the Co^{+++} ion, in the "activated" Co^{+++} ion, and in Co(III) in the hexamminecobalt(III) ion.

Figure 31-9

Schematic representation of dsp^2 hybridization

(a) The $_{28}Ni$ atom (has two unpaired electrons). (b) The Ni^{++} ion in the ground state (still has two unpaired electrons). (c) The Ni^{++} ion in the bonding state. (d) Ni(II) in the $[Ni(CN)_4]^{--}$ ion (no unpaired electrons).

	1s	2s	2p	3s	3p	3d	4s	4p
(a)	↑↓	↑↓	↑↓ ↑↓ ↑↓	↑↓	↑↓ ↑↓ ↑↓	↑↓ ↑↓ ↑↓ ↑ ↑	↑↓	○ ○ ○
(b)	↑↓	↑↓	↑↓ ↑↓ ↑↓	↑↓	↑↓ ↑↓ ↑↓	↑↓ ↑↓ ↑↓ ↑ ↑	○	○ ○ ○
(c)	↑↓	↑↓	↑↓ ↑↓ ↑↓	↑↓	↑↓ ↑↓ ↑↓	↑↓ ↑↓ ↑↓ ↑↓ ○	○	○ ○ ○
(d)	↑↓	↑↓	↑↓ ↑↓ ↑↓	↑↓	↑↓ ↑↓ ↑↓	↑↓ ↑↓ ↑↓ ↑↓ ↑↓	↑↓	↑↓ ↑↓ ○

(last 3d, 4s and two 4p orbitals in (d) boxed: dsp^2 hybridization)

Figure 31-10

Schematic representation of d^2sp^3 hybridization

(a) The $_{27}Co$ atom. (b) The Co^{+++} ion. (c) The Co^{+++} ion in the bonding state. (d) Co(III) in the $[Co(NH_3)_6]^{+++}$ ion.

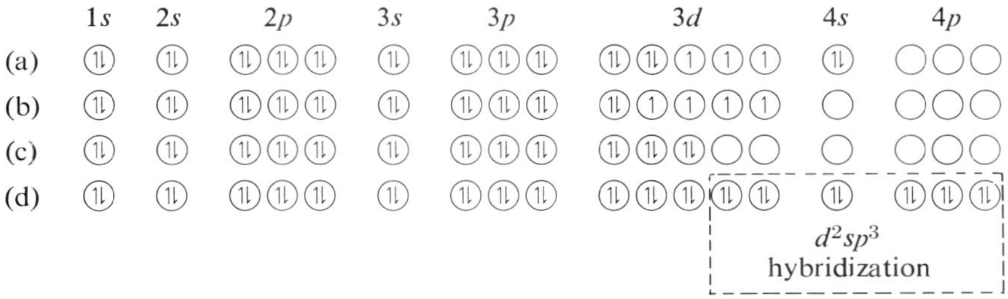

In recent years a quite different approach, known as *ligand field theory*, has been developed to explain certain properties of transition metal complexes that were not understood in terms of the valence bond theory. The ligand field theory assumes that the bonding of the ligands to the central metal ion is primarily electrostatic rather than covalent. In other words, the ligands and the metal are held together by electrical attraction between the positively charged metal ion and the electrons associated with the ions or polar molecules that are acting as ligands. The maximum coordination number of an ion is limited by the electrical repulsions between the ligands and by the space available for them around the central metal ion.

Because the ligands are so oriented that their negative ends point toward the central ion, there will be an electrostatic repulsion between the electrons of the ligands and the electrons of the central metal ion. The five *d* orbitals of the central ion are affected differently by the electrostatic field of the ligand so that they are no longer equivalent as they are in the isolated ion.

In order to discuss the ligand field theory further, we must consider the shapes and orientations of *d* orbitals, as we did earlier for *s* and *p* orbitals (Figure 4-1). The shapes of the five *d* orbitals are shown in Figure 31-11. These orbitals are more complicated than *s* and *p* orbitals. It is important to note that two of the *d* orbitals point along the coordinate axes, whereas the other three are directed in between the coordinate axes.

Each of the three *d* orbitals that are directed in between the coordinate axes consists of four lobes arranged like a four-leafed clover or, better, like crossed hourglasses. A fourth orbital has this same four-lobed shape with one pair directed along the *x* axis and the other along the *y* axis. The fifth orbital has two lobes on the *z* axis, and a ring, which may be likened to a doughnut, in the *xy* plane.

In an isolated atom or ion, an electron in any of the five *d* orbitals has the same energy. For example, the one *d* electron in an isolated Ti^{+++} ion has no preference as to which *d* orbital it will occupy. In an isolated Mn^{++} ion the five *d* electrons will each occupy one of the five *d* orbitals.

Figure 31-11

Shapes of *d* orbitals

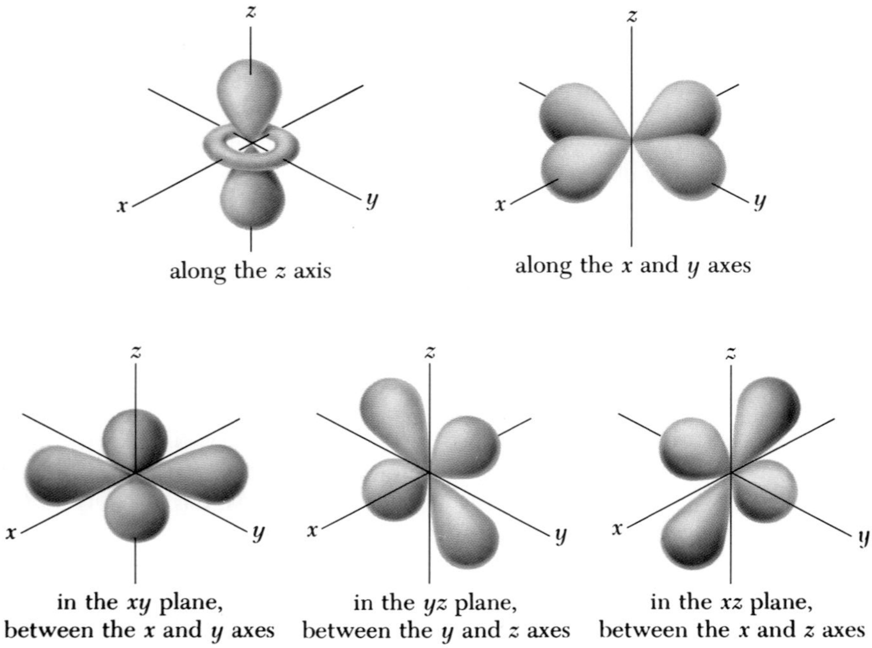

Now let us see what happens when six ligands are arranged symmetrically about a central ion to form an octahedral complex. Let us assume that the six ligands approach the central ion from both directions along each of the three coordinate axes x, y, and z. Two of the d orbitals will point their electrons (if they have any) directly at the ligands, but the electrons in the other three d orbitals will point away from the ligands. Consequently, any electrons that may be present in the five d orbitals will not have the same energy. All the d orbital energies will be raised by electrostatic repulsion, but the energies of the two orbitals that point directly at the negative charges of the ligands will be raised more than the energies of the three orbitals that are directed toward points between the negative charges of the ligands. The relative effect of the electrostatic repulsion on electrons in the two sets of d orbitals is shown in Figure 31-12. The change in the energies of some d orbitals relative to other d orbitals is known as *ligand field splitting*.

In octahedral complexes of an ion containing one d electron, one of the orbitals of the low-energy set will be occupied preferentially by the electron. However, by picking up a quantum of radiation of the required frequency the electron can move into one of the orbitals of the high-energy set. Thus, the absorption of radiation depends on the presence of some d electrons and is a function of the type of electron-repelling groups (ligands). The number of d electrons must be less than ten for this mechanism of light absorption to operate.

Figure 31-12

Splitting of *d* orbitals in an octahedral crystal field

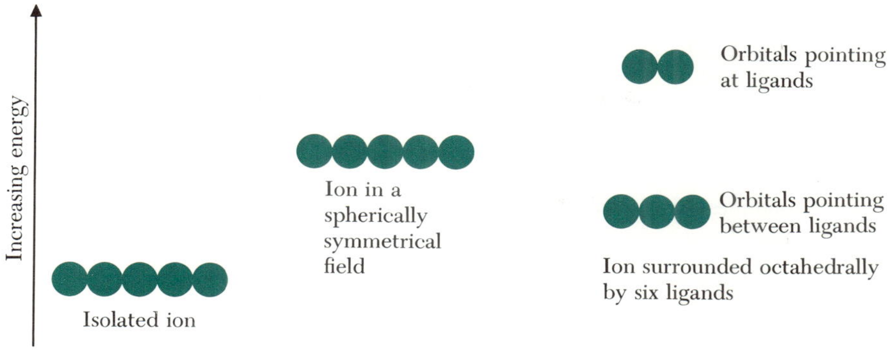

There can be no shifting of electrons from one *d* orbital to another if each of the five orbitals is already filled with a pair of electrons. This is the explanation of the tendency of all the first-row transition metals except zinc to form colored compounds. One of the deficiencies of the valence bond theory is that it does not explain the wide variety of colors characteristic of so many transition metal complexes.

New Terms

Chelate: a complex containing a ligand that occupies two or more positions in the coordination sphere of a single metal atom or ion.

Complex: an ion or molecule composed of two or more species capable of existing independently of each other.

Coordination number: the number of electron pairs coordinated to the central atom or ion in a complex.

Formation constant: the equilibrium constant for the formation of a complex ion from its components.

Ligands: the groups coordinated to the central atom or ion in a complex.

Sequestration: the suppression of a property or a reaction of a metal ion without its actual physical removal from solution, resulting from its conversion to a complex with a large formation constant.

Exercises

31-1 Define: coordination compound, coordination number, ligand, chelate, and sequestration.

31-2 Explain the differences in the electrical conductivities of $CoCl_3 \cdot 6NH_3$, $CoCl_3 \cdot 5NH_3$, $CoCl_3 \cdot 4NH_3$, and $CoCl_3 \cdot 3NH_3$.

31-3 Name the following compounds and ions:

(a) $K[Cr(NH_3)_2Cl_4]$
(b) $[Pt(NH_3)_2Cl_2]$
(c) $[Co(NH_3)_6]Cl_3$
(d) $[Co(H_2NCH_2CH_2NH_2)_3]Cl_3$
(e) $[Cr(NH_3)_6][Co(CN)_6]$
(f) $[Co(NH_3)_6][Cr(CN)_6]$
(g) $[Zn(NH_3)_4]^{++}$
(h) $[Zn(H_2O)_4]^{++}$

31-4 Show the electronic configurations of the silver atom, the silver(I) ion, and the diamminesilver(I) ion. Explain what is meant when the diamminesilver ion is described as linear.

31-5 What determines whether a 4-coordinate complex will be tetrahedral or square coplanar?

31-6 Show that there are two five-membered rings in the bis(ethylenediamine)-copper(II) ion.

31-7 The formation constant for the $[Ag(CN)_2]^-$ ion is 1×10^{20}. Calculate the concentration of Ag^+ ion in a 0.2 M solution of $Na[Ag(CN)_2]$.

31-8 Discuss the coordination chemistry involved in the separation of silver(I) from mercury(I) in qualitative analysis.

31-9 Explain the reaction of amphoteric hydroxides with strong bases in terms of coordination chemistry.

31-10 Sometimes a precipitate forms upon adding A to B but no precipitate forms upon adding B to A. How is this possible?

31-11 How would the following compounds respond to tests for chloride ions? For bromide ions?

(a) $[Pt(NH_3)_4Cl_2]Br_2$
(b) $[Pt(NH_3)_4Br_2]Cl_2$

References

Basolo, F., and Johnson, R. C., *Coordination Chemistry*, Benjamin, New York, 1964 (paperbound). A volume in the General Chemistry Monograph Series.

Martin, D. F., and Martin, B. B., *Coordination Compounds*, McGraw-Hill, New York, 1964 (paperbound). A small volume written for beginning students.

Murmann, R. K., *Inorganic Complex Compounds*, Reinhold, New York, 1964 (paperbound). A volume in the Selected Topics in Modern Chemistry Series.

32

The Noble Gases

32-1 Properties of the Noble Gases

The noble gases are the elements with atomic numbers 2, 10, 18, 36, 54, and 86. They are called the *noble gases* because of their lack of chemical reactivity. Prior to 1962 the members of this family were thought to be incapable of forming bonds with other elements and for this reason they were known as the *inert gases*. In the belief that zero is the only oxidation state in which they can exist, the group made up of these elements was designated Group 0 in the periodic table. Their apparent inertness provided the key to the explanation of the chemical activity of the other elements.

Figure 32-1

Electronic configurations of the noble-gas atoms

	1	2	3	4	5	6
$_{2}$He	s^2					
$_{10}$Ne	s^2	s^2p^6				
$_{18}$Ar	s^2	s^2p^6	s^2p^6			
$_{36}$Kr	s^2	s^2p^6	$s^2p^6d^{10}$	s^2p^6		
$_{54}$Xe	s^2	s^2p^6	$s^2p^6d^{10}$	$s^2p^6d^{10}$	s^2p^6	
$_{86}$Rn	s^2	s^2p^6	$s^2p^6d^{10}$	$s^2p^6d^{10}f^{14}$	$s^2p^6d^{10}$	s^2p^6

Table 32-1

Properties of the noble gases

	Boiling point (°C)	*Melting point* (°C)	*Density at boiling point* (g/ml)	*First ionization energy* (kcal/mole)	*Abundance in atmosphere* (parts per million, by volume)
$_{2}He$	−268.9	−272.2 (26 atm)	0.125	566	5
$_{10}Ne$	−246	−249	1.21	497	12
$_{18}Ar$	−186	−189	1.40	363	9400
$_{36}Kr$	−153	−157	2.41	322	1
$_{54}Xe$	−107	−112	3.06	279	0.08
$_{86}Rn$	−62	−77	4.4	247	variable

The extraordinary stability characteristic of these elements is the result of the electronic structure of their atoms (Figure 32-1). It is for this reason also that their molecules are monatomic. Being monatomic and symmetrical, they are nonpolar. The resulting weakness of the van der Waals forces explains the low melting points and low boiling points, all of which are far below room temperature (Table 32-1). With increasing atomic number, the increasing number of electrons results in greater intermolecular attraction (See Section 9-16). For this reason the heaviest noble gases have the highest melting and boiling points.

The gases are colorless, odorless, and tasteless. Liquefaction produces colorless liquids, and freezing gives snow-white solids. As would be expected, the heats of fusion and the heats of vaporization are small.

There is a regular increase in the sizes of the atoms as the atomic numbers increase. However, the addition of each outer shell of electrons is accompanied by a contraction of the inner shells resulting from the increasing nuclear charge. When we compare the radii of the atoms of the noble-gas elements with the radii of the simple ions with which they are isoelectronic (Figure 32-2), we see clearly that the effect of increasing nuclear charge is to reduce the size.

32-2 Discovery of Argon

When Mendeleev and Meyer constructed the periodic table, the noble gases were still unknown. Their lack of chemical reactivity causes them to occur in nature only as free elements. As minor components of the earth's atmosphere, they escaped early detection. When first discovered they therefore were called the *rare gases*.

Aside from such variable components as water vapor and carbon dioxide, the atmosphere consists chiefly of nitrogen and oxygen. The composition of the atmosphere was investigated in 1785 by Henry Cavendish. He mixed extra

Figure 32-2

Sizes of noble-gas atoms and isoelectronic ions

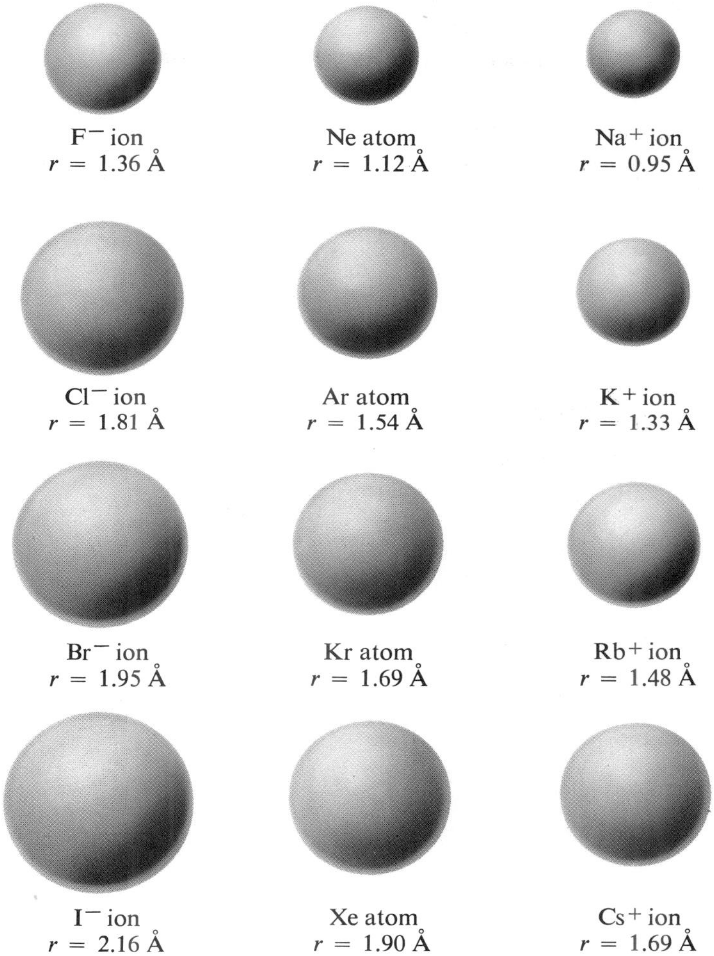

oxygen with a sample of air and passed an electric spark through the mixture, causing nitrogen to combine with oxygen to form a compound, NO_2, which was dissolved in water. After there was no further decrease in volume, he removed the excess oxygen from the residual gas by absorbing it in another solution. He found that only a small bubble of air remained unabsorbed, its volume being not more than $\frac{1}{120}$ of that of the original air. He concluded that *if* there is any part of our atmosphere that is neither nitrogen nor oxygen, it is not more than $\frac{1}{120}$ of the whole. Cavendish did not claim to have proved the existence of another substance.

In 1894, Lord Rayleigh was engaged in research that required careful measurements of the densities of hydrogen, oxygen, and nitrogen gases. Nitrogen obtained from air was found to have a density of 1.2572 g per liter at standard conditions; but nitrogen obtained by decomposition of any of several nitrogen compounds was found to have density 1.2505 g per liter at standard conditions. Sir William Ramsay collaborated with Lord Rayleigh in an investigation of this discrepancy. Essentially, they repeated the experiment of Cavendish and found that the residual gas that did not combine with oxygen is nearly 50 percent denser than nitrogen. They showed by spectroscopic analysis that the residual gas was not nitrogen but a new element, which they named *argon* (Greek, inactive). Ramsay began intensive research on this minor constituent of the atmosphere (less than 1 percent by volume), and with his students he isolated five other elements. Thus, the original argon, obtained in 1894, was actually a mixture containing small quantities of other noble gases still awaiting discovery.

32-3 Discovery of Helium

The spectroscope was first used to study the sun's corona in 1868. A bright yellow line that did not belong to the spectrum of any known element was observed, and it was concluded that the line was caused by a new element, which was named *helium* (Greek: *helios*, sun).

In 1889 it was found that certain minerals containing uranium evolve small quantities of a gas when heated. Six years later Ramsay investigated the gas to see if it might be argon. From observations of its spectrum he concluded that the gas was helium—at last helium was proved to exist not only on the sun but also on the earth. Later, Ramsay found helium to be present in small amounts in the earth's atmosphere.

The scarcity of helium in our atmosphere can be attributed to the lightness of its molecules. The mean velocity of helium molecules at 0°C is 1.3 km per sec. The velocity that a molecule, or any other moving body, needs to escape the earth's gravitational field is about 11 km per sec. It has been calculated that if the mean velocity of the molecules of a gas is as great as one third the "escape velocity," i.e., 3.7 km per sec, one half the gas will escape within a few weeks. It can be seen that helium molecules will escape rather rapidly (although not as rapidly as hydrogen molecules). If we do not conserve our terrestial supply of helium, we shall suffer an irretrievable loss.

The presence of helium in radioactive minerals was not explained until after the discovery of radium in 1898. It was then found that helium is a stable product of the disintegration of several radioactive elements, an alpha particle being the nucleus of a helium atom. It is probable that the helium in the atmosphere has been formed in this manner.

The name helium is unfortunate for a noble gas, for the ending *-ium* is characteristic of names of metallic elements. When helium was discovered on the sun, it was surmised that this new element might be a metal.

32-4 A New Periodic Group

Mendeleev provided no space in his periodic table in which either argon or helium could be placed. The discovery of these elements suggested the possible existence of an entire group of unreactive elementary gases. Ramsay therefore thought that there ought to exist a member of this new group with an atomic weight intermediate between helium and argon, and with the assistance of Morris Travers he began a search for the missing element.

At about this time in England and in Germany methods were developed for the bulk production of liquid air. The various components of the atmosphere could now be separated by fractional distillation of the liquid air. A fraction of the air that was chemically inert was found to have a spectrum unlike that of any element previously known. The new element was named *krypton* (Greek: *kryptos*, hidden). From the density of the gas it was found that the element belonged between bromine and rubidium in the periodic table. Although krypton was a new element in Group 0, it was not the one for which Ramsay had been looking! Continued searching for the lighter gas led to the discovery of another new element, characterized by the crimson lines in its spectrum. This new element was named *neon* (Greek: *neos*, new).

With a larger liquid air machine, helium was isolated and still another noble gas was discovered and named *xenon* (Greek: *xenos*, stranger).

While engaged in their research on radium, Pierre and Marie Curie observed that air which has come into contact with radium compounds is itself radioactive. The explanation of this phenomenon was provided in 1900 by Friedrich Dorn, who showed that the disintegration of radium produces a radioactive gaseous element. Ramsay isolated this element in 1908 and determined its atomic weight. It is the heaviest of the noble gases and the only one that is not a normal component of the atmosphere. Since 1923 it has been called *radon* to show its origin from radium and its membership in the group of noble gases.

32-5 Production and Uses of the Noble Gases

In 1907 Hamilton P. Cady made the important discovery that a small percentage of helium is present in natural gas from certain wells in Kansas. Most of the world's supply of helium comes from natural gases obtained in the southwestern United States. These gases sometimes contain as much as 5 percent helium, but 1 percent is more usual.

Helium is used for filling balloons and blimps. For this purpose it is much safer than hydrogen because it does not burn. The lifting power of helium is nearly 93 percent the lifting power of hydrogen, even though the density of helium is twice that of hydrogen. This is because the lifting power of a gas is determined by the *difference* between its density and the density of air.

Helium is less soluble in water than nitrogen. When it is inhaled it dissolves in the bloodstream in smaller quantities than nitrogen. For this reason a mixture of helium and oxygen is used in place of natural air for divers and others who work under high air pressure. Breathing compressed air causes considerable nitrogen to dissolve in the blood. When a diver is brought up rapidly and begins to breathe air at normal pressure again, much of the nitrogen that dissolved under high pressure comes out of solution, forming bubbles that block the circulation of the blood. This painful and dangerous condition is known as "the bends" and is avoided by the use of the helium-oxygen mixture.

Helium is used in industry to provide an inert atmosphere for melting and welding metals that are easily oxidized and in chemical processes that must be carried out in the absence of air. The chief scientific use for helium, which has the lowest boiling point of any known substance, is in cryogenics, which is the science and art of producing very low temperatures.

Since argon makes up almost 1 percent of the atmosphere, it is really an abundant element even though it is a member of the family originally referred to as the rare gases. Argon is used in light bulbs to prolong the life of the filament by retarding the vaporization of the tungsten. Its greater chemical inertness and lower thermal conductivity make argon superior to nitrogen for this purpose.

Neon is used in the familiar "neon signs." The brilliant red glow is caused by the passage of electric current through neon gas under low pressure. The color of the light given off can be changed by mixing mercury vapor and argon with the neon.

Lamps filled with krypton are used at some airports as approach lights since their light can penetrate dense fog unusually well. In 1960 the meter was redefined by international agreement to be equal to 1,650,763.73 times the wavelength of the orange-red line in the spectrum of krypton-86.

Xenon emits an instant, intense light when present in discharge tubes. This property of xenon is utilized in high-speed electronic flashbulbs used by photographers.

Neon, argon, krypton, and xenon are obtained by fractional distillation of liquid air. Helium is also obtained in small quantities from this source. Radon is collected from radium salts and is used in medicine as a source of gamma rays. The gas is sealed in small capsules which are implanted in the body to destroy malignant growths.

32-6 Remarkable Properties of Liquid Helium

Liquid helium has some remarkable properties. At 1 atmosphere pressure it is a normal liquid from its boiling point at 4.18°K to 2.18°K and is designated He-I. Below 2.18°K it is a liquid of very unique properties and is designated He-II. This strange form of liquid helium has no measurable viscosity, and it cannot be confined in an open container because it creeps over the walls and flows down the outside. It is referred to as a superfluid. Its conductivity for heat and electricity is several hundred times as great as that which metallic copper has at room temperature.

No fully satisfactory theory has been proposed for explaining the unusual properties of He-II.

At atmospheric pressure helium remains liquid at the lowest temperatures yet obtained. It is not known whether it would freeze without being under pressure if it were possible to cool it to absolute zero.

32-7 Compounds of Noble Gases

The noble gases have higher ionization energies than any of the other elements in their respective periods of the periodic table. However, if we compare the ionization energies of the *heaviest* noble gases with the ionization energies of other elements in general, we note that they are of similar magnitude to those of nitrogen, oxygen, and chlorine. It would not be unreasonable to suppose that a highly electronegative element like fluorine might react with a heavy noble gas such as xenon. Nevertheless, no compounds of the noble gases were known until 1962, when Neil Bartlett discovered that xenon reacts with PtF_6 to form a yellow solid believed to have the formula $XePtF_6$. For chemists, this was a discovery of unusual importance. It led to the discovery that xenon could combine directly with fluorine, the most electronegative of all elements. At 400°C the two elements combine to form XeF_4, which exists in the form of colorless crystals below its melting point, 90°C.

Compounds subsequently prepared include XeF_2, XeF_6, RnF_4, and KrF_4. Hydrolysis of xenon hexafluoride produces xenon trioxide:

$$XeF_6 + 3H_2O \rightarrow XeO_3 + 6HF$$

The xenon trioxide is a white crystalline solid that is highly explosive.

Since the ionization energies of the noble gases become less with increasing atomic number, we may expect xenon to be the most reactive member of the group, excepting radon, and we may expect helium to be the least reactive. Helium, neon, and probably argon will likely continue to be known as inert gases. Of the chemical elements known, these three are certainly the most inert.

New Terms

Inert gas: a gas that does not react with other substances. Prior to 1962 the elements of Group 0 were thought to be inert gases.

Noble gas: a gas that is extraordinarily stable in comparison with other gases. The elements of Group 0 are now called noble gases.

Rare gas: any of the elements that are minor components of the atmosphere. The noble gases were once called rare gases, but the name is inappropriate because argon constitutes almost 1 percent of the atmosphere.

Exercises

32-1 The alkali metals have low ionization energies and low electron affinities. The halogens have high ionization energies and high electron affinities. What may be said of the ionization energies and electron affinities of the noble gases?

32-2 Why is it to be expected that helium and neon should be the most inert of the noble gases?

32-3 Why were the noble gases originally called rare gases? Why were they later known as inert gases?

32-4 What was the origin of the octet rule? Show how it is incompatible with the existence of such compounds as SF_6 and XeF_6.

32-5 Calculate the weight at standard conditions of 1 liter of (a) helium, (b) hydrogen, (c) radon, and (d) air.

References

Asimov, I., *The Noble Gases*, Basic Books, New York, 1966. A treatment of the historical, scientific, and practical significance of the noble gases.

Chernick, C. L., "The Noble Gas Compounds," *Chemistry*, **37** (1), 6 (1964). Written by a chemist who is active in fluorine-noble gas chemistry.

Frey, J. E., "Discovery of the Noble Gases and Foundations of the Theory of Atomic Structure," *J. Chem. Educ.*, **43** (7), 371 (1966). Historical.

Huntoon, R. D., "Status of the National Standards for Physical Measurement," *Science*, **150,** 169 (1965). A discussion of procedures employed in establishing and revising standards for units of measurement, such as the international standard meter.

Hyman, H. H., "The Chemistry of the Noble Gases," *J. Chem. Educ.*, **41** (4), 174 (1964). The challenges of noble-gas chemistry to the theoretical chemist. Based on a talk presented at the 1963 summer conference of the California Association of Chemistry Teachers.

Selig, H., Malm, J. G., and Claassen, H. H., "The Chemistry of the Noble Gases," *Sci. Am.*, May 1964, p. 66. An account of developments since the first noble-gas compound was synthesized in 1962.

33

Carbohydrates

33-1 The Classification of Carbohydrates

Carbohydrates are polyhydroxy aldehydes, or polyhydroxy ketones, or substances which yield such compounds, and only such compounds, upon hydrolysis. The term *carbohydrate* originated from superficial consideration of the molecular formulas of many compounds of this class, which permit them to be represented as hydrates of carbon. Examples are glucose, which is $C_6H_{12}O_6$ or $C_6(H_2O)_6$, and sucrose, $C_{12}H_{22}O_{11}$ or $C_{12}(H_2O)_{11}$. Structurally, the substances of this class are in no sense hydrates of carbon. Furthermore, carbohydrates are now known in which the ratio of hydrogen to oxygen is not 2:1.

Carbohydrates occur in all plants and animals. They constitute one of the important classes of foods for animals. In plants they are produced from carbon dioxide and water by photosynthesis. In animal bodies the oxidation of carbohydrates, forming carbon dioxide and water, serves as a source of energy in the amount of about 4 kcal per g. The metabolism of sugars and starches in an organism occurs in successive stages of oxidation under the control of enzymes.

A carbohydrate that cannot be hydrolyzed to give simpler carbohydrates is known as a *monosaccharide*. A carbohydrate is classified as a *disaccharide* if the hydrolysis of one of its molecules results in the formation of two monosaccharide molecules. From the equations for the hydrolysis of sucrose and maltose,

$$\underset{\text{sucrose}}{C_{12}H_{22}O_{11}} + H_2O \rightarrow \underset{\text{glucose}}{C_6H_{12}O_6} + \underset{\text{fructose}}{C_6H_{12}O_6}$$

$$\underset{\text{maltose}}{C_{12}H_{22}O_{11}} + H_2O \rightarrow \underset{\text{glucose}}{C_6H_{12}O_6} + \underset{\text{glucose}}{C_6H_{12}O_6}$$

we can see that sucrose and maltose are disaccharides. Since they share the same molecular formula, sucrose and maltose are isomers. Similarly, glucose and fructose are isomers. Since neither glucose nor fructose can be hydrolyzed to give simpler carbohydrates, they are monosaccharides.

Oligosaccharides are carbohydrates that yield only a few monosaccharide molecules upon hydrolysis. Practically all the oligosaccharides that occur in nature are disaccharides, trisaccharides, or tetrasaccharides. *Polysaccharides* are carbohydrates of high molecular weight that can be hydrolyzed to give, ultimately, a large number of monosaccharide molecules.

Sugars are colorless crystalline carbohydrates, more or less sweet, and soluble in water. They char when heated:

$$C_{12}H_{22}O_{11} \xrightarrow{\text{heat}} 12C + 11H_2O$$

The sugars are the simpler carbohydrates, usually monosaccharides and disaccharides.

Individual carbohydrates frequently have names ending in *-ose*. They are structurally characterized by the presence of an "ose group,"

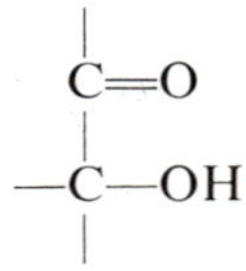

or of a structure that gives rise to an ose group upon hydrolysis. If a hydrogen atom is attached to the carbonyl carbon of a monosaccharide, the compound is a hydroxy aldehyde and is known as an *aldose*. A monosaccharide in which the carbonyl carbon is bonded to two other carbon atoms is a hydroxy ketone and is known as a *ketose*.

On the basis of the number of carbon atoms it contains, a monosaccharide is classified as a triose, tetrose, pentose, etc. For example, a ketopentose is a five-carbon monosaccharide that contains a keto group. An aldohexose is a six-carbon monosaccharide that contains an aldehyde group.

Free and combined sugars occur throughout the living world. For example, sucrose is the product of the photosynthetic action of plants. The sucrose is transported throughout the plants and is stored in some of them. Better known is the storage of *starch* in plants and of *glycogen* in animals. Starch and glycogen are polysaccharides. The most abundant of all polysaccharides is *cellulose*, which contributes to the structural strength of plants.

To attain any significant comprehension of the nature of carbohydrates it is necessary to consider the spatial or steric arrangement of the atoms within the molecules. This requires an understanding of the nature of optical isomerism (Sections 19-13 and 19-14).

33-2 Trioses: The Simplest Carbohydrates

The simplest of all compounds that fit the definition we have given for carbohydrates (Section 33-1) are dihydroxyacetone, $HOCH_2 \cdot CO \cdot CH_2OH$, and glyceraldehyde, $HOCH_2 \cdot CHOH \cdot CHO$. Both of these could be theoretically obtained by the controlled oxidation of glycerol. Commercially, dihydroxyacetone is indeed obtained by the oxidative fermentation of glycerol, using the organism *Acetobacter suboxydans:*

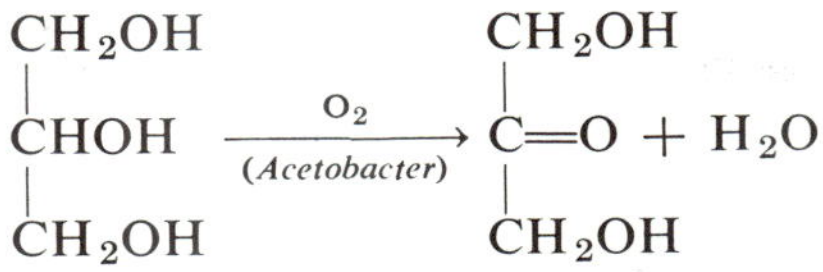

This interesting compound reacts with proteins to form a brown product and is used in lotions for producing synthetic suntan.

Glyceraldehyde is prepared from glycerol in the chemical laboratory by a series of reactions whose net effect is the selective oxidation of one of the primary alcohol groups:

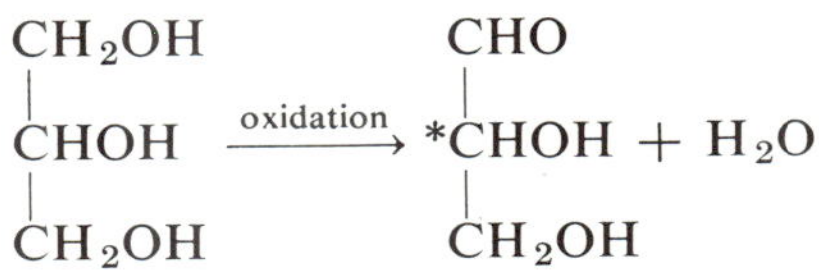

Attention is called to the presence of an asymmetric carbon atom in glyceraldehyde by marking it with an asterisk. As prepared by chemical synthesis the glyceraldehyde is optically inactive because equal numbers of molecules of the two mirror-image forms are obtained. Upon resolution, dextrorotatory glyceraldehyde is found to have a specific rotation of $+13.5°$. The specific rotation of levorotatory glyceraldehyde is $-13.5°$.

Dextrorotatory compounds can be indicated by writing (+) as a part of the name and levorotatory compounds, by writing (−). Thus, dextrorotatory glyceraldehyde can be designated (+)glyceraldehyde, and the levorotatory isomer, (−)glyceraldehyde.

33-3 D and L Configurations

The two mirror-image forms of glyceraldehyde can be represented by *perspective formulas* designed to give a three-dimensional perspective of the arrangement of the four different groups attached to the asymmetric carbon atom. In one style of perspective formula (Figure 33-1) the central carbon atom and the

Figure 33-1

Perspective formulas for the isomers of glyceraldehyde

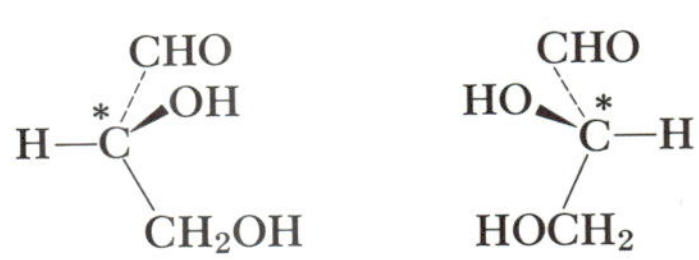

I II

bonds represented by solid lines are regarded as being in the plane of the paper, bonds represented by dotted lines lie behind the plane of the paper, and bonds represented by wedges are extending outward from the plane of the paper toward the reader. The difference between the two configurations is merely that II is the mirror image of I. If you hold I in front of a mirror, the image you see in the mirror is identical in configuration with the formula you see as II in Figure 33-1.

The representation of steric configurations is simplified by projecting the space formulas into the plane of the paper:

D-glyceraldehyde L-glyceraldehyde

In reference to these projection formulas we say that the —OH group is "on the right" in D-glyceraldehyde and "on the left" in L-glyceraldehyde. In Figure 33-1 the perspective formula for D-glyceraldehyde is I and that for L-glyceraldehyde is II.

In 1951, it was established by X-ray analysis of related compounds that the configuration designated by D is the configuration of (+)glyceraldehyde. The configuration designated by L is the configuration of (−)glyceraldehyde. The complete names for these isomeric aldotrioses are, therefore, D(+)glyceraldehyde (pronounced "dee plus glyceraldehyde") and L(−)glyceraldehyde (pronounced "ell minus glyceraldehyde"). These two compounds are called *enantiomers* (a contraction of "enantiomorphic isomers"), as are any two compounds whose only difference is direction of rotation. Enantiomers are nonsuperposable mirror images of each other.

Other compounds can be related configurationally to one or the other of the glyceraldehydes by reactions that do not involve breaking any of the bonds to an asymmetric carbon atom. Compounds that are related configurationally to D-glyceraldehyde are given the designation D, whether they are dextrorotatory or levorotatory. Compounds related to L-glyceraldehyde configurationally are given the designation L regardless of direction of rotation.

The oxidation of D(+)glyceraldehyde gives D(−)glyceric acid:

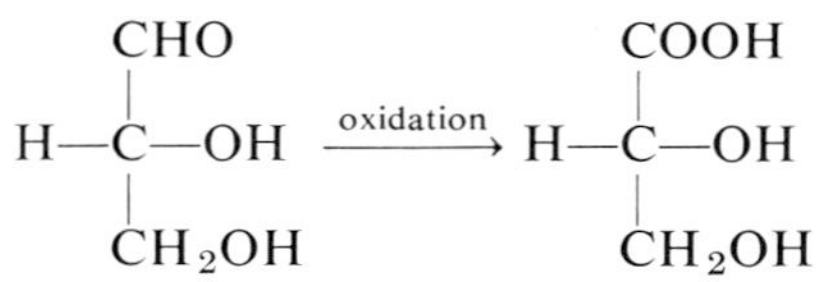

D(+)glyceraldehyde D(−)glyceric acid

Although D-glyceraldehyde is dextrorotatory, D-glyceric acid is levorotatory. However, in the conversion of glyceric acid to lactic acid the direction of rotation is not changed.

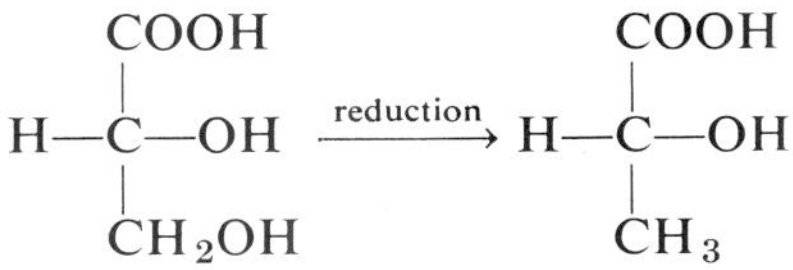

D(−)glyceric acid D(−)lactic acid

The specific rotation of D-lactic acid is different from that of D-glyceric acid. Most reactions of optically active compounds are accompanied by a change in specific rotation. Even the direction of rotation changes in some reactions, as we have just noted in the oxidation of glyceraldehyde to glyceric acid. In general we cannot predict whether a compound with a D configuration will be found to be dextrorotatory or levorotatory. To do so we would need to know the effect exerted by each of the different groups attached to the asymmetric carbon atoms.

33-4 The Aldotetroses

The aldotetroses share the formula $CHO \cdot CHOH \cdot CHOH \cdot CH_2OH$. Each has two asymmetric carbon atoms. Consequently, there are four optical isomers (Figure 33-2).

In an aldose the configuration of the asymmetric carbon atom farthest from the aldehyde group determines whether the molecule belongs to a D or an L family. If the asymmetric carbon most remote from the aldehyde group has the same configuration as D(+)glyceraldehyde (—OH group on the right in the projection formula), the compound is defined as a member of the D family. If this carbon has the same configuration as L(−)glyceraldehyde, the compound belongs, by definition, to the L family.

The two enantiomers of erythrose (Figure 33-2) have physical properties that are identical except for optical activity. Threose differs from erythrose not only in optical activity but also in other physical properties. Although threose and erythrose are optical isomers, the one is *not* the mirror image of the other. Optical isomers that are not mirror images are called *diastereoisomers*. Dia-

Figure 33-2

Configurations of the aldotetroses

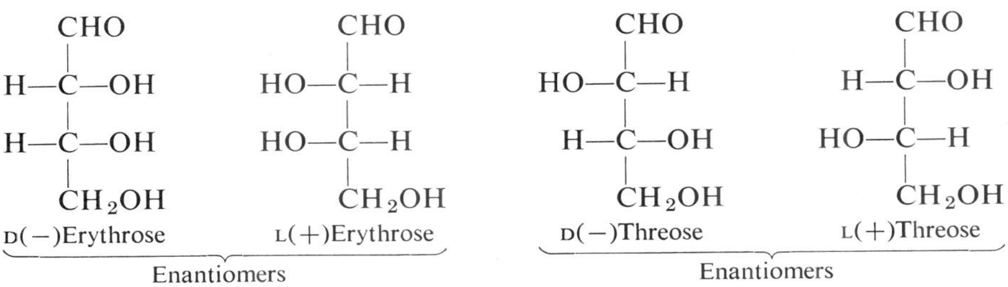

stereoisomers have different melting points, boiling points, densities, etc., in addition to having different specific rotations.

Sugars have names ending in *-ose*. The preceding portion of the name frequently refers to the source of the sugar. Erythrose received its name from the fact that it is formed by oxidation of a polyhydric alcohol named erythritol. Erythrose is related to erythritol in the same way that glyceraldehyde is related to glycerol.

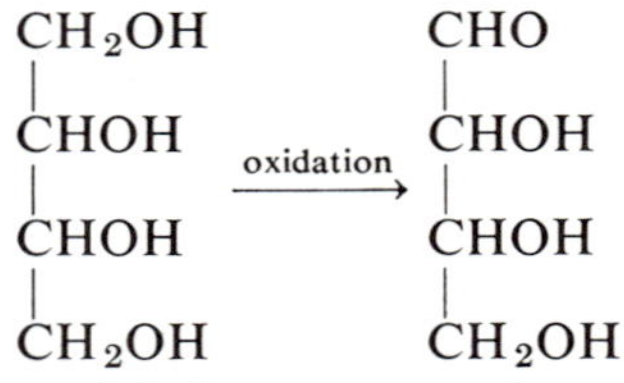

erythritol erythrose

Threose is an artificial sugar that has not been found in natural products. Its name was formed by selecting and rearranging some of the letters of "erythrose."

33-5 The Aldopentoses

Just as the aldotetroses may be considered as being built up from glyceraldehyde, as they actually have been experimentally, so can the aldopentoses be built up from the aldotetroses (Figure 33-3). Ribose, arabinose, xylose, and lyxose are diastereoisomers.

Figure 33-3

Configurations of the D family of aldotetroses and aldopentoses

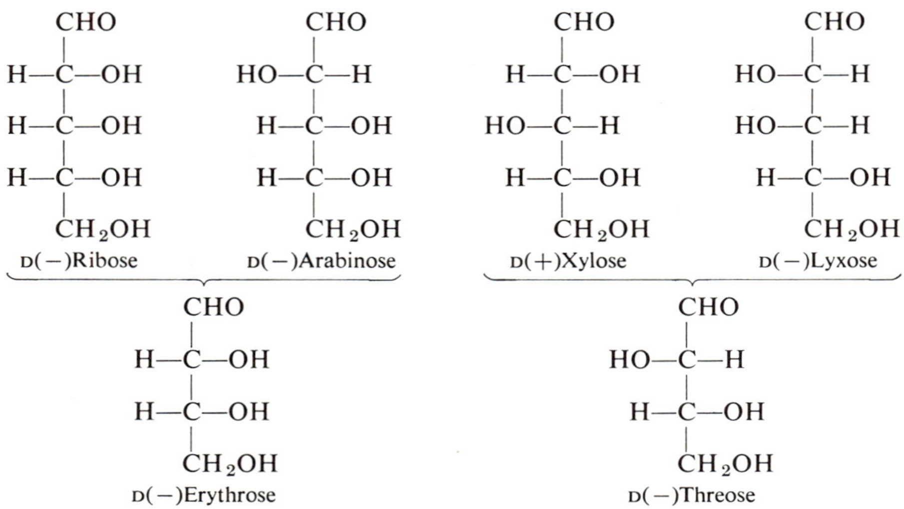

Xylose (Greek: *xylon*, wood) is also known as wood sugar. It is obtained by boiling wood, bran, corncobs, etc. with dilute acids. Only D(+)xylose occurs in nature. It is not utilizable by many animals. Arabinose is obtained, with xylose, by boiling gum arabic with dilute acids. Both enantiomorphs of arabinose occur in nature. Lyxose is a synthetic sugar. In nature (−)ribose is of great importance because of its presence in combination in ribonucleic acids (the RNA's), which are present in all parts of living cells and are involved in the biosynthesis of proteins.

33-6 The Aldohexoses

Eight D-aldohexoses can be built up from the four D-aldopentoses, and eight L-aldohexoses from the four L-aldopentoses. If you compare the configurations of the D-aldohexoses (Figure 33-4) with those of the D-aldopentoses (Figure 33-3) you will note that allose and altrose are built up from ribose, glucose and mannose from arabinose, gulose and idose from xylose, and galactose and talose from lyxose. Diastereoisomers that differ only in the configuration of the

Figure 33-4

Configurations of the D family of aldohexoses

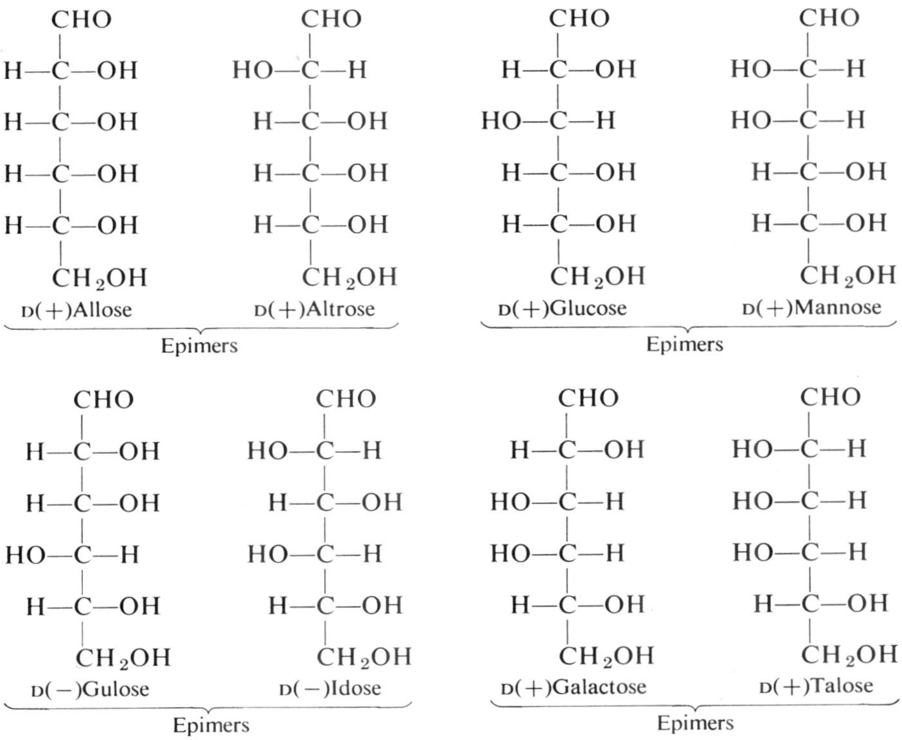

D(+)Allose, D(+)Altrose — Epimers; D(+)Glucose, D(+)Mannose — Epimers; D(−)Gulose, D(−)Idose — Epimers; D(+)Galactose, D(+)Talose — Epimers

carbon atom next to the aldehyde group are called *epimers*. Thus, allose and altrose are epimers; glucose and mannose are epimers; etc.

The only common aldohexoses are D(+)glucose, D(+)mannose, and D(+)galactose. The most important is D(+)glucose, known commercially as "dextrose" (a contraction of "dextro-glucose"). The word *glucose* without any indication of configuration or sign of rotation always means D(+)glucose because L(−)glucose does not occur naturally. Such is the biochemical importance of molecular configuration that L-glucose has no food value.

Glucose is also known as *blood sugar*. It is present to the extent of about 1 mg per ml in the blood of normal mammals. Persons afflicted with diabetes have as much as 3 mg per ml in their blood. Glucose is also called *grape sugar*. It is found in grapes and other fruits, in plant juices, and in honey. It has been called *corn sugar* because it can be obtained by the hydrolysis of corn starch. Large quantities are produced for the candy industry by this method. It is also used for intravenous feeding of medical patients who are unable to obtain their nourishment orally. Glucose is a major component of many oligosaccharides (including sucrose and lactose) and polysaccharides (e.g., cellulose, starch, and glycogen).

In American industry the term "glucose" is used as a synonym for *corn syrup*, a product obtained by the incomplete hydrolysis of cornstarch. It contains dextrins, maltose, and D(+)glucose.

Mannose received its name from the fact that it was obtained by the oxidation of mannitol, a polyhydroxy alcohol that is the chief constituent of manna, the sweetish exudate of certain trees. Mannose is most readily obtained by the hydrolysis of vegetable ivory. Shavings obtained as by-products of the manufacture of buttons are used for this purpose. (Vegetable ivory is the hard endosperm of the ivory nut, seed of the ivory palm of South America. Chemically, it is in no way related to the ivory composing elephants' tusks, which is a fine-grained dentine.)

D(+)Galactose is produced, with glucose, by the hydrolysis of lactose (milk sugar). Large amounts of D(+)galactose in the diet of chickens have been found to result in violent spasms and eventual death. Similar effects are suffered by humans when they are unable to assimilate the galactose portion of lactose in milk. Infants with congenital galactosemia lack one of the enzymes required for the metabolism of galactose.

Talose, gulose, idose, allose, and altrose are synthetic sugars.

33-7 Other Aldoses

The length of the carbon chain in monosaccharides has no theoretical limit. Aldoheptoses, octoses, nonoses, and decoses have been synthesized in the laboratory, and aldoheptoses have been obtained by the hydrolysis of certain polysaccharides of bacterial origin.

Some sugars occurring in nature have a hydrogen atom in place of one or more of the —OH groups. Such sugars are called *deoxy sugars*. Examples

Figure 33-5

Examples of deoxy sugars

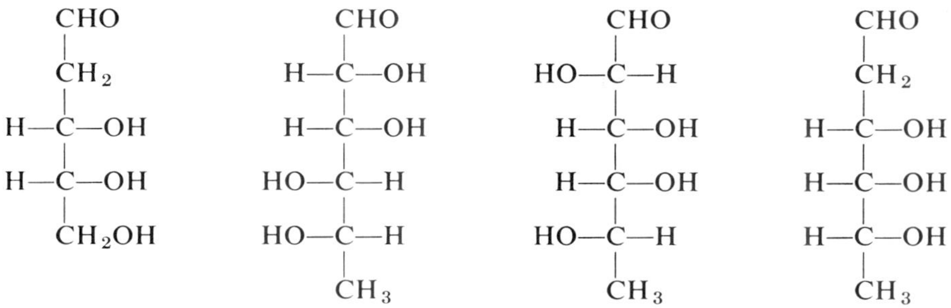

2-Deoxy-D-ribose   L(+)Rhamnose   L(−)Fucose   Digitoxose

(Figure 33-5) include deoxyribose, rhamnose, fucose, and digitoxose. Of great biochemical importance is the 2-deoxy-D-ribose, which is present in combination in deoxy-ribonucleic acids (the DNA's), which play an important role in heredity. The most common deoxy sugar occurring naturally is L(+)rhamnose, which is 6-deoxy-L-mannose. Rhamnose is obtained by hydrolysis of compounds present in buckthorn berries. It has been detected in various polysaccharides of bacterial origin. Hydrolysis of the materials comprising the cell walls of marine algae yields fucose as one of the products. Fucose is 6-deoxy-L-galactose. Digitoxose, which is 2,6-dideoxy-D-allose, occurs in combination in digitalis, which is used as a heart stimulant. None of the deoxy sugars have formulas of the type $C_x(H_2O)_y$ ordinarily associated with carbohydrates.

Amino sugars are sugars that contain one or two $—NH_2$ groups instead of —OH groups. Two amino sugars that are widely distributed as constituents of polysaccharides and other substances of biochemical origin are D-glucosamine and D-galactosamine (Figure 33-6).

Figure 33-6

Examples of amino sugars

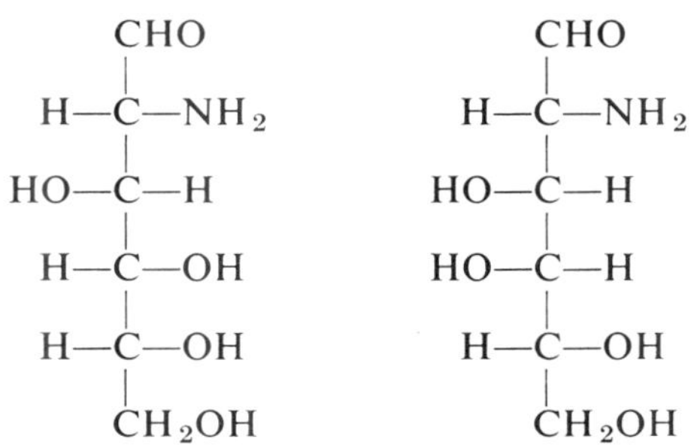

D-Glucosamine   D-Galactosamine

Figure 33-7

Examples of branched chain sugars

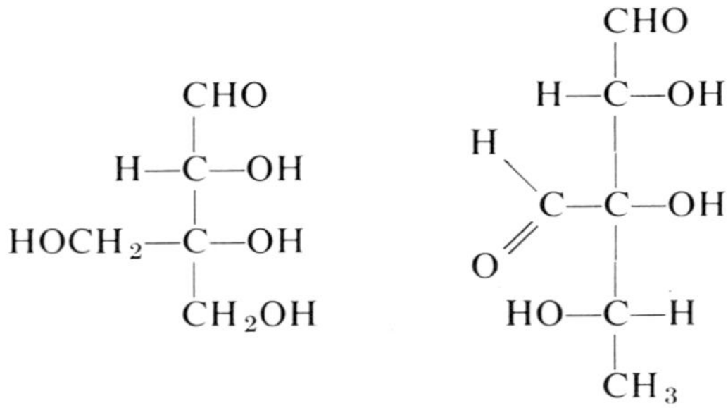

Apiose Streptose

Figure 33-8

Configurations of the D family of ketoses

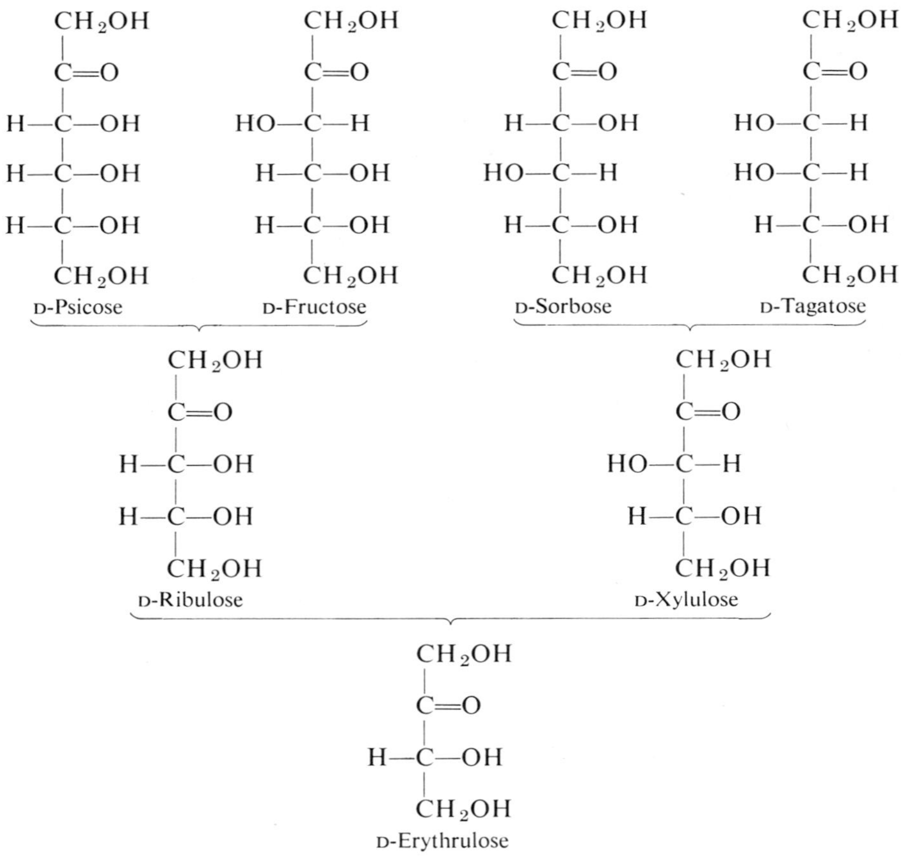

Sugars with branched carbon chains also occur in nature, but they are not common. Examples (Figure 33-7) are apiose, which occurs in combination in parsley and celery, and streptose, which constitutes a portion of the molecule of the antibiotic streptomycin.

33-8 Ketoses

Ketoses are isomeric with the aldoses that have the same number of carbon atoms. Structurally, a ketose differs from an aldose in that the carbonyl group never contains a terminal carbon atom. In the common ketoses the carbonyl group contains the second carbon atom so that the general formula for these sugars is $HOCH_2 \cdot CO \cdot (CHOH)_n \cdot CH_2OH$.

The simplest ketose is dihydroxyacetone (Section 33-2), which contains no asymmetric carbon atom. There is only one ketotetrose, which is called erythrulose. In the systematic naming of sugars the ending *-ulose* is used for ketoses. The ketopentoses may be considered as being built up from erythrulose, and the four ketohexoses from the two ketopentoses (Figure 33-8). The ketopentoses are commonly referred to by their systematic names, ribulose and xylulose, but systematic names are not commonly used for the ketohexoses.

The most abundant ketose is D(−)fructose, which occurs with glucose and sucrose in fruit juices and honey. The older name "levulose" is due to the fact that this sugar always occurs in nature as the levorotatory isomer. Fructose is the sweetest of all sugars. (Several synthetic compounds are many times sweeter than fructose, but they are not sugars and have no food value.)

The only other readily available ketose is L(−)sorbose, which is obtained by bacterial oxidation of sorbitol, a polyhydric alcohol occurring in the sorb apple (serviceberry). The same organism, *Acetobacter suboxydans*, that is used for the production of dihydroxyacetone from glycerol is used for the production of sorbose from sorbitol. Sorbose is important for the commercial synthesis of vitamin C. The sorbitol for the production of the sorbose is manufactured by the catalytic hydrogenation of glucose:

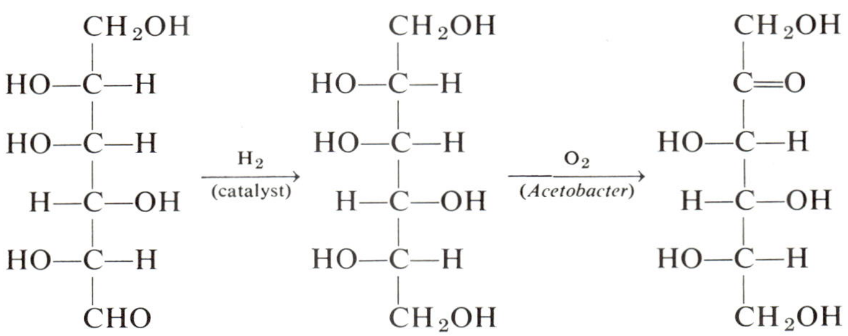

D(+)glucose sorbitol L(−)sorbose

D-Tagatose has been obtained by hydrolysis of a gum exudate of certain tropical trees. D-Ribulose has been found to play a part in the photosynthesis of carbohydrates.

The only monosaccharide known to be of biological significance that has more than six carbon atoms is the seven-carbon ketose, D-sedoheptulose, which is an intermediate in photosynthesis:

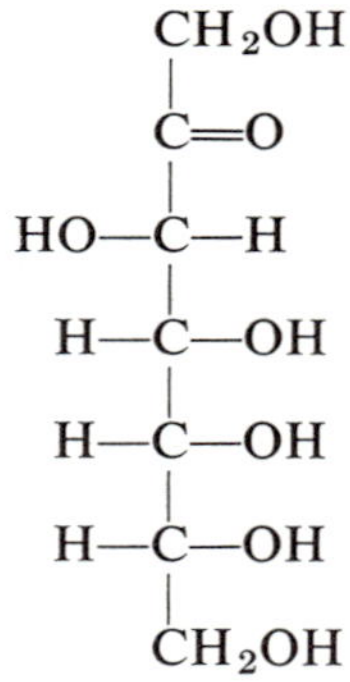

33-9 Ring Structures in Carbohydrates

An aldehyde in alcoholic solution is in equilibrium with a *hemiacetal*:

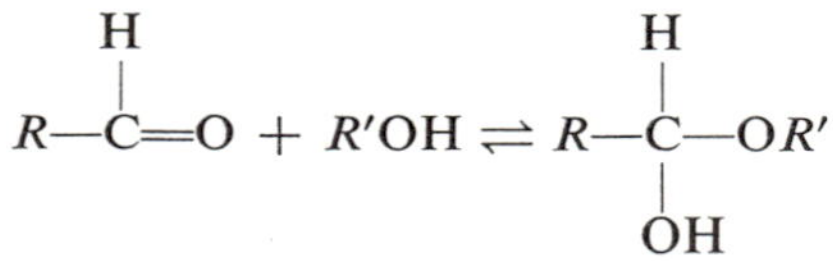

aldehyde alcohol hemiacetal

A hemiacetal is both an ether and an alcohol. In most cases hemiacetals are so unstable that they cannot be isolated.

In the presence of an acid a hemiacetal reacts with more of the solvent alcohol to form an *acetal:*

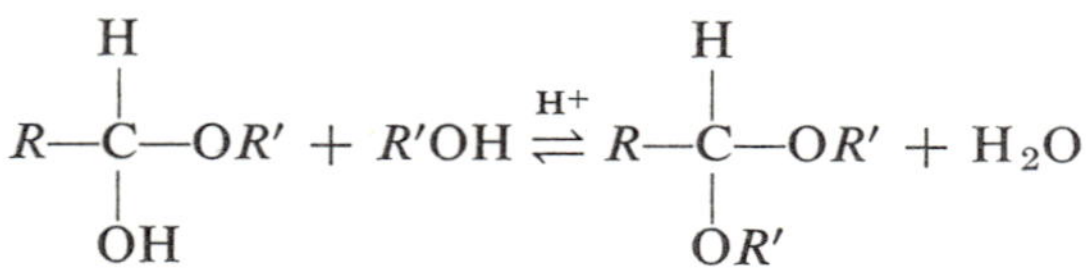

hemiacetal alcohol acetal

Acetals have the structure of ethers, but they differ from ethers in the ease and rapidity with which they react with water, in the presence of acid, to form aldehyde and alcohol:

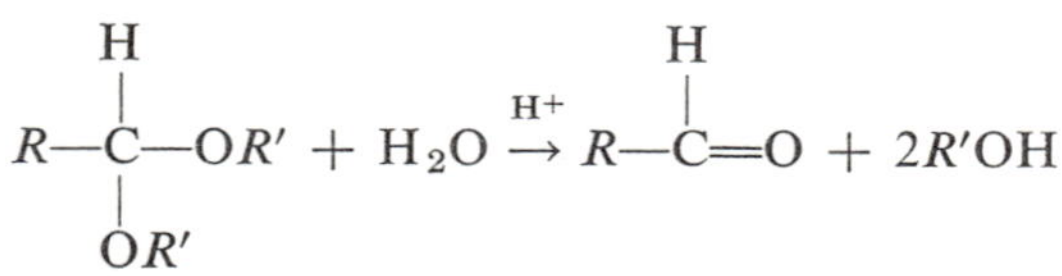

acetal aldehyde alcohol

Figure 33-9

The cyclic structure of D(+)glucose

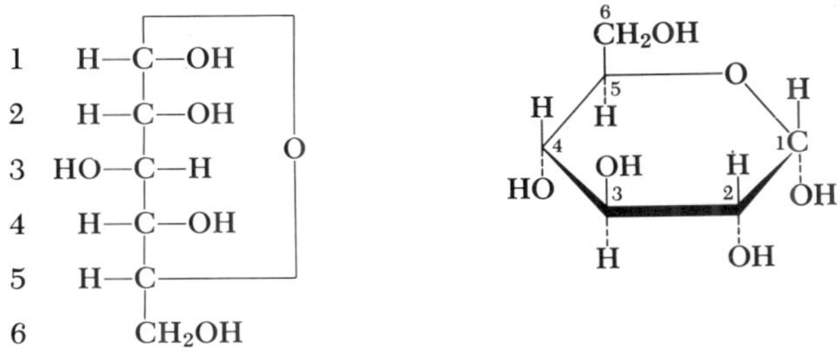

(a) (b)

α-D(+)glucose, m.p. 146°C, [α] = +112°

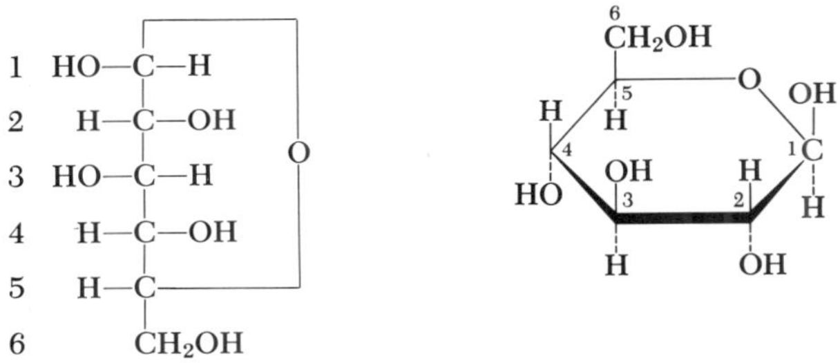

(c) (d)

β-D(+)glucose, m.p. 150°C, [α] = +19°

Carbohydrates have alcoholic hydroxyl groups and a carbonyl group in the same molecule. Consequently, there can be intramolecular interaction between the carbonyl group and one of the hydroxyl groups. Of necessity, the reaction tesults in a ring; i.e., the product is a cyclic hemiacetal. Thus, D(+)glucose forms a six-membered ring by interaction of its aldehyde group with the C-5 hydroxyl group. The resulting cyclic structure is represented crudely in Figure 33-9 a and c to show the relationship of the cyclic structure to the open-chain aldehyde structure. The cyclic structure is represented more accurately by the *perspective formulas* in Figure 33-9 b and d. Here the ring is shown as a hexagon slightly tilted out of the plane of the paper, with thick lines representing the bonds nearest the viewer and thin lines representing those that are farthest away (see Section 25-1). Ring carbon atoms are omitted with the exception of C-1. Solid lines indicate positions above the ring, and broken lines indicate positions below the ring. The groups that appear on the right in the open-chain formula are written below the ring in the ring formula.

When the open-chain aldehyde is converted to the cyclic hemiacetal structure, C-1 becomes asymmetric. For this reason there are two isomeric hemiacetal structures, known as *anomers*. The designation α is given to the anomer in which the —OH group of C-1, called the *anomeric carbon*, is on the right in the projection formula, and below the ring in the perspective formula, for members of the D family. Similarly, β-D-glucose has the —OH group of C-1 on the left in the projection formula and above the ring in the perspective formula. (For members of the L family, α designates the anomer that has the —OH group of the anomeric carbon *above* the ring.)

It should be noted that α-D-glucose and β-D-glucose are diastereoisomers, not enantiomers. Both are dextrorotatory, but to different degrees. Techniques have been developed for obtaining the two anomers separately in crystalline form. The α-D(+)glucose precipitates from aqueous solution upon addition of ethanol below 30°C. The β-D(+)glucose crystallizes from aqueous solutions evaporated at temperatures above 98°C. The α crystals melt at 146°C; the β crystals melt at 150°C.

Freshly prepared aqueous solutions of α-D-glucose have a specific rotation of +112°, but the specific rotation gradually drops to +52.7°. Freshly prepared aqueous solutions of β-D-glucose have a specific rotation of +19°, but the value gradually rises to +52.7°. The final value corresponds to an equilibrium mixture of the α and β forms. Since +52.7° is not the average of +19 and +112°, we know that the equilibrium mixture does *not* contain equal amounts of the two anomers. There is, of course, no reason for assuming that it should. (Can you calculate the relative amounts of the α and β forms in the equilibrium mixture?) The change in rotation of each solution to the equilibrium mixture is called *mutarotation*. It is caused by the ready opening and closing of the hemiacetal ring in both the α and β forms. The opening of the ring of either form gives the open-chain aldehyde structure, which by intramolecular interaction can give either the original form or the opposite form. Thus, in aqueous solution either form is converted, through the open-chain form, into an equilibrium mixture containing both cyclic forms. All evidence, chemical and spectroscopic, indicates that the amount of open-chain aldehyde in solution, if present at all, is very small.

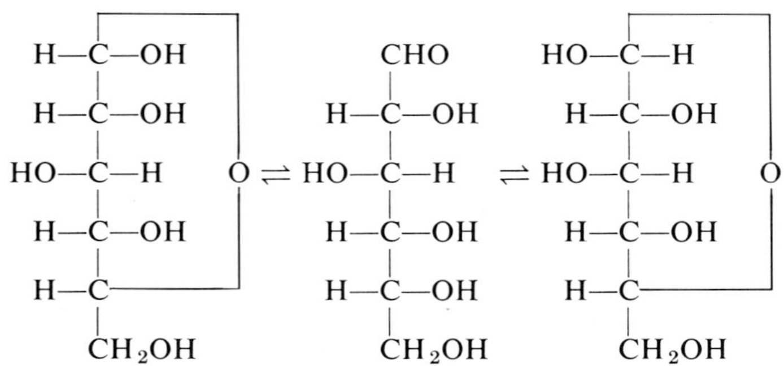

α-D(+)glucose β-D(+)glucose

When we attempt to prepare an acetal from glucose and methanol, in the presence of acid, we obtain two isomeric cyrstalline products which behave

Figure 33-10

Formation of methyl glucosides

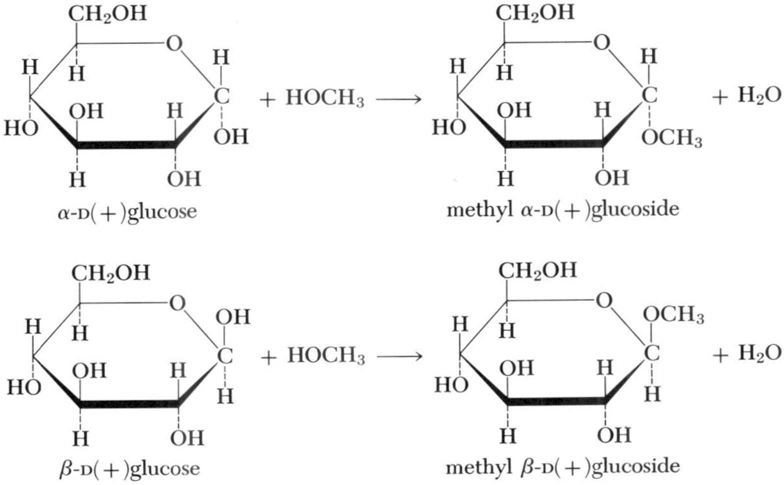

like acetals, but analysis shows that only one methyl group has been introduced (even though a molecule of water has been eliminated). This behavior can be explained in terms of the cyclic structure of glucose. When it forms an acetal it does so with only one molecule of alcohol because it is itself a hemiacetal.

The acetals formed by sugars are called *glycosides*. The glycosides formed by glucose are called glucosides; those formed by fructose are called fructosides; etc. Most natural glycosides are glucosides.

The reaction of methyl alcohol with α-D-glucose gives methyl α-D-glucoside, and its reaction with β-D-glucose gives methyl β-D-glucoside (Figure 33-10).

The six-membered ring that we have shown for glucose is known as a pyran ring because *pyran* is the name of a hypothetical cyclic compound whose ring consists of five carbon atoms and one oxygen atom. The pyran ring is also present in many other carbohydrates. Any carbohydrate containing a six-membered ring therefore is called a *pyranose* and its glycosides are called *pyranosides*. Thus, the size of the ring in β-D-glucose and in methyl β-D-glucoside can be specified by naming these compounds β-D-glucopyranose and methyl β-D-glucopyranoside, respectively.

Five-membered ring structures are also common in sugars. A carbohydrate containing a five-membered ring is called a *furanose* because *furan* is the name of a cyclic compound whose ring consists of four carbon atoms and one oxygen atom. The glycosides of a furanose are called *furanosides*.

The rings present in carbohydrates are not planar. If they were, the bond angles would be considerably larger than the natural bond angle of 109.5°. The rings are buckled so as to permit this bond angle to be maintained.

33-10 Oligosaccharides

Carbohydrates that yield only a few monosaccharide molecules on hydrolysis are called *oligosaccharides* (Greek: *oligos*, few, small). Examples are sucrose, lactose, and maltose, which are disaccharides, $C_{12}H_{22}O_{11}$, and raffinose, which is a trisaccharide, $C_{18}H_{32}O_{16}$.

Sucrose (French: *sucre*, sugar) is our common table sugar. The most important sources of sucrose are sugar cane and the sugar beet. It also occurs in the sap of the sugar maple tree. Indeed, sucrose is present in all photosynthetic plants. It was isolated from cane juice in essentially pure crystalline form as early as 300 A.D. in India. Sucrose was discovered in the juice of the sugar beet in 1747.

Hydrolysis of sucrose occurs in the presence of acids or of the enzyme *sucrase* (also called *invertase*). In the hydrolysis reaction each sucrose molecule yields one molecule of glucose and one molecule of fructose. The specific rotation of sucrose is +66.5°. The specific rotation of glucose is +52.7° and that of fructose is −92.4°. The equimolar mixture of fructose and glucose therefore is levorotatory. The hydrolysis of sucrose is commonly known as the *inversion* of sucrose because of the change in sign of rotation that occurs during the reaction. The glucose-fructose mixture is called *invert sugar*. Invert sugar has about the same sweetening power as sucrose but has much less tendency to crystallize. For this reason it is used in the manufacture of candies and syrups. Honey is chiefly invert sugar.

Unlike glucose and many other sugars, sucrose does not show the reducing action that would indicate the presence of an aldehyde group. Neither does it form methyl glycosides. These and other facts indicate that the fructose and glucose portions are linked through their anomeric carbon atoms. Since sucrose

Figure 33-11

The inversion of sucrose

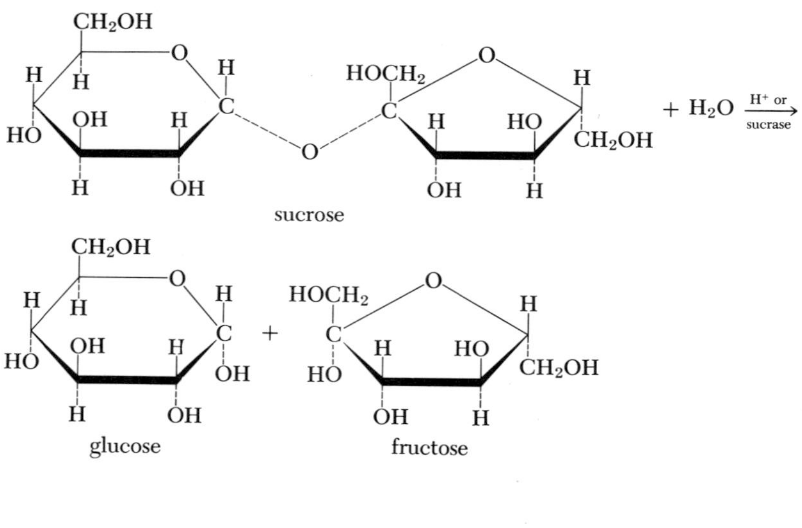

Figure 33-12

The structure of lactose

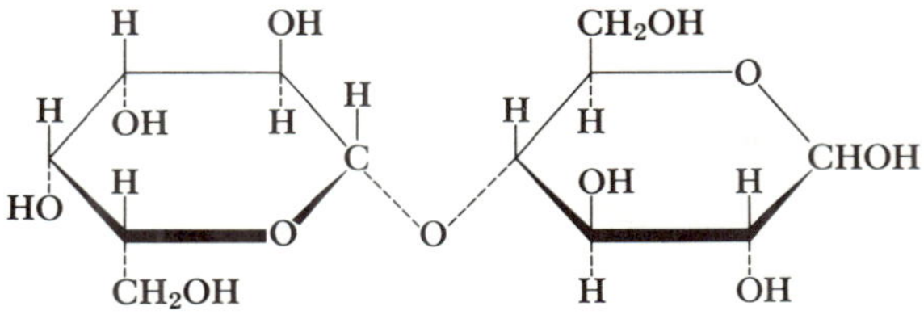

is hydrolyzed by sucrase, an enzyme that hydrolyzes β- but not α-fructofuranosides, the anomeric carbon of the fructose portion is assigned the β configuration. Since sucrose is also hydrolyzed by α-glucosidase, an enzyme that attacks any sugar containing an α-linked glucose, but is not hydrolyzed by β-glucosidase, an enzyme that attacks only sugars containing β-linked glucose, the anomeric carbon of the glucose portion of the sucrose molecule is assigned the α configuration. This structure has been confirmed by X-ray analysis. Thus, sucrose is both a β-D-fructoside and an α-D-glucoside. (Its complete chemical name can be given as α-D-glucopyranosyl β-D-fructofuranoside or as β-D-fructofuranosyl α-D-glucopyranoside.) The structure of sucrose is shown in Figure 33-11, where its inversion is represented.

Lactose (Latin: *lac*, milk) is commonly known as milk sugar. It makes up about 5 percent of the milk of mammals. The souring of milk consists of the conversion of lactose into lactic acid. In addition to lactose, the milk of nonruminants (including human milk) contains tri-, tetra-, penta-, and hexasaccharides that are not present in the milk of cows, goats, or sheep. Commercially, lactose is a by-product of cheese-making; it is recovered from whey, which is the aqueous solution remaining after the milk proteins have been coagulated. It is used for preparing "modified milk" for infants.

Lactose is a reducing sugar. On hydrolysis (by acid or by enzyme), each lactose molecule yields one molecule of glucose and one molecule of galactose. It has been established that the glucose unit contains the reducing portion of the molecule. Thus, it is a galactoside, not a glucoside. Its structure is indicated in Figure 33-12. It exists in α and β forms.

Maltose, also known as malt sugar, is formed by the hydrolysis of starch through the action of the enzyme diastase during the germination of cereal grains, as in the preparation of malt. Commercially it is obtained by the partial hydrolysis of cornstarch by acid:

$$\underset{\text{starch}}{(C_6H_{10}O_5)_n} + \frac{n}{2}H_2O \xrightarrow[\text{or } H^+]{\text{diastase}} \underset{\text{maltose}}{\frac{n}{2}C_{12}H_{22}O_{11}}$$

Maltose is present in food for infants and in malted milk. It is a reducing sugar. On hydrolysis (by acid or by the enzyme maltase), each maltose molecule

Figure 33-13

The structure of maltose

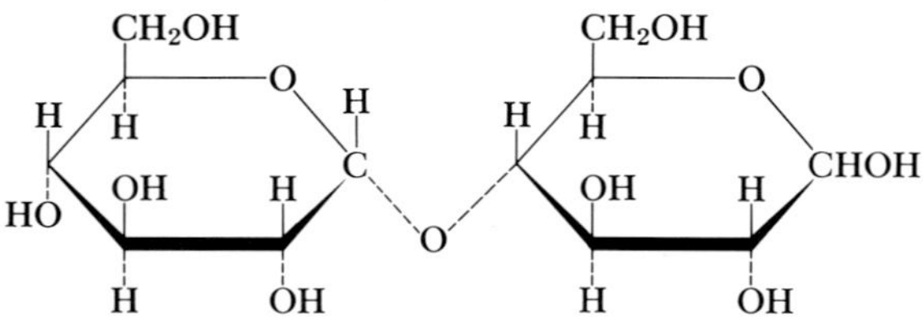

yields two molecules of glucose. Its structure is indicated in Figure 33-13. It exists in α and β forms.

Raffinose is present in many plants and is obtained as a by-product of the refining of beet sugar. Its complete hydrolysis yields equimolar quantities of glucose, fructose, and galactose.

33-11 Polysaccharides

Polysaccharides consist of hundreds of monosaccharide units and have average molecular weights varying from tens of thousands to hundreds of millions. Polysaccharides are not molecularly homogeneous, i.e., the polymeric molecules of a given substance do not all have the same molecular weight.

Another term with the same meaning as *polysaccharide* is *glycan*. A polysaccharide built up from monosaccharide units that are all of the same kind are sometimes named by taking the root of the sugar name and adding *-an*. Thus, xylans are polymers of xylose, and a mannan is a polymer of mannose. Some polysaccharides were given names ending in *-in* early in the history of chemistry and are still referred to by these names because of uninterrupted usage. Examples are pectin and dextrin.

The most important polysaccharides are cellulose and starch. Both are produced in plants from CO_2 and H_2O by photosynthesis and both are made up of D(+)glucose units.

Starch is the form in which reserve food is stored in the roots, tubers, and seeds of plants. Commercial sources of starch include corn, potatoes, wheat, rice, tapioca, and sago. Most starch produced in the United States is obtained from maize (corn).

Starch is essentially a polymer made up of α-glucose units, but some of the —OH groups are esterified with phosphoric acid. Furthermore, starch consists of two distinct glucosans (glucose polymers) of very different molecular weights. One of these, called *amylose*, is composed of several hundred glucose units in a continuous chain. The other, called *amylopectin*, has molecular weights measured in the millions and consists of highly branched molecules. The two polysaccharides differ substantially in their properties. Amylose is water-soluble; amylopectin is insoluble in cold water but forms a paste with hot water. It is the amylose that reacts with iodine to give the blue color reaction typical of starch.

Most starches contain from three to four times more amylopectin than amylose. Genetic experiments have produced mutant varieties of grains ("waxy corn" etc.) that contain only amylopectin and a mutant strain of pea that contains mostly amylose.

Stepwise hydrolysis of starch by acid gives soluble, gummy carbohydrates known as dextrins, then maltose, and finally glucose. The dextrins have adhesive properties that make them useful on the backs of stamps and labels.

The chief use of starch is as food. It is hydrolyzed stepwise in the mouth by the action of the enzyme ptyalin. There are no starch-digesting enzymes in the stomach, and ptyalin is inactivated by the hydrochloric acid of the stomach, which is itself too dilute to catalyze the hydrolysis of starch. However, the digestive action of ptyalin persists in the stomach for a while because the swallowed food masses are penetrated only slowly by the gastric juice. Starch digestion is resumed in the intestines, where the hydrochloric acid of the gastric juice is neutralized and the enzyme called amylopsin, contained in the pancreatic juice, completes the hydrolysis of starch to maltose. The intestinal juice contains three enzymes, maltase, lactase, and sucrase, which hydrolyze maltose, lactose, and sucrose to the monosaccharides (glucose, fructose, and galactose). These pass through the intestinal walls into the bloodstream.

Glycogen (animal starch) is the reserve carbohydrate of animals and is thus the counterpart of starch in plants. It is distributed throughout the protoplasm, available upon metabolic demand. Excess glycogen is stored in the liver where it is readily converted into glucose if the blood sugar (glucose) level falls below normal. The structure of glycogen is similar to that of amylopectin but it is even more highly branched. It has a higher average molecular weight and a shorter average chain length than amylopectin.

Cellulose is the chief component of plant fibers and is the most abundant organic substance in the world. Cotton consists of cellulose in nearly pure form. It has been determined by X-ray analysis that cellulose consists of unbranched chains of great length which lie side by side in bundles. Hydrogen bonds are responsible for holding the chains together. The bundles are twisted in ropelike fashion. The fibers that are visible are composed of groups of the ropelike structures.

Complete hydrolysis of cellulose by acid yields only one monosaccharide, D(+)glucose. In this respect cellulose resembles starch. Both are glucosans of high molecular weight. The molecular weight of cellulose is even greater than that of starch. A more important difference between starch and cellulose is indicated by the results of their *partial* hydrolysis. Incomplete hydrolysis of starch gives maltose, but when the hydrolysis of cellulose is carried to the same stage of degradation the product is a different disaccharide called *cellobiose*. The hydrolysis of either maltose or cellobiose gives two molecules of glucose, but different enzymes are required to bring about the hydrolysis. Whereas maltose has an α-D-glucosidic linkage, cellobiose has a β-D-glucosidic linkage. Thus, starch may be regarded as a polymer of maltose, and cellulose as a polymer of cellobiose. The same idea can be expressed by characterizing starch as a polymer made up of α-glucose units, and cellulose as a polymer composed of β-glucose units.

Unlike starch and glycogen, which readily undergo digestion, cellulose passes through the human digestive tract without degradation. Most animals do not have the enzymes necessary for hydrolysis of cellulose and, consequently, it merely provides bulk to the diet. However, herbivorous animals can utilize cellulose as food because the symbiotic microorganisms that inhabit their digestive tracts produce enzymes, called cellulases, that are capable of hydrolyzing cellulose. Termites are able to digest wood because of the enzymatic activity of the microorganisms present in their alimentary tract.

The rigidity of wood results from the fact that the cellulose fibers are embedded in lignin, giving a structure that can be compared to that of reinforced concrete. Lignin is a complex polymeric material whose origin and structure have not yet been established. In the manufacture of paper from wood it is necessary to remove the lignin by dissolving it with such reagents as calcium bisulfite.

A polysaccharide called *chitin* makes up the hard shells of insects and crustaceans (lobsters, shrimps, etc.). Chitin is a polymer of acetylglucosamine. The structure of chitin is identical with that of cellulose except that the acetylamino group, CH_3CONH—, replaces the —OH group at C-2. (See Figure 33-6 for the structure of glucosamine.) Chitin is fairly resistant to acid and alkaline hydrolysis but is readily broken down by an enzyme called chitinase, found in certain microorganisms and snails.

New Terms

Acetal: any compound that fits the general formula $RCH(OR')_2$.

Aldose: a polyhydroxy aldehyde.

Amino sugars: sugars that have —NH_2 groups in place of one or two —OH groups.

Anomers: diastereoisomers of the type illustrated by α-D-glucose and β-D-glucose.

Carbohydrate: a polyhydroxy aldehyde (aldose), polyhydroxy ketone (ketose), or any substance that yields such compounds (and only such compounds) upon hydrolysis.

Deoxy sugars: sugars that have —H in place of one or more —OH groups.

Diastereoisomers: optical isomers that have different melting points, boiling points, etc., as well as different specific rotations.

Disaccharide: a carbohydrate that can undergo hydrolysis to give two monosaccharide molecules, e.g., sucrose, which hydrolyzes to give glucose and fructose.

Enantiomers: a pair of compounds whose only difference is in the direction of rotation of polarized light. Their molecules are nonsuperposable mirror images.

Epimers: diastereoisomers that differ only in the configuration of the carbon atom next to the aldehyde group.

Furanose: any carbohydrate containing a five-membered ring.

Glycan: another name for polysaccharide.

Glycosides: the acetals formed by sugars.

Hemiacetal: any compound that has an —OH group and an —O*R* group attached to the same carbon atom.

Ketose: a polyhydroxy ketone.

Monosaccharide: a carbohydrate that cannot be hydrolyzed to give simpler carbohydrates.

Mutarotation: the change of optical rotation of a solution with time.

Oligosaccharide: a carbohydrate that yields only a few monosaccharide molecules upon hydrolysis.

Polysaccharide: a carbohydrate of high molecular weight. A polysaccharide can be hydrolyzed to give, ultimately, a large number of monosaccharide molecules.

Pyranose: any carbohydrate containing a six-membered ring.

Triose: a carbohydrate that contains three carbon atoms in each molecule.

Exercises

33-1 Show how the meaning of the term *carbohydrate* has changed from its original meaning.

33-2 Give definitions for the following terms: sugars, monosaccharides, oligosaccharides, and polysaccharides.

33-3 Show why there are only two isomeric glyceraldehydes but four isomeric aldotetroses.

33-4 What are diastereoisomers? Epimers? Anomers?

33-5 List, and account for, several different names for D(+)glucose.

33-6 Are α-D-glucose and β-D-glucose optical isomers? Explain.

33-7 What is a pyranose? A furanose?

33-8 Discuss the inversion of sucrose as to meaning, origin of the term, and similarity to other reactions.

33-9 Define and explain *mutarotation.*

33-10 Describe the basic composition and structure of starch, cellulose, and chitin.

33-11 The formula for formaldehyde may be written CH_2O. Does this mean that it is a carbohydrate? Explain your answer.

References

Ihde, A. J., *The Development of Modern Chemistry*, Harper & Row, New York, 1964. Chapter 13 deals with the structure of natural products, and Chapter 17 discusses the development of biological chemistry.

Jellinck, P. H., *Biochemistry: An Introduction*, Holt, Rinehart & Winston, New York, 1963. Written at an elementary level for nonspecialists. Chapter 7 is an outline of carbohydrate chemistry, and Chapter 8 deals with carbohydrate metabolism.

Noller, C. R., *Chemistry of Organic Compounds*, 3rd ed., Saunders, Philadelphia, 1965. Chapter 17 deals with optical isomerism, and Chapter 18 discusses carbohydrates.

34

Lipids

34-1 Types of Lipids

Lipids comprise the fats and a variety of other substances whose properties are similar in some degree to those of fats. Among these properties are solubility in the "fat solvents" such as acetone, alcohol, ether, and chloroform; insolubility in water; and a greasy feel. Lipids include, among others, the following types of substances:

1. Waxes: esters of fatty acids with monohydric alcohols.
2. Fats: esters of fatty acids with glycerol.
3. Phospholipids: substituted fats containing phosphorus and nitrogen.
4. Glycolipids: compounds containing both carbohydrate and lipid portions.
5. Lipoproteins: complex substances in which lipids are associated with proteins.

34-2 Waxes

The simplest lipids are the waxes, which are simple esters, $RCOOR'$, derived from long-chain monohydric alcohols and long-chain fatty acids. Both the parent acid and the parent alcohol have 16 or more carbon atoms in the natural waxes. Actually, any natural wax is a mixture of different esters; other types of compounds, such as long-chain alcohols and long-chain hydrocarbons, are frequently present also. Beeswax, the material from which bees build honeycomb, is chiefly myricyl palmitate, $C_{15}H_{31}COOC_{30}H_{61}$.

Commercially, the most important natural wax is carnauba wax, which occurs as a coating on the leaves of the Brazilian wax palm. Because of its hardness and its imperviousness to water, it is used in floor polishes, automobile polishes, and carbon paper coatings. Its chief ingredients are esters of C_{24} and C_{28} acids with C_{32} and C_{34} alcohols. Also present are considerable quantities of polymeric esters of ω-hydroxy acids, $HO(CH_2)_nCOOH$, where n is 17–29. (Omega, ω, the last letter of the Greek alphabet, is used to indicate that the —OH group is attached to the carbon atom furthest removed from the —COOH group.) The unique physical properties of carnauba wax are probably due to the presence of high molecular weight esters of this type.

34-3 Fats and Oils

One of the important classes of foodstuffs is the group of substances known as the fats. These organic compounds occur naturally in the bodies of animals and in the fruits and seeds of plants. A fat that is liquid at ordinary temperatures is called an oil.

The fats in our diet are obtained from such sources as cream, butter, lard, tallow, corn oil, and cottonseed oil. In addition to their importance in foods, some substances of this type are used as drying oils in paints and varnishes (e.g., linseed oil). Some vegetable oils have uses as special lubricants and plasticizers (Section 18-21). Of great economic significance is the utilization of fats and oils in the manufacture of soaps and synthetic detergents.

The vegetable oils that are produced in greatest quantity (more than 1 billion pounds a year) are coconut oil, cottonseed oil, soybean oil, linseed oil, olive oil, palm oil, and peanut oil. Other vegetable oils produced in large quantity (more than 200 million pounds a year) are corn oil, sesame oil, castor oil, and sunflower seed oil.

34-4 The Chemical Composition of Fats and Oils

The fats and oils used as foods are esters of monocarboxylic acids. All are derived from the same trihydric alcohol, glycerol, and are known as *glycerides*. All edible oils are of this type and are not to be confused with the hydrocarbon fuel oils and lubricating oils that are obtained from petroleum.

Differences between different fats are due to the fact that they are derived from different acids. With very few exceptions the acids are straight-chain compounds. Only those containing an even number of carbon atoms occur in significant amounts. Some of the acids are saturated; others contain one or more double bonds.

The most important saturated acids occurring in fats are the following:

$CH_3(CH_2)_{10}COOH$	lauric acid
$CH_3(CH_2)_{12}COOH$	myristic acid
$CH_3(CH_2)_{14}COOH$	palmitic acid
$CH_3(CH_2)_{16}COOH$	stearic acid

The most important unsaturated acids have 18 carbon atoms, with a double bond located at the middle of the chain. If other double bonds are present they are more distant from the carboxyl group:

$$CH_3(CH_2)_7CH{=}CH(CH_2)_7COOH \qquad \text{oleic acid}$$
$$CH_3(CH_2)_4CH{=}CHCH_2CH{=}CH(CH_2)_7COOH \qquad \text{linoleic acid}$$
$$CH_3CH_2CH{=}CHCH_2CH{=}CHCH_2CH{=}CH(CH_2)_7COOH \qquad \text{linolenic acid}$$

Since the glycerol molecule contains three —OH groups, it can be esterified with either one, two, or three molecules of acid:

$$\begin{array}{cccc}
\begin{array}{c} HOCH_2 \\ | \\ HOCH \\ | \\ HOCH_2 \end{array} &
\begin{array}{r} HOCH_2 \\ | \\ R\text{COOCH} \\ | \\ HOCH_2 \end{array} &
\begin{array}{r} HOCH_2 \\ | \\ R\text{COOCH} \\ | \\ R\text{COOCH}_2 \end{array} &
\begin{array}{r} R\text{COOCH}_2 \\ | \\ R\text{COOCH} \\ | \\ R\text{COOCH}_2 \end{array} \\
\text{glycerol} & \text{a monoglyceride} & \text{a diglyceride} & \text{a triglyceride}
\end{array}$$

The fats and oils are *triglycerides*. *Monoglycerides* and *diglycerides*, obtained by reacting triglycerides with glycerol, are useful as emulsifying agents in oleomargarine and in cosmetic creams.

A *simple triglyceride* is one in which all three —OH groups have been esterified with the same acid. Examples are glyceryl tristearate (stearin) and glyceryl trioleate (olein):

$$\begin{array}{cc}
\begin{array}{r} CH_3(CH_2)_{16}COOCH_2 \\ | \\ CH_3(CH_2)_{16}COOCH \\ | \\ CH_3(CH_2)_{16}COOCH_2 \end{array} &
\begin{array}{r} CH_3(CH_2)_7CH{=}CH(CH_2)_7COOCH_2 \\ | \\ CH_3(CH_2)_7CH{=}CH(CH_2)_7COOCH \\ | \\ CH_3(CH_2)_7CH{=}CH(CH_2)_7COOCH_2 \end{array} \\
\text{stearin} & \text{olein}
\end{array}$$

Historically, simple triglycerides were given trivial names based on their sources, e.g., *stearin* (Greek: *stear*, tallow). Common names of fatty acids were derived from the names of the fats, e.g., stearic acid.

A *mixed triglyceride* is one in which the —OH groups of the glycerol molecule have been esterified with two or three different acids.

All fats are mixtures of different triglycerides. Furthermore, most of the triglyceride molecules in a fat are of the mixed glyceride type. The different *R*COO— groups in a fat appear to be distributed at random. Unless more than two thirds of the *R*COO— groups are of one kind, simple glycerides are not present in quantity.

34-5 Chemical Composition and the Properties of Fats

The melting points of the saturated straight-chain monocarboxylic acids increase with chain length from $-8°C$ for butyric acid (four carbon atoms) to 70°C for stearic acid (18 carbon atoms). Double bonds tend to lower the melting point. Thus, oleic acid (18 carbon atoms, one double bond) melts at 14°C; linoleic acid (18 carbon atoms, two double bonds) melts at $-5°C$; and the melting point of linolenic acid (18 carbon atoms, three double bonds) is $-111°C$.

The effects of double bonds on the melting points of the fatty acids are reflected in the melting points of the glycerides formed from them. In general, unsaturated acid chains predominate in the vegetable fats and saturated groups in animal fats. The melting points of glycerides are also affected by molecular weight. Butter contains only small amounts of unsaturated acids, but it is nevertheless a low-melting fat. This is due in part to the low molecular weights of some of the acids from which butter fats are formed. Butter softens gradually over a range of temperature because it contains a large number of different glycerides.

Coconut oil and palm oil are examples of fats that are referred to as oils despite the fact that they are solids in the temperate zones. However, they are liquid in the tropics where they are obtained.

The composition and properties of animal fats are affected by the diet of the animals. As an example, lard from hogs fed on corn has a higher melting point than lard from hogs fed on peanuts. However, beef tallow has a higher melting point than hog lard, regardless of the diet of the animals. The first stomach (rumen) of a cud-chewing animal (ruminant) has an enzyme that reduces the double bonds of unsaturated fats to single bonds.

34-6 Hydrogenation of Oils

The hydrogenation of oils is a large-scale technical operation carried out for the purpose of raising the melting point. An example is the conversion of olein to stearin:

$$\begin{array}{c} CH_3(CH_2)_7CH{=}CH(CH_2)_7COOCH_2 \\ | \\ CH_3(CH_2)_7CH{=}CH(CH_2)_7COOCH \\ | \\ CH_3(CH_2)_7CH{=}CH(CH_2)_7COOCH_2 \\ \text{olein} \end{array} + 3H_2 \xrightarrow[\text{catalyst}]{\text{nickel}} \begin{array}{c} CH_3(CH_2)_{16}COOCH_2 \\ | \\ CH_3(CH_2)_{16}COOCH \\ | \\ CH_3(CH_2)_{16}COOCH_2 \\ \text{stearin} \end{array}$$

Hydrogen gas is bubbled through the hot oil containing finely divided nickel in suspension. Any desired consistency in the product may be obtained by controlling the amount of hydrogen added. It is usually desirable to hydrogenate only part of the double bonds.

The original reason for converting vegetable oils to solid fats, a process often referred to as the "hardening" of fats, lay in the unsuitability of the natural oils for use as less expensive substitutes for butter and lard. Since 1953 the production of oleomargarine has exceeded that of butter. Also, solid fats are preferred for making soap. The rancidity that oils develop is due mostly to atmospheric oxidation at the site of the double bonds. This, of course, is prevented by hydrogenation.

Vegetable oils are more resistant to the development of rancidity through atmospheric oxidation than are animal oils because of the presence of small amounts of natural antioxidants. Rancidity results from the formation of volatile acids and aldehydes with unpleasant odors. In some cases the rancidity is brought about by microorganisms.

Glycerides of fatty acids containing two or more double bonds are known as *drying oils* because of the polymerization they undergo in the presence of oxygen, forming a tough film. Examples of natural products containing such glycerides are linseed oil and tung oil, which are important constituents of paints and varnishes. In the "drying" of paint the evaporation of a volatile solvent (such as turpentine) is only a minor part of the process. More important is the chemical reaction in which the unsaturated oil polymerizes to form a protective coating over the painted surface. Neither the polymerization process nor the structure of the waterproof film is well understood.

Waste or rags containing unsaturated oils are a fire hazard unless air is excluded or enough ventilation is provided to remove the heat generated by the oxidation of the oils. If the temperature rises, the rate of oxidation increases. This in turn causes a further rise in temperature. When the kindling temperature is reached the material bursts into flame. Such *spontaneous combustion* can be prevented by spreading oily rags out flat or by confining oily waste within an air-proof container.

Since the susceptibility of an oil to polymerization and to atmospheric oxidation depends on the presence of double bonds, the extent of unsaturation is an important characteristic of each natural oil. A traditional procedure for determining empirically the degree of unsaturation of an oil is to ascertain the number of grams of iodine which will add to 100 g of the oil. This is known as the *iodine value* of the oil.

34-7 Hydrolysis of Fats

As esters, the glycerides can be hydrolyzed to give the parent acids and the parent alcohol (glycerol):

$$\begin{array}{c} RCOOCH_2 \\ | \\ R'COOCH \\ | \\ R''COOCH_2 \\ \text{a mixed glyceride} \end{array} + 3H_2O \rightarrow \begin{array}{c} \left\{ \begin{array}{c} RCOOH \\ R'COOH \\ R''COOH \end{array} \right\} \\ \text{fatty acids} \end{array} + \begin{array}{c} CH_2OH \\ | \\ CHOH \\ | \\ CH_2OH \\ \text{glycerol} \end{array}$$

To occur at a substantial rate, this reaction must be catalyzed with acid, with alkali, or with certain digestive enzymes called lipases. With alkali, the metal salts of the acids are obtained. Since salts of fatty acids are *soaps*, the reaction of alkali with a glyceride is called *saponification* (Latin, "soapmaking"). The term saponification has been broadened to include the reaction of alkali with any ester. The following is an example of saponification of a glyceride:

$$\begin{array}{c} CH_3(CH_2)_{16}COOCH_2 \\ | \\ CH_3(CH_2)_{16}COOCH \\ | \\ CH_3(CH_2)_{16}COOCH_2 \\ \text{stearin (a fat)} \end{array} + 3Na^+OH^- \rightarrow \begin{array}{c} 3C_{17}H_{35}COO^-Na^+ \\ \text{sodium stearate} \\ \text{(a soap)} \end{array} + \begin{array}{c} CH_2OH \\ | \\ CHOH \\ | \\ CH_2OH \\ \text{glycerol} \end{array}$$

Different fats require different amounts of alkali for saponification because the fatty acids have different molecular weights and because a natural fat may contain certain substances that do not react with alkali (e.g., hydrocarbons and alcohols). The amount of alkali actually required for the saponification of a given weight of fat varies with different fats and is characteristic of them. The quantity known as the *saponification value* of a fat is arbitrarily defined as the number of milligrams of KOH required to saponify 1 g of fat. The determination of saponification values is useful for detecting the adulteration of butter and other fats used in the food industry.

34-8 Soaps

Ordinary soap is a mixture of sodium salts of long-chain monocarboxylic acids. It is a mixture because it is made from fats that are mixtures of different glycerides. Important components of ordinary soap are sodium palmitate, sodium stearate, and sodium oleate. Hard soaps are made from fats that are mostly saturated; and soft soaps, from highly unsaturated fats.

Each individual alkali metal salt of a carboxylic acid is called a soap, provided it contains from 10 to 18 carbon atoms. If the chain is less than 10 carbon atoms long, the salt is unable to emulsify oil in water. If the chain contains more than 18 carbon atoms, the salt is not sufficiently soluble in water to act as an effective detergent. Soaps of low molecular weight are most soluble in water. Potassium soaps are more soluble than sodium soaps (but also more costly to make).

The anions of the salts present in soaps, e.g., the stearate ion,

$$CH_3CH_2CH_2CH_2CH_2CH_2CH_2CH_2CH_2CH_2CH_2CH_2CH_2CH_2CH_2CH_2COO^-$$

have long hydrocarbon "tails" and polar "heads." The polar ends, the carboxylate groups, tend to make the anions water-soluble, and the nonpolar chains tend to make them oil-soluble. Although "oil and water don't mix" (i.e., they cannot dissolve in each other to form a true solution), oil can be emulsified by being shaken with an aqueous solution of soap. The soap anions surround the oil droplets with their hydrocarbon chains oriented inward and their polar heads extending outward into the surrounding water. The resulting emulsion is not a true solution, for the oil is not molecularly dispersed in the water.

Soap solutions have a surface tension much less than that of water, and this makes the water "wetter," i.e., the wetting power of a soap solution is greater than that of water alone. Soap solutions therefore are able to envelop grease and dirt particles and remove them from the surface to which they are attached. Thus, soaps serve as detergents through a combination of wetting action and emulsifying action.

Soaps cannot be used in acidic solution because the water-insoluble fatty acids precipitate out as a scum:

$$C_{17}H_{35}COO^- + H^+ \rightarrow C_{17}H_{35}COOH \downarrow$$

The soap anions are also precipitated by the calcium, magnesium, and iron(II) ions present in hard water:

$$2C_{17}H_{35}COO^- + Ca^{++} \rightarrow (C_{17}H_{35}COO)_2Ca \downarrow$$

This is the cause of the bathtub "ring" and other undesirable effects of using soap in hard water.

34-9 Synthetic Detergents

Detergent properties are shown not only by soaps but also by other long-chain compounds that have a polar group at one end. An example of a detergent that is not a soap is sodium lauryl sulfate, $NaC_{12}H_{25}SO_4$.

The making of soap is one of the oldest of the chemical arts, but the synthetic detergents ("syndets") have been developed since 1930. Since 1954 the sale of syndets has exceeded that of soaps.

The fat most commonly used for the production of detergents is coconut oil. The first step in the process is the reduction of the fat. The reduction of glycerides, like the reduction of other esters, is brought about either with hydrogen and a catalyst or with sodium metal and alcohol (this now being one of the major uses of sodium metal):

$$\begin{array}{c} RCOOCH_2 \\ | \\ R'COOCH \\ | \\ R''COOCH_2 \end{array} \xrightarrow{\text{reduction}} \left\{ \begin{array}{c} RCH_2OH \\ \\ R'CH_2OH \\ \\ R''CH_2OH \end{array} \right\} + \begin{array}{c} CH_2OH \\ | \\ CHOH \\ | \\ CH_2OH \end{array}$$

a glyceride — mixture of primary alcohols — glycerol

Note that glycerol is a by-product of this first step in the manufacture of syndets. Even so, the soap and syndet industries combined are unable to supply all the glycerol required for its many uses.

Reduction of coconut oil yields a high proportion of lauryl alcohol. It is first converted to lauryl hydrogen sulfate, which is an ester of sulfuric acid:

$$\underset{\text{lauryl alcohol}}{CH_3(CH_2)_{10}CH_2OH} + \underset{\text{sulfuric acid}}{HOSO_2OH} \rightarrow \underset{\text{lauryl hydrogen sulfate}}{CH_3(CH_2)_{10}CH_2OSO_2OH} + H_2O$$

Treatment of lauryl hydrogen sulfate with sodium hydroxide yields sodium lauryl sulfate:

$$CH_3(CH_2)_{10}CH_2OSO_2OH + Na^+OH^- \rightarrow CH_3(CH_2)_{10}CH_2OSO_3^-Na^+ + H_2O$$

Sodium lauryl sulfate acts in essentially the same manner as soap when it is used as a detergent. The lauryl sulfate ion, like the stearate ion, has a long hydrocarbon "tail" and a polar "head." The advantage of sodium alkyl sulfates over sodium carboxylates is that they do not form precipitates with the calcium, magnesium, and iron(II) ions present in hard water.

Sodium lauryl sulfate was the first of the synthetic detergents to be placed on the market. A large number of compounds useful as detergents have since been developed. Syndets vary considerably in their chemical structure. Unfortunately, those with highly branched chains are not readily destroyed by microorganisms. The wide use of detergents of this type has resulted in their appearance in domestic water supplies and has caused serious foaming problems in sewage disposal plants. Since 1965 the manufacture of detergents has been limited to *biodegradable* compounds, i.e., those with unbranched hydrocarbon chains.

34-10 Fats in Food

When oxidized to carbon dioxide and water, fats produce more than twice as much energy as an equal weight of carbohydrate or protein (9.5 kcal per g compared to 4 kcal per g). When stored in the body, fats serve as thermal insulators, and they protect the body and its vital organs against mechanical injury from blows and falls.

Fats in their ordinary form consist of large particles that are insoluble in water. In this form they cannot be absorbed through the intestinal walls. They are acted upon in the intestine by fat-splitting enzymes (lipases), which partially hydrolyze them, forming diglycerides and monoglycerides, which act as emulsifying agents upon the unhydrolyzed fat. Emulsification is also brought about by the bile salts secreted by the liver.

Only about one third of the fat is broken down by the lipases; the remainder, upon emulsification, is absorbed from the intestine as unchanged triglyceride. The emulsified fats are of sufficiently small particle size to be able to pass directly through the villi of the intestinal wall into the lymphatic system. From there they enter the general circulation through the thoracic duct and are transported throughout the body. Some are oxidized for energy; some are stored in fat depots from which they may be taken when required for future use; and some are converted into phospholipids, glycolipids, lipoproteins, etc.

The fats in the diet must supply either linoleic or linolenic acid to maintain a healthy condition of the skin. It is also claimed that inclusion of unsaturated glycerides in the diet lowers the cholesterol content of the blood. There has been much interest recently in the replacement of saturated fats in the diet by corn oil and safflower oil because of the apparent correlation between high cholesterol content of the blood and the incidence of heart disease and hardening of the arteries.

With the exception of the unsaturated fatty acids mentioned above, fats are not essential in the diet. It is well known that starches, when present in the diet in excess of the needed amounts, are converted into fats by the animal body. The fat synthesis takes place inside the cells, where glucose is degraded into active two-carbon units which are used to make fatty acids. By a different type of degradation, glucose gives rise to glycerol. Esterification of the fatty acids with the glycerol produces fats, which are transported to fat depots and stored like fats contained in the diet.

34-11 Phospholipids, Glycolipids, and Lipoproteins

The phospholipids, also known as phosphatides, are present in all living cells, both plant and animal. They are abundant in the brain and spinal cord, in eggs, and in soybeans. Large quantities of soybean lecithin are used to stabilize emulsified fats, e.g., oleomargarine and mayonnaise.

The dry matter of the human brain is about half protein and half lipid. The most abundant of the lipids present are phospholipids, glycolipids, and cholesterol.

The simplest of the phospholipids are esters of the phosphatidic acids, which are themselves fatty acid esters of glycerophosphoric acid.

glycerophosphoric acid	a phosphatidic acid
$HOCH_2$	$RCOOCH_2$
|	|
$HOCH$	$RCOOCH$
|	|
$(HO)_2POOCH_2$	$(HO)_2POOCH_2$

Glycolipids are compounds formed between sugars and lipids. They are often called cerebrosides because they occur in the brain and nervous tissue. The sugar unit is usually galactose.

Lipoproteins are complex substances consisting of protein linked with lipid. They occur in blood serum, brain tissue, eggs, etc. Much of the lipid in blood plasma is transported in the form of loose lipoprotein complexes.

New Terms

Fats (glycerides): monocarboxylic acid esters of glycerol.

Glycolipids: compounds formed between sugars and lipids.

Lipases: fat-splitting enzymes.

Lipids: the group of substances that comprise the fats and a variety of other substances whose properties are similar in some degree to those of fats.

Lipoproteins: complex substances consisting of protein linked with lipid.

Phospholipids: substituted fats containing phosphorus and nitrogen.

Soaps: salts obtained by reaction of strong base with fats; alkali metal salts of long-chain monocarboxylic acids.

Waxes: simple esters derived from long-chain monohydric alcohols and long-chain fatty acids.

Exercises

34-1 Give the chemical composition of five types of lipids.

34-2 What are glycerides? How do monoglycerides and diglycerides differ from triglycerides?

34-3 What is the distinction between simple glycerides and mixed glycerides?

34-4 The saponification of a sample of a fat produced sodium stearate, sodium oleate, sodium palmitate, and glycerol. Write the formulas for all of the glycerides that might be present in the fat.

34-5 Corn oil is nutritious but diesel oil is not. Explain.

34-6 Why are vegetable oils hydrogenated in the food industry? Why has there been a new emphasis in recent years on leaving the vegetable oils partially unsaturated?

34-7 What type of chemical structure is necessary in drying oils?

34-8 Explain why glycerol is a by-product of both the soap and syndet industries.

34-9 What becomes of the fats and carbohydrates we eat?

34-10 Why is phosphorus one of the elements essential to life?

References

Bennett, T. P., and Frieden, E., *Modern Topics in Biochemistry*, Crowell Collier & Macmillan, New York, 1966 (paperbound). Chapter 8 deals with lipids.

Rafelson, M., and Binkley, S. B., *Basic Biochemistry*, Macmillan, New York, 1965 (paperbound). Chapter 3 is a discussion of lipids, and Chapter 9 deals with lipid metabolism.

35

Proteins

35-1 The Composition of Proteins

Proteins are complex combinations of amino acids. Hence, all of them contain carbon, hydrogen, oxygen, and nitrogen. They may also contain sulfur, phosphorus, and traces of other elements, including iron, copper, iodine, manganese, and zinc.

The proteins are essential to life. Protoplasm itself has proteins as its basis. Accordingly, proteins are present in all plant and animal cells. Each type of cell makes its own specific kinds of proteins. Nerve and muscle tissue, skin, hair, nails, feathers, wool, horns, and hoofs are all composed of proteins. Hemoglobin, antibodies, viruses, enzymes, and many of the hormones are proteins.

Even the simplest of the proteins are far more complicated than sugars or fats. The molecular weights of proteins vary from several thousand upward, depending upon the source (Table 35-1). Upon hydrolysis, proteins are broken down into α-amino acids. Thus far, twenty-six different amino acids have been obtained by hydrolysis of proteins from various sources. No single protein gives all twenty-six of them, but each protein gives several different amino

Table 35-1

Molecular weights of typical proteins

Protein	Molecular weight
Beef insulin	6,000
Pepsin	36,000
Hemoglobin	66,000
Urease	480,000

Table 35-2

Amino acids present in proteins

(Names of the ten essential amino acids are indicated by asterisks.)

Name	*Abbreviation*	*Formula*
I. *Neutral Amino Acids*		
1. Glycine	Gly	CH_2COOH NH_2
2. (+)Alanine	Ala	$CH_3CHCOOH$ NH_2
3. (+)Valine*	Val	$(CH_3)_2CHCHCOOH$ NH_2
4. (−)Leucine*	Leu	$(CH_3)_2CHCH_2CHCOOH$ NH_2
5. (+)Isoleucine*	Ileu	CH_3 $CH_3CH_2CHCHCOOH$ NH_2
6. (−)Serine	Ser	OH $CH_2CHCOOH$ NH_2
7. (−)Threonine*	Thr	OH $CH_3CHCHCOOH$ NH_2
8. (−)Cysteine	CySH	SH $CH_2CHCOOH$ NH_2
9. (−)Cystine	CyS—SCy	$HOOCCHCH_2S—SCH_2CHCOOH$ NH_2 NH_2
10. (−)Methionine*	Met	$CH_3SCH_2CH_2CHCOOH$ NH_2
11. (−)Phenylalanine*	Phe	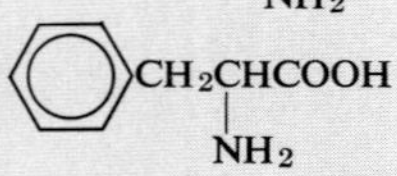
12. (−)Tyrosine	Tyr	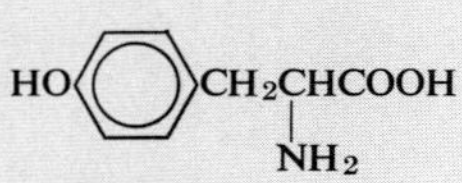
13. (+)3,5-Dibromotyrosine		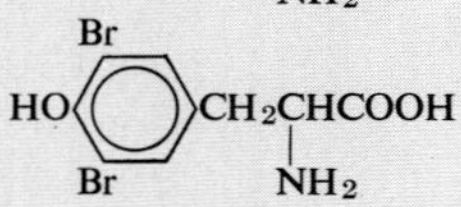

Name	Abbreviation	Formula
14. (+)3,5-Diiodotyrosine		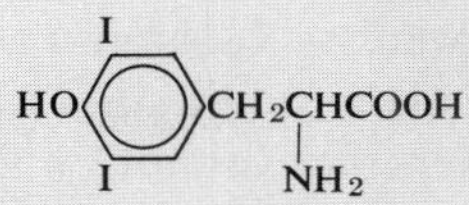
15. (−)Proline	Pro	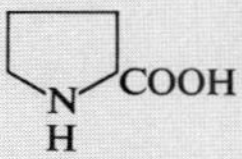
16. (−)Hydroxyproline	Hypro	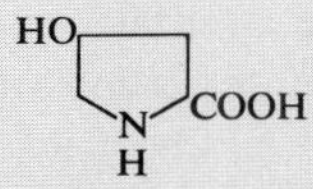
17. (−)Tryptophan*	Try	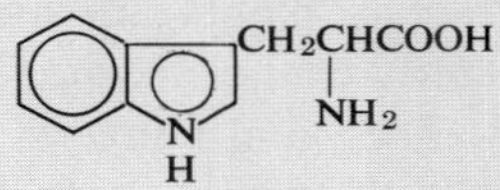
18. (+)Thyroxine		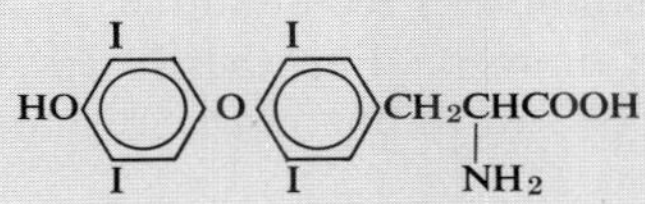
II. *Acidic Amino Acids*		
19. (+)Aspartic acid	Asp	$HOOCCH_2\underset{NH_2}{CH}COOH$
20. (−)Asparagine	Asp(NH₂)	$H_2NCOCH_2\underset{NH_2}{CH}COOH$
21. (+)Glutamic acid	Glu	$HOOCCH_2CH_2\underset{NH_2}{CH}COOH$
22. (+)Glutamine	Glu(NH₂)	$H_2NCOCH_2CH_2\underset{NH_2}{CH}COOH$
III. *Basic Amino Acids*		
23. (+)Lysine*	Lys	$H_2NCH_2CH_2CH_2CH_2\underset{NH_2}{CH}COOH$
24. (−)Hydroxylysine	Hylys	$H_2NCH_2\underset{OH}{CH}CH_2CH_2\underset{NH_2}{CH}COOH$
25. (+)Arginine*	Arg	$H_2N\underset{\overset{\|}{NH}}{C}NHCH_2CH_2CH_2\underset{NH_2}{CH}COOH$
26. (−)Histidine*	His	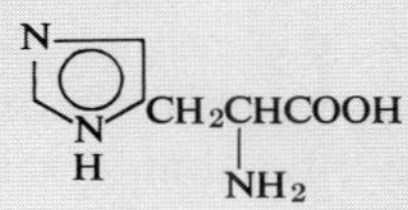

acids. One of the smallest and simplest protein molecules is the insulin molecule, which contains only fifty-one amino acid units; even so, its hydrolysis yields seventeen different amino acids.

Of the twenty-six α-amino acids present in the proteins of the human body, all but ten can be synthesized within the body from other materials available in food. The ten that cannot be synthesized within the body must be present in the food we eat; for this reason, they are referred to as the *essential* amino acids. Casein, the principal protein of milk, contains all ten of the essential amino acids.

In general, the animal organisms obtain their supply of essential amino acids by eating plants, other animals, and animal products (e.g., milk). Therefore, the ultimate source of all proteins is plants and bacteria. In the synthesis of amino acids, plants are assisted by microorganisms present in the soil. Nitrite bacteria change ammonia to nitrite, and nitrate bacteria change nitrites to nitrates. The plant converts nitrates and ammonia into α-amino acids and then converts the amino acids into proteins.

Commercial synthesis of several of the essential amino acids has been developed in recent years so that they are available as dietary supplements.

35-2 Amino Acids Present in Proteins

The twenty-six amino acids that have been found to occur in proteins are listed in Table 35-2 by name and formula. Abbreviations are given for most of them, since these are commonly used as a space-saving convenience. The names of the ten essential amino acids are indicated by asterisks. Those amino acids with (+) included in their names are dextrorotatory, and those that are levorotatory have (−) included with their names.

The α carbon atom is asymmetric in all the α-amino acids except glycine. It has been shown, through chemical interrelationships, that the configuration of the α carbon atom is the same in all amino acids derived from plant and animal proteins. This configuration is regarded as being the same as in (−)glyceraldehyde, if in the formula RCH(NH_2)COOH we consider the —COOH, —NH_2, —H, and —R groups to correspond, respectively, to the —CHO, —OH, —H, and —CH_2OH groups of glyceraldehyde. By analogy to the nomenclature of the carbohydrates, the α-amino acids derived from proteins therefore are said to belong to the L family.

The so-called "neutral" amino acids are those that have an equal number of basic —NH_2 groups and acidic —COOH groups. The so-called "acidic" amino acids have an additional carboxyl group, as in aspartic and glutamic acids, or a potential carboxyl group in the form of the hydrolyzable carboxamide group (—$CONH_2$), as in asparagine and glutamine. The "basic" amino acids have an additional basic group, which is an amino group in lysine, a guanidino group in arginine, and an imidazole ring in histidine.

In Table 35-2 the amino acids found in proteins are arranged in order of increasing complexity within each of the three categories. In Table 35-3 the

Table 35-3

Chemical names for amino acids

Common name	*Chemical name*
Alanine	α-aminopropionic acid
Arginine	α-amino-δ-guanidinovaleric acid
Asparagine	α-aminosuccinamic acid
Aspartic acid	aminosuccinic acid
Cysteine	α-amino-β-mercaptopropionic acid
Cystine	bis(2-amino-2-carboxyethyl) disulfide
3,5-Dibromotyrosine	α-amino-β-(3,5-dibromo-4-hydroxyphenyl)propionic acid
3,5-Diiodotyrosine	α-amino-β-(3,5-diiodo-4-hydroxyphenyl)propionic acid
Glutamic acid	α-aminoglutaric acid
Glutamine	α-aminoglutaramic acid
Glycine	aminoacetic acid
Histidine	α-amino-β-(5-imidazolyl)propionic acid
Hydroxylysine	α, ϵ-diamino-δ-hydroxycaproic acid
Hydroxyproline	4-hydroxy-2-pyrollidinecarboxylic acid
Isoleucine	α-amino-β-methylvaleric acid
Leucine	α-aminoisocaproic acid
Lysine	α, ϵ-diaminocaproic acid
Methionine	α-amino-γ-(methylthio)butyric acid
Phenylalanine	α-amino-β-phenylpropionic acid
Proline	2-pyrollidinecarboxylic acid
Serine	α-amino-β-hydroxypropionic acid
Threonine	α-amino-β-hydroxybutyric acid
Thyroxine	3,5,3′,5′-tetraiodothyronine
Tryptophan	α-amino-β-(3-indolyl)propionic acid
Tyrosine	α-amino-β-(4-hydroxyphenyl) propionic acid
Valine	α-aminoisovaleric acid

acids are arranged in a single alphabetical listing for ready reference in determining their descriptive chemical names.

35-3 Amphoteric Properties of Amino Acids

Amino acids that have the same number of $—NH_2$ groups as —COOH groups are called "neutral" amino acids, but when they are dissolved in water, amino acids of this type form solutions that are slightly acidic instead of neutral. In their reactions, amino acids act both as bases and as acids. In the reaction

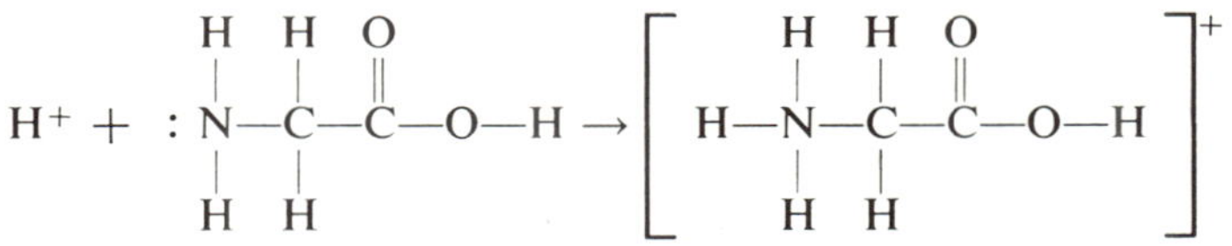

the H^+ ion joins the glycine molecule through the unshared electron pair on the nitrogen atom. In the presence of a base, the proton from the carboxyl group neutralizes OH^- ions:

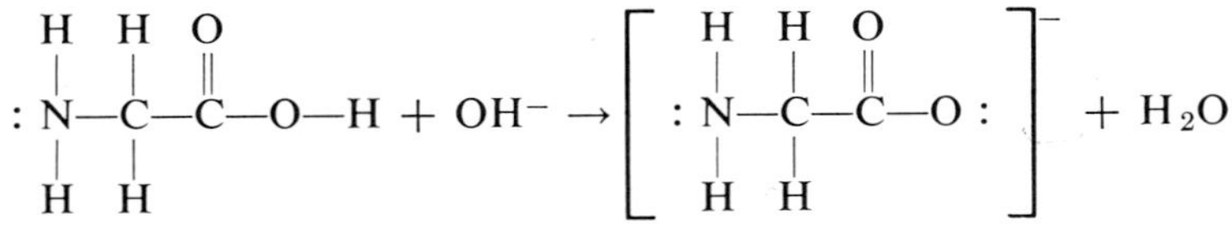

If glycine is in solution alone, the most abundant species present is *none of the above structures* but the one that results from the reaction that occurs within the amino acid itself—the transfer of a proton from the carboxyl group to the amino group:

$$H_2NCH_2COOH \rightarrow {}^+H_3NCH_2COO^-$$

The species resulting from this proton transfer is called a *zwitterion* or a *dipolar ion*. It is not really an ion, for its net charge is zero. It is actually a highly polar molecule; the charges are written to emphasize its polar character.

In contrast to amines and carboxylic acids, which have low boiling points, the amino acids are nonvolatile crystalline solids with fairly high melting points. Furthermore, they are insoluble in nonpolar solvents, such as benzene or ether. In brief, their properties are the properties characteristic of salts. Having a zwitterion structure in the pure state, they are *inner salts*. Viewed from one end, an amino acid zwitterion appears to be a substituted ammonium ion; from the other end, it looks like a substituted acetate ion.

Since they are amphoteric electrolytes, amino acids exist in aqueous solution as an equilibrium mixture whose composition depends upon the pH (Figure 35-1). Upon addition of sufficient alkali to establish a pH of about 11, the amino acid is present predominantly in the form of negative ions. When acid is added to make the pH about 1, the predominant species is a positive ion.

The behavior of a solution of an amino acid when placed between two oppositely charged electrodes depends on the pH of the solution. Since the

Figure 35-1

Equilibria in a solution of a "neutral" amino acid

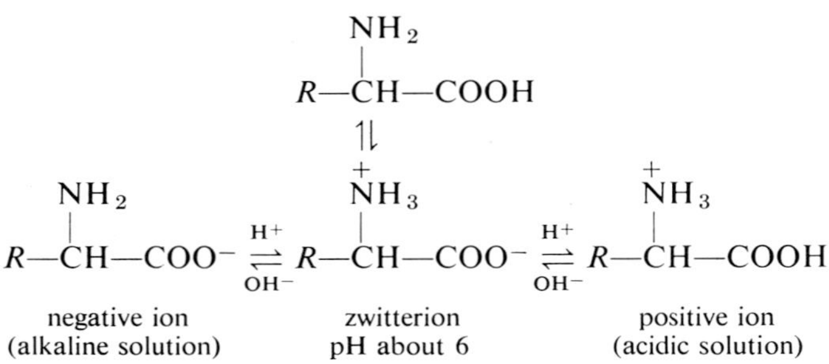

particles in a solution of low pH are positively charged, they migrate toward the negative electrode. In a solution of high pH the particles are negatively charged and migrate toward the positive electrode. At a certain intermediate pH the amino acid is present almost entirely in zwitterion form and does not migrate toward either electrode. The pH at which there is no migration is called the *isoelectric point* of the amino acid and may be designated as pI.

Each amino acid has a slightly different isoelectric point. This point usually is not the neutral point (pH 7). All amino acids of the so-called neutral class have pI values in the neighborhood 6–7. With the general formula $R\text{CH}(\text{NH}_2)\text{COOH}$, α-amino acids have a basic group (—NH_2), an acidic group (—COOH), and (in all cases except glycine) an organic group represented by R attached to the same carbon atom. The isoelectric point depends on the basic strength of the —NH_2 group and the acid strength of the —COOH group, but both of these vary with the nature of R. The pI value is also affected directly by the presence of other functional groups that are themselves basic or acidic; e.g., the phenolic —OH group in tyrosine. The pI value tends to be lower, of course, for amino acids that have an excess of —COOH over —NH_2 groups; and it tends to be higher for those that have an excess of —NH_2 over —COOH groups.

35-4 The Peptide Linkage

The reaction between the amino group of an amino acid molecule and the carboxyl group of another amino acid molecule forms a type of amide known as a *peptide*. The

$$\text{—}\overset{\overset{\displaystyle \text{O}}{\|}}{\text{C}}\text{—}\overset{\overset{\displaystyle \text{H}}{|}}{\text{N}}\text{—}$$

group, which is present in peptides, is called a *peptide linkage*. The reaction between two glycine molecules forms the *dipeptide* glycylglycine (Gly · Gly):

$$\underset{\text{glycine}}{\text{H}_2\text{N—CH}_2\text{—}\overset{\overset{\displaystyle \text{O}}{\|}}{\text{C}}\text{—OH}} + \underset{\text{glycine}}{\text{H—}\overset{\overset{\displaystyle \text{H}}{|}}{\text{N}}\text{—CH}_2\text{—COOH}} \rightarrow$$

$$\underset{\text{glycylglycine (Gly · Gly)}}{\text{H}_2\text{N—CH}_2\text{—}\overset{\overset{\displaystyle \text{O}}{\|}}{\text{C}}\text{—}\overset{\overset{\displaystyle \text{H}}{|}}{\text{N}}\text{—CH}_2\text{—COOH}} + \text{H}_2\text{O}$$

Similarly, the reaction between two alanine molecules forms the dipeptide alanylalanine:

$$\underset{\text{alanine}}{\text{H}_2\text{N}\overset{\overset{\displaystyle \text{CH}_3}{|}}{\text{C}}\text{HCOOH}} + \underset{\text{alanine}}{\text{HNH}\overset{\overset{\displaystyle \text{CH}_3}{|}}{\text{C}}\text{HCOOH}} \rightarrow \underset{\text{alanylalanine (Ala · Ala)}}{\text{H}_2\text{N}\overset{\overset{\displaystyle \text{CH}_3}{|}}{\text{C}}\text{HCONH}\overset{\overset{\displaystyle \text{CH}_3}{|}}{\text{C}}\text{HCOOH}} + \text{H}_2\text{O}$$

Glycylglycine, like glycine itself, has a terminal —NH_2 group capable of reacting with the —COOH group of another amino acid molecule. It also has a terminal —COOH group capable of reacting with the —NH_2 group of another molecule. If either of these reactions occurs, the result is a *tripeptide*. A peptide containing a large number of units is called a *polypeptide*. The units making up a peptide are amino acids less the elements of water and are called *amino acid residues*.

Proteins are peptides of high molecular weight. The structure of a protein, or any other peptide, can be represented conveniently by the use of standard abbreviations to represent the amino acid residues that it contains. We have used this method of representing the structure of peptides when we symbolized glycylglycine as Gly · Gly and alanylalanine as Ala · Ala.

The complete hydrolysis of a peptide breaks all the peptide linkages and converts each amino acid residue into the corresponding amino acid. For example, the hydrolysis of glycylglycine gives two molecules of glycine:

$$H_2NCH_2CONHCH_2COOH + H_2O \rightarrow H_2NCH_2COOH + H_2NCH_2COOH$$

A peptide often yields more than one kind of amino acid upon hydrolysis. This means, of course, that the peptide chain contains more than one kind of amino acid residue. Such a peptide would be formed by the reaction of a molecule of glycine with a molecule of alanine. Actually, two different dipeptides could be formed by this reaction, depending on whether the amino group of the glycine reacts with the carboxyl group of the alanine or the amino group of the alanine with the carboxyl group of the glycine:

$$\underset{\text{glycine}}{H_2NCH_2COOH} + \underset{\text{alanine}}{HNH\overset{\overset{\displaystyle CH_3}{|}}{C}HCOOH} \rightarrow \underset{\text{glycylalanine (Gly · Ala)}}{H_2NCH_2CONH\overset{\overset{\displaystyle CH_3}{|}}{C}HCOOH} + H_2O$$

$$\underset{\text{alanine}}{H_2N\overset{\overset{\displaystyle CH_3}{|}}{C}HCOOH} + \underset{\text{glycine}}{HNHCH_2COOH} \rightarrow \underset{\text{alanylglycine (Ala · Gly)}}{H_2N\overset{\overset{\displaystyle CH_3}{|}}{C}HCONHCH_2COOH} + H_2O$$

Hydrolysis of either glycylalanine or alanylglycine gives one molecule of glycine and one molecule of alanine. To distinguish the one dipeptide from the other by hydrolysis alone is impossible.

A procedure that will serve to distinguish glycylalanine from alanylglycine must depend upon the fact that the residues at the two ends of a peptide are different. One, called the N-terminal residue, contains a free amino group; the other, called the C-terminal residue, contains a free carboxyl group. Several methods are now available for "tagging" the terminal amino group. For example, phenyl isothiocyanate reacts with the free —NH_2 group of a peptide to give a substituted thiourea. Treatment of the substituted thiourea with dry hydrogen chloride selectively removes the N-terminal residue as a heterocyclic compound whose identification reveals which amino acid residue was the N-terminal residue in the original peptide:

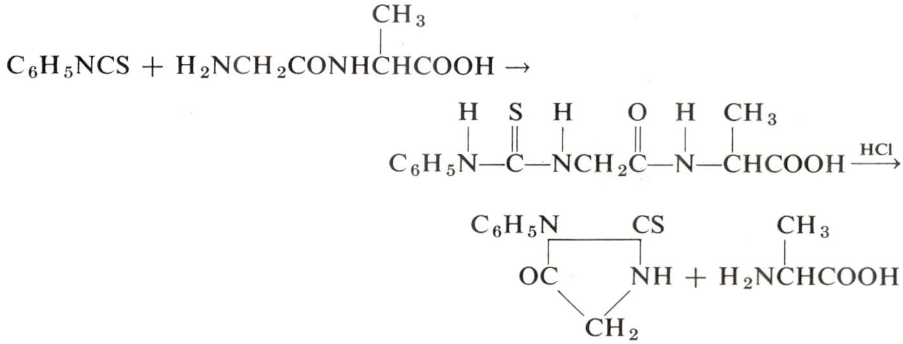

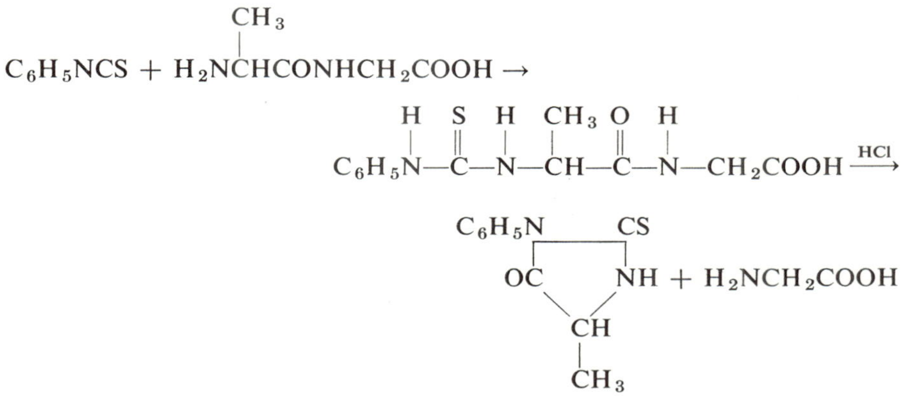

35-5 The Separation of Amino Acids and Proteins

Since individual amino acids have different isoelectric points, they behave differently at any particular pH in an electric field. A mixture of amino acids therefore can be separated by varying the pH of a solution situated between oppositely charged electrodes. The movement of charged particles in an electric field is called *electrophoresis*. The separation of amino acids by electrophoresis is useful not only for their identification but also in their isolation and purification.

Proteins, like the amino acids from which they are built, are amphoteric. Usually a protein is least soluble at its isoelectric point but will dissolve in solutions of markedly higher or lower pH than the pI value. The pI values vary from one protein to another, depending upon the kinds and relative amounts of amino acids that they contain.

Electrophoretic analysis of a mixture of proteins in aqueous solution involves establishing a series of different pH values and observing the behavior of the proteins in an electric field at each pH value. In general, such a mixture at a particular pH will contain certain proteins that migrate toward the negative electrode, each at its characteristic rate; other proteins will migrate toward the positive electrode, some more rapidly than others; and a protein will not migrate at all if the pH of the solution is the same as its isoelectric point. When a different pH is established in the solution, the electrophoretic behavior of the individual proteins will change. Thus, it is possible to separate the proteins of a mixture by electrophoresis.

Another method of separating proteins consists in adjusting the pH of the solution to the pI value of the protein to be precipitated and then adding a protein precipitant. The common protein precipitants are salts, such as sodium chloride, ammonium sulfate, and magnesium sulfate. Nonaqueous solvents that are miscible with water but are not solvents for proteins may also be used to aid the precipitation. Examples are alcohol and acetone.

35-6 Determining the Structure of Peptides

A peptide, whether classified as a polypeptide or a protein, is a polymer consisting of a chain of amino acid residues,

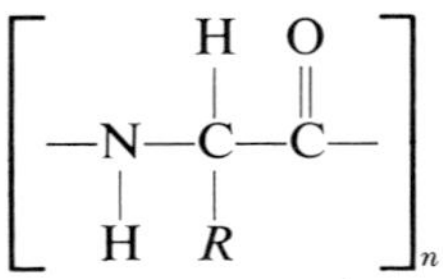

joined to an N-terminal residue,

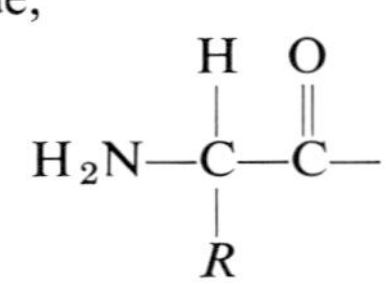

at one end and a C-terminal residue,

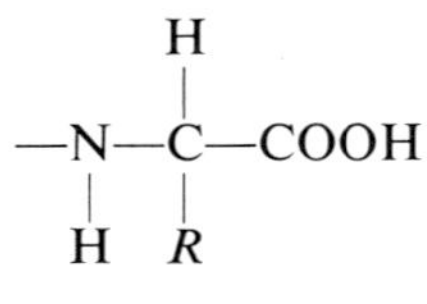

at the other end. To determine the structure of a peptide, we must ascertain the identity of the *R* group in each terminal residue and in each successive residue in the chain.

The first step is to determine which amino acid residues are present in the chain and to establish their relative amounts. This is done by subjecting a sample of the peptide to complete hydrolysis, breaking all peptide linkages. The products of the hydrolysis must be individually identified and their amounts determined by analysis. Knowing the molecular weights of the individual amino acids, we can calculate the number of moles of each acid obtained from the sample. From this result we then compute the relative numbers of each type of amino acid residue in the peptide molecule. As a simple example, hydrolysis of glutathione, a component of most living cells, yields three amino acids in equimolar amounts: glycine (Gly), glutamic acid (Glu), and cysteine (CySH). The "empirical formula" for glutathione therefore can be written as Gly, Glu, CySH. We use commas in the "empirical formula" to indicate that we have not yet established the sequence of the amino acid residues in the chain.

The next step in determining the structure of a peptide is to measure its molecular weight so that we will know *how many* residues are present in the chain. The molecular weights of simple peptides can be determined from the lowering of the freezing point, but the molecular weights of proteins are too large to be determined by this method. Molecular weights of proteins have been determined by the measurement of osmotic pressure, rate of diffusion, rate of settling in an ultracentrifuge, and the scattering of light.

The use of molecular weight information in determining the structure of a peptide can be illustrated by considering again the simple example of glutathione. Its molecular weight can be determined to be approximately 300. Knowing already that its "empirical formula" is Gly, Glu, CySH, we can compute the sum of the weights of the three residues as 307. Clearly, there is only one glycine residue, one glutamic acid residue, and one cysteine residue in a molecule of glutathione. Thus, glutathione is a tripeptide and its "molecular formula," like its "empirical formula," is Gly, Glu, CySH. We again use commas in the "molecular formula" to indicate that we still do not know the sequence of the residues in the chain.

The third step in determining the structure of a peptide is to determine the sequence of the units in the chain. For proteins this is a difficult and laborious task. The English biochemist Frederick Sanger received the 1958 Nobel prize in chemistry for developing the method and for using it to determine the amino acid sequence of insulin. The method involves partial hydrolysis and terminal residue analysis.

The structure of a simple peptide such as glutathione can be determined by partial hydrolysis alone. Even in the case of glutathione there are six ways in which the three residues could conceivably be arranged. However, partial hydrolysis gives only two dipeptides, glutamylcysteine (Glu · CySH) and cysteinylglycine (CySH · Gly). This establishes the position of the cysteine residue in the middle of the glutathione chain, with glutamic acid as the N-terminal residue and glycine as the C-terminal residue. The sequence, therefore, is Glu · CySH · Gly. Periods (or hyphens), rather than commas, are used in the "structural formulas" for peptides, since they are designed to show the sequence of residues in the chain. By convention, the N-terminal residue is written at the left end and the C-terminal residue at the right end. Thus the "formula" Glu · CySH · Gly is shorthand for the structure

$$\mathrm{H_2N\underset{\underset{\displaystyle COOH}{|}}{C}HCH_2CH_2CONH\underset{\underset{\displaystyle CH_2SH}{|}}{C}HCONHCH_2COOH}$$

With proteins, partial hydrolysis must be supplemented by terminal residue analysis. Phenyl isothiocyanate is among the reagents used for labeling the

N-terminal residue. The C-terminal residue is determined by an enzymatic method. The enzyme *carboxypeptidase* (from the pancreas) breaks only peptide linkages adjacent to free α-COOH groups in peptide chains. A C-terminal residue therefore is removed selectively by this enzyme.

35-7 The Structure of Proteins

The unique character of individual proteins cannot be accounted for completely in terms of the length and composition of the peptide chains. In recent years much attention has been given to the ways in which the peptide chains can be arranged to give the most stable conformations.

Hydrogen bonding can occur between an amide hydrogen and any carbonyl oxygen which it can approach at the proper distance. In many peptides and proteins a highly favorable arrangement is a coiled conformation in which the peptide chain forms a helix like a spiral staircase or a corkscrew. The helix is coiled in such a way that hydrogen bonding occurs between amide hydrogens and carbonyl oxygens that are four peptide linkages apart. There are 3.7 amino acid residues along each turn of the helix, and the spacing between turns is about 5.4 Å. The hydrogen bonds are approximately parallel to the long axis of the coil. Only right-handed helixes have been found in proteins.

The structure of proteins can be considered at different levels. At the lowest level, the *primary structure* is the order of linkage of atoms in the chain by covalent bonding, i.e., the specific sequence of amino acid residues in the polypeptide chain. The *secondary structure* is the way in which the chain is arranged in space, e.g., the helix described above. The *tertiary structure* is the way the coils are arranged relative to one another—wound into a ball, aligned beside each other in fibrils, or twisted together like the strands of a rope.

The properties of proteins are sometimes changed markedly by agents which do not break any of the peptide linkages. Such changes are called *denaturation* because they profoundly change the nature of the proteins. For example, the characteristic biological activity of enzymes is usually destroyed. Denaturation involves changes in the secondary structure of proteins. One of the most familiar examples of denaturation is the coagulation of egg white by heat.

35-8 Classification of Proteins

The two broad classes of proteins are *globular proteins* and *fibrous proteins*. Globular proteins are soluble in water or in aqueous solutions of acids, bases, or salts. Fibrous proteins are insoluble in aqueous media.

The names of the two classes give some indication of their physical structure. Fibrous proteins consist of long, threadlike particles which lie side by side; the particles may be coiled molecules, or they may be ropelike structures consisting of parallel coiled molecules twisted or braided together. The insolubility of fibrous proteins is due to the strong intermolecular forces between molecules that are held together at many points by hydrogen bonds or other types of attraction. Globular proteins consist of coiled molecules that are folded into compact units that may more or less approach spherical shape. The hydrogen bonding occurs between different points within the same molecule so that forces between molecules are relatively weak. Solutions of globular proteins are colloidal because the molecules are so large that they are of colloidal size. Only the globular proteins undergo denaturation.

Fibrous proteins serve as structural materials in animal tissues. They include collagen, elastin, keratin, fibroin, and myosin. *Collagen* is the main constituent of connective tissue and occurs in cartilage, bone, tendon, and skin. Collagen has an unusually high content of hydroxyproline and hydroxylysine. The prolonged action of boiling water on bones, skin, and connective tissue converts collagen to soluble proteins called gelatin. *Elastin* is similar to collagen but is more elastic. It occurs in the ligaments, in the walls of arteries, in the lungs, and in other tissues where elasticity is important. Elastin is not converted into gelatin. *Keratin* is the main constituent of hair, wool, nails, hoofs, horns, scales, and feathers. It has a high content of cystine and is resistant to digestive enzymes. *Fibroin* is the chief constituent of silk. *Myosin* is the fibrous protein in muscle.

The globular proteins are traditionally classified on the basis of solubility behavior. *Albumins* are proteins that are soluble in water and in dilute salt solution but are coagulated by heat. Examples include ovalbumin from egg white, lactalbumin from milk, serum albumin from blood serum, and legumelin from legumes. *Globulins* are insoluble in water but are soluble in dilute salt solutions. They are coagulated by heat. Examples are ovoglobulin in eggs and lactoglobulin in milk. Blood plasma contains at least four globulins, called fibrinogen and α, β, and γ serum globulins. *Prolamins* are insoluble in water or in absolute ethanol but are soluble in 70–80 percent aqueous ethanol. They are obtained mainly from the seeds of cereals. Zein from corn is an example. *Glutelins* do not dissolve in any of the solvents mentioned above but dissolve readily in dilute acid or alkali. They are present in gluten, the viscid material which gives adhesiveness to dough made from wheat flour.

On the basis of chemical composition, proteins have been traditionally classified into two major groups. Those that yield α-amino acids exclusively when they undergo complete hydrolysis are called *simple proteins*. The term *conjugated proteins* refers to those that yield, in addition to α-amino acids, one or more substances of another kind. The groups of a nonprotein nature present in conjugated proteins are referred to as conjugates or as *prosthetic groups*.

In one type of conjugated protein the prosthetic group is a carbohydrate, which may be simple or complex. Proteins that yield high proportions of carbohydrate may be referred to as *mucoproteins*. Those that yield small proportions of carbohydrate are designated *glycoproteins*.

In *lipoproteins* the prosthetic group is a lipid. In *phosphoproteins* phosphoric acid is linked as an ester to the hydroxyl groups of hydroxy amino acids. Examples of phosphoproteins are casein of milk and vitellin in egg yolk. *Chromoproteins* are characterized by a colored prosthetic group containing a metallic element such as iron, magnesium, or copper. An example is hemoglobin, the oxygen-carrying chromoprotein present in mammalian blood. In hemoglobin the prosthetic group is heme, a chelate complex containing iron. The globin portion of hemoglobin has a relatively high content of histidine and lysine and a small amount of isoleucine. There are several types of human hemoglobin, resulting from differences in the globin portion of the molecule.

The most complex conjugated proteins are the *nucleoproteins*, which are present in all living cells. In these materials the prosthetic groups are nucleic acids.

35-9 Metabolism of Proteins

When proteins are ingested by an animal organism, they undergo progressive hydrolysis in the digestive tract, giving peptides and amino acids. The hydrolysis in the stomach, where the pH is 1–2, is catalyzed by the enzyme *pepsin*; in the intestines, where the pH is 6–8, the acting enzymes are *chymotrypsins* and *peptidases*.

The amino acids pass through the intestinal walls into the bloodstream, which carries them to the body tissues. There the amino acids are reconstituted into proteins characteristic of the particular tissues. The liver converts amino acids into plasma proteins, which can also be used as a source of amino acids for the synthesis of tissue proteins.

It is remarkable that body chemistry is so well regulated as to enable an organism to synthesize a particular protein for each purpose. This involves selecting the necessary amino acids from the bloodstream and putting them together in the proper sequence.

Organisms are also able to convert some amino acids into others. Furthermore, some of the amino acids can be synthesized in the body from ammonium salts and α-keto acids derived from carbohydrates. The amino acids that can be formed in the body in either of these ways need not be present in the diet of the organism. However, those that cannot be so formed *must* be present in the diet. These are the so-called "essential" amino acids. The term "essential" means only that it is essential that these amino acids be supplied in the diet (usually in the form of proteins containing them). These particular amino acids are *not* more important than the others.

By means of studies made with the aid of isotopes it has been discovered that the proteins of the body are being continuously built up and torn down so that there is a fairly rapid turnover of the amino acids. The average "half-life" of proteins in the body of the adult human has been calculated to be about eighty days. This is the average time required for half of the original amino acid units in the body proteins to be replaced by other amino acid units.

35-10 Enzymes

When chemical reactions that take place within living organisms are duplicated in the laboratory, high temperatures or very strong reagents are usually required for them to occur at reasonable speeds. In general, neither of these conditions is compatible with life. Living cells produce substances whose function is to accelerate these reactions at a temperature compatible with life and under conditions no more extreme than mildly acidic or moderately alkaline. These biocatalysts, once called "ferments," are now known as *enzymes*.

Enzymes are made only by living cells, and all living cells contain them. Many have been extracted from their sources and utilized for catalyzing reactions in test tubes. It has been repeatedly demonstrated that an enzyme will selectively catalyze a particular reaction to the exclusion of other reactions that might take place between the same materials. Their biochemical importance is a consequence of their ability to direct the reactions that occur within an organism and of the specificity of their directive effect.

A similar specificity of catalytic powers is well known among inorganic catalysts. For example, platinum catalyzes the reaction of carbon monoxide with hydrogen to form methane and water, but zinc chromate catalyzes the combination of these two gases to form methanol.

Reactions catalyzed by enzymes, like other chemical reactions, lead to a state of chemical equilibrium, even though the reactions often appear to go to completion because the equilibrium is so far over to one side. Actually, the reactions can go either way, depending on concentrations. If the living cell excretes a product, or stores it in a separate compartment, the product is removed from the equilibrium system so that the enzyme-catalyzed reaction can continue in the one direction—either the direction of synthesis or the direction of decomposition, as the case may be.

Living cells often have means of disposing of the end products of a reaction: they may diffuse away or they may be stored in a special place or they may be immediately converted into something else by another reaction. In such cases enzymes may appear to promote a reaction in only one direction. In the laboratory, many substances have been synthesized by the action of the same enzymes that would have hydrolyzed them under other concentration conditions. The same equilibrium point has been reached from both directions. Although nothing appears to be happening at the equilibrium point, both synthesis and hydrolysis are in reality taking place—at exactly the same rate.

Many enzymes have been obtained in crystalline form. In every case the crystals have been proved to be proteins. It is now considered probable beyond a reasonable doubt that all enzymes are proteins. The general properties of enzymes, such as their behavior toward acid, alkali, and heat, indicate that they have a protein nature.

The rates of enzyme-catalyzed reactions are affected by temperature, by pH, by the concentration of the enzyme, and by the concentration of the substance undergoing chemical change under the influence of the enzyme. The generalization that the speeds of chemical reactions increase with rising temperatures holds true, within limits, for reactions catalyzed by enzymes. However, being proteins, enzymes are inactivated by denaturation at high temperatures. There is, therefore, an optimum temperature for each enzyme at which its catalytic activity is greatest.

Enzymes belong to different classes of proteins. Thus, carboxypeptidase is a globulin, but the pancreatic proteinases behave like albumins. Molecular weights of different enzymes have been found to vary over a wide range. Although some enzymes appear to be simple proteins, others are conjugated. The prosthetic groups in enzyme proteins are often related to one of the vitamins. It may be that one reason for the necessity of including vitamins in the diet is their ability to enable the organism to synthesize necessary enzymes.

The substance undergoing accelerated chemical change under the effect of an enzyme is called its *substrate*. Thus, in the hydrolysis of fat under the influence of a lipolytic enzyme, the substrate of the enzyme is the fat. Enzyme specificity is often described in terms of an analogy by which substrate and enzyme are compared to a lock and a key. The analogy seems particularly appropriate when it is considered that enzymes exhibit stereochemical specificity. As an example, natural L-leucylglycine is hydrolyzed by intestinal dipeptidase, but the synthetic D-leucylglycine is not affected by this enzyme. Thus, an enzyme can distinguish between two molecules whose only structural difference is that one is the mirror image of the other. From this it is to be inferred that the enzyme itself has a complex molecular architecture—and that its catalytic activity and specificity is a consequence of its structural complexity.

Enzymes were originally given whatever names pleased their discoverers. Among the old names that have persisted are ptyalin, pepsin, trypsin, and rennin. It was later agreed that enzymes would be named by adding the suffix *-ase* to the name of the substrate. Thus, carbohydrases hydrolyze carbohydrates, lipases hydrolyze fats, etc.

At the present time enzymes are classified mainly on the basis of function. Most enzymes fall into one of four groups: enzymes of hydrolysis (*hydrolases*), enzymes of oxidation (*oxidases*), enzymes of transfer (*transferases*), and enzymes of molecular rearrangement (*mutases*).

Enzymes of hydrolysis include the carbohydrases, lipases, and proteinases. An example of the enzymes of oxidation is *lactic dehydrogenase*, which is present in muscle; it catalyzes the oxidation of L-lactic acid (but not of D-lactic acid) to pyruvic acid:

$$\underset{\text{lactic acid}}{CH_3CHOHCOOH} \underset{+2H}{\overset{-2H}{\rightleftharpoons}} \underset{\text{pyruvic acid}}{CH_3COCOOH}$$

Transaminases are biochemically important examples of enzymes of transfer. A simple case is the transaminase-catalyzed transfer reaction between glutamic acid and pyruvic acid, producing α-ketoglutaric acid and alanine:

$$\underset{\text{glutamic acid}}{HOOCCH_2CH_2\overset{\overset{NH_2}{|}}{C}HCOOH} + \underset{\text{pyruvic acid}}{CH_3COCOOH} \rightleftharpoons$$

$$\underset{\alpha\text{-ketoglutaric acid}}{HOOCCH_2CH_2COCOOH} + \underset{\text{alanine}}{CH_3\overset{\overset{NH_2}{|}}{C}HCOOH}$$

Mutases are known to be involved in the metabolism of carbohydrates.

Certain substances are known to increase the action of enzymes to the extent that the enzymes are ineffective in their absence. A *coenzyme* is an organic molecule of low molecular weight that attaches itself to the enzyme molecule and, in so doing, makes the enzyme active. Substances that combine with enzymes to make them ineffective are called *inhibitors*. Many toxic substances exert their effect on the living organism by acting as enzyme inhibitors. Hydrogen cyanide is lethal because it inhibits an enzyme necessary for the normal consumption of oxygen in the respiratory metabolism of the cells. So-called nerve gases combine with an enzyme whose proper functioning is necessary for the transmission of nerve impulses.

35-11 Nucleoproteins

Nucleoproteins are present in all living cells. As the name suggests, they were once believed to occur only in the cell nucleus, but it is now known that they are also present in other parts of the cell.

The nucleoproteins are conjugated proteins in which the prosthetic groups are high polymers known as nucleic acids. They are essentially saltlike combinations of simple proteins with the nucleic acids. Much of their biochemical importance is due to their association with the chromosomes of cell nuclei. They are believed to play an important role in the genetic mechanism of life. They are also present in the cytoplasm, where they are involved in the biosynthesis of proteins.

The structures of nucleic acids are studied in much the same way as protein structures, e.g., by hydrolysis and by X-ray analysis. The nucleic acids are chemically quite different from proteins, but the two classes of substances resemble each other in consisting of long chains to which side groups are attached.

The progressive hydrolysis of a nucleoprotein occurs in the following steps:

1. Nucleoprotein → Nucleic acid + Protein
2. Nucleic acid → Nucleotides
3. Nucleotide → Nucleoside + H_3PO_4
4. Nucleoside → Organic base + Sugar

The sugars present in nucleic acids are D-ribose and D-2-deoxyribose. The nucleic acids containing ribose are known as ribonucleic acids (the RNA's), and those containing deoxyribose are called deoxyribonucleic acids (the DNA's).

The sugar in a nucleic acid is always present in cyclic form (the furanose ring). Each ring is attached through C-1 to a nitrogen atom of a heterocyclic organic base. The combination of one molecule of sugar with one molecule of base is called a *nucleoside*. An important example is *adenosine* (Figure 35-2), in which ribose is combined with adenine.

The bases present in the RNA's are adenine, guanine, cytosine, and uracil (Figure 35-3). Those present in the DNA's are adenine, guanine, thymine, cytosine, and 5-methylcytosine.

Figure 35-2

Structure of adenosine

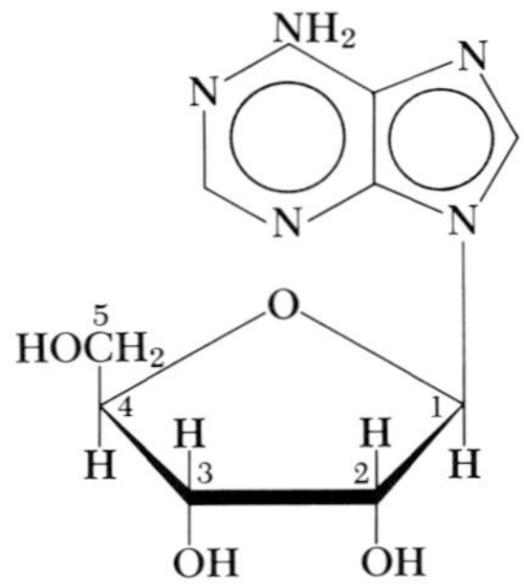

The ester formed between phosphoric acid and a hydroxyl group of a nucleoside sugar ring is called a *nucleotide*. An important example is adenylic acid (Figure 35-4), also known as adenosine monophosphate (AMP).

Nucleic acids are polynucleotides. In the same sense that proteins are polyamides composed of amino acid units, nucleic acids are polyesters composed of nucleotide units. Each sugar ring is joined to phosphate through the C-3 and C-5 hydroxyl groups (Figure 35-5).

The difference between one nucleic acid and another is in the properties of the different organic bases attached to the sugar rings and the sequence in which they occur along the polynucleotide chain. It is now believed that particular sequences of bases are likely responsible for particular hereditary characteristics. This is a prime reason for the current emphasis on "molecular biology." Structural theory is now the basis of the science of chemistry, both organic and inorganic, and it aspires to become the basis for our understanding of life.

Figure 35-3

Heterocyclic bases present in nucleic acids

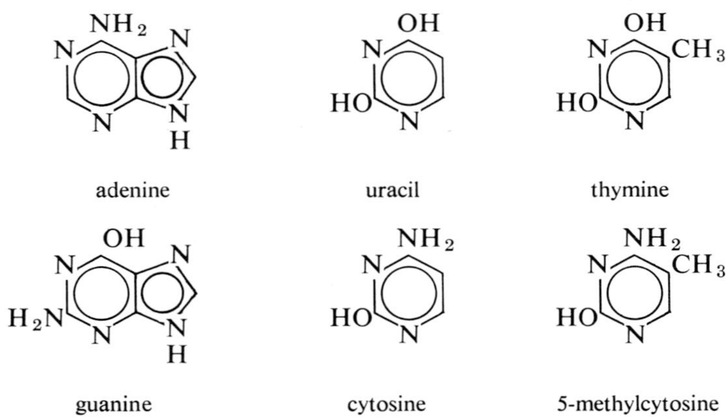

Figure 35-4

Structure of adenylic acid

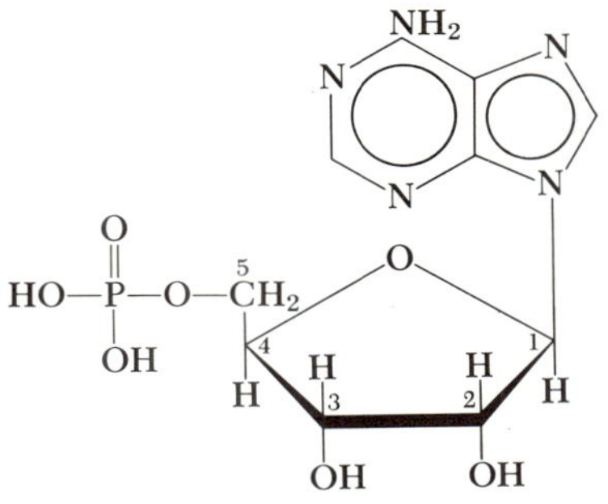

New Terms

Electrophoresis: the movement of charged particles in an electric field.

Isoelectric point: the pH at which an amino acid or protein does not migrate in an electric field.

Proteins: the complex combinations of amino acids that are the essential constituents of all living cells.

Zwitterion: a highly polar molecule produced by the transfer of one of its own protons from an acidic site to a basic site within the molecule, as when the proton from the carboxyl group of a glycine molecule transfers to the $—NH_2$ group.

Figure 35-5

Portion of chain of RNA

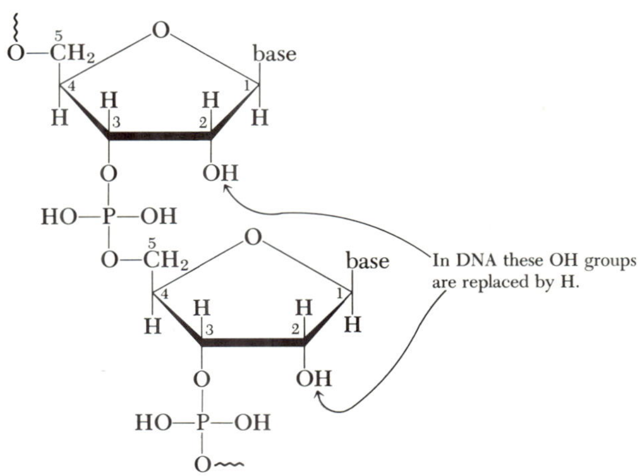

Exercises

35-1 Which elements are present in all proteins? What are some of the other elements that occur in some proteins?

35-2 Explain why the simplest proteins are far more complicated than sugars or fats.

35-3 How are the structures of proteins restricted as to (a) the positions of the $—NH_2$ groups, (b) the steric configurations of the alpha carbon atoms, and (c) the direction of the spiral in the helical conformation?

35-4 Explain the existence of amino acids in solution as zwitterions. Explain how the charge of an amino acid in solution may be altered by changing the pH.

35-5 Explain why amino acids are crystalline substances with fairly high melting points.

35-6 Define isoelectric point and itemize some of the factors that affect the pI value.

35-7 Define electrophoresis and discuss its applications in protein chemistry.

35-8 Define the following: peptide, peptide linkage, dipeptide, tripeptide, polypeptide, amino acid, amino acid residue, and protein.

35-9 Describe how the sequence of amino acid residues in a polypeptide chain is determined.

35-10 Define the following: denaturation, conjugated proteins, prosthetic groups, substrate, enzyme, coenzyme, nucleoprotein, nucleic acid, nucleotide, and nucleoside.

35-11 Outline the differences between globular proteins and fibrous proteins and give examples of each class.

35-12 List some factors that affect the rates of enzyme-catalyzed reactions.

35-13 Show how a polymer might result from the reaction between adipic acid, $HOOC(CH_2)_4COOH$, and $H_2N(CH_2)_6NH_2$.

References

Bennett, T. P., and Frieden, E., *Modern Topics in Biochemistry*, Macmillan, New York, 1966 (paperbound). Chapter 3 discusses proteins; Chapter 4, enzymes; and Chapter 9, nucleic acids.

Fairley, J. L., and Kilgour, G. L., *Essentials of Biological Chemistry*, Reinhold, New York, 1963. Chapter 2 discusses proteins, amino acids, and peptides; Chapter 5, nucleic acids and nucleotides; and Chapter 6, enzymes.

Jellinck, P. H., *Biochemistry: An Introduction*, Holt, Rinehart & Winston, New York, 1963. Chapter 3 discusses proteins; Chapter 4, enzymes; Chapter 5, nucleic acids; and Chapter 10, protein metabolism.

Rafelson, M., and Binkley, S. B., *Basic Biochemistry*, Macmillan, New York, 1965 (paperbound). Chapter 4 discusses proteins; Chapter 5, nucleic acids and nucleoproteins; and Chapter 6, enzymes.

36

Transmutation of the Elements

36-1 Discovery of Radioactivity

In 1896 Henri Becquerel, a French physicist, made the chance observation that potassium uranyl sulfate, $K_2SO_4 \cdot UO_2SO_4 \cdot 2H_2O$, emits penetrating radiation spontaneously and continuously. The radiation was found to affect photographic plates even when they were wrapped in lightproof black paper. Systematic investigation revealed that the radiation is emitted by all compounds containing uranium. Mme. Marie Sklodowska Curie gave to this remarkable property the name *radioactivity*.

Becquerel's discovery of radioactivity occurred within a few months after Röntgen's discovery of X rays. Becquerel found that the radiation from radioactive materials is similar to X rays in penetrating materials that are opaque to ordinary light, in affecting photographic plates in much the same manner as light, and in ionizing the molecules of air in the immediate vicinity, as demonstrated by the discharge of an electroscope.

In 1898 Mme. Curie undertook to make reliable quantitative comparisons of the relative activities of various radioactive materials. She found that the activities of all compounds of uranium are proportional to the quantities of uranium they contain. She also found that all compounds containing thorium are radioactive. It was fortunate that she discovered so early that radioactivity is not a property that is limited to one element. Soon she was to need this information to understand her remarkable observation that two uranium minerals, pitchblende and carnotite, are more radioactive than uranium itself. She postulated that these minerals contain another element which is more radioactive than uranium. Pierre Curie joined his wife in the search for this element.

36-2 Radioactive Elements

The Curies carried out a series of chemical separations which finally yielded a small quantity of a material 400 times as radioactive as uranium. From the nature of the chemical reactions involved in isolating this material, the Curies concluded that it was a compound of a previously unknown metal. They named the element *polonium* in honor of Mme. Curie's native Poland (Latin: *Polonia*). Continuing their work, they isolated, in December 1898, another substance 900 times as radioactive as uranium. This substance, they concluded, was a compound of another hitherto unknown metal, to which they gave the name *radium* in recognition of the high intensity of its radiation.

To confirm their discoveries of the two elements, the Curies obtained a ton of pitchblende residues from which most of the uranium had been extracted. From this Mme. Curie isolated 0.1 g of radium chloride, $RaCl_2$, of sufficient purity for the determination of the atomic weight of radium. This culminated what has been termed one of the supreme efforts of faith and perseverance on the part of one individual in the history of science. For this work Mme. Curie received the 1911 Nobel prize in chemistry. The Curies and Becquerel had previously shared the 1903 Nobel prize in physics for the discovery and early investigation of radioactivity.

By 1911 various scientific workers had discovered about forty different radioactive species. At the time there were only twelve positions available in the periodic table to accommodate them, and it was apparent that several radioactive species would have to occupy a single position in the table. This was equivalent to saying that different radioactive species may be different kinds of atoms of the same element, for in the periodic table there is one place for each element and presumably only one element for each place. Thus was born the concept of isotopes (Greek: *iso* + *topos*, same + place).

36-3 Three Types of Rays

After the announcement of Becquerel's discovery of radioactivity, the nature of the radiation itself had to be investigated. When a small sample of radioactive material is placed in a deep recess in a block of lead, the emerging radiation travels in a straight line as a pencil of parallel rays, all other rays being absorbed within the lead. If the rays are permitted to pass through an electric field and then impinge upon a photographic plate (Figure 36-1), it is found that the rays are of three types. These rays have been named *alpha* (α), *beta* (β), and *gamma* (γ) *rays*. This experiment was first performed by Ernest Rutherford (later Lord Rutherford) while he was professor of physics at McGill University in Montreal.

From the directions in which they are deflected in the electric field, it is apparent that alpha rays are positively charged, that beta rays are negatively charged, and that gamma rays have no charge at all. Rutherford found that the alpha radiation consists of *particles* which are doubly charged helium ions. The beta radiation also consists of particles; beta rays are high-speed electrons and

Figure 36-1

Behavior of alpha, beta, and gamma rays in an electric field

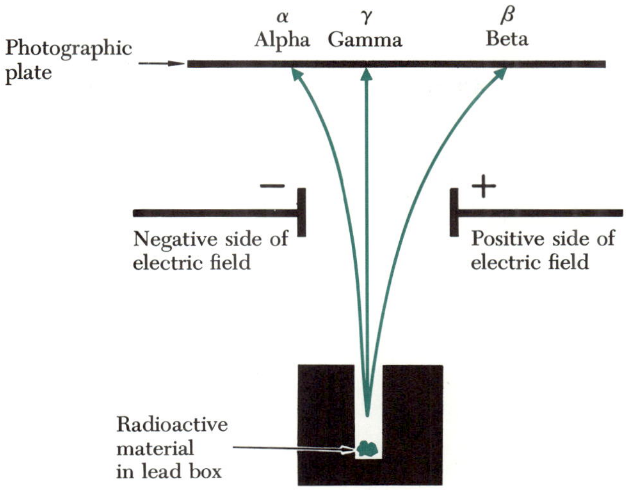

are thus similar in character to cathode rays (Section 3-5). The gamma rays are not particles but are a type of electromagnetic radiation like X rays. Gamma rays have even greater penetrating power than X rays because of their very short wavelengths, 10^{-8} to 10^{-10} cm. The properties of the three types of rays are summarized in Table 36-1. From this table it will be noted that the mass number (i.e., mass to the nearest integer) is 4 for the alpha particle and 0 for the beta particle.

Table 36-1

Rays emitted from radioactive substances

Ray	*Character*	*Charge*[a]	*Mass*[b]	*Velocity*	*Penetration*
α	He^{++}	+2	4.0026	About 0.1 speed of light	Stopped by paper or human skin
β	Electron	−1	0.000548	Up to 0.9 speed of light	Several mm in human tissue
γ	Short X rays	0	0	Speed of light[c]	Human body or several feet of concrete

[a] The unit of charge employed here is the amount of charge of the electron itself.
[b] The mass is expressed on the atomic-weight scale, defined as C^{12} = 12 exactly.
[c] The velocity of light in a vacuum is 3.00×10^{10} cm per sec.

Seldom does one radioactive species emit all three types of radiation. Nevertheless, most specimens of natural radioactive materials emit at least some of all three types, for such materials are usually mixtures of alpha-emitters and beta-emitters, and gamma radiation usually accompanies emission of beta particles.

The intensity of the radiation from a sample is not affected by raising or lowering the temperature, and it is independent of the oxidation state which the radioactive element is in. The activity depends only on the identity of the radioactive element and the quantity that is present. The intensity decreases with time, in some cases very rapidly and in other cases so slowly as to be imperceptible by direct measurement.

36-4 Disintegration of Atoms

Rutherford and his colleague Frederick Soddy proposed a theory of radioactive disintegration to account for the many surprising facts that had come to light since Becquerel's discovery. They suggested that the atoms of radioactive species undergo spontaneous disintegration with the emission of alpha or beta particles and form atoms of a new element. This bold suggestion was directly opposed to the concept that atoms are indestructible; according to Dalton's atomic theory, such *transmutation* of one element into another is not to be expected. However, it can be demonstrated with a spectroscope that after a period of time elements are present in a sample of radioactive material that were not originally present. In most cases the new element is itself radioactive.

The rates at which radioactive disintegrations occur vary from very slow to very rapid. Half of any sample of a radioactive element disintegrates in a definite period of time known as the *half-life* of that element. For radium the half-life is 1590 years. This means that, on the average, half of the radium atoms contained in a particular sample will disintegrate in 1590 years, that half of the remaining atoms will disintegrate in the next 1590 years, etc. Half-life periods vary from fourteen billion years for one of the isotopes of thorium to less than one billionth of a second for one of the isotopes of polonium. The importance of radium is a consequence of its half-life, which is neither so long as to make its radiation feeble nor so brief as to give the element only a fleeting existence.

By 1913 it was established that the nature of the emitted radiation is the factor that determines the identity of the new element. The relationship was stated by Soddy in the *group displacement law:*

1. Emission of an α particle reduces the mass of the atom by four atomic mass units, and the daughter element is located two groups to the left of its parent in the periodic table.
2. Emission of a β particle causes little change in the mass of the atom, but the daughter element is located one group to the right of its parent in the periodic table.

The nuclear theory of the atom (Section 3-7) and the atomic number concept (Section 3-8) made Soddy's group displacement law understandable and made it possible for the existence of isotopes to be reconciled with the periodic table. The order of increasing atomic numbers placed cobalt and nickel, tellurium and iodine, potassium and argon, and thorium and protactinium in their correct positions in the periodic table, although the order of increasing atomic weights did not do so. The number of possible lanthanide elements could now be foretold, and the periodic-table location of elements not yet discovered was definitely known for the first time. Although the recognition that radioactive disintegration involves transmutation had made the distinction between compounds and elements somewhat ambiguous, it gave rise to the concept of isotopes, which made it possible to define an element simply as a substance all of whose atoms have the same atomic number (nuclear charge).

36-5 Radioactive Disintegration Series

The radioactive disintegration of a radium nucleus consists in the ejection of an alpha particle, which is itself a helium nucleus. The radium nucleus is thereby transformed into a radon nucleus:

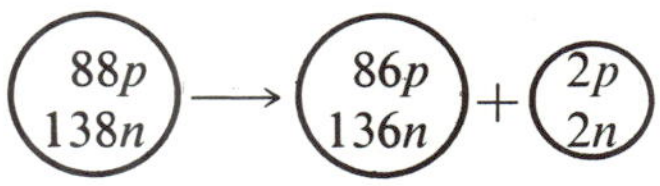

The transmutation can be represented by a *nuclear equation:*

$$^{226}_{88}\text{Ra} \longrightarrow {}^{222}_{86}\text{Rn} + {}^{4}_{2}\text{He}$$

In this kind of equation the symbol for an element represents the nucleus of one atom of that element. The number at the lower left is the atomic number, i.e., the number of protons in the nucleus; the number at the upper left is the mass number, i.e., the mass of the nucleus expressed to the nearest whole number, which is the number of protons plus the number of neutrons. Since the daughter element produced by alpha emission has an atomic number two units less than the parent, it must occupy the position in the periodic table two places to the left of the parent, as stated in Soddy's group displacement law.

The disintegration of a thorium-234 nucleus consists in the ejection of a beta particle, which has a charge of -1 unit but has such a small mass (0.000548 on the atomic-weight scale) that its mass number (mass to the nearest integer) is 0. The thorium-234 nucleus is therefore converted to a protactinium-234 nucleus:

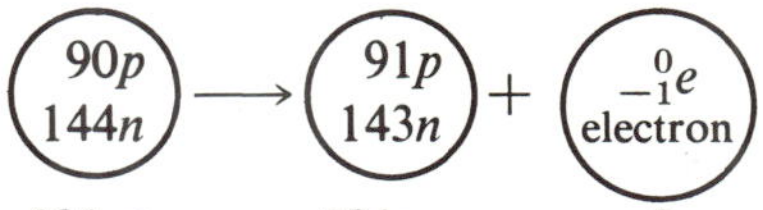

$$^{234}_{90}\text{Th} \longrightarrow {}^{234}_{91}\text{Pa} + {}^{0}_{-1}e$$

Whenever a beta particle is ejected from a nucleus, the daughter nucleus has one more proton and one less neutron than the parent; evidently a neutron in the nucleus is converted to an electron which is ejected at high velocity and a proton which remains in the nucleus:

$$\underset{\text{neutron}}{{}^{1}_{0}n} \rightarrow \underset{\text{proton}}{{}^{1}_{1}\text{H}} + \underset{\text{electron}}{{}^{0}_{-1}e}$$

The nucleus which is formed by the emission of an alpha or a beta particle may itself be radioactive, and usually is. Thus, the radon nucleus produced by the alpha decay of a radium nucleus emits an alpha particle and forms a polonium-218 nucleus:

$${}^{222}_{86}\text{Rn} \rightarrow {}^{218}_{84}\text{Po} + {}^{4}_{2}\text{He}$$

Similarly, the protactinium-234 produced by the beta decay of thorium-234 emits a beta particle and forms uranium-234, which, in turn, by alpha decay forms thorium-230, and so on.

Uranium-238 is the starting point for a series of radioactive disintegrations that finally ends in the formation of a nonradioactive isotope of lead, ${}^{206}_{82}\text{Pb}$. The successive "generations" in this *uranium disintegration series* are given in Table 36-2. Every known ore of uranium contains radium, and the uranium-radium ratio is always the same, this being a consequence of the brevity of the half-life of radium compared with that of uranium. All uranium ores also contain nonradioactive lead. This lead probably has originated as the end product of the disintegration series, for it has an atomic weight (approximately

Table 36-2

The uranium disintegration series

Element	*Atomic number*	*Mass number*	*Particle emitted*	*Half-life period*
Uranium	92	238	α	4.5×10^9 yr
Thorium	90	234	β	24 days
Protactinium	91	234	β	1.2 min
Uranium	92	234	α	2.5×10^5 yr
Thorium	90	230	α	8.0×10^4 yr
Radium	88	226	α	1590 yr
Radon	86	222	α	3.8 days
Polonium	84	218	α	3.05 min
Lead	82	214	β	27 min
Bismuth	83	214	β	20 min
Polonium	84	214	α	1.6×10^{-4} sec
Lead	82	210	β	19 yr
Bismuth	83	210	β	5.0 days
Polonium	84	210	α	138 days
Lead	82	206	Stable	——

206) which differs from the atomic weight of lead from most other sources (207.19). The atomic weight of an element is merely the average mass of its different isotopes.

From the known half-life periods of the elements involved, the time has been calculated that would be required for the formation of the lead in uranium ores. The result is of the order of three billion years. Presumably the earth is at least this old.

The uranium series is one of four *radioactive disintegration series*. The *thorium series* starts with thorium-232 and ends in the formation of nonradioactive lead-208. Since all members of this series have mass numbers that are divisible by four, the thorium series is known as the $4n$ series. The uranium series is the $4n + 2$ series. A third series is the $4n + 3$ or *actinium series*, which begins with uranium-235 and ends with lead-207. The fourth series is the $4n + 1$ or *neptunium series*, which involves man-made elements and was discovered during World War II in connection with the American atomic energy project. This series begins with plutonium-241 and ends with bismuth-209. In recent years both $^{241}_{94}Pu$ and $^{209}_{83}Bi$ have been found to be present in nature in small amounts.

36-6 Artificial Transmutation

With transmutation taking place spontaneously in nature, man was challenged to produce it artificially. In 1919 Lord Rutherford observed that discharging alpha particles into a space containing nitrogen gas results, on rare occasions, in a collision between an alpha particle and a nitrogen nucleus that results, in turn, in the formation of a proton and an oxygen-17 nucleus:

$$^{4}_{2}He + ^{14}_{7}N \rightarrow ^{17}_{8}O + ^{1}_{1}H$$

This was the first instance of deliberate *artificial transmutation* of one element to another. Most of the elements from boron to potassium were subsequently found to emit protons when bombarded with alpha particles. Frequently the energy of the protons was found to be higher than that of the alpha particles.

Bombardment of beryllium with alpha particles from polonium gives a highly penetrating radiation, which was proved by James Chadwick in England (1932) to consist of particles having no charge but a mass approximating that of the proton. The newly discovered particle was named the *neutron*. Its lack of charge accounts for its high penetrating power and for the difficulty of detecting it (and thus for the long delay in its discovery). The neutron is now regarded as a fundamental unit of atomic structure.

In 1934 Irene Curie (daughter of Pierre and Marie Curie) and her husband, Frederic Joliot, found that bombardment of magnesium with alpha particles causes emission of both neutrons and *positrons* (positive electrons). The emission of neutrons ceases when the source of alpha particles is removed, but positrons continue to be produced for some time. These observations were accounted for by the assumption that the transmutation of magnesium produced a radioactive isotope of silicon:

$$^{24}_{12}\text{Mg} + {}^{4}_{2}\text{He} \rightarrow \underset{\text{radioactive}}{{}^{27}_{14}\text{Si}} + {}^{1}_{0}n$$

$$\hookrightarrow {}^{27}_{13}\text{Al} + \underset{\text{positron}}{{}^{0}_{1}e}$$

The first reaction ceases as soon as alpha particle bombardment is ended, but the *artificial radioactivity* continues as long as the silicon-27 is spontaneously disintegrating to form aluminum. The Curie-Joliot discovery was significant, for it meant that radioactive isotopes of common elements could be obtained artificially for use in research. Radioactive carbon, nitrogen, and phosphorus have proved very useful as tracer atoms in biochemical research.

Artificial transmutations that occur instantaneously cause emission of a proton, a neutron, or some particle composed of protons and neutrons such as a deuteron (deuterium nucleus) or an alpha particle. Electrons and positrons are produced as secondary products from the spontaneous disintegration of the artificially produced radioactive isotopes. Just as electrons (β particles) are emitted as products of the decomposition of neutrons, so positrons are attributed to the disintegration of protons:

$$\underset{\text{proton}}{{}^{1}_{1}\text{H}} \rightarrow \underset{\text{neutron}}{{}^{1}_{0}n} + \underset{\text{positron}}{{}^{0}_{1}e}$$

The positron has the same mass and the same quantity of electric charge as an electron; however, the charge is positive instead of negative. (The positron was discovered in 1932 by the American physicist C. D. Anderson.)

Artificial transmutation has been achieved in recent years through the use of cyclotrons and other powerful particle-accelerator machines. Such instruments as the huge synchrotron at Brookhaven National Laboratory have been built in order to use strong magnetic and electric fields to impart tremendous energies to protons, deuterons, alpha particles, and electrons. For example, bombardment of carbon-12 nuclei with protons accelerated to high velocities in a cyclotron produces radioactive nitrogen-13:

$$^{12}_{6}\text{C} + {}^{1}_{1}\text{H} \rightarrow \underset{\text{radioactive}}{{}^{13}_{7}\text{N}} + \text{energy}$$

$$\xrightarrow[\text{10 min}]{\text{half-life}} \underset{\text{stable}}{{}^{13}_{6}\text{C}} + {}^{0}_{1}e$$

Since neutrons have no charge, they are not repelled by atomic nuclei. For this reason they are extremely effective as atomic artillery. They can be obtained by allowing alpha particles to bombard beryllium. The following are examples of nuclear reactions brought about by streams of neutrons:

$$^{200}_{80}\text{Hg} + {}^{1}_{0}n \rightarrow {}^{201}_{80}\text{Hg} + \text{energy } (\gamma \text{ rays})$$
$$^{40}_{20}\text{Ca} + {}^{1}_{0}n \rightarrow {}^{40}_{19}\text{K} + {}^{1}_{1}\text{H}$$

36-7 Man-made Elements

Artificial transmutation has been used not only to change many of the known elements one into another and to obtain radioactive isotopes of all known elements, but to form elements not known to occur in nature. Of the 103 elements now known to exist, fourteen were first obtained by artificial transmutation.

In the middle 1930's four elements with atomic numbers less than 92 (uranium) were still unknown, and no elements beyond uranium were known or believed necessarily possible of existence. The four elements missing from the periodic table were those with atomic numbers 43, 61, 85, and 87.

Element 43 was first obtained in 1937 by bombardment of molybdenum with deuterons:

$$^{96}_{42}\text{Mo} + ^{2}_{1}\text{H} \rightarrow ^{97}_{43}\text{Tc} + ^{1}_{0}n$$

It was given the name *technetium*, symbol Tc (Greek: *technetos*, artificial), because it was the first element not known on earth to be made artificially.

Element 61, the missing lanthanide, was found to be a product of uranium fission in connection with the atomic energy project during World War II. The name *promethium*, symbol Pm, which was given to the element was derived from the mythological character Prometheus who brought fire from heaven for the use of man, much as nuclear fission made atomic energy available.

Element 85, the missing halogen, was first obtained in 1940 by bombardment of bismuth with alpha particles (Section 24-13).

Element 87, the missing alkali metal, was discovered in 1939 (in France) and was named *francium*, symbol Fr. It occurs in nature in extremely small quantities; its longest-lived isotope has a half-life of only 21 minutes, which makes francium the most unstable of any element occurring naturally. Francium isotopes have been identified among products of artificial transmutation reactions.

The first transuranium elements were obtained in 1940 by bombarding uranium with high-velocity deuterons:

$$^{238}_{92}\text{U} + ^{2}_{1}\text{H} \rightarrow \underset{\text{radioactive}}{^{239}_{92}\text{U}} + ^{1}_{1}\text{H}$$

$$^{239}_{92}\text{U} \rightarrow \underset{\text{radioactive}}{^{239}_{93}\text{Np}} + ^{0}_{-1}e$$

$$^{239}_{93}\text{Np} \rightarrow ^{239}_{94}\text{Pu} + ^{0}_{-1}e$$

Element 93 was named *neptunium*, symbol Np, after Neptune, the planet directly beyond Uranus in the solar system. Element 94 was named *plutonium*, symbol Pu, after Pluto, the planet beyond Neptune which was discovered in 1930.

Elements 95–103 have all been subsequently obtained by artificial means. The actinide series was completed with the production of a very small amount of element 103, named *lawrencium*, in 1961. Its half-life is 8 seconds. A group of Russian scientists have recently announced the preparation of element 104 (Section 3-14) by bombardment of Pu-244 with Ne-20 ions.

36-8 Nuclear Fission

Before 1939 atomic scientists had little hope that atomic energy could be used for practical purposes. Although many of the artificial transmutation reactions then known released large amounts of energy, the bombardment procedures employed to produce the reactions consumed even more energy. Both the targets (atomic nuclei) and the projectiles (protons, etc.) in such procedures are so small that there is no way to aim the projectile or even to see the target. "Scoring a hit" under such circumstances is an improbable event of infrequent occurrence. Most of the energy invested in the operation is wasted on projectiles that go wide of their mark. In 1937, the year of his death, Lord Rutherford held the opinion that the outlook for gaining useful energy by artificial processes of transmutation was not promising.

In 1938 Otto Hahn and Fritz Strassmann in Germany unexpectedly discovered that bombardment of uranium with neutrons results in the formation of barium. This was impossible to account for in terms of the transmutation reactions previously known, for the atomic number of barium is only 56, i.e., thirty-six less than the atomic number of uranium. Early in 1939 Lise Meitner (a refugee from Nazi Germany, who had formerly worked with Hahn in Berlin), in association with her nephew, Otto Frisch, postulated that the impact of a neutron upon a uranium nucleus sometimes causes the latter to disintegrate into portions of somewhat similar mass. The phenomenon was given the name *nuclear fission*. The following equation represents one of the many ways in which a uranium atom may undergo fission:

$$^{235}_{92}\mathrm{U} + ^{1}_{0}n \rightarrow \underset{\text{radioactive}}{^{141}_{56}\mathrm{Ba}} + \underset{\text{radioactive}}{^{92}_{36}\mathrm{Kr}} + 3^{1}_{0}n + \text{energy}$$

The discovery of nuclear fission immediately changed the attitude of atomic scientists toward the atomic energy question because the process is accompanied by the liberation of large amounts of energy and because the fission, which is brought about by neutrons, is accompanied by the liberation of neutrons. The analogy with fire is clear. A fire is a self-sustaining reaction because the heat liberated by the reaction (combustion) maintains the temperature required for the reaction to continue. In similar fashion it was assumed that nuclear fission might be made self-sustaining by having neutrons liberated in the process initiate fission in other nuclei. If this could be accomplished, the practical utilization of nuclear energy would be realized.

Natural uranium is a mixture of isotopes having the approximate composition 99.3 percent U-238 and 0.7 percent U-235. Only the U-235 nuclei undergo fission. Neutrons absorbed by U-238 atoms are unavailable for bringing about fission of U-235 atoms. The utilization of the nuclear fission phenomenon in the so-called atomic bomb required that the U-235 be separated from most of the U-238 with which it occurs.

Unexpectedly, it was discovered that the absorption of neutrons by U-238 results in the formation of another fissionable species of atom, Pu-239:

$$^{238}_{92}\text{U} + {}^{1}_{0}n \rightarrow {}^{239}_{92}\text{U}$$

$$^{239}_{92}\text{U} \rightarrow {}^{239}_{93}\text{Np} + {}_{-1}^{0}e$$

$$^{239}_{93}\text{Np} \rightarrow {}^{239}_{94}\text{Pu} + {}_{-1}^{0}e$$

Although the plutonium-239 is itself radioactive, its half-life is of the order of 25,000 years.

36-9 Nuclear Energy

Weights of the individual isotopes of an element are expressed relative to $^{12}_{6}\text{C} = 12.0000$ exactly. Precision measurements show that the mass of each isotope is very nearly a whole number, but C-12 is the only one whose mass is exactly a whole number. The exact value for the mass of an isotope is called the *atomic mass* or *isotopic weight*, and the nearest integer is called the *mass number*. As examples, the isotopic weight of Ne-20 is 19.999 and that of Ne-22 is 21.998. The helium atom (mass number, 4) has the isotopic weight 4.0026. The U-235 atom has an isotopic weight slightly in excess of 235. Atoms at the beginning and at the end of the periodic table have isotopic weights in excess of their mass numbers. Atoms in the middle of the periodic table have isotopic weights less than their mass numbers.

Nuclear fission of elements near the end of the periodic table, such as uranium, produces elements nearer the middle of the table, such as barium and krypton. The loss of mass that accompanies the fission is responsible for the formation of such enormous amounts of energy. Albert Einstein, in 1905, had derived the equation

$$E = mc^2$$

which relates the magnitude of the mass that is lost to the quantity of energy produced. In this equation, m represents the mass in grams, E is the energy in ergs, and c is the velocity of electromagnetic radiation (light, etc.) in a vacuum, which is 3.00×10^{10} cm per second. Thus, conversion of 1 g of mass produces 9×10^{20} ergs. In other words, destruction of a small quantity of matter produces a tremendous amount of energy. In the nuclear fission of U-235 or Pu-239, the masses of the fission products are not much less than the masses of the starting materials, but even so the amount of energy produced by the fission of a small quantity of material is tremendous.

Nuclear reactors have been constructed for the utilization of nuclear energy for purposes other than explosions. In all such devices the energy is produced by fission.

The converse process, whereby nuclei of small mass undergo *nuclear fusion* to form nuclei of atoms nearer the middle of the periodic table, provides another means for the practical utilization of nuclear energy. The fusion of deuterium and tritium nuclei produces helium:

$$^{2}_{1}\text{H} + {}^{3}_{1}\text{H} \rightarrow {}^{4}_{2}\text{He} + {}^{1}_{0}n$$

The sum of the masses of the helium nucleus and the neutron is sufficiently less than the sum of the masses of the deuteron and the triton to account for the formation of enormous amounts of energy by this process. All that is required to initiate the fusion reaction is an extremely high temperature. For this reason the process is called a *thermonuclear reaction*. The only known method for producing the required temperature of many millions of degrees is the explosion of a fission-type atomic bomb. Therefore, an atomic bomb (A-bomb) is used to detonate a hydrogen bomb (H-bomb). There is no theoretical limit to the possible size of an H-bomb.

Controlled fusion for peacetime purposes has proved much more difficult to achieve than controlled fission and is still in the process of development.

New Terms

Artificial radioactivity: radioactivity arising from the formation, through artificial transmutation, of an atomic species that is radioactive and does not occur in nature.

Artificial transmutation: the transmutation of one element into another as the result of human action and intent (as distinguished from the "natural" spontaneous type of transmutation familiar as radioactivity).

Half-life: the time required for half of any sample of a radioactive element to disintegrate.

Nuclear equation: an equation representing a nuclear "reaction," i.e., transmutation. In a nuclear equation the chemical symbol of an element represents the nucleus of one atom of that element; the atomic number and the mass number are represented by a subscript and a superscript, respectively.

Nuclear fission: a type of transmutation characterized by the disintegration of an atomic nucleus into portions of somewhat similar mass.

Nuclear fusion: a type of transmutation in which atomic nuclei of somewhat similar mass unite to form a nucleus of greater mass than either of them.

Thermonuclear reaction: same as nuclear fusion (so called because nuclear fusion is initiated by extremely high temperature).

Exercises

36-1 If you have a 16-mg sample of iodine-128 at 1:00 P.M. and only 0.25 mg remains at 3:30 P.M., what is its half-life?

36-2 How do nuclear reactions differ from chemical reactions?

36-3 If the nucleus of a Bi-209 atom captures a neutron and the product undergoes beta decay, what is the final product?

36-4 What is the difference between radioactivity and artificial transmutation?

36-5 What distinguishes nuclear fission from nuclear fusion?

36-6 Why does uranium-235 not undergo self-sustaining nuclear fission in nature?

36-7 Why are the atomic weights of the elements not whole numbers? (Give two reasons.)

36-8 Complete the following equations:

(a) $^{27}_{13}Al + ? \rightarrow ^{24}_{11}Na + ^{4}_{2}He$

(b) $? + ^{2}_{1}H \rightarrow ^{12}_{6}C + ^{4}_{2}He$

(c) $^{23}_{11}Na + ^{2}_{1}H \rightarrow ^{24}_{11}Na + ?$

36-9 When would you say the Atomic Age began?

References

Glasstone, S., *Sourcebook on Atomic Energy*, 3rd ed., Van Nostrand, Princeton, N.J., 1967. A source of basic atomic energy information for readers with many different interests. Sponsored by the U.S. Atomic Energy Commission, this book has had worldwide acceptance.

Hewlett, R. G., and Anderson, O. E., Jr., *The New World, 1939/1946*, Pennsylvania State University Press, University Park, Pa., 1962. A complete account of the step-by-step discovery of the nature of nuclear fission and the first utilization of atomic energy. Volume I in *A History of the U.S. Atomic Energy Commission.*

Seaborg, G. T., *Man-made Transuranium Elements*, Prentice-Hall, Englewood Cliffs, N.J., 1963. An intellectually stimulating account of the discovery and investigation of the eleven transuranium elements. Written for use with the Chemical Education Material Study program.

Smyth, H. D., *Atomic Energy for Military Purposes*, Princeton University Press, Princeton, N.J., 1945. Complete title: "A general account of the development of methods of using atomic energy for military purposes under the auspices of the U.S. Government, 1940–45." The first book published on the subject, the Smyth report was released six days after the first use of an atomic bomb in warfare.

Appendix

1 Alphabetical List of Symbols of the Elements

Symbol	*Name*	*Atomic number*	*Symbol*	*Name*	*Atomic number*
Ac	Actinium	89	F	Fluorine	9
Ag	Silver	47	Fe	Iron	26
Al	Aluminum	13	Fm	Fermium	100
Am	Americium	95	Fr	Francium	87
Ar	Argon	18	Ga	Gallium	31
As	Arsenic	33	Gd	Gadolinium	64
At	Astatine	85	Ge	Germanium	32
Au	Gold	79	H	Hydrogen	1
B	Boron	5	He	Helium	2
Ba	Barium	56	Hf	Hafnium	72
Be	Beryllium	4	Hg	Mercury	80
Bi	Bismuth	83	Ho	Holmium	67
Bk	Berkelium	97	I	Iodine	53
Br	Bromine	35	In	Indium	49
C	Carbon	6	Ir	Iridium	77
Ca	Calcium	20	K	Potassium	19
Cd	Cadmium	48	Kr	Krypton	36
Ce	Cerium	58	La	Lanthanum	57
Cf	Californium	98	Li	Lithium	3
Cl	Chlorine	17	Lu	Lutetium	71
Cm	Curium	96	Lw	Lawrencium	103
Co	Cobalt	27	Md	Mendelevium	101
Cr	Chromium	24	Mg	Magnesium	12
Cs	Cesium	55	Mn	Manganese	25
Cu	Copper	29	Mo	Molybdenum	42
Dy	Dysprosium	66	N	Nitrogen	7
Er	Erbium	68	Na	Sodium	11
Es	Einsteinium	99	Nb	Niobium	41
Eu	Europium	63	Nd	Neodymium	60

Symbol	Name	Atomic number	Symbol	Name	Atomic number
Ne	Neon	10	Sc	Scandium	21
Ni	Nickel	28	Se	Selenium	34
No	Nobelium	102	Si	Silicon	14
Np	Neptunium	93	Sm	Samarium	62
O	Oxygen	8	Sn	Tin	50
Os	Osmium	76	Sr	Strontium	38
P	Phosphorus	15	Ta	Tantalum	73
Pa	Protactinium	91	Tb	Terbium	65
Pb	Lead	82	Tc	Technetium	43
Pd	Palladium	46	Te	Tellurium	52
Pm	Promethium	61	Th	Thorium	90
Po	Polonium	84	Ti	Titanium	22
Pr	Praseodymium	59	Tl	Thallium	81
Pt	Platinum	78	Tm	Thulium	69
Pu	Plutonium	94	U	Uranium	92
Ra	Radium	88	V	Vanadium	23
Rb	Rubidium	37	W	Tungsten	74
Re	Rhenium	75	Xe	Xenon	54
Rh	Rhodium	45	Y	Yttrium	39
Rn	Radon	86	Yb	Ytterbium	70
Ru	Ruthenium	44	Zn	Zinc	30
S	Sulfur	16	Zr	Zirconium	40
Sb	Antimony	51			

2 Names and Charges of Important Cations

Name	Formula and charge	Name	Formula and charge
Aluminum	Al^{+++}	Lithium	Li^{+}
Ammonium	NH_4^{+}	Magnesium	Mg^{++}
Antimony	Sb^{+++}	Manganese(II), manganous	Mn^{++}
Antimonyl	SbO^{+}	Mercury(I), mercurous	Hg_2^{++}
Barium	Ba^{++}	Mercury(II), mercuric	Hg^{++}
Bismuth	Bi^{+++}	Methylammonium	$CH_3NH_3^{+}$
Bismuthyl	BiO^{+}	Nickel	Ni^{++}
Cadmium	Cd^{++}	Potassium	K^{+}
Calcium	Ca^{++}	Silver	Ag^{+}
Chromium(III), chromic	Cr^{+++}	Sodium	Na^{+}
Cobalt(II), cobaltous	Co^{++}	Strontium	Sr^{++}
Copper(II), cupric	Cu^{++}	Tetramethylammonium	$(CH_3)_4N^{+}$
Dimethylammonium	$(CH_3)_2NH_2^{+}$	Tin(II), stannous	Sn^{++}
Hydrogen, hydronium	H^{+}, H_3O^{+}	Trimethylammonium	$(CH_3)_3NH^{+}$
Iron(II), ferrous	Fe^{++}	Uranyl	UO_2^{++}
Iron(III), ferric	Fe^{+++}	Zinc	Zn^{++}
Lead	Pb^{++}		

3 Names and Charges of Important Anions

Name	Formula and charge
Acetate, CH_3COO^-	$C_2H_3O_2^-$
Aluminate	$Al(OH)_4^-$
Amide	NH_2^-
Arsenate	AsO_4^{---}
Azide	N_3^-
Benzoate, $C_6H_5COO^-$	$C_7H_5O_2^-$
Bromate	BrO_3^-
Bromide	Br^-
Bromite	BrO_2^-
Carbonate	CO_3^{--}
Hydrogen carbonate, bicarbonate	HCO_3^-
Chlorate	ClO_3^-
Chloride	Cl^-
Chlorite	ClO_2^-
Chromate	CrO_4^{--}
Dichromate	$Cr_2O_7^{--}$
Cyanate	OCN^-
Cyanide	CN^-
Ferricyanide, hexacyanoferrate(III)	$Fe(CN)_6^{---}$
Ferrocyanide, hexacyanoferrate(II)	$Fe(CN)_6^{4-}$
Fluoaluminate	AlF_6^{---}
Fluoride	F^-
Fluosilicate	SiF_6^{--}
Formate, $HCOO^-$	CHO_2^-
Hydroxide	OH^-
Hypobromite	BrO^-
Hypochlorite	ClO^-
Hypoiodite	IO^-
Iodate	IO_3^-
Iodide	I^-
Lactate, $CH_3CHOHCOO^-$	$C_3H_5O_3^-$
Metaperiodate	IO_4^-
Molybdate	MoO_4^{--}
Nitrate	NO_3^-
Nitride	N^{---}
Nitrite	NO_2^-
Oxalate	$C_2O_4^{--}$
Hydrogen oxalate, binoxalate	$HC_2O_4^-$
Oxide	O^{--}
Paraperiodate	IO_6^{5-}
Perchlorate	ClO_4^-
Permanganate	MnO_4^-
Peroxide	O_2^{--}
Phosphate	PO_4^{---}
Monohydrogen phosphate	HPO_4^{--}
Dihydrogen phosphate	$H_2PO_4^-$
Salicylate, $HOC_6H_4COO^-$	$C_7H_5O_3^-$
Selenate	SeO_4^{--}
Selenide	Se^{--}
Silicate	SiO_3^{--}
Sulfamate	$SO_3NH_2^-$
Sulfate	SO_4^{--}
Hydrogen sulfate, bisulfate	HSO_4^-
Sulfide	S^{--}
Hydrogen sulfide, bisulfide	HS^-
Sulfite	SO_3^{--}
Hydrogen sulfite, bisulfite	HSO_3^-
Tartrate	$C_4H_4O_6^{--}$
Telluride	Te^{--}
Tetraborate	$B_4O_7^{--}$
Thioarsenate	AsS_4^{---}
Thiocyanate	SCN^-
Thiosulfate	$S_2O_3^{--}$
Tungstate	WO_4^{--}
Uranate	UO_4^{--}
Zincate	$Zn(OH)_4^{--}$

4 Important Acids and Their Ionization Constants, Listed in Order of Decreasing Strength

Name	*Formula*	*Ionization constant at 25°C*
Perchloric acid	$HClO_4$	Completely ionized
Hydriodic acid	HI	Completely ionized
Hydrobromic acid	HBr	Completely ionized
Sulfuric acid	H_2SO_4	Completely ionized
Hydrochloric acid	HCl	Completely ionized
Nitric acid	HNO_3	Completely ionized
Iodic acid	HIO_3	2.0×10^{-1}
Oxalic acid	$H_2C_2O_4$	5.4×10^{-2}
Sulfurous acid	H_2SO_3	1.2×10^{-2}
Hydrogen sulfate ion	HSO_4^-	1.2×10^{-2}
Phosphoric acid	H_3PO_4	7.5×10^{-3}
Ferric ion	$Fe(H_2O)_6^{+++}$	6.0×10^{-3}
Hydrotelluric acid	H_2Te	2.3×10^{-3}
Hydrofluoric acid	HF	6.7×10^{-4}
Nitrous acid	HNO_2	5.1×10^{-4}
Formic acid	$HCHO_2$ ($HCOOH$)	1.8×10^{-4}
Lactic acid	$HC_3H_5O_3$	1.4×10^{-4}
Hydroselenic acid	H_2Se	1.7×10^{-4}
Benzoic acid	$HC_7H_5O_2$ (C_6H_5COOH)	6.6×10^{-5}
Hydrogen oxalate ion	$HC_2O_4^-$	5.4×10^{-5}
Acetic acid	$HC_2H_3O_2$ (CH_3COOH)	1.8×10^{-5}
Carbonic acid	H_2CO_3	4.4×10^{-7}
Hydrosulfuric acid	H_2S	1.1×10^{-7}
Dihydrogen phosphate ion	$H_2PO_4^-$	6.3×10^{-8}
Hydrogen sulfite ion	HSO_3^-	5.6×10^{-8}
Hypochlorous acid	$HClO$	3.2×10^{-8}
Ammonium ion	NH_4^+	5.7×10^{-10}
Hydrocyanic acid	HCN	4.0×10^{-10}
Hydrogen carbonate ion	HCO_3^-	4.7×10^{-11}
Hydrogen peroxide	H_2O_2	2.4×10^{-12}
Monohydrogen phosphate ion	HPO_4^{--}	1.7×10^{-12}
Hydrogen sulfide ion	HS^-	1.0×10^{-14}

5 Solubility Product Constants

Name	*Formula*	K_{sp} *at 25°C*
Aluminum hydroxide	$Al(OH)_3$	1.9×10^{-33}
Barium carbonate	$BaCO_3$	8.1×10^{-9}
Barium chromate	$BaCrO_4$	2.4×10^{-10}
Barium fluoride	BaF_2	1.7×10^{-6}
Barium sulfate	$BaSO_4$	1.1×10^{-10}

Name	*Formula*	K_{sp} *at 25°C*
Calcium carbonate	$CaCO_3$	4.8×10^{-9}
Calcium fluoride	CaF_2	3.9×10^{-11}
Calcium sulfate	$CaSO_4$	2.4×10^{-5}
Copper(II) sulfide	CuS	8.7×10^{-36}
Iron(III) hydroxide	$Fe(OH)_3$	1.1×10^{-36}
Lead carbonate	$PbCO_3$	1.5×10^{-13}
Lead chromate	$PbCrO_4$	1.8×10^{-14}
Lead sulfate	$PbSO_4$	1.8×10^{-8}
Lead sulfide	PbS	8.4×10^{-28}
Magnesium carbonate	$MgCO_3$	1.0×10^{-5}
Magnesium hydroxide	$Mg(OH)_2$	1.5×10^{-11}
Silver bromide	$AgBr$	5.0×10^{-13}
Silver chloride	$AgCl$	1.6×10^{-10}
Silver chromate	Ag_2CrO_4	2.4×10^{-12}
Silver iodide	AgI	1.5×10^{-16}
Strontium carbonate	$SrCO_3$	9.4×10^{-10}
Strontium sulfate	$SrSO_4$	2.8×10^{-7}
Zinc sulfide	ZnS	1.1×10^{-20}

6 Vapor Pressure of Water

Temperature (°C)	*Pressure, torrs* (mm Hg)	*Temperature* (°C)	*Pressure, torrs* (mm Hg)
0	4.6	28	28.4
5	6.5	29	30.0
10	9.2	30	31.8
11	9.8	31	33.7
12	10.5	32	35.7
13	11.2	33	37.7
14	12.0	34	39.9
15	12.8	35	42.2
16	13.6	36	44.6
17	14.5	37	47.1
18	15.5	38	49.7
19	16.5	39	52.4
20	17.5	40	55.3
21	18.6	50	92.5
22	19.8	60	149.4
23	21.1	70	233.7
24	22.4	80	355.1
25	23.8	90	525.8
26	25.2	100	760.0
27	26.7	110	1074.6

7 Units of Measurement

Any measurement is expressed by a number and a unit. The size of the unit used for a particular measurement is a matter of choice. It is desirable to express measurements with numbers that are neither fantastically large nor absurdly small. Thus, in any system of weights and measures there are multiples and submultiples of the fundamental units.

In the United States both the English system and the metric system are used. The English system employs such units as the inch for length, the acre for area, and the pound for weight. The disadvantage of the English system lies in the lack of systematic interrelationships among units. For example, the units of length are not related in a consistent manner: 1 foot = 12 inches, 1 yard = 3 feet, etc.

The metric system, less than two hundred years old, derives its name from its basic unit of length, the *meter* (abbreviation: m). In this system the units are related by powers of 10. The most frequently used metric units of length are the following:

UNITS OF LENGTH (DISTANCE)

1 kilometer (km) = 1000 m (10^3 m)

1 centimeter (cm) = 0.01 m (10^{-2} m)

1 millimeter (mm) = 0.001 m (10^{-3} m)

1 micron (μ) = 0.000001 m (10^{-6} m)

1 Ångström (Å) = 0.0000000001 m (10^{-10} m. 10^{-8} cm)

The meter was originally intended to be one ten-millionth of the distance from the earth's equator to the North Pole (measured on a meridian passing through Paris). After this distance was measured (as accurately as possible), two fine lines were scribed on a metal bar made of an alloy of the noble metals platinum and iridium. Since the middle of the nineteenth century this metal bar has been kept in a vault at the International Bureau of Weights and Measures (at Sèvres, near Paris), and the distance between the two lines became the international standard of length. In 1875 the meter was adopted as the official standard of length by the United States, and the standard yard was defined as $\frac{3600}{3937}$ meter (from which it follows that 1 inch is approximately 2.54 cm). The requirements of present-day science and technology have necessitated increased accuracy in measurements; accordingly, the standard meter was redefined by international agreement in 1960 in terms of spectroscopic measurements. (The meter is now defined as 1,650,763.73 times the wavelength of the orange-red line of krypton-86.)

The basic unit of mass in the metric system is the *gram* (abbreviation: g), but the primary standard of mass is that of the international prototype kilogram, a body of platinum-iridium alloy kept at the International Bureau of Weights and Measures.

UNITS OF MASS (WEIGHT)

1 kilogram (kg) = 1000 g (10^3 g)

1 metric ton = 1000 kg (10^3 kg, 10^6 g)

1 milligram (mg) = 0.001 g (10^{-3} g)

1 microgram (μg or γ) = 0.000001 g (10^{-6} g, 10^{-3} mg)

Approximately, 1 pound is 453.59 g, 1 ounce is 28.35 g, and 1 kilogram is 2.20 lb.

The basic unit of volume in the metric system is the *liter*. In October 1964, the Twelfth General Conference on Weights and Measures defined the liter to equal a cubic decimeter, i.e., 1000 cubic centimeters. Formerly (since 1901) the liter was defined as the volume occupied by 1 kilogram of water at 4°C. This made the liter equal to 1000.028 cm^3, i.e., slightly larger than a cubic decimeter. The new definition makes the milliliter the same as the cubic centimeter.

UNITS OF VOLUME

1 liter = 1 cubic decimeter

1 milliliter (ml) = 0.001 liter

1 liter = 1000 cm^3 (cc)

1 ml = 1 cm^3

Approximately, 1 liter is 1.057 quarts (liquid measure), and 10 liters are 2.64 United States gallons or 2.20 British Imperial gallons.

8 Significant Figures

It is possible to count discrete objects without introducing an error, but measurements are never completely free from error because the accuracy of the procedure and the precision of the measuring instruments are limited. In the technical reporting of measurements (and calculations based on them), the convention used is to write down all digits definitely known plus one additional digit. The known digits plus the doubtful one constitute the *significant figures* of the number. Thus, a reported length of 23.42 cm is to be interpreted as being 23.42 $\pm$ 0.01 cm; i.e., the first three digits are assumed to be definitely known, with the true value of the length lying between 23.41 and 23.43 cm.

The number of significant figures used to record a measurement depends on the calibration of the measuring instrument. For example, if a ruler calibrated in millimeters is used for a measurement, the result is expressed to the nearest millimeter, and an estimate is made to the nearest 0.1 mm. If a quantity is measured only to the nearest tenth, it should not be reported to the nearest hundredth, e.g., 32.1 cm as against 32.10 cm. On the other hand, a length measured as 32.1000 cm should not be recorded as 32.1 cm.

Zeros may or may not be significant figures, depending on how they are used. A zero used only to locate a decimal point is not a significant figure. Thus, there is no difference between 27 mm, 2.7 cm, 0.027 m, and 0.000027 km, and each has two significant figures. However, the zeros in such numbers as 328.0 and 2.037 are counted as significant figures.

When we use numbers in a computation, we must make sure that we do not throw precision away nor appear to add to it. When a set of measurements is added or subtracted, the accuracy of the sum or difference is determined by the least accurate measurement in the set. For example, let us compute the sum of 32.1000 cm + 23.42 cm + 9.752 cm:

32.1000 cm	32.1000 cm
23.42	23.42??
9.752	9.752?
65.2720 cm	65.27?? cm

On the left we have set up the problem in the conventional fashion for adding and have performed the addition in the naive manner of a person who regards each recorded number as a quantity known with absolute exactness. On the right we have set up the problem with question marks to indicate our lack of information about certain digits; this shows why the sum should be reported as 65.27 cm.

Sometimes the answer to a computation must be subjected to the procedure known as rounding off. For example, let us add 14.23 g and 8.649 g:

$$\begin{array}{r} 14.23\ \ \text{g} \\ 8.649\phantom{\ \text{g}} \\ \hline 22.879\ \text{g} \end{array}$$

Clearly, there should be only four digits in the reported sum, but the number to be reported is 22.88 g. The rule in rounding off is to increase the last retained digit by 1 if the following digit to be discarded exceeds 5 (as in our example). If the following digit to be discarded is less than 5, the last retained digit is left unchanged. Thus,

1.423 rounds off to 1.42
1.427 rounds off to 1.43

If the digit to be discarded is 5, the standard procedure is to increase the preceding digit if it is odd and to leave it unchanged if it is even. Thus,

1.425 rounds off to 1.42
1.435 rounds off to 1.44

When you are multiplying or dividing measured quantities, there can be only as many significant figures in the answer as the least number of significant figures in any of the quantities used in the calculation. This means that multiplication of a number with two significant figures by a number with three significant figures gives an answer that has only two significant figures. For example, let us compute the product

$$40.2 \text{ cm} \times 0.11 \text{ cm}$$

The multiplication of 40.2 by 0.11 gives 4.422, but you should report the result as 4.4 cm^2.

There is a special problem connected with exact numbers that are not the result of measurement. Suppose 775 g of sugar is divided into 30 equal piles; how much sugar is in each pile? Because 30 is an exact number here, it has an infinite number of significant figures; i.e., it could be written as 30.00000 . . . . Since the least precisely known quantity in this problem is 775 g, you are justified in retaining three significant figures in the answer (25.8 g).

9 Exponential Notation

Numbers obtained in scientific measurements and computations range from very large to extremely small. Such numbers can be expressed and manipulated most easily by the use of exponential notation, i.e., by expressing them as multiples of powers of 10. For example, the mass of the earth is about 6,000,000,000,000,000,000,000,000,000 g, a quantity more conveniently written in exponential notation as 6×10^{27} g.

Numbers greater than 10 are usually written with positive powers of 10. Thus,

$$\begin{aligned} 14.52 &= 1.452 \times 10 \\ 145.2 &= 1.452 \times 10^2 \\ 1452 &= 1.452 \times 10^3 \\ 14520 &= 1.452 \times 10^4 \\ 145200 &= 1.452 \times 10^5 \end{aligned}$$

In each case the exponent of 10 corresponds to the number of places the decimal point is moved to the left to obtain a number between 1 and 10.

Usually, numbers less than 1 are written with negative powers of 10. For example, the mass of an electron, which is approximately 0.000,000,000,000,000,000,000,000,000,9 g, is more conveniently written as 9×10^{-28} g. This use of negative powers of 10 can be illustrated as follows:

$$\begin{aligned} 0.1452 &= 1.452 \times 10^{-1} \\ 0.01452 &= 1.452 \times 10^{-2} \\ 0.001452 &= 1.452 \times 10^{-3} \\ 0.0001452 &= 1.452 \times 10^{-4} \\ 0.00001452 &= 1.452 \times 10^{-5} \end{aligned}$$

In each case the negative exponent of 10 indicates how many places the decimal point must be moved to the right to obtain a number between 1 and 10.

With a number such as 143,000 the zeros may or may not be significant figures. To avoid such ambiguity, use exponential notation. Writing 143,000 as 1.43×10^5 indicates three significant figures; as 1.430×10^5, four significant figures; as 1.4300×10^5, five significant figures; as 1.43000×10^5, six significant figures.

Exponential numbers cannot be added or subtracted unless the powers of 10 are the same.

EXAMPLE Add 3.00×10^2 and 4.00×10^3.
First, write 4.00×10^3 as 40.0×10^2. Then,

$$\begin{aligned}(3.00 \times 10^2) + (40.0 \times 10^2) &= 43.0 \times 10^2 \\ &= 4.30 \times 10^3\end{aligned}$$

An alternate solution is to first write 3.00×10^2 as 0.300×10^3. Then,

$$(0.300 \times 10^3) + (4.00 \times 10^3) = 4.30 \times 10^3$$

It is better to report the answer as 4.30×10^3 than as 43.0×10^2 because 4.3 is between 1 and 10.

To multiply exponential numbers, multiply the digit terms in the customary manner and also multiply the powers of 10 (by adding their exponents).

EXAMPLE Multiply 4.8×10^5 by 2.0×10^{-7}.

$$\begin{aligned}(4.8 \times 10^5)(2.0 \times 10^{-7}) &= (4.8 \times 2.0)(10^5 \times 10^{-7}) \\ &= 9.6 \times 10^{-2}\end{aligned}$$

When one exponential number is divided by another, the digit term of the numerator is divided by the digit term of the denominator, and the powers of 10 are divided by subtracting the exponent of the denominator from the exponent of the numerator.

EXAMPLE Divide 4.2×10^{-8} by 7×10^{-3}.

$$\begin{aligned}\frac{4.2 \times 10^{-8}}{7 \times 10^{-3}} &= \frac{42 \times 10^{-9}}{7 \times 10^{-3}} \\ &= \frac{42}{7} \times \frac{10^{-9}}{10^{-3}} \\ &= 6 \times 10^{-6}\end{aligned}$$

To raise an exponential number to a higher power, raise the digit term to the stated power by multiplying it by itself the necessary number of times, and raise the power of 10 to the higher power by multiplying the exponent by the appropriate number.

EXAMPLE Square the number 9.0×10^3.

$$\begin{aligned}(9.0 \times 10^3)^2 &= (9.0)^2 \times (10^3)^2 \\ &= 81 \times 10^6 \\ &= 8.1 \times 10^7\end{aligned}$$

EXAMPLE What is the cube of 4.0×10^7?

$$\begin{aligned}(4.0 \times 10^7)^3 &= (4.0)^3 \times (10^7)^3 \\ &= 64 \times 10^{21} \\ &= 6.4 \times 10^{22}\end{aligned}$$

In taking the square root of an exponential number, first make the power of 10 divisible by 2; then extract the square root of the digit term and divide the exponential term by 2.

EXAMPLE Find the square root of 2.5×10^{-7}.

$$\begin{aligned}2.5 \times 10^{-7} &= 25 \times 10^{-8} \\ \sqrt{25 \times 10^{-8}} &= \sqrt{25} \times \sqrt{10^{-8}} \\ &= 5.0 \times 10^{-4}\end{aligned}$$

In taking the cube root, proceed in similar fashion except that the power of 10 is made divisible by 3.

EXAMPLE What is the cube root of 8,000,000?

$$\begin{aligned}8{,}000{,}000 &= 8 \times 10^6 \\ \sqrt[3]{8 \times 10^6} &= \sqrt[3]{8} \times \sqrt[3]{10^6} \\ &= 2 \times 10^2\end{aligned}$$

10 Logarithms

Natural numbers	0	1	2	3	4	5	6	7	8	9	Proportional parts								
											1	2	3	4	5	6	7	8	9
10	0000	0043	0086	0128	0170	0212	0253	0294	0334	0374	4	8	12	17	21	25	29	33	37
11	0414	0453	0492	0531	0569	0607	0645	0682	0719	0755	4	8	11	15	19	23	26	30	34
12	0792	0828	0864	0899	0934	0969	1004	1038	1072	1106	3	7	10	14	17	21	24	28	31
13	1139	1173	1206	1239	1271	1303	1335	1367	1399	1430	3	6	10	13	16	19	23	26	29
14	1461	1492	1523	1553	1584	1614	1644	1673	1703	1732	3	6	9	12	15	18	21	24	27
15	1761	1790	1818	1847	1875	1903	1931	1959	1987	2014	3	6	8	11	14	17	20	22	25
16	2041	2068	2095	2122	2148	2175	2201	2227	2253	2279	3	5	8	11	13	16	18	21	24
17	2304	2330	2355	2380	2405	2430	2455	2480	2504	2529	2	5	7	10	12	15	17	20	22
18	2553	2577	2601	2625	2648	2672	2695	2718	2742	2765	2	5	7	9	12	14	16	19	21
19	2788	2810	2833	2856	2878	2900	2923	2945	2967	2989	2	4	7	9	11	13	16	18	20
20	3010	3032	3054	3075	3096	3118	3139	3160	3181	3201	2	4	6	8	11	13	15	17	19
21	3222	3243	3263	3284	3304	3324	3345	3365	3385	3404	2	4	6	8	10	12	14	16	18
22	3424	3444	3464	3483	3502	3522	3541	3560	3579	3598	2	4	6	8	10	12	14	15	17
23	3617	3636	3655	3674	3692	3711	3729	3747	3766	3784	2	4	6	7	9	11	13	15	17
24	3802	3820	3838	3856	3874	3892	3909	3927	3945	3962	2	4	5	7	9	11	12	14	16
25	3979	3997	4014	4031	4048	4065	4082	4099	4116	4133	2	3	5	7	9	10	12	14	15
26	4150	4166	4183	4200	4216	4232	4249	4265	4281	4298	2	3	5	7	8	10	11	13	15
27	4314	4330	4346	4362	4378	4393	4409	4425	4440	4456	2	3	5	6	8	9	11	13	14
28	4472	4487	4502	4518	4533	4548	4564	4579	4594	4609	2	3	5	6	8	9	11	12	14
29	4624	4639	4654	4669	4683	4698	4713	4728	4742	4757	1	3	4	6	7	9	10	12	13
30	4771	4786	4800	4814	4829	4843	4857	4871	4886	4900	1	3	4	6	7	9	10	11	13
31	4914	4928	4942	4955	4969	4983	4997	5011	5024	5038	1	3	4	6	7	8	10	11	12
32	5051	5065	5079	5092	5105	5119	5132	5145	5159	5172	1	3	4	5	7	8	9	11	12
33	5185	5198	5211	5224	5237	5250	5263	5276	5289	5302	1	3	4	5	6	8	9	10	12
34	5315	5328	5340	5353	5366	5378	5391	5403	5416	5428	1	3	4	5	6	8	9	10	11
35	5441	5453	5465	5478	5490	5502	5514	5527	5539	5551	1	2	4	5	6	7	9	10	11
36	5563	5575	5587	5599	5611	5623	5635	5647	5658	5670	1	2	4	5	6	7	8	10	11
37	5682	5694	5705	5717	5729	5740	5752	5763	5775	5786	1	2	3	5	6	7	8	9	10
38	5798	5809	5821	5832	5843	5855	5866	5877	5888	5899	1	2	3	5	6	7	8	9	10
39	5911	5922	5933	5944	5955	5966	5977	5988	5999	6010	1	2	3	4	5	7	8	9	10
40	6021	6031	6042	6053	6064	6075	6085	6096	6107	6117	1	2	3	4	5	6	8	9	10
41	6128	6138	6149	6160	6170	6180	6191	6201	6212	6222	1	2	3	4	5	6	7	8	9
42	6232	6243	6253	6263	6274	6284	6294	6304	6314	6325	1	2	3	4	5	6	7	8	9
43	6335	6345	6355	6365	6375	6385	6395	6405	6415	6425	1	2	3	4	5	6	7	8	9
44	6435	6444	6454	6464	6474	6484	6493	6503	6513	6522	1	2	3	4	5	6	7	8	9
45	6532	6542	6551	6561	6571	6580	6590	6599	6609	6618	1	2	3	4	5	6	7	8	9
46	6628	6637	6646	6656	6665	6675	6684	6693	6702	6712	1	2	3	4	5	6	7	7	8
47	6721	6730	6739	6749	6758	6767	6776	6785	6794	6803	1	2	3	4	5	5	6	7	8
48	6812	6821	6830	6839	6848	6857	6866	6875	6884	6893	1	2	3	4	4	5	6	7	8
49	6902	6911	6920	6928	6937	6946	6955	6964	6972	6981	1	2	3	4	4	5	6	7	8
50	6990	6998	7007	7016	7024	7033	7042	7050	7059	7067	1	2	3	3	4	5	6	7	8
51	7076	7084	7093	7101	7110	7118	7126	7135	7143	7152	1	2	3	3	4	5	6	7	8
52	7160	7168	7177	7185	7193	7202	7210	7218	7226	7235	1	2	2	3	4	5	6	7	7
53	7243	7251	7259	7267	7275	7284	7292	7300	7308	7316	1	2	2	3	4	5	6	6	7
54	7324	7332	7340	7348	7456	7364	7372	7380	7388	7396	1	2	2	3	4	5	6	6	7

Natural numbers	0	1	2	3	4	5	6	7	8	9	Proportional parts								
											1	2	3	4	5	6	7	8	9
55	7404	7412	7419	7427	7435	7443	7451	7459	7466	7474	1	2	2	3	4	5	5	6	7
56	7482	7490	7497	7505	7513	7520	7528	7536	7543	7551	1	2	2	3	4	5	5	6	7
57	7559	7566	7574	7582	7589	7597	7604	7612	7619	7627	1	2	2	3	4	5	5	6	7
58	7634	7642	7649	7657	7664	7672	7679	7686	7694	7701	1	1	2	3	4	4	5	6	7
59	7709	7716	7723	7731	7738	7745	7752	7760	7767	7774	1	1	2	3	4	4	5	6	7
60	7782	7789	7796	7803	7810	7818	7825	7832	7839	7846	1	1	2	3	4	4	5	6	6
61	7853	7860	7868	7875	7882	7889	7896	7903	7910	7917	1	1	2	3	4	4	5	6	6
62	7924	7931	7938	7945	7952	7959	7966	7973	7980	7987	1	1	2	3	3	4	5	6	6
63	7993	8000	8007	8014	8021	8028	8035	8041	8048	8055	1	1	2	3	3	4	5	5	6
64	8062	8069	8075	8082	8089	8096	8102	8109	8116	8122	1	1	2	3	3	4	5	5	6
65	8129	8136	8142	8149	8156	8162	8169	8176	8182	8189	1	1	2	3	3	4	5	5	6
66	8195	8202	8209	8215	8222	8228	8235	8241	8248	8254	1	1	2	3	3	4	5	5	6
67	8261	8267	8274	8280	8287	8293	8299	8306	8312	8319	1	1	2	3	3	4	5	5	6
68	8325	8331	8338	8344	8351	8357	8363	8370	8376	8382	1	1	2	3	3	4	4	5	6
69	8388	8395	8401	8407	8414	8420	8426	8432	8439	8445	1	1	2	2	3	4	4	5	6
70	8451	8457	8463	8470	8476	8482	8488	8494	8500	8506	1	1	2	2	3	4	4	5	6
71	8513	8519	8525	8531	8537	8543	8549	8555	8561	8567	1	1	2	2	3	4	4	5	5
72	8573	8579	8585	8591	8597	8603	8609	8615	8621	8627	1	1	2	2	3	4	4	5	5
73	8633	8639	8645	8651	8657	8663	8669	8675	8681	8686	1	1	2	2	3	4	4	5	5
74	8692	8698	8704	8710	8716	8722	8727	8733	8739	8745	1	1	2	2	3	4	4	5	5
75	8751	8756	8762	8768	8774	8779	8785	8791	8797	8802	1	1	2	2	3	3	4	5	5
76	8808	8814	8820	8825	8831	8837	8842	8848	8854	8859	1	1	2	2	3	3	4	5	5
77	8865	8871	8876	8882	8887	8893	8899	8904	8910	8915	1	1	2	2	3	3	4	4	5
78	8921	8927	8932	8938	8943	8949	8954	8960	8965	8971	1	1	2	2	3	3	4	4	5
79	8976	8982	8987	8993	8998	9004	9009	9015	9020	9026	1	1	2	2	3	3	4	4	5
80	9031	9036	9042	9047	9053	9058	9063	9069	9074	9079	1	1	2	2	3	3	4	4	5
81	9085	9090	9096	9101	9106	9112	9117	9122	9128	9133	1	1	2	2	3	3	4	4	5
82	9138	9143	9149	9154	9159	9165	9170	9175	9180	9186	1	1	2	2	3	3	4	4	5
83	9191	9196	9201	9206	9212	9217	9222	9227	9232	9238	1	1	2	2	3	3	4	4	5
84	9243	9248	9253	9258	9263	9269	9274	9279	9284	9289	1	1	2	2	3	3	4	4	5
85	9294	9299	9304	9309	9315	9320	9325	9330	9335	9340	1	1	2	2	3	3	4	4	5
86	9345	9350	9355	9360	9365	9370	9375	9380	9385	9390	1	1	2	2	3	3	4	4	5
87	9395	9400	9405	9410	9415	9420	9425	9430	9435	9440	0	1	1	2	2	3	3	4	4
88	9445	9450	9455	9460	9465	9469	9474	9479	9484	9489	0	1	1	2	2	3	3	4	4
89	9494	9499	9504	9509	9513	9518	9523	9528	9533	9538	0	1	1	2	2	3	3	4	4
90	9542	9547	9552	9557	9562	9566	9571	9576	9581	9586	0	1	1	2	2	3	3	4	4
91	9590	9595	9600	9605	9609	9614	9619	9624	9628	9633	0	1	1	2	2	3	3	4	4
92	9638	9643	9647	9652	9657	9661	9666	9671	9675	9680	0	1	1	2	2	3	3	4	4
93	9685	9689	9694	9699	9703	9708	9713	9717	9722	9727	0	1	1	2	2	3	3	4	4
94	9731	9736	9741	9745	9750	9754	9759	9763	9768	9773	0	1	1	2	2	3	3	4	4
95	9777	9782	9786	9791	9795	9800	9805	9809	9814	9818	0	1	1	2	2	3	3	4	4
96	9823	9827	9832	9836	9841	9845	9850	9854	9859	9863	0	1	1	2	2	3	3	4	4
97	9868	9872	9877	9881	9886	9890	9894	9899	9903	9908	0	1	1	2	2	3	3	4	4
98	9912	9917	9921	9926	9930	9934	9939	9943	9948	9952	0	1	1	2	2	3	3	4	4
99	9956	9961	9965	9969	9974	9978	9983	9987	9991	9996	0	1	1	2	2	3	3	3	4

11 Answers to Selected Exercises

1-3 Present century: nylon, DDT, synthetic rubber, Corfam, etc. Since prehistoric times: wood, copper, water, stone, etc.

1-6 The components of air can be separated without resorting to chemical change (e.g., by fractional distillation of liquid air). Chemical decomposition is necessary to produce hydrogen and oxygen from water.

2-1 45; 3; 200

2-3 One fourth as heavy

2-5 14

2-7 2.14 g; 1.97 g

2-9 28.8

2-11 Same

2-13 (a) $CaBr_2$
(b) Mn_3S_4
(c) $K_2Cr_2O_7$
(d) KH_2PO_4
(e) Na_3AlF_6

3-2 KNO_3; $Be(NO_3)_2$; BF_3; SiS_2; H_2S; CF_4

3-3 Heaviest isotopes of argon and lightest isotopes of potassium are most abundant.

4-2 18; 32; 8

4-4 VIIA structure is ns^2np^5;
VIIB structure is $(n - 1)d^5ns^2$.
IA structure is ns^1;
IB structure is $(n - 1)d^{10}ns^1$.
IVA structure is ns^2np^2;
IVB structure is $(n - 1)d^2ns^2$.
0 structure is ns^2np^6;
VIII structure is $(n-1)d^{6,7,8}ns^2$.

5-1 Li^+, I^-; K^+, Br^-; Na^+, S^{--}; Ca^{++}, O^{--}; Al^{+++}, S^{--}; Ca^{++}, F^-

5-4 Mg_3O_2, no; Mg_2O_3, no; PO, no; KCl_2, no; RbBr, yes; CoNe, no; BaSe, yes; NaI, yes

6-1 32.064 g; 30.028 g; 0.552 g; 2.346×10^{-11} g

6-3 28.01 g; 90.08 g; 3.84×10^{-4} g

6-5 1063 moles; 2126 moles

6-7 2.92 moles

6-9 3.338×10^{-22} g

7-6 $KClO_3$

7-10 160 g

8-2 $LiAlH_4$

8-3 2.88×10^3 kcal; 8.16×10^4 kcal

8-4 Carbon is oxidized from the -4 state to the $+4$ state; oxygen is reduced from the 0 state to the -2 state.

9-1 $-40°$

9-5 320 ml

9-7 1.322 liters

9-9 50

9-13 Cl_2, Ar, O_2, N_2, Ne, NH_3, He, H_2; Cl_2 diffuses most slowly, H_2 most rapidly.

9-15 35 liters

10-1 98.9°C; 100.6°C; 100.0°C

10-7 2.09 g

10-9 25.8 g

11-3 1.82×10^4 cal

11-4 Melted iodine: a nonconductor, because it consists of molecules (not ions).
Melted sodium: an excellent conductor, like solid sodium and the other metals.
Melted sodium iodide: a good conductor, since it consists of ions that are free to move.

12-2 51.16%

12-4 $Na_2B_4O_7 \cdot 10H_2O$

12-12 48.21%

13-3 0.200*m*; 187

13-5 100.204°C

13-7 (a) 174; (b) $C_6H_{12}O_6$

13-9 (a) -2.8°C; (b) -0.93°C; (c) -14°C

13-11 24.8 ml

14-1 (a) 289,470 coulombs;
(b) 192,980 coulombs;
(c) 771,920 coulombs;
(d) 5,789,400 coulombs

14-3 80 hr

14-5 0.204 amp

15-1 (a) 342.3 g; 4108 g; 51.35 g; 256.7 g; 0.4279 g

(b) 58.44 g; 701.3 g; 8.766 g; 43.83 g; 0.07305 g
(c) 246.5 g; 2958 g; 36.97 g; 184.9 g; 0.3081 g

15-3 0.51

16-1 6.8×10^{-4}

16-3 1.3×10^{-11}

16-5 (a) 2, 12; (b) 10.5, 3.5; (c) 3.9, 10.1; (d) 2.5, 11.5

16-7 9.0×10^{-12}

17-1 (a) −103.69 kcal;
(b) −56.52 kcal;
(c) −150.85 kcal;
(d) −41.69 kcal;
(e) −74.43 kcal;
(f) −600.18 kcal

17-2 −92 kcal

17-4 +27.44 kcal

18-10 Skeletal isomerism: the two butanes
Positional isomerism: 1-butene and 2-butene
Geometric isomerism: *cis*-2-butene and *trans*-2-butene

18-11 The six carbon-carbon bonds are of equal length.

19-1 No real difference, for every carbon compound can be derived from a hydrocarbon.

19-10 The lactic acid molecule contains a carbon atom to which four different groups are attached. The citric acid molecule contains no such atom.

20-4 (a) H_2 and NaOH at cathode, O_2 and HF at anode
(b) Ag at cathode, O_2 and HNO_3 at anode
(c) Cu at cathode, Cl_2 at anode (O_2 and HCl if solution is dilute)
(d) H_2 at cathode, O_2 at anode
(e) H_2 at cathode, O_2 at anode

20-5 Yes, yes, no, yes, no

20-6 (a) 3, 5, 2, 3, 5;
(b) 1, 4, 2, 1, 2, 2;
(c) 1, 2, 1, 1, 2;
(d) 1, 8, 1, 4, 4;
(e) 3, 8, 3, 3, 2, 4;
(f) 2, 5, 6, 2, 5, 8;
(g) 2, 5, 6, 2, 5, 8

21-5 Strong acids: $HClO_4$, HI, HBr, H_2SO_4, HCl, HNO_3, H_3O^+
Their conjugate bases (all very weak bases): ClO_4^-, I^-, Br^-, HSO_4^-, Cl^-, NO_3^-, H_2O

21-6 Strong bases: CH_3^-, H^-, CH_3NH^-, NH_2^-, $C_2H_5O^-$, OH^-
Their conjugate acids (all very weak acids): CH_4, H_2, CH_3NH_2, NH_3, C_2H_5OH, H_2O

22-2 The Solvay process yields $CaCl_2$ as a by-product.

22-5 Na_2CO_3, KOH, KNO_3, $LiAlH_4$, C_2H_5Li

22-8 0.55 g

23-3 Air, water, coal, limestone, salt, sulfur, natural gas, petroleum

23-4 980 lb

23-8 The sodium sulfate causes the formation of insoluble $BaSO_4$, which is not poisonous.

24-5 Cl_2, $COCl_2$, $(ClC_2H_4)_2S$, Cl_3CNO_2

24-7 32 tons NaOH, 0.81 ton H_2, 28 tons Cl_2

25-1 (a) NaOH, H_2, Cl_2
(b) NaClO, H_2

25-2 (a) Digit furthest to right is number of F atoms: digit next to right is *one more than* number of H atoms; third digit (if there is one) is *one less than* number of C atoms.
(b) CCl_3F, CCl_2F_2, $CHClF_2$, $CClF_2 \cdot CClF_2$

26-1 −2: H_2O, H_2S, H_2Se, H_2Te, H_2Po, CaO, NaOH, Na_2S, NaHS, etc.
+4: SO_2, $SeBr_4$, TeI_4, etc.
+6: SO_3, H_2SO_4, SeF_6, H_6TeO_6, etc.

26-2 In a peroxide, two oxygen atoms are linked by a covalent bond; in a polysulfide, sulfur atoms are linked in a chain by covalent bonds.

27-2 Inactivity is due to great strength of triple bond in N_2 molecule.

27-5 In water, acids furnish H_3O^+ ions and bases furnish OH^- ions. In liquid ammonia, acids furnish NH_4^+ ions and bases furnish NH_2^- ions.

27-9 N_2O; NH_4HCO_3; glyceryl trinitrate; nitroglycerin taken up in porous material, e.g., wood pulp; cellulose nitrate; trinitrotoluene.

28-1 (a) N atom too small
(b) I atom too large
(c) Fluorine is the strongest oxidizing agent

28-2 $(CH_3)_2AsH$, $(CH_3)_4PI$, $(CH_3)_2As—As(CH_3)_2$, Na_2HPO_4, Na_3SbS_4

28-6 Aqueous solution, 82% H_3PO_4; As_4O_6; trisodium phosphate mixture, 2 moles of $CaSO_4$ to 1 mole of $Ca(H_2PO_4)_2$.

29-3 Graphite, Si, and Ge show decrease in resistance with rising temperature; Sn and Pb show an increase.

29-4 H_2NCOOH, $CO(NH_2)_2$, $COCl_2$, Pb_3O_4, $Pb(OH)_2 \cdot PbCO_3$

30-2 Phosphorus and the elements of Groups IA, IIA, IIIA, IVA (except carbon), and VIIA.

30-3 B and Al; Ga, In, and Tl; Ga, In, and Tl

31-3 (a) potassium tetrachlorodiamminechromate(III)
(b) dichlorodiammineplatinum
(c) hexamminecobalt(III) chloride
(d) tris(ethylenediamine) cobalt(III) chloride
(e) hexamminechromium(III) hexacyanocobaltate(III)
(f) hexamminecobalt(III) hexacyanochromate(III)
(g) tetramminezinc ion
(h) hexaquozinc ion

31-7 $8 \times 10^{-8}\ M$

31-11 (a) Negative for chloride, positive for bromide
(b) Negative for bromide, positive for chloride

32-1 They have higher ionization energies and lower electron affinities than the other elements in their respective periods.

32-5 (a) 0.18 g; (b) 0.090 g; (c) 10 g; (d) 1.3 g

33-3 Glyceraldehyde has only one asymmetric carbon atom, therefore two isomers, D and L. An aldotetrose has two asymmetric carbon atoms; configuration of either of them can be D or L, giving four possible combinations: D, D; L, L; D, L; and L, D.

33-6 They are optical isomers but not enantiomers. They are diastereoisomers.

34-2 Glycerides (fats) are esters of fatty acids with glycerol. Monoglycerides and diglycerides contain unesterified —OH groups; in a triglyceride all —OH groups have been esterified.

34-3 Simple triglyceride has all three —OH groups esterified with same acid. Mixed triglyceride has —OH groups esterified with two or three different acids.

34-10 Phospholipids, which contain phosphorus, are essential components of all living cells.

35-1 (a) Carbon, hydrogen, oxygen, and nitrogen
(b) Sulfur, phosphorus, iron, copper, iodine, manganese, zinc

35-12 Temperature, pH, concentration of enzyme, concentration of substrate

36-1 25 min

36-3 Polonium-210

36-8 (a) ${}^{1}_{0}n$; (b) ${}^{14}_{7}N$; (c) ${}^{1}_{1}H$

Index

Italicized page numbers refer to the "New Terms" sections.

9 0 D 1 E 2 F 3 G 4 H 5 I 6 J 7

( ) indicates the most stable or best-known isotope.

	Symbol	Atomic Number	Atomic Weight
Actinium	Ac	89	(227)
Aluminum	Al	13	26.9815
Americium	Am	95	(243)
Antimony	Sb	51	121.75
Argon	Ar	18	39.948
Arsenic	As	33	74.9216
Astatine	At	85	(210)
Barium	Ba	56	137.34
Berkelium	Bk	97	(247)
Beryllium	Be	4	9.0122
Bismuth	Bi	83	208.980
Boron	B	5	10.811
Bromine	Br	35	79.909
Cadmium	Cd	48	112.40
Calcium	Ca	20	40.08
Californium	Cf	98	(249)
Carbon	C	6	12.01115
Cerium	Ce	58	140.12
Cesium	Cs	55	132.905
Chlorine	Cl	17	35.453
Chromium	Cr	24	51.996
Cobalt	Co	27	58.9332
Copper	Cu	29	63.54
Curium	Cm	96	(247)
Dysprosium	Dy	66	162.50
Einsteinium	Es	99	(254)

	Symbol	Atomic Number	Atomic Weight
Erbium	Er	68	167.26
Europium	Eu	63	151.96
Fermium	Fm	100	(253)
Fluorine	F	9	18.998
Francium	Fr	87	(223)
Gadolinium	Gd	64	157.25
Gallium	Ga	31	69.72
Germanium	Ge	32	72.59
Gold	Au	79	196.967
Hafnium	Hf	72	178.49
Helium	He	2	4.002
Holmium	Ho	67	164.930
Hydrogen	H	1	1.007
Indium	In	49	114.82
Iodine	I	53	126.904
Iridium	Ir	77	192.2
Iron	Fe	26	55.847
Krypton	Kr	36	83.80
Lanthanum	La	57	138.91
Lawrencium	Lw	103	(257)
Lead	Pb	82	207.19
Lithium	Li	3	6.939
Lutetium	Lu	71	174.97
Magnesium	Mg	12	24.312
Manganese	Mn	25	54.938
Mendelevium	Md	101	(256)